The Routledge Companion to Rural Planning

The Routledge Companion to Rural Planning provides a critical account and state of the art review of rural planning in the early years of the twenty-first century.

Looking across different international experiences – from Europe, North America and Australasia to the transition and emerging economies, including BRIC and former communist states – it aims to develop new conceptual propositions and theoretical insights, supported by detailed case studies and reviews of available data. The Companion gives coverage to emerging topics in the field and seeks to position rural planning in the broader context of global challenges: climate change, the loss of biodiversity, food and energy security, and low carbon futures. It also looks at old, established questions in new ways: at social and spatial justice, place shaping, economic development, and environmental and landscape management. Planning in the twenty-first century must grapple not only with the challenges presented by cities and urban concentration, but also grasp the opportunities – and understand the risks – arising from rural change and restructuring. Rural areas are diverse and dynamic. This Companion attempts to capture and analyse at least some of this diversity, fostering a dialogue on likely and possible rural futures between a global community of rural planning researchers.

Primarily intended for scholars and graduate students across a range of disciplines, such as planning, rural geography, rural sociology, agricultural studies, development studies, environmental studies and countryside management, this book will prove to be an invaluable and up-to-date resource.

Mark Scott is Professor of Planning at University College Dublin, Ireland.

Nick Gallent is Professor of Housing and Planning and Head of the Bartlett School of Planning, University College London, UK.

Menelaos Gkartzios is Senior Lecturer in Planning and Development at Newcastle University's Centre for Rural Economy, UK.

The Routledge Companion to Rural Planning

Edited by Mark Scott, Nick Gallent and Menelaos Gkartzios

Routledge
Taylor & Francis Group

LONDON AND NEW YORK

First published 2019
by Routledge
2 Park Square, Milton Park, Abingdon, Oxon OX14 4RN

and by Routledge
605 Third Avenue, New York, NY 10017

First issued in paperback 2020

Routledge is an imprint of the Taylor & Francis Group, an informa business

British Library Cataloguing-in-Publication Data
A catalogue record for this book is available from the British Library

Library of Congress Cataloging-in-Publication Data
A catalog record has been requested for this book

Typeset in Bembo
by Swales & Willis Ltd, Exeter, Devon, UK

ISBN 13: 978−0−367−73196−0 (hbk)
ISBN 13: 978−1−138−10405−1 (hbk)
ISBN 13: 978−1−315−10237−5 (ebk)

Contents

Contents

Contents

Figures

Tables

Boxes

Contributors

Loka Ashwood is an assistant professor at Auburn University. She earned her PhD in sociology from the University of Wisconsin-Madison. She is author of the book *For-Profit Democracy: Why the Government is Losing the Trust of Rural America* (forthcoming).

Stuart Auckland is the programme coordinator for community health development at the Centre for Rural Health, University of Tasmania. He has extensive experience in rural community development and has undertaken a range of community based projects and participatory research initiatives in both the natural resource management and rural health sectors.

Jo Barton is a senior lecturer in sports and exercise science at the University of Essex. Her research expertise focuses on the relationship between the environment and human health, specifically 'green exercise'. She is interested in how nature-based interventions can be used to facilitate physical activity, enhance physiological and mental health and initiate behaviour change.

Jonathan Bell is policy manager at Northern Ireland Environment Link (a leading environmental NGO in Northern Ireland). Jonathan has extensive experience in environmental research, teaching, policy and practice. His research interests cover environmental policy and governance, sustainable development and rural planning.

Michael M. Bell is Vilas Distinguished Achievement Professor of Community and Environmental Sociology and director of the Center for Integrated Agricultural Systems at the University of Wisconsin-Madison. He is the author or editor of ten books, three of which have won national awards. His most recent book, *City of the Good: Nature, Religion, and the Ancient Search for What Is Right*, was published in 2018.

Mark Bevan is senior research fellow in the Centre for Housing Policy, University of York. He has maintained a keen interest in housing issues in rural areas throughout his career, including work on the private rented sector, the role of social housing, the housing and support needs of older people, the impact of second and holiday homes on rural communities and rural homelessness.

Bettina B. Bock is a professor at the rural sociology group at Wageningen University and the cultural geography group at Groningen University, both in the Netherlands. Most of her research regards questions around inclusive rural development, the quality of life in depopulating rural areas, and interrelation between urbanisation and rural marginalisation. She is the editor-in-chief of *Sociologia Ruralis*.

Andrew Butt is associate professor in sustainability and urban planning in the School of Global, Urban and Social Studies (GUSS) and the Centre for Urban Research (CUR) at RMIT, Australia. His current research and supervision is in the area of land use change and planning policy associated with regional Australia, food systems and peri-urban development.

Luís Camarero is professor at the National Distance Education University (Madrid, Spain). His research work is focused in the analysis of rural living conditions, demographic processes, gender inequalities, labour markets, and the role that spatial mobilities play in the transformation of social structure in rural areas.

Paul Cowie is a faculty research fellow working in the Centre for Rural Economy at Newcastle University. His research interests are in rural innovation and entrepreneurship particularly in the non-traditional economic sectors. He is also interested in the use of theatre as research and engagement method.

Julie Crawshaw is lecturer in material culture at Northumbria University. She has a background in visual arts and holds a PhD in planning from University of Manchester. Her research interests are classical and feminist pragmatism, the role of art and anthropology in rural and urban planning, and arts administration and cultural management.

Andreas Culora is a postgraduate researcher in the School of Geography and Environment at Loughborough University. His main research interests relate to housing in multiple occupation, housing markets and population change, socio-economic inequalities and housing, internal and international migration, and studentification.

Hemalata C. Dandekar is professor of city and regional planning, California Polytechnic State University, San Luis Obispo, Planning Commissioner City of San Luis Obispo, and a licensed architect, State of California. She is the editor of *The Planner's Use of Information* (Routledge). Her current research examines housing affordability and access.

Tom Daniels is a professor in the Department of City and Regional Planning at the University of Pennsylvania, Philadelphia, Pennsylvania, USA. His research interests include metropolitan growth management, farmland preservation, environmental planning, and rural and small town planning.

Christopher DeGeer is a graduate student at the University of Waterloo School of Planning, Waterloo, Canada.

Jennifer Dickie is a lecturer in environmental geography at the University of Stirling. Her main research interests relate to socio-environmental interactions of the energy landscape, including public perceptions and reaction to energy transitions, rural sustainability, energy justice and the social challenges associated with the development of the shale gas industry.

Petra L. Doan is professor of urban and regional planning at Florida State University in Tallahassee Florida. She has edited two volumes on planning and LGBTQ communities and written a number of articles which explore the critical intersections of planning and the LGBTQ community.

Michael Drescher is an associate professor in the University of Waterloo School of Planning, Waterloo, Canada, and director of the Heritage Resources Centre. His research interests include environmental conservation and natural resource use on private lands, climate change and land use effects on temperate and boreal forests, alternative knowledge sources and their applications in natural resource management and planning.

Mozart Fazito is lecturer at the Tourism Department in the Federal University of Rio Grande do Norte, Brazil. His research interests include regional planning, geography of development, socio-environmental conflicts, and tourism and leisure studies.

Robert Feick is an associate professor in the University of Waterloo School of Planning, Waterloo, Canada. His research focuses broadly on the application of spatial information technology to assist decision making and public participation in land management and planning.

Linda Fox-Rogers is a lecturer in environmental planning in the School of Natural & Built Environment at Queen's University Belfast. Her research is guided by her interest in understanding how issues of power and politics permeate the planning system.

Kathryn I. Frank is an associate professor in the Department of Urban and Regional Planning at the University of Florida, USA. Her research, teaching, and service integrate the fields of rural planning, environmental planning, governance, and community development.

Nick Gallent is professor of housing and planning at University College London and was, for eight years, head of the Bartlett School of Planning. His research is concerned mainly with planning for housing and with rural communities' engagement with planning and development processes. He is the author or editor of numerous books on these subjects. The most recent include *Introduction to Rural Planning* (with Iqbal Hamiduddin, Meri Juntti, Sue Kidd and Dave Shaw; 2nd edition 2015), *Rural Planning and Development* (edited with Mark Scott; 2017) and *New Money in Rural Areas* (with Iqbal Hamiduddin, Meri Juntti, Nicola Livingstone and Phoebe Stirling; 2019).

Menelaos Gkartzios is senior lecturer in planning and development at Newcastle University's Centre for Rural Economy. His research has focused on mobilities and social change, rural housing, the relationship between art and local development, and international comparative analysis. He has published extensively in these areas in, *inter alia*, the *Journal of Rural Studies*, *Regional Studies*, *Population, Space & Place*, *Geoforum* and *Land Use Policy*. He sits on the editorial board of *Sociologia Ruralis* and has been a Visiting Associate Professor at the University of Tokyo where he taught a rural planning course and conducted fieldwork on art festivals in regional towns in Japan.

Valerie Gladwell is a senior lecturer in sports and exercise science based in the School of Sport, Rehabilitation and Exercise Sciences at the University of Essex. Her research interests include physical activity and health, including 'green exercise', combining physical activity in a natural environment.

Matthew Gorton is professor of marketing at Newcastle University Business School. His main research interests relate to small businesses and rural development, food marketing and agri-food supply chains. He is the co-ordinator of the EU Horizon 2002 project Strength2Food.

Lise Herslund is associate professor at the Institute for Geosciences and Natural Resource Management at the University of Copenhagen. Her research focuses on capacities for collective action and community-led social innovation within a broad field of themes including climate change adaptation and regional and rural development.

Peter Hetherington is past chair of the Town and Country Planning Association, Britain's oldest housing and planning charity, and former regional affairs editor of *The Guardian*. He is the author of several books, including *Whose Land is Our Land? The Use and Abuse of Britain's Forgotten Acres* (2015) and a RIBA 'Shaping the City' book on Newcastle–Gateshead (2010). He lives in Northumberland.

Michael Hibbard is a professor emeritus in the School of Planning, Public Policy and Management at the University of Oregon. His research focuses on sustainable regional development and the social impacts of economic change on small towns and rural regions.

George C. Homsy is an assistant professor of public administration at Binghamton University and directs the university's sustainable communities programme. His research focuses on the factors that shape local sustainability policies as well as land use and economic development planning and citizen participation.

Julia Horsley is a research fellow in geography at the University of Western Australia, and has undertaken research on a range of themes related to regional development. Her most recent research is focused on the relationships between government policy and private firms in regional development.

Daniel P. Hubbard is a professional planner with the Florida Department of Economic Opportunity. He is a pre-doctoral student in urban and regional planning at Florida State University and plans to write a dissertation on planning and queer spaces.

Gertrud Jørgensen is professor of spatial planning at the University of Copenhagen. She works on a broad variety of planning issues, including urban transformation, sustainable urban development, urban liveability, peri-urban development and strategic planning in rural areas. Her research interest lies in the relation between planning tools, processes and planning outcomes in terms of spatial quality, sustainability and liveable environments.

Meri Juntti is a senior lecturer at Middlesex University, Department of Law and Politics. Her research focuses on environmental governance and sustainable rural development, in particular, on the role of different knowledges and representations of the environment in rural environmental policy and development initiatives and in the delivery of urban greenspace.

Karita Kan is assistant professor of social policy and administration in the Department of Applied Social Sciences, Hong Kong Polytechnic University. Her research interests include the political economy of rural transformation in China, Chinese land and property rights reform, contentious politics related to land disputes, as well as state building and social governance at the rural–urban grassroots.

Sue Kilpatrick is professor of education in the Faculty of Education, University of Tasmania, Australia. Until 2015 she was pro vice-chancellor (students), University of Tasmania where her responsibilities included access and outreach programs. She was formerly pro vice-chancellor (rural and regional) at Deakin University, Australia, and director of the University Department of Rural Health, Tasmania.

Chloe Kinton is a research associate, Department of Geography and Environment, Loughborough University. Her research interests include student geographies, in particular processes of (de-)studentification; geographies of social and population change; rural/urban change, in particular processes of gentrification and urban abandonment and decline; and geographies of higher education.

Michiel Köhne is assistant professor in anthropology of law and development and works at the Sociology of Development and Change group of Wageningen University. His areas of expertise are: land conflict, resistance, extraction, oil palm, energy practices, and anthropology of law, involving empirical studies in Bolivia, Indonesia, Australia and The Netherlands.

Tamara Krawchenko is a policy analyst in the Department of Regional Development Policy in the Organisation for Economic Co-operation and Development (OECD). Her research has included studies of metropolitan governance, public budgeting, inter-generational equity and more recently, the governance of land use. This includes the OECD's recent work on comparative land use governance across OECD countries.

Mark B. Lapping is the distinguished university professor emeritus at the University of Southern Maine in Portland, Maine. He is the author/co-author or editor of nine volumes, including the American Planning Association's *Small Town Planning Handbook*.

Mick Lennon is lecturer in planning and environmental policy in the School of Architecture, Planning and Environmental Policy, University College Dublin, having previously worked as a lecturer in spatial planning in Cardiff University. His research interests centre on interpretative analysis, particularly in how the meaning-making process influences approaches to environmental policy and planning.

Philip Lowe is emeritus professor and founding director of the Centre for Rural Economy at Newcastle University. He has been a leading figure in the development of interdisciplinary rural studies in the UK and internationally.

Katherine MacTavish is an associate professor at Oregon State University. She earned her PhD in human and community development from the University of Illinois–Champaign/Urbana. She is the co-author of *Singlewide: Chasing the American Dream in a Rural Trailer Park*.

Chris McDonald is a policy analyst in the Regional Development and Tourism Division, OECD. Chris has worked across a number of roles including with the Victorian State Government, Monash University, and the OECD. Chris's undergraduate qualification is in urban planning and he has a PhD which focused on the governance of regional development from Monash University, Australia.

Amanda McMillan Lequieu is a graduate student in sociology at the University of Wisconsin-Madison. Her research interrogates how place-based, working-class communities adapt to globalising economies and changing environments over time – from farmers and consumers, to urban steelworkers and rural iron miners.

Ian Mell is a senior lecturer in environmental and landscape planning at the University of Manchester. His main area of research examines the development, financing and functionality of green infrastructure in the UK and internationally. He is the author of *Global Green Infrastructure* (2016) and *Green Belts: Past, Present and Future* (with John Sturzaker; 2016).

Tom Moore is a lecturer in planning at the University of Liverpool. His main research interests are in housing policy and practice and community planning. He is an Associate Editor of the *International Journal of Housing Policy*.

Michael Murray retired in 2016 as reader in spatial planning at Queen's University Belfast, where he continues to teach on field trips and maintain his research interests related to international planning, rural development and planning law.

Lucy Natarajan is a research associate at the Bartlett School of Planning, University College London. Her research covers a wide range of policy and decision-making subjects within the field planning, and is focused on the role of relationships between the public and government. Her most recent work focuses on the role of 'publics and evidence' in decision-making on renewable energy infrastructure.

Jesús Oliva is a tenured lecturer of sociology at the Public University of Navarra (Pamplona, Spain). His lines of research focus on rural change, mobilities, labour processes, urban mobility, rural sustainability and regional planning.

Apostolos G. Papadopoulos is professor of rural sociology and geography in Harokopio University, Athens. His main research interests include rural transformation, class and social stratification, social and spatial mobility, rural migration and local food. He is member of the advisory board of CIHEAM 2017–2021.

Chris Paris, FAcSS, is emeritus professor of housing studies, Ulster University. He has held senior academic posts in the UK and Australia, most recently as a research fellow in the Centre for Housing, Urban & Regional Planning at Adelaide University. He has published over 30 books, monographs and research reports and over 100 journal articles on urban studies, housing, planning and demography, including *Affluence, Mobility and Second Home Ownership* (2011).

Arthur Parkinson is a conservation architect and lecturer in planning and urban design at University College Dublin. His research lies at the interface between these areas, examining the politics of built heritage, conservation planning, and design in historic contexts, and in particular the competing discourses that frame decision-making and conflicts. He is co-author of the forthcoming book *Urban Heritage and Contested Planning: Making Use of Ireland's Built Past* (with Mark Scott and Declan Redmond).

John Pendlebury is a town planner and urban conservationist. He is professor of urban conservation in Newcastle University and undertakes research on heritage, conservation and planning. Principal publications include *Conservation in the Age of Consensus* (2009) as well as the edited collections *Valuing Historic Environments* (with Lisanne Gibson; 2009) and *Alternative Visions of Post-War Reconstruction: Creating the Modern Townscape* (with Erdem Erten and Peter Larkham; 2015).

Martin Phillips is professor of human geography in the School of Geography, University of Leicester. He has research interests in the areas of rural social and cultural geography, historical geography, society/environment relations, and philosophy in geography. He is currently part of an international research team (iRGENT), funded by ESRC, NSF and ANR, studying international rural gentrification in UK, USA and France.

Jeremy Phillipson is professor of rural development at the Centre for Rural Economy at Newcastle University. He has research interests on the development needs of rural economies and fishing communities, processes of expertise exchange within rural land management, and on the integration of social and natural sciences in resource management.

Jules Pretty OBE is is deputy vice-chancellor and professor at the University of Essex, School of Biological Sciences. He was head of the Department of Biological Sciences from 2004–2008, pro-vice-chancellor (science and engineering) from 2010–2012, then deputy vice-chancellor from 2012. His research interests include connections to nature and place, agricultural sustainability/sustainable intensification, green minds and health benefits of nature, and social capital and natural resources. He received an OBE in 2006 for services to sustainable agriculture.

Elisabet Dueholm Rasch is assistant professor working in the Sociology of Development and Change group of Wageningen University. Her research interests focus on mining, activism, citizenship, indigenous peoples, development, and ethnography.

Dalton Richardson is a Masters student in the Department of Agricultural Economics and Rural Sociology in the College of Agriculture at Auburn University. His research is concerned with understanding the intersections between the environment, the law, and industrial agriculture.

Guy M. Robinson is a human geographer who works on rural development and environmental management. He is the editor of the journal *Land Use Policy,* departmental associate in the Department of Land Economy, University of Cambridge, adjunct professor at the University of Adelaide, and guest professor in the Chinese Academy of Sciences, Beijing.

Mike Rogerson is a researcher and associate lecturer in sports and exercise science based in the School of Sport, Rehabilitation and Exercise Sciences at the University of Essex. His primary research interests are in the area of exercise and environment, including green exercise; physical activity in dementia cohorts; health by stealth in older and rural populations, and virtual reality exercise.

Paula Russell is a lecturer and director of graduate planning programmes in the School of Architecture, Planning and Environmental Policy, University College Dublin. Her main area of research relates to the role of civil society in the planning process, looking at issues of engagement and influence. She is a member of the board of the European Urban Research Association (EURA) and an associate editor of the journal *Urban Research and Practice*.

Madhu Satsangi has worked at the University of Glasgow, Heriot-Watt University and the University of Stirling. His main research interests relate to rural housing and planning and land reform. With Nick Gallent and Mark Bevan, he wrote *The Rural Housing Question* (2010).

Alister Scott is professor of environmental geography and planning in University of Northumbria. He is also a Natural Environment Research fellow in Green Infrastructure. His main research interests relate to spatial planning and governance, urban greenspace and ecosystem science.

Mark Scott is professor of planning in University College Dublin. His research is primarily focused on rural planning and development, environment and sustainability, and planning and heritage. He has published widely in these issues including over 60 papers in peer review journals and 20 book chapters. He recently edited *Rural Planning and Development* (with Nick Gallent; 2017). In addition to undertaking funded research projects, he has acted as an advisor or consultant on spatial planning to the OECD, the Oireachtas (Irish Parliament), the Northern Ireland Assembly and various local and regional planning authorities. He is an editor of the leading international journal, *Planning Theory and Practice*, and also is on the editorial board of *Town Planning Review*.

Robert Shipley was an associate professor in the University of Waterloo School of Planning, Waterloo, Canada, until his recent retirement. His research interests have focused on heritage conservation planning and the economics of heritage, strategic planning and the use of visioning, and decision-making processes and public participation

Sally Shortall is the Duke of Northumberland Chair of Rural Economy at Newcastle University. Her research interests are the role of women on farms and in rural development, social changes in farming practice and the links between evidence and rural development policy. She is currently president of the European Society for Rural Sociology, and first vice-president of the International Rural Sociology Association.

Darren P. Smith is professor of geography in the School of Geography and Environment, Loughborough University. He is a population social geographer with research interests in contemporary migration and population change, and how new social relations and conflicts are created. In a rural context, his work has investigated processes of rural gentrification and is currently part of an international research team (iRGENT), funded by ESRC, NSF and ANR, that is studying international rural gentrification in UK, USA and France. He is co-editor of the journal *Population, Space and Place*, and also on the editorial board of the *Journal of Rural Studies*.

Max Spoor is professor emeritus at the International Institute of Social Studies, The Hague, part of Erasmus University Rotterdam, and research professor at IBEI, University Pompeu Fabra in Barcelona. His research interests are rural development and environment, and the issue of water stress, particularly focused on Central Asia and China. He published in, among others, *Land Use Policy*, *China Economic Review*, and *Journal of Peasant Studies*.

Aileen Stockdale is professor in environmental planning at Queen's University Belfast, Fellow of the Academy of Social Sciences, and co-editor of the journal *Population, Space and Place*. Her research interests cover all aspects of rural policy and planning but specifically rural society and economy.

John Sturzaker is a senior lecturer in Civic Design and Planning in the Department of Geography and Planning, Liverpool University. His research interests focus on how cities grow in more 'sustainable' ways, with an understanding of sustainability that considers all those involved, including those with less power and influence in the planning and development systems.

Laura E. Taylor studies exurbia and ideologies of nature in planning. She is associate professor, Faculty of Environmental Studies, York University, Toronto, Ontario Greenbelt Council member, and co-editor of *A Comparative Political Ecology of Exurbia* (2016) and *Landscape and the Ideology of Nature* (2013).

Anne Tietjen is associate professor of urban design and planning at the University of Copenhagen. She specialises in the transformation, preservation and development of existing built environments and landscapes. Her work draws on relational design and planning theory, specifically based on new materialism and actor-network theory. Substantially her work focuses on strategic rural planning, urban/rural public space, heritage in planning and research-by-design.

Pattanapong Tiwasing is a research associate at Newcastle University's Centre for Rural Economy. He holds a PhD on agricultural economics. His research interests are in the areas of rural development, nutrition and food security, and small business economics.

John Tomaney is professor of urban and regional planning in the Bartlett School of Planning, University College London. Previously he was Henry Daysh Professor of Regional Development and director of the Centre for Urban and Regional Development Studies (CURDS), Newcastle University. He has published over 100 books and articles on questions of local and regional development including *Local and Regional Development* (2nd edition 2017) and *Handbook of Local and Regional Development* (2011) co-authored with Andy Pike and Andrés Rodríguez-Pose.

Matthew Tonts is professor of geography at the University of Western Australia (UWA), and has longstanding interests in regional development and policy. Matthew is presently executive dean of the Faculty of Arts, Business, Law and Education (UWA).

Oane Visser is associate professor at the International Institute of Social Studies, The Hague, part of Erasmus University Rotterdam, and was visiting researcher at the University of Oxford, University of Toronto and Cornell University. He studies farmland investment, financialisation, digital farming, smallholder farming and alternative food networks. He edited special issues in *Journal of Agrarian Change* and *Agriculture and Human Values*.

Mildred E. Warner is a professor of city and regional planning at Cornell University. Her research focuses on local government finance, service delivery, economic development and environmental and social service policies.

Jessica Woodroffe is currently the academic course coordinator of the University of Tasmania's Connections Program (UCP), High Achiever Program (HAP) and oversees the University Students in Schools (USIS) programme, schools engagement grants, college masterclasses and leads research and evaluation efforts for the area. She is currently collaborating on a number of cross-disciplinary research and evaluation projects in the areas of community learning, educational program evaluation, social inclusion and health and wellbeing.

Michael Woods is professor of human geography at Aberystwyth University. He has worked extensively on change in the contemporary countryside, rural policy and protests, and globalisation in rural areas. He is author of *Rural Geography* (2005) and *Rural* (2010), and editor of the *Journal of Rural Studies*.

Acknowledgements

We have received a great deal of support during the production of this Companion. The team at Routledge, Andrew Mould and Egle Zigaite, gave encouragement to this project in its early stages and have been our guides during the production process. We would also like to thank Sandra Mather at Liverpool University, who redrew the excellent illustrations, maps and figures that now adorn many of the chapters. Front covers are often an afterthought, but not so with this Companion. Gemma Burditt, artist in residence at Berwick Visual Arts and Newcastle University's Centre for Rural Economy, produced the fantastic design on the front board – with only the lightest of steers from the editors. We are also hugely indebted to our chapter contributors, all of whom delivered well-polished, engaging chapters ahead of schedule – making our task, as editors, far easier than it might otherwise have been. The penultimate acknowledgement goes to the small army of rural researchers with whom we've engaged over the years. Conversations at conferences and seminars have provided much of the inspiration needed to produce this Companion and the thoughts shared have shaped much of our thinking on the issues covered.

The final thank you must go to family and friends. Mark would particularly like to thank Karen and the kids, Ada and Lucas. Nick is, as always, grateful for the love and support of Manuela, Marta and Elena. And Menelaos would like to thank Mark Dunningham, as well as his family of colleagues and friends at both Newcastle University and University of Tokyo, for their support while working on this book.

New horizons in rural planning

Mark Scott, Nick Gallent and Menelaos Gkartzios

Introduction

While rural concerns were central to the emergence of planning in the last century, during the second half of the twentieth century planning theory and practice have been dominated by urban challenges, with an increasingly unimaginative rural planning regime driven largely by a dominant agricultural agenda. This productivist agenda continues to relegate potentially progressive rural planning debates behind farmland preservation, amenity protection, and a minimal approach to socio-economic issues (Lapping, 2006). However, the continued impacts of urbanisation, demographic changes, the decline of the traditional rural economic base, the emergence of multi-functional rural landscapes, and deeply contested new demands for rural space suggests an urgent need to reinvent rural planning for the twenty-first century.

While we increasingly live in an 'urban age' – characterised, according to Brenner and Schmid (2014), by an uneven yet planetary process of urbanisation – 'the rural' still matters, and through this Companion we call for planning theory and practice to engage proactively with rural regions and localities and not simply to cast these places as residual or as scenic backdrops to growing urbanisation. A quarter of the population across Organisation for Economic Co-operation and Development (OECD) countries can be classified as living in predominantly rural regions, defined in terms of low population density and the absence of major urban centres (OECD, 2018; see also OECD, 2011 for methodology). A further 26% of the population across the OECD live in intermediate regions, characterised largely as near-urban rural areas or the rural–urban fringe – these regions are often undergoing rapid and contested social, economic and environmental changes in the face of urbanisation processes and changing demands for land resources and rural amenities. Moreover, the OECD (2018) highlight that 83% of land across OECD countries are within predominantly rural regions. Managing land-use change and mediating between competing interests in the use of land is central to the rural planning challenge, particularly given threats to natural resources and importance of balancing global challenges with local demands and needs.

The effective management of rural resources and land use is central to key global challenges that require critical advancement in order to reposition rural planning as a key framework to address, *inter alia*, climate change mitigation (e.g. transitions to post carbon landscapes, carbon sinks), climate change adaptation (e.g. ecosystem services for managing flood risks), biodiversity loss (e.g. site specific conservation to ecological networks), energy security (e.g. roll-out of renewable energy infrastructure), global food security, contested resource extraction (from minerals to 'fracking' debates), and the siting of key strategic infrastructure (e.g. interurban or international transportation networks). These global concerns overlap and interact with local and more traditional challenges surrounding housing supply, sustainable and inclusive communities, protecting valued landscapes and rural heritage, or addressing rural decline. In this context, the rural should also be understood as a social space (a living and working countryside) and not simply as a resource base.

This Companion sets out to provide a comprehensive statement and reference point for rural planning in an international context. The Companion aims to provide critical reviews of the current state-of-the-art of conceptual, theoretical and empirical knowledge and understanding of rural planning, addressing the multi-dimensional nature of the field. Moreover, the Companion aims to push the boundaries of rural planning in three ways. *First*, the Companion gives coverage to emerging topics in rural planning so as to reposition rural planning within the context of global challenges in the twenty-first century. *Second*, the Companion addresses topics that currently represent significant gaps in the rural planning literature – such as a queer perspective of rural planning, rural planning and environmental justice, artistic methodologies in rural planning research, and so-called green and land grabbing – and by addressing these topics, the Companion aims to set new agendas for rural planning research. *Third*, the Companion includes emerging geographical frontiers of rural planning. Whereas rural planning research is dominated UK and US literature, the Companion seeks to foster an international dialogue with contributors from throughout Europe, Canada, Australasia, and importantly from emerging and transition economies representing a unique feature of the Companion – primarily BRIC economies and former communist societies. Therefore, the contributing authors have been tasked with reviewing the international 'state-of-the-art' in relation to their topic and to incorporate regional/local case study material to illustrate these wider trends when appropriate. A key position for us in editing this book was to avoid marginalising specific contexts by addressing them collectively as 'international case studies', while presenting others as the norm. We recognise that rural planning knowledge and research carries its own asymmetries; the literature is very much shaped by mature and western economies, but we have aimed to include emerging and transition economy cases and commentary as much as possible. Through this approach therefore, the Companion includes empirical and case study material from the US, Canada, UK, Norway, Ireland, Greece, France, Poland, Sweden, Finland, Netherlands, Denmark, Spain, Japan, Australia, Brazil, Russia, India, China, South Africa and Uganda.

Understanding rurality and rural change

What is meant by the word 'rural'? How is rural space differentiated globally, and is there an observable, universal taxonomy of rural types? The rural studies literature has grappled with these questions at least since the 1960s. Within the Companion, Gallent and Gkartzios (Chapter 2) discuss in more detail the various conceptions of 'rural' and how these relate to planning debates. Each new turn in theoretical and methodological approaches raises different sets of questions for rural planning theory and practice.

Gallent and Gkartzios chart the constant development in both definitions and classifications of 'the rural' over recent decades. This includes attempts to define the 'rural' as functional 'positivist' spaces based on measurable and observable data (e.g. population density, settlement, landcover, employment and labour market characteristics and so on; see, for example, Cloke, 1977; Cloke and Edwards, 1986). While this approach remains popular within policy debates, by the mid 1990s, Chapter 2 outlines how positivist accounts of rurality had increasingly been displaced by cultural approaches to understanding the rural, and by political economy approaches seeking to describe the shift towards the so-called post-productivist countryside. The 'cultural turn' within rural studies shifted the methodological focus to include feminist approaches, ethnographies and discourse analyses, which attempt to capture rurality as a phenomenon which is socially and culturally constructed, and therefore contested (Halfacree, 1993; Cloke, 1997). A rich vein of research emerged illustrating the significance of discourse, social representations and cultural symbolism in constructing the rural, stressing multiple and redefinable social spaces, rather than set, rural geographical spaces (see, for example, Jones, 1995; Halfacree, 1995; Pratt, 1996; Cloke, 1997; Phillips, 1998; van Dam et al., 2002). This provided a new understanding of rural conflicts as researchers increasingly highlighted that the very notion of rurality had become deeply contested, underpinned by latent social conflicts – termed by Woods (1998, 2003) as the 'politics of the rural'.

While there has been much focus in rural studies in conceptualising the rural, there has also been much debate concerning (uneven) processes of rural change or restructuring and its impact on redefining the role of rural areas and in underpinning different development trajectories across rural space. In this context, analysis of the shifting role and function of agriculture has provided a key departure point for examining rural transformations. For example, an extensive body of literature emerged in the 1990s (Marsden et al., 1993; Lowe et al., 1993; Munton, 1995; Marsden, 1998; Murdoch et al., 2003) charting the demise of productivist agricultural models, opening opportunities for a substantial growth in demand for new uses for rural space (e.g. amenity, recreation, conservation, residential) and creating new conditions for actors to pursue their demands both in the market place and in the political system. In contrast to the cultural turn in rural studies, this work placed more emphasis on the politics of place and power struggles to examine rural change, with two key inter-related themes emerging from this research: (1) that the dominant concerns of productivist agriculture are declining in the face of competing consumption interests; and (2) new patterns of diversity and differentiation are emerging across rural space in relation to governance and regulative processes – the so-called 'differentiated countryside'. The key argument underpinning this approach to understanding rural change is outlined by Lowe et al.:

> It is our argument that fundamental changes in the productivist regime have, firstly 'opened the door' for various (non agricultural) economic interests to assert claims on rural land and, secondly, created a vacuum in rural policy making which competing policy 'communities' . . . have sought to fill. This has resulted in uneven processes of economic development and uneven processes of regulation – both between sectors *and* places.
>
> *(Lowe et al., 1993: 218)*

Since the 1990s, key themes within the rural restructuring literature have included the shifting nature of agriculture (Wilson, 2008) and the related and contested emergence of new uses for rural space and competing societal demands for the consumption of 'rurality' (e.g. Woods, 2005, 2006). A significant body of literature has also examined the emergence of a new set of rural social geographies associated with diverse processes of rural in-migration. These include

3

extensive studies of counterurbanisation (e.g. Halfacree, 2001; Mitchell, 2004), rural gentrification (e.g. Smith and Phillips, 2001; Phillips, 2004, 2010; Phillips and Smith, 2018), international migration (e.g. Halfacree, 2011), international second home-ownership (e.g. Wong and Musa, 2014), and retiring to the countryside (e.g. Stockdale, 2011). Increasing attention has also been given to the influence of external actors in shaping rural localities including capital, consumers and regulating bodies from processes of economic globalisation (e.g. Brunori and Rossi, 2007) and the increased significance of neoliberal ideas, policies and practices to the unfolding of social and spatial life in rural areas (e.g. Dibden et al., 2009; Shucksmith and Rønningen, 2011). In essence, these studies have charted a series of 'radical breaks and ruptures within rural societies' (Smith, 2007: 275) leading to a fundamental reconfiguration of rural housing and land markets, local economies, rural resource use, spatial mobilities, and rural politics (Marsden, 2009; Bell and Osti, 2010).

While this literature has provided rich accounts of the changing rural condition, other researchers have questioned the use of *rural* to explain changing spatial dynamics. Hoggart (1988), for example, questions the notion of rural restructuring as being blind to the causal processes that transcend any rural–urban divide. This led Hoggart (1990) to argue for abandoning rural as an analytical construct – which he suggests is too often deployed as a term of convenience but without justification as a causal category. Hoggart contends that causal forces in rural areas are not distinctive (e.g. from some urban areas), nor are they uniform within or across rural areas. Hoggart's perspective, moreover, challenges those interested in rural planning to consider if there is something distinctive about planning in rural contexts – is rural planning fundamentally different to urban planning? Is 'rural' an explanatory factor in understanding planning outcomes?

In a similar vein to Hoggart, recent scholarship has also questioned the usefulness and universality of 'rural' as a ubiquitous concept. Brenner and Schmid (2014), for example, unpack the usefulness of 'urban' and 'rural' as opposing analytical concepts, suggesting that there is limited analytical value in dividing human settlement into urban and rural containers in light of their structuration by wider political-economic processes. These themes are also addressed (but from a rural perspective) by Lichter and Brown (2011) who examine rural restructuring through exploring the rural–urban relationship and the influence of urbanism and globalisation on rural life. They argue that there is a greater interpenetration of rural and urban life and a blurring of rural–urban spatial and social boundaries – and that the pace of change is accelerating. They conclude that rural–urban relations are increasingly symmetrical rather than asymmetrical and, therefore, it is becoming more difficult to discuss rural and urban change without acknowledging the other. However, despite a dominant discourse of global urbanisation, Woods's ongoing research into the *global countryside* (see Chapter 53) re-emphasises that rural areas retain a distinctiveness, both functionally and politically, and therefore it is vital that the rural agenda is not overlooked, or indeed merged with urban planning agendas (Gallent et al., 2015). More recently, Gkartzios and Remoundou (2018) challenged the role of monolingual academia in constructing the rural, with diverse and nuanced meanings of the rural often missed from international research when they are disseminated in the language of English.

Overall, within this Companion we take a pragmatic and inclusive approach to rurality, allowing each author and chapter to take forward their own approach or interpretation of 'the rural'. This approach has a number of advantages. First, any attempt to define the rural and present an international and workable definition would merit an entire Companion on its own – thus, we recognise the existence of a diverse range of countrysides and rural experiences. Second, encouraging the various contributors to highlight their own approaches provides a much richer and more diverse account of rural studies, which also better reflects the wide range of approaches to understanding how the rural is conceptualised within planning theory and practice. Third, any understanding of 'the rural' is inherently context specific and sensitive, and varies considerably across international borders, linguistic traditions and cultural imaginations. Given these factors, we hope that the collection of chapters in this Companion reflects the wide diversity of rural experiences, including near urban and the rural–urban fringe, from the rural within mature urbanised countries to the rural condition within rapidly urbanising places, and the experiences of more remote, sparsely populated or depopulating rural regions. Within these different contexts, the various chapters provide a detailed account of how rural planners have conceptualised and dealt with the wide range of planning issues they face.

Understanding rural planning

In taking a broad approach to understanding rurality, the Companion also takes a pragmatic approach to defining 'what is' rural planning. Definitions of planning have changed over time and vary considerably across the world and inevitably reflect specific governance and institutional traditions. On a basic level, Healey (2010) refers to planning activity as being about 'making better places', which applies equally to rural as urban localities. Early accounts of rural planning tended to focus on the impacts of physical change, often resulting from the perceived threat of urbanisation on important landscapes and agricultural lands (see, for example, Abercrombie, 1926), with change managed through state-centred 'command-and-control'. In this context, the rural planning literature traditionally focused on the notion of planning as an independent, neutral arbitration for competing demands for resources in the countryside (Cloke and Little, 1997). This perspective has, however, been largely dismissed since the late 1970s as rural planning processes have increasingly been debated within the wider context of rural 'power' (see for example, Newby, 1986; Buller and Hoggart, 1986; Cloke and Little, 1997). These analyses not only revealed power struggles and new politics, but also emphasised the relationship between different levels of state activity together with the interaction of the public, voluntary and private spheres.

Nowadays, with the retreat of top-down, command-and-control intervention, it is more common to think of planning in terms of 'steering' or as a coordinator of place-making and place-shaping, implemented through more flexible planning arrangements. For the purposes of this book, planning is interpreted broadly as policy, place-making, regulation and design interventions that shape rural places. This invariably involves multiple actors (e.g. central and local state, private sector, civil society), is multi-level operating at and connecting with different spatial scales, and involves coordinating activities across economic, environmental and social dimensions of rurality. In this context, rural planning goes well beyond land-use planning, comprising a matrix of place-making strategies, spatial coordination and mediation, land-use regulation and development control, landscape management and design, protecting, restoring and enhancing ecosystem services and natural capital, and community action. Rural planning, therefore, takes various forms depending on context and the challenges that confront it – global and local. These issues are explored in more depth in Chapter 2 (Gallent and Gkartzios), which outlines the scope of rural planning, and also Chapter 3 (Lapping and Scott) and Chapter 4 (Dandekar), which chart the evolution of rural planning and policy in the Global North and within India as an illustrative example of an emerging economy, respectively.

Is rural planning distinct from urban planning? In terms of state-led planning and land-use regulation processes, there may be limited differences between planning in urban and rural communities. For example, strategic planning or managing new development may, in process terms, be similar across urban and rural space. However, the substantive focus and context is very different, often leading to very diverse and differentiated planning outcomes across urban and rural places. For example, rural planning with its traditional focus on preservation is disconnected from wider rural development debates and economic development in contrast to the way that urban planning and regeneration appear more closely aligned.

As highlighted above, a key departure point for much analysis of the rural condition is the shifting nature of agriculture, and how non-agricultural actors and new, often contested, uses for rural space have become increasingly central to rural futures. The emergence of potential new functions for rural places (summarised in Table 1.1) is central to understanding the future rural planning challenge and planning outcomes as competing actors exert power and influence. We summarise some of the potential new rural functions.

This Companion is concerned with policy, planning and design interventions to enhance the resilience and sustainability of differentiated rural places (from remoter, peripheral rural areas to near-urban metropolitan fringes). The Companion offers an analytical overview of the forces of rural restructuring and uneven development and the resultant planning response and policy outcomes. It provides detailed coverage of key topics using original contributions that either offer entry points to those topics, divergent perspectives, theoretical frameworks or in-depth case studies drawing on empirical research. The collection is intended to be a resource – brought together in one place – for students and researchers attempting to make sense of the field of rural planning.

The book is 'pan-national' in its coverage. We choose the phrase 'pan-national' rather than international as the field of rural planning is primarily a Global North project, with rural development being the primary rural policy goal in the Global South. However, in an era of globalisation and urbanisation, rural planning concerns are crossing borders into new emerging contexts. Therefore, while many of the contributions explore rural planning in relation to mature, advanced economies in the Global North, the Companion also unpacks rural planning challenges in the context of rapidly urbanising and transitioning economies, exploring new frontiers for rural planning research, in BRIC and post-communist transition economies.

Table 1.1 Emerging functions of the rural.

The rural as . . .	
A playground . . .	. . . as a place of consumption for new second home-owners, tourists, food consumers
A dumping ground . . .	. . . for controversial unwanted land uses (e.g. waste incinerators, prisons)
A post carbon landscape . . .	. . . a site for the (often contested) deployment of renewable energy – wind farms, solar farms, biomass
A resource sink . . .	. . . a site for extracting resources, often with short term 'boomtown' effects and limited long term re-investment in rural futures
A cultural heritage repository . . .	. . . tied into perceptions of rural 'authenticity' and nostalgia, cultural landscapes and the commodification of place
A food basket . . .	. . . as agriculture remains as a dominant land use and function of rural places
A provider of ecosystem services . . .	. . . whereby ecosystems provide functions and services essential to human wellbeing, from recreation to flood alleviation or carbon storage
A social space . . .	. . . at the scale of everyday life, where people live and interact, often characterised by strong place attachment

Source: Gallent and Scott (2017)

The Companion contains a number of distinctive features. First, the Companion comprises new contributions from the leading and emerging voices in the field that reflect the diversity and variety of conceptual, theoretical and practical approaches to thinking about and 'doing' rural planning, with international applicability. Second, we place an emphasis on exploring the relationship between the theory *and* practice of rural planning through advancing our conceptual understanding of rural planning challenges while critically reviewing policy, planning and design interventions. The Companion develops the theoretical and methodological foundations of rural planning theory and practice as well as the fields of spatial planning and rural studies more generally. Third, and crucially, the Companion provides a critical (re)framing of rural planning within the context of key global challenges, including climate change, biodiversity loss, energy security, food security, resource extraction, and social justice. Given the centrality of land use to these issues, we argue that spatial planning should focus beyond the urban fix to understand the role of planning and governance in shaping these wider resource debates. Fourth, as highlighted above, the Companion sets out to connect global debates to local planning challenges in diverse rural contexts with contributions from Europe, Australasia, North America and BRIC economies – these latter insights explore the challenge of managing rural places in the context of rapidly urbanised and globalised economies and transitioning societies. Through pushing the conceptual, substantive and geographical boundaries of rural planning, the Companion seeks to reinvent rural planning for future generations of students, researchers and practitioners.

Organisation of the Companion

The Companion itself is divided into nine parts, each with a more detailed introduction from the editors setting out the focus, scope and thematic areas covered. *Part I* sets out the key *concepts and foundations* of rural planning with the aim of charting the terrain of rural planning and providing a clear conceptual foundation for further developing the field. This begins with three key foundation chapters, with the first chapter outlining the (often contested) meaning of rurality and how this relates to the scope of contemporary rural planning theory and practice, followed by two chapters exploring the evolution of the rural planning field in an international context. The following chapters provide an understanding of key concepts that underpin analysis of the rural condition and planning outcomes in rural localities, namely sustainability and resilience; production, consumption and protection; land and property; and environmental justice. Together these chapters provide an understanding of the competing interests shaping rural places and a basis for reappraising of the productive, consumption and environmental forces steering rural planning debates and outcomes.

Part II contextualises rural planning debates through exploring *the state and rural governance*, and examines the shifting nature of rural governance alongside an understanding of the range of policy instruments employed to manage rural space. This begins with examining rural planning within its multi-level governance context and the balance between strategic, top-down planning and bottom-up initiatives. Rural planning is then considered against the back-cloth of the neoliberal countryside examining the entrenchment of roll-out and roll-back neoliberalism as the dominant political project shaping rural places. The next two chapters examine two contrasting instruments for shaping rural planning processes and outcomes. First, is a chapter dealing with the use of market-based instruments within rural planning, which draws on the US experience of land preservation. Second, is a chapter on the role of direct community ownership of rural assets, specifically examining the role of land trusts to address affordable housing supply in rural localities. While many of the chapters throughout the Companion highlight and explore the importance of community engagement, the final chapter in Part II examines the 'dark side' of community and the translation of perceived local interests into clientelist and corrupt politics and planning outcomes.

Part III examines *planning for the rural economy*, and aims to address the 'disconnect' between traditional rural and land-use planning concerns and the rural development literature whilst exploring the role of rural spatial planning in underpinning sustainable rural livelihoods. Part III begins with a chapter reviewing rural development approaches, providing an understanding of the evolution of exogenous to endogenous development models, before highlighting a hybrid approach of neo-endogenous rural development, combining local action with a recognition of the importance of extra-local agents and networks. The next chapter positions 'the rural' within a regional development context, questioning the appropriate scale for economic development policy and how rural localities connect with urban centres. The next chapter then addresses the challenge and legacy of the global financial crisis of 2008 within rural localities, including the politics of austerity and its impact on rural planning and development pathways. Following this are three thematic chapters focused on various dimensions of the new rural economy including: innovation and small business development; the creative class and entrepreneurship; and payments for ecosystem services. Part III concludes with a reflection on the role of spatial planning in promoting the rural economy.

Part IV is focused on *social change and planning*, and aims to develop a socially progressive agenda (along with Part V below) for rural planning to complement traditional emphasis on environmental and landscape protection. The part begins by exploring demographic shifts in the differentiated countryside, examining the underlying population dynamics that underpin housing markets and rural services. This chapter will examine processes of rural in- and out-migration across different types of rural contexts and gentrification concerns. This is linked directly to planning in the following chapter which focuses on contested housing debates in rural areas and how this links to sustainable rural communities. Housing markets are further explored in the next chapter on second home ownership in rural localities and the challenges and opportunities that housing investment generates. We then have two chapters that examine two essential building blocks for rural social sustainability: firstly, is a chapter examining community health planning in rural contexts, followed by a chapter on mobilities and accessibility in rural localities. The last chapter in Part IV examines the role of art as a means of rural planning research that unpacks processes of rural change.

Part V explores *planning the inclusive countryside*. This begins with an overview chapter which examines sectional inclusions and identities and how they relate to rural space, planning practice and inequality. We then focus selectively on three key themes (recognising that other perspectives exist) to address different dimensions of planning and inclusivity, both in terms of advancing theoretical frameworks and examining inclusionary/exclusionary planning practices. These include chapters relating to gendering rural planning debates, queerying rural planning, and planning for an ageing countryside.

Part VI explores *rural settlement, planning and design*. This part starts with an overview of rural infrastructures that support people, communities and settlements. Moving beyond the traditional focus on rural services, this chapter examines rural infrastructures in terms of both soft and hard supports, from community infrastructure to roads and broadband, and how these can support rural wellbeing. This is followed by three chapters on planning for rural settlements, particularly at the village scale: Chapter 32 discusses settlement strategies for rural places, which revisits the role of key rural settlement policy, the merits or downsides of concentration and implications for social outcomes. This is followed by a chapter on participatory rural planning at the village scale and then a chapter on preserving or enhancing place distinctiveness through effective village design. Building on this theme of distinctiveness, we then have a chapter on heritage and rural planning, which explores both built and natural (landscape) heritage. Part VI finishes with two chapters which examine rural settlement challenges in very different rural contexts, starting with rural planning challenges in a rapidly urbanising context (China), followed with managing rural decline or shrinkage in more remote rural places (drawing on Denmark).

Part VII moves from the built environment to examine *landscape, amenity and the rural environment*. Rural landscape protection has dominated rural planning practice for much of the twentieth century, and this part aims to critically examine this legacy while reimagining landscape and conservation planning for twenty-first-century needs. Part VII starts with three chapters that examine a traditional concern of rural planning – the preservation of rural landscapes. This begins with an exploration of National Parks, perhaps one of the longest established methods of protecting 'outstanding' rural landscapes and for countryside management. This is followed by a chapter focused on cultural heritage landscapes, exploring a range of community engagement methods for identifying and prioritising local values in landscape planning. Then, the future of green belts as a strategy for growth management is reconsidered in the following chapter, specifically in the context of housing market pressures and environmental challenges. Building on this discussion of green belts, the next chapter examines in more detail the often 'messy' arena of the rural–urban fringe, arguing that often neglected fringe spaces represent a hybrid opportunity space for rural planning. The next few chapters consider alternative approaches to conservation planning. This includes chapters exploring the landscape scale for green infrastructure planning, framing landscape within health and wellbeing debates, and the potential of 'rewilding' for rural land management.

The focus of *Part VIII* is *energy and resources*, a deeply contested aspect of rural planning and management as rural places become sites for energy production, infrastructure and extraction. The part begins by examining post-carbon ruralities and how post-carbonism will increasingly shape rural futures. This is developed in the following two chapters which will explore the rural planning dimension of renewable energy deployment and the emergence of 'fracking' as a new land-use conflict. The next chapter examines the impact of mineral and energy resource extraction on rural transformation, focusing on conflicts surrounding mineral extraction and managing fragile landscapes in Brazil. Related to this, the following chapter examines land-use governance in the context of global food security and how agricultural land use competes with energy needs (e.g. biofuels), housing and recreation, and the impact of further food production intensification on landscapes and economies. Part VIII concludes with a chapter on land grabbing and displacement of small-scale farmers by larger and often external capital interests in the context of Russia.

Finally, *Part IX* concludes the Companion by reconceptualising rural planning through a series of *reflections and futures* chapters. The first chapter in this part examines the potential of climate change concerns to reshape rural planning and to reposition rural planning as a key framework for developing holistic and territorial-based approaches to rural climate change mitigation and adaptation, currently dominated by sectoral-based interventions. This is followed by a final reflection on rural governance and power structures, particularly uneven power between local and external actors. The next chapter then examines the future of rural places, particularly exploring the global countryside thesis. The concluding chapter draws together the key themes presented in the Companion to map out future directions for reinventing rural planning theory, research and practice.

Conclusion

The chapters included in this Companion provide a foundation for readers in seeking a holistic understanding of rural planning. While we have attempted to provide broad coverage of the field, undoubtedly additional topics could have been included and new areas of interest will invariably emerge in coming years. We hope this collection of chapters from leading researchers in our field captures the nature and at least some of the scope of rural planning, but also charts 'what might be' as new societal, economic and environmental imperatives emerge to re-shape rural planning theory and practice. Within this context, understanding the balance between bottom-up and strategic initiatives, the blending of the local with globalisation, and the blurring of urban/rural boundaries will provide a key interface for future research.

References

Abercrombie, P., 1926. The preservation of rural England, *Town Planning Review, 12*(1), pp. 5–56.

Bell, M. and Osti, G., 2010. Mobilities and ruralities: An introduction, *Sociologia Ruralis, 50*, pp. 199–204.

Brenner, N. and Schmid, C., 2014. The 'urban age' in question, *International Journal of Urban and Regional Research, 38*(3), pp. 731–755.

Brunori, G. and Rossi, A., 2007. Differentiating countryside: Social representations and governance patterns in rural areas with high social density: The case of Chianti, Italy, *Journal of Rural Studies, 23*, pp. 183–205.

Buller, H. and Hoggart, K., 1986. Nondecision-making and community power: Residential development control in rural areas, *Progress in Planning, 25*, pp. 131–203.

Cloke, P.J., 1977. An index of rurality for England and Wales, *Regional Studies, 11*(1), pp. 31–46.

Cloke, P., 1997. Country backwater to virtual village? Rural studies and 'the cultural turn', *Journal of Rural Studies, 13*, pp. 367–375.

Cloke, P. and Edwards, G., 1986. Rurality in England and Wales 1981: A replication of the 1971 index. *Regional Studies, 20*(4), pp. 289–306.

Cloke, P. and Little, J., 1997. Introduction: Other countrysides?, in P. Cloke and J. Little (eds), *Contested Countryside Cultures, Otherness, Marginalization and Rurality*, Routledge, London, pp. 1–18.

Dibden, J., Potter, C. and Cocklin C., 2009. Contesting the neoliberal project for agriculture: Productivist and multifunctional trajectories in the European Union and Australia, *Journal of Rural Studies, 25*, pp. 299–308.

Gallent, N., Hamiduddin, I., Juntti, M., Kidd, S. and Shaw, D., 2015. *Introduction to Rural Planning: Economies, Communities and Landscapes*, Routledge, London.

Gallent, N. and Scott, M., 2017. Introduction, in N. Gallent and M. Scott (eds), *Rural Planning and Development*, Routledge, London, pp. 1–28.

Gkartzios, M. and Remoundou, K., 2018. Language struggles: Representations of the countryside and the city in an era of mobilities, *Geoforum, 93*, pp. 1–10.

Halfacree, K., 1993. Locality and social representation: Space, discourse and alternative definitions of the rural, *Journal of Rural Studies, 9*, pp. 1–15.

Halfacree, K., 1995. Talking about rurality: Social representations of the rural as expressed by residents of six English parishes, *Journal of Rural Studies, 11*, pp. 1–20.

Halfacree, K., 2001. Constructing the object: Taxonomic practices, 'counterurbanisation' and positioning marginal rural settlement, *International Journal of Population Geography, 7*, pp. 395–411.

Halfacree, K., 2011. Heterolocal identities? Counter-urbanisation, second homes, and rural consumption in the era of mobilities, *Population, Space and Place, 18*(2), pp. 209–224.

Healey, P., 2010. *Making Better Places: The Planning Project in the Twenty-First Century*, Macmillan, London.

Hoggart, K., 1988. Not a definition of rural, *Area, 20*, pp. 35–40.

Hoggart, K., 1990. Let's do away with rural, *Journal of Rural Studies, 6*(3), pp. 245–257.

Jones, O., 1995. Lay discourses of the rural: Developments and implications for rural studies, *Journal of Rural Studies, 11*, pp. 35–49.

Lapping, M., 2006. Rural policy and planning, in P. Cloke, T. Marsden and P. Mooney (eds), *Handbook of Rural Studies*, Sage, London, pp. 104–122.

Lichter, D. and Brown, D., 2011. Rural America in an urban society: Changing spatial and social boundaries, *Annual Review of Sociology, 37*, pp. 565–592.

Lowe, P., Murdoch, J., Marsden, T., Munton, R. and Flynn, A., 1993. Regulating the new rural spaces: The uneven development of land, *Journal of Rural Studies, 9*, pp. 205–222.

Marsden, T., 1998. New rural territories: Regulating the differentiated rural spaces, *Journal of Rural Studies*, *14*, pp. 107–117.

Marsden, T., 2009. Mobilities, vulnerabilities and sustainabilities, *Sociologia Ruralis*, *49*, pp. 113–131.

Marsden, T., Murdoch, J., Lowe, P., Munton, R. and Flynn, A., 1993. *Constructing the Countryside*, UCL Press, London.

Mitchell, C., 2004. Making sense of counterurbanization, *Journal of Rural Studies*, *20*, pp. 15–34.

Munton, R., 1995. Regulating rural change: Property rights, economy and environment – a case study from Cumbria, UK, *Journal of Rural Studies*, *11*, pp. 269–284.

Murdoch, J., Lowe, P., Ward, N. and Marsden, T., 2003. *The Differentiated Countryside*, Routledge, London.

Newby, H., 1986. Locality and rurality: The restructuring of rural social relations, *Regional Studies*, *20*(3), pp. 209–215.

OECD, 2011. *Regional Typology*, OECD, Paris.

OECD, 2018. *National Area Distribution (Indicator)*, OECD, Paris.

Phillips, M., 1998. The restructuring of social imaginations in rural geography, *Journal of Rural Studies*, *14*, pp. 121–153.

Phillips, M., 2004. Other geographies of gentrification, *Progress in Human Geography*, *28*(1), pp. 5–30.

Phillips, M., 2010. Counterurbanisation and rural gentrification: an exploration of the terms, *Population, Space and Place*, *16*(6), 539–558.

Phillips, M. and Smith, D.P., 2018. Comparative approaches to gentrification: Lessons from the rural, *Dialogues in Human Geography*, *8*(1), pp. 3–25.

Pratt, A., 1996. Discourses of rurality: Loose talk or social struggle?, *Journal of Rural Studies*, *12*, pp. 69–78.

Shucksmith, M. and Rønningen, K., 2011. The uplands after neoliberalism? The role of the small farm in rural sustainability, *Journal of Rural Studies*, *27*(3), pp. 275–287.

Smith, D., 2007. The changing faces of rural populations: '"(Re)fixing" the gaze' or 'eyes wide shut'?, *Journal of Rural Studies*, *23*, pp. 275–282.

Smith, D.P. and Phillips, D.A., 2001. Socio-cultural representations of greentrified Pennine rurality, *Journal of Rural Studies*, *17*(4), pp. 457–469.

Stockdale, A., 2011. A review of demographic ageing in the UK: Opportunities for rural research, *Population, Space and Place*, *17*, pp. 204–221.

Van Dam, F., Heins, S. and Elbersen, B., 2002. Lay discourses of the rural and stated and revealed preferences for rural living: Some evidence of the existence of a rural idyll in the Netherlands, *Journal of Rural Studies*, *18*, pp. 461–476.

Wilson, G., 2008. From 'weak' to 'strong' multifunctionality: Conceptualising farm-level multifunctional transitional pathways, *Journal of Rural Studies*, *24*, pp. 367–383.

Wong, K.M. and Musa, G., 2014. Retirement motivation among 'Malaysia My Second Home' participants, *Tourism Management*, *40*, pp. 141–154.

Woods, M., 1998. Researching rural conflicts: Hunting, local politics and actor networks, *Journal of Rural Studies*, *14*, pp. 321–340.

Woods, M., 2003. Conflicting environmental visions of rural: Windfarm development in mid-Wales, *Sociologia Ruralis*, *43*, pp. 272–288.

Woods, M., 2005. *Rural Geography*, Sage, London.

Woods, M., 2006. Redefining the 'rural question': The new 'politics of the rural' and social policy, *Social Policy and Administration*, *40*, pp. 579–595.

Part I

Concepts and foundations

The aim of this part is explore the foundations and evolution of rural planning, and also to introduce key concepts that cut across discussions throughout this Companion, such as rurality, resilience, justice and land. The first chapter of this collection, by Gallent and Gkartzios, critically reviews the development and theorisation of the term rurality, and introduces different approaches to rural planning systems internationally. The authors here draw attention to definitional debates of rurality across, primarily, positive functionalist and social constructivist approaches as well as their fusion more recently, notwithstanding problems of internationalising such debates across different cultural and linguistic contexts. Gallent and Gkartzios also draw attention to works of spatial analysis and, subsequently, the emergence of rural typologies by identifying the key players or change drivers in the development of rural places. The authors point out that meanings of rurality are critical in planning, dictating in many cases the potential of rural areas and the remit for intervention. Following this, the authors provide a summary of the nature of rural planning across different cultural contexts and forms of expertise. In particular, the authors summarise five main components with different scalar and governance characteristics: public-led land-use regulation; vision-focused spatial or territorial planning; community action and planning activity; countryside management; and the broader programmes and projects orchestrated by pan-national or national bodies. As with the concept of rurality, all these different forms of rural planning are characterised by the emergence of new rural politics and the various ways planning is expected to address different and sometimes conflicted interests.

The section continues with an overview of rural planning traditions across the Global North by Lapping and Scott. The authors here start by discussing the experience of the USA and Canada which share some similarities – contexts that lack a national land-use planning system, but with innovative local rural policy programmes and projects (such as Quebec's rural policy in Canada or Vermont's and Maine's local food strategies). This is contrasted by the European experience with a more institutionalised rural development policy towards endogenous, territorial and integrated local approaches (for example, through the LEADER programme). The experiences of two particular European countries are discussed, offering unique insights on

European rural development: Ireland, due to its large, politically important rural population and the development of a laissez-faire planning system, and Sweden with its strong rural policy focus to support the economic diversification of the countryside, build civil society and enhance local governance through inclusive public/private partnerships, technological investments and more recently towards a national food strategy. The authors then review the evolution of spatial planning and spatial governance discourses in Europe and critique, from a rural perspective, the disconnect between rural development and rural planning programmes. Finally, the Japanese rural planning experience is presented within the country's characteristic rural ageing and agricultural abandonment problems and emerging LEADER-like policy interventions towards multifunctional rural economies.

The experience of rural planning in an emergent economy is discussed in detail by Dankekar, focusing on India. India is an insightful context to examine rural planning, because of the complex interplay of exogenous and endogenous influences and also its leading positionality in a globalised market economy. For example, India's post-independence experience is characterised by prioritising agriculture and extension services drawing on the US model, but, unlike the US laissez-faire economic development approach, Indian plans explicitly framed goals within equity and justice concerns. Dankekar discusses the evolution of rural planning throughout the development decades that followed, and explores how traditional goals (such as agricultural development, access to education and housing) are continuously challenged in the quest for economic development in an increasingly globalised economy and the emergence of new policy priorities – such as climate change and implementation of sustainable strategy goals. Dankekar is particularly critical of the prevailing neoliberal ideology and how private sector support for economic development (which is supported across both governmental and non-governmental actors) has in practice resulted in significant budget cuts in rural poverty programmes. She recognises how globalisation has erased some of the traditional gaps across urban and rural areas, as well as diversifying the rural economy. However, she raises concerns as the goals of rapid urbanisation and globalisation are becoming a priority at the expense of social justice and participatory governance – which are even more critical in the rural context.

The concept of resilience within rural and agricultural development discourses – employed in many chapters throughout this Companion – is then discussed by Robinson. The author first reviews the evolution of resilience as a concept from science to social science debates, and its potential application across 'bounce back' and 'bounce forward' approaches. The literature on the resilience of communities is critically reviewed in the context of 'adaptive capacity' and 'resourcefulness' as well as how such terms can be approached in policy circles within a neoliberal logic, while ignoring questions of power. In the remainder of the chapter, Robinson approaches resilience through a focus on economic and social development linked to agricultural policies as 'a fundamental basis for rural development'. First, he discusses resilience in the context of multifunctionalism, with the examples of the European Union's Rural Development Regulation towards more diverse rural economies, and the creation of policies that support products bearing unique local or regional properties towards more ethical food markets. Second, he looks at the policy interplay between resilience and sustainability in the agricultural sector, which has favoured a focus on organic farming practices and land conservation projects (such as the Farm Bills in the USA and the Sloping Land Conversion Program in China). Conclusions are then developed highlighting not only aspects of participation and local responsibility towards understandings of resilience, but, in line with the neo-endogenous thesis, also the role of governments and external to the community actors.

The concept of multifunctionality employed in the previous chapter, is further contextualised in the chapter by Frank and Hibbard within rural planning concerns. The authors here in particular introduce the 'multifunctional transition' as a shift from a narrow focus on rural landscapes for production to multifaceted and often overlapping uses, including for consumerist purposes as well as for conservation. The implications for rural planning are discussed through a series of interviews with rural specialists in North American contexts, which highlight understandings of rural areas as human-landscape systems and the role multi-scalar governance across public, private and non-profit sectors for their development strategies. Frank and Hibbard further discuss how the multifunctional framework can inform rural planning concerns, particularly as regards responses to globalisation and environmental degradation, learning to negotiate across increasingly opposing views and across the goals of economic development and social equity, and recognising urban-rural interlinkages. As in the previous chapter, the authors here position these challenges within conceptualisations of resilience and sustainability and point out how, due to all these synergies, rural areas can become laboratories for new and alternative planning practices.

Continuing on key concepts within planning debates, *Guardian* columnist Peter Hetherington provides a succinct account on the land question, drawing on the British experience. The chapter is written polemically, highlighting in many cases how different attitudes to land ownership and land development have prevailed in Britain compared to continental Europe. Hetherington provides a review of the history of land ownership (covering the Enclosures, Irish land reform and revolution, and the 'quiet revolution' as regards the transfer of land ownership in Britain) – as well as the symbolic and cultural importance of land in the construction of national identity and pride. He identifies a few interrelated and sometimes conflicted challenges that need addressing in any discussion about land: food provision; land trading and investment; housing provision; flood management; and tenancy rights. As Hetherington acknowledges, England is unique in certain aspects: for example land ownership 'seems stuck in an early-twentieth-century time-warp'. And while only less than 10 per cent is built on, at the same time, the public believes that England is over urbanised. An obvious exception within the British context is Scotland with its twenty-first-century land reform which has resulted in shifting ownership (although moderately) to community trusts. This has not been without criticisms, particularly as regards the (non-)regulation the amount of land elite classes can own, and arguments about breeching landowners' human rights. In the context of a looming Brexit, Hetherington makes a series of suggestions to compensate for the lack of any institutional approach to addressing all these land pressures and towards an all embracing ministry of land.

Spatial and environmental justice is the next key concept of this section, explored by Ashwood, MacTavish and Richardson. The authors first position their chapter within understandings of injustice, drawing primarily on race and class. Then the authors focus on rural American contexts to discuss two iterative processes of spatial and environmental injustice: dispossessing those who still possess property rights; and, dispossessing those already dispossessed; The first ('dispossessing the possessed') refers to the establishment of 'Right to Farm' laws and their impact on stripping rural populations of their property rights, while putting residents at health risks. The second ('dispossessing the dispossessed') refers to rural trailer parks, which function as a form of informal affordable housing in the USA, and their legal treatment often results in further exclusion and elimination. Both cases enable rural exploitation and the authors examine how 'the power of the rhetoric' legitimises such processes. For example, the Right to Farm draws on positive images surrounding the need to grow food, rather than corporate agribusiness; trailer parks although as 'characteristic to rural America as agriculture', are not a legitimate representation of the rural idyll, and are further marginalised and continuously stereotyped as 'white trash'. Despites these instances of rural injustice, the authors also reviews cases in favour of rural rights, demonstrating the multiple ways the law can support spatial justice and ultimately sustainability.

The contributions in this section aim to provide informed accounts on some of the key concepts used throughout this book, such as rurality, resilience, sustainability, land and justice. While planning covers a wide range of instruments and functions internationally, as discussed by Gallent and Gkartzios, certain issues presented in this section appear almost universal. For example the role of 'politics of place' in planning and mediating across different (and contrasting) views about the future of rural areas; the need to support a combination of environmental, social and economic goals, while understanding that rural areas have been drastically changed and affected by global and local changes; and, the focus on enhanced governance and the role of rural communities (identified in multiple ways) in the process of policy development.

2

Defining rurality and the scope of rural planning

Nick Gallent and Menelaos Gkartzios

Introduction

Definitions of rurality have shifted considerably. Statistical measures focused on the structure of rural economies and land use have given way to experiential and cultural accounts of what makes a place rural. While important land use and economic characteristics persist – and reveal a need for a particular approach to land use planning and development strategy – a more nuanced understanding of the countryside and rural society has taken root, which suggests a necessary extension of planning for rural areas beyond the standard concerns of farming, housing, conservation, infrastructure and economic development. Rural areas are shown to be extremely diverse and complex, although a 'rural pathology' usually prevails in popular imagination. Moreover, experience of the rural is subjective, culturally contingent, gendered and dependent on market position. Likewise, the countryside is both a place to escape to, and from – being a context in which people enjoy wealth and advantage but also endure poverty and inequality.

Most planning practitioners and scholars recognise that there is no ubiquitous rural; however, the need for planning outside of metropolitan areas is evidenced by ongoing debates about the future of rural areas globally. The days of 'rural planning' being viewed as a remedial action are long gone. If the term planning is to have any relevance and currency, it needs to work with and through different actors, connect in some way with the complexity of the countryside and be a multi-sectoral activity, sometimes dealing with broad structural challenges, but more usually concerned with subjective wellbeing in particular places. As well as providing a sketch of the complexity of rural places, this chapter offers a statement on the nature of rural planning. That nature, alongside the broader purpose of planning, is then unpacked in the chapters that follow.

The chapter has two parts. In the first part, detail is added to the brief sketch of definitional debates provided above. In the second part, the evolving and current scope of rural planning is analysed. The purpose of this chapter is to provide a conceptual anchor for later contributions. Not everything that is said will ring true in all contexts and for all scholars, but the conclusions reached in this chapter – and the general pictures of rurality and planning presented – will be reflected upon at the end of this book; the aim being to elicit a more nuanced account of rural spaces and the scope and role of planning therein.

Thinking about the nature of the 'rural'

Rural spaces extend from the near-urban to remote wildernesses. They differ in form and content from one place, and one country, to the next. Lowland agricultural landscapes – punctuated with farms and villages – in western Europe contrast with the emptier uplands of Scandinavia. Around the world, fragmented peripheral areas mixing urban and rural uses (Gallent et al., 2006), near-urban hinterlands, and well-connected areas dedicated to arable farming provide one identifiable rural space. This is the 'country-side' abutting or enveloping larger or smaller settlements. Further out, such landscapes give way to more sparsely populated regions, less directly influenced by the daily patterns of movement between urban and rural places. These are variably badged as 'remote', 'wild' (or at least 'semi-natural') or as 'backwaters' and are often less developed because of topographical or climatological impediment – they are uplands or mountainous, or comprise undrained fenland, semi-desert, bush or outback. For a hundred years, if any thought was given to the constitution of rural areas, it was in the form of very broad characterisation: areas were distinguished by a dominant economic activity – usually farming – and by a sparsity of habitation. The rurality of such areas had been guaranteed by the more limited intrusion of development and, underpinning this, weaker market interest. In many places, patterns and levels of market interest have waxed and waned. The declining importance of farming in some industrialising nations – which turned to foreign sources for basic foods – triggered depopulation and the hollowing out of rural areas.

But later on, the growing middle classes of those same nations chose – either occasionally, seasonally or permanently – to return to the countryside. This prompted new interest and new economic activity, transforming former productive spaces into places of consumption and mass tourism. It is arguably that return to the countryside in 'the west' – and the collision between urban and rural ideas and values – that prompted new interest in the nature and basic characteristics of rural areas. For a long time, the preoccupation of planning practice and research was with industrialisation and its urban outcomes. However, this general claim should be tempered with the acknowledgement that the British 'town and country' planning system placed an early emphasis on the perceived need of rural areas to be protected from urban encroachment; an emphasis that was not shared in other countries, which tended not to see rural areas as a space for planning intervention (Gkartzios and Shucksmith, 2015).

Counter-urbanisation during the second half of the twentieth century re-ignited interest in rural spaces, which were now host to a range of social, economic, environmental and political transformations. Greater effort was suddenly invested in trying to unpack the range of meanings conveyed by the word 'rural'. Basic definitions were soon followed by complex typologies, which attempted to capture something of the variety of rural places.

The rural studies literature has grappled with the basic questions of definition and typology at least since the 1960s. There has been constant development in both definitions and classifications of 'the rural' since that time, which have gradually moved from a functional 'positivist' delimitation of rural areas to a 'post-rural' social construction of space, characterised by the decoupling of function from the values and meanings assigned to particular places. In short, concern with the 'materiality' (and objects) of rural places has given ground to a search for the 'immaterial' (and subjective) basis of rurality.

This shift can be read in the British-focused works of Cloke (1977, 2006), Hoggart (1988) and Halfacree (1993). Cloke's (1977) attempt to move beyond previous 'subjective ideas' of the rural (based on a broad and loosely defined characterisation) towards a concise statement of rural–urban differentials aimed to provide 'an aid to the possible standardisation of planning solutions in areas with similar problems', using key indicators to reveal whether areas 'inclined' towards being rural or urban. This overtly positivist view sought to demonstrate rurality from datasets for employment, population, migration, housing conditions, land use and spatial remoteness. These data were used to quantify rurality against measures of density, engagement in primary industry and outward commuting proxies – using multi-variate analysis. A decade later, Cloke and Edwards (1987) repeated the exercise only to be rebuked for their 'unidimensional' account of outward rural attributes (Hoggart, 1988). Hoggart was among the first researchers to question the 'discredited' assumptions – relating to outward patterns of economic and social behaviour – underpinning the quantitative profiling of rural areas. He accused such profiling of being blind to the causal processes that transcend any urban–rural divide, arguing that it is the relative importance of competitive, monopoly and state sectors that drive observable spatial outcomes rather than surface issues such as low population density. It was Halfacree (1993) who picked up the gauntlet thrown down by Hoggart, beginning his development of new perspectives on rurality with three points of agreement with Hoggart. He agreed, firstly, that statistical descriptions provide only shallow insight into the nature of rurality. He agreed, secondly, that socio-cultural traits cannot be tied to rural and urban spaces in a simple way (referencing Pahl, 1966) because of the manner in which space is produced by (a counter-urbanising) society and thereafter produces other space (causing a mixing of urban and rural society and space). And lastly, he agreed that economic causal processes do not respect area types. But from these three points of agreement, Halfacree argued for a need to look beyond the 'causal' to 'social representation': the rural must be read through discourse, including the lay discourse ignored in causal analysis. Representational definitions of the rural require new gender, ethnic and class specificities. This shift was then summarised by Cloke (2006) who charted the way in which positivist tendencies (including his own) had given way, for the most part (except in agricultural economics) to new conceptualisations of the rural. Citing his own 1977 paper, Cloke drew attention to the flaws of positivist functionalism before making the critical claim that while the starting point for understanding the rural is (undoubtedly) its attendant values and meanings, these meanings have become increasingly detached from function, giving rise to the notion of 'post rurality'. Although Cloke was ready to embrace the transition in his own field, he reserved some criticism of this functional detachment and a 'cultural turn' that has failed to gain traction in rural studies largely because it has lost sight of the 'politics of place'. By the mid-2000s, there remained too little focus in rural studies on power relationships, social practice and politics, prompting Cloke's call for a new theoretical pragmatism and some fusion of the cultural turn – with its inherent richness – with political and economic materialism.

These debates positioned British academia (and British human geography) at the forefront of rural research, resulting in a degree of sustained 'Anglo-centricity' (Lowe, 2012) and also extending British perspectives into international debates on the nature of rurality (Gkartzios, 2013). These perspectives are certainly present in the development of continuums and rural typologies, which broadly address material, social, political and economic change. Antrop's (2004) focus on urbanisation in Europe has drawn attention to the way that this process affects and creates a taxonomy of rural landscapes and situations. Urban pressure has produced patterns of change that are 'very different for the countryside near major cities, for metropolitan villages and for remote rural villages' (see opening discussion) and 'planning and designing landscapes for the future requires that this is understood' (ibid.: 9). While Antrop's focus was on physical attributes and the 'long making of the landscape', he strayed into numerous debates around the urbanisation of the rural 'way of life', accepting that urbanisation produces social as well as material and functional change.

Another important contribution, *inter alia*, is that of Marsden and colleagues (1993), who dealt directly with Cloke's 'politics of place', examining the way in which economic shifts accompanied by in-migration (again, movement and mixing) alter the practised politics of rural areas. Their own classification of rural space was researched and written before the cultural turn (in rural studies) and its 'functional detachment'. These authors argued that the politics 'practised' in rural areas is differentiated and determined by contrasting power sources, meaning that spaces can be more or less traditional and subject to continuity or change depending, largely, on the degree to which they have been affected by economic, social and cultural shifts. Looking again at England, four political rural spaces were identified: an anti-development 'preserved' countryside, in which the middle classes dominate decision-making; a 'contested' countryside beyond commuter catchments, in which landowners and farmers continue to play a leading and conflicting political role; a 'paternalistic' countryside dominated by landed estates and the monopoly power afforded by landed status; and a 'clientelist' countryside in which relative isolation and peripherality (generating welfare concerns) meets strong agricultural corporatism.

Their work is now well known for its attempt to dissect the power and political relationships of rural areas, underpinned by socio-economic changes in the latter part of the twentieth century. And despite contrasting experiences of rural places and politics internationally (Shortall and Alston, 2016), the idea of the 'differentiated countryside' has proven surprisingly mobile, providing an analytical frame for the study of power relations in many contexts. This is perhaps odd given that the constellation of employed concepts – including stewardship, paternalism and gentrification – are so specific to the English experience (Lowe, 2012) making translation, let alone application, extremely difficult. Indeed, such concepts are deeply embedded within Britain's landed-class history. Their use elsewhere can lead to confusion and misunderstanding. The risk of this – or 'language politics' – is now becoming a focus of concern for rural researchers (Gkartzios et al., 2017).

More broadly, the word 'rural' itself – derived from the Latin *rus* (countryside) – seems to have resonance in some European languages (Woods, 2011), while remaining meaningless, or at least cryptic, in others. This has led some research to abandon the rural label, referencing instead the 'non-metropolitan', the 'periphery', 'region' or 'province' (Laschewski et al., 2002). Where there is a need to be more specific, 'land' or 'agricultural' concerns are linked to peripheral or non-metropolitan discourses. Moreover, linguistic associations with 'rural' can result in very different perspectives and understandings. While in England, it has become a framing for middle-class and national identity (Newby, 1985), in other contexts to label something 'rural' is to be dismissive or emphasise its backwardness (UN, 1980).

Yet despite such sensitivities, spatial analyses of rural transition remain as popular as ever, though often focused on territorial change and avoiding deeper connotation. Lowe and Ward (2009) have offered a more recent economic classification of rural areas and a base on which to project future change. Drawing, in part, on Marsden et al. (1993) – which Phillip Lowe contributed to – they argue that rural England – as their illustrative case – divides into seven area types: settled commuter, dynamic rural, dynamic commuter, transient rural, retirement retreats, deep rural and peripheral amenity. Each area type has its associated 'change drivers' and the future mix of these areas will depend on whether more people move into the countryside, whether planning policy becomes more or less restrictive, and whether rural policy in general becomes more or less oriented to economic growth goals. Lowe and Ward build on the idea of 'political spaces' to develop a typology that is mainly economic in its perspective but also embraces social and policy-induced change (the latter being locally tied to in-migration, especially in the 'preserved countryside'). Adopting a similar approach, Verburg et al. (2010) link change drivers to future patterns of urbanisation, agricultural expansion and abandonment across Europe under four scenarios. These imagine different balances between the scaling of neoliberal economies and the scaling of planning. Hence, they envisage a global economy scenario in which competition drives spatial outcomes; a continental market scenario in which blocks of countries club together to compete; a global cooperation scenario in which future development is planned at a pan-national level; and a 'regional communities' scenario prioritising a form of localism in which communities become key to development trajectories. In both of these works, a range of spatial attributes – material, social, political and economic – are brought together as a basis for rural futurology. And in both, the socio-economic challenges of the countryside – to be navigated by the market and by pan-national, national and local policy – are set out.

The importance of this body of work – and many other studies dealing with the material characteristics and immaterial dimensions of rural places – relates, at least in this Companion, to the emerging rural planning challenge. These spatial discussions of the nature of rurality flag important aspects of the economic, landscape, built and socio-political context with which planning must connect. Selective meanings and terms of rurality chosen and used by policy makers, planners and academics, in varying contexts, dictate their development trajectories: what rural places are for and what they are not for. They illustrate the way in which economic and social transformations during the twentieth century have generated a mixed rural politics and governance processes that steer planning – as an instrument of that governance – onto different development or preservationist paths. In the remainder of this chapter, we turn to consider the nature of rural planning and its potential to operate effectively in rural locations.

Thinking about the nature of 'rural planning'

Internationally, 'planning' embodies different traditions and knowledges, situated across disciplines such as architecture, engineering, and (more recently) social sciences, and different systems of governance. Yet almost everywhere, planning remains a highly contested political activity because of its attempts to deliver against contested public goals and conflicting concerns. Planning has in the past been narrowly defined – being seen purely as a public-sector activity (e.g. the control of land use) intended to regulate change or deliver a specific public policy goal, for example relating to the siting of new housing or key infrastructure. But how organisations and individuals plan for, and manage, change is always broader than that, especially in the countryside where there are built places, extensive 'natural' environments and farmed spaces to attend to.

Rydin (2011: 12) has defined planning as 'a means by which society collectively decides what urban change should be like and tries to achieve that vision by a mix of means'. There is no reason why this same definition cannot be applied to rural change, and the idea of achieving a vision by a 'mix of means' – many well outside the traditional boundaries of land-use control – has particular resonance for rural areas. Planning, wherever undertaken, was once very much a public sector project. Many European countries established comprehensive planning systems after the Second World War and charged the local state with responsibility to orchestrate and deliver reconstruction on a grand scale. But this model has since given ground to greater 'interaction between the public and private sectors' (ibid.) and more indirect means of steering and facilitating development agendas, which are often set locally and embrace the visions and priorities of different communities and stakeholders. There has been a backlash against public planning and professionalism in recent decades, and also against the distant administration of urban and rural change and development. Rittel and Webber (1973: 155) observed that 'we've been hearing ever-louder public protests against the professions' diagnoses of the clients' problems, against professionally designed governmental programmes, against professionally certified standards for the public services'. These protests have arisen because definitions of the 'public good', defined centrally by experts, are often disputed.

Gallent et al. (2015: 12) suggest that modern rural planning comprises a composite of core public planning functions (primarily the control of development), spatial or territorial planning that brings together a wider range of public and private stakeholders (but remains public sector led; see Albrechts, 2004), community action and planning in various forms, countryside management (which deals with the spaces beside or between settlements and is often led by farming interests) and the programmes and projects spearheaded by government departments and agencies or by local groups (Bishop and Phillips, 2004: 4). It is further suggested that rural planning, wherever it happens, is broad in scope; it is undertaken by numerous agents; it comprises a range of initiatives, interventions and local actions; and it goes well beyond mere land-use planning, particularly when ideas of countryside management and the role of rural areas delivering multifunctional benefits are added into the mix.

The components of rural planning listed in Table 2.1 provide a mix of means through which change is managed and delivered. Throughout this Companion, these different means are shown to produce particular outcomes and to interact in specific ways. Moreover, each operates across a specific band of scales. Planning is also a product of the 'politics of place', legitimising certain development interests while rendering others illegal, undesirable and even 'unsustainable' (Murdoch and Lowe, 2003; Sturzaker and Shucksmith, 2011). For example, in relation to housing provision in rural areas, very different (often opposing) outcomes are judged beneficial in different places. Where the conservation ethic is dominant (underpinned by a particular rural meaning and discourse) intolerance to housing development may be viewed as a 'good'

Table 2.1 The scope of rural planning.

Components	Functions
Public land-use planning	• National policy • Strategic planning for infrastructure and housing • Development (settlement) planning • Land-use control and other regulatory functions
Spatial or territorial planning	• Area visioning • Co-ordination of service investments • Co-ordination of all public/private and third sector initiatives
Community action and planning	• Campaigning and lobbying • Voluntary control of services • Support for community development and social infrastructure • Community visioning • Interfacing with public and spatial planning activity
Countryside management	• Farming and stewardship • Strategies and actions that focus on the spaces besides or between physical development • Strategies for renewable energy, mineral extraction or 'fracking'
Other projects and programmes	• Governmental and pan-national directives and programmes • Departmental or agency-based (sectoral) projects around health, education, transport and so on • Development agency interventions • Private sector (industrial) programmes and initiatives

Source: Gallent et al. (2015: 12)

planning outcome. Elsewhere, a permissive approach to development might be celebrated as a route to greater social justice, bringing much needed development to hitherto 'backward' rural areas. It is difficult to judge whether one system, or one outcome, is better than another. Each is the product of complex socio-political conditions. It is impossible therefore – in this Companion or elsewhere – to catalogue good or bad planning practice. It is, however, reasonable to analyse how planning systems have treated rural space, and consider whose interests and needs have been prioritised or ignored and for what reason.

It was noted above that many Western countries introduced comprehensive systems of land-use planning during the twentieth century. These were often seeded much earlier, and their development responded to critical urban growth challenges and sometimes to the urbanisation of hitherto rural areas. Indeed, the pressure to exert greater political control over 'market processes' began to build in the middle of the nineteenth century as rapid industrialisation and urbanisation across Europe, North America and elsewhere generated a mix of problems centred on poor public transport, uncoordinated infrastructure provision, bad housing, and declining public health. Many systems were centrally and strategically coordinated and built on the notion that general principles and ordinances should be agreed that would determine future development outcomes in specified zones. Zonal or compliance-based land-use planning systems set core parameters for development, to be more or less rigidly applied in many European countries, North America and Southeast Asia. A different approach took root in the United Kingdom where intervention in 'private' development matters had been a topic of heated political debate since the early years of the nineteenth century.

The regulation of land-use change is only one part of the broader activity of planning in either urban or rural areas. Land-use and broader socio-economic planning components are brought together in some planning systems such as comprehensive planning in the UK. Other planning systems had maintained a broad spatial or territorial focus so it is perhaps unfair to present a clear divide between matters of land use and wider socio-economic planning. However, this distinction does generally separate the regulatory (zonal, ordinance- or compliance-based) components from the visioning function of planning. Spatial or territorial planning treats regulation – including the regulation of land use – as one means, among many, of delivering broader development goals. Others could include the co-ordinated investments of development agencies and their private partners, or the decisions over the siting of schools or medical centres taken by other bodies. Building on the earlier work of Kunzmann (2000) and Healey (1997), Albrechts (2004: 747) notes that while spatial planning is generally public-sector-led, it should be seen as a broader 'socio-spatial process through which a vision, actions, and means for implementation are produced that shape and frame what a place is, and may become'. Spatial, strategic-spatial or territorial planning might also be conceived as a wider process of place-governance that connects to, and delivers through, multiple partners. This has been the experience of many different forms of spatial planning across Europe.

Another very important component of rural planning occurs at the community level and involves a mix of informal activism and formalised very local planning processes. 'Community level' can infer different scales of activity. The formal activities of a town council might be viewed as community level and so too might the collective actions of a hundred or so residents of a village who grasp the opportunity to form a residents' association or some other type of forum. But below these levels, sub-groups may take specific interest in key community concerns – around older people, the needs of children, or the protection of a local asset – and engage in some kind of activism designed to achieve specific goals. Such community level action plays its part in shaping local places. It generates new capacities, contributes to the preservation of essential assets and infrastructure and combats the occasional sense of disillusionment felt in some rural areas faced with the remoteness of public authorities. Indeed, community groups often form to grapple seemingly 'forgotten' or overlooked issues that are below the radar of private enterprise or public sector concern / capacity. Hence, communities might take charge of some services or provide the soft infrastructure of community support for more vulnerable groups. A key concern here has been the social inclusion of certain groups and enhanced local governance. The forms that community action and planning can take are analysed by Gallent and Ciaffi (2014). Numerous examples are given from across Europe, North America and elsewhere – some rural and some much more urban. In the US, community action has often been epitomised through the work of residents' associations in the major cities, which have worked to improve the housing conditions of poorer sections of society. In rural Australia, community support has played a critical role in substituting for missing healthcare infrastructure and helped compensate for the isolation of some outback communities. In Europe, rural residents and formal 'community planning' structures have roles in service provision, are leading on the planning of new housing development and are contributing to participatory budgeting processes and disaster relief. One of the key challenges for higher levels of public planning has been how to connect to the sometimes sporadic and fragmented activities of community groups without undermining representative democracy (Owen et al., 2007). This presents integrated planning with two challenges: first, how to achieve a good fit between local actions and higher level strategy; and second, how to avoid a situation in which most decisions are taken by 'self-identifying groups of articulate residents' while other groups are left feeling disenfranchised. In order to avoid the second pitfall, there must be an effective means of dealing with the first challenge – so that democratic practice at a community level is inclusive and builds on the mandate achieved by representative democracy. The very local level is important in rural planning but also fraught with risk – principally, the risk that self-serving cliques pursue narrow and negative private interest agendas.

The components listed above focus, principally, on human, settlement or physical aspects of rural (or urban) areas. Equally important for these areas is countryside management, dealing with the spaces on the edges or in between settlements. Environmental and landscape changes affecting rural areas have a variety of global, European and national policy drivers, with policy frequently attempting to frame the countryside as a multi-functional space, in which opportunities to enhance biodiversity, deliver climate change adaptation and mitigation, improve water quality and alleviate flood risk should go hand-in-hand with other forms of development. At the same time, the traditional landscapes of the countryside, with their inherent biodiversity, are viewed as a precious resource for future generations. Exploiting and protecting 'nature' have become the twin goals of countryside management, though this aspect of rural planning often struggles to reconcile beneficial exploitation – from conventional farming to energy crops and from wind farms to fracking – with protection aimed at maintaining current amenity: e.g. access to rural land, aesthetic value and quiet enjoyment. It is countryside management, in broad terms, that needs to bring together different interest and users of rural spaces, often attempting to reconcile projects such as onshore wind farms – that have a huge climate change mitigation potential – with community and societal interests in maintaining countryside character. The role that the countryside can play in addressing global challenges makes it an increasingly contested space in which global goals and local ambitions conflict. Today, a broad range of actors are involved in countryside management, from communities and local interest groups, to farmers and institutional land-owners, all the way to government agencies and supra-national bodies which set the directives and frameworks that determine the nature of that management. More traditionally, farmers have been credited with a stewardship role in the countryside by virtue of their control over extensive land areas. A flow of new investment money into rural areas and the arrival of institutional landowners, seeking to transform productive income into financial income, is altering traditional patterns of landownership. The new 'stewards' of many rural areas are investment banks and other financial service providers which are acquiring and consolidating land holdings, leasing them back to tenant farmers, and deriving income from rural land assets (first from rental income and ultimately from land value appreciation) which they are able to package as investment funds (Gallent et al., 2018). New landowners are often actively engaged in countryside management. Some bring with them new ideas and values and many are pro-active in their engagements with nearby communities, viewing good relations and broader contributions as part of their corporate social responsibility (ibid.).

The final component of rural planning, linking strongly to countryside management, comprises a mix of programmes and projects (and framing directives), initiated at different scales by a broad range of actors. In the US, these might comprise Federal Government initiatives. In the Global South, programmes and projects may be orchestrated by the UN's Food and Agriculture Organization (FAO). At the level of the European Union (EU), various grant schemes operated under the auspices of the Common Agricultural Policy have had a profound impact on the development of rural areas, focusing initially on agricultural production and later on a wider set of environmental goals. The EU's 'INTERREG' (inter-regional) Structural Programmes have channelled large sums of money into rural areas judged to need significant investment in their infrastructure. These programmes (and the projects taken forward) have been very much top-down, but other EU initiatives – including various rounds of the LEADER (Liaison Entre Actions de Développement de l'Économie Rurale) programme – have lent support to local initiative, funding partnerships that promote 'integrated, high-quality and original strategies for sustainable development'. It is also the case that 'Europe' frames, for example, aspects of national planning and conservation policy in the different parts of Europe through various EU Directives and Regulations. Some EU programmes support projects that are defined and delivered by a range of national ministries and agencies. But these national bodies also run their own rural initiatives. Therefore, another level of programmes and projects, under the pan-national level, are those taken forward by national governments.

It is extremely difficult to set fixed boundaries around rural planning. Decisions and actions are being taken at many levels and by a broad array of actors. There are many influences on the trajectories of development and change in the countryside, though the separation of planning into five main areas seems to capture the breadth of this influence.

Conclusions

In this chapter, we have addressed two basic questions: what delimits ostensibly rural areas from urban ones and what is the function and scope of 'planning' in those areas? The answers to these questions have been necessarily brief. Rural areas can be defined by their assemblage of material assets (patterns of land use, economic activity, built form etc.) and immaterial qualities (their particular social life and the subjective experiences of being in a rural place). Over time, positive functionalism (emphasising objective measures of material rurality) has given ground first to the cultural turn in rural studies (and its emphasis on the cultural specificity of rural experience) and then to a recognition of the functional detachment of the cultural turn from social and political materialities – although the application of these approaches in policy making has remained elusive. There is today a broad concern with both the material and the immaterial and with the transformation (or 'restructuring') of rural areas from the late 20th century onwards: in particular, with the ways in which counter-urbanisation have generated a new rural politics built on social change and diverse expectations of what the rural should be.

That new rural politics often plays out in the planning arena. Planning in this chapter has been characterised as the coordination and management of change through a mix of means and by an array of different public, private and community interests. Five components of rural planning were listed: public led land-use regulation; vision-focused spatial or territorial planning; community action and planning activity; countryside management (and the strengthening role of institutional land investors); and the broader programmes and projects orchestrated by pan-national or national bodies. Rural planning is multi-scale and multi-actor. It centres on the mediation of different interests – and seeks to address potential conflicts arising from contestation around the function and nature of rural places. That mediation – between sometimes conflicting visions – plays out in public planning and regulation, in community-level debates, in countryside management and in broad rural programmes. The different levels of planning often reach their own conclusions as to the right path for rural development or preservation, which may conflict with other levels. Public planning, for example, may have development aspirations that are out of kilter with community ambition or pan-national directives. The interplay between the complexities of what the rural is and what planning does, at different levels, has created a rich context for rural planning research, as evidenced by the extraordinary range of chapters contained in the rest of this Companion.

References

Albrechts, L. (2004) Strategic (spatial) planning re-examined. *Environment and Planning B: Planning and Design*, 31(5), 743–758.

Antrop, M. (2004) Landscape change and the urbanization process in Europe. *Landscape and Urban Planning*, 67(1), 9–26.

Bishop K. and A. Phillips (2004) Then and now: planning for countryside conservation. In K. Bishop and A. Phillips (eds), *Countryside Planning: New Approaches to Management and Conservation*, 1–15. London: Earthscan.

Cloke, P. J. (1977) An index of rurality for England and Wales. *Regional Studies*, 11(1), 31–46.

Cloke, P., (2006) Conceptualizing rurality. In P. Cloke, T. Marsden and P. Mooney (eds), *Handbook of Rural Studies*, 18–28. Oxford: Sage.

Cloke, P., and Edwards, G. (1987) Rurality in England and Wales 1981: a replication of the 1971 index. *Regional Studies*, 20(4), 289–306.

Gallent, N. and Ciaffi, D. (eds) (2014) *Community Action and Planning: Contexts, Drivers and Outcomes*. Bristol: Policy Press.

Gallent, N., Andersson, J. and Bianconi, M. (2006) *Planning on the Edge*. London: Routledge.

Gallent, N., Hamiduddin, I., Juntti, M., Kidd, S. and Shaw, D. (2015) *Introduction to Rural Planning*, 2nd edition. London: Routledge.

Gallent, N., Hamiduddin, I., Juntti, M., Livingstone, N. and Stirling, P. (2018) *New Money in Rural Areas*. London: Palgrave.

Gkartzios, M. (2013) 'Leaving Athens': narratives of counterurbanisation in times of crisis. *Journal of Rural Studies*, 32, 158–167.

Gkartzios, M., Remoundou, K. and Garrod, G. (2017) Emerging geographies of mobility: The role of regional towns in Greece's 'counterurbanisation story'. *Journal of Rural Studies*, 55, 22–32.

Gkartzios, M. and Shucksmith, M. (2015) 'Spatial anarchy' versus 'spatial apartheid': rural housing ironies in Ireland and England. *Town Planning Review*, 86(1), 53–72.

Halfacree, K. H. (1993) Locality and social representation: space, discourse and alternative definitions of the rural. *Journal of Rural Studies*, 91(1), 23–37.

Healey, P. (1997) *Collaborative Planning: Shaping Places in Fragmented Societies*. Basingstoke: Macmillan.

Hoggart, K. (1988) Not a definition of rural. *Area*, 20, 35–40.

Kunzmann, K. (2000) Strategic spatial development through information and communication. In W. Salet and A. Faludi (eds), *The Revival of Strategic Spatial Planning*, 259–265. Amsterdam: Royal Netherlands Academy of Arts and Sciences.

Laschewski, L., Teherani-Krönner, P. and Bahner, T. (2002) Recent rural restructuring in East and West Germany: experiences and backgrounds. In K. Halfacree, I. Kovach and R. Woodward (eds), *Leadership and Local Power in European Rural Development*, 145–172. Aldershot: Ashgate.

Lowe, P. (2012) The agency of rural research in comparative context. In M. Shucksmith, D. L. Brown, S. Shortall, J. Vergunst and M. E. Warner, (eds), *Rural Transformations and Rural Policies in the US and UK*, 18–38. New York: Routledge.

Lowe, P. and Ward, N. (2009) England's rural futures: a socio-geographical approach to scenarios analysis. *Regional Studies*, 43(10), 1319–1332.

Marsden, T., Murdoch, J., Lowe, P., Munton, R. and Flynn, A. (1993) *Constructing the Countryside: An Approach to Rural Development*. London: Routledge.

Murdoch, J. and Lowe, P. (2003) The preservationist paradox: modernism, environmentalism and the politics of spatial division. *Transactions of the Institute of British Geographers*, 28, 318–322.

Newby, H. (1985) *Green and Pleasant Land? Social Change in Rural England*. London: Penguin.

Owen, S., Moseley, M. and Courtney, P. (2007) Bridging the gap: an attempt to reconcile strategic planning and very local community-based planning in rural England. *Local Government Studies*, 33, 49–76.

Pahl, R. E. (1966) The rural-urban continuum. *Sociologia Ruralis*, 6(3), 299–329.

Rittel, H. and Webber, M. (1973) Dilemmas in a general theory of planning. *Policy Sciences*, 4, 155–169.

Rydin, Y. (2011) *The Purpose of Planning*. Bristol: Policy Press.

Shortall, S. and Alston, M. (2016) To rural proof or not to rural proof: A comparative analysis. *Politics and Policy*, 44(1), 35–55.

Sturzaker, J. and Shucksmith, M. (2011) Planning for housing in rural England: Discursive power and spatial exclusion. *Town Planning Review*, 82(2), 169–193.

UN. (1980) *Patterns of Urban and Rural Population Growth*. Population Studies no. 68. New York: Department of International Economic and Social Affairs, United Nations.

Verburg, P. H., van Berkel, D. B., van Doorn, A. M., van Eupen, M. and van den Heiligenberg, H. A. (2010) Trajectories of land use change in Europe: a model-based exploration of rural futures. *Landscape Ecology*, 25(2), 217–232.

Woods, M. (2011) *Rural*. London: Routledge.

3

The evolution of rural planning in the Global North

Mark B. Lapping and Mark Scott

Introduction

In the previous chapter, Gallent and Gkartzios outlined the nature and scope of rural planning and how this relates to our understanding or conceptualisation of rurality. Building on their assessment, this chapter will examine the evolution of rural policy and planning practice within the context of the Global North. As recorded by Lapping (2006), many of the early pioneers of planning, particularly within the Anglo-American tradition, held an avowed interest in rural people and places, with many foundational planning ideas focused on rural challenges. In the US, this included the early 'regionalists' and the tradition of rural mega-projects for regional development, with the 1930s depression era Tennessee Valley Authority's hydroelectric schemes providing the exemplar. In rapidly urbanising and industrialising European contexts, the early decades of the twentieth century witnessed new planning policies designed to protect rural places from urban encroachment and sprawl, seeking to preserve the unique qualities of rural places. For example, as far back as 1926, Patrick Abercrombie raised concerns with people 'colonising' the English countryside and the need to preserve the countryside on aesthetic grounds as England's greatest national monument (Abercrombie, 1926). However, despite the centrality of 'rural' concerns to the early planning project, the post-Second World War era was marked by an increasing focus on urban debates within planning, and rural planning began to play a more marginal role within both the planning profession and academy.

In this chapter, we examine how rural policy and planning have evolved in the Global North, selectively focusing on the North American (USA and Canada), some European and Japanese experiences. Accordingly, the chapter is structured as follows: firstly, the US experiences of rural development and planning are outlined, followed by a review of the evolution of rural planning in Canada. The chapter then compares this evolution to the European experience, by examining firstly, the emergence of the European approach to endogenous rural development – with two illustrative examples of Ireland and Sweden – followed by a brief overview of spatial planning in rural Europe. We then examine the how rural planning and policy have evolved in Japan, a highly urbanised country with an ageing rural population. While place-based rural development and rural planning are both *spatial* approaches to policy and managing rural change, the chapter suggests that these two policy domains are often poorly integrated in practice. The chapter concludes by calling for more imaginative and innovative responses to planning for the rural environment, economy and communities.

Rural policy and planning in the USA

In a recently published assessment of the practice of rural planning in North America, several findings point to the changes occurring both in rural regions as well as the very nature and characteristics of planning in such places (Frank and Hibbard, 2016). The study sums up the current rural situation thus:

> The rural areas of North America are at a vital transition point in the early years of the twenty-first century. They are experiencing fundamental shifts in their economies. Agriculture and natural-resource production continue to become more and more capital intensive; many rural places are shifting from being sites of production to consumption, driven by their natural amenities; and others are losing their economic raison d'etre altogether.
>
> *(Frank and Hibbard, 2016: 299)*

Long viewed by the larger national policy nexus as essentially residual areas, Frank and Hibbard continue that 'rural areas in the US have been a bathtub that's draining, extracting people and resources' for the emerging globalised post-industrial world (ibid.: 302). While seen by many as an accurate picture of much of rural North America, such a generalisation masks the reality that different rural areas, such as those on the urban/rural fringe, as well as those with high amenity values, are witnessing growth often accompanied by the in-migration of non-traditional populations. Perhaps the most accurate portrayal of rural planning in North America is one that places less stress on land use, once the sine qua non of rural planning. Land use issues are now but a part of a larger and more comprehensive rural development thrust. What has been emerging is a broader development emphasis stressing locally based and locally focused sustainability planning – in its broadest sense – as an appropriate response to post-industrialisation, neo-liberalism and globalisation along with profound environmental challenges, notably related to climate change. Much the same may be said of other rural areas throughout the Global North.

Despite some attempts in the 1970s and 1980s to craft federal responses to rural planning problems, there is no national land use planning policy in the United States. Instead, land use has historically been understood as essentially 'local' in nature and hence has been the domain of states, municipalities and communities to enact land use policies and programmes. Indeed, commentators during those decades talked about a 'quiet revolution' in land use among many states that attempted to develop aggressive and far-sighted planning systems (Bosselman and Callies, 1972). In terms of rural areas, the focus was often on preserving productive agricultural and grazing lands. Over the years, many of these efforts have been eroded in a piecemeal fashion. In terms of actual policies to preserve farmland as a state-wide priority, many of the most progressive states, like Oregon, Florida, California, Vermont and Hawaii, among others, have seen their efforts stalled, at best, and rolled back in the most extreme case. The result has been the on-going loss of prime agricultural lands. The private sector, in the form of non-government organisations (NGOs) such as the American Farmland Trust and any number of local and state-wide land trusts, have attempted with some success to fill the gap that has been created by the retrenchment in public sector programmes. However, because in most cases the focus has been on preserving farmland rather than preserving and enhancing farming *per se*, these efforts often missed the mark. Further, given that agriculture defines the economic base and rationale in only a dwindling number of America's rural counties and communities – and tend to be concentrated in distinct regions, like the Great Plains – many rural problems have remained fundamentally unaddressed over the past several decades.

In term of federal involvement in rural planning and rural development, the US Department of Agriculture (USDA) has been and remains the 'lead' agency since the 1980s. This reflects a long-held view that agriculture equated rural and that rural areas were essentially populated by farm families. Over the years the majority of USDA funding has been devoted to increasing farm incomes and price stability in commodity markets, especially on large and corporate farms. While evidence suggests that subsidies and income supports for farming have helped to increase farm incomes and productivity, such programming has also had the perverse impact of pushing more and more rural labour out of such regions and into metropolitan areas (USDA, 2007).

The agency has an undersecretary for rural development, and its task has essentially been to advocate for resources as well as to coordinate services for rural America across all other federal agencies – including those for housing and urban development, health and human services and interior – which have programming that touches the lives of many rural Americas. Each presidential administration has attempted to find programme efficiencies and increases in programme impact, but sometimes ends-up just reducing federal financial support to localities. The 1990s saw the emergence of a number of state level rural development councils with a federally funded director whose job was to act as a liaison between that state's governor and various federal agencies to provide greater coordination of programming. Some of these councils remain active though the federal financial support for these entities has largely disappeared. All too often these councils failed to garner genuine local or grass-roots support as they were perceived as top-down rather than bottom-up entities. Also in the 1990s the Rockefeller Report, named after former Arkansas governor Winthrop Rockefeller who chaired the commission that generated the study, issued a report to the president, entitled *Revitalizing Rural America* (in Honadle, 2001). While the report made many recommendations, the most substantial one may well have been the need to 'target' and concentrate federal programmes in specific places rather than to distribute them broadly across rural America. This utterly failed to take hold given that it is the normative political tendency of public officials to bestow federal largess among as many constituents and places as possible. This is a reality that continues to the present day. Each new administration, so it seems, pledges reform, seeks change, promulgates policies, reorganises within and between the USDA and other agencies, and reimagines, repackages, and renames existing programmes. The underlining issue may well be that for contemporary rural America, the USDA is less relevant than in past decades. The assumption that the USDA must remain the key federal agency in rural development increasingly seems a dubious proposition, for as Freshwater and Scorsone (2002) rightly observe, farm policy has lost much of its ability to shape realities in rural America. Overall, as one of the most astute students of American rural policy, Beth Honadle, has written, '[i]t seems that most people no longer expect the Federal government to make a major rural policy pronouncement' (Honadle, 2001: 106).

If there has been an absence of an avowed American rural policy, numerous state and local programmes and projects, often supported with modest federal funding, have attempted to address outstanding issues. Unfortunately, too few states have funded their own projects and programmes preferring to utilise pass-through financial support from the federal government. For the longest time these programmes have tended to embrace a place-based or a territorially defined approach. Borrowing from the experiences of urban communities, place-based strategies, like enterprise zones, have sought to attract economic activity into rural communities.

As in some lagging metropolitan regions, place-based rural strategies have been focused on underperforming places by supporting infrastructure investments, industrial parks, tax-increment financing, 'Main Street' or downtown renewal projects, and other initiatives. The goal of such programmes has been to make such places more attractive to private sector investment. Public investments targeted to rural areas have a substantial history in the United States tracing back to the Tennessee Valley Authority which sought, during the Great Depression, to energise the middle South through massive investments in hydroelectric dams that provided cheap power to bring manufacturing to the region (Martin, 1967; Hargrove and Conklin, 1983). Behind such approaches lies the concept of agglomeration which, over time, creates defined industrial clusters that generate numerous spillover effects that, in turn, broaden and deepen wealth creation. Overall the effectiveness of such place-based focused programming in rural areas has been mixed at best and has increasingly been challenged (OECD, 2006; Partridge and Olfert, 2009).

There has developed over the past decade or so considerable criticism of this approach. Some have argued that the roots of the criticism of place-based strategies reflects a 'managerial turn' toward a more market-oriented approach that discounts the value of place distinctiveness and meaning (Padt and Luloff, 2009). Nevertheless the orientation is changing and instead of place-based approaches many advocate a people-based or people-oriented approach that focuses greater attention on enhancing population quality through greater access to education and training, health and human services investments, social capital formation and reinforcement, tourism development, local leadership cultivation and, most especially, entrepreneurship development. The state-wide small business development centres, funded in part by the US Department of Commerce, have been especially active in terms of the latter type of programming. Likewise, the Cooperative Extension Service, components of the land-grant university system, have been aggressive in nurturing change in rural areas with this people-oriented and bottom-up development approach, often referred to as asset-based development. Given the American emphasis on labour mobility, the fear is, of course, that such investments will ultimately lead to the further out-migration of labour from rural communities. Perhaps to counter this concern, entrepreneurship initiatives have been emphasised as a means both to create new sources of wealth and employment as well as helping in the intergenerational transfer of local assets, such as grocery stores and other key local businesses, due to retirements and/or relocations. Preventing the 'hollowing out' of small town and community business and commercial districts has been an important objective. Yet, one of the great obstacles facing existing and new business start-ups in rural areas has historically been the limited access to adequate financing, as well as the absence of a national commitment to provide appropriate digital communications technologies, such as high-speed broadband as well as transportation infrastructure, such as upgrades to rural airports. If there is a growing 'digital divide' in the United States, it is not only between the rich and the poor but also between metropolitan regions and many rural areas

In contrast, some rural places are not facing problems of the out-migration of the young and educated, isolated rapidly ageing populations, or the stagnation of economic activity. Communities in the urban/rural fringe within commuting distance of metropolitan employment and service opportunities continue to grow across the American landscape. While there has been considerable discussion about the 'urban comeback' and indeed, many cities are witnessing in-migration and growth, the urban fringe continues to be a 'contested landscape' loaded with land use conflicts, local political turmoil and the infrastructure and housing demands generated by growth (Furuseth and Lapping, 1999). And still other more rural places and small towns are facing problems generated by growth. Invariably these are places with intrinsically high amenity value which have been attracting population and service sector employment growth linked to second-home development, leisure facilities and sites (e.g. ski areas), and emerging rural retirement opportunities. These new rural residents are part of a trend of amenity migrants who seek out leisure and retirement housing. Some argue that this development reflects 'newer attitudes towards mobility, wealth, and the construction of identity in global society' rather than a return to the rural roots of the nation (Esparza, 2011). These new rural residents tend to be highly selective, better educated and wealthier adults who are attracted to 'rural areas that offer scenic beauty (e.g. mountains, beaches, lakes, and river) and those places near or adjacent to urban centers' (Longino and Hass, 1993: 18). Lower living costs, a fear of urban crime, and favourable climates also contribute to the 'pull factors' generating rural amenity-based in-migration. While many new rural residents generate demands for services and stimulate new employment opportunities, conflicts sometimes emerge between long-time residents and new-comers largely around issues of the cost and provision of public services and infrastructure, rising housing costs and taxes. For long-time elderly residents who depend on transfer payments, like pensions and Social Security, being priced out of the housing market has become a genuine concern. And for those who work in the service sector, rising housing prices may force them ever further away from their place of work thus requiring longer commutes. This can translate into additional change in nearby communities which now host service workers as they become 'gateway communities' for the more amenity-rich locations. As Marcouillier et al. (2011) conclude, exurban amenity-driven growth has created both rural 'haves' and 'have nots'.

Several other planning developments warrant note. The 'Smart Growth' movement has a strong urban emphasis that tries to rebuild many cities, better integrate mass transit with housing and commercial development, enhance community 'walkability' and a reduction in vehicular use and dependence – with Portland, Oregon leading the way. The smart growth movement has an important theoretical linkage with 'new urbanism' and form-based zoning and planning. In rural areas and small towns smart growth has manifested itself in greater attention being paid to downtown and Main Street renewal, walkability, mixed-use zoning and, most importantly, the preservation of open space (Daniels and Lapping, 2005). Likewise, the rise of the local food movement is having potentially significant consequences for small communities and rural areas. Growing consumer demand for locally produced foods as well as organic and non-GMO commodities has led to something of a renaissance in small scale agriculture across the country. More and more communities, urban and rural, are establishing programmes to accommodate and nurture local foods, re-establish the once declining fortunes of farmer's markets, and support mechanisms promoting direct farm-to-consumer sales. Maine has been in the forefront of these developments having recently passed into law 'food sovereignty' legislation that allows such farm-based direct sales to consumers with minimal overview by state agencies (Bayly, 2017). The State of Vermont is leading the way in developing an integrated state-wide approach to local food production, processing and distribution through its 'Farm-to-Plate' strategy (Vermont Sustainable Jobs Fund, 2018). These developments are taking place in two of America's most rural states where family farming is facing substantial problems. Finally, many communities are seeking to help with the rapid ageing of rural America through instituting in a wide array of approaches to support ageing- in-place. These programmes are aimed at permitting rural elders to remain in the rural communities and places where they raised their families and lived for years and where they can maintain their social connections and support systems.

Canadian rural policy and planning

Like the United States, Canada has a federal form of government. Inherent in this framework is a level of tension between the federal government and provincial governments. This makes intergovernmental relationships a little more difficult in Canada than in the US context. If one accepts a very broad definition of rural, then the creation in 1999 of a new federal territory, Nunavut, surely ranks as one of the most interesting developments in rural policy. A polar territory slightly smaller than Mexico with approximately 37,000 largely aboriginal inhabitants spread across a vast wilderness, Nunavut depends on resource extraction, including gold mining, traditional Inuit livelihoods and self-provisioning, and government supports.

Canada faces a good many rural problems which are not dissimilar from those in other developed economies of the Global North. In terms of the preservation of agricultural land, the Province of British Columbia has led the way in addressing the problem as early as the late 1970s. Of its pioneering Agricultural Land Reserve Programme, which protects approximately 5 million acres, two leading researchers have written that 'It is impossible not to be impressed by the qualities of the political act which grasped the farmland nettle . . . it is skilful, logical, bold and strong' (Wilson and Pierce, 1982: 17). Yet, as in so many regions in the Global North, urban sprawl is eroding some of the most valuable farmland in the province and the nation (Hume, 2013). More generally, however, as one recently published report notes:

> The problems that rural Canada face include: social and economic restructuring; decline in the manufacturing sector; demographic ageing as young people leave their home communities; and the diminishing of the social safety net because of the decline of the Canadian welfare state and the rise of the Canadian neoliberal state.
>
> *(Lauzon et al., 2017: 2–3)*

Canadian rural policy has also historically focused on agriculture. At the federal level there has been a recognition that this is no longer entirely relevant and appropriate, and slowly a sectoral approach focused on agriculture is gradually being replaced. At the federal level, a rural secretariat was established to bring greater continuity and coordination to rural programming among all ministries. While agricultural policies remain quite important in the Prairie provinces, other regions, particularly those with high levels of chronic unemployment, like the Maritimes, require programmes that emphasise business and entrepreneurial development. Some of these efforts have had a positive impact (Savoie, 1992). These policies reflect the fact that traditionally for much of Canada, as Freshwater (1991) has noted, regional policy *is* rural policy. Canadian public policy has concentrated for well over two generations or more on regional disparities (Savoie, 2000). This orientation is underscored, for example, by the establishment of the Atlantic Canada Opportunities Agency (ACOA) which funds projects and programmes throughout the Maritime provinces of New Brunswick, Prince Edward Island, Nova Scotia and Newfoundland and Labrador. While the approach of ACOA might work in the Maritime provinces, there is a recognition that other regions and areas of Canada require different approaches. In western Canada the Western Economic Diversification programme has reached out to traditionally under-represented groups to help address the problem of a rural skills gap. The Organisation for Economic Co-operation and Development (OECD) recently highlighted the programme's mobile laboratory initiative which brings training in carpentry, auto services, and machining to rural areas as a 'best practice' (OECD, 2011). All of these programmes and more are buttressed by the national Community Futures Network of Canada which, through nearly 300 local organisations, provides training and loans to support small business formation and growth (see http://communityfuturescanada.ca for details). In other words, where regional policy is rural policy, Canada has recognised that one size does not fit all.

Something of a hybrid between place-based and people-based development paradigms is emerging in the Province of Quebec. It is place-based in the sense that, as Reimer and Markey (2008: 4) argue, specific places are 'where people's assets are situated, that is where services are delivered, that is where governance takes place'. But it is also people-based in that it seeks to strengthen local social capacity through enhancing social capital, better local government performance in economic development, and civil society more broadly. The OECD (2010: 1) has gone so far as to say that 'Quebec has developed one of the most advanced rural policy approaches in the OECD . . . as suggested in the OECD's New Rural Paradigm'. Enshrined in its 'Politique nationale de la ruralite' (PNR), provincial policy tries to focus programmes and investments in socio-economic growth and development through the enhancement of social capital and civil society, asking local governments to become more directly involved in local entrepreneurial growth through public/private partnerships, more non-government development agencies and by assuming greater responsibility for self-development (OECD, 2011). Further, it supports collaboration between rural communities and municipalities and attempts to reduce programme and investment duplication. Bruno Jean has written, relative to Quebec, that 'the paradox of our post-modern world is that it is going in two opposite directions: globalization being one direction, and the rediscovery of the virtues of local communities being the other' (Jean, 2006: 65). The success or failure of Quebec's approach to rural development may well determine if it is possible for rural people and places to live in both worlds at one and the same time.

Rural policy and planning in the European Union territory

Rural development

As discussed in detail in Chapters 9 and 14, rural development policy in Europe has witnessed considerable convergence, leading some commentators to suggest that a Europeanisation of rural development has taken place, whereby EU promoted policy and practices have been mainstreamed within member states' rural development policies and into local development strategies (Shortall and Shucksmith, 2001; Ward and Lowe, 2004). European approaches to rural issues echo much of the evolution of rural development in North America with a distinctive shift from a rural modernisation paradigm, viewing agriculture as synonymous with rural, towards a focus on endogenous, bottom-up rural development, focused on territorial and integrated development usually at the local scale.

As outlined by Woods (2011), rural modernisation gained momentum in Europe after the Second World War – particularly focused on enhancing food security on a continent scarred by food shortages during and in the aftermath of the war. National policies prioritised agricultural modernisation through a transition from subsistence farming towards commercial food production through the mechanisation and industrialisation of farming, farmer training and extension programmes, the application of new agri-chemicals and biotechnologies, farm consolidation and specialisation, and the reorganisation of the agri-food sector. This drive towards productivism was often supported through state support, and with the emergence of the European Economic Community (later the EU) in 1957, the establishment of the Common Agricultural Policy (CAP) for agricultural subsidies and price support.

However, by the 1980s, CAP was increasingly criticised, particularly in relation to budget resources and as further consolidating regional disparities across Europe. Moreover, while CAP represented the EU's primary policy programme for supporting rural areas (and EU's most ambitious cross-national project), CAP had a singular focus on agriculture that failed to recognise other dimensions of the rural economy and society and, as a top-down centrally defined programme, neglected the diversity of rural Europe. In this context, in both the EU and member states, a policy discourse emerged in the 1990s which envisaged a fundamental shift in rural policy from sectoral support policies (predominantly agriculture) to territorial development and spatial approaches (Moseley, 1997, 2000; Shortall and Shucksmith, 2001). This policy shift recognised that spatial strategies and policies can integrate sectoral dimensions to public policy delivery (agriculture, housing, employment creation, transport, etc.) and offer holistic approaches to balancing the economic, social and environmental processes which shape Europe's rural areas. Throughout the 1990s, rural development theory and practice was firmly focused on local development and bottom-up approaches to face the challenge of the continued re-structuring of the agricultural industry, exemplified by the EU's LEADER programme (see, for example, Amdam, 1995; Moseley, 1997, 2000; Ploeg et al., 2000; Ray, 2000, 2002; Scott, 2002; Shucksmith, 2000).

Established in 1991, the Liaison Entre Actions de Développement de l'Économie Rurale (LEADER) programme represented a venture by the European Commission's DGVI (Agriculture) into a participatory, 'bottom-up' approach to rural development. The LEADER initiative aspired to be geared to local requirements and local origin, making use of available organisational capacity and expertise (CEC, 1991). The programme aimed to establish a network of rural development action groups enjoying a substantial degree of flexibility in implementing at a local level the initiatives financed by 'global grants'. From its inception, the LEADER approach was defined by the EC more as a set of principles than through 'pre-ordained', technocratic, sectoral measures (Ray, 1998). Participation at the local level meant an emphasis on community-based strategies. Having designed a LEADER plan and had it approved by the EC, a local action group would gain access to a block grant from the Structural Funds. As Ray highlights, the combination of a small 'grass roots' organisation, the flexibility of a block grant system of funding and the guiding objective to search for local innovative development projects marked the LEADER programme as potentially a very new policy style. The essential elements of this approach to rural development are identified by Moseley (1997) and Ray (2000) and include: a territorial and integrated focus; an endogenous development accent; the use of local resources; and local contextualisation through active public participation (see also the review in this Companion by Gkartzios and Lowe, Chapter 14). These themes are further elaborated through a brief discussion of rural policy in Ireland and Sweden.

Among all OECD countries no nation has a higher percentage of its population living in rural areas than Ireland. As recorded at the last Census in 2016, approximately forty per cent of the Irish population lives in rural areas, defined as living in settlements of less than 1500 people (CSO, 2016). However, the issue of rural population decline has provided *the* central narrative for rural policy in Ireland, with most western counties experiencing population loss for most of the last century, either through emigration (typically to US, UK or Australian cities) or through out-migration to eastern counties or to large urban centres in Ireland. In more recent times, the Irish economy has become more concentrated in the capital city, Dublin, leading to significant regional imbalances and following the Irish economic crash and international financial crisis of 2007/2008, rural Ireland has been deeply affected by a range of austerity measures (Murphy and Scott, 2014). While public and private sector services have been withdrawing for many years from rural Ireland, this process intensified during the post-crash years with businesses, hospital services, post offices, police stations and banking services closed or rationalised into larger urban centres – representing symbolic decline of many rural places.

Rural underdevelopment and measures for its alleviation have a long history in the Republic of Ireland, dating back, at least, to the efforts of the Congested Districts Board at the end of the nineteenth century (Cawley and Keane, 1999). Moreover, this interest in rural issues was often underpinned by local and community involvement, exemplified by the Muintir Na Tire (National Association for the Promotion of Community Development in Ireland movement) movement in the 1930s, and the 'Save the West' campaign of the 1960s (see Varley and Curtin, 1999). Although rural policy in the 1960s and 1970s was dominated by the modernisation of the agricultural sector (particularly in the wake of joining the European Economic Community in 1973), by the mid- to late 1980s, increasingly rural policy looked beyond agriculture resulting from the increased recognition given to rural areas by the EU and the growing realisation that the Common Agricultural Policy and top-down policies were not reducing regional disparities (McDonagh, 2002).

Throughout the 1990s, Ireland's rural areas, as elsewhere, increasingly embraced local action and local development solutions to face the challenge of the continued re-structuring of the agricultural industry, exemplified by the EU's LEADER programme. This approach to local development has been widely adopted in the Irish Republic, with a proliferation of rural development partnerships throughout the State, and increasingly the local partnership model has become mainstreamed into national rural development programmes. Reflecting and in parallel to European policy discourses, a suite of territorial rural development initiatives have been advanced in partnership with rural communities since the early 1990s focusing on local development, such as the Pilot Programme for Integrated Rural Development (1988–1990); the EU LEADER Programmes (I, II and LEADER+) (since 1991); area-partnerships for social inclusion (since 1991); and three successive National Rural Development Programmes (adopted since 2000). The current thrust of rural development policy is outlined in the Government's Action Plan for Rural Ireland (Government of Ireland, 2017), which represents a continuity with LEADER-style measures (aligned with local government reform) while also providing a significant new focus on small town and village physical regeneration.

In contrast to this widespread interest in rural development, the appetite for land-use planning or spatial planning in rural Ireland has been more limited. Historically, the fate of smaller settlements and rural areas received less than significant attention from physical planners, with rural areas often perceived as 'scenic backdrops to the drama of urban based investment in infrastructure, industry and services' (Greer and Murray, 1993: 3). This focus on the urban was reinforced by a lack of development pressures on the countryside due to historically high levels of out migration from rural communities, and also by the common perception of the rural arena as agricultural space. This standpoint has led to the operation of a liberal planning system in rural Ireland, described as one of the more lax rural planning regimes in Europe (Duffy, 2000) leading to relaxed attitudes towards building new single, dispersed dwellings in the open countryside (McGrath, 1998). This has led Gallent et al. (2003: 90) in a comparative European study of rural housing to classify rural planning in Ireland as a laissez-faire regime, suggesting that: 'the tradition of a more relaxed approach to regulation, and what many see as the underperformance in planning is merely an expression of Irish attitudes towards government intervention'.

Sweden

Few nations have so embraced the structure for rural development established by the European Union (EU) and the OECD as has Sweden. Throughout the years and the many changes that EU policy has seen –whether through the Mansholt Plan of 1968, the McSharry reforms of 1992 or the CAP and its sub components—Swedish programming has had a consistent sectoral orientation toward agriculture as the focus of national rural development policy (Ilkonen and Knobblock, 2007). But as the CAP itself has changed and matured, so has Sweden's policy system to reflect a broader conception of the relationship between agriculture and the rural (Granvik et al., 2012). What has emerged has been a more nuanced strategy that has sought increased competitiveness in both the agricultural and forestry sectors, improvement in the environmental quality of rural places, a commitment to economic diversification in the countryside and, using the LEADER approach, a strong push to build civil society and stronger local governance through inclusive public/private partnerships (Lindberg et al., 2012). There is some evidence that, after a modest start, such partnerships – especially where government has initially assumed a role in establishing such organisations – are taking hold in the Swedish countryside, especially as they concern the utilisation of natural resources (Bjarstig, 2017; Bjarstig and Sandstrom, 2017).

This more sophisticated approach still sees a multifunctional agriculture as one of the essential elements of Swedish rural policy, with food production combined with value-added activities to enhance the rural economy (UN, 2005). While recognising that those farms located near to population centres tend to be economically stronger, the policy provides that those in 'less favoured' regions are to be provided with compensation to maintain the agricultural landscape that is necessary for biodiversity and to support the growth of tourism. Additionally, a policy thrust to enhance Sweden's culinary traditions and innovations has buttressed agriculture in many places. As Rytkonen has observed, '[t]he reemergence of farm dairies, the establishment of a wine sector, the intensification of farm tourism, the constant increase of farm stores is just some of the many expressions of the new Swedish rurality' (Rytkonen, 2014: 1196).

Most recently the Swedish government has reinforced this multifunctional approach to agriculture by proposing a national food strategy (Ministry of Enterprise and Innovation, 2017). The strategy is a comprehensive plan that looks at all facets of the Swedish food value chain. It sets targets to be reached by 2030. The plan stresses the need for greater food self-sufficiency, a more competitive system in the global marketplace, greater organic and local food production, substituting biofuels for fossil fuels, and animal welfare and environmental stewardship consistent with resilience and sustainability principles. The plan calls for new investments in research and scientific innovation, technology transfer within the entire value chain, increases in the amount of pastureland and direct support for farms in more marginally fertile areas. Further, the plan calls for changes in the CAP and some of its mechanisms (Ernkrans, 2017). While ambitious, if the strategy is instituted and fully funded, its impact on rural Sweden can be transformative.

To assist rural economic development and technology transfer in all sectors, rural broadband has been expanded. Recognising that the private sector would not undertake this investment outside of population centres, the national government has used its funding to do so. The result has been that the Fibre to the Home Council Europe (FTTH) considers Sweden to be a 'showcase' for rural broadband (Babaali, 2013). As one OECD study has noted, 'The Internet offers the possibility to provide services in rural areas and for providers in rural areas to offer services outside their immediate territory' (OECD, 2011: 125). Enhanced broadband coverage is but one element of a larger rural infrastructure commitment.

Spatial planning and rural Europe

While rural development within the EU has witnessed a convergence of approaches across its member states, spatial planning for rural regions across the EU is much more diverse. Firstly, spatial planning is not a competency of the EU and therefore spatial planning remains a concern for individual member states with significantly less influence from European wide institutions compared to other policy areas, despite efforts to introduce a common spatial vocabulary as discussed further below, and a gradual shift from merely regulating land use change to enhanced 'spatial' governance. Secondly, and relatedly, spatial planning traditions, governance and legal systems vary widely across the European territory (Knaap et al., 2015). In this context, various commentators have attempted to group European planning systems into broad typologies. Nadin and Stead (2008), for example, identify three governance types in Europe: a Nordic model, an Anglo-Saxon model and continental/corporatist models. In turn, these have influenced a variety of planning cultures including comprehensive integrated planning (e.g. Netherlands, Germany), land-use regulation and management (e.g. traditionally the approach in the UK and Ireland), regional economic planning (e.g. France) and urbanism (closely aligned with urban design, more common in southern Europe). More recently, these traditions have become more blurred in practice.

While European planning systems remain rooted in contrasting governance traditions, the EU has nevertheless sought to exert influence on spatial development through regional development initiatives, environmental directives, and structural and cohesion funding to foster coordination and integration (Faludi, 2015). Moreover, for almost two decades the EU has attempted to influence national and regional planning through a growing interest in strategic spatial planning, as illustrated with its publication of the European Spatial Development Perspective (ESDP) (CSD, 1999) and subsequent development of the EU's 'Territorial Agenda', culminating in the publication of the Territorial Agenda of the EU 2020 (CEC, 2011). Through this EU spatial planning process, a new discourse of European spatial development has taken shape, with the definition of a new policy discourse, new knowledge forms and knowledge transfer, and new policy options (Richardson, 2000). ESDP promoted concepts – such as polycentric urban development, balanced spatial development, a new urban–rural relationship and transnational planning – were increasingly translated and applied into individual member state's national and regional policies and strategies throughout the 2000s (Healey, 2004; Lambregts and Zonneveld, 2004; Krätke, 2001).

The publication of the ESDP in 1999 also supplied a new vocabulary for *rural* spatial planning and has enabled policy-makers in member states an opportunity to 'rethink' spatial policies for rural Europe. In contrast to the emphasis on *local* development within rural development debates, the ESDP proposed tying rural areas much more into their urban and regional contexts and 'transform the countryside both physically and socially into images and identities of those who consume rural resources' (Hadjimichalis, 2003: 108). In this regard, the ESDP called for the strengthening of the partnership between urban and rural areas to overcome 'the outdated dualism between city and countryside' (CSD, 1999: 19) and to provide an integrated approach to regional problems. As Tewdwr-Jones and Williams (2001) argue, this focus on core-periphery (or urban–rural) relations necessitates an analysis of territory, rather than periphery, urban or rural alone.

However, both this policy direction and the ESDP's construction of rurality are contested. Hadjimichalis (2003) argues that from a rural perspective, a strong urban bias is evident in the ESDP. First, he suggests that cities are constructed as the sole driving forces and motors of regional development, which could lead to further agglomeration and a widening gap between urban and rural areas. This raises questions relating to the future of rural territory and space, particularly for population movement, if non-urban areas are constructed as areas of agriculture, green tourism and environmental protection (Richardson, 2000). Second, Hadjimichalis is critical of the ESDP's polycentric urban model, which may be relevant in flat, economically mature north-central EU countries like north-central France, Germany, Benelux and Denmark, but it marginalises rural and peripheral geography such as south-eastern Europe and Nordic countries. And third, Hadjimichalis criticises the urban–rural partnership for submerging small rural towns into their large counterparts, as urban needs predominate. As Richardson observes:

> The construction of rurality in this discourse raises many concerns for those with specific interests in rural development. Within this discourse, rurality is partly defined in a relational way – in relation to the urban – in partnership, but also in a way that subsumes the rural into a new European regional political economy. Cities and regions are the principal units of implementation of the ESDP's policies.
>
> *(Richardson, 2000: 66)*

As this critique suggests, the very notion of rurality is often deeply contested within planning policy, which is also reflected in contrasting approaches to how Europe's individual member states' conceive and frame the role and function of rural places within planning systems across the European territory. In broad terms, in more urbanised EU countries, planning in rural places has tended to prioritise protecting rural landscapes – related to aesthetic preferences, preventing urban sprawl and amenity protection – and protecting agricultural land for productive purposes. These countries, such as the Netherlands and Britain, tend to have mature and highly regulated planning systems. In contrast, in more predominately rural regions, or more peripheral regions, promoting development has often been prioritised over protection, favouring local development interests over the perceived preferences of external interests (e.g. environmental groups). In these cases, countries such as Greece, Spain and Ireland, development is less regulated, often characterised by laissez-faire approaches or by market enabling planning regimes (see, for example, Economou, 1997; Gkartzios and Shucksmith, 2015; and Chapter 13, this volume). In both highly urban societies and predominantly rural regions, local spatial planning and rural development policies are often poorly integrated (Gkartzios and Scott, 2014; Scott and Murray, 2009), with planning policy focused on managing physical change in the countryside and rural development concerned primarily with promoting economic and community development. This fragmentation reflects conflicting and competing constructions of rurality and is reinforced by local governance arrangements: formulating agreed rural sustainable development policies at a local level, therefore, remains an elusive and contested policy goal.

Rural policy and planning in Japan

Japan may well be a harbinger of things to come in many countries of the global North. With falling fertility rates and very little in-migration, Japan's population is rapidly ageing and declining. Nowhere is this more apparent than in the nation's rural areas. Rural Japan is, quite simply, hollowing out (Tanaka and Iwasawa, 2010).

Starting with the Meiji Restoration in 1869 through to the post-Second World War II period, Japan undertook significant reforms that profoundly altered land-use and land management as well as land tenure from a near feudal system to a unique small holdership pattern devoted to the intensive production of rice (Sorenson, 2010). The agricultural landscape of Japan is defined by man-made rice paddies connected through extensive irrigation systems. Farmed and settled land, the latter typified by high density living in village communities, contrasts with mountainous, forested landscapes. Forests cover marginally less Japanese land than in Scandinavian countries, and wilderness areas, either in the form of national parks, other protected lands, or privately owned tracts, play a crucial role in protecting the headwaters of rivers that provide much of the water for hydroelectric facilities and the country's irrigations systems, so vital to rice production.

In both its planning and agricultural/rural policy there has been a certain consistency and continuity in that Japan's policy system is heavily concentrated and top-down. It has been shaped by the almost uninterrupted political dominance of the conservative Liberal Democratic Party since the end of the Second World War. Agricultural land ownerships are quite small and were reduced in size even further to provide support for returning soldiers after the war. It has been the policy of Japan's federal government to favour agriculture in certain ways, not the least of which has been a pricing strategy for rice production that has kept farm incomes at or near par with those of urban workers. This reflected a growing concern over the disparity between urban–rural incomes (Hashimoto and Nishi, 2016). This has had the result, along with some other factors of a cultural nature, of reducing the willingness of farmers to convert land to more intensive uses. Indeed, even large-scale farmers have found it difficult to expand their operations by any means other than leaseholds. A National Land Agency monitors land markets but has little power to intervene in transaction. Most farm families secure substantial income from off-farm work with one or more members of the household working in nearby villages or in cities. As Tabayashi and Iguchi (2005) point out, contemporary Japanese agriculture is defined by part-time farm households with ageing and female farmers. Further, Hebbert (1989: 141) has documented the highly mechanised and efficient nature of contemporary Japanese farming thus:

> With mechanized planters, monorail or cable conveyors, lightweight sprinklers, vinyl greenhouses and artificial fertilizers, crops have increased as effort has diminished. The productivity of Japanese farming per unit area is the highest in the world . . .
>
> *(Hebbert, 1989: 141)*

As Japan industrialised and urbanised at unprecedented rates in the post-Second World War era, the pressure to convert farmland to other uses intensified despite popular concerns. A strategy to develop industry and land in and around growth-pole metropolitan areas was instituted as part of a national approach to development. Invariably it saw farmland near urban areas as potential development sites for industries and housing. In the early 1950s, The Agricultural Land Act was implemented to give prefecture (the equivalent of states or provinces) governors greater control over land conversion. Later in that decade, the central government's Ministry of Agriculture established a set of standards for agricultural land conversion. Neither of the two measures proved to be particularly effective. By 1969 the situation grew so dire that the Agricultural Promotion Areas Act was passed into law, which gave the prefectural governors the power to control potential conversions. In the year prior to the establishment of agricultural zones under the promotions act, a complex national city planning law was instituted that essentially tried to create green belts and urban growth boundaries as constraints against sprawl and development. These approaches failed to constrain land conversions. In the case of Tokyo's greenbelt, it was 'abolished so soon mainly because of strong development pressure and fierce political opposition from property owners and developers' (Ding and Zhao, 2011: 914). Despite the creation of a National Land Use Plan instituted in 1978, pressure on farmland continued apace. Additionally, some farmland, especially in mountainous and other marginal areas, was simply abandoned because no one was left in the community to farm it (Hashiguchi, 2014). Various other plans to conserve forests and conservation areas have been implemented but essentially village or rural planning in Japan lacks coherence and direction. As one assessment notes, 'rural landscapes are an important conservation challenge in Japan because they are being lost rapidly' (Natori et al., 2011: 285).

If turning around Japan's demographic decline is proving difficult to achieve, attempts to make the country's rural areas more resilient and sustainable has a somewhat longer history. The post-Second World War land reform included a programme that focused on enhancing the lives of rural women. Known as the Rural Livelihood Improvement Programme (R-LIP), women were introduced to concepts and methods of community leadership, alternative modes of income generation and family health enhancement. These efforts were similar to the work of the American Cooperative Extension Service and other programmes within some OECD nations (Mizuno, 2012). Another important effort to assist farmers has been the development of food cooperatives that connect producers with consumers. These function as networks that bring together women as producers with women as consumers, and as food safety issues have emerged over the decades, so have these institutions. Such programmes have the benefit of supporting local or what Odagiri (2011) has called Japan's 'tiny economies'. Other EU LEADER-like programmes have begun to be introduced in Japan, including the conceptualisation in policy of farming as a multifunctional activity and greater urban–rural linkages.

Various government ministries have, over the years, invested in rural infrastructure projects such as roads, sewage and water systems, and other sanitations projects (Hashimoto and Nishi, 2016). Some of these projects were geared to stimulate tourism in rural Japan. Given the nature of Japanese culture, returning to rural areas has long had a certain appeal for urbanities and governments at all levels have seized upon tourism as a way to diversify rural economies (Creighton, 1997; see also Japan Times, 2018). As in cases elsewhere, some of these investments have increased land prices in rural areas and brought with them damage to the local agricultural economy and negative externalities in the environment (Goto, 2008). Other programmes and policies, such as providing mobile food and medical services, have also been implemented. Yet, as in some other countries of the Global North, it is hard to see how Japan will surmount some of its demographic and economic challenges to produce better outcomes for its rural communities and people.

Conclusion

While rural development policy has slowly evolved from a focus on agricultural modernisation to a broader focus on territorial development, over the past half-century the landscape for rural planning has remained largely static. As Lapping (2006: 118) suggests, 'rural planning and policy has demonstrated an amazing consistency and lack of imagination in terms of their focus and orientation. In many national and international contexts an emphasis on agriculture as the rural persists.'

Despite rapidly changing demographic, social, environmental and economic realities, rural planning policy and practice often continue to emphasise the primacy of agriculture, which in turn can produce policies that hamper the wider development of the rural economy or society. In the so-called 'urban age', rural planning has become more marginal within the planning profession and planning academia, as cities are framed as the key drivers of innovation and economic growth, with the rural constructed as the scenic backdrop or 'breadbasket'. Rural areas increasingly become sites for urban infrastructure, often mega projects, to connect cities and create new scale-economies. However, as the remaining chapters in this book demonstrate, it is now timely to reimagine and reinvigorate rural planning to move beyond narrow development versus preservation debates to imagine more sustainable rural futures. In addition to identifying some of the limitations of rural planning through examining its evolution, the chapter has also highlighted some examples of better practice across North America and the EU. This includes experimenting with rural governance arrangements, the importance of 'community' as a building block for local policy formulation, transitioning towards multifunctional agriculture and rural places, and new infrastructural needs (particularly broadband roll-out) to support rural communities and as well as diversified and creative economies. In this context, there is considerable scope for more comparative research for cross-learning and practice-transfer.

References

Abercrombie, P. (1926) The preservation of rural England, *Town Planning Review*, 12 (1), 5–56.

Amdam, J. (1995) Mobilisation, participation and partnership building in local development planning: experience from local planning on women's conditions in six Norwegian communes, *European Planning Studies*, 3, 305–332.

Babaali, N. (2013) *Sweden: A Showcase for Rural FTTH*, retrieved from www.ftthcouncil.eu/documents/ Opinions/2013/Rural_FTTH_Nordics_Final.pdf (accessed 12 April 2018).

Bayly, J. (2017) Food Sovereignty Bill passes in Maine House, Senate, *Bangor Daily News*, 3 June, retrieved from https://bangordailynews.com/2017/06/03/news/state/food-sovereignty-bill-passes-in-maine-house-senate (accessed 12 April 2018).

Bjarstig, T. (2017) Does collaboration lead to sustainability? A study of public-private partnerships in the Swedish mountains, *Sustainability*, 9 (10), 1–22.

Bjarstig, T. and Sandstrom, C. (2017) Public-private partnership in a Swedish rural context: a policy tool for the authorities to achieve sustainable rural development? *Journal of Rural Studies*, 49, 58–59.

Bosselman, F. and Callies, D. (1972) *The Quiet Revolution in Land Use Control*, Washington, DC: President's Council on Environmental Quality.

Cawley, M. and Keane, M. (1999) Current issues in rural development, in J. Davis (ed.), *Rural Change in Ireland*, 143–160. Belfast: Institute of Irish Studies, Queen's University Belfast.

CEC. (1991) LEADER, Commission notes to member states, *Official Journal of the European Communities*, OJ C73 (19 March), 33–37.

CEC. (2011) *Territorial Agenda of the EU 2020*, Brussels: Commission of the European Communities.

Creighton, M. (1997) Consuming rural Japan: the marketing of tradition and nostalgia in the Japanese travel industry, *Ethnology*, 36, 239–254.

CSD. (1999) *European Spatial Development Perspective: Towards a balanced and sustainable development of the territory of the EU*, Luxembourg: Commission of the European Communities.

CSO. (2016) *Census 2016*, Dublin: Central Statistics Office.

Daniels, T. and Lapping, M. (2005) Land preservation: an essential ingredient in smart growth, *Journal of Planning Literature*, 19 (3), 316–329.

Ding, C. and Zhao, X. (2011) *Urbanization in Japan, South Korea and China: Policy and Reality*, retrieved from http://smartgrowth.umd.edu/assets/urbanization_japan_southkorea_china.pdf (accessed 12 April 2018).

Duffy, P. (2000) Trends in nineteenth and twentieth century settlement, in T. Barry (ed.), *A History of Settlement in Ireland*, 206–227, London: Routledge.

Economou, D. (1997) The planning system and rural land use control in Greece: a European perspective, *European Planning Studies*, 5 (4), 461–476.

Ernkrans, M. (2017) *Which CAP Can Support our New Swedish Food Strategy? Chair Report*, Swedish Parliamentary Committee on Environment and Agriculture, retrieved from www.ksla.se/wp-content/ uploads/2017/02/Matilda-Ernkrans.pdf (accessed 12 April 2018).

Esparza, A. X. (2011) The exurbanization process and rural housing markets, in D. Marcouiller, M. Lapping and O. Furuseth (eds), *Rural Housing, Exurbanization, and Amenity-Driven Development: Contrasting the 'Haves' and the 'Have Nots'*. Aldershot: Ashgate.

Faludi, A. (2015) The European Union context of national planning, in G. Knaap, Z. Nedović-Budić and A. Carbonell (eds), *Planning for States and Nation-states in the US and Europe*, 259–290, Cambridge, MA: Lincoln Institute of Land Policy.

Frank, K. and Hibbard, M. (2016) Rural planning in the twenty-first century: context-appropriate practices in a connected world, *Journal of Planning Education and Research*, 37, 299–308.

Freshwater, D. (1991) Canadian rural policy: mostly a regional matter, *Rural Development Perspectives*, 7, 13–18.

Freshwater, D. and Scorsone, E. (2002) The search for an effective rural policy: an endless quest or an achievable goal, paper presented at the Rural Matters RUPRI Policy Conference, Nebraska City, Nebraska.

Furuseth, O. and Lapping, M. (1999) *Contested Countryside: The Rural Urban Fringe in North America*, Aldershot: Ashgate.

Gallent, N., Shucksmith, M. and Tewdwr-Jones, M. (2003) *Housing in the European Countryside, Rural Pressure and Policy in Western Europe*, London: Routledge.

Gkartzios, M. and Scott, M. (2014) Placing housing in rural development: exogenous, endogenous and neo-endogenous approaches, *Sociologia Ruralis*, 54 (3), 241–265.

Gkartzios, M. and Shucksmith, M. (2015) 'Spatial anarchy' versus 'spatial apartheid': rural housing ironies in Ireland and England, *Town Planning Review*, 86 (1), 53–72.

Goto, J. (2008) Evolution of rural and agricultural policy in Japan, 1945–2002, in L. Apedaile and N. Tsuboi (eds), *Revitalization: Fate and Choice*, n.p., Brandon, Manitoba: Brandon University Rural Development Institute and the Canadian Rural Revitalization Foundation.

Government of Ireland. (2017) *Realising Our Rural Potential, Action Plan for Rural Development*, Dublin: Government of Ireland.

Granvik, M., Lindberg, G., Stigzelius, K. A., Fahlbeck, E. and Surry, Y. (2012) Prospects of multifunctional agriculture as a facilitator of sustainable rural development: Swedish experience of Pillar 2 of the Common Agricultural Policy (CAP), *Norsk Geografisk Tidsskrift-Norwegian Journal of Geography*, 66 (3), 155–166.

Greer, J. and Murray, M. (1993) Rural Ireland – personality and policy context, in M. Murray and J. Greer (eds) *Rural Development in Ireland, A Challenge for the 1990s*, 1–18, Avebury: Aldershot.

Hadjimichalis, C. (2003) Imagining rurality in the new Europe and dilemmas for spatial policy, *European Planning Studies*, 11, 103–113.

Hargrove, E. and Conklin, P. (1983) *TVA: Fifty Years Grass-Roots Bureaucracy*, Urbana, IL: University of Illinois Press.

Hashiguchi, T. (2014) Current status of agriculture and rural areas in Japan and prospect of new policy framework: comparisons with the direct payment system in Japan and Europe, paper presented at the European Agricultural Economics Congress, Ljubljana, Slovenia.

Hashimoto, S. and Nishi, M. (2016) Policy evolution and land consolidation and rural development in postwar Japan, *Geomatics, Landmanagement and Landscape*, 3 (3), 57–75.

Healey, P. (2004) The treatment of space and place in the new strategic spatial planning in Europe, *International Journal of Urban and Regional Research*, 28, 45–67.

Hebbert, M. (1989) Rural land-use planning in Japan, in P. Cloke (ed.), *Rural Land-Use Planning in Developed Nations*, 130–151, London: Unwin, Hyman.

Honadle, B. W. (2001) Rural development policy in the United States: beyond the cargo cult mentality, *Journal of Regional Analysis and Policy*, 31, 93–108.

Hume, M. (2013) Developers winning out over farmland preservation, planner says, *Toronto Globe and Mail*, 15 November 2013, retrieved from www.theglobeandmail.com/news/british-columbia/devel opers (accessed 4 December 2017).

Ilkonen, R. and Knobblock, E. (2007) An overview of rural development in Sweden, in A. K. Copus (ed.), *Continuity or Transformation? Perspectives on Rural Development in the Nordic Countries*, 90–110, Stockholm: Nordregio Report 4.

Japan Times. (2018) New group to promote agricultural tourism in Rural Japan, *The Japan Times*, 9 February, retrieved from www.japantimes.co.jp/news/201802/09/national/news-group-promote-nohaku-agricultural-tourism-rural-japan (accessed 11 February 2018).

Jean, B. (2006) The study of rural communities in Quebec: from the 'folk society' monograph approach to the recent revival of community as place-based rural development, *Journal of Rural and Community Development*, 1 (2), 56–68.

Knaap, G., Nedović-Budić, Z. and Carbonell, A. (eds) (2015) *Planning for States and Nation-states in the US and Europe*, Cambridge, MA: Lincoln Institute of Land Policy.

Krätke, S. (2001) Strengthening the polycentric urban system in Europe: conclusions from the ESDP, *European Planning Studies*, 9, 105–116.

Lambregts, B. and Zonneveld, W. (2004) From Randstad to Deltametropolis: changing attitudes towards the scattered metropolis, *European Planning Studies*, 12, 299–321.

Lapping, M. B. (2006) Rural policy and planning, in P. Cloke, T. Marsden and P. Mooney (eds), *Handbook of Rural Studies*, 106–107, London: Sage Publications.

Lauzon, A, Bollman, R. and Ashton, B. (2017) *State of Rural Canada Report, 2017* Canadian Rural Revitalization Foundation.

Lindberg, G., Copus, A., Hedstrom, M. and Perjo, L. (2012) CAP rural development policy in the Nordic countries: what can we learn about implementation and coherence? Working Paper 6, Stockholm: Nordregio.

Longino, C. F. Jr. and Hass, W. H. (1993) Migration and the rural elderly, in C. N. Bull (ed.), *Aging in Rural America*, Newbury Park, CA: Sage Publishers.

Marcouiller, D., Lapping, M. and Furuseth, O. (2011) *Rural Housing, Exurbanization and Amenity-Driven Development: Contrasting the 'Have' and the 'Have nots'*, Aldershot: Ashgate.

Martin, J. R. (1967) *The Economic Impact of the TVA*, Knoxville, TN: University of Tennessee Press.

McDonagh, J. (2002) *Renegotiating Rural Development in Ireland*, Aldershot: Ashgate.

McGrath, B. (1998) Environmental sustainability and rural settlement growth in Ireland, *Town Planning Review*, 3, 227–290.

Ministry of Enterprise and Innovation (2017) *A National Food Strategy for Sweden: More Jobs and Sustainable Growth Throughout the Country*, Stockholm: Government Office of Sweden.

Mizuno, M. (2012) Rural development – the role of rural livelihood improvement, in T. Toyada et al. (eds), *Economic and Policy Lessons from Japan to Developing Countries*, 134–142, Basingstoke: Palgrave Macmillan.

Moseley, M. (1997) New directions in rural community development, *Built Environment*, 23, 201–209.

Moseley, M. (2000) Innovation and rural development: some lessons from Britain and western Europe, *Planning Practice and Research*, 15, 95–115.

Murphy, E. and Scott, M. (2014) 'After the crash': life satisfaction, everyday financial practices and rural households in post Celtic Tiger Ireland, *Journal of Rural Studies*, *34*, 37–49.

Nadin, V. and Stead, D. (2008) European spatial planning systems, social models and learning, *disP The Planning Review*, 44 (172), 35–47.

Natori, Y., Silbernagel, J. and Adams, M. (2011) Biodiversity conservation planning in rural landscape in Japan: integration of ecological and visual perspectives, in I. Pavlinov (ed.), *Research in Biodiversity: Models and Applications*, 285–306, London: Intech Publishers.

Odagiri, T. (2011) *Rural Regeneration in Japan*, Rural Economy Research Report, 56, Newcastle Upon Tyne: Centre for Rural Economy, University of Newcastle Upon Tyne.

OECD. (2006) *OECD Policy Brief, Reinventing Rural Policy*, Paris: OECD.

OECD. (2010) *Rural Policy Reviews: Quebec, Canada*, Paris: OECD.

OECD. (2011) *New rural policy: linking up for growth*, Paris: OECD.

Padt, F. J. and Luloff, A. E. (2009) An institutional analysis of rural policy in the United States, *Community Development*, 40 (3), 232–246.

Partridge, M. D. and Olfert, M. R. (2009) Lessons from the evaluation of Canadian and US rural policy, paper presented to the Expert Consultation on Aiding the Process of Agricultural Policy Reform, OECD, Paris.

Ploeg, J. van der, Renting, H., Brunori, G., Knickel, K., Mannion, J., Marsden, T., Roest, K. de, Sevilla-Guzmán, E. and Ventura, F. (2000) Rural development: from practices and policies towards theory, *Sociologia Ruralis*, 40, 391–408.

Ray, C. (1998) Territory, structures and interpretation: two case studies of the European Union's LEADER I programme, *Journal of Rural Studies*, 14, 79–87.

Ray, C. (2000) Endogenous socio-economic development in the European Union – issues of evaluation, *Journal of Rural Studies*, 16, 447–458.

Ray, C. (2002) A mode of production for fragile rural economies: the territorial accumulation of forms of capital, *Journal of Rural Studies*, 18, 225–231.

Reimer, B. and Markey, S. (2008). *Place-Based Policy: A Rural Perspective*, report for Human Resources and Social Development Canada, retrieved from www.crcresearch.org/files-crcresearch_v2/files/ReimerMarkeyRuralPlaceBasedPolicySummaryPaper20081107.pdf (accessed 12 April 2018).

Richardson, T. (2000) Discourses of rurality in EU spatial policy: the European spatial development perspective, *Sociologia Ruralis*, 40, 53–71.

Rytkonen, P. (2014) Constructing the new rurality: challenges and opportunities of a recent shift in Swedish rural policies, *International Agricultural Policy*, 2, 7–19.

Savoie, D. (1992) *Regional Economic Development: Canada's Search for Solutions*, Toronto: University of Toronto Press.

Savoie, D. (2000) *Community Economic Development in Atlantic Canada*, Moncton, NB: Canadian Institute for Research on Regional Development.

Scott, M. (2002) Delivering integrated rural development: insights from Northern Ireland, *European Planning Studies*, 10, 1013–1025.

Scott, M. and Murray, M. (2009) Housing rural communities: connecting rural dwellings to rural development in Ireland, *Housing Studies*, 24, 755–774.

Shortall, S. and Shucksmith, M. (2001) Rural development in practice: issues arising in Scotland and Northern Ireland, *Community Development Journal*, 36, 122–133.

Shucksmith, M. (2000) Endogenous development, social capital and social inclusion: perspectives from LEADER in the UK, *Sociologia Ruralis*, 40, 208–218.

Sorenson, A. (2010) Land, property rights and planning in Japan: institutional design and institutional change in land management, *Planning Perspectives*, 25, 279–302.

Tabayashi, A. and Iguchi, A. (2005) Changing agricultural and farm successors in Japan, *Studies in Human Geography*, 29, 85–134 (in Japanese with English abstracts).

Tanaka, K. and Iwasawa, M. (2010) Aging in rural Japan: limitations in the current social care policy, *Journal of Aging Social Policy*, 22, 394–406.

Tewdwr-Jones, M. and Williams, R. (2001) *The European Dimension of British Planning*, London: Spon Press.

UN. (2005) *Sweden: Rural Development*, New York: UN, retrieved from www.un.org/esa/agenda21/natlinfo/countr/sweden/ruralDevelopment.pdf (accessed 12 April 2018).

USDA. (2007) Sources and levels of operator household income, structure and finances of US farms: family farm report, retrieved from www.ers.usda.gov/publications/eib24/eib24e.pdf (accessed 24 February 2018).

Varley, T. and Curtin, C. (1999) Defending rural interests against Nationalists in 20th century Ireland: a tale of three movements, in J. Davis (ed.), *Rural Change in Ireland*, Belfast: Institute of Irish Studies, Queen's University Belfast.

Vermont Sustainable Jobs Fund. (2018) *Plan: Vermont Farm to Plate: Strengthening Vermont's Food System*, Montpelier: Vermont Sustainable Jobs Fund.

Ward, N. and Lowe, P. (2004) Europeanizing rural development? Implementing the CAP's second pillar in England, *International Planning Studies*, 9 (2–3), 121–137.

Wilson, J. W. and Pierce, J. T. (1982) The agricultural land commission of British Columbia. *Environments*, 14 (3), 11–20.

Woods, M. (2011) *Rural*, London: Routledge.

4

Rural planning in an emerging economy

India

Hemalata C. Dandekar

Introduction

Following Indian Independence in 1947, India embarked on five-year national development plans, the first from 1951–1956. Under the leadership of Prime Minister Jawaharlal Nehru, national plans took inspiration from Soviet five-year plans, first implemented in 1928–1932. India embraced the Soviet ideal of rapid industrialisation as a way to strengthen the national economy. However, in its own approach to rural planning, it rejected the Soviet approach of replacing individual peasant farming with a system of collective farming in which peasant property and entire villages were incorporated into the state. It thus stood in contrast to China, which in 1953 encouraged farmers to form small cooperatives to increase farm yields, but, beginning in 1958, turned to Soviet style large-scale collective farming policies.

Rather than follow this collective farming approach, India turned to the US for inspiration, looking at its experiments with inducing agricultural changes based on scientific experimentation and a government-financed extension system to disseminate best practices. The formation of farm cooperatives in rural US were also of interest. But rural planning in India, shaped by the complex specificities of its rural areas, took its own trajectory. As a key member of the 'Third World' block of countries whose leaders included India's Prime Minister Nehru and Egypt's President Nasser, it adopted rural planning policies that struck an intermediate and distinct path, differing from the US and the Soviet Union, and explicitly embraced goals of equity and redistributive justice.

With its long tradition of settled agriculture dating back to 3,000 BC and even 10,000 BC (Krishna and Morrison, 2009), India entered the post-Independence development period with complex and entrenched socio/cultural traditions which complicated the adoption of the US science based approach. The outcomes, particularly with respect to farm size, farm employment, and urbanisation have been quite different. Established traditions in India had enabled populations to cultivate land and improve the quality of human existence in balance with the area's natural resource base and land quality – the foundation of a traditional village economy (Dandekar, 1986). The legacy of heavily populated newly independent India in the mid-twentieth century, was an agriculture-based economy and a predominantly agrarian based population with long entrenched, hierarchical, societies, histories and traditions. Although India took a significantly different track than the US in framing and implementing rural planning, it emulated its indirect market-driven approach, tempering it to attain social justice by redistributing productive assets and thus their benefits. This was motivated both by considerations of ideology and political efficacy. Strategies encompassed both the development of agriculture and improving the access of rural people to infrastructure, education, housing and amenities, and, reducing income disparities.

The specific foci of rural government policy in India and the level of resources committed has varied over successive five-year plans reflecting changes in social and economic conditions and the political and ideological climate (Muthusamy, 2012; Olfert et al., 2010). They have addressed: land ownership and inheritance; cultivation practices including crop types, rotation and tilling cycles; protective practices and rights in shared resources such as forest areas and grazing lands; water sharing and irrigation practices; and, division of labour and gender responsibilities. A broader definition of rural planning as encompassing policy that both explicitly and implicitly impacts rural conditions – economic, social, and physical – better serves to capture the sum total of what is happening currently to, and in, rural places in India. Legislation, policy and programmes in closely related sectors such as agriculture, transportation and communications, education, health, and social welfare, at national or state levels, have far reaching and profound impacts on rural conditions. They shape the availability of infrastructure and jobs and determine the range and quality of services accessible to rural populations. In contrast, rural development plans, at state and national levels, are largely defined by social welfare agendas to address inequalities in access and to achieve redistributive justice.

In the twenty-first century the focus of rural planning is shifting. Although still concerned with addressing conflicting goals of economic development and resource conservation and developing and protecting physical and human capital, the specific issues and emphasis within the rural sector also respond to the effects of globalisation on the local economy. The traditional goals, development of agriculture along with improving the access of rural people to infrastructure, education, housing and amenities persist. But the drive for economic development in the context of global competition and extreme international fragmentation of supply side production and distribution has raised concerns. Some echo those that have been encountered by planners working in urban contexts such as: the need to reduce income disparities and spatial segregation; remediate for factors inducing climate change; reduce greenhouse gas emissions; and increase access to good quality housing. And the more traditional focus on agriculture and resources has shifted to embrace the rhetoric, if not the practice, of sustainable agriculture and the judicious use of natural resources particularly with respect to water conservation and use of renewable forms of energy. This evolution of rural planning in India since the development decades of the mid twentieth century to its current emphasis in the context of an emerging, global, economy is described here to highlight issues of the rural in a rapidly emerging economy.

Rural planning in the development decades (1940s to 1980s)

India attained independence from British colonial rule at the end of the Second World War and self-identified as 'Third World' – not to categorise itself along the Rostowian (1956) 'stages of development' but rather on grounds of political ideology – as centrist rather than left or right. In the five or so development decades which followed the Second World War, from the 1940s to the 1990s, this ideology, to a greater or lesser extent, shaped the country's attitudes and approaches to planning in general, and rural development in particular. Newly liberated, India embarked on ambitious, centralised, national development plans. Given a predominantly rural population, planning for agricultural and overall rural development was initially perceived as an integral aspect of the national plans. The rural sector, one of several interrelated sectors of the national plan, had a decidedly spatial, locational (non-urban, non-city) connotation, one in which practices and objectives were many and varied. The key governmental concern was about agriculture, its efficiency, diversity, and its ability to sustain the rural population, produce surpluses to support urban residents, and, earn needed foreign exchange through exports. Rural contributions to economic growth were key indicators of success and rural development policies were designed to reshape traditional societal structures and change rural people's attitudes and capacities 'to make them modern'.

During the development decades following the Second World War, approaches to rural planning in the US, with its emphasis on the modern farm, which evolved with assistance from research at land grant universities and extension services funded by the US government, were emulated in India. They included, particularly in areas of irrigated agriculture, the use of high yield hybrids, fertilisers, pest controls, and extensive mechanisation to serve national and international markets. But, as one of the founding countries of the 'Third World' bloc, India attempted to forge a path that differed from US-style capitalism and Soviet-style communism. Its national constitution endorsed centralised government planning to attain modernisation by facilitating rapid industrialisation of both the private and public sectors. Rural planning was a key element of this development model, with policies intended to increase production, but also to redistribute wealth on social justice and human welfare grounds. The latter provided legitimacy for the new government of a country with a significant population that was impoverished and rural. India aimed to use agricultural surplus to provide capital for industrialisation and infrastructure development for both urban and rural populations. Rapid socioeconomic change was concentrated in industrialisation at chosen metropolitan centres on the assumption that the benefits would 'trickle down' through the economy and spread to rural areas.

Promoting complementary rural planning to reduce existing and anticipated urban–rural disparities in wealth was recognised, and the desire to increase agriculture production was tempered with a broader objective – to bring about improvements in the living standards and quality of life of economically disadvantaged rural populations. This emphasis was an acknowledgement that the majority of India's populations was rural, asset poor, involved in subsistence production, and, included members of lower strata (caste) groups who were oppressed by social structures that were feudal in character. India evolved a planning model for 'integrated rural development' that recognised gender and class (caste) differences and attempted to address social inequalities in access to rural resources. It was an attempt to balance between agricultural and economic development/growth and social welfare and redistributive justice.

At Indian Independence in 1947, rural areas constituted some 80 per cent of the national population. Rural planning's social equity with redistributive justice paradigm focused on strategies that included: passage of 'land to the tiller' legislation to redistribute land from large absentee landowners to active farmers; building rural infrastructure; building roads to improve access; increasing irrigation of land to help productivity; and, enhancing education systems and access to them to build the capacity of the rural work force. Existing indigenous school systems were expanded to create government-run, secular schools with required universal primary school education for both sexes, a radical departure from traditional schools where education was based on religious texts and imparted largely to upper-caste male children. Health care services were expanded and the supply of safe, treated, drinking water and improvements in the quality of housing of the rural poor were promoted as a way to reinforce preventive health care.

In agricultural production, planners also closely studied the approach that had proven so successful in the US a century earlier. The goal was to increase agricultural productivity by disseminating scientific research and promoting implementation in farming practices. The preferred model for dissemination was the US extension system – formalised in 1914, but with roots back to the Morrill Act of 1862 that established land-grant universities to educate farmers and develop a scientific approach to agriculture (USDA, 2014). The government was to support these efforts by developing essential rural infrastructure.

A summary of the approach to rural planning in the Indian first five-year plan (1951–1956) leads with the following statement:

> The 'Community Development Programme' (CDP) was launched on 2 October 1952 by Prime Minister Pandit Jawaharlal Nehru through which emphasis was given to the development of agriculture, irrigation, energy and power, industry and minerals, village small scale industry, transport, employment etc. The National Extension Service Programme, Mettur Dam, Hirakud Dam, and Bhakra Nangal Dam were established as irrigation programme during the plan.
>
> *(Bhende and Yatanoor, 2017: 23)*

Mosher, a US international agricultural development expert, recruited as a consultant and advisor to the Government of India, describes Indian rural planning as seeking moral and social transformation, economic efficiency, redistribution and social justice, and, active participation in decision-making (Mosher, 1976). He summarises the phases of Indian rural development planning describing its common characteristics as:

- Recognising that although economic development is important, it is not an essential precondition to a social transformation that is considered the central objective of a rural development strategy.
- Increasing efficiency with the adoption of fertilisers and high yield varieties of cereals. The attempt was to boost both food grain production, increase commercial crops and promote production of commodities for which there was an international market.
- Developing programmes that address equity and asset redistribution.
- Enhancing rural citizen's participation in the evolution of locally appropriate planning programmes.

Rural planning measures in India, conspicuously emulating the US model of agricultural development, included: investments in irrigation projects; the promotion of scientifically developed cash crops through access to improved seeds, equipment and fertiliser (the so-called Green Revolution approach to making agriculture productive); providing credit for agricultural investments to stimulate agricultural production; and, establishing an extension system to introduce scientific farming techniques at the village level. Government also regulated markets to stimulate trade in agricultural commodities; invested in roads and communications; and founded rural cooperatives to buy and sell agricultural products and facilitate the marketing of goods in rural areas. This system had worked effectively in the land-rich colonisation of the US, but was, in some respects, an uncomfortable overlay on the stratified, caste-based occupational structure of the Indian village and the small size of average land holdings throughout the country (Dandekar, 1986).

The consequences of these national policies and investments in industry and commercial agriculture were, in hindsight, predictable. They served to increase the gap between those involved in the corporate formal sectors of the economy, urban and rural, and those in the informal, traditional, subsistence sectors. Rural landless and land poor, squeezed out of traditional positions in village society, migrated to the city in unprecedented numbers. Cities were unable to accommodate the massive influx of displaced, often unskilled, rural labourers and urban environments deteriorated as slum settlements of the poor proliferated. Responding to this phenomenon inspired a new set of rural planning policies. Stemming migration to the city by improving rural life and diversifying the rural economy emerged as prime planning objectives. Actions to improve rural living conditions included provision of postal service, healthcare facilities, clean drinking water, and expansion of the electrical grid into the countryside. As this approach evolved, its objectives also expanded to explicitly include increasing equity and distributing the benefits of development to the poor. Planning measures included placing a ceiling on individual land holdings, reforming tenancy to enhance the rights to the land of those who cultivated it, initiating programmes to support traditional artisans with better tools, establishing credit and marketing networks to allow them to compete with industry or at least sustain them until alternative jobs were created. Finally, the Indian government put greater emphasis on free, mandatory elementary school education throughout rural areas to improve skills the poor needed to compete for development benefits.

The shifts in Indian rural planning over the years are reflected in how the sector is addressed in the five-year national plans. In the second plan (1956–1961) a greater emphasis is placed on economic development. This is reflected in the rural sector by programmes such as promotion of Khadi and Village Industries, creation of Intensive Agricultural Districts and Rural Housing and Tribal Area development. In the third plan (1961–1966), during border wars with both China and Pakistan, the emphasis was on defence industries. States were charged with expanding electricity grids in rural areas and improving primary and secondary education. The fourth plan (1969–1974), developed under the leadership of Prime Minister Indira Gandhi, emphasised alleviating poverty and included numerous programmes to help the small and marginal farmers such as providing assistance to dig wells, loans for animal husbandry and other efforts to raise on-farm incomes. The fifth plan (1974–1979) was a period when Mrs. Gandhi imposed a 'State of Emergency' on the country and suspended democratic rights. The plan included rural programmes such as training youth for employment. The sixth plan (1980–1985) featured a one-child policy and programmes were introduced which aimed at improving conditions for women and children and increasing rural employment with a particular focus on the landless unemployed. These and other anti-poverty and job creation programmes were continued and expanded in the seventh plan (1985–1990). The period from 1992 to 1997 was one of economic instability and only annual plans were adopted which had an emphasis on population control and poverty eradication.

Economic liberalisation and the emerging economy of the twenty-first century

The disintegration in December 1991 of the Soviet Union into fifteen separate countries, celebrated in the west as a triumph of democracy over totalitarianism and evidence of the superiority of capitalism over socialism, blurred the international philosophical differences that had characterised various national development planning approaches in the development decades. Ideological distinctions between and within political blocks blurred, and a shift to capitalist modes of production was ushered in throughout the world. By the end of the twentieth century engaging in the global market place became the universally embraced path to economic development. Economic liberalisation, instituted by the late 1990s in most countries of the developing world, significantly shifted production toward global markets. Rural planning too reflected this and encompassed policies giving greater importance to market mechanisms, particularly in the agriculture and resource sectors. And thus the diverse objectives of rural planning once promulgated in countries such as Tanzania, Egypt, India, and China under Mao were effectively erased. In addition, the balance between populations that lived in areas that were categorised as urban and those that were rural shifted in favour of the urban. World urban population grew, increasing to 54 per cent in 2014 and projected to be 66 per cent in 2050 representing a dramatic change from 29 per cent in 1950 (United Nations, 2014).Consequently, urban issues and concerns received an increasingly larger share of government attention and resources.

In addition to the shift of populations from rural to urban, economic systems were themselves changing. Liberalisation of the economy became the governing paradigm and helped spur the trend towards engagement in a globally integrated market. This shift and an embrace of the efficiency-driven model of development was aided and enabled by the ubiquitous and pervasive influence of the world-wide-web. It provided access to information, to know-how to global consumer markets and had a significant impact on national economies. India was one of the countries at the leading edge of this globalisation which manifested the following trends.

Primacy of the free market

There emerged an international acceptance of the ideal of free global trade and capital flows. Policy and social attitudes favouring trade, exchange and the primacy of the market place, which was an underpinning of countries in the first world, became the dominant paradigm of those in the second and thirds world too. The almost universal engagement in freeing up global trade and capital flows was accomplished by the creation of economic blocks – the European Union's (EU) Single Market in 1993, North American Free Trade Agreement (NAFTA) in 1994, ASEAN Free Trade Area (AFTA) in 1992, South Asian Free Trade Area (SAFTA) in 2004. A loosening up of national boundaries and a scaling up of migration and immigration flows between participating nations followed. For rural areas the emphasis on global trade has served to incentivise an increase in larger scale, often corporate, agriculture and an increase in the scale and corporatisation of natural resource extraction.

Primacy of urban areas

The world's population increasingly moved to reside in areas designated as urban. By the turn of the century it was clear that a majority of the world population would be living in cities and urban concerns received greater government attention and an increasing share of national resources. In the earlier periods of their post-Independence histories, rural planning oriented to improving physical infrastructure and the quality of life for rural populations was an important policy agenda in developing countries. As urbanisation rates rise in these countries and cities experience hyper growth (despite earlier efforts to stem rural migration and population increases which by and large are not central in rural policy today) rural planning is a lower priority endeavour, but remains on the national planning agenda, and is primarily aimed at poverty alleviation.

Diminished friction of distance

The evolution and spread of digital technology, particularly the Internet, which has made communications easier and cheaper, served, if not to erase the friction of distance, certainly to diminish it significantly. It reduced global, inter and intra-country barriers to information flows. The world grew smaller and the power to communicate inexpensively devolved down to individuals. As the ability to make connections around the world became inexpensive and broadly accessible, businesses in formerly Second and Third World countries have been able to better compete globally. Communication flows between urban and rural areas and between rural and rural areas have improved. In places like India this has strengthened the reach of agricultural extension services and the farmers' and rural residents' ability to communicate with experts and obtain timely advice to help them compete in the market place. With the increasing access to the world wide net by way of inexpensive mobile hand-held devices, the technology has also enabled the access to larger markets for even individual small-scale producers (Basu and Banerjee, 2011). And, it has also created a potential for reducing the unequal access to education and training for people living in rural places.

The knowledge economy and inequality

The evolution and growth of national economies has been spurred by, and privileged, those who contribute to the so-called 'knowledge economies' and possess the technical expertise that makes for their success. It is a globally integrated system, one in which the characteristic attributes of First and Third Worlds (as defined by quality of life rather than political ideology) are juxtaposed within a country. In addition to a long standing lack of equity and opportunity in rural places, there has emerged an overriding concern about the phenomenon of a general increase in income inequality both urban and rural, most clearly reflected in access, or lack thereof, to affordable housing.

Rural planning in India in the twenty-first century

During this transition, attending to rural planning continued to be significant in the Indian five-year plans. Although the urban population in India is estimated to have almost doubled from 17 per cent in 1950 to 33 per cent in 2015, two-thirds of India's total population of 1.21 billion (Census of India, 2011) still live in rural areas. A guiding principle stated in the 2013–2014 annual report of the Ministry of Rural Development indicates that, at the very least in its articulation of current objectives, the ministry adheres to the objectives embraced in the development decades.

> The mission of the Indian Ministry of Rural Development is 'sustainable and inclusive growth of rural India through a multi-pronged strategy for eradication of poverty by increasing livelihood opportunities, providing social safety net and developing infrastructure for growth and improvement of quality of life in rural India'. The Ministry has devised different programmes to meet primary needs of rural population such as employment, infrastructural development, social assistance etc.
>
> *(Ministry of Rural Development, 2014: vii)*

The efforts to attain these objectives include: job creation in public works such as road building which had offered minimum wage work for rural poor; and supporting production in small cottage industry with training, market networks and infrastructure. The approach and attention to social welfare and poverty alleviation is implemented in programmes described on the official web site of the Ministry of Rural Development (2013–2014). It prioritises those for rural employment, livelihood and connectivity, followed by social assistance and caring for differentially abled. An emphasis on employment and welfare is clearly articulated and poverty alleviation programmes such as public works related jobs-creation to provide a minimum income to underpin poverty are supported (Muthusamy, 2012).

But the investments in these efforts are relatively modest. The notion that it is the private sector or public/private partnerships that will yield employment and result in economic development has gained ground as the nation-wide development strategy has shifted to market mechanisms and free trade. Even the non-governmental sector has embraced this shift. Examples of this approach include small-scale production cooperatives formed in tribal areas for developing indigenous products such as cultivation of herbal medicines (Torri, 2010) and, production of other local products that require a presence and quality control on the ground. But as Singh (2014) notes, under the newly elected pro-market government of Prime Minister Narendra Mody, the 2014 budget of Finance Minister Arun Jaitley cuts, sometimes to half of the 2013 budget, all allocations for rural poverty reduction programmes such as rural housing and safe drinking water, an indication that the new government will rely on market forces and market mechanisms to promote rural change.

The World Bank has made significant investments in rural India to support the following three top priority areas:

1   raising agricultural productivity per unit of land;
2   reducing rural poverty through a socially inclusive strategy that comprises both agriculture as well as non-farm employment; and
3   ensuring that agricultural growth responds to food security needs (World Bank, 2012).

The level of World Bank and Indian Government investments in rural areas is contextualised as follows:

> With some $5.5 billion in net commitments from both IDA and IBRD, and 24 ongoing projects, the World Bank's agriculture and rural development program in India is by far the Bank's largest such program worldwide in absolute dollar terms. This figure is even higher when investments in rural development such as rural roads, rural finance and human development are included. Nonetheless, this amount is relatively small when compared with the Government's – both central and state – funding of public programs in support of agriculture. Most of the Bank's agriculture and rural development assistance is geared towards state-level support, but some also takes place at the national level.
>
> *(World Bank, 2012: n.p.)*

The World Bank critiques the regulation of domestic agricultural trade, and interventions in the land, labour and credit markets, which are residues of earlier government efforts to equalise access and support during the socialised programmes of the development decades. It advocates for deregulation and public-private partnerships a framework, which resembles the US approach.

Issues of sustainability

The industrialisation of agricultural cultivation has disconnected the historically immediate and tangible imperative traditional rural Indian communities had to conserve, land and rural resources. Significant among these was the notion of long-term sustainability – social, economic and environmental. As the long-term negative impacts on the environment of the 'Green

Revolution' technologies have surfaced (i.e. degradation of soil quality; depletion of ground-water tables; the salinisation of land, and; pollution of water systems from high fertiliser use and agricultural run-off), the long-term sustainability of modern agricultural cultivation on prime land is under question. The goal is for sustainable practices to keep production in balance with replacement, maintain the rural resource base, strive for equity, and allow an adaptation to climate change. In India one finds such awareness in both the non-profit and government approaches. A representative example of a pro-sustainability approach is the non-profit Aga Khan Foundation's Rural Support Program in India, which designates sustainability-oriented goals for rural support programmes in tribal and drought prone areas of three Indian states (Gujarat, Madhya Pradesh and Bihar). The investment areas include economic and social development, basic service provision and improved governance. Programmes include:

- strategies for food security and increased income;
- soil and water conservation;
- forest conservation and management;
- climate resilience;
- alternative energy;
- potable water and sanitation;
- community organisation; and
- savings and credit.

There is congruence around the notion that the key challenge of rural planning is to grow and diversify the rural economy for the new global marketplace, but also sustain rural communities, rural assets and reduce inequity. Salient concerns include, in addition to concerns about the disparity of opportunity between urban and rural contexts, sensitivity to the increasing economic gap *within* rural contexts, between the affluent and those in jobs that pay minimum wages. In India differences in income and wellbeing between rural residents who own the enterprises that supply urban markets and those who provide the labour for these enterprises are significant and problematic. The lowered investment in poverty alleviation and increased investment in market production and market driven incentives by the new Indian government, therefore warrants continued examination.

Globalisation in India has erased some of the traditional gaps between urban and rural locales, and rural economies have diversified and responded to the vagaries of urban and international markets. An improved standard of life has accrued for rural people, facilitated by government investments in water supply, sanitation, roads, markets, cooperatives, health care and education, and served to raise the rural standard of living. And, the isolation of rural areas has been further reduced with the proliferation of television, the Internet, and cell phones (Basu and Banerjee, 2011). Farmers with adequate land holdings or those engaged in rural business have prospered and utilised these technologies to optimise market returns.

Income disparities between those in the formal sectors of the economy and those in subsistence agriculture, small-scale services, and the informal economy continue to warrant governmental attention. The need is recognised to: reduce inequity and access to shelter and work; provide new services that help guide sustainable utilisation of land and natural resources; and frame an integrated, inter-sectoral approach to rural planning. The demographic reality of a significantly large rural population provides political pressure for maintaining government investments in rural planning. Currently in the context of economic transformation in India, rural planning has lower priority to programmes that address rapid urbanisation, globalisation and gaining global market share. Programmes related to social justice, participatory decision-making, sustainability, and the environment take second place to transformation of rural production. If inequality rises in rural areas, there will be increased public pressure to ameliorate it. And in that effort, the struggle to forge a model of rural development that is respectful and attentive to the sustainability of the natural resource base and human-social capital is going to be needed, one that balances economic, social and environmental sustainability.

References

Basu, T. and Banerjee, S., 2011, Impact of internet on rural development in India: a case study, *Amity Journal of Media and Communications Studies*, 1(2), 12–17.

Bhende, B. S. and Yatanoor, C. M., 2017, A study of five years plans of rural development programmes in India, *International Journal of Academic Research and Development*, 2(4), 23–26.

Census of India, 2011.

Dandekar, H. C., 1986, *Men to Bombay, women at home: urban influence on village life in Deccan Maharashtra, India, 1942–82*, Ann Arbor, MI: University of Michigan Press.

Krishna, K. R. and Morrison, K. D., 2009, History of South Indian agriculture and agroecosystems in South Indian agroecosystems: nutrient dynamics and productivity, retrieved from www.kathleen morrisonlab.com/storage/Krishna%20Morrison%20-%202009%20-%20History%20of%20South%20 Indian%20Agriculture%20and%20Agroecosystems.pdf (accessed 2 January 2017).

Ministry of Rural Development, 2014, *Government of India, Annual Report (2013–14)*, New Delhi: Ministry of Rural Development, Government of India.

Mosher, A. T., 1976, *Thinking About Rural Development*, New York: Agricultural Development Council.

Muthusamy, R., 2012, The role of Mahatma Gandhi National Rural Employment Guarantee Scheme in poverty alleviation in India, *International Journal of Research in Commerce, Economics and Management*, 2(11), 119–123.

Olfert, M. R. and Partridge, M., 2010, Best practices in twenty-first-century rural development and policy, *Growth and Change*, 41(2), 147–164.

Rostow, W. W., 1956, The take-off into self-sustained growth, *The Economic Journal*, March, 25–48.

Singh, N., 2014, Union budget: a pro-poor and rural development-centric plan?, *International Business Times*, 10 July, retrieved from www.ibtimes.co.in/union-budget-pro-poor-rural-development-centric-plan-604136 (accessed 2 January 2018).

Torri, M.-C., 2010, Community-based enterprise: a promising basis towards an alternative entrepreneurial model for sustainability enhancing livelihoods and promoting socio-economic development in rural India, *Journal of Small Business and Entrepreneurship*, 23(2), 237–248.

United Nations, 2014, *World Urbanization Prospects: The 2014 Revision, Highlights* (ST/ESA/SER.A/ 352), retrieved from http://esa.un.org/unpd/wup/Highlights/WUP2014-Highlights.pdf (accessed 3 October 2014).

USDA, 2014, History of extension, retrieved from www.csrees.usda.gov/qlinks/extension.html (accessed 4 October 2014).

World Bank, 2012, India: issues and priorities for agriculture, *World Bank News*, May, retrieved from www.worldbank.org/en/news/feature/2012/05/17/india-agriculture-issues-priorities (accessed 7 October 2014).

5

Sustainable and resilient ruralities

Guy M. Robinson

Introduction: the concept of resilience

Resilience is a concept originating in physics, mathematics and ecology. It usually refers to the ability of a system or material to recover its shape after encountering a displacement or disturbance (Norris et al., 2008), when being bent, compressed or stretched. In ecology it describes the persistence of natural systems in the face of significant change caused by fires, floods and human interventions (Folke, 2006; Maguire and Hogan, 2007). Engineering resilience focuses on the time taken for a system to return to equilibrium or steady state after a shock (Gunderson 2000), sometimes termed 'bounce back' (Zolli and Healy, 2012), whereas the term ecological resilience is applied to a system's capacity to reorganise under change to reach a new equilibrium while retaining the same essential functions (Holling, 2001). For over twenty years there has been a transfer of these ideas about resilience into the social sciences, especially relating resilience to the development of human communities and the role of resilience in attaining sustainable development (Adger, 2000; Berkes and Folke, 1998), often linked to urban and regional development (Hill et al., 2008; Martin and Sunley, 2015).

The derivation of ideas from ecology has produced a conceptualisation of human society in which individuals and families live and function in a community shaped by systemic influences, both internal and external. However, there is a recognition that human systems may never reach a state of equilibrium, but rather they evolve as complex systems that constantly adapt to sustain their development paths (see Table 5.1; Pike et al., 2010; Scott, 2013). This introduces another term to describe the resilience of human systems, namely 'adaptive resilience', or 'the ability of the system to undergo anticipatory or reactionary reorganization of form and/or function so as to minimize impact of a destabilizing shock' (Martin, 2012: 5). There is a psychological dimension to resilience if individuals are considered, referring to an individual's ability to adapt to stress and adversity. So, the resilience of a community reflects the cumulative well-being of its individual members (Werner, 1989). In this approach an individual's resilience reflects their ability to respond to adversity by presenting positive adaptability to change (e.g. Luthar and Cicchetti, 2000; Luthar et al., 2000), which is closely linked to the socio-cultural context in which individuals operate, and hence their particular economic and social circumstances (e.g. Chaskin, 2008; Flora and Flora, 2008; Ungar, 2008). It is also closely linked to the notion of vulnerability, which some view as the opposite of resilience, with risks/shocks rendering communities vulnerable, while some have resilience to overcome or adapt to the risk (Freshwater, 2015).

Conceptions of resilience usually include the idea that human communities possess an adaptive capacity in response to change, expressed as 'community capacity', defined as 'the combined influence of a community's commitment, resources and skills that can be deployed to build on community strengths and address community problems and opportunities' (Aspen Institute, 1996: 17). So, adaptation refers to a dynamic social process and how well the community can exist with or respond to change, i.e. bounce back, as a proactive or reactive process, which may be unintentional (Adger, 2006). Certain types of resource-dependent rural communities have often been described as resilient in the face of economic or environmental crises, with labels such as adaptive, flexible, proactive and deliberative with respect to future development strategies (Steinführer, 2013). Wilson (2015) contends that the inbuilt (social) memory of a community helps shape resilience pathways. In rural communities this memory refers to rites, traditions and social learning processes. Thus, the variability within 'rural' will predispose some rural communities to be more resilient than others, e.g. depending on the combinations they possess of various rural characteristics (Halfacree, 2007).

Magis (2010: 401) contends that 'communities can develop resilience by actively building and engaging the capacity to thrive in an environment characterized by change, and that community resilience is an important indicator of social sustainability'. She noted that the most appropriate community responses to change or disruption can vary from maintenance to adaptation to transformation so that a healthy community may be sustained. However, given that a key characteristic of human communities is their unpredictability, and as there is no single best or most moral path to resilience, this limits the effectiveness of ecological approaches to resilience (Robinson and Carson, 2016: 118). Such approaches can also privilege established social structures which may need to be changed to order to effect desirable transformation.

Table 5.1 Key features of equilibrium and evolutionary approaches to resilience.

Equilibrium resilience	Evolutionary resilience
'Bounce-back' resilience	'Bounce-forward' resilience
The ability of a system to accommodate disturbances without experiencing changes to the system	The ability of a system to respond to shocks and disturbances by adaptation and adaptability
Emphasises a return to a steady-state after disturbance – 'business as usual'	Emphasises transformation or path creation in response to disturbances – 'do something different'
Short-term response to shocks and disturbances	Long-term response, emphasising adaptive capacity
Prominent in the literature surrounding disaster management, managing geo-environmental hazards	Prominent in the literature surrounding regional economic development, spatial planning
Conservative approach, naturalising man-made crises and depoliticising responses	Recognises the politics of resilience, involving normative and value judgements
A reactionary tool, reinforcing existing power structures	A critical tool, enabling reform

Source: Scott (2013: 601)

Other limitations regarding the conceptualisation of resilient communities include the fundamental issue of whether resilience is a normative concept: is it possible to define it prescriptively and is it a concept that can be subjected to empirical testing? In addressing these questions, MacKinnon and Derickson (2013) provide a strong critique of the use of resilience within public policy, arguing that communities cannot be expected to develop adaptive capacity as self-contained systems that are divorced from national and global flows of capital and power. Hence, they argue that policy is frequently misplaced in terms of spatial scale 'since the processes which shape resilience operate primarily at the scale of capitalist social relations' (ibid.: 253). Yet, policy that seeks to develop community resilience is commonly linked to notions of community participation in decision making, the creation of an inclusive and creative culture, a local economy based on sound environmental principles and supportive inter-community links (Featherstone et al., 2012). This has been termed 'apolitical inclusive localism' (Mason and Whitehead, 2012), acknowledging that policy tends to promote acceptance of the 'shocks' imposed by globalisation, maintaining and legitimising forms of social hierarchy and control, in part by producing a political agenda in which cuts in public expenditure are allied to calls for greater community resilience, but which in reality recreate unequal social relations (Neal, 2013). Nevertheless, resilience frequently requires communities to display 'resourcefulness' in which local autonomous action generates broader development of social justice through local political expression to release resources, skills sets and local knowledge, use of indigenous and 'folk' knowledge, and cultural recognition (MacKinnon and Derickson, 2013: 266). This is essentially a practical prescription for local action that can 'develop resilience', and it is repeated by various policies directed to rural communities, in which rural resilience is defined as 'the capacity of a rural region to adapt to changing external circumstances in such a way that a satisfactory standard of living is maintained' (Heijman et al., 2007: 383). Hence, local 'capital' is often seen as the key to generating resilience. This draws upon the ideas of Pierre Bourdieu (1986) who recognised several principal forms of capital, notably economic, symbolic, cultural and social. Others have developed these ideas, often adding a physical/environmental element to the list.

There are close links between sustainability and resilience, especially if the former is regarded as the continuation of an entity over time because of the presence of conditions enabling development to be maintained, as in the case of a resilient community bouncing back after experiencing a shock. Thus, according to Folke (2006), emphasising the capacity for renewal, re-organisation and development is essential for the sustainability discourse. This chapter focuses on resilience as part of this broader perspective of sustainability, but primarily in the context of addressing economic and social development as opposed to dealing with resilience in response to 'natural' disasters such as floods and earthquakes or adaptation/mitigation in the face of climate change.

Resilience and multi-functionalism

The concept of rural community resilience has been linked to multi-functionality, with communities that possess high economic, social and environmental capacity said to exhibit both strong multi-functionality and resilience, often associated with agricultural systems possessing positive attributes (e.g. creating employment, a stable food supply, environmental benefits and contributing to increased social, cultural and institutional capital) (Renting et al., 2009; Zasada, 2011). There are links to community capacities, so that multi-functionality can take on a global character (Dibden and Cocklin, 2009) whereby rural communities can be situated on a spectrum from highly developed to weakly developed capacities, with some exhibiting resilience and some not (Cutter et al., 2008; Rigg et al., 2008). In this conceptualisation, Wilson (2012: 368) acknowledges that 'resilience can be scaled down to the household and individual level, and it is the totality of economic, social and environmental actions/responses of individuals and households within a rural community that shape a community's overall resilience'.

Scaling up from the individual to a community and beyond has involved identifying the components and determinants of community resilience in specific spatial localities, including both rural and city regions (Christopherson et al., 2010; Simmie and Martin, 2010; Scott and Gkartzios, 2015). Hence there is a body of literature on resilient cities and regions (Newman et al., 2009; Polèse, 2009), but in terms of resilient rural places, one basis for resilience might be the presence of diverse, pluriactive or multifunctional agriculture from which agribusiness may form the basis of a vibrant rural economy supporting strong rural services. This may also refer to the capacity of agriculture to produce various non-commodity outputs in addition to food (Wilson, 2007), but it has also been used in conceptualising changes extending beyond agriculture into rural society.

In policy terms, there have been various attempts to promote multi-functional agriculture as part of rural development initiatives. Measures are usually operational at an individual farm level (e.g. subsidies to farmers to help develop farm-based tourism or to enhance on-farm environmental attributes). At a community level, therefore, the impacts of these measures depend on how well farming is embedded within broader rural society. Key individual farmers and their advisors can act as leaders for measures affecting the wider community (Raymond and Robinson, 2013). A good example is the Landcare programme in Australia where farmers and the community voluntarily tackle environmental degradation, usually at the scale of individual river catchments (Tennent and Lockie, 2013). However, multi-functionality can operate at a regional level, as in the case of the European Union (EU)'s Structural Funds, which have provided multi-functional opportunities since 1988, and its LEADER programme, operating from 1991 to support rural development projects initiated at the local level to revitalise rural areas and create employment. Across four different iterations, LEADER developed seven principles for local development that can be readily recognised as part of resilience discourse: based in small socially cohesive areas, a bottom-up (locally determined) process, public–private partnership, innovative, integrated not sectoral, with development of community networks and fostering co-operation (Ray, 2000).

Rural development policy in the European Union (EU) has focused on multifunctionality as a means of producing sustainable development. Within this, there is a growing concern for a 'new rural paradigm' in which some communities and regions are pursuing 'place-based' initiatives to develop new trajectories for sustainable development (Horlings and Marsden, 2014). Two in particular may be observed: one focused on the bio-economy and the other on the eco-economy. The latter refers to management of rural resources to enhance the local and regional ecosystem. The former relies on largely corporate-controlled production of biomass and biofuels, and related elements (such as biotechnology, genomics, chemical engineering and enzyme technology) (Marsden, 2010) to capture the latent value in biological processes and renewable bio-resources.

These two economies challenge globalising and homogenising processes by emphasising the need for a focus on sustainable development. In so doing, the aim is 'to capture the interrelations between six conceptual domains or dimensions in regional rural development: endogeneity, novelty, production, social capital, market governance, new institutional arrangements and sustainability' (Horlings and Marsden; 2014: 7; see Table 5.2). Strategies include niche innovation (new product–market combinations, mostly related to food, such as better marketing, new products, new markets; new interfaces (new arrangements are formed that enable new forms cooperation between actors through shorter food chains, co-operative marketing,

Table 5.2 Domains of rural development.

Domain	Characteristics
Endogeneity	The degree to which rural economies are (a) built upon local resources; (b) organised according to local models of resource combination; and (c) strengthened through the distribution and reinvestment of produced wealth within the local/regional constellation
Novelty	New insights, practices, artefacts and/or combinations (of resources, technological procedures, bodies of knowledge) that carry the promise that specific constellations function better
Social capital	Networks of relationships among people who live and work in a particular society, enabling that society to function effectively or, more specifically, the ability of individuals, groups, organisations or institutions to engage in networks, cooperate and employ social relations for common purpose and benefit
Market governance	Institutional capacities to control and strengthen existing markets and/or to construct new ones
New institutional arrangements	New institutional constellations that solve coordination problems and support cooperation among rural actors
Sustainability	Existence of social and ecological conditions necessary to support human life at a certain level of well-being through future generations

Source: based on Horlings and Marsden (2014: 7)

new rural–urban linkages; and reorientation of territorial capital, which involves valorisation of local/regional assets or is rooted in social–cultural notions of regional identity. The latter involves a gradual shift from agricultural diversification to a more integrative rural and regional development.

Policies supporting these new forms of rural development in the EU include the creation of Protected Designation of Origin (PDO), whereby only items produced in a specific area in a certain way may bear that label in the European market, and Protected Geographical Indicator (PGI), whereby products must be traditionally and at least partially manufactured within a specific region and thus acquire unique properties. This association with a place/region has been referred to as a 'relocalisation' of the agri-food system (Ilbery and Maye, 2015). Introduced in 1992, these labels are designed to protect (and brand) foods that are unique to specific regions of Europe, especially rural areas. There are now over 1,000 protected products within the EU, dominated by cheese and drink products (e.g. gorgonzola, feta, camembert, champagne, Armagnac, Herefordshire cider) and often strongly motivated by protectionism. However, the latter has helped producers obtain higher prices for their goods, fostering the creation of ethical food markets that fix capital in certain areas while narrowing competition. It also confers rights on certain groups of producers and shifts responsibility from those introducing the regulation (the EU) to the consumers purchasing the goods and who may be willing to pay a premium for food with a PGI or PDO (Lang, 2010).

In their Australian case studies of resilient farming communities, McManus et al. (2012) identified key aspects of resilience, including community spirit (and leadership), participation, education, housing supply, health care, employment, crime and safety, sense of belonging, retail, recreation and leisure, and quality of the physical environment. It is a set of characteristics that combines aspects for which appropriate policies can be implemented (e.g. focusing on services and facilities), but with other aspects drawing on a mixture of inherent local capital and external support. However, it must be acknowledged that all too often policies formulated in urban centres and applying urban standards disadvantage rural regions, as illustrated by school closures in rural Nova Scotia, Canada, based on threshold pupil numbers based primarily on metropolitan concerns (Corbett, 2014). Such developments may be countered by more place-based education and training, as in the example of nurses and social workers trained specifically to deal with the special needs of rural and remote communities (Bourke et al., 2012).

Resilience and sustainability

Resilience has only featured strongly and explicitly in policy discourse since 2000. Its appearance in policy has often been linked to governments finding ways of promoting community action to develop desirable economic, environmental and social capabilities. For example, there are an increasing number of community support measures linked to developing resilience in the face of climate change, such as the Australian Government's AU$4.5 million climate change adaptation funding stream for local government, 'Building resilience of coastal communities' (Commonwealth of Australia, 2011). Reference to resilience now appears in many local government climate change strategies and plans worldwide, often linking resilience to risk management in scenarios whereby 'resilient communities' can either withstand and adapt to climate change or take steps to reduce risk.

As globalisation has increasingly impacted on farmers across the world, rural communities have been placed at a disadvantage to many urban areas in terms of investment and economic growth. Yet, various characteristics of rural places have enabled some to remain resilient and sustainable in face of major changes to the economic and social base, such as fewer but larger farms and ongoing rural outmigration. External measures by the state to support sustainable rural development have often centred on improvements to infrastructure, linking rural places to larger service centres, but much of the resilience exhibited has come from endogenous sources, such as local entrepreneurs, regional trade associations and clusters of small businesses. In the United States research suggests that if endogeneity gives rural people more control over the local economy rather than being reliant on inputs from external sources such as government or multi-national corporations, then there is a greater chance of socio-economic well-being and rising civic welfare (Blanchard et al., 2011). Yet, there may be significant limits to how the local level can shape and influence different development path creation processes, especially given the limited capabilities of many small villages and the power of globalising forces.

Scott (2013) suggests that resilience provides an alternative policy narrative for rural development practice, especially if policy can be set in the context of generating evolutionary resilience. This entails a focus on capacity building within rural communities alongside more traditional policy aims such as job creation. Other key policy aspects that he identifies are the need to embed environmental and ecological considerations into rural policy (e.g. generating low-carbon rural futures and clusters of eco-businesses, and blending the local and the global in searching for new pathways for local economies). One of the contradictions in rural policies explicitly aimed at creating resilient communities is the link established between increased self-reliance and the parallel removal of government support for rural services. For example, the LEADER initiative is set alongside state removal of rural services and restructuring of the Common Agricultural Policy (Scott, 2004).

If sustainability is regarded as an inherent trait of resilience, then policies to promote sustainable agriculture should also be viewed as contributing to resilient rural development. Such policies have taken various forms, but in terms of promoting an ecocentric view of sustainability, the focus has been on organic farming. This is a production system largely excluding the use of synthetic compounded fertilisers, pesticides, growth regulators and livestock feed additives. Organic farming systems rely upon crop rotations, crop residues, animal manures, legumes, green manures, off-farm organic wastes and aspects of biological pest control to maintain soil productivity and tilth, to supply plant nutrients, and to control insects, weeds, and other pests. They are supported by national certification schemes, which have contributed to 43 million ha recognised worldwide (Willer and Lernoud, 2017). Certification, like PGIs and PDOs, works as a branding system that gives consumers confidence in the product. This is increasingly important given there is a growing concern over food quality and healthy eating in some segments of the market. The certification schemes are often accompanied by financial assistance for farmers to enable them to pay for the costs of conversion, including compensation for loss of profits during conversion.

In addition to official governmental designations, there are increasingly numerous farm, food and rural tourism quality assurance schemes (QAS) developed by regional groups, retailers and representative organisations. These tap into growing public awareness of ethical, environmental and safety implications of industrial-style farming systems while linking QAS-backed products to perceptions of good taste and sophistication. This linkage between regulation and quality can be seen in the intersection between agro-ecology and branding of Californian wine grapes (Warner, 2007).

More direct attempts to foster both greater sustainability and multifunctionality have grown in popularity in the EU since 1986 when agri-environmental schemes (AES) were first introduced to promote environmentally friendly practices on farms. Subsequently these have been extended, with various farm supports from 2005 onwards predicated on farmers meeting certain environmental standards, termed cross-compliance, though the AES are usually voluntary. AES account for 22 per cent of EU expenditure for rural development. They include results-based schemes focusing on payments that reward improvements in farmland biodiversity rather than just paying farmers to participate in an activity, e.g. The MEKA programme B4 in south-west Germany, established in 2000, offers small payments (€60 per hectare) to 4,800 participating farmers who manage species-rich grassland containing at least four key plant indicator species (Troost et al., 2015). Many of these schemes now take the form of market-based instruments, termed payment for ecosystem services (PES), and other forms of conditional social transfers (CSTs) intended to improve ecosystems and alleviate poverty.

Some similar environmentally friendly measures have been adopted in the United States since the 1930s, initially through the paid diversion of erodible land into conservation uses. A similar idea was embodied in the 1956 Agricultural Act, creating a national Soil Bank. In 1985 the Conservation Reserve Program (CRP) was signed into law in that year's Farm Bill, becoming the country's largest private-lands conservation programme, with a long-term goal to re-establish valuable land cover to improve water quality, prevent soil erosion, and reduce loss of wildlife habitat. In exchange for a yearly rental payment, farmers who enrol in the programme agree to remove environmentally sensitive land from agricultural production. On this land they plant species that will improve environmental health and quality. Contracts for land enrolled in CRP are ten to fifteen years in length. Over 10 million ha have been enrolled, with total funding of approximately $2 billion annually (Stubbs, 2014). Measurable benefits include a reduced loss of soil from erosion (325 million tons saved), 0.8 million ha in wetlands, the same amount in buffers, and as new wildlife habitat. There has been approximately 52 million metric tons net reduction in CO_2 from sequestration, reduced fuel use and nitrous oxide emissions avoided from no fertiliser use, and a net reduction of about 607 million pounds of nitrogen and 122 million pounds of phosphorus (ibid.: 14).

Similar principles relating to removing land from production have been adopted on a grand scale in China through the Sloping Land Conversion Program. Running since 1999 this has expended US$69 billion, affecting 32 million households across 25 provinces. On average, payments have accounted for about 14 per cent of per capita income of participating households, with 15.31 million ha afforested by conversion from arable land and grassland. Estimates are that this has reduced silt run-off by a quarter and sequestered at least 225 million tons of CO_2 (Leshan et al., 2017: 4). Participants are compensated directly through subsidies and seedlings, and indirectly through promotion of off-farm employment, aimed at reducing income inequality. The programme is now in its fourth stage, with a further 2.8 million ha of cropland to be converted to forest or grassland by 2020. This has had relatively little effect on continued large-scale rural outmigration, with at least 200 million rural migrants now working in the major Chinese cities alone (Pieke and Mallee, 2014). So, while environmental gains are evident, the phenomenon of 'hollow' villages is occurring, with villages becoming the preserve of the elderly and very young (Liu et al., 2010). This rural depopulation has offered opportunities for land consolidation measures to increase agricultural productivity (Long, 2014), while more resilient communities, especially within hinterlands of major cities, have prospered through farm diversification and participation in new markets (Song et al., 2017). To counter the loss of farmland created by land conversion, a requisition–compensation balance of farmland has been introduced by central government aimed at replenishing farmland loss caused by urban expansion and environmental conversion. It is not clear, though, whether the top-down policy will be effective as it conflicts with local interests, since land conversion from agriculture to construction is a prime means by which local governments attract investment and raise fiscal revenue (Shen et al., 2017).

Conclusion

Policies designed to promote resilient rural communities have emphasised different aspects of resilience, drawing upon both equilibrium and evolutionary conceptualisations (as illustrated in Table 5.1). There has been a strong emphasis on releasing endogenous capabilities to respond to changed circumstances, but with different external inputs that can help transform local capacity. This chapter has concentrated on two broad sets of such policies, linked to agriculture as a fundamental basis for rural development.

In focusing on multifunctional agriculture, and ultimately multifunctional rural regions, policies such as the EU's LEADER programme have sought to harness rural community resources to create more diverse rural economies. Notable successes have occurred when there has been potential to diversify from agriculture into tourism and culturally related initiatives (Bell, 2015; Cawley and Gillmor, 2008). Place-based initiatives in the EU have also helped generate new developments through enabling measures focused on the bio- and eco-economies. Similar developments in Australia and Vietnam are the support for oyster and fish farming in declining small coastal communities (Pierce and Robinson, 2013). Measures that have enabled rural producers to gain a price premium, e.g. through PGIs and PDOs, are other examples of providing not only more income to the producers but also increasing employment opportunities within communities by generating new or expanding old markets. The potential for 'relocalisation' and a place-based focus has engendered growth in some rural communities, while recognising that there are various 'communities of place' (Skerratt, 2013), in which a rural community actually comprises various different 'communities' sharing a particular place. Harnessing the capacity of individuals and communities can bring about positive outcomes, although many rural places still suffer from the downward spiral created by declining population, loss of services and negative impacts of broader structural trends in farming and other rural industries.

The second set of policies considered herewith has been those that link resilience to sustainability in the agricultural sector. Broadly, these are measures to address the environmental base, providing encouragement to farmers not only to redress negative externalities of industrial farming methods, but also to adopt environmentally friendly measures that contribute to production of attractive landscapes. The growing recognition of the importance of ecosystem services has contributed to the range of schemes generated worldwide: from ones promoting organic production methods (e.g. for conversion and certification) to grandiose projects aimed at large-scale environmental improvements (as in the case of the CRP and Grain for Green) to PES and broader landscape stewardship schemes (Bieling and Plieninger, 2017). These largely aim to improve the environmental dimension of sustainability while facilitating new sources of income to rural households.

Ultimately, policy intending to create more resilient futures for rural communities has to ask questions about who and what is to be made resilient, in what manner, and for what purposes. In face of growing concerns about impacts of climate change, it is likely that increased emphasis will be placed on measures linked to disaster preparedness and response. However, in common with policies aimed at strengthening rural economies, these too have tended to be part of the trend in the Western world towards 'local integrated place-based outcomes' (Coaffee, 2013: 244). This is part of a broad neo-liberal approach to resilience that emphasises local participation and responsibility. Nevertheless, many of the policy examples cited here suggest that inputs by government at all levels and contributions by external agencies can be vital in mobilising and releasing local capabilities to produce positive local outcomes. Hence it is the judicious marriage of internal and external that can perhaps best promote both rural community resilience and sustainability.

References

Adger, W. N. (2000) 'Social and ecological resilience: are they related?' *Progress in Human Geography* 24, 347–364.

Adger, W. N. (2006) 'Vulnerability', *Global Environmental Change* 16, 268–281.

Aspen Institute (1996) *Measuring community capacity building: a workbook in progress for rural communities*, Washington, DC: Aspen Institute.

Bell, D. (2015) 'Cottage economy: the "ruralness" of rural cultural industries', in K. Oakley and J. O'Connor (eds), *The Routledge companion to the cultural industries*, 222–231. New York: Routledge.

Berkes, F. and Folke, C. (eds) (1998) *Linking social and ecological systems: management practices and social mechanisms for building resilience*, Cambridge: Cambridge University Press.

Bieling, C. and Plieninger, T. (eds) (2017) *The science and practice of landscape stewardship*, Cambridge: Cambridge University Press.

Blanchard, T. C., Tolbert, C. and Mencken, C. (2011) 'The health and wealth of US counties: how the small business environment impacts alternative measures of development', *Cambridge Journal of Regions, Economy and Society* 5(1), 149–162.

Bourdieu, P. (1986) 'The forms of capital', in J. G. Richardson (ed.), *Handbook for theory and research for the sociology of education*, 241–258. New York: Greenwood.

Bourke, L., Humphreys, J. S., Wakerman, J. and Taylor, J. (2012) 'Understanding rural and remote health: a framework for analysis in Australia', *Health & Place* 18(3), 496–503.

Cawley, M. and Gillmor, D. A. (2008) '"Culture economy", "integrated tourism" and "sustainable rural development": evidence from Western Ireland', in G. M. Robinson (ed.), *Sustainable rural systems: sustainable agriculture and rural communities*, 145–160. Burlington, VT: Ashgate.

Chaskin, R. J. (2008) 'Resilience, community, and resilient communities: conditioning contexts and collective action', *Child Care in Practice* 14, 65–74.

Christopherson, S., Michie, J. and Tyler, P. (2010) 'Regional resilience: theoretical and empirical perspectives', *Cambridge Journal of Regions, Economy and Society* 3, 3–10.

Coaffee, J. (2013) 'Rescaling and responsibilising the politics of urban resilience: from national security to local place-making', *Politics* 33, 240–252.

Commonwealth of Australia (2011) *Coastal communities building resilience to climate change*, Canberra: Department of Climate Change and Energy Efficiency, Commonwealth of Australia.

Corbett, M. J. (2014) 'We have never been urban: modernization, small schools, and resilient rurality in Atlantic Canada', *Journal of Rural and Community Development* 9(3), 1–13.

Cutter, S. L., Barnes, L., Berry, M., Burton, C. G., Evans, E., Tate, E. and Webb, J. (2008) 'A place-based model for understanding community resilience to natural disasters', *Global Environmental Change* 18, 598–606.

Dibden, J. and Cocklin, C. (2009) 'Multifunctionality: trade protectionism or a new way forward?' *Environment and Planning A* 41, 163–182.

Featherstone, D., Ince, A., MacKinnon, D., Cumbers, A. and Strauss, K. (2012) 'Progressive localism and the construction of political alternatives', *Transactions of the Institute of British Geographers* 37, 177–182.

Flora, J. L. and Flora, C. B. (2008) *Rural communities, legacy and change*, 3rd edition, Boulder, CO: Westview Press.

Folke, C. (2006) 'Resilience: the emergence of a perspective for social-ecological system analyses', *Global Environmental Change* 16, 253–67.

Freshwater, D. (2015) 'Vulnerability and resilience: two dimensions of rurality', *Sociologia Ruralis* 55(4), 497–515.

Gunderson, L. H. (2000) 'Ecological resilience – in theory and application', *Annual Review of Ecology and Systematics* 31, 425–439.

Halfacree, K. (2007) 'Trial by space for a "radical rural": introducing alternative localities, representations and lives', *Journal of Rural Studies* 23(3), 125–141.

Heijman, W., Hagelaar, G. and Heide, M. (2007) 'Rural resilience as a new development concept', in D. Tomić and M. M. Šervalić (eds), *Development of agriculture and rural areas in Central and Eastern Europe: 100th seminar of the EAAE*, 383–396. Novi Sad, Serbia: EAAE.

Hill, E. W., Wial, H. and Wolman, H. (2008) *Exploring regional economic resilience*, working paper 204, Berkeley, CA: Berkeley Institute of Urban and Regional Development.

Holling, C. S. (2001) 'Understanding the complexity of economic, ecological and social systems', *Ecosystems* 4, 390–405.

Horlings, L. G. and Marsden, T. K. (2014) 'Exploring the "New Rural Paradigm" in Europe: eco-economic strategies as a counterforce to the global competitiveness agenda', *European Urban and Regional Studies* 21(1), 4–20.

Ilbery, B. W. and Maye, D. (2015) 'The changing dynamics of alternative agri-food networks: a European perspective', in G. M. Robinson and D. A. Carson (eds), *Handbook on the Globalisation of Agriculture*, 425–445. London: Edward Elgar Publishing.

Lang, T. (2010) 'From "value-for-money" to "values-for-money"? Ethical food and policy in Europe', *Environment and Planning A* 42(8), 1814–1832.

Leshan, J., Porras, I., Lopez, A. and Kazis, P. (2017) *Sloping Lands Conversion Programme*, People's Republic of China, London: IIED.

Liu, Y., Liu, Y., Chen, Y. and Long, H. (2010) 'The process and driving forces of rural hollowing in China under rapid urbanization', *Journal of Geographical Sciences* 20(6), 876-888.

Long, H. (2014) 'Land consolidation: an indispensable way of spatial restructuring in rural China', *Journal of Geographical Sciences* 24(2), 211–225.

Luthar, S. S. and Cicchetti, D. (2000) 'The construct of resilience: implications for interventions and social policies', *Development and Psychopathology* 12, 857–885.

Luthar, S. S., Cicchetti, D. and Becker, B. (2000) 'The construct of resilience: a critical evaluation and guidelines for future work', *Child Development* 71, 543–562.

MacKinnon, D. and Derickson, K. D. (2013) 'From resilience to resourcefulness: a critique of resilience policy and activism', *Progress in Human Geography* 37, 253–270.

McManus, P., Walmsley, J., Argent, N., Baum, S., Bourke, L., Martin, J., Pritchard, B. and Sorensen, T. (2012) 'Rural community and rural resilience: what is important to farmers in keeping their country towns alive?' *Journal of Rural Studies* 28(1), 20–29.

Magis, K. (2010) 'Community resilience: an indicator of social sustainability', *Society and Natural Resources* 23, 401–416.

Maguire, B. and Hogan, P. (2007) 'Disasters and communities: understanding social resilience', *The Australian Journal of Emergency Management* 22(2), 16–20.

Marsden, T. (2010) 'Mobilising the regional eco-economy: evolving webs of agri-food and rural development in the UK', *Cambridge Journal of Regions, Economy and Society* 3, 225–244.

Martin. R. (2012) 'Regional economic resilience, hysteresis and recessionary shocks', *Journal of Economic Geography* 12, 1–32.

Martin, R. and Sunley, P. (2015) 'On the notion of regional economic resilience: conceptualization and explanation', *Journal of Economic Geography* 15, 1–42.

Mason, K. and Whitehead, M. (2012) 'Transition urbanism and the contested politics of ethical place making', *Antipode* 44, 493–516.

Neal, S. (2013) 'Transition culture: politics, localities and ruralities', *Journal of Rural Studies* 32, 60–69.

Newman, P., Beatly, T. and Boyer, H. (eds) (2009) *Resilient cities: responding to peak oil and climate change*, Washington, DC: Island Press.

Norris, F. H., Stevens, S. P., Pfefferbaum, B., Wyche, K. F. and Pfefferbaum, R. L. (2008) 'Community resilience as a metaphor, theory, set of capacities, and strategy for disaster readiness', *American Journal of Community Psychology* 41, 127–150.

Pieke, F. N. and Mallee, H. (eds) (2014) *Internal and international migration: Chinese perspectives*, New York: Routledge.

Pierce, J. and Robinson, G. M. (2013) 'Oysters thrive in the right environment: the social sustainability of oyster farming in the Eyre Peninsula, South Australia', *Marine Policy* 37, 77–85.

Pike, A., Dawley, S. and Tomaney, J. (2010) 'Resilience, adaptation and adaptability', *Cambridge Journal of Regions, Economy and Society* 3, 59–70.

Polèse, M. (2009) *The wealth and poverty of regions: why cities matter*, Chicago, IL: University of Chicago Press.

Ray, C. (2000) 'The EU LEADER program: rural development laboratory', *Sociologia Ruralis* 40(2), 163–171.

Raymond, C. M. and Robinson, G. M. (2013) 'Factors affecting rural landholders' adaptation to climate change: insights from formal institutions and communities of practice', *Global Environmental Change* 23(1), 103–114.

Renting, H., Rossing, W. A. H., Groot, J. C. J., Van der Ploeg, J. D., Laurent, C., Perraud, D., Stobbelaar, D. J. and Van Ittersum, M. K. (2009) 'Exploring multifunctional agriculture: a review of conceptual approaches and prospects for an integrative transitional framework', *Journal of Environmental Management* 90, S112–S123.

Rigg, J., Veeravongs, S., Veeravongs, L. and Rohitarachoon, P. (2008) 'Reconfiguring rural spaces and remaking rural lives in central Thailand', *Journal of Southeast Asian Studies* 39, 355–381.

Robinson, G. M. and Carson, D. A. (2016) 'Resilient communities: transitions, pathways and resourcefulness', *Geographical Journal* 182(2), 114–122.

Scott, M. (2004) 'Building institutional capacity in rural Northern Ireland: the role of partnership governance in the LEADER II programme', *Journal of Rural Studies* 20(1), 49–59.

Scott, M. (2013) 'Resilience: a conceptual lens for rural studies', *Geography Compass* 7, 597–610.

Scott, M. and Gkartzios, M. (2015) 'Rural housing: questions of resilience', *Housing and Society* 41(2), 247–276.

Shen, X., Wang, L., Wu, C., Lv, T., Lu, Z., Luo, W. and Li, G. (2017) 'Local interests or centralized targets? How China's local government implements the farmland policy of requisition–compensation balance', *Land Use Policy* 67, 716–724.

Simmie, J. and Martin, R. (2010) 'The economic resilience of regions: towards an evolutionary approach', *Cambridge Journal of Regions, Economy and Society* 3, 27–43.

Skerratt, S. (2013) 'Enhancing the analysis of rural community resilience: evidence from community land ownership', *Journal of Rural Studies* 31, 36–46.

Song, B., Robinson, G. M. and Zhou, Z. (2017) 'Agricultural transformation and ecosystem services: A case study from Shaanxi Province, China', *Habitat International* 69, 114–125.

Steinführer, A. (2013) 'Vulnerability and resilience: new perspectives on rurality?', paper presented at the Conference on 'Rurality: new perspectives and themes', Universität Bamberg, 15–16 November.

Stubbs, M. (2014) *Conservation Reserve Program (CRP): status and issues*, Washington, DC: Congressional Research Service.

Tennent, R. and Lockie, S. (2013) 'Vale Landcare: the rise and decline of community-based natural resource management in rural Australia', *Journal of Environmental Planning and Management* 56(4), 572–587.

Troost, C., Walter, T. and Berger, T. (2015) 'Climate, energy and environmental policies in agriculture: simulating likely farmer responses in Southwest Germany', *Land Use Policy* 46, 50–64.

Ungar, M. (2008) 'Resilience across cultures', *British Journal of Social Work* 38, 218–235.

Warner, K. D. (2007) 'The quality of sustainability: agroecological partnerships and the geographic brand-ing of California wine-grapes', *Journal of Rural Studies* 23, 142–155.

Werner, E. E. (1989) *Vulnerable but invincible: a longitudinal study of resilient children and youth*, New York: McGraw-Hill.

Willer, H. and Lernoud, J. (2017) 'Organic agriculture worldwide 2017: current statistics', retrieved from http://orgprints.org/31197/1/willer-lernoud-2017-global-data-biofach.pdf

Wilson, G. A. (2007) *Multifunctional agriculture: a transition theory perspective*, Wallingford: CABI.

Wilson, G. A. (2012) *Community resilience and environmental transitions*, London: Routledge.

Wilson, G. A. (2015) 'Community resilience and social memory', *Environmental Values*, 24(2), 227–257.

Zasada, I. (2011) 'Multifunctional peri-urban agriculture – a review of societal demands and the provision of goods and services by farming', *Land Use Policy* 28, 639–648.

Zolli, A. and Healy, A. M. (2012) *Resilience: why things bounce back*, New York: Free Press.

6

Production, consumption and protection

The multifunctional transition in rural planning

Kathryn I. Frank and Michael Hibbard

Introduction

A wide range of activities have emerged in the last 20 years or so – watershed restoration, community forestry, sustainable agriculture, value-chain differentiated products, eco-tourism and adventure tourism, and management for ecosystem services, to mention several examples – that point toward a re-thinking of the human uses of rural space. There is an emerging shift from the dominant use of rural landscapes for resource production toward a more complex and often overlapping mix of uses that scholars have termed the multifunctional transition (McCarthy, 2005; Holmes, 2006).

Multifunctionality does not entail abandonment of agriculture and natural resource extraction, but rather a condition in which rural landscapes and their communities concurrently serve production, consumption, and protection functions. For example, a forest managed to mimic natural conditions of biodiversity, regeneration capacity, and vitality can yield logs and wood chips (production) and also provide adventure tourism (consumption) – and riparian restoration (protection).

The rise of global markets and supply chains, new regulatory regimes, local economic restructuring, and related forces are at work broadly to shape multifunctional rural landscapes. As Hall and Stern (2009) have noted, they combine in multiple ways in different locales and institutional contexts to produce different actions, both locally and at a distance. In this chapter, we explore the significance of the multifunctional transition for rural planning and development in North America by means of our recent work on the topic. We incorporate the results from a series of interviews with a purposively selected set of ten rural specialists in the United States and Canada (Frank and Hibbard, 2017), a comprehensive literature review (Frank and Reiss, 2014), and an analysis of the multifunctional transition and its implications for planning and development (Frank and Hibbard, 2016). The rural specialists we interviewed are practitioners and academics with deep practice experience. Two are currently based in Canada and eight in the US, though four of the latter have extensive Canadian experience. All can be characterised as rural generalists. However, they fall roughly into specialty areas – two each in rural policy, economic development, environmental planning, land use, and design. In discussing the implications, we provide numerous quotes from the interviews to adequately represent their concerns and emerging approaches.

Next we discuss the changing nature of rurality and the rise of multifunctionality. We then connect the evolution of rurality to current and future changes in rural planning and development principles and practices. We conclude with a discussion of the importance of attending to the multifunctional transition in rural places.

Multifunctionality in rural places

Rural restructuring

Rural regions across the industrialised world have been experiencing 'rural restructuring' since the 1980s. Globalisation, the emergence of the neo-liberal state, the rise of automated production in agriculture and natural resource extraction, and amenity-driven gentrification are among the forces that have been affecting the rate and nature of rural change (Hoggart and Paniagua, 2001; McCarthy, 2008). The result has been the outmigration of many long-term residents no longer able to support themselves, while affluent urbanites have moved into high amenity rural areas. Further complicating things, international migrants are also settling in rural areas to take advantage of both the available primary production jobs as well as jobs emerging in the service sector because of rural gentrification (Nelson et al., 2010; Sandoval, 2013). So, population change entails a multifaceted and dynamic transformation in the rural economy, associated with a socio-cultural transformation as people continuously move in and out (Hibbard et al., 2011; Stauber, 2001). As well, rural regions are experiencing a complex range of issues including climate change, biodiversity loss, alternative resource development, land-use conflict, and food security (Morrison et al., 2015; Fitzgibbon, 2010). Restructuring challenges the abilities of rural regions to respond in traditional ways.

The declining significance of primary production as a source of jobs and wealth for rural communities, despite its continuing importance as a source of raw materials and basic products for the global economy, has led to awareness that the 'monofunctional' landscape is a product of the twentieth century. For most of human history, landscapes were multifunctional; it is only with the emergence of commodity production that monofunctionality became the norm. The drive to maximise material outputs led to the prioritisation of production values over all other ways of thinking about the landscape (Holmes, 2006). Previously, land was seen as 'a central medium through which all aspects of life are mediated, and economic considerations are merely part of an intimate, immediate, fundamentally holistic relationship' (Strang, 1997: 84). In the twenty-first century, after decades of rural restructuring and its attendant issues, contemporary global society has created multiple demands on rural landscapes, and there is a trend toward managing for a more variable, multifunctional rural landscape.

The multifunctional transition

Multifunctionality involves managing the landscape for a variety of environmental and socio-economic values in addition to production. Looking at the multifunctional transition in forest management, Kelly and Bliss (2012) point to a shift in goals among private industrial forestland owners, from maximising timber revenue to include aims of conservation, maintaining public access, and restoring degraded forest conditions (see also Klein and Wolf, 2007). Similar shifts are occurring across rural landscapes (Buttel, 2003; McCarthy, 2006).

Initially, multifunctionality focused on agriculture – producing farm products while simultaneously protecting biodiversity, generating employment, and contributing to the vitality of rural communities (McCarthy, 2006). A number of observers have argued for 'liberating the multifunctionality concept from its restrictive application solely to agriculture and recognizing [it] . . . as a pivotal component in the reconstitution of rural space' (Holmes, 2012: 252). Consistent with this, the core concepts of multifunctionality are implicated in a widening range of activities, variously termed 'working landscapes', 'working forests', 'working waterfronts', and the like (Hibbard et al., 2015).

The multifunctional transition from the dominant use of rural landscapes for production toward a multifaceted and often overlapping mix of commodity and non-commodity uses entails conventional *production* along with *consumption*, using the landscape without using it up, as in tourism and recreation, and *protection*, maintaining, conserving, and restoring the landscape. They do not occur sequentially or in different parts of the landscape, as in 'multiple use', but are often simultaneously present in a given place. Simultaneous engagement in production, consumption, and protection means that 'economic activity may have multiple outputs and, by virtue of this, may contribute to several societal objectives at once. . . . [It is] an activity-oriented concept that refers to specific properties of the production process and its multiple outputs' (OECD, 2001: 11). Drawing on the resource base for other uses, in addition to commodity production, enables a more diverse rural economy, as well as supports the social and environmental health of the local community.

Implications for rural planning and development

The challenge for rural planning and development is to understand the economic, social, and environmental processes shaping the countryside, their inter-relationships, and the possibilities for action in specific localities. Planning is fundamentally concerned with optimising the organisation and management of land and resources. As well, it has a future-seeking dimension, aiming to improve the circumstances of human existence. And, as a process, it has a problem-solving focus that seeks to mediate among diverse claimants (Hibbard and Lane, 2004). These mutually reinforcing characteristics of planning take on new meaning in the shaping of rural areas under the multifunctional transition.

Interviews with rural specialists (see also Frank and Hibbard, 2017) revealed planning and development responses to the multifunctional transition within two themes:

1   as a characteristic of rural settings, multifunctionality compels viewing the objects of planning as *human–landscape systems*, consisting of people and interacting productive, consumptive, and protective subsystems; and
2   multifunctional systems require and empower varied stakeholder involvement in public decision making, making diverse *governance* a key theme.

In introducing these themes, it is useful to note the close alignment between *planning* and *development*. Many of our interviewees made no distinction between them:

> 'Rural development operates at the intersection between the natural and built environment to create prosperity, maintain and enhance healthy ecosystems, and provide a high quality of life.'

> 'Development is about increasing opportunities and life choices in rural places . . . [through] the interaction and mutual influence of physical and socio-economic development in support of community resilience and sustainability.'

Human–landscape systems

The rural specialists we interviewed approached rural areas as human–landscape systems interacting at multiple scales. As one respondent stated, 'Rural is about the larger landscape. Even small towns are objects within the larger landscape.' Another said, 'Rural development recognises that people are embedded in and depend on the environment.' One interviewee reasoned that 'agricultural management goes beyond asking what crop to grow to include issues of the nutrient runoff and environmental conservation, which lead to regulation and planning activities', concluding that 'many planning topics translate to rural areas, such as businesses/economic development, residential land use, water and sewer'. Several interviewees emphasised the importance of integrating different planning sectors, especially physical planning (land use, environmental management, and design of the built environment) and economic development. As one respondent said, 'I see planning as beyond land use to include environmental and resource issues . . . There is not "a" definition of rural development. It is as much social as physical.'

The rural specialists' holistic, integrative perspectives led them to suggest that rural planning and development should be oriented towards general problem solving rather than strict disciplinary boundaries: 'When I went to college, I was told I could not be a generalist or entrepreneur, that I should have a narrow perspective. Now I see that there are [multiple] forms of assets, and we should work towards building each without undermining the others. We need to build systems to do that.'

A respondent critical of planning's silos said, 'I usually run screaming from planners . . . still wedded to the old paradigm . . . [They] said we were too hands on . . . [They are] not open to these ideas.' Another advocated 'using design as a problem-solving process, also known as "design thinking" . . . The new "Extension Reconsidered" model is about multidisciplinary innovation, creativity, and entrepreneurship, i.e., design. Everyone, including planners, become educated about the design process.' The respondents had more to say: 'The rural planner has to be a jack of all trades. S/he works on any and all issues from land use to economic development to infrastructure to natural resource management and the environment.' Another added, '[Rural planning is] engagement of communities, and tackling of issues not normally addressed by planners – immigration, food sector, environmental standards, agricultural sector.'

Several rural specialists discussed the power of the geographic perspective for paradoxically distinguishing and linking rural and urban areas. One respondent pointed out that the foundational social science in Europe and Canada is geography; this provides greater support for rural planning than in the United States, where the foundational social science is economics. In essence, 'rural planning concerns the rural geography'. Another respondent described a Canadian study that identified several rural geographic-economic regions and inter-regional interactions. A common sentiment was that 'we must look at rural and urban areas together, and work together'. Another respondent thought the rural–urban distinction was a distraction:

> My job isn't to revitalise rural planning; it is to better understand reciprocity and relationships between urban and rural. To do that, we need to know the full scope of current markets and politics. We are in desperate need of a new paradigm . . . If rural is seen as a separate entity, we are not going to get there.

The rural specialists connected the planning field and the geographic perspective through their concerns for place and community:

'People are beginning to recognise the common interests, and the importance of sense of place.'

'Planning and design can help citizens understand the aspects of a place and what is great about it.'

'[Rural planning initiatives] are concentrating on quality of place.'

'There's a conscious effort to create community by building community gathering places.'

Governance

Multifunctional landscapes are associated with diverse and multiscalar stakeholders, agencies, and constituencies. Their differing goals and expectations from the rural landscape make new competing, and sometimes conflicting, demands on rural planning. The literature's discussion of multifunctional rural areas elaborates on the governance view – networking of public agencies, private firms, and the nonprofit sectors, in planning, policy making, and implementation (Klein and Wolf, 2007; Morrison et al., 2015). Our respondents upheld that theme.

The interviewees elaborated various justifications for governance of multifunctional rural areas across organisational boundaries. Several observed that a legitimate role of government is to help realise the full value of rural areas by complementing the market: 'Rural areas have some public goods, some private. Some activities supported [by government policies] are not economically profitable in themselves but are a part of the larger good.' Within government institutions, respondents noted the need for rural planning to 'coordinate across policies' at the local, state, and federal levels. However, respondents reported that the common situation in rural areas of 'declining local institutional and financial capacity to deal with issues' can motivate local government restructuring to increase efficiencies, and recruitment of governance partners, that is, greater involvement of the private/nongovernmental sectors. In some cases, nongovernmental organisations (NGOs) may perform similar coordinating functions as government: 'A good role of nonprofits is to hold the big picture and facilitate, to discover shared interest.' One rural specialist suggested, 'It might be helpful to make [planners] aware of alternative tracks – in NGOs, for example – where there is more opportunity to be an agent of change.'

As a result of rural governance trends, a respondent characterised rural planning as 'evolving from "lines on a map" to brokering, facilitating, resolving'. Several respondents noted a new challenge of reconciling the growing influence of local stakeholders with national public and international private sector interests: 'When planning for rural agricultural areas, planning must acknowledge the major multinational players, such as Dow Chemical.' A respondent described Canada, where citizens are 'less oriented towards property rights', as experiencing a 'push-back to centralised government . . . by local residents over wind turbines', which raised the philosophical issue of 'different public interests [planners] must work through' during global changes. A second respondent from Canada stated:

> There are debates in rural areas in Canada about ownership, land claims and treaties, around who owns, inherits, and plans the land and resources. In the coming decade, these debates will become a major part of the discourse about what rural is and who has obligations and authority to plan.

A US-based respondent welcomed the debates: 'Planning needs to be involved with ownership and control. . . . there is no point in creating wealth if it doesn't stick.'

All respondents discussed the need to continue to build local governance capacity:

> 'There are political barriers [to planning] more at the local level than at the provincial level; people don't understand the value of planning.'

> 'The political system lags the public consciousness.'

> 'Unfortunately, the political old guard of county commissioners like the way things are currently done, which is "back door" . . . and tend not to embrace new ideas.'

This leadership gap is a challenge for rural planners, requiring them to be simultaneously technically proficient and politically savvy.

Many respondents advocated citizen participation as a strategy to increase support among the 'old guard', build governance capacity, and provide information to planning processes. In Canada, which is seeing more conflicts across scales of interests, a respondent reported 'community engagement' as an 'innovative' response. Another described citizen participation as a key component of 'design thinking':

> [Design thinking] works with rural citizens, policy makers, institutions, and foundations, and leads groups to think beyond their own borders towards regional perspective . . . How processes are conducted makes all the difference. We do design 'workshops', not 'charrettes'. The workshops lead to citizen-selected solutions.

The human–landscape perspective and diverse governance approach enable planners to be effective within multifunctional rural areas. The next section examines what planners can do to further advance the multifunctional transition so that rural areas can better support communal and long-term goals of social, economic, and ecological sustainability and resilience.

Planning for multifunctionality, sustainability and resilience

Multifunctional places by definition reflect diverse values, and *sustainability and resilience* are closely related ethical frameworks suited to balancing and integrating values. Sustainability and resilience are the leading normative perspectives on human–landscape systems in planning and development (see also Chapter 5 of this Companion). These perspectives in turn highlight the need for system diversity and interaction, and they inspire strategies for multifunctionality and associated guidelines for professional practice in rural settings (Klein and Wolf, 2007; Lovell and Johnston, 2008; Musacchio, 2009; Scott, 2013). Sustainability, resilience, and multifunctional strategies to support these goals figured prominently in our interview results, and we present them below according to observed sub-themes.

Responding to globalisation

When asked about the major planning issues facing rural areas, the respondents described dual crises of globalisation and degradation. According to one, rural areas suffer from 'neo-liberal myths about the highest, best use . . . [leading to] the extraction of resources, people, and money of a rural place . . . and the centralisation of everything'. Echoing that sentiment, another remarked on 'corporatisation – the consolidation of public services, privatisation of health care, and so on'. Across the interviews, the resulting problems facing rural areas covered environmental, social, and economic aspects:

'Exploitation, misuse, and lack of protection for natural resources – especially water.'

'Economic instability because of population change [i.e. loss and aging].'

'Under-investment – in infrastructure, social capita.'

'We have become more of a developing country in the rural areas.'

'[There is] stereotyping of rural people and places . . . no recognition of rural diversity.'

'Unwanted' land development is due to 'outward expansion of metro areas into urban fringe, and growth pressures in high amenity areas'.

One respondent summed up the situation: 'There are enormous changes occurring . . . to which rural areas are still trying to adjust.'

Despite the losses, rural people and places have significant assets and, through appropriate planning and development, potentials for supporting sustainability and resilience from local to global scales. A respondent remarked, 'The system is dysfunctional without feedback loops . . . Planning should get honest, and make visible how the status quo is costly.' There was also a thread of social equity, especially as it related to obsolescence and marginalisation from globalisation. One rural specialist described the aim of rural planning and development as to 'resist external control and maintain a separate local identity'.

Combining diverse goals

The multifunctional transition introduces an abundance of stakeholders – for example, commodity producers, recreationists, environmentalists, Indigenes, and retirees – whose viewpoints, even if they are conflicted with each other, have to be heard. The human–landscape sustainability literature, both rural and urban, advocates simultaneous planning for environmental, economic, social equity, and cultural goals (Morrison and Lane, 2006; Musacchio, 2009; Wu, 2013). In the planning literature, sustainability is seen as a means of achieving resilience (Jun and Conroy, 2014), however, the magnitude of global issues (e.g. climate change) has led to recent debates about whether responses should focus on resilience or sustainability, and some authors have declared 'the end of sustainability' (Benson and Craig, 2014). Scholars are identifying how planners can negotiate between the two, and recent literature explores ways to move beyond complementarity toward synergies that directly relate to the concept of multifunctionality (Dymén and Langlais, 2013; Duguma et al., 2014).

The interviewees expressed similar views. For example, a respondent said, 'The goals of rural development are economic security, healthy ecosystems, and social inclusion'. The rural specialists explicitly referenced sustainability and resilience frameworks, such as saying 'we need to reverse the paradigm . . . [towards] sustainability and inclusivity', using the term 'sustainable economic development', and promoting 'a new goal of fostering resilient communities'. The respondents also conveyed holistic sustainability goals, such as to 'enable a decent quality of rural life' and to '[increase] life choices so people can derive better satisfaction'.

Fostering social equity through community-based economic development

The interview responses included specific objectives and strategies for sustainability and resilience in social-economic and physical realms. Many of the respondents were concerned with 'ways to do economic development without subsidies in deep rural areas'. In these places, community objectives entailed interrelated strategies of business retention, workforce training, and endogenous economic development, such as 'finding a new economic reason to exist as [the] old economy disappears'; 'preparing people for the jobs of tomorrow'; and answering 'how do we facilitate growth, provide labor retraining, develop and retain the agricultural sector?'. Strategies for endogenous economic development included fostering community-based entrepreneurialism focused on local assets:

> 'There is rising interest in rural entrepreneurship.'

> 'Rural areas need to transition away from the traditional paradigm of buffalo hunting [also known as smokestack chasing], toward development from within, identifying and building on assets/strengths.'

> 'Development has historically meant sending people out to rural areas to help with technology, knowing the land and what it will grow; it was top-down. Now there is more emphasis on self-help and the local economy, but there is still a provincial role.'

Several respondents believed that communities had significant local power to organise and meet their needs due to their 'culture of self-help, self-reliance', which has resulted in regional programs such as 'a local food movement, local special service districts, and formal and informal helping networks'. One noted that fostering resilient communities 'is about capacity building, assets not deficits, working forward, leadership'.

The respondents viewed local economic development as a means to an end – social equity – and this shaped the recommended strategies. A respondent advocated that rural communities 'attract investment that they own and control'. As one respondent explained:

> The root of poverty is isolation from the mainstream economy, healthcare, and wealthier neighborhoods . . . Local people can gain more power by starting on the demand side of the market. Ask: who has a self-interest in seeing the local community succeed, and what could local people bring to the table to make the investment succeed? . . . Rural areas need to make the case, 'If you invest with us, such as healthy food, you will get a return on investment.'

Relating rural and urban areas

In contrast to the concern with declining deep rural areas far from cities, several respondents discussed strategies for supporting sustainability and resilience in rural areas experiencing urbanisation. These strategies tended to fall into the physical realm, especially land-use planning:

> 'How do we protect farmland, increase density, create appropriate standards, and control sprawl?'

> 'In growing areas, planning should focus on preserving what has made the place attractive and special, and not allowing growth to overwhelm the character of the place.'

> Protection occurs 'through rural conservation design and greenway planning'.

> 'There's a renewed commitment to efficient infrastructure and physical development'.

Most respondents recognised the strategy of linking rural and urban concerns together via economic development and land use strategies:

> We have been seeking to reposition rural areas to be seen as valued and essential parts to urban and suburban areas . . . We need to make clear that people in cities need food, fiber, and recreation, energy; things provided by rural areas . . . No one is an island. We need to identify the self- and shared interests of each and develop an economy in sustainability.

Such a perspective can blur the line between rural and urban:

> The [Ontario] Greenbelt Act recently won an APA award; it focused on the countryside, but with significant aspects of urban . . . The provincial growth belt is an innovation, but is it urban or rural planning?

Barriers to rural planning and the multifunctional transition

Despite the rural needs and innovations in practice, the rural specialists reflected the uphill battle faced by rural planning and development:

> 'We don't get a hell of a lot of respect in the [planning] profession.'

> 'There is no program in rural design anywhere in the world. I can't even get my own college to create one.'

> 'Unfortunately, community development extension is fading out.'

> 'There is neglect of rural policy by the federal government.'

> 'I was surprised by the degree to which the strength of markets makes local voice very difficult.'

The emergence of multifunctionality requires a dramatic shift in disciplinary thought away from reductionism and towards holism, as well as sophisticated coordination, sharing, and intellectual demands on planning and governance, which are difficult barriers to overcome:

> 'Rural policy is still to an alarming degree seen as synonymous with agriculture.'

> 'Nonprofits don't think as networks. Everything has to be a project; they want to own it.'

One respondent described even the traditional rural development field as too narrowly conceived: 'Community development is about voice, economic development is about jobs and money, international development is rights-based; it would be powerful to combine these perspectives.' Another saw hope in the fact that 'planning becomes possible in times of stress'.

Conclusion

The multifunctional transition is leading towards a transformation in rural planning thought and practice. The emerging perspective, with emphases on human–landscape systems, governance, and sustainability and resilience, necessitates the integration of actors and sub-disciplines, such as land use planning and economic development, it elevates ethics, and it focuses on planning as problem-solving, design, and entrepreneurship.

The emerging perspective contrasts with the North American planning field's dominant orientations towards urban areas and the modern economic, social, policy, and scientific paradigms that have underpinned globalisation. In particular, the multifunctional perspective is the opposite of the planning field's inclination to separate ostensibly 'incompatible' uses. At a time when the dominant paradigms are being called into question, rural areas have become breeding grounds for alternative planning thought and practices.

References

Benson, M.H. and Craig, R.K. 2014. The end of sustainability. *Society and Natural Resources*. 27(7), 777–782.

Buttel, F.H. 2003. Communities and disjunctures in the transformation of the U.S. agro-food system. In: Brown, D.L. and Swanson, L.E. eds. *Challenges for Rural America in the 21st Century*. University Park, PA: Pennsylvania State University Press, 177–189.

Duguma, L.A., Minang, P.A. and van Noordwijk, M. 2014. Climate change mitigation and adaptation in the land use sector: From complementarity to synergy. *Environmental Management*. 54(3), 420–432.

Dymén, C. and Langlais, R. 2013. Adapting to climate change in Swedish planning practice. *Journal of Planning Education and Research*. 33(1), 108–119.

Fitzgibbon, J. 2010. Resources, environment, and community: Conflict and collaboration. In: Douglas, D.J.A. ed. *Rural Planning and Development in Canada*. Toronto: Nelson Education, 151–178.

Frank, K. and Hibbard, M. 2016. Production, consumption, and protection: Perspectives from North America on the multifunctional transition in rural planning. *International Planning Studies*. 21(3), 245–260.

Frank, K. and Hibbard, M. 2017. Rural planning in the 21st century: Context-appropriate practices in a connected world. *Journal of Planning Education and Research*. 37(3), 299–308.

Frank, K.I. and Reiss, S.A. 2014. The rural planning perspective at an opportune time. *Journal of Planning Literature*. 29(4), 370–385.

Hall, P.V. and Stern, P. 2009. Reluctant rural regionalists. *Journal of Rural Studies*. 25(1), 67–76.

Hibbard, M. and Lane, M. 2004. By the seat of your pants: Indigenous action and state response. *Planning Theory and Practice*. 5(1), 97–104.

Hibbard, M., Seltzer, E., Weber, B. and Emshoff, B. 2011. *Toward One Oregon: Rural-Urban Interdependence and the Evolution of a State*. Corvallis: Oregon State University Press.

Hibbard, M., Senkyr, L. and Webb, M. 2015. Multifunctional rural regional development: Evidence from the John Day watershed in Oregon. *Journal of Planning Education and Research*. 35(1), 51–62.

Hoggart, K. and Paniagua, A. 2001. What rural restructuring? *Journal of Rural Studies*. 17(1), 41–62.

Holmes, J. 2006. Impulses towards a multifunctional transition in rural Australia: Gaps in the research agenda. *Journal of Rural Studies*. 22(2), 142–160.

Holmes, J. 2012. Cape York Peninsula, Australia: A frontier region undergoing a multifunctional transition with indigenous engagement. *Journal of Rural Studies*. 28(3), 252–265.

Jun, H.J. and Conroy, M.M. 2014. Linking resilience and sustainability in Ohio township planning. *Journal of Environmental Planning and Management*. 57(6), 904–919.

Kelly, E.C. and Bliss, J. 2012. From industrial ownership to multifunctional landscapes: Tenure change and rural restructuring in Central Oregon. *Society and Natural Resources*. 25(11), 1085–1101.

Klein, J.A. and Wolf, S.A. 2007. Toward multifunctional landscapes: Cross-sectional analysis of management priorities in New York's Northern Forest. *Rural Sociology*. 72(3), 391–417.

Lovell, S.T. and Johnston, D.M. 2008. Creating multifunctional landscapes: How can the field of ecology inform the design of the landscape? *Frontiers in Ecology and the Environment*. 7(4), 212–220.

McCarthy, J. 2005. Rural geography: Multifunctional rural geographies – reactionary or radical? *Progress in Human Geography*. 29(6), 773–782.

McCarthy, J. 2006. Rural geography: Alternative rural economies – the search for alterity in forests, fisheries, food, and fair trade. *Progress in Human Geography*. 30(6), 803–811.

McCarthy, J. 2008. Rural geography: Globalizing the countryside. *Progress in Human Geography*. 32(1), 129–137.

Morrison, T. and Lane, M. 2006. The convergence of regional governance discourses in rural Australia: Enduring challenges and constructive suggestions. *Rural Society*. 16(3), 341–357.

Morrison, T.H., Lane, M.B. and Hibbard, M. 2015. Planning, governance and rural futures in Australia and the USA: Revisiting the case for rural regional planning. *Journal of Environmental Planning and Management.* 58(9), 1601–1616.

Musacchio, L.R. 2009. The scientific basis for the design of landscape sustainability: A conceptual framework for translational landscape research and practice of designed landscapes and the six Es of landscape sustainability. *Landscape Ecology.* 24(8), 993–1013.

Nelson, P.B., Oberg, A. and Nelson, L. 2010. Rural gentrification and linked migration in the United States. *Journal of Rural Studies.* 26(4), 343–352.

OECD. 2001. *Multifunctionality: Towards an Analytical Framework.* Paris: OECD Publication Service.

Sandoval, G. 2013. Shadow transnationalism: Cross-border networks and planning challenges of transnational unauthorized immigrant communities. *Journal of Planning Education and Research.* 33(2), 176–193.

Scott, M. 2013. Resilience: A conceptual lens for rural studies? *Geography Compass.* 7(9), 597–610.

Stauber, K.N. 2001. Why invest in rural America – and how? A critical public policy question for the 21st century. *Economic Review – Federal Reserve Bank of Kansas City.* 86(2), 57–87.

Strang, V. 1997. *Uncommon Ground: Cultural landscapes and Environmental Values.* New York: Berg.

Wu, J. 2013. Urban ecology and sustainability: The state-of-the-science and future directions. *Landscape and Urban Planning.* 125(May), 209–221.

7

Land, property and reform

Peter Hetherington

Valuing land: the British case

Are the London-based ruling class largely indifferent towards the use, and the abuse, of a land mass that defines Britain and, in many ways, underpins a distinct identity for citizens of three nations: Scots, Welsh and English?

Senior British ministers running the grandly titled Department of the Environment, Food and Rural affairs, or Defra – an English-only ministry now that Scotland and Wales are largely self-governing – lie well down the Whitehall pecking order. Not so in Scotland, where ownership of land, and its varied use, has a higher priority in a devolved government.

But in England, ministers in charge of Defra come and go with regularity, although the latest incumbent – Michael Gove, a leading 'Brexiteer' – has shown more interest, than a succession of short-lived predecessors, in a key role of his department: namely, the environment. He thinks 'Brexit' provides an opportunity to strengthen countryside stewardship and wildlife protection. Farmers, on the other hand, are clearly anxious. According to a report from an advisory body to which many of them subscribe through a levy, incomes could halve after 'Brexit' from a current average level of £38,000 to £15,000 (Financial Times, 2017). And even with an EU–UK trade deal, there's little indication of how £3bn annually in agricultural support from the EU's Common Agricultural Policy will be replaced – indeed, how it can be in full – with competing demands on the government's diminishing revenues?

Nevertheless, we have much to celebrate. Our small island is glorious in both its diversity and in the variety of its landscapes – from the Lake District, to the Scottish Highlands and Islands, flatlands of the English east, downlands of the south and the mountain and coastal splendour of Wales.

But who cares beyond the self-interest of the farming lobby? The answer should be quite a lot. When membership of environmental and countryside groups is totted up – such as the National Trust, its Scottish equivalent, the Royal Society for the Protection of Birds (RSPB), the Ramblers, and countless others – you can easily head towards the 10 million mark. And that is before the estimated 110 million who visit our national parks each year are thrown in, albeit often on multiple trips.

Yet with our island facing so many pressures – not least feeding, watering and housing its three nations in the face of myriad challenges – any debate surrounding the compelling case for an active land policy, through a new ministry adopting a 'whole systems approach' (Hetherington, 2015), has been muted. The ruling class, across the political divide, has had other priorities.

Are we, then, in danger of wasting our most precious resource through a combination of government inaction and collective indifference – in sharp contrast to a combination of active government and progressive civil society earlier in the twentieth century and beyond to the late 1940s?

Among the British public, only one area, it seems, has generated considerable anger, albeit with more heat than light: housing the largest nation of Britain (particularly the south eastern part of it). Below lurid headlines of a country threatened by bulldozer, brick and breeze-block – the threat of being 'concreted over' is a favourite headline – middle-England is uneasy. Surveys show (e.g. Ipsos MORI, 2012) that many think between a quarter and a half of England is urbanised. In fact, while clearly overcrowded in London and the south east, only 8.8 per cent of England is built on – and much less than that is occupied by housing (Hetherington, 2015) – while almost three-quarters is devoted to farmland and 14.5 per cent classed as 'natural', embracing mainly mountain, moor, heath and wild grassland.

However, perception is all. In 2017, after national BBC news headlined the findings of a detailed survey of land use, an irate viewer in his eighties from Windsor and Maidenhead, west of London, phoned the academic behind the exercise (Professor Alasdair Rae, University of Sheffield) in disbelief (BBC, 2017). His quality of life had deteriorated so markedly in recent years, with infrastructure seemingly unable to cope with new housing, roads constantly gridlocked and trains overcrowded that he simply could not believe the statistics.

That, of course, is a relatively small part of a country – one of the most populous in Europe – where perception sometimes collides with reality. But consider this: with so much arable and pasture land, particularly in England, Britain has more areas actively farmed than most larger European nations which have just 50 per cent devoted to livestock and crops – and Norway a mere five per cent (Taylor, 2016).

While much of our land is undoubtedly 'green and pleasant', those campaigning to protect their perception of a timeless countryside they regard as 'unchanging' sometimes forget it has changed beyond recognition in relatively recent history. As the historian Howard Newby (1988) has noted:

> The countryside may be the repository of our 'heritage', but it is often the countryside of tower silos (beside farms) . . . uprooted hedgerows, ploughed-up moorland, burning stubble, pesticides, factory farming . . . sometimes our idyllic image stops us seeing reality.

Yet the countryside – surely, *our* land morally, if not legally – is facing a variety of pressures which, frankly, are barely being addressed: providing more home-grown food, additional water supplies, and adequate housing for the 66 million+ population of our nations for a start. We seem to be drifting aimlessly without an over-arching vision embracing all land use, let alone developing a wide-ranging policy to oversee our most precious resource.

One key challenge, among several, emerges: while much of the country is devoted to farming, which directly produces just 0.7 per cent of GDP, we are well short of self-sufficiency in basic crops. And our most productive acres lie in eastern England, where the most fertile land is at or below sea level, and only protected from rising sea levels by often inadequate flood defences.

On one level, we draw so much from our land, physically and mentally. It nourishes the inner soul, providing solace, inspiration, exercise and recreation. The British romanticise it through great artists, poets, composers: Wordworth, Burns, Constable, the school of Scottish Colourists, Delius, Vaughan Williams, Ewan McColl's 'Manchester Rambler' ('free man on Sunday, wage slave on Monday'), and a derivation of Woody Guthrie's classic 'This Land is Your Land', adapted by Chris Ellis and Rosie Toll in a version associated with the musician and radical activist Billy Bragg:

> This land is your land
> This land is my land
> From the Downs to the Western Highlands
> From the oak wood forest to the Lakeland waters
> This land was made for you and me.

But here is the question: do we value that most basic resource – our land – as much as we should beyond the obscene monetary gain from both trading, and hoarding our most basic resource which, for the super-rich, has become a more valuable commodity than both gold and prime central London property (Financial Times, 2015)? The moneyed elite, from Britain and overseas, have certainly taken Mark Twain's oft-quoted advice to heart: *Buy land, they're not making it any more.*

And, courtesy of the Office for National Statistics (ONS), we thankfully can place more of a real value on land. Last year, it helpfully published (Cornish, 2017) so-called 'ecosystem accounts' which placed a value on natural resources benefiting people in three areas: farmland, freshwater habitats (such as rivers and lakes) and woodland. As national 'assets' it concluded that, in 2015, they were collectively worth £177.7 billion (ibid.).

Even if that was a notional figure, it represents a good starting point for the areas we should, as a nation, be addressing – and, it certainly has been calculated from far more detailed research, and analysis, than anything passing as an 'accurate' valuation of, say, farmland and the considerable tracts with planning permission for housing which benefit from a huge development uplift in value from planning permission which, invariably, is captured by developers rather than by the local community.

So how should the country begin to address the myriad pressures, the challenges, and the opportunities facing our land? From the outset five glaring issues urgently need addressing:

Home-grown food

According to the National Farmers Union (NFU, 2015), we are lucky to produce 62 per cent of the food we are capable of growing – down by a fifth since 1980. While it thinks the country has the potential to reach 85 per cent self-sufficiency, it argues that inadequate productivity and under-investment in research and development has placed the country behind mainland European counterparts. And unless we are careful, it warns that in 25 years' time self-sufficiency will fall by a further 10 per cent. Now, with the prospect of Brexit, we at least have a chance to redistribute farming subsidies – almost £3 billion annually – towards genuine food production rather than mindlessly subsidising the biggest landowners regardless of how they use, or misuse the land. As it is, around a third of our food is imported from the EU.

Land trading

In Britain, particularly England, land has become the safest investment for those with a few spare millions to offload. Prices were up by a staggering 277 per cent in a decade according to one survey of UK land by Savills (Financial Times, 2015). In the 1994–2004 decade they grew by 41 per cent. Prime London property, by contrast in the 2004–2014 period rose by a mere 127 per cent, although thankfully they are now falling. Our prime farmland was considered more valuable than gold. Institutions, and the rich, are still queuing up to get their hands on the country's most basic resource.

Why? Two reasons. Farmland is exempt from inheritance tax if it is actively farmed after two years. And additional relief allows the sale of a farming asset to be rolled into a new business, and capital gains tax is thus deferred until the sale of the asset. If you want to avoid tax just keep selling, buying, rolling and deferring endlessly. The other reason, perhaps the main attraction, is that farm-land has been recession-proof. Agriculture is the last great subsidised industry, operating outside the norms of the market. Regardless of the state of the national economy, or the efficiency or the prosperity of farmers – be they 'barley barons' in the east of England, well-endowed landowners or modest Pennine hill farmers up against it – the subsidy cheques keep rolling in courtesy of the EU's CAP. But for how much longer?

Will the prospect of Brexit – presumably with a UK government offering reduced, and more targeted agricultural support – bring down the cost of farm land and shake-out the industry? As it is, many outside 'investment' buyers are not interested in farming, preferring to hand its management over to contractors so that 'the new owner enjoys the fiscal benefits of being an "active farmer" while leaving the responsibilities to (others)' (Taylor, 2016: n.p.). And unlike much of mainland Europe, there is neither regulation, not limitation of foreign investment in either land or real estate.

Moreover, although all land in England is required to be registered by the Land Registry – a state agency – following any significant change in its title, around 20 per cent, most of it rural, still remains unregistered because it has not changed hands for a considerable time. This means that accurate statistics on the nature of land ownership are difficult to produce, although on some estimates two thirds of UK land is owned by 189,000 families.

Homes for all

Are farmers and landowners gaining handsomely by keeping land off the market? Is the most basic resource – *our* land – being rationed in the face of a housing crisis with landowners awaiting a windfall as land prices rocket into the valuation stratosphere? These might be open questions. But an uplift in value from planning permission encourages a high level of land trading, rather than development, resulting in windfalls for the developer rather than the local community. On some estimates, land now accounts for two thirds of the cost of a home.

Late in 2014, the Lyons Housing Review Commission (Lyons, 2014: 6) warned that an artificial scarcity of land was distorting the housing market, limiting building and 'incentivising the acquisition and trading of land'. It spoke of six firms of land agents alone holding strategic land banks of 23,000 acres – enough perhaps for 400,000 homes. It expressed particular concern that 'non-developers' – farmers, landowners, institutions – were holding onto land either under an option for development or with planning permission. Although they had no intention of currently building, Lyons said they 'may be motivated by speculating on future land values' (ibid.: 62).

What is the consequence, then, of this trading, whether for farming or for building? Certainly, at least at the margins, it does appear that food production is of secondary importance to tax 'efficiency'. One interviewee of mine – farmer of the year in 2014 – volunteered that he was appalled by the state of some recently traded land, farmed not by owners but by contractors.

In this opaque, unregulated market – with land for housing at a premium – one casualty is the size of homes: they are getting smaller. The other casualty is the small building firm, squeezed out by high land prices: they are becoming fewer. And the third, of course, is the aspiring home owner, priced out of the market. The use and abuse of our land takes no account of the nation's housing needs. In England it is widely acknowledged that we should be building approximately 250,000 houses annually to meet rising household formation. The figure has barely reached half that level in recent years.

Flood remediation and resilience

It is a rarely discussed issue which only – literally – surfaces when floods, sometimes following tidal surges, affect large tracts of England. Briefly there is a public outcry, then the issue is largely forgotten. Yet we should all be concerned by the consequences of rising sea levels, induced by climate change. And we have been warned – not least by tidal surges along the east coast of England late in 2013. The Foresight Land Futures report (2010) pulled no punches: 57 per cent of our best Grade 1 farm land, and 13 per cent of Grade 2, occupies flood plains – an important asset for national food security. And it put the capital value of 'at risk' land at £15 billion. Here is the choice it put: either governments provide high levels of flood protection, or they abandon large tracts to the sea. In 2014, the National Audit Office, the official public spending watchdog went further, warning millions of acres are at risk because Government flood defence spending is spread too thinly (NAO, 2014). About half of the country's defences, 1400 schemes, are only being maintained to a minimal level.

But far from rising, flood defence spending is flat-lining. Flood resilience should be a matter of the most intense focus in government. Local government is rightly infuriated by cuts in flood defence funding. The House of Commons Environment, Food and Rural Affairs Committee (2016) has called for a complete overhaul of flood defence and resilience policy, arguing that responsibility for it should be separated from the Environment Agency, a state body. And it wants a much bigger experiment in land management to develop a post-Brexit system of farm payments that subsidises land use as a weapon against downstream flooding.

Land denied

While tenants farm at least a third of the agricultural acres of England and Wales, on some estimates the figure is probably 10 per cent higher, although disguised by agreements both formal and informal. Britain needs young, ambitious farmers if it is to ramp-up food production and replace EU imports. But there is a problem: current, short-term tenancy agreements, while providing little security for farmers also offer no incentive for investment in land and essential equipment to farm and maintain it.

Many landowners prefer short-termism. Almost half of all agricultural tenancies in England and Wales are now let (thanks to 1995 legislation) for only four years on average under so-called 'farm business tenancies'. For George Dunn (2017), energetic chief executive of the Tenant Farmers' Association, this serves the country extremely badly. He has an answer: change the taxation framework for agricultural land, which leaves farms and buildings largely free from inheritance tax with significant relief from capital gains tax, to make it contingent on longer farming tenancies and, hence, a commitment to nurture the land.

That so much land remains in the hands of big institutions, the new rich and the old aristocracy is a testament to the influence, and – yes – sometimes the guile of the old landed class alongside the incentives for 'investment' in land delivered by our taxation system. It rewards the wealthy with both a safe haven to offload spare millions and protection for family trusts and, hence, safeguard the big tenanted estates which form the bedrock of the aristocracy's continuing influence in swathes of England.

To be fair, the amount of tenanted land has fallen significantly, since the early twentieth century when 90 per cent of farm land was tenanted – including a significant part of what is now the Republic of Ireland, where most is now owned by those who farm it. This followed considerable agitation and political unrest in the late nineteenth century, forcing the British government to make concessions in an attempt to appease Irish nationalism. By 1914, for instance – eight years before Irish independence – 75 per cent of farmers had acquired the farms they tenanted, after holdings had been compulsory purchased by an Irish Land Commission (only abolished in 1999).

Significantly, alongside the insurrection in Ireland, mainland Britain was about to experience a quiet revolution in land ownership. A combination of heavy taxation (death duties increased to 40% on the biggest estates) and the deaths of young aristocratic heirs to estates in battle, resulted in the biggest shift in land ownership since the dissolution of the monasteries in the sixteenth century. As a result, by 1939, over half of all our agricultural acres were held by owner-occupied farmers. Nevertheless, by selling outlying portions of estates, many aristocrats held onto core businesses. Today they are often thriving, and expanding into wider property development. As Arthur Marwick (cited in Hetherington, 2015: 22) notes in his study of British society after the First World War:

> They emerged . . . still in residence in their country seats with their territorial empires considerably reduced but with their incomes . . . probably much healthier than they had been for many years.

Progression – or regression?

To the relative outsider examining land use, and abuse, it doubtless appears remarkable that in the light of so many social changes, and upheavals since the early to mid-twentieth century, ownership of our most basic resource in England, Wales and, to some extent in Scotland (where land reform is beginning to take root) sometimes seems stuck in an early-twentieth-century time-warp.

England has rarely experienced agitation on a scale of, say, the Irish Land League in the late nineteenth century, campaigning against landlordism and for tenants to own the land they farmed – or, for that matter, the powerful emotions aroused by a dark period in Scottish history 200 years ago, the Highland Clearances. That involved the brutal removal of families 'cleared' from the land to make way for sheep, and it unleashed a lingering bitterness still deeply etched in the collective psyche of some Scots – much stronger than the emotional fall-out from the Enclosures, largely in England, one of the most turbulent periods of rural history. Through mainly the eighteenth and nineteenth centuries, millions of acres of open fields were appropriated and enclosed in a series of parliamentary acts, forcing the poor off the land. According to the author and campaigner Kevin Cahill (2002), the Enclosures were particularly invidious because the aristocracy took from a weak monarchy 'rights which were really those of the common people . . . at a time when the common people had neither representation nor power'. In one estimate six million acres, a quarter of all cultivated acreage were appropriated from the second quarter of the eighteenth century to the first quarter of the nineteenth. Thus, by the 1870s, after the final Enclosure Act, the idea of collective land rights in England had been largely extinguished.

The repercussions from these 'legal' land grabs undoubtedly influenced a reforming Liberal chancellor of the exchequer (minister of finance), and subsequent prime minister, David Lloyd George – a passionate advocate of land reform – in the early twentieth century. Land became a defining political issue in his 1910 'People's Budget'. It hit the aristocracy with land taxes to help pay for an emerging welfare state and, after the First World War, further Lloyd George reforms bore down even harder on the landed class.

In truth, the quiet revolution in land transfers from estates to tenant farmers, in the 1920s–1930s (outlined previously), was short-lived. Successive and progressive post-war legislation, such as the 1947 Town and Country Planning Act (which briefly 'nationalised' the right to develop land, seen as a precursor to full public ownership of land) failed to address wider ownership. As the historian David Kynaston (2007) has pointed out, full land nationalisation was soon off the political agenda. Instead, with a new Agriculture Act in 1947 guaranteeing prices for farmers – precursor to the EU's Common Agricultural Policy (CAP) – it became 'jackpot time' for many farmers. Agriculture thus emerged as a powerful lobby with Labour as its friend.

To be sure, other legislation, delivering a string of national parks in England and Wales – and, earlier this century, two in Scotland – and guaranteeing a right to walk along 14,000 miles of footpaths, with the same legal force as any highway – opened up the countryside. But as the last century progressed, the reformist zeal of Lloyd George was sadly diminished. And, if anything, subsequent years, far from being progressive, have proved regressive – certainly in the insecurity brought by short-term farming tenancies.

Until relatively recently, young farming aspirants could sometimes count on one main avenue into agriculture: county council farms, another legacy of the progressive reforms adopted by Lloyd George. His 1919 Housing and Town Planning Act provided subsidies for councils to build tens of thousands of homes for rent – establishing a principle that housing is a social need – while a Land Settlement Act in the same year encouraged councils to provide land for a new generation of farmers. This resulted, over the years, in the creation of a substantial, local government-owned farming enterprise spread across England. From North Yorkshire to Staffordshire, Norfolk to Gloucestershire, Cambridgeshire to Devon, county councils bought large tracts of farmland – and let them to a new breed of tenant farmers. It proved a remarkably successful initiative, providing a first step on the farming ladder for many – that is, until the mid-1980s when councils began selling them off. Although still covering under 200,000 acres, well over half have been sold from a county council estate once covering at least 340,000 acres.

Charles Coats, who managed Gloucestershire's county farms for 27 years, and has become a national authority on a system once labelled a gateway into agriculture, laments that the aims of subsequent legislation in 1970 – laying down a general aim for councils to acquire more land and requiring them to submit annual plans to the government for scrutiny – were never met: 'They were a discretionary, not a mandatory service, which meant they became very vulnerable' (interviewed by the author).

As it is, Britain offers an open market in land. As a leading rural surveyor at a large farming consultancy notes (Charles Coats, interviewed by the author), in most of Europe authorities rely much more heavily on restrictions on land ownership to influence the use of land and to protect rural communities, although this seems to fly in the face of the Treaty of Rome. Nevertheless governments justify the controls on another article of the EU constitution setting out objectives to ensure a 'fair standard of living' for agricultural communities. Only the Netherlands, it seems, has an open land market similar to that in the UK with all other EU countries controlling land ownership in some form.

Reformist Scotland

Scotland might share the same island as England, yet the ideological and philosophical divide between two nations is considerably wider than – say – the Solway Firth estuary separating the two nations at the west of the 100-mile Anglo-Scottish border.

It was not always so. Before Scotland was granted self-government over a broad domestic agenda – including farming, rural affairs and, hence, land management – after the 1997 UK general election, a Westminster (London-based) government effectively oversaw Scotland. In the early 1960s, a British Labour government created a powerful Highlands and Islands Development Board to improve the economic fortunes of much of the Scottish land mass. Land reform was high on its agenda; a former board chairman, an economics professor, even railed about the 'unacceptable face of feudalism' in the Highlands. But the subsequent election of a Conservative government put paid to such radical sentiments. Reform was quietly sidelined. Then self-government arrived, albeit with Westminster controlling the Treasury and foreign affairs. A Labour–Liberal coalition government, elected from a new Scottish parliament in 2000, put land reform high on its agenda. One statistic, above all – quoted by a land reform activist, Andy Wightman – highlighted the issue: namely that only 432 owners, a combination of the old aristocracy and a new rich, often from overseas, owned 50 per cent of privately owned land. By 2003 the Scottish Parliament had passed a Scottish Land Reform Act giving communities the first right of refusal when a large estate, or land holding, came on the market. Although rarely invoked, the legislation proved a spur: today 562,230 acres – albeit representing 2.9 per cent of the Scottish land mass – are owned by community trusts, often covering large tracts of the highlands and, particularly, much of the 130-mile long Western Isles archipelago. Now a 2016 Land Reform Act, pursued by a Scottish National Party government, has ostensibly given ministers the power to compel the sale of an estate, or holding, to a community if they believe it is not being managed in a sustainable way – in other words, it does not require a willing seller. The legislation has also created a Scottish Land Commission – created in April, 2017 – tasked with creating nothing short of a new order where (to quote a promotional video) 'everybody benefits from the ownership, management and use of the nation's land and buildings'. The legislation now applies to urban, as well as rural areas – thus giving groups in towns and cities, theoretically, the power to press for government intervention if parcels of land fail the 'sustainability' test. It is clearly a big ask.

Significantly, the act provides for more immediate action: slapping business rates on 'sporting' estates (i.e. huge areas of, mainly, the highlands and islands, where stalking red deer, salmon fishing and associated shooting 'sports' mainly for game birds such as grouse, provide valuable income). Money raised from business rates will boost a Scottish Land Fund with the aim of doubling community ownership to one million acres.

The legislation has not been without criticism: owners of large estates, such as the Duke of Buccleuch and Queensberry (240,000 acres), maintain that the act breaches landowners' human rights. And campaigners for land reform, such as Andy Wightman, argues that the legislation has both failed to restrict the amount of land a wealthy individual can own and prevent the use of offshore havens as a vehicle for tax evasion.

Conclusions: moving forward

In September 2015, Defra announced it was developing a 25-year plan for the environment, a Conservative Party manifesto commitment, although no publication date has been announced. If the prospect of Scottish-style land reform is remote in England, it is perhaps significant that voices calling for measures to curb speculation in land prices – alongside the case for an all-embracing ministry of land – are getting marginally louder. George Dunn, of the Tenant Farmers' Association, is certainly not alone in calling for a new taxation regime to curb land speculation and, hence, bring down land prices.

Against the background of our land facing multiple pressures, one issue stands out: the absence of any national, holistic strategy to address the pressures, challenges and opportunities facing the British land mass. So, what to do in this unregulated, unregistered free-for-all in land trading and its impact on farming and housing the nation? First, the case for ending breaks on inheritance tax and capital gains tax is so overwhelming that measures should be taken to bring a sense of order to the land market. It delivers both windfalls, courtesy of tax 'efficiency' for a few – in reality, a substantial state subsidy – at the expense of the many unable either to get on the farming ladder or secure a home to rent or to buy.

Second, a diminished Defra must be given the tools, the funds and the co-ordination role its name implies to address the crisis of both inadequate flood defences, particularly in eastern England, diminishing domestic food production, and the need to adequately house the nation. It is a broad agenda which would take this ministry far beyond what many see as its current focus: farming, and related issues.

As it is, another revolution is looming, through Brexit, which will assuredly determine the use of land in Britain and the billions of pounds in public subsidies currently supporting farming and allied environmental programmes. It could be as profound, and far-reaching, as the raft of post-war social and economic programmes in the late 1940s. For a start, it is unlikely that hill farming on marginal land, and the widespread production of animals – sectors dependent on subsidies for at least half of farming incomes – will survive on such a large scale outside the EU. The implications for upland Britain, with the prospect of limited sheep farming at best, are clearly profound. The future of farming as we know it has rarely been so uncertain.

We are, then, approaching a perfect storm – increasing demand for food, energy, housing and water – without any vision, still less a plan, for the land mass of Britain, particularly England. Striking a balance between efficient farming on the one hand, and managing the land and its fragile ecosystems on the other, is not easy; providing the billions needed to safeguard our best farm land from more extreme weather will involve borrowing to invest and some hard choices; reforming the tax system to ensure landowners pay their fair share of tax, rather than avoiding it, will take political courage. But we cannot continue as we are.

References

BBC (2017). How much of your land is built on? Retrieved from www.bbc.co.uk/news/uk-41901294.

Cahill, K. (2002). *Who Owns Britain and Ireland*. London: Canongate.

Cornish, C. (2017). Mother Nature worth £177bn to UK. *Financial Times*, 25 July. Retrieved from www.ft.com/content/535aa22a-713a-11e7-93ff-99f383b09ff9.

Dunn, G. (2017). Tenant Farmers' Association (TFA) media release MR17/36. Retrieved from www.tfa.org.uk/tfa-media-release-mr1736 (accessed 5 December 2017).

Financial Times (2015). UK land returns more than Mayfair. *Financial Times*, 18 February. Retrieved from www.ft.com/content/e8efa742-b3a8-11e4-a6c1-00144feab7de.

Financial Times (2017). UK farmers risk seeing incomes halve after Brexit. *Financial Times*, 17 October. Retrieved from www.ft.com/content/cd9323b8-ad0e-11e7-beba-5521c713abf4.

Foresight Land Use Futures Project (2010). *Final Project Report*. London: The Government Office for Science.

Hetherington, P. (2015). *Whose Land Is Our Land? The Use and Abuse of Britain's Forgotten Acres*. Bristol: Policy Press.

House of Commons Environment, Food and Rural Affairs Committee (2016). Future flood prevention. Retrieved from www.parliament.uk/business/committees/committees-a-z/commons-select/environ ment-food-and-rural-affairs-committee/inquiries/parliament-2015/inquiry (accessed 2 November 2016).

Ipsos MORI (2012). Survey for British Property Federation – public attitudes towards development. Retrieved from www.ipsos.com/sites/default/files/migrations/en-uk/files/Assets/Docs/Polls/SRI_ IpsosMORIBPFtopline_080512.PDF.

Kynaston, D. (2007). *Austerity Britain 1945–51*. London: Bloomsbury.

Lyons, M. (2014). The Lyons Housing Review: Mobilising across the nation to build the homes our children need. Retrieved from www.policyforum.labour.org.uk/uploads/editor/files/The_Lyons_ Housing_Review_2.pdf.

Marwick, A. (2006). *The Deluge, British Society and the First World War* (2nd edition). Basingstoke: Palgrave Macmillan.

NAO (2014). Strategic flood risk management. Retrieved from www.nao.org.uk/report//strategic-flood-risk-management-2.

Newby, H. (1988). *The Countryside in Question*. London: Hutchinson.

NFU (2015). Backing British farming in a volatile world. Retrieved from www.nfuonline.com/ assets/43617.

Taylor, M. (2016). Who owns the land? Retrieved from www.barbers-rural.co.uk/blog/who-owns-land (accessed 9 February 2016).

8

Legal enforcement of spatial and environmental injustice

Rural targeting and exploitation

Loka Ashwood, Katherine MacTavish and Dalton Richardson

Introduction

Traditionally, two steadfast vanguards have prevailed as explanatory predictors of environmental injustices: race and class. Their efficacy, to an extent, is inarguable. Those bearing the brunt of the most extractive and hazardous industries within Europe, North America, and Australia are minorities of race and ethnicity, and possess a minority of capital (Bullard, 1990, 1993; Cole and Foster, 2001; Davidson and Anderton, 2000; Downey, 2007). Globally, the worst of toxic waste and the least regulated of production practices play out in India, China, and Nigeria, to name only a few nation-state examples (Bunker, 1984; Hornborg, 1998; Jorgenson and Clark, 2009; Anand, 2004). Documentation of hazardous and toxic working and living conditions are well known in the poorest and most rapidly industrialising countries. On the aggregate, then, it seems race and class are quite sufficient descriptors to understand environmental injustice.

Yet, scholars warn that the monolithic treatment of race as colour without context, along with overly simplistic victim–perpetrator framings of environmental injustices, misses much of what's happening (Pulido, 1996; Pellow, 2000). Injustices practiced upon the most vulnerable are not just intersectional, but they systematically compile. That compilation happens as part of the internal mechanisms of the state (Pellow, 2014), even in democracies that increasingly are governed by profit, rather than people. The utilitarian orientation that governs states, what Ashwood (2018) calls for-profit democracy, leaves those fewest in number and possessing the least money not just socially vulnerable, but legally targeted. The state can justify its exploitation of such populations because such people have fewer votes and less money, which naturalises the siting of the most dangerous industries as a positive for those who come to neighbour it.

Until recently, the rural has yet to be studied as an explicit factor that helps explain environmental burden under the demands of the utilitarian state (Ashwood and MacTavish, 2016; Pellow, 2016). In part, the naturalisation of the rural as big on resources, short on education, high on working class ethics, and low on innovation, has historically aided the unquestioning treatment of the rural as a dumping ground and a site of extraction. The ideological stigmatisation of the rural space coincides with the structural demands that rural people resign their rights.

Consequently, the law often provides the enforcement tool that leaves rural people with limited standing (Pruitt and Sobczynski, 2016).

We turn our attention to the United States as a key example of the minority-majority dynamic estranging rural people in the context of utilitarian states (Tocqueville, 1969). We focus on zoning in the context of trailer parks and industrial animal facilities to demonstrate environmental burden and the resignation of rights in rural America through the law. We do so in three main parts. First, we examine the meaning of rights in the rural context by examining how those who possess property rights are currently being dispossessed in the context of 'right to farm' (RTF) laws, and how those who do not possess property rights are perpetually dispossessed in the context of rural trailer parks. We reveal how legal confusion, whether over the operator of an industrial animal facility or the ownership of a trailer park, enables rural exploitation. We explore a series of means to counter the stigmatisation and exploitation of the rural. We focus on ways in which reframing and renaming of privileged operations, like corporate agribusinesses that call themselves farmers, along with countering the stereotypes applied to the most vulnerable, like the stigmatisation of those who live in trailers as white trash, can help gain rights for rural people.

The meaning of rights: justice for whom?

The legal treatment of trailer parks and agricultural operations provide two compelling examples of how spatial and environmental injustices are enacted and sustained in rural communities. Specifically, we use RTF laws and the legal treatment of rural trailers as unique cases through which to examine the role of property rights and zoning laws in targeting and exploiting rural residents by dispossessing them of rights that the constitution ensures: namely, the right to property and protection from deprivation. As we illustrate below, what results is a system that privileges the rights of the powerful and often extra-local at the cost of justice for rural residents and local communities.

Dispossessing the possessed

The rapid urbanisation of the United States over the last century inspired some state legislatures to pass RTF laws under the presumption that they could protect farms from urban sprawl (Grossman and Fischer, 1983). Specifically, these laws emerged in response to demetropolitanisation in the 1970s, wherein urban populations left city centres and developed suburban hubs within rural areas (Satterthwaite et al., 2010). Since the establishment of RTF laws, agricultural production has drastically contracted, including a reduction in farmland and farmers, simultaneous to significant increases in the scale and intensity of production (Lobao and Meyer, 2001). In action, RTF laws have enabled these changes by protecting corporate industrial farming operations rather than farmland, the latter of which is more synonymous with existing small and medium sized farms, and the former with rapidly expanding concentrated animal feeding operations (CAFOs) (Goeringer and Goodwin, 2013; Hamilton, 1998; DeLind, 1995). RTF laws have done so by employing a technocratic, legal language that depersonalises the farmer within the agricultural frame and favours vertical and horizontally integrated industrial production (Ashwood et al., 2014).

Farming operations have traditionally been managed through zoning laws that identified how land could be used within a municipality (Maantay, 2002). RTF laws initially complemented zoning laws by demarcating the conditions under which farming operations could be labelled a nuisance. Since their inception, however, the role of RTF laws has changed from protecting sole-proprietor and family operations to protecting industrial operations from nuisance litigation (Smart, 2016). Moreover, such laws often supersede local governance, leaving rural communities without the ability to zone out unwanted operations. Today, RTF laws offer broad protections to farms that meet certain criteria, such as length of time in operation and the type of agricultural product produced (Walker, 2017).

Right to farm laws now consequently play a key role in shaping who possesses and is dispossessed of their rights. Most explicitly, this encompasses property, as nuisance suits are at a basic level about one's capacity to enjoy one's property. Via the proxy of property, the dispossession of rights is also about health and environment, which can be threatened by industrial scale operations, especially large scale animal production. As CAFO utilisation has increased, so too has the utilisation of open-air manure lagoons to store the prodigious amounts of manure produced by livestock (Hinrichs and Welsh, 2003). The RTF laws in a number of states contain clauses that allow agricultural operations to expand their size after a certain amount of time, so long as the operation was not considered a nuisance within the time before it expanded (Walker, 2017). Furthermore, most states do not protect the first established land use, meaning intergenerational family farms or homesteaders are disenfranchised to the benefit of corporate operations. These clauses allow operations to encroach on residential areas, exposing residents to ammonia, hydrogen sulfide, methane, and endotoxins that can facilitate asthma, severe coughs, loss of smell, respiratory impairment, aerobic metabolism, headaches, fatigue, weakness, chest tightness, and the development of antibiotic-resistant bacteria (Wing et al., 2008). The inability to enforce local zoning ordinances that conflict with a state's RTF legislation leaves rural communities and residents with little capacity to use the law to stop the building or operation of such facilities. These legal interactions are unique between RTF laws and zoning ordinances, and work to strip rural populations of their right to demand property protection in court.

Dispossessing the dispossessed

While farms and farming evoke sentimental notions of the rural, a rural trailer park regularly brings up less idyllic images (Isenberg, 2016). Yet trailer parks, a housing form stigmatised even today by the toxic slurs 'trailer trash' and 'white trash', are as characteristic to rural America as agriculture (Salamon and MacTavish, 2017). Across the US, trailer parks, which currently function as the leading source of unsubsidised affordable housing, shelter some 12 million individuals. With three quarters of the nation's estimated 50,000–60,000 parks located in the countryside, trailer parks are a decidedly rural solution to the affordable housing need (Housing Assistance Council, 2012). As an affordable housing option, parks densely concentrate a community's younger and older, poorer, and less educated households on property privately held by a park owner, most often on the edge of town (Salamon and MacTavish, 2017). That stigmatised identity, neighbourhood demographic, spatial location, and a status as *landless* tenants, links trailer park residence to a litany of environmental, economic, and social injustices that compile and add to the vulnerabilities of rural households (MacTavish et al., 2006). These injustices are reinforced by the legal exclusion and elimination of trailer parks. Such treatment naturalises and is naturalised by notions of who does and who does not belong in rural (Salamon, 2003). In the end, the legal treatment of trailer parks further dispossesses those already dispossessed from access to rights assured in the US National Housing Act of 1949 that promises, 'a decent home in a suitable living environment for every American'. Mitigating these injustices demands that we reconsider the legal treatment of trailer parks in the rural context.

Zoning practices that relegate trailer parks to the remote and rural reflect long standing legal efforts to keep this housing type away from more conventional residential areas (Baker et al., 2011; Genz, 2001). Exclusionary zoning that restricts trailers to parks and parks to the margins arises largely out of 'NIMBY' (not in my back yard) concerns around appearance, safety, quality, occupants, and threats to the tax base and property values (Mandelker, 2016). Such policies push trailer parks to cheap commercial and industrial land on the edges of town invariably adjacent to flood zones, tornadoes alleys, train tracks, and waste treatment facilities (Beamish et al., 2001; MacTavish, 2007). Such placement brings with it increased exposure to environmental hazards, but it also dispossesses residents of access to educational and social services and employment opportunities critical to individual and family wellbeing (Baker et al., 2011; Pierce et al., 2018). Processes termed 'municipal underbounding' reinforce exclusion when small towns selectively choose not to incorporate a trailer park even as town boundaries expand (Lichter et al., 2007). The transportation tax of a remote location along with out-of-town fees for library use and recreation strain already limited household budgets (Salamon and MacTavish, 2017). The perceived impacts of family social and economic vulnerability on small town schools, social services, and police protection reinforce local resentments toward trailer park and the households who live there (MacTavish, 2007).

Beyond restrictive zoning and underbounding, the legal reinforcement of trailer parks as a private solution to a public need for affordable rural housing isolates park households from community resources and protections. With US federal investments in rural affordable housing on a steep decline for decades, the very private solution offered by trailer parks has prevailed (Salamon and MacTavish, 2017). As a privately run business, the provision of basic services like safe drinking water, adequate sanitation, and the maintenance of roads and infrastructure falls in the domain of the trailer park owner rather than the local municipality. The industry argues that such an approach may actually save local governments money, but resulting environmental quality issues are well documented. Trailer park residence brings an increased risk for contaminated drinking water, failing sewer systems, inadequate trash pickup, and ill maintained streets (Cutter et al., 2003; Moore and Matalon, 2011; Peirce and Jimenez, 2016; Zwerdling, 2016). The shabby park conditions that ensue when owner upkeep is lax significantly impact park life and draw 'the focus of the entire community's wrath' (Salamon and MacTavish, 2017: 20). The hands-off or complaint-only approach small towns take in dealing with such issues, if only for lack of fiscal resources, reinforces a power hierarchy privileging the landowner (MacTavish, 2007). Park residents can do little, as voicing concerns too loudly can lead to eviction and, given the prohibitive cost of relocating a trailer, moving means leaving a home behind.

The most egregious form of dispossession emerging within this system comes in limited legal protections against the closure of trailer parks. For trailer park residents, the threat of dispossession from the land and their home when a park sells for what is termed a higher and better use, is a threat increasingly present where rural land values are rising (Zwerdling, 2016). Park closures, combined with zoning restrictions on new park development, fuel a national shortage of sites on which a household might relocate their trailer home (Erenfeucht, n.d.). When a park closes, a small town is left to absorb the housing needs of those displaced, whether elderly on fixed incomes or working-class families critical to the local economy (Hartman, 2003).

The forcible resignation of property, health, and environmental rights by rural people in favour of industrial agriculture and the industrialisation of trailer park points to a crucial mismatch in power and purpose of governance. Namely, those with a mind of living upon the land, drinking its water, and breathing the air possess fewer rights than those who use land, water, and environment to generate profit. Altogether, rural people who formally hold a legal deed (whether to land or a trailer home) nonetheless lose parallel rights that accompany ownership elsewhere. Thus, the constitutional protection of property ownership, based in common law, is revoked from rural people outside the industrial agricultural frame, leading to the dispossession of the possessed and further dispossession of those already dispossessed.

Confusing justice by confusing ownership

The mechanisms we describe as confusing justice through the excluded and privatised nature of trailer parks and the co-opted use of RTF laws emerge from a similar force – that of the industrialisation of operations through extra-local corporate ownership. The primary issue in agriculture is that farming operations maintain a local face through the use of contract farming. The contracted farmers, however, do not actually own the hogs on the farms nor do they own the operations as the corporate entity has the ability to renew or decline to renew their contract (Ashwood et al., 2014). When there are local complaints about the types of technologies utilised for farming operations, they often fall on deaf ears as the farmers are constrained by the types of technologies provided by the contractor, and the contractor can avoid local regulations because they are extra-local.

In the case of trailer parks, that shift to corporate investor ownership accelerated in the 1990s when the significant investment potential of trailer parks was recognised (Williams, 2017). The more benevolent 'mom and pop' trailer park operations originally developed on family farmland became part of investment portfolios bought and sold as packaged real estate deals in the same way as strip-malls. Transferring ownership from a local individual to a far-off corporation added complexity to efforts aimed at addressing injustices. It also brought with it the legal resources of a powerful industry lobby. In the wake of such change, park residents and small-town communities can do little to manage the injustices that are sustained and legally reinforced.

Preventing rural spatial and environmental injustices

In the previous sections, we articulated significant connections between the role of the law and the establishment and continuation of spatial and environmental injustices within rural communities. How is the status quo maintained? How is it that rural people continue to be dispossessed of what they have, while the expropriation from the already landless continues? While it is certainly important to identify the legal factors that facilitate injustices in order to refute them, it is also necessary to situate the law within a broader social context. After all, Weber identified the law as in large part a reflection of social norms then codified by the state. While we do not want to overstretch this approach, as states are easily co-opted by powerful elites, it is nonetheless crucial to identify the ways in which dispossession is justified by those who exert it and those who have to live with it. In the following section, we discuss the power of rhetoric, or more specifically a name given to expropriative processes that normalise them or guise them. Normalisation or legal veiling of identities can mold or even deflate the conflicts often seen between trailer park tenant and owner rights, as well as nuisance litigation and RTF laws. We end this section by discussing potential solutions to the current issues posed by the legal treatment of housing and agricultural operations.

What's in a name?

Names are an important, yet often overlooked, aspect of the law. Perhaps this is because names are so commonplace among all aspects of daily life that an assumption reigns they are value neutral. Despite this assumption, names communicate particular ideas about an object so-named (Bloom, 2002). The importance of naming a law cannot be understated. Discussions of laws often are informed first by its name, as many people lack the necessary legal education with which to discuss the particular nuances of statutes.

Consider the name 'right to farm'. The title suggests to the casual listener or reader a benign necessity – a law that gives farms just authority that they may otherwise lack. And the results are potentially disastrous. Without such a law, how then could farmers grow food or people have enough food to eat? The power of the 'right to farm' name further benefits from its timing. The title is the first thing a listener hears. Any consequent legal explanation fumbles to explain itself against the farm-rights-first context. Its simplistic, yet broad, nature prompts unbounded and personalised interpretations of what constitutes a farm, which are often enough melancholy ones. Farming operations enjoy a uniquely personal face in rural and urban communities, a narrative cultivated early on in children's books, with farmers and animals and all the joys in between. Farmers thus are akin to a beloved neighbour or a community luminary. Such a farmer may not actually exit, at least as a labourer with a small herd working a few animals in an open pasture. Those driven to utilise the RTF defence are often investors or managers operating CAFOs. Yet an investor behind an amiably named corporate hog facility, such as Shamrock Acres LLC, can ride on the coattails of the farmer imaginary propped up by corporate laws that do not require divulging identities (Ashwood et al., 2014). The notion of farm continues to beckon a farming idyll, not corporate investment, to the benefit of those who hold the most capital. As a result, movements against industrial operations are not understood to be against extra-local corporate entities, but against community members, neighbours, and even friends. These interpretations are further reinforced by the historical importance given to agriculture within rural communities, thus ensuring that industrial operations are protected despite how harmful and disruptive they often are to the surrounding community and environment.

The importance of a name works in an alternative way for rural trailer parks. The parks that hold trailers are not chronicled in stacks of children's books. Nor are they celebrated as affordable living places with ample green landscapes and open pastures. They, like RTF laws, distract from the villain. But rather than projecting animosity away from the condition at hand, they volte-face that wrath upon those it names directly. And with much force. Trailer park jokes remain one of the last unquestioned relics of political incorrectness in the US (Salamon and MacTavish, 2017). Commonplace jokes make cultural assumptions that the people who live in trailer parks are simple-minded, lawless, reckless with fertility, and indifferent to their behaviours. All this culminates in the label 'trailer trash'. In the meantime, the problem becomes about the people who live there, but not the owners who extract rents and exacerbate the conditions within the park. The 'trailer trash' brand works to stigmatise an entire category of people so as to marginalise them from mainstream society. Trailer park dwellers have few advocates. Faced with the prospect of a rundown park, the wrath of a small town can easily focus on the residents of a park rather than the profit-making owner (MacTavish, 2007). And so the landlords benefit, while the renters suffer the economic and cultural consequences.

Right to farm laws and trailer parks then face up against one another, at least in terms of who they serve: in the first case, the powerful guised in an amiable image, and the second, the less powerful guised in condemnation. Yet they carry the same torch forward, one that serves the fewest at the expense of the rural many.

Effective lawsuits and action

Despite the daunting ideological and material conditions rendered in these two cases, there have been important wins that favour rural rights more generally. In 1989, the Farm Environmental Defense Foundation of Jackson County, Michigan, reached an out-of-court settlement against Sand Livestock Systems, Inc., for the operation of 'hog hotels' and open-air manure lagoons that were damaging local farms and the everyday lives of residents (DeLind, 1995). In 2002, fifteen Missourian farmers filed a lawsuit against Premium Standard Farms for the operation of a 4,300-acre hog finishing farm located outside the town of Berlin that contained 80,000 hogs.

The suit awarded the plaintiffs $11.05 million in damages and is the largest monetary award against a hog farm for odor nuisance (Weiss, 2010). While these successful suits have primarily occurred in the Midwest, currently there are 26 ongoing nuisance lawsuits in North Carolina against Murphy-Brown LLC that have over 500 plaintiffs and are set to proceed to trial in the first half of 2018 (Sorg, 2017).

Such success plays out in an environment where many states have strengthened their RTF laws to discourage future litigation. After the settlement in Michigan, the state legislature strengthened protections for operations by limiting monetary awards in a successful nuisance suit against an agricultural operation (DeLind, 1995). Moving beyond their original legislative actions, the state of Missouri held a public referendum to enshrine the RTF in the state constitution, and passed with 50.1 per cent of the vote (Kennedy, 2016). Similar legislative actions have occurred in North Carolina (Smart, 2016) and public votes concerning constitutional RTF amendments have occurred in North Dakota and Oklahoma.

Our national debate about the utility of trailer parks as a housing solution for low income rural families will no doubt continue. Yet a number of recent legal actions and lawsuits evidences that the dialogue is beginning to move forward and focus on how we might preserve and improve this housing sector so essential for our nation's low income rural families. In 1984, the non-profit New Hampshire Community Loan Fund (NHCLF) assisted in the conversion of a land-lease park to a resident-owned community or ROC (Bradley, 2007). In a ROC, collective resident ownership of the land removes many of the social and economic injustices that come from investor ownership (Ward, French and Giraud, 2006). By 2010, NHCLF had helped close to 100 investor-owned parks make that conversion with other non-profits joining the movement since. Beginning in the early 2000s, states began implementing 'right of first refusal' protections critical to ROC conversions (Goronstein, 2012). States like Washington, Oregon and California, at the urging of mobile home park homeowner groups, also implemented mobile home park tenant rights laws and protections against displacement through the sale of parks. Recent and substantial legal settlements around park owner negligence stand as powerful examples of the courts reinforcing the rights of trailer park residents (Harris, 2017; Kaplan, 2014, 2017). In 2015, a state supported 'Oregon Solutions' project developed a 'tool kit' for a proactive housing and social services response to the threat of park closure (Mobile Home Park Solutions Collaborative, 2016). Together, these legal instances provide a source of hope that the outdated stigma against trailer parks is beginning to fade.

Reformulating the utilitarian state

It seems that no matter the advancements made in combating spatial and environmental justices that manifest from dispossessing rural people, the state often restructures and reinforces the law to circumvent future action. The utilitarian state's many, either the many of profit or the many in number, regularly justifies the marginalisation of the rural poor and their expropriation in favour of park owners or industrial agriculture extraction. What then is to be done regarding the current state of RTF laws or the privatisation of mobile home parks across the country?

Increasingly public awareness about RTF laws and trailer parks, as well as the true identities of those who benefit from their most exploitative practices, helps to demystify the complicated and specific legal terms that serve to protect corporate interests. Even as courts contract to limit rural people's legal standing, some key rulings continue to give momentum to RTF reformation in favour of common law property right protections. To stall dispossession requires a larger reworking of the way academics and the public conceive of property rights. Not all property rights are equal, as RTF and laws regulating trailer parks suggest. The exertion of property rights remains beholden to the marketisation of those rights, whether the selling of parks upon which mobile homes rest, or the consolidation of markets farmers that farmers rely on to maintain their livelihoods. We call for researchers concerned with rural wellbeing to focus on the ways in which the law can push back against spatial injustice by protecting justly distributed property, and supporting laws that are wedded to home, stewardship, place, and sustainability. There are other options for a property rights regime outside of the one encouraged by the utilitarian state, which regularly takes from the rural to serve the urban, as it takes from those with less to consolidate more in the hands of the few.

Note

Ashwood and MacTavish shared first authorship responsibilities for this chapter.

References

Anand, R. (2004) *International Environmental Justice: A North–South Dimension.* London: Routledge.

Ashwood, L. (2018) *For-Profit Democracy: Why the Government is Losing the Trust of Rural America.* New Haven, CT: Yale University Press.

Ashwood, L., and K. MacTavish. (2016) Tyranny of the Majority and Rural Environmental Injustice. *Journal of Rural Studies* 47, 271–277.

Ashwood, L., D. Diamond, and K. Thu. (2014) Where's the Farmer? Limiting Liability in Midwest Industrial Hog Production. *Rural Sociology* 76(1), 2027.

Baker, D., K. Hamshaw and C. Beach. (2011) A Window into Park Life: Findings From a Resident Survey of Nine Mobile Home Park Communities in Vermont. *Journal of Rural and Community Development* 6(2), 53–70.

Beamish, J., R. Goss, J. Atiles and Y. Kim. (2001) Not a Trailer Anymore: Perceptions of Manufactured Housing. *Housing and Policy Debate* 12(2), 373–392.

Bloom, P. (2002) Mindreading, Communication and the Learning of Names for Things. *Mind & Language* 17(1–2): 37–54.

Bradley, P. (2007) Gaining Ground. Retrieved from https://shelterforce.org/2007/04/23/gaining_ground/

Bullard, R. D. (1990) *Dumping in Dixie: Race, Class, and Environmental Quality.* Boulder, CO: Westview Press.

Bullard, R. D. (1993) Anatomy of Environmental Racism and the Environmental Justice Movement. In *Confronting Environmental Racism: Voices from the Grassroots*, ed. Robert D. Bullard. Cambridge, MA: South End Press, 15–39.

Cole, L. and S. Foster (2001) *From the Ground up: Environmental Racism and the Rise of the Environmental Justice Movement.* New York: New York University Press.

Cutter, S., B. Buroff and W. Shirley (2003) Social Vulnerability to Environmental Hazards. *Social Sciences Quarterly* 84(2), 242–261.

Davidson, P. and D. Anderton (2000) Demographics of Dumping II: A National Environmental Equity Survey and the Distribution of Hazardous Materials Handlers. *Demography* 37(4), 461–466.

DeLind, L. (1995) The State, Hog Hotels, and the 'Right to Farm': A Curious Relationship. *Agriculture and Human Values* 12(3), 34–44.

Downey, L. (2007) US Metropolitan-Area Variation in Environmental Inequality Outcomes. *Urban Studies* 44, 953–977.

Erenfeucht, R. (n.d.) Moving Beyond the Mobile Myth: Preserving Manufactured Home Communities. Retrieved from https://issuu.com/groundedsolutionsnetwork/docs/moving_beyond_the_mobile_myth

Genz, R. (2001) Why Advocates Need to Rethink Manufactured Housing. *Housing Policy Debate* 12(2), 393–414.

Goeringer, L. P. and H. L. Goodwin (2013) An Overview of Arkansas's Right-to-Farm Law. *Journal of Food Law and Policy* 9, 1.

Goronstein, D. (2012) Home Sweet Mobile Home: Coop Deliver on Ownership. Retrieved from www.npr.org/2012/05/02/151863518/home-sweet-mobile-home-co-ops-deliver-ownership

Grossman, M. and T. Fischer. (1983) Protecting the Right to Farm: Statutory Limits on Nuisance Actions Against the Farmer. *Wisconsin Law Review* 95–165.

Hamilton, N. D. (1998) Right-to-Farm Laws Reconsidered: Ten Reasons Why Legislative Efforts to Resolve Agricultural Nuisances May be Ineffective. *Drake Journal of Agriculture* L. 3, 103.

Harris, M. (2017) Court Approves Class Action Settlement Between Residents. Retrieved from www.vcstar.com/story/news/local/communities/conejo-valley/2017/09/19/court-approves-class-action-settlement-between-residents-owner-t-o-mobile-home-park/681522001

Hartman, C. (2003) Evictions: The Hidden Housing Problem. *Housing Policy Debate* 14(4), 461–501.

Hinrichs, C. and R. Welsh. (2003) The Effects of the Industrialization of US Livestock Agriculture on Promoting Sustainable Production Practices. *Agriculture and Human Values* 20(2), 125–141.

Hornborg, A. (1998) Towards an Ecological Theory of Unequal Exchange: Articulating World System Theory and Ecological Economies. *Ecological Economics* 30(25), 127–136.

Housing Assistance Council (2012) *Taking Stock: Rural People, Poverty and Housing in the 21st Century.* Washington, DC: Housing Assistance Council.

Isenberg, N. (2016) *Poor White Trash: The 400-Year Untold History of Class in America.* New York: Viking.

Jorgenson, A. and B. Clark. (2009) The Economy, Military, and Ecologically Unequal Exchange Relationships in Comparative Perspective: A Panel Study of the Ecological Footprints of Nations, 1975–2000. *Social Problems* 1 56(4), 621–646.

Kaplan, T. (2014) Jury Award Record $111 Million to Trailer Park Residents. Retrieved from www.mercurynews.com/2014/04/20/jury-awards-record-111-million-to-trailer-park-residents/

Kaplan, T. (2017) Trailer Park Wins: A Negligent Owner, $10 Million Settlement and the San Jose Residents Who Prevailed. Retrieved from www.mercurynews.com/2017/03/05/4434205

Kennedy, A. (2016) Sustainable Constitutional Growth: The Right to Farm and Missouri's Review of Constitutional Amendments. *Missouri Law Review* 81: 205–250.

Lichter, D., M. Parisi, S. Grice and M. Taquino. (2007) 'Municipal Underbounding': Annexation and Racial Exclusion in Small Southern Towns. *Rural Sociology* 72(1), 47–68.

Lobao, L. and K. Meyer (2001) The Great Agricultural Transition: Crisis, Change, and Social Consequences of Twentieth Century US Farming. *Annual Review of Sociology* 27,103–124.

Maantay, J. (2002) Zoning Law, Health, and Environmental Justice: What's the Connection? *The Journal of Law, Medicine & Ethics* 30(4), 572–593.

MacTavish, K. (2007) The *Wrong* Side of the Tracks: Social Inequality and Mobile Home Park Residence. *Community Development* 38, 74–91.

MacTavish, K., M. Eley, and S. Salamon (2006) Housing Vulnerability among Rural Mobile Home Park Residents. *Georgetown Journal of Poverty Law and Policy* 13, 95–117.

Mandelker, D. (2016) Zoning Barriers to Manufactured Housing. *Urban Lawyer* 48(2), 233–278.

Mobile Home Park Solutions Collaborative (2016) Manufactured Home Park Solutions Collaborative Local Agency Toolkit Online. Retrieved on 23 October 2018 from www.oregon.gov/ohcs/CRD/mcrc/docs/Manufacture-Home-Park-Solutions-Collaborative-Local-Agency-Toolkit.pdf.

Moore, E. and E. Matalon (2011) *The Human Cost of Nitrate-Contaminated Drinking Water in the San Juaquin Valley.* Oakland, CA: Pacific Institute.

Peirce, G. and S. Jimenez (2016) Unreliable Water in US Mobile Home Parks: Evidence from the American Housing Survey. *Housing Policy Debate* 25(4), 739–753.

Pellow, D. (2000) Environmental Inequality Formation: Toward a Theory of Environmental Injustice. *American Behavioral Scientist* 43(4), 581–601.

Pellow, D. (2014) *Total Liberation: The Power and Promise of Animal Rights and the Radical Earth Movement.* Minneapolis, MN: University of Minnesota Press.

Pellow, D. (2016) Environmental Justice and Rural Studies: A Critical Conversation and Invitation to Collaboration. *Journal of Rural Studies* 47(38), 1e386.

Pierce, G., C. J. Gabbe and S. R. Gonzalez. (2018) Improperly-zoned, Spatially-marginalized, and Poorly-served? An Analysis of Mobile Home Parks in Los Angeles County. *Land Use Policy* 76, 178–185.

Pruitt, L., and L. Sobczynski. (2016) Protecting People, Protecting Places: What Environmental Litigation Conceals and Reveals About Rurality. *Journal of Rural Studies* 47, 326–336.

Pulido, L. (1996) A Critical Review of the Methodology of Environmental Racism Research. *Antipode* 28(2), 143–159.

Salamon, S. (2003) From Hometown to Nontown: Rural Community Effects of Suburbanization. *Rural Sociology* 68(1), 1–24.

Salamon, S. and K. MacTavish. (2017) *Singlewide: Chasing the American Dream in a Rural Trailer Park.* Ithaca, NY: Cornell University Press.

Satterthwaite, D., G. McGranahan, and C. Tacoli. (2010) Urbanization and its Implications for Food and Farming. *Philosophical Transactions of the Royal Society B* 365(1554), 2809–2820.

Smart, C. M. (2016) The 'Right to Commit Nuisance' in North Carolina: A Historical Analysis of the Right-to-Farm Act. *North Carolina Law Review* 94(6), 2098–2154.

Sorg, L. (2017) In a Setback to Murphy-Brown, Hog Nuisance Suits Can Go On, Federal Judge Rules. North Carolina Policy Watch, November 13. Retrieved 29 January 2018 from http://pulse. ncpolicywatch.org/2017/11/13/setback-murphy-brown-hog-nuisance-suits-can-go-federal-judge-rules/#sthash.NGiDlH07.dpbs

Tocqueville, A. (1969) *Democracy in America.* New York: Anchor Books.

Walker, F. (2017) Right to Farm Laws: A Thematic Analysis. MS thesis, Department of Agricultural Economics and Rural Sociology, Auburn University.

Ward, S., C. French and K. Giraud (2006) Residents Ownership of New Hampshire's Mobile Home Parks: A Report on Economic Outcomes. Carsey School of Public Policy, University of New Hampshire. Retrieved from https://carsey.unh.edu/publication/resident-ownership-new-hampshires-mobile-home-parks-report-economic-outcomes-revised

Weiss, S. (2010) $11 Million Verdict – Premium Standard Farms. Retrieved 17 February 2017 from www. seegerweiss.com/news/hog-farm-victory

Williams, C. (2017) The Quiet Moneymaker: Trailer Park REITs are Averaging 27% Annual Return. Retrieved from www.forbes.com/sites/bisnow/2017/08/04/the-quiet-moneymaker-trailer-park-reits-are-averaging-27-annual-returns/#3effd8102116

Wing, S., R. Horton, S. Marshall, K. Thu, M. Tajik, L. Schinasi, and S. Schiffman. (2008) Air Pollution and Odor in Communities Near Industrial Swine Operations. *Environmental Health Perspectives* 116(10), 1362–1368.

Zwerdling, D. (2016) When Residents Take Ownership a Mobile Home Community Thrives. Retrieved from www.npr.org/2016/12/27/503052538/when-residents-take-ownership-a-mobile-home-community-thrives

Part II

The state and rural governance

Part II contextualises rural planning debates through exploring the state and rural governance. The various contributions to this section examine the shifting nature of rural governance alongside an understanding of the range of policy instruments employed to manage rural space. Rural policy and politics have become increasingly complex, reflecting a growing shift from government to governance, and characterised by economic, social and political tranformations, which have transformed the manner in which policy is made and delivered. These governance shifts place an emphasis on involving multiple stakeholders (including active citizens and active communities), partnership arrangements or new 'soft spaces' for policy-making, 'flexibility' and government 'steering' (Murdoch and Abram, 1998). Policy-making and plan-making increasingly involve more complex interactions between politicians, public officials, state agencies, voluntary sector organisations, and interest groups. Within this interaction, 'the State', today, is rarely conceived as a neutral arbitrator among competing demands for rural space and resources. Instead, the dominance of neoliberalism has hollowed out the state, while state actors simultaneously re-regulate in favour of market interests. As Allmendinger (2016: 1) argues, planning policy and actors have 'shifted incrementally but perceptively away from an area of public policy that was an arena where . . . issues could be determined in the public interest to one that legitimizes state-led facilitation of growth and development by superficially involving a wider range of interests and issues'.

This part addresses these themes through examining broad shifts in the governance of rural regions and localities. Two key themes emerge: firstly, is the dominance of neoliberalism in shaping rural life and the subsequent role-out of market-oriented policies, including the increased use of market-based instruments. Secondly, and related, is the increased role given to community organisations in not only the design of rural policy but also in the delivery of rural services or the direct ownership of rural assets. This is driven both by neoliberal models of state withdrawal from rural service provision, but also by demands from community actors to be actively involved and mobilise local resources for development from within.

In the first chapter in this section, Bock provides an overview of the rural dimension of multi-level governance, placing rural debates within wider theoretical discussions on the nature of power and governance. This begins with a discussion concerning the shift in power from government to governance highlighted in the literature during the 1990s, which in turn framed much debate concerning new endogenous models of rural development that became widely established at that time (such as the European Union's LEADER programme). While these bottom-up models offered new opportunities to identify development pathways based on local resources, Bock also notes the tensions within this approach, particularly its legitimacy vis-à-vis representative democracy and the danger of 'capture' by local elite groups rather than more deep-seated community empowerment. Despite these limitations, the governance of rural areas has largely continued in this direction, underpinned by the continuous withdrawal of the state from public services provision, consolidated by the politics of austerity following the financial crisis of 2008. While driven by a lack of public funds, Bock highlights that the withdrawal of the state is presented as an 'opportunity' for citizen empowerment, representing a form of 'double devolution', from national to localism, and from the state outsourcing tasks to citizens. Within this context, Bock calls for more research to 'better understand the potential and significance of the shift of public responsibilities towards civic community organisations', particularly in marginal rural areas where community organisations are now a central actor in local governance. Many of the other sections in this Companion include chapters that emphasise the importance of community participation in place-making. However, the centrality of community actors has now moved beyond collaborative models of decision-making or policy-making, shifting towards the *co-production* of services in which many community organisations have adopted responsibility for managing service delivery. Bock suggests that this shifting role is not only a reaction to changing government policy, but often results from a disappointment in government that fuels a desire for local autonomy (and generally inspired by commitment, pride and a sense of service) – an important 'defence against abandonment' in rural localities.

Many of these themes are further extended by Tonts and Horsley's analysis of the 'neoliberal countryside' in the following chapter. The political-economic ideas of neoliberalism have become deeply entrenched within public administration in most parts of the world with profound sociospatial consequences as increasingly the belief that the market should discipline politics at a variety of spatial scales has become adopted as orthodoxy. Tonts and Horsley examine the spatial and temporal logics of neoliberalisn and examine the ways in which this has shaped rural economies, societies and cultures. Complementing Bock's analysis, this chapter outlines the shift from state-led to market oriented approaches towards service delivery and economic development which has penetrated most aspects of rural life. The chapter begins with an overview of neoliberalism, including 'roll back' and 'roll out' neoliberal modes of working and the role of the state in re-regulation in the interests of business. Noting the uneven geographies of neoliberalism, Tonts and Horsley outline the importance of examining 'actually existing neoliberalisms', which emphasise the importance of understanding particular institutional frameworks, policy regimes, regulatory practices, and socio-cultural norms. Within this context, focusing on rural geographies contributes further to understanding how the wider neoliberal project works. Tonts and Horsley provide an overview of rural scholarship on neoliberalism and also chart the evolution of neoliberalising practices in rural places. This analysis starts with agriculture, including the growing competitiveness agenda, the withdrawal of state support (noting the exception of the EU) and the liberalisation of global markets. The authors move on to examine the intersection of neoliberalism with globalisation, drawing on Michael Woods' conceptualisation of the 'global countryside' (examined by Woods in Part IX). This draws our attention to the importance of global networks in shaping the contemporary countryside, both in terms of productive processes (e.g. liberalisation of agricultural trade) and consumption processes (e.g. the commodification of rural landscapes for tourism and amenity migration). Tonts and Horsley conclude by stating that 'the reconstitution of rural places under "neoliberalised globalisation" is complex, non-linear and spatially contingent' and that 'neoliberalism, as a key driver of globalisation, proceeds by hybridisation, fusing and mingling the local and the extra-local to produce new formations', adding new complexities to uneven rural development processes.

While these two chapters develop a robust critique of market oriented approaches, in the next chapter Daniels highlights the potential of harnessing markets to maximise environmental benefits through examining the use of market-based instruments in planning and agricultural land preservation. Although less common in a European context, Daniels highlights the importance and growing popularity of financial incentives in the US for the protection of rural lands for farmland and forestland preservation, wildlife habitat protection, and to secure recreational lands. Although rural planning regulation is generally weak in the US, Daniels provides a comprehensive account of how voluntary, market-based instruments are deployed within US rural planning approaches, most commonly through the purchase and donation of conservation easements and the transfer of development rights. Daniels suggests that market-based incentives appear most effective when combined with comprehensive plans and land use regulations to manage growth. Regulations – such as zoning and urban growth boundaries – are less expensive than market-based instruments, but are vulnerable to political change: 'the acquisition of conservation easements provides long-term protection from incompatible development through a binding legal contract'; however, the cost of acquiring conservation easements can be as high as 90 per cent of the fair market value of the property. Daniels argues that the transfer of development rights has achieved some noteworthy success, but is more difficult to implement than a purchase of conservation easements programme. Daniels concludes by discussing the wider potential of applying market-based tools within rural planning practice. Specifically, the chapter outlines the potential of reducing greenhouse gas emissions based on land management practices that sequester carbon, which could expand significantly if the federal government were to adopt a national cap-and-trade programme with carbon offsets.

In contrast to harnessing markets to capture positive rural planning outcomes, Moore in Chapter 12 examines the potential of direct community ownership of rural assets as a form of place-making. This extends Bock's earlier analysis of how community organisations have moved from being involved in policy design to playing much more direct roles in rural governance and delivery. Moore provides an assessment of rural community ownership of assets, which represents an emerging local response to economic and social change in rural areas, including the loss of local amenities and, particularly in England, the lack of affordable rural housing. As a case study, the chapter examines the role of community land trusts in rural areas, tracing their development, application, and governance to understand the potential of community asset ownership as a governance model that can effectively respond to the local concerns and changes within rural communities. Although examined here in relation to affordable housing provision, similar experiments have been applied to other areas, such as renewable energy projects and woodland preservation. Moore argues that community land trusts have had notable successes, particularly in overcoming traditional rural planning problems associated with accommodating new development in rural localities. For example, while locals often resist renewable energy projects or proposals for new housing development, community land trusts are able to use 'narratives of community, local benefit, and local need to mitigate opposition and build support'. However, Moore also notes that support is needed for community land trusts to help reduce the risks associated with relying on voluntary action for service delivery, while the 'scaling-up' of interventions may risk distancing community land trusts from their very local community-based roots.

The final chapter in this section explores the 'dark side' of community within rural planning practice. In the 1990s, seminal work on the 'differentiated countryside' by Marsden and colleagues developed a four-fold classification of the English countryside to examine the regulation of land and development (Marsden et al., 1993; Marsden 1998). This included: *the preserved countryside*, characterised by strong anti-development and preservationist attitudes and decision-making, *the contested countryside*, where the development process is marked by increasing conflict between old and new social groups, *the paternalistic countryside*, referring to areas where large private estates and large farms still dominate, and *the clientelistic countryside*: likely to be found in remote rural regions where agriculture and its associated political institutions still have power. Processes of rural development in these areas have traditionally been dominated by farming, landowning, local capital and state agencies, usually working in close corporatist relationships. Local politics tends to be dominated by employment concerns and the welfare of the rural 'community'.

Fox-Rogers extends this analysis of the clientelist countryside through drawing on the experiences of rural planning in Ireland, where local government has long been characterised as operating along clientelist lines and which experienced major planning corruption scandals in the 1980s and 1990s. Fox-Rogers begins her analysis by noting that while the term 'community' is synonymous with the very essence of planning, 'making planning decisions in line with perceived local or community interests can often revolve around informal processes, clientelist politics and the domination of farming, local capital and landowning interests'. In contrast to the extensive planning literature on public engagement and collaboration, this chapter examines and critically reflects on the uneven power dynamics that exist in rural communities that can lead to opaque planning practices within 'the shadows of the planning system's formal structures'. It begins by identifying the conditions which have been identified in the literature as being central to understanding the planning system's vulnerability to clientelism and corruption. Fox-Rogers then goes on to explore how clientelist exchanges arise and operate within the context of rural communities before analysing how they play out in terms of specific social, economic and environmental outcomes. The chapter concludes by exploring approaches to enhancing the transparency of planning decision-making. While institutional reforms and new legal obligations have been implemented in various countries to reduce the potential for corruption, Fox-Rogers also highlights the limitations of this approach including the ability of 'clientelist networks [to] develop more innovative channels of influence that are more difficult to detect'. As Fox-Rogers argues, more research is needed to examine how power relations develop and play out within rural communities 'in ways which distort the democratic nature of planning practice'.

References

Allmendinger, P. (2016) *Neoliberal Spatial Governance*. London: Routledge.

Marsden, T. (1998) New rural territories: regulating the differentiated rural spaces. *Journal of Rural Studies*, 14: 107–117.

Marsden, T., Murdoch, J., Lowe, P., Munton, R. and Flynn, A. (1993) *Constructing the Countryside*. London: UCL Press.

Murdoch, J. and Abram, S. (1998) Defining the limits of community governance. *Journal of Rural Studies*, 14(1): 41–50.

9

Rurality and multi-level governance

Marginal rural areas inciting community governance

Bettina B. Bock

Introduction

In the late 1990s the shift of power from government to governance was a critical subject of academic debate. In the rural context, studies focused on the role of governance in endogenous models of development. The discussion centred on the opportunities for development from the bottom-up and based on local resources, the danger of elite groups' dominance and the legitimacy of rural governance within a society organised as a representative democracy. Changes in the governing of rural areas have continued ever since and proceeded in the same direction. They reflect the continuous withdrawal of the state from public tasks in the aftermath of the financial crisis and as part of welfare state reforms. Lack of public funds is one driver of this development, yet this is – again – presented as an opportunity for citizen empowerment. Marginal rural areas are affected in a particular way; because of the diminishing support from central governments, active citizens play an increasingly important role in maintaining vital services. In a way they may, hence, be considered as natural laboratories for testing and understanding the opportunities and limits of multilevel governance. This chapter looks into the development of thought about rural governance in relation to rural development. It sketches how research and theory developed since the 1990s, appointing and discussing the most important strands of thought in rural governance literature.

It then focuses on current research in marginal rural areas and the problems that result from the further withdrawal of the state, welfare state reforms, and the ongoing process of urbanisation reflected in rural depopulation and the reorientation of public funds towards urban centres. As a result, in particular, the more remote rural areas are confronted with the closure of essential public and private services. The actions with which rural community groups confront and resist marginalisation and endeavour to take service provision into their own hands are the subject of the following section with specific attention given to initiatives in Scotland and The Netherlands. The concluding section returns to rural governance literature and discusses if and to what extent these initiatives reveal the emergence of novel forms of community governance, in which the take-over of local welfare provision reflects a desire and aspiration for local autonomy incited by the acknowledgement of state withdrawal.

Shifts in rural governance

The rural governance literature deals with the management and regulation of rural areas, with the concept of governance already implying that the state is not the only actor and agency involved. In the 1990s, it was this shift from government to governance which was an important focus of academic debate (Jessop, 1997; Rhodes, 1996; Stoker, 1998) together with its specific effects on the rural (Goodwin, 1998). Part of the discussion centred on defining what governance is. Following Cheshire:

> . . . governance can be understood to represent a new mode of governing that is no longer enacted solely through the formal, coercive powers of the nation state, but is exercised through a range of government and non-governmental actors and entities operating at different spatial scales and across different actors.
>
> *(Cheshire, 2016: 596)*

In practice, governance includes the upward devolvement of national states' functions and authorities to supra-national bodies such as the European Union, as well as the downward delegation and decentralisation to regional and local bodies.

This vertical shift of power and responsibility is accompanied by a horizontal one through the involvement of a variety of public and private stakeholders in new institutional arrangements (Ward and McNicholas, 1998) that are collaboratively engaged in the formulation and implementation of policy. Among the main actors involved are representatives from the state, business and civil society (Swindal and McAreavey, 2014). This new type of governance is characterised as multi-actor governance as it should function through partnerships that involve multiple stakeholders; it is expected to be multi-sectoral through the representation and integration of private, public and third-sector actors, and multi-level as it should include participation from the national to the local level. The general discussion often centres on the idea of governance as a tool for improving the efficiency and effectiveness of government and its potential to reinforce democracy while rendering policymaking and delivery more inclusive through the building of community capacity and the fostering of social engagement. The EC's White Paper on European Governance promoted working in partnerships as one of the pillars of 'good governance' (CEC, 2001).

The shift from government to governance has taken place in society at large but has profound impacts on rural areas (Cheshire, 2016; Goodwin, 1998), resulting in numerous studies and ongoing academic debate ever since the 1990s. An important part of rural governance debates coincides with a discussion on area-based, territorial development, as the delegation of governmental authority to the local level should reinforce the power of 'the local community to identify and pursue their own development goals' (Macken-Walsh, 2016: 615). At the same time, however, it also stressed the communities' responsibility for achieving those goals (Winter, 2002; Ray, 2000).

Involving public and private stakeholders fits the approach of endogenous development (see also Chapter 14 of this Companion), in which local residents and organisations are perceived as the main development actors (van der Ploeg and Long, 1994), and which has been popular among academics as well as policymakers ever since the publication of *The Future of Rural Society* by the European Communities in 1988 (CEC, 1988) and the first Cork Declaration on Rural Development in 1996. The central idea is that local actors are best positioned to identify the strong points and resources of their localities, to assemble various local stakeholders and promote their collaboration for the sake of local development. The public authorities are expected to shift authority to local communities and become a facilitator of bottom-up development, for example, through the provision of funding, promotion of training and capacity building. The reality proves to be different and the space for local citizen action as rather limited, as the rural development policies formulated and implemented at national and European level predefine goals and actions and identify the issues to be at the core of rural development action. Following Swindal and McAreavey (2014: 283) 'it is the state's imaginaries of rural space that define the subjects and objects of development, [and] the state also shapes the strategies and actions that emerge'. The arenas of rural development and rural governance are, hence, not really 'popular spaces', in their view, but 'invited spaces' conceptualised by the state (Cornwall, 2004).

The EU promotes the idea of endogenous local and regional development through its regulations for the delivery of structural funds (Geddes, 2000; Ward and McNicholas, 1998; Dax, 2015) and introduced LEADER (Links between Actions for the Development of the Rural Economy) as its flagship bottom-up development programme in 1991, initially as an experiment to test the opportunities a new rural development approach offered. Local endogenous development has remained the dominant approach within the EU, expressed in a varying vocabulary – territorial, place-based development and currently community-led local development (CLLD) as bottom-up, integrated, multi-actor inclusive development (EC, 2014). While initially targeted at specific, particularly vulnerable areas, the LEADER approach has now been translated to all regions. National governments incorporated the idea of multi-level area-based development into their own programmes to different extents and in different ways with distinctions in national political culture and structure as well as institutional capacity playing an important role (Douglas, 2016). Local democracy was, for instance, already strongly anchored in the political arena of many Scandinavian countries, which supported the implementation of community led development (Bjärstig and Sandström, 2017).

Research follows this development with sympathy for the new approach but also a critical eye for 'the practices and relations of power that play out in governance arrangements' (Cheshire, 2016: 297). Many studies demonstrate that the partnerships involved in policymaking are far from inclusive as they tend to involve established interest groups, organised through well-known traditional organisations (Woods, 1997; Osti, 2000) and, hence, exclude less well-organised groups such as women and youth (Derkzen et al., 2008). Research reveals that it is often the local elite and those who are skilled in project management and political lobbying (Kovách and Kučerova, 2006) who dominate local development groups and that less powerful groups, or actors with divergent ideas of development, risk becoming side-lined. This also affects the content of rural development programmes with economic competitiveness as a prioritised goal and activities that reinforce traditional (gender) identities (Pini, 2006; Bock, 2015; Shortall and Bock, 2015). In doing so, scientists also criticise the idea of one consentient and harmonious community as the basis of territorial development as fictitious and exclusionary (Shortall, 2004; Derkzen et al., 2008).

This again leads to discussions about the democratic legitimacy and accountability of rural governance particularly vis-à-vis traditional representative democracy (e.g. Connelly et al., 2006). Although the involvement of stakeholders is promoted as utterly democratic, critics point to the selective engagement of stakeholders, the inaccessibility of many partnerships to groups without institutional support, which enhances the power and dominance of the local elites (Shortall, 2008; Esparcia et al., 2015; Connelly et al., 2006). Research, moreover, reveals that resourceful and well connected communities (Fischer and McKee, 2017) are more successful in promoting local development because they are more embedded in networks with significant external actors and agencies (Bock, 2016a; Bosworth et al., 2016; Shucksmith, 2010). Others focus on the process of development and point at shortcomings in the organisation of development programmes, and in particular the LEADER programmes: the hesitance of public authority to let go of their traditional role and tendency to continue their control of local action groups (Furmankiewicz and Macken-Walsh, 2016; Fałkowski, 2013); the need to submit elaborate proposals and business plans, short project cycles which do not allow for the gradual development of capacity and capability, which advantage those groups that are experienced in formulating projects for funding (Kovách and Kuçerová, 2006; Shortall and Shucksmith, 1998). In a similar vein, studies criticise the instrumentalisation of local agency in policy programmes, their disciplining through programme regulations and the stigmatisation of communities which do not manage to develop themselves (Herbert-Cheshire and Higgins, 2004; Young, 2016). Others stress that the state has a crucial enabling role to play in particular in rural areas, to initiate, financially support and regulate public-private partnerships and to organise the necessary conditions for regional network engagement (Bjärstig and Sandström, 2017; Morrison, 2014) or in other words 'meta-governance' (Jessop, 1997) or 'the regulation of self-regulation' (Wilson et al., forthcoming). Following the latter this includes the strategic design of the governance network (including institutional structuring, framing, selective activation and resourcing) as well as process management in order to promote positive interaction of network actors, including mediation and facilitation. Both approaches, however, are therefore criticised, as having either too much or too little state involvement, reflecting also a different interpretation of governance as co-governance or self-governance.

In recent years governance is discussed in connection with advancing neoliberalism in the context of the financial crisis which gave a new impetus to state withdrawal as a result of public budget cuts and the privatisation, marketisation and individualisation of risks formerly buffered by public programmes (Young, 2016). The reduction of public spending is an important motive although these changes are still promoted through discourses of community empowerment (Bock, 2016a). Both are highly relevant in the context of rural marginalisation as will be discussed in the following section.

Rural marginalisation

Processes of rural differentiation that have been going on for decades have intensified in recent years and have contributed to a polarisation between affluent and poor rural regions (Bock et al., 2016). Many rural areas, in particular those in reach of the bigger cities, are prospering as the inflow of urban residents and recreational dwellers strengthens the local economy (Halfacree, 2012), supporting private businesses and cost effectiveness of public services. It also encourages investment in transport and infrastructure which facilitates travelling and commuting to urban centres. Furthermore, the perceived amenity value of rural areas is an important factor encouraging urbanites to visit and move to specific rural areas, for instance upon retirement (Brown and Glasgow, 2008; Arnason et al., 2009).

The situation is quite different in the more remotely located areas that experience depopulation and run the risk of a continuous decline in living conditions. The attraction exerted by the city among young people is one factor, next to the loss of industry and with it employment, and the closure of public and private services (Kühn, 2015; Bock et al., 2016; Steiner and Teasdale, 2018). Continuous depopulation undermines the cost-effectiveness of private and public services, but the restructuring of welfare systems in combination with public budget cuts and austerity measures is a pertinent driver as well as it results in the centralisation and diminution of social welfare services (Richter, 2018; Bock, 2016b; Shucksmith and Brown, 2016). Following Young (2016), the financial crisis has also renewed states' commitment to austere forms of neoliberalism which emphasise regional competitiveness and local entrepreneurialism, legitimating the devolution of key responsibilities of the nation state to local communities. In addition, many national governments in Europe have reoriented public funding towards urban centres, city regions and their hinterlands with little attention for isolated rural areas (Shortall and Warner, 2014). This goes along with a reduction of support for rural and less populated areas and calls into question earlier ideals of universal service provision (Skerrat, 2016; Farmer et al., 2011). Hodges et al. (in Steiner and Teasdale, 2018: 2) conclude that rural areas 'have been disproportionately affected by an increasing withdrawal of physical public services'.

The downturn of the economic development and loss of services threatens the quality of life in these rural areas. At the same time selective outmigration and demographic transitions impact on the vitality of many rural communities; it is in particular the young women and more ambitious, well-educated and well paid citizens who leave, which results in an over representation of elderly and low educated citizens in many rural areas (Champion and Brown, 2014; Meijer and Syssner, 2017: 59). This affects community cohesion, citizens' capacity to act and address problems of economic downturn, loss of services and the deteriorating quality of life.

It is important to acknowledge these risks and account for the enormous impact that public budget cuts and welfare state reforms have on rural areas in general and in particular on the more vulnerable marginal areas threatened by peripheralisation (Kühn, 2015). At the same time, it is important to keep an open eye for the active resistance of rural communities and their engagement in the development of novel forms of service delivery and community governance. In the editorial to the *Routledge International Handbook of Rural Studies*, Shucksmith and Brown (2016) refer to the rural as a site of resistance to neoliberalism, the development of real utopias (Wright, in Shucksmith and Brown, 2016) and post-neoliberal alternatives through 'the remobilisation, recognition and valuation of multiple, local forms of development, rooted in local cultures, values and movements' (Peck, in Shucksmith and Brown, 2016: 15). State withdrawal, austerity measures and welfare state reforms pose a real threat to rural communities; nevertheless, it also calls for resistance and motivates rural communities 'to spur the post neoliberal imagination' (Peck et al., in Shucksmith and Brown, 2016: 16) and pushes them to reach for local autonomy and self-governance. The following section looks into these initiatives and their significance for the rural governance debate.

Rural governance today

Rural governance has remained an important subject of research and academic debate since the 1990s, and its limitations in terms of participation, legitimacy and effectiveness are still important themes within the literature. In addition a number of new issues have emerged in recent years that address the effects of the financial crisis and the threats they pose for the quality of rural life. This research sheds also new light on rural governance and brings more practice oriented hands-on versions of community governance to the fore.

As explained in more detail above, a growing number of rural areas are experiencing marginalisation as a result of depopulation, public budget cuts and reform of welfare state policies which threatens the maintenance of employment as well as essential public and private services. Overall, more and more rural residents experience the quality of life of their rural home to be degrading and fear for its future. The reaction to these threats differs. In some places they are met by local action and resistance whereas others seem to passively await what happens. Research demonstrates that the likelihood of successfully contesting marginalisation through local action may be explained by access to resources such as social capital, materialisation of charismatic leadership and embeddedness in relevant networks (Horlings and Padt, 2013; Beer, 2014; Fischer and McKee, 2017). Simply said, it is generally the better situated villages and regions with higher levels of linking capital (Meijer and Syssner, 2017) which are more active and more successful compared to other, geographically and socio-economically more peripheral places. Not all rural communities, hence, manage to resist neoliberalism.

The actions undertaken by the more active communities include the design and enactment of local policies in the sense of plans and programmes, with objectives and instruments. These are, however, non-governmental in nature, rolled out by citizen groups and not local governments. The emergence of this type of community governance centres around the provision of services that public authorities, professional organisations and private businesses are no longer providing and which local communities adopt to maintain and improve their delivery.

In the UK context they are often indicated as 'social enterprises' (Steiner and Teasdale, 2018), in other countries other modes of organisation prevail. In The Netherlands they are often organised as cooperatives or voluntary associations. In the remainder of the chapter, therefore, the term 'community organisations' is used. What the different modes of organisation have in common is that non-governmental actors adopt or partake in the delivery of formerly *publicly and formally* organised tasks, such as transport, health and social care, education and energy production (Richter, 2018). Their start is motivated by the desire to maintain or regenerate the quality of life in their place by filling the gaps that occur through the withdrawal of the state, business and professional organisations. The existence of a 'culture of self-help' (Steinerowski and Steinerowska-Streb, 2012: 173) plays an important role as well, combined with a 'sense of community and solidarity' (Farmer et al., 2008: 455).

The ways in which these community organisations function differ across countries and types of activities (Best and Myers, 2018). Some community organisations use the right to challenge to take over services formerly delivered through the state in return for (part) of the budget that was intended for these services (Skerrat, 2010). It allows community organisations to create employment, redesign and reorganise the delivery of services according to local needs and to reinvest realised profits in the community. Others start from local needs and step by step weave a web of interrelated community activities, such as social activities for children and the elderly, transport services, sports and training, the adoption of vacant buildings and their restoration and use for a village shop, child care centre and/or mobile health care centre; it may even include the construction, sale or rent of new housing (Slee, 2017; Bock, 2016a). Most of the community organisations collaborate in some way or other with public authorities; some because they need financial support (Richter, 2018), others because the adoption of public services requires contractual agreement and permissions (Meijer and Syssner, 2017).

Collaboration with policymakers is arguably more important when community organisations adopt or coproduce (formerly) public services such as health and social care, which are heavily regulated by law. Here, research points at the emergence of innovative forms of collaboration between community and professional health care organisations, which may again take various forms, such as social enterprises or firms, cooperatives, and community development organisations (Best and Myers, 2018). Central to these initiatives is that they search to develop multistakeholder partnerships in order to collaboratively design and deliver services in a way that fits the rural context (Steiner and Teasdale, 2018). This requires what Meijer and Syssner (2017: 61) call linking social capital – the 'formation of vertical ties between non-governmental and governmental actors'.

Research in the Netherlands among health cooperatives, for example, underlines their success in addressing the clients' desires for personalised care, in which the client's voice is decisive and which takes place outside health care institutions. The health cooperatives are also successful in the sense of delivering care more cost-efficiently, which is partly possible through the engagement of volunteers. Research, however, appoints problems as well; for the health care cooperative they often result in stringent health care regulations that fit badly with health care delivery outside institutions, and which complicates collaboration with health care professionals and health insurances (Bokhorst, 2015). Research into community initiatives more generally points to problems due to internal conflicts or conflicts with public authorities which have difficulty in letting go of traditional role divisions (Ubels et al., forthcoming a).

Research into the endeavour of community organisations in Scotland sketches a similar image of problems and promises. In rural Scotland, social enterprises play an important role in maintaining services in remote rural areas and the creation of employment and in doing so importantly contribute to the viability of rural areas (Steiner and Teasdale, 2018). They do, however, experience problems in their collaboration with public authorities (Fischer and McKee, 2017). And there are more problems as Skerrat (2016: 22) underlines: only the already competent communities manage to maintain services, 'with the potential of exclusion from service provision of whole communities as well as individuals within communities'. Nevertheless Scottish social enterprises are recognised by policymakers as playing 'a central role in the inclusive growth strategy' (Steiner and Teasdale 2018: 3) and the government continues in devolving responsibility in many policy areas to social enterprises (Alcock, in Steiner and Teasdale, 2018: 1).

Conclusion: towards community governance

More research needs to be done to better understand the potential and significance of the shift of public responsibilities towards civic community organisations. Current studies indicate that community organisations play a pertinent role in the maintenance of services that are essential for the quality of life in rural areas and, hence, their future. By accepting the responsibility to adopt or co-produce essential public services they enter the field of rural governance as an important actor.

One might argue that this reinforces their position and renders '*governance more conducive towards co-production*' (Pestoff et al., in Steiner and Teasdale, 2018: 5) Following Davoudi and Madanipour (in Meijer and Syssner, 2017) this development can be described as 'double devolution' – from national to local and from local government outsourcing tasks to citizens. Meijer and Syssner (2017: 67) argue in a similar vein that 'community *initiatives have now become part of a local government policy*' as many governments are actively inviting communities to participate in the delivery of services and where outsourcing public services is becoming the rule.

Such a development may be captured in the concept of collaborative government, although this has so far mainly been discussed in terms of the engagement of multiple stakeholders in decision-making about 'laws and rules for the provision of public goods' (Ansell and Gash, 2007: 545). What is described above goes beyond joint decision-making as the design and implementation of services is taken over from government. It is collaborative as the government and other agencies are generally involved in some way, yet this can be at quite a distance in those cases in which community organisations have adopted the responsibility and authority for designing and managing service provision. The extent to which community organisations do so differs a lot and affects the extent to which power structures may be transformed (McKee, 2015). Ansell and Gash (2007) underline the importance of interdependence as a core contingency for successful collaborative governance. In the cases described above, community organisations depend on the government for funding and permissions, yet the contrary is also true – in rural areas local governments increasingly rely on community organisations for the delivery of public tasks.

Whether or not this is a good development is questionable and open for debate. How to answer that question depends also very much on the perspective chosen. It seems unfair that residents of rural areas no longer have access to essential services that are financed by taxes paid by them too (Farmer et al., 2011). It is also questionable if essential services should be outsourced at all, be it to private business or civic community groups. Douglas (2016: 606), for instance, remarks that the delegation of service provision by the state may also be interpreted as 'offloading' and underlines that 'the collaborative process, where it has existed, and the controlled degrees of participatory process, where they have been accommodated, have generally been subservient to centre's hegemonic agenda' (ibid.: 611). Young (2016) warns us that it is a way to absolve senior governments to protect rural livelihoods, services and environments, and allows them to transfer responsibility to communities 'for the (inevitable) failures that afflict many places' (ibid.: 646). Others underline that the community groups in charge are not necessarily inclusive or represent the whole community (Skerrat and Steiner, 2013; Ubels et al., forthcoming b). This points again at the risk of an inconsistent landscape of welfare provision, without warranted accessibility for all. Another risk regards the continuity of service provision, as dependence on volunteers to maintain services is fragile. All this is certainly true and needs to be taken into account. At the end of the day, however, many rural communities are confronted with a situation in which very few services would be provided if community organisations did not step in.

In terms of governance it is an interesting development. Scholars temper our expectation of local autonomy by stressing the influence of extra-locals and the ability of higher levels of governance to '"steer" the self-governing process of small rural communities' (O'Toole and Burdess, in Macken-Walsh, 2016: 618). They point at the limits of self-governance depending on the nature and extent of services provided (Douglas, 2016). Nevertheless, scholars also refer to this development as a step towards local autonomy and community governance (e.g. Steiner and Teasdale, 2018; Slee, 2017). The adoption of public tasks plays an essential role here and results in a rather hands-on practical governance of community affairs, in terms of implementation as well as design. It focuses on what needs to be done in order to manage the community and take care of its needs in terms of health and social care, energy provision, transport and education. It is a reaction to changing governmental policy and very much influenced by it – a defence against abandonment one could argue. And it is often disappointment in the government, professional organisations and private businesses, which fuels a desire for local autonomy. Nevertheless the ambition for autonomous community governance is also inspired by love and care for rural life, pride, and confidence of being able to serve the community well.

References

Ansell C. and A. Gash. 2007. Collaborative governance in theory and practice. *Journal of Public Administration Research and Theory*. 18: 543–571.

Arnason A., M. Shucksmith and J. Vergunst. 2009. *Comparing Rural Development: Continuity and Change in the Countryside of Western Europe*. Aldershot: Ashgate

Beer A. 2014. Leadership and the governance of rural communities. *Journal of Rural Studies*. 34: 254–262.

Best S. and J. Myers. 2018. Prudence of speed: Health and social care innovation in rural Wales. *Journal of Rural Studies*. https://doi.org/10.1016/j.rurstud.2017.12.004.

Bjärstig T. and C. Sandström. 2017. Public–private partnerships in a Swedish context – a policy tool for the authorities to achieve sustainable rural development? *Journal of Rural Studies*. 49: 58–68.

Bock B.B. 2015. Gender-mainstreaming and rural development policy: The trivialisation of rural gender issues. *Gender, Place and Culture*. 22(5): 731–745.

Bock B.B. 2016a. Rural marginalisation and the role of social innovation: A turn towards nexogenous development and rural reconnection. *Sociologia Ruralis*. 56(4): 552–573.

Bock B.B. 2016b. Introduction: Inequality in rural areas: a territorial and relational perspective. In M. Shucksmith and D. Brown (eds), *Routledge International Handbook for Rural Studies*. New York: Routledge, 427–432.

Bock B.B., G. Osti and F. Ventura. 2016. Rural migration and new patterns of exclusion and integration in Europe. In M. Shucksmith and D. Brown (eds), *International Handbook for Rural Studies*. New York: Routledge, 71–84.

Bokhorst M. 2015. De koers van zorg coöperaties; samenwerken zonder te verworden tot paradepaard of werkpaard van de participatiesamenleving. *Bestuurskunde*. 24(2): 27–39.

Bosworth G., I. Annibal, T. Carroll, L. Price, J. Sellick and J. Shepherd. 2016. Empowering local action through neo-endogenous development; the case of LEADER in England. *Sociologia Ruralis*. 56(3): 427–449.

Brown D. and N. Glasgow (eds). 2008. *Rural Retirement Migration*. Berlin: Springer.

CEC. 1988. *The Future of Rural Society*. COM(88) 501. Brussels: European Commission.

CEC. 2001. *European Governance: A White Paper*. COM(2001) 428. Brussels: European Commission.

Champion T. and D.L. Brown. 2014. Migration and urban–rural population redistrubution in the UK and US. In M. Shucksmith, D.L. Brown, S. Shortall, J. Vergunst and M.E. Warner (eds), *Rural Transformations and Rural Policies in the US and UK*. New York: Routledge, 39–56.

Cheshire L. 2016. Power and governance: empirical questions and theoretical approaches for rural studies. In M. Shucksmith and D. Brown (eds), *Routledge International Handbook of Rural Studies*. New York: Routledge: 593–601.

Connelly S., T. Richardson and T. Miles. 2006. Situated legitimacy: deliberative arenas and the new rural governance. *Journal of Rural Studies*. 22: 267–277.

Cornwall A. 2004. Introduction: New democratic spaces? The politics and dynamics of institutionalised participation. *IDS Bulletin*. 35(2): 1–10.

Dax T. 2015. The evolution of European rural policy. In A. Copus and P. De Lima (eds), *Territorial Cohesion in Rural Europe: The Relational Turn in Rural Development*. New York: Routledge, 35–52.

Derkzen P., B.B. Bock and A. Franklin. 2008. The equality claim dismantled: power struggle as signifier for successful partnership working. A case study from Wales. *Journal of Rural Studies*. 24: 458–466.

Douglas D.J.A. 2016. Power and politics in the changing structures of rural local government. In M. Shucksmith and D. Brown (eds). *Routledge International Handbook of Rural Studies*. New York: Routledge, 601–614.

EC 2014. Community-led local development. Retrieved from http://ec.europa.eu/regional_policy/sources/docgener/informat/2014/community_en.pdf (accessed 26 January 2018).

Esparcia J., J. Escribano and J.J. Serrano. 2015. From development to power relations and territorial governance: increasing the leadership role of LEADER Local Action Groups in Spain. *Journal of Rural Studies*. 42: 29–42.

Fałkowski J. 2013. Political accountability and governance in rural areas: some evidence from the Pilot Programme LEADER+ in Poland. *Journal of Rural Studies*. 32: 70–79.

Farmer J., A. Steinerowski and S. Jack. 2008. Starting social enterprises in remote and rural Scotland: best or worse of circumstances? *International Journal of Entrepreneurship and Small Business*. 6: 450–464.

Farmer J., A. Nimegeer, J.H. Farrington and G. Rodger. 2011. Rural citizens' rights to accessible health services: an exploration. *Sociologia Ruralis*. 52(1): 134–144.

Fischer A. and A. McKee. 2017. A question of capacities? Community resilience and empowerment between assets, abilities and relationships. *Journal of Rural Studies*. 54: 187–197.

Furmankiewicz M. and A. Macken-Walsh. 2016. Government within governance? Polish rural development partnerships through the lens of functional representation. *Journal of Rural Studies*. 46: 12–22.

Geddes M., 2000. Tackling social exclusion in the European Union? The limits to the new orthodoxy of local partnership. *International Journal of Urban and Regional Research*. 24(4): 782–800.

Goodwin M. 1998. The governance of rural areas: some emerging research issues and agendas. *Journal of Rural Studies*. 14(1): 5–12.

Halfacree K. 2012. Heterolocal identities? Counter-urbanisation, second homes, and rural consumption in the era of mobilities. *Population, Space and Place*. 18(2): 209–224.

Herbert-Cheshire L. and V. Higgins. 2004. From risky to responsible: expert knowledge and the governing of community-led rural development. *Journal of Rural Studies*. 20(3): 289–302.

Horlings I. and F. Padt. 2013. Leadership for sustainable regional development in rural areas: bridging personal and institutional aspects. *Sustainable Development*. 21(6): 413–424.

Jessop B. 1997. The entrepreneurial city: re-imaging localities, redesigning economic governance, or restructuring capital? In N. Jewson and S. MacGregor (eds), *Transforming Cities: Contested Governance and New Spatial Divisions*. London: Routledge, 28–41.

Kovách I. and E. Kuçerova. 2006. The project class in central Europe: the Czech and Hungarian cases. *Sociologia Ruralis*. 46(1): 3–21.

Kühn M. 2015. Peripheralization: Theoretical concepts explaining socio-spatial in equalities. *European Planning Studies*. 23(2): 367–378.

McKee K. 2015. Community anchor housing associations: Illuminating the contested nature of neo-liberal governing practices at the local scale. *Environment and Planning C: Government and Policy*. 33(5): 1076–1091.

Macken-Walsh A. 2016. Governance, partnerships and power. In M. Shucksmith and D. Brown (eds), *Routledge International Handbook of Rural Studies*. New York: Routledge: 615–625.

Meijer M. and J. Syssner. 2017. Getting ahead in depopulating areas – how linking social capital is used for informal planning practices in Sweden and The Netherlands. *Journal of Rural Studies*. 55: 59–70.

Morrison T.T. 2014. Developing a regional governance index: the institutional potential of rural regions. *Journal of Rural Studies*. 35: 101–111.

Osti G. 2000. LEADER partnerships: the case of Italy. *Sociologia Ruralis*. 40(2): 172–180.

Pini B. 2006. A critique of 'new' rural local governance: the case of gender in a rural Australian setting. *Journal of Rural Studies*. 22: 396–408.

Ray C. 2000. The EU LEADER programme: rural development laboratory. *Sociologia Ruralis*. 40(2): 163–171.

Rhodes, R.A.W., 1996. The new governance: governing without government. *Political Studies*. XLIV: 652–667.

Richter R. 2018. 2018. Rural social enterprises as *embedded intermediaries*: The innovative power of connecting rural communities with supra-regional networks. *Journal for Rural Studies*. https://doi.org/10.1016/j.rurstud.2017.12.005.

Shortall S. 2004. Social or economic goals, civic inclusion or exclusion? An analysis of rural development theory and practice. *Sociologia Ruralis*. 44(1): 109–123.

Shortall S. 2008. Are rural development programmes socially inclusive? Social inclusion, civic engagement, participation and social capital: exploring the differences. *Journal of Rural Studies*. 24(4): 450–457.

Shortall S. and B.B. Bock. 2015. Rural women in Europe: the impact of place and culture on gender mainstreaming the European Rural Development Programme. *Gender, Place and Culture*. 22(5): 662–669.

Shortall S. and M. Shucksmith. 1998. Integrated Rural Development: issues arising from the Scottish experience. *European Journal of Planning Studies*. 6(1): 73–88.

Shortall S. and M.E. Warner. 2014. Rural transformations: conceptual and policy issues. In M. Shucksmith, D.L. Brown, S. Shortall, J. Vergunst and M.E. Warner (eds). *Rural Transformations and Rural Policies in the US and UK*. New York: Routledge, 3–17.

Shucksmith, M. 2010. Disintegrated rural development? Neo-endogenous rural development, planning and place-shaping in diffused power contexts. *Sociologia Ruralis*. 50(1): 1–14.

Shucksmith M. and D. Brown. 2016. Framing rural studies in the Global North. In M. Shucksmith and D. Brown (eds). *Routledge International Handbook of Rural Studies*. New York: Routledge: 1–26.

Skerratt, S. 2010. Hot Spots and Not Spots: addressing infrastructure and service provision through combined approaches in rural Scotland. Invited paper to 'Human populations in remote areas', special edition of *Journal Sustainability*. 2(6): 1719–1741.

Skerrat S. 2016. The power of rhetoric in re-shaping health and social care: implications for the under-served. *Sociologia e Politiche Sociali*. 19(3): 9–28.

Skerrat S. and A. Steiner. 2013. Working with communities-of-place: complexities of empowerment. *Local Economy*. 28(3): 320–338.

Slee B. 2017. Revitalising rural services through social innovation. Presentation at the 2nd Meeting of Thematic Group on 'Smart Villages', Brussels, 7 December. Retrieved from https://enrd.ec.europa.eu/sites/enrd/files/tg2_smart-villages_social-innovation_slee.pdf.

Steiner A. and S. Teasdale. 2018. Unlocking the potential of rural social enterprise. *Journal of Rural Studies*. http://doi.org/10.1016/j.jrurstud.2017.12.021.

Steinerowski A. and I. Steinerowska-Streb. 2012. Can social enterprise contribute to creating sustainable rural communities? *Local Economy*. 27: 167–182

Stoker G. 1998. Governance as theory: five propositions. *International Social Science Journal*. 50(155): 17–28.

Swindal M.G. and R. McAreavey. 2014. Rural governance: participation, power and possibilities for action. In M. Shucksmith, D.L. Brown, S. Shortall, J. Vergunst and M.E. Warner (eds). *Rural Transformations and Rural Policies in the US and UK*. New York: Routledge, 269–286.

Ubels H., B.B. Bock and T. Haartsen. Forthcoming a. Experimental collaborative governance arrangements and role shifts between municipalities and residents in population decline areas. *Environment and Planning C*.

Ubels H., T. Haartsen and B.B. Bock. Forthcoming. Social innovation and community focussed civic initiatives in the context of rural depopulation: for everybody by everybody? The case of Ulrum 2034. *Journal for Rural Studies*.

Van der Ploeg J.D. and A. Long (eds). 1994. *Born from within. Practice and perspectives of endogenous rural development*. Assen: van Gorcum.

Ward N. and K. McNicholas. 1998. Reconfiguring rural development in the UK: objective 5b and the new rural governance. *Journal of Rural Studies*. 14(1): 27–39.

Wilson C., T. Morrison and J.A. Everingham. Forthcoming. A case study of steering collaborative social policy in Australia's unconventional gas regions. *Sociologia Ruralis*.

Winter D.M. 2002. *Rural policy: new directions and new challenges. Research to identify the policy context on rural issues in the South West*. Exeter: Centre for Rural Research University of Exeter.

Woods M. 1997. Researching rural conflicts: hunting, local politics and actor–networks. *Journal of Rural Studies*. 14(3): 321–340.

Young N. 2016. Responding to rural change: adaptation, resilience and community action. In M. Shucksmith and D. Brown (eds), *Routledge International Handbook of Rural Studies*. New York: Routledge, 638–649.

10

The neoliberal countryside

Matthew Tonts and Julia Horsley

Introduction

Neoliberalism has often been depicted as a pervasive structural force that has transformed urban and rural spaces through the latter tweintieth and early twenty-first centuries (Perkins et al., 2015; Peck et al., 2013). Critical studies of neoliberalism have articulated a powerful set of narratives that emphasise the multi-scalar and multidimensional nature of neoliberalism (Woods, 2007, 2011). Indeed, one of the common themes within this body of work is that neoliberalism is dynamic, highly contextualised, and spatially uneven (Peck et al., 2013). This is particularly evident in rural spaces, where neoliberalism is a central and geographically complex dynamic underpinning the transformation of economies, environments, politics, and socio-cultural conditions. Yet, its uneven, contextual and adaptive nature means that rarely is interpreting the logics and implications of the 'neoliberal countryside' straightforward.

This chapter examines the spatial and temporal logics of neoliberalism, and the ways in which this has underpinned much of the economic, social and cultural transformation witnessed across the 'global countryside'. It argues that while neoliberalism has become a pervasive influence in almost all areas of rural life, the ways in which it has been interpreted, implemented and reshaped has varied enormously across spatial contexts and scales. Moreover, as a political 'project', neoliberalism has proven to be dynamic and adaptive, and is routinely being reinterpreted and recast in the face of changing assemblages of economic, political and socio-cultural circumstances and relationships.

Grand narratives and actually existing neoliberalisms

As a political project, neoliberalism comprises a series of pro-market values, ideas and policy settings that are designed to improve national and international competitiveness via a reorientation of the roles of government and private enterprise (Lockie and Higgins, 2007). Peck and Tickell (2002) have distinguished between what they term 'roll back' and 'roll out' neoliberalism. 'Roll back' neoliberalism emerged in the 1980s in countries such as the USA, UK and Australia and was associated with the dismantling of state institutions, and the reduction of public benefits associated with the Keynesian welfare state (Tonts and Haslam-Mckenzie, 2005). These forms of government intervention were held to interfere with or distort market forces and thereby reduce competitiveness and the efficient allocation of resources (Head, 1988). In contrast, in the mid-1990s, there was a subtle shift in the agenda with the emergence of what has been described as 'roll out' of neoliberalism via the creation of new institutions and policies whereby the state actively intervenes (re-regulates) where it considers it can more directly serve the interests of business, improve competition and foster community responsibility (Peck and Tickell, 2002).

Of course, the reality is that the global 'imposition' of neoliberalism has been highly uneven, both socially and geographically, and its institutional forms and socio-political consequences have varied significantly across spatial scales (Beer et al., 2005). However, these unevenly developed geographies are not mere variations around an emergent norm, but rather evidence of the co-constitutive nature of the project reflecting expressions of local histories, socio-institutional difference, contests, and compromises (Peck et al., 2013). Indeed, in contrast to a simplistic assumption in which market forces operate according to immutable laws no matter where they are 'unleashed', research into 'actually existing neoliberalisms' has emphasised the importance of understanding particular institutional frameworks, policy regimes, regulatory practices, and socio-cultural norms (Brenner and Theodore, 2002). An understanding of actually existing neo-liberalisms must therefore explore the contextually specific interactions between the local and the global and a broad range of geographical scales in between.

In the context of rural space, the past three decades has witnessed a deep and ongoing scholarly commitment to understanding the ways in which neoliberalism has played out across multiple spatial scales and localities. While much of the early work tended to emphasise the transformation of agri-food systems as part of a more globalised agriculture (see, for example, Goodman and Watts, 1997; McMichael, 1994; Lawrence, 1987), there was also an increasing appreciation of the ways in which neoliberalism was penetrating other areas of rural life. A vibrant research agenda emerged that sought to understand how various forms of neoliberalism were operating at multitude of sub-national scales to reshape, *inter alia*, land use, service provision, environmental regulation, social wellbeing and rural policy (Perkins et al., 2015; Higgins, 2014b; Argent, 2011; Wilson, 2008).

The emerence of a neoliberal rural agenda

Across much of the developed world, the three decades following the Second World War were characterised by a relatively high levels of state support for agriculture and rural industries more generally (Blandford and Hill, 2006). Agriculture existed under a protective mantle that typically included some combination of price support mechanisms for agricultural commodities, state investment in rural infrastructure, generous credit provisions, tax incentives, state-sponsored research and extension services, emergency assistance, tariffs and import restrictions. In broad economic management terms, agriculture was simply part of a Keynesian economic management 'consensus' in which the state took significant responsibility for ensuring economic stability and growth (Wilson, 2007). There was, of course, considerable diversity in how different nations provided support under this model of Keynsianism (Peck and Tickell, 2002).

Yet, support for rural industries was not simply linked to pragmatic concerns about economic management. In many rural areas, particularly in Europe, the United States and to a lesser extent Australia, political sensibilities were also critical in shaping rural policy. Various national and sub-national electoral configurations saw considerable power in the hands of rural voters which, in turn, served to promote policy frameworks that were supportive of agrarian interests (Brett, 2011; Woods, 2003). This was often justified on the grounds that agricultural industries were critical to national economic fortunes. Indeed, agriculture contributed around 40 per cent of all world merchandise trade in 1950 (GATT Secretariat, 1993), and in many countries was was critical to export earnings, public revenue, and domestic economic activity. For example, in Australia agricultural exports contributed more than 85 per cent of export earnings in the early 1950s and nearly a quarter of gross domestic product (Lawrence, 1987). This helped underpin the notion of 'countrymindedness'; a political logic that argued agriculture was more 'deserving' of support than urban industries on the basis of its economic contribution (Aitken, 1985).

Alongside the support provided to agriculture was a broader commitment to addressing spatially uneven development. Between the 1950s and 1980s, governments around the world invested heavily in services and infrastructure, and were particularly focused on narrowing social and economic disparities between rural and urban localities (Beer et al., 2005). While there were considerable variations in approaches within and between nations, there was nevertheless a consistent focus on improving health, education and social service outcomes (Furuseth, 1998). In addition, investments in infrastructure were a critical part of regional development strategies, particularly in terms of roads, rail, ports and telecommunication (Head, 1988).

By the 1980s, however, the tide had begun to shift. In the wake of the economic shocks of the 1970s, and the inability of policies based on Keynesian demand management to resolve the crisis, the principles of neoliberalism were viewed as central to unlocking economic growth and prosperity (Peck and Tickell, 2002). The focus on increasing competitiveness through market forces saw the systematic deregulation of national economic systems and a greater focus on free trade. The reduction in government spending in order to reduce the tax encumbrances on economies also resulted in major shifts in service and infrastructure delivery. In many cases, privatisation, service rationalisation, and even service withdrawal were common outcomes (Tonts and Jones, 1997). In the case of agriculture, the parallel 'forces' of neoliberalisation and globalisation placed pressure on traditional state support structures. As a result there was a gradual but inexorable liberalisation of domestic agricultural sectors. Subsidies, bounties, market support schemes, concessional financing programmes, and state-led research and extension programmes accreted during the post War era were stripped away (Gray and Lawrence, 2000).

Yet the changes experienced in agriculture were highly varied. This was particularly evident in the international debates about the role of agricultural protection through subsidies and import restrictions. During the 1980s the Cairns Group, a coalition of nineteen countries that were highly dependent on the export agricultural goods sought to exert international pressure to ensure that the liberalisation of agricultural trade remained a high priority in trade talks. This group of countries were active in opening up their own agricultural industries to competition through deregulation, and at the same time led the global critique of the trade 'distorting' impact of agricultural subsidies in the context of successive World Trade Organization-sponsored trade rounds (Higgot and Cooper, 1990). The Cairns Group countries advocated strongly for the reduction in protectionist policies (particularly farm subsidies) in Europe and North America, which they argued flew in the face of the principles of free trade (Higgins, 2014b).

Australia's Bureau of Agricultural Economics took the lead on behalf of the Cairns coalition, not only in compiling and analysing the technical details of the costs of subsidies and protection in agriculture, but also in widely publicising this information to marshal the arguments against the supposed short-term gains of protectionist policies as opposed to the long-term goal of a fairer trading system (Higgot and Cooper, 1990). Indeed, the research was utilised explicitly as a diplomatic tool both in the bilateral context and in the multilateral arena. Following the Australian initiative, the Organisation for Economic Co-operation and Development (OECD) took steps to institutionalise the concept of producer subsidy equivalents and use it as the index of government support for agriculture (Higgot and Cooper, 1990).

However, even among countries and regions that have otherwise fully subscribed to the general neoliberal project, such as the UK, Europe and the United States, the decrease in agricultural protection policies has not been followed through consistently. The agricultural sectors of Australia and New Zealand 'led the charge' in reducing subsidies, and continue to be among the least assisted of the member countries of the OECD. As part of the steady decline, for example, in 2016 their aggregate agricultural producer support estimates were, respectively, 1.95 per cent and 0.86 per cent, compared with the OECD average of 18.8 per cent (OECD, 2018). This is in contrast to the European Union (EU) and the United States where state assistance remains high at 21 per cent and 8.7 per cent, respectively. In the EU, an exceptionalist case continues to be made that the provision of state assistance to economically marginal producers is needed to retain a countryside occupied and managed by family farmers (Higgins, 2014a). This defence of government subsidies is partly based on the argument that agriculture has important environmental and social as well as economic functions that need to be underwritten by the state (Dibden et al., 2009).

In reality, however, this claim regarding agriculture's wider values represents a complex set of political and social dynamics. While the Cairns Group countries argued in favour of a neoliberal agenda on the grounds of industry efficiency and competitiveness, in Europe and the United States, distinctive agrarian cultures and political forces saw arguments emerge around the need to protect heritage, landscape attributes and a 'way of life' (Bunce, 1994). Moreover, the political implications of a radical shift in agricultural policy in the United States and France in particular meant that the application of neoliberal principles in agriculture would be mediated by the threat of electoral backlash.

In many respects, this provides a rural counterpart to Peck et al.'s (2013: 1093) observation that 'cities . . . are not merely at the "receiving end" of neoliberalism, imposed unilaterally from above'. The authors go on to note that processes of neoliberalism are actively consistuted and contested at a range of scales. In the context of rural environments, the arguments of farming interests in Europe and the United States might be constructed as a form of resistance to neoliberalisation, given the defence of income support and environmental protection within economically marginal rural areas (Higgins, 2014a). Indeed, critics have argued persistently that this policy position is a thinly veiled form of protectionism for European and American agriculture (Dibden et al., 2009).

There is considerable evidence to suggest that the impact of neoliberal policies in rural regions have been significant at the local level (Argent, 2011). In those countries where agriculture experienced rapid deregulation and the reductions in support, rates of farm amalgamation and expansion increased rapidly as farmers sought to increase economies of scale (Adams et al., 2014; Wilson, 2007). Indeed, in countries like Australia government schemes were put in place to 'adjust' smaller farmers out of the sector to increase opportunities for larger operators to expand. Farm debt often spiralled as farmers sought to expand and invest in new technologies or expand their holdings (Smailes, 1996). The expansion of farms typically led to outmigration by farm families, contributing to what Sorensen (1993) described as a vicious cycle of decline for rural communities. The loss of farm families typically reduced expenditure in small towns and resulted in a contraction of local economic activity, job losses and further outmigration (Argent and Tonts, 2015). At the same time, services and infrastructure came under increasing pressure as part of a general shift towards a more 'economically rational' allocation of services that took less account of issues associated with socio-spatial equity and focussed on the financially efficient allocation of resources (Beer et al., 2005; Tonts and Jones, 1997).

There is of course an important caveat to the above argument, in that there was evidence that the declining global terms of trade for agriculture had been leading to adjustment across the farm sector well before the 1980s (Blandford and Hill, 2006). Cost-price pressures had seen farm amalgamations and investments in new labour saving technology increase rapidly from the 1960s (Smailes, 1996). Yet, it is clear that neoliberalism added a new set of pressures that intensified these processes (Dibden et al., 2009). It is also important to emphasise that the experience and outcomes of neoliberalism were not uniform. Indeed, there is also evidence that increasing competitive pressures resulted in increased farm productivity through the introduction of new technology, new farm and environmental management practices, commodity diversification, and increasing off-farm income (Tamásy and Diez, 2016; Blandford and Hill, 2006). Moreover, a more diverse rurality is emerging as regions adapt to a policy environment that favours competition, adaptability and integration with global economic processes.

Neoliberal ruralities and the 'global countryside'

Much of the recent debate on neoliberalism in the context of rural space has focussed on the increasing interaction between assemblages of broad global processes and more localised settings. Of particular interest is the ways in which the neoliberalism and globalisation have combined in ways that have contributed to an increasingly 'differentiated countryside' comprising new hybrid and dynamic production and consumption ruralities (Marsden et al., 1993). In contrast to the prevailing wisdom of the late 1980s that rural restructuring would lead to the creation of homogenised landscape of 'factory farms' and a deep social malaise across the countryside (Lawrence, 1987), the reality has been quite different. In integrating rural economies into wider and increasingly complex global circuits of capital, neoliberalism has contributed to a fundamental reshaping of territorial development.

In 2007, Michael Woods posited the notion of the 'global countryside' as a hypothetical space within which globalising tendencies are fully realised in the transformation of rural place (Woods, 2007; see also Chapter 53 of this Companion). Rather than viewing rural change as being 'determined' by neoliberalism and globalisation, Woods (2007, 2011) draws on a relational perspective to develop insights into the remaking of rural places within the context of these intertwined processes. Importantly, Woods's (2007) notion of the global countryside is a heuristic device for exploring and investigating the realisation of global 'tendencies' in rural space; that is, as a framework for identifying the partial and hybrid articulation of these globalising tendencies in real, present-day rural localities (Box 10.1).

Box 10.1   The global countryside: defining characteristics

1   Primary and secondary sectors inserted into spatially extensive commodity networks.

2   Sites of 'increasing corporate concentration and integration, with corporate networks organised on a transnational scale' (Woods, 2007: 492).

3   Both a supplier and an employer of international migrant labour.

4   Select high amenity sites attract flows of the globally mobile population.

5   Host to expanding and expansive flows of foreign direct investment, including into local and regional property markets.

6   Established local approaches to human/non-human relations and their management increasingly challenged by global discourses.

---

7   Landscapes bear the material imprint of globalisation processes.

8   Increasing social polarisation at the local scale due to the combined effect of 1–4.

9   Prone to the effects of re-scaling and re-siting of political authority and governance.

10 As a corollary of 8 and 9, increasingly a site of contestation.

Source: Woods (2007) as summarised in Argent and Tonts (2015: 142)

As Argent and Tonts (2015) point out, how the characteristics summarised in Box 10.1 are expressed in rural space is largely dependent on highly contextual local and regional histories and geographies. Not surprisingly, therefore, there has been considerable interest in understanding precisely how, and to what extent, these globalising tendencies are actually experienced and mediated in different rural contexts (see Tamásy and Diez, 2016; Šťastná et al., 2013; Argent, 2011; Lockie and Higgins, 2007). Collectively this body of work provides a rich set of insights into the increasing diversity of experiences of globalisation and neoliberalism as expressed in land use, economic activity, demographic change, rural politics and socio-economic conditions. One means of making some sense of this diversity is through Holmes's (2006) notion of the multifunctional rural transition. Holmes suggests that the increasingly differentiated, or multi-functional, countryside is characterised by the intersecting forces of protection, consumption and production.

Protection in this context refers to the growing societal concern for ethical and sustainable management of natural and other rural resources. In this context, there is a strong alignment with aspects of roll-out neoliberalism, with the focus on farmer-led initiatives on sustainable agriculture, community based environmental management, and regulatory frameworks emphasising ethical treatment of animals (Argent and Tonts, 2015; Lockie and Higgins, 2007). Production, of course, draws attention to the ongoing importance of food and fibre production in the context of a globalised and liberalised agriculture. Consumption draws attention to the growing commodification of rural landscapes and heritage (Taylor and Tonts, 2016). This has been particularly important in driving amenity migration into selected rural areas and the growth of tourism and leisure economies (Perkins et al., 2015). As Holmes (2006) points out, these forces rarely play out in isolation, and are often overlapping and competing. Further, as Marsden et al. (1993) noted more than two decades ago, the transformation of the countryside under conditions of globalisation and neoliberalism are making it not only increasingly differentiated, but also highly contested.

The notion of an increasingly contested rural space/place is indeed critical to Woods's (2007) conceptualisation of the global countryside. Increasing integration into global commodity networks, the rising presence of corporate interests, social polarisation and amenity migration have been accompanied by rescaling of and re-siting of political scale and governance (Argent, 2011). This is consistent with the neoliberal 'hollowing out' of state territoriality (Brenner, 1999), or what is sometimes referred to as scaling down, that is, a shift toward decision-making at more localised governance scales, and scaling out, that is, increased participation of extra-governmental actors in decision-making (Cohen and McCarthy, 2015).

This transformation in established relationships between local communities and relevant tiers of government, together with the private and non-governmental sectors, manifests in a variety of forms and combinations across space, within the broader spatial unevenness of the shift from a productivist structured coherence to a multifunctional agricultural regime (Wilson, 2007; Holmes, 2006). The upbeat discourse of 'scaled down' governance in neoliberalism places emphasis on local community empowerment, partnership, capacity building and social capital.

Together with the production imperatives of farmers and local communities, who are increasingly immersed in globalised economic relations, this new form of localised governance and regulation is arguably aimed at developing the entrepreneurial conduct of subjects. Farmers are expected to become entrepreneurial and 'active' agents who improve their productivity and competitiveness without government interference. On the other hand, however, with the scaling back of government intervention and regulation, farmers are often expected to place community interests before their own by providing long-term social and environmental benefits. This is particularly evident in terms of the growing expectations that farmers lead initiatives related to environmental stewardship, landscape preservation, and ecosystem rehabilitation (Dibden et al., 2009).

Beyond the local scaling down, non-government organisations representing a wide variety of sectoral interests have similarly risen to prominence in rural affairs as part of a scaling out process. Across a range of policy domains, including environmental management, regional economic development and social services, an increasingly complex array of non-government actors are now enmeshed in rural governance (Argent, 2011). This scaling out process has added diversity and in some cases dissonance to rural policy and planning as a broader array of social values, cultures and political sensibilities become engaged in debates about the future of the countryside (Dibden et al., 2009; Tonts and Greive, 2002; Murdoch and Marsden, 1994; Marsden et al., 1993).

Rural development and neoliberalism

The rescaling of political authority and governance within the context of the global countryside is particularly significant for rural development policy and planning. The global countryside is characterised by a new, highly variegated geography of economic activity, settlement and land use, producing many different 'rurals' each with varying development capacities. Moreover, neoliberal policy approaches themselves have continued to adapt and evolve, sometimes in response to their unintended or more socially contentious outcomes (Tonts and Haslam-McKenzie, 2005). This has included a re-engagement by the state, albeit on a limited scale, in regional development, service provision and infrastructure development. The approach has typically remained commited to neoliberal principles, emphasising the importance of civic responsibility and self-help. Consistent with the notions of 'scaling down' and 'scaling out' there is an emphasis on local responses and the engagement of a broad array of actors across business, non-government organisations, and government agencies (Beer et al., 2005).

This emerging paradigm is particularly evident in the ways in which decisionmakers are now approaching the question of rural policy planning (see Ward and Brown, 2009). While earlier models of rural development favoured state support for agriculture through various macro-economic policy interventions and regional land-use regulation, newer perspectives emphasise strategic investment across multiple sectors and engagement of multiple actors (see Table 10.1). In line with the principles of neoliberalism, central to contemporary rural development is identifying existing or potential local competitive advantages. These competitive advantages might include intensive agricultural production, environmental amenities, cultural industries, niche manufacturing, or high value food processing. Rather than direct state support for the development of these advantages, attention is increasingly placed on building local leadership and entrepreneurial capacity, private sector investment, and in some cases public–private partnerships.

The emphasis on territorial approaches to development are driving increasing economic, social and land-use heterogeneity across the countryside, consistent with Holmes's (2006)

Table 10.1 The new rural development paradigm.

	State support paradigm	Neoliberal development paradigm
Objective	Farm productivity, stable farm incomes, spatial equity, coordinated and centralised policymaking	Rural competitiveness, valorisation of local assets, exploitation of underutilised resources
Target sectors	Agriculture and other natural resources	Diversity of sectors (manufacturing, tourism, recreation, agriculture, creative industries etc.)
Policy tools	Subsidies, state expenditure, horizontal fiscal equalisation	Private sector investment, lightly funded government supported capacity building, public–private partnerships
Key actors	National and state/provincial governments, farmer interest groups	Diversity of stakeholders across all levels of government, non-government organisations, and private capital

Source: adapted from Ward and Brown (2009)

notion of a multifunctional transition. Rural policy and planning now centres on identifying and developing tailored place-based strategies that seek to maximise competitive advantages and integrate rural places into global circuits of capital. While the specific policy and planning strategies are diverse, Woods (2011) has identified eight broad responses:

- *Global resource providers* – rural localities that are able to capitalise on mineral and energy resources that are required as raw materials by the global economy. The policy challenge in this context is ensuring local economic benefits and the social dislocation often associated with resource economies is minimised.
- *Branch plant economies* – rural localities that are engaged in attracting inward investment from national and/or transnational firms. Regional policy in these circumstances is oriented towards investment attraction, maintaining local competitiveness, and ensuring local endogenous business development.
- *Global playgrounds* – rural localities with particularly high amenity-values with capacity to attract international tourists and amenity migrants. Key policy considerations include protecting amenity assets from excessive development, land-use conflicts, and the potential for rising social inequality.
- *Niche innovators* – rural localities that are able to engage with the global economy by producing and exporting goods to serve global niche markets, or attracting international visitors to niche events and attractions.
- *Trans-border networkers* – intensification of networking with neighbouring regions across conventional national borders, which can serve to subvert established core-periphery dynamics.
- *Global conservators* – rural localities that have been integrated – willingly or unwillingly – into international structures for nature conservation and environmental protection, informed by global environmental discourses.
- *Re-localisers* – rural localities reaffirming and consolidating local aspects of economic and social life.

- *Structurally marginalised regions* – rural localities where there is limited international activity by local businesses and which are net exporters of labour to the global economy, with precarious economies and are vulnerability to market and policy changes; individual entrepreneurs and agencies may attempt to engage with global economic networks, but without particular success, frustrated by structural factors such as geographical location, the absence of valuable natural resources, and political-economic culture.

The responses outlined above are not mutually exclusive to a single region, and indeed multiple responses can often be found within a region, sometimes complementing each other, at other times in conflict with each other. Moreover, as regions stand in quite diverse relations to 'the global', so the capacity to shape future development will vary. Many rural regions have found that their opportunities are constrained to a greater or lesser degree by structural factors, from the presence of natural resources, to geographic location, to deep path dependent histories of economic and social marginalisation. Indeed, the focus on localised responses also carries with it the danger that development outcomes can be highly uneven depending on business and political acumen, availability of local resources, human capital and socio-economic conditions (Dibden et al., 2009). Thus, extant patterns of uneven development may be reinforced or reconsistuted within this neoliberal development paradigm.

Conclusion

The reconstitution of rural places under 'neoliberalised globalisation' is complex, non-linear and spatially contingent. Indeed, as Woods (2007: 495) has suggested 'There is no pre-existent stable and uniform rural place upon which "globalisation" can act, but then neither is there a single, unidirectional force of globalisation' (Woods, 2007: 495). Neoliberalism, as a key driver of globalisation, proceeds by hybridisation, fusing and mingling the local and the extra-local to produce new formations. Neoliberalism has both 'rolled back' much of the Keynesian statist approach to regional development practised by federal and state governments and subsequently 'rolled out' a new model of so-called locally led, bottom-up entrepreneurialism and community development as the panacea to regional inequality (Beer et al., 2005) and underdevelopment (Argent, 2011). Building on Polanyi's analysis of liberalism, several scholars writing from a political economy perspective have argued that neoliberalism, like nineteenth century liberal forms of capitalism, is characterised by a 'double movement' in which accelerating social and environmental degradation produces social resistance to market liberalisation: thus 'the immersion of all things into the marketplace [is] countered by predictable calls for regulation and restraint' (McCarthy, 2004: 335). Yet, rather than any significant retreat from the principles of neoliberalism there has been an ongoing process of adaptation and reinvention that continues to emphasise the underlying libertarian principles. Within this unfolding geography of neoliberalism, numerous different constructions of the relationship between markets, agriculture, rural society and the environment have emerged, adding yet more complexity and unevenness to the global countryside.

References

Adams, M., Brown, N. and Wickes, R. 2014. *Trading Nation: Advancing Australia's Interests in World Markets*. Sydney: University of New South Wales Press.

Aitken, D. 1985. Countrymindedness: the spread of an idea. *Australian Cultural History* 4: 34–41.

Argent, N. 2011. Australian agriculture in the global economic mosaic. In M. Tonts and M.A. Siddique (eds), *Globalisation, Agriculture and Development: Perspectives from the Asia-Pacific*, 7–28. Cheltenham: Edward Elgar.

Argent, N. and Tonts, M. 2015. A multicultural and multifunctional countryside? International labour migration and Australia's productivist heartlands. *Population Space and Place* 21: 140–156.

Beer, A., Clower, T., Haughtow, G. and Maude, A. 2005. Neoliberalism and the institutions for regional development in Australia. *Geographical Research* 43(1): 49–58.

Blandford, D., and Hill, B. 2006. *Policy Reform and Adjustment in the Agricultural Sectors of Developed Countries: From Pioneer to Policy*. Wallingford: CABI.

Brenner, N. 1999. Globalisation as reterritorialisation: the re-scaling of urban governance in the European Union. *Urban Studies* 36(3): 431–451.

Brenner, N. and Theodore, N. 2002. Cities and the geographies of actually existing neoliberalisms. In N. Brenner and N. Theodore (eds), *Spaces of Neoliberalism*, 2–33. Oxford: Blackwell Publishing.

Brett, J. 2011. Fair share: country and city in Australia. *QE* 42: 1–67.

Bunce, M. 1994. *The Countryside Ideal: Anglo-American Images of Landscape*. London: Routledge.

Cohen, A. and McCarthy, J. 2015. Reviewing rescaling: strengthening the case for environmental considerations. *Progress in Human Geography* 39(1): 3–25.

Dibden, J., Potter, C. and Cocklin, C. 2009. Contesting the neoliberal project for agriculture: productivist and multifunctional trajectories in the European Union and Australia. *Journal of Rural Studies* 25: 299–308.

Furuseth, O. 1998. Service provision and social deprivation. In B. Ilbery (ed.), *The Geography of Rural Change*, 233–256. Harlow: Longman.

GATT Secretariat 1993. *International Trade Statistics 91–92: Statistics*. Geneva: General Agreement on Tariffs and Trade Secretariat.

Goodman, D. and Watts, M. 1997. *Globalizing Food: Agrarian Questions and Global Restructuring*. London: Routledge.

Gray, I. and Lawrence, G. 2000. *A Future of Regional Australia: Escaping Global Misfortune*. Cambridge: Cambridge University Press.

Head, B. 1988. The Labor government and economic rationalism. *Australian Quarterly* 60: 466–477.

Higgins, V. 2014a. Neoliberalising rural environments. *Journal of Rural Studies* 36: 386–390.

Higgins, V. 2014b. Australia's developmental trajectory: neoliberal or not? *Dialogues in Human Geography* 4(2): 161–164.

Higgot, R. and Cooper, A. 1990. Middle power leadership and coalition building: Australia, the Cairns Group, and the Uruguay round of trade negotiations. *International Organization* 44(4): 589–632.

Holmes, J. 2006. Impulses towards a multifunctional transition in rural Australia: gaps in the research agenda. *Journal of Rural Studies* 22: 142–160.

Lawrence, G. 1987. *Capitalism and the Countryside: The Rural Crisis in Australia*. Sydney: Pluto Press.

Lockie, S. and Higgins, V. 2007. Roll-out neoliberalism and hybrid practices of regulation in Australian agri-environmental governance. *Journal of Rural Studies* 23: 1–11.

Marsden, T., Murdoch, J., Lowe, P., Munton, R. and Flynn, A. 1993. *Constructing the Countryside*. London: UCL Press.

McCarthy, J. 2004. Privatizing conditions of production: trade agreements as neoliberal environmental governance. *Geoforum* 35(3): 327–341.

McMichael, P. 1994. *The Global Restructuring of Agro-food Systems*. Ithaca, NY: Cornell University Press.

Murdoch, J. and Marsden, T. 1994. *Reconstituting Rurality: Class, Community and Power in the Development Process*. London: University College London Press.

OECD. 2018. Agricultural support. Retrieved from https://data.oecd.org/agrpolicy/agricultural-support.htm (accessed 25 February 2018).

Peck, J., Theodore, N. and Brenner, N. 2013. Neoliberal urbanism redux? *International Journal of Urban and Regional Research* 37(3): 1091–1099.

Peck, J. and Tickell, A. 2002. Neoliberalising space. In N. Brenner and N. Theodore (eds), *Spaces of Neoliberalism*. Oxford: Blackwell.

Perkins, H.C., Mackay, M. and Espiner, S. 2015. Putting pinot alongside merino in Cromwell District, Central Otago, New Zealand: rural amenity and the making of the global countryside, *Journal of Rural Studies* 39: 85–98.

Smailes, P. 1996. Entrenched farm indebtedness and the process of agrarian change. In D. Burch, R. Rickson and G. Lawrence (eds), *Globalization and Agri-food Restructuring: Perspectives from the Australasia Region*, 301–322. Aldershot: Avebury.

Sorensen, A. 1993. The future of the country town: strategies for local economic development. In A. Sorsensen and R. Epps (eds), *Prospects and Policies for Rural Australia*, 201–240. Melbourne: Longman Cheshire.

Šťastná, M., Toman, F., Vaishar, A., Woods, M. and Vavrouchová, H. 2013. European rural regions in the era of globalization: the South Moravia case study. *Innovation: The European Journal of Social Science Research* 26(4): 354–364.

Tamásy, C. and Diez, J. (eds). 2016. *Regional Resilience, Economy and Society: Globalising Rural Places.* London: Routledge.

Taylor, M. and Tonts, M. 2016. Agriculture in chains: farms, firms and contracts. In C. Tamásy and J. Diez (eds), *Regional Resilience, Economy and Society: Globalising Rural Places.* London: Routledge.

Tonts, M. and Greive, S. 2002. Commodification and creative destruction in the Australian rural landscape: the case of Bridgetown, Western Australia. *Australian Geographical Studies* 40(1): 58–70.

Tonts, M. and Haslam-McKenzie, F. 2005. Neoliberalism and changing regional policy in Australia. *International Planning Studies* 10: 183–200.

Tonts, M. and Jones, R. 1997. From state paternalism to neoliberalism in Australian rural policy. *Space and Polity* 1: 171–190.

Ward, N. and Brown, D. 2009. Placing the rural in regional development. *Regional Studies* 43(10): 1237–1244.

Wilson, G. 2007. *Multifunctional Agriculture: A Transition Theory Perspective.* Wallingford: CABI.

Wilson, G. 2008. From 'weak' to 'strong' multifunctionality: conceptualising farm-level multifunctional transitional pathways. *Journal of Rural Studies* 24: 367–383.

Woods, M. 2003. Deconstructing rural protest: the emergence of a new social movement. *Journal of Rural Studies* 19(3): 309–325.

Woods, M. 2007. Engaging the global countryside: globalization, hybridity and the reconstitution of rural place. *Progress in Human Geography* 31(4): 485–507.

Woods, M. 2011. Regions engaging in globalisation: a typology of regional responses in rural Europe. Paper presented to the Anglo-American-Canadian Rural Geographers Quadrennial Conference, July, Manitoba.

11

Market-based instruments and rural planning in America

Tom Daniels

Introduction

Rural planning in the United States has featured efforts by the federal government, individual states, specific regions, local governments, and non-governmental organisations. The federal government owns and manages about 240 million hectares of the nation's nearly 900 million hectares (Daniels, 2014). The large majority of this land is in the western states and includes national parks, national forests, wilderness areas, rangeland, desert, and wildlife refuges. About 60 per cent of the United States is privately owned. Farmers and ranchers own 366 million hectares or slightly more than half of all private land (US Department of Agriculture, 2014). Private forest landowners hold about 170 million hectares (Daniels, 2014).

Rural planning for private land varies considerably among the 50 states. Some states, such as Maryland and Oregon, have state level planning agencies and state plans aimed at requiring local governments to adopt plans that direct growth and protect working farm and forest landscapes. In general, however, county and municipal governments in America have the primary authority for rural planning and land-use regulation. A few fragile ecological regions have their own planning and regulatory authority including the Adirondack Park of New York State, the New Jersey Pinelands, and greater Lake Tahoe in both California and Nevada. A wide range of rural planning and regulatory efforts exists among America's local governments. Some counties, as in California and Oregon, have used large minimum lot size zoning (e.g. one house per 16 hectares to 32 hectares) to protect private farmland and forests. On the other extreme, in Texas, counties have no zoning authority. Furthermore, there are more than 500 private non-profit land trusts that list rural land protection as a major part of their mission (Land Trust Alliance, 2011). Although private land trusts do not have regulatory powers or create public land-use plans, they do undertake internal planning aimed at keeping certain lands undeveloped as natural areas, working forests, or active farmland. Land trusts offer landowners market-based financial incentives either alone or jointly with government agencies to protect rural lands.

America's rural planning and regulation are generally weak. Private land in rural areas is usually a household's largest asset, and attempts to reduce the value of that asset through land-use regulations are typically resisted with fervour. Regulations, such as zoning, are legitimate if the land subject to the regulations still has a reasonable economic use. However, a regulation can 'go too far' and result in the 'taking' of private property if the regulation removes all reasonable economic use. In such a case, the government would have to pay compensation to the landowner or rescind the regulation. This tension between government regulation as a potential taking of private property and rural landowners' resistance to regulation has compelled governments to seek market-based financial incentives to shape landowner behaviour for public benefit. A second reason for financial incentives is to stem the loss of farm and forest lands to higher valued commercial and residential development. For example, recently, farmland has been converted to other uses at a rate of about 800,000 acres a year (NRCS, 2015), and about half of this land is prime farmland.

What are market-based instruments?

Market-based instruments are financial incentives that governments and non-governmental organisations (NGOs) offer rural landowners to induce desirable land-use outcomes. A market-based instrument differs from a subsidy in that the landowner places greater restrictions on the use of his or her property in return for a cash payment or tax savings. Also, market-based instruments are voluntary: landowners decide which, if any, market-based instruments to employ. The success of any one instrument depends on how many landowners participate. Adequate government and/or private sector funding sources are also crucial for the success of market-based instruments. If funding is low or non-existent, then few landowners will be able to participate in market-based programmes. Market-based instruments in America include the purchase or donation of conservation easements, the transfer of development rights, and payments for environmental services, discussed in the following sections.

Purchase or donation of conservation easements

American farmers and foresters face three main challenges: (1) earning a living from the land; (2) passing the farm or forest to the next generation; and (3) resisting the temptation to sell land for house lots and commercial sites. For example, the average age of farmland owners is 58 years old (US Department of Agriculture, 2014). Tens of millions of hectares of agricultural land will change hands over the next few decades, which could impact farming across the United States.

A market-based incentive known as the purchase or donation of conservation easements can aid farmers and foresters by providing needed capital to strengthen their operations, providing capital for a retirement nest egg which can facilitate the transfer of the farm or forest to the next generation, and offering an alternative to selling land for development. Other rural landowners have property with important ecological values, such as wildlife habitat, water re-charge areas, riparian buffers, scenic vistas, and recreation opportunities. Many of these landowners want to keep the land in their family and do not want to see it developed. The purchase or donation of a conservation easement can provide cash and/or tax benefits to enable a family to keep the property intact and provide an attractive option instead of selling land for development.

A landowner in America legally owns a bundle of rights to the land. These include water rights, mineral rights, air rights, the right to pass the land to heirs, the right to use the land, the right to lease or sell the land, and the right to develop the land. Each right in this bundle can be separately severed and sold or donated. For example, rural landowners have often sold mineral rights to energy and mining companies to allow these companies to search for oil, gas, coal, and minerals underneath the property. The purchase or donation of a conservation easement is a voluntary act in which a landowner severs the right to develop the property and retires it through a legally binding document known as a Deed of Easement. The Deed of Easement restricts the uses of the land to farming, forestry, natural areas, or open space. The landowner may choose to sell or donate a conservation easement to a government agency or to a qualified private, non-profit land trust. Both the landowner and the government agency or land trust sign the Deed of Easement, which is then recorded in the land records at the local county court-house. The conservation easement runs with the land, so if the land is sold or passed on to heirs, the land-use restrictions of the conservation easement continue to apply all future landowners.

A conservation easement may exist for a specified number of years, known as a term ease-ment, or in perpetuity. The large majority of the more than 113,000 conservation easements created so far are perpetual, and must be perpetual for the donation of a conservation to be eligible for a federal income tax or estate tax deduction (IRC, 2017). The land remains private property, usually with no right of public access, and the landowner may sell the property, sub-ject to the conservation easement. The holder of the conservation easement, whether a govern-ment agency or a private land trust, must monitor and inspect the property (usually once a year) to ensure that the landowner is meeting the terms of the conservation easement. The easement holder must also enforce the easement to protect the conservation values of the property.

The value of a conservation easement is determined by an appraisal conducted by a quali-fied appraiser. The appraiser first estimates the fair market value of the property, based on sales of comparable properties in the area. Then, the appraiser determines the value of the property restricted by a conservation easement to farming, forestry, or open space, based on comparable sales or an income approach. The difference between the fair market value and the restricted value of the property is the value of the conservation easement (see Table 11.1).

The landowner in Table 11.1 may have the choice of selling the conservation easement for cash, donating the conservation easement to receive income tax and/or estate tax benefits, or completing a bargain sale easement of part cash and part donation with either a land trust or a government agency. The advantage of the sale of a conservation easement is that it is taxed as a capital gain, at a lower rate than as ordinary income, and the income tax benefits of a dona-tion of a conservation easement depend on the income of the landowner (Daniels and Bowers, 1997). Many rural landowners are land rich and cash poor with relatively low annual income. The donation of a conservation easement is much less attractive to them than the sale of a con-servation easement. For example, if in the Table 11.1 example the landowner is a farmer and earns $50,000 a year in the year of the donation and for each of the next six years, the value of the income tax benefits is $31,500. By contrast, the return on the sale of the conservation ease-ment, even after paying a 15 per cent capital gains tax, is $297,500. For a bargain sale of the con-servation easement, let's assume that the farmer receives $150,000 in cash and makes a $200,000 donation. In this case, the donation offsets the cash portion of the bargain sale as well as farmer's $50,000 annual income, so the farmer owes no federal tax for one year and has $150,000 in cash.

Conservation easements have proven to be durable, with fewer than 20 being overturned by the courts (Anella and Wright, 2004). A government agency may physically take land

Table 11.1 Conservation easement value example.

150-acre farm – appraisal	Value
Fair market value	$950,000
Restricted value	$600,000
Conservation easement value	$350,000

subject to a conservation easement for a public purpose, but must compensate the landowner and easement holder. Private land trusts have acquired conservation easements on 6.7 million hectares, mainly in rural areas (Land Trust Alliance, 2016). Federal, state, and local governments have acquired conservation easements on more than 4.3 million hectares of rural land. These include about 1.4 million hectares of agricultural land (NRCS, 2017a, 2017b; American Farmland Trust, 2016a, 2016b), 1.1 million hectares of forest land (US Forest Service, 2016), and 1.8 million hectares of wetlands (NRCS, 2017c).

The purchase of conservation easements has been popular in the Northeastern states, California, Colorado, and Ohio. Although 28 states have created purchase of conservation easement programmes for farmland, only seven states – Colorado, Delaware, Maryland, New Jersey, New York, Pennsylvania and Vermont – have each acquired conservation easements on more than 40,000 hectares (see Table 11.2). Land trusts have been successful in acquiring conservation easements on more than one million hectares of agricultural land, especially on nearly 900,000 hectares of western ranchlands in California, Colorado, Kansas, Montana, Oregon, Texas, and Wyoming. Much of the forest land preserved through conservation easements has been in the Northeastern states, Upper Midwest, and West (see Table 11.3).

Since 1996, the federal government has made $2 billion in grants to state and local governments and land trusts to acquire conservation easements on more than 500,000 hectares of agricultural land (NRCS, 2012, 2017b). Since 1990, the federal Forest Legacy Program has made $756 million in grants for the acquisition of conservation easements on slightly more than one million hectares of forest land (US Forest Service, 2016); and since 1985, the federal government has spent about $3 billion for conservation easements to protect almost 1.2 million hectares of wetlands (NRCS, 2017b, 2017c, 2017d). The federal funding has become increasingly

Table 11.2 Leading states in farmland under conservation easements, 2016.

State	Area of land protected in hectares	Cost
Colorado	732,919	$241,691,708
Pennsylvania	555,894	$992,394,321
Maryland	528,608	$1,389,360,688
New Jersey	231,935	$1,187,875,080
Vermont	153,034	$73,545,140
Delaware	116,223	$150,104,160
New York	100,117	$678,383,309

Source: American Farmland Trust (2016a)

Table 11.3 Leading states in forestland under conservation easements, 2016.

State	Area of land protected in hectares	Cost
Maine	296,135	$141,545,681
New Hampshire	105,882	$65,583,841
Montana	89,514	$120,211,326
New York	79,283	$54,284,826
Michigan	62,618	$57,370,187
Minnesota	58,973	$23,463,222
California	45,096	$98,908,425

Source: US Forest Service (2016)

important because of budget cuts that many state and local government programmes suffered during and after the recession of 2008–2010.

The strength of conservation easements is that they offer landowners a way to get cash or tax benefits out of the land without having to sell any land. The landowner can use the financial benefits to re-invest in a farm or forestry operation or maintain a family property. The sale of a conservation easement can enable older landowners to create a retirement nest egg and then sell the land subject to the conservation easement at a reduced price to a family member. The more landowners in an area who participate in selling or donating conservation easements, the more likely the area will remain in farming, forestry, or open space uses. The purchase of conservation easements typically costs much less to a government agency or a private land trust that the outright purchase of the land. Also, private landowners bear the primary burden for managing the property. For example, in some states and with any federally funded conservation easement, a landowner is required to have a soil and water conservation plan on the property at the time of the sale of the conservation easement. Finally, conservation easements can enable governments to better direct future growth, protect environmental quality, and maintain working rural lands.

The shortcomings of acquiring conservation easements are: (1) the purchase of conservation easements can be expensive, as much as 90 per cent of the fair market value, and more than $25,000 per hectare; (2) public and private funds may not be available to purchase conservation easements; (3) the donation of a conservation easement does not provide attractive income tax benefits to most farmers and foresters who have relatively low incomes; (4) if only a small number of landowners sell or donate conservation easements, the protected land may be scattered. In rural areas with weak zoning, protected lands can attract developers who want to build next to a 'permanent view' which can cause conflicts between newcomers and neighbouring farm and forestry operations; (5) in remote rural areas, conservation easements are likely to have little value, a reflection of the limited development potential; and (6) selling a conservation easement takes time. For instance, several state and local farmland protection programmes have a long backlog of applicants interested in selling a conservation easement. If funding is not adequate and the wait is many years, some applicants may decide to pursue development options instead.

Transfer of development rights

The transfer of development rights is a market-based tool that local governments in the United States have used to protect rural land while also promoting development. A transferable development right is not included in a landowner's bundle of rights. Instead, a local government creates transferable development rights (TDRs), based on state enabling legislation. TDRs are permitted in 25 states, mainly in the Northeast, West Coast, Midwest and Florida (Nelson et al., 2012).

To create a TDR programme, a county, city, or township government must first revise their comprehensive plan to designate one or more sending areas (such as agricultural lands) that the local government would like to protect and to identify one or more receiving areas where the local government would like to encourage more intensive development. The local government then gives TDRs to landowners in the sending area(s), such as one TDR per two hectares, and requires developers in the receiving areas to purchase TDRs to build at a higher density than normally allowed in the local zoning regulations, such as one TDR for each additional dwelling unit in a residential zone. When a landowner in a sending area sells TDRs, the local government places and holds a permanent conservation easement on the land. No additional development is permitted. TDRs, like any market-based instrument, are voluntary and depend on willing buyers and sellers. Landowners are not required to sell their TDRs and developers are not compelled to purchase TDRs.

The local government creates a market in TDRs. The overall supply of TDRs and the demand for TDRs determine the price. Thus, if the supply of TDRs is large relative to the demand for TDRs, the price will tend to be low. If there are few TDRs available relative to the demand for TDRs, then the price will tend to be high. Yet, as in any real estate transaction, a seller of TDRs (a landowner in a sending area) and a buyer (a developer in a receiving area) negotiate the price of a TDR. To add liquidity, some local governments have established a TDR bank to enable the local government to purchase TDRs from willing sellers in the sending areas and then sell them to buyers for use in the receiving areas. A TDR bank can also set a floor price for TDRs.

There are two types of TDR programmes: voluntary and mandatory (Johnston and Madison, 1997). In a voluntary TDR programme, the landowner can decide how much of his or her property to develop under the zoning regulations and whether to sell no TDRs, some TDRs, or all of the TDRs. For example, a landowner owns a 40-hectare farm; the property is zoned at one house per 2 hectares; and the landowner has 20 TDRs. The landowner could sell 15 TDRs (protecting 30 hectares) and then subdivide the remaining property into five lots of 2-hectares each. A leading voluntary TDR programme is Montgomery County, Maryland where nearly 20,000 hectares of agricultural land have been protected through TDRs (Daniels, 2017).

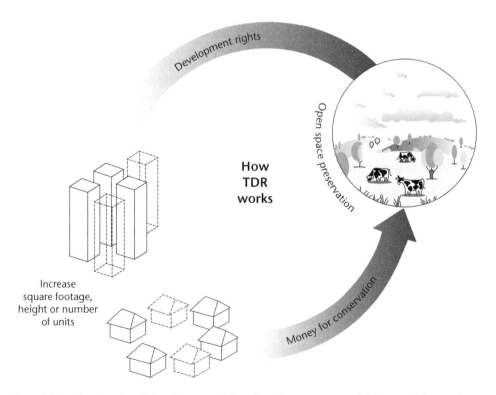

Figure 11.1   The transfer of development rights. Development potential is moved from a farm in the TDR sending area to a development site in the TDR receiving area.

Source: King County (2017); used with permission

In a mandatory TDR programme, the landowners in the sending area decide whether or not to sell TDRs, but they cannot subdivide lots for residential or commercial development. The landowners can sell TDRs instead of developing their property, and buyers of TDRs can use the TDRs in separate designated receiving area where growth is desired. The TDR programme of the Tahoe Regional Planning Agency (in California and Nevada) is a leading example of a mandatory TDR programme (Barrett, 2011). Since the late 1960s, about 75 local governments have created TDR programmes to protect agricultural or forest land, with thousands of TDR transactions and more than 60,000 hectares of agricultural land protected (Nelson et al., 2012). TDRs have been used less to protect forestland, but, notably, King County, Washington has protected more than 55,000 hectares of forestland through its TDR programme (King County, 2017).

The strengths of TDR programmes include:

1   The preservation of rural lands in metropolitan counties where extensive contiguous areas of agricultural or forest land still exist and it is possible to move development potential from these rural areas to designated growth areas;
2   Providing rural landowners with financial compensation in return for land-use regulations, and thus reducing the likelihood of a successful challenge of government 'taking' private property;
3   Leveraging private dollars for rural land protection, and decreasing the need for public funding;
4   Managing growth through the use of receiving areas designated for growth and sending areas for land protection (Daniels and Lapping, 2005). Thus, population growth and economic development continue while rural lands are maintained as working landscapes or open space; and,
5   Keeping farm land and forest land affordable for farmers and foresters.

The weaknesses of the transfer of development rights technique are:

1   TDRs are complex and difficult to establish, including identifying distinct sending and receiving areas and making changes to the comprehensive plan and zoning regulations;
2   TDRs do not work well in remote rural areas where there is typically a large supply of TDRs and little demand from developers. Hence, TDR prices are low, attracting few sellers in the sending areas;
3   A TDR programme requires a well-trained staff and careful record-keeping of the status of TDRs on all parcels in the sending and receiving areas. If the local government operates a TDR bank, the local government must keep track of the number of TDRs in the bank, the proceeds from the sale of TDRs, and the amount of public funds spent on purchasing TDRs; and,
4   Local governments may be reluctant to acquire conservation easements through a TDR programme because the local government must take on the expense of monitoring and enforcing the terms of the conservation easements.

Payments for environmental services: carbon offset credits

The United States has not enacted a national cap-and-trade programme aimed at reducing greenhouse gas (GHG) emissions (Daniels, 2010). In 2006, however, the California state legislature passed the Global Warming Solutions Act (Assembly Bill 32 or AB 32), which mandated a state-wide reduction in GHG emissions to 1990 levels by the year 2020. In 2012, the California Air Resources Board (CARB), the state agency responsible for implementing AB 32, launched a cap-and-trade programme to reduce GHG emissions. The programme has capped allowances of GHG emissions for a wide array of business operations, which produce an estimated 85 per cent of California's GHG emissions (CARB, 2014).

The cap-and-trade programme allows GHG emitters to meet up to eight per cent of their allowed emissions through the purchase of carbon-offset credits. The offsets are mainly derived from forests in the lower 48 states that meet requirements for additional, verifiable increases in carbon sequestration through improved forest management. Forest operators have so far provided offset credits involving more than 500,000 hectares of forestland (CARB, 2017). By comparison, as of 2015, forest carbon projects had protected an estimated 24 million hectares around the world (Goldstein and Ruef, 2016).

An important factor is the initial $10 a ton floor price that California imposed on greenhouse gas allowances, which the state periodically sells to GHG emitters at auction. The offsets sell at about a ten per cent discount below the allowances and involve negotiations between the buyers and sellers as in any market. Forests outside of California have generated most of the offset credits; and strong potential exists for additional offsets both within California and in other states. Most forest owners in the offset programme hold more than 5,000 acres, which reflects the high cost of applying for and verifying the offset credits (CARB, 2017). The offset programme has shown the potential to mesh well with the federal forest legacy programme, which pays forest owners not to develop their forests. Thus, forest operators can receive income from three streams: the sale of a conservation easement to keep the forest in forest use; the sale of offsets to California greenhouse gas emitters, and the sale of timber from harvests in accordance with a forest management plan. The California offset programme has begun to make owners of agricultural land eligible to sell carbon credits as well.

Conclusion

America's experience with market-based incentives for rural land-use planning has shown uneven and overall modest success. Governments and private land trusts have used the purchase and donation of conservation easements to protect nearly 10 million hectares of privately held rural land. This represents 1 per cent of America's land area, or an area equal to the State of Maine, the 39th largest state. The transfer of development rights involves buying the right to develop the property and acquiring a conservation easement, and then moving that right from the restricted property to enable additional development on another property within the local government's receiving area. It is unlikely that new large TDR programmes will be created in the near future. Although there are some noteworthy TDR successes, the purchase and donation of conservation easements are easier techniques to implement and have protected much more rural land.

The federal government has provided important grant funding for the purchase of conservation easements through the farmland, grassland, and wetlands programmes within the Agricultural Conservation Easement Program (ACEP) and the Forest Legacy Program. Whether funding for these programmes will continue and, if so, at what level, are critical questions for the future of conservation easement acquisitions in rural areas. State and local governments have reduced their funding for purchasing conservation easements to farmland since the Great Recession of 2008–2010. Again, future funding levels will determine the robustness of these programmes. Land trusts rely mainly on donations from individuals, companies, and other NGOs. There are now more than 500 land trusts that list rural land protection as a major part of their mission. Future funding from the private sector will be crucial to further rural land protection by land trusts.

Payments for environmental services provided by owners of rural land are in their infancy in America. California's carbon offset programme as part of the state's cap-and-trade programme has shown that forestland can sequester additional amounts of carbon and while providing landowners with three streams of income: (1) the sale of a conservation easement to ensure the forestland cannot be developed for non-forest uses; (2) the sale of carbon offset credits to GHG emitters; and (3) the sale of timber harvested according to a forest management plan.

Finally, land-use planning and regulation should be viewed as short- to medium-term arrangements. Conservation easements provide more lasting protection for farmland, forestland, and natural areas from conversion to developed uses, and can help to direct growth to appropriate locations.

References

American Farmland Trust. 2016a. Farmland Information Center: Status of State PACE Programs. Retrieved from www.farmlandinfo.org/sites/default/files/State_Purchase_of_Agricultural_Conservation_Easement_Programs_2016_AFT_FIC_09-16.pdf (accessed 8 August 2017).

American Farmland Trust. 2016b. Farmland Information Center: Status of Local PACE Programs, January 2016. Retrieved from www.farmlandinfo.org/sites/default/files/Local_Purchase_of_Agricultural_Conservation_Easement_Programs_2016_AFT_FIC_0.pdf (accessed 8 August 2017).

Anella, Anthony and John B. Wright. 2004. *Saving the Ranch: Conservation Easement Design in the American West*. Washington, DC: Island Press.

Barrett, Gordon. 2011. Restoring the Tahoe Region with Comprehensive Regional Planning. In Carelton Montgomery, ed. *Regional Planning for a Sustainable America*. New Brunswick, NJ: Rutgers University Press.

CARB. 2014. California's Compliance Offset Program. Retrieved from www.thepmr.org/system/files/documents/ARB%20Offsets%20PMR%20webinar%202014%20FINAL_Mexico.pdf (accessed 15 August 2017).

CARB. 2017. ARB Offset Credits Issued January 11, 2017. Retrieved from www.arb.ca.gov/cc/capandtrade/offsets/issuance/arb_offset_credit_issuance_table.pdf (accessed 15 August 2017).

Daniels, Thomas L. 2010. Integrating Forest Carbon Sequestration Into a Cap-and-Trade Program to Reduce Net CO_2 Emissions. *Journal of the American Planning Association*, 76 (4): 463–475.

Daniels, Tom. 2014. *The Environmental Planning Handbook*, 2nd edn. Chicago, IL: APA Planners Press.

Daniels, Tom. 2017. Montgomery County's Agricultural Reserve. *Planning*, 83 (5): 38–39.

Daniels, Thomas L. and Mark B. Lapping. 2005. Land Preservation: An Essential Ingredient in Smart Growth. *Journal of Planning Literature* 19: 316–329.

Daniels, Tom and Deborah Bowers. 1997. *Holding Our Ground: Protecting America's Farms and Farmland*. Washington, DC: Island Press.

Goldstein, Allie and Franziska Ruef. 2016. View from the Understory: Forest Carbon Finance 2016, Overview. Washington, DC: Forest Trends Ecosystem Market Place. Retrieved from www.forest-trends.org/documents/files/doc_5498.pdf (accessed 14 August 2017).

IRC. 2017. 26 US Code Section 170. Retrieved from www.law.cornell.edu/uscode/text/26/170 (accessed 10 August 2017).

Johnston, Robert A. and Mary E. Madison. 1997. From Landmarks to Landscapes: A Review of Current Practices in the Transfer of Development Rights. *Journal of the American Planning Association*, 63: 365–379.

King County, Washington. 2017. Program Overview – Transfer of Development Rights. Retrieved from www.kingcounty.gov/services/environment/stewardship/sustainable-building/transfer-development-rights/overview.aspx (accessed 10 August 2017).

Land Trust Alliance. 2011. *2010 National Land Trust Census Report*. Washington, DC: Land Trust Alliance.

Land Trust Alliance. 2016. *2015 National Land Trust Census Report*. Washington, DC: Land Trust Alliance. Retrieved from www.landtrustalliance.org/about/national-land-trust-census (accessed 2 August 2017).

Nelson, A. C., Rick Pruetz, and Doug Woodruff. 2012. *The TDR Handbook*. Washington, DC: Island Press.

NRCS. 2017a. Farm and Ranch Lands Protection Program. Washington, DC: Natural Resources Conservation Service. Retrieved from www.nrcs.usda.gov/Internet/NRCS_RCA/reports/fb08_cp_frpp.html (accessed 1 August 2017).

NRCS. 2017b. Agricultural Conservation Easement Program. Washington, DC: Natural Resources Conservation Service. Retrieved from www.nrcs.usda.gov/Internet/NRCS_RCA/reports/srpt_cp_acep.html (accessed 1 August 2017).

NRCS. 2017c. Wetlands Reserve Program. Washington, DC: Natural Resources Conservation Service. Retrieved from www.nrcs.usda.gov/Internet/NRCS_RCA/reports/fb08_cp_wrp.html (accessed 1 August 2017).

NRCS. 2017d. Wetlands Reserve Program Maps. Washington, DC: Natural Resources Conservation Service. Retrieved from www.nrcs.usda.gov/Internet/NRCS_RCA/maps/cp_wrp_maps.html

NRCS. 2015. Natural Resource Inventory Summary Report. Washington, DC: Natural Resources Conservation Service. Retrieved from www.nrcs.usda.gov/Internet/FSE_DOCUMENTS/nrcseprd396218.pdf (accessed 25 August 2017).

US Department of Agriculture. 2014. *2012 Census of Agriculture*. Washington, DC: USDA.

US Forest Service. 2016. Forest Legacy Program, Funded and Completed Projects. Retrieved from www.fs.fed.us/cooperativeforestry/images/flp_completed_tracks.jpg (accessed 14 August 2017).

12

Community ownership of rural assets

The case of community land trusts

Tom Moore

Introduction

While other chapters in this Companion explore the potential of public engagement, this contribution explores the role of 'communities' in not only participating in policy-making, but in playing a more direct role through the ownership of community assets. Community ownership can be applied to a wide range of rural assets (such as local energy schemes) and in this chapter will be examined through the lens of community land trusts to critique the potential of community ownership as a governance mode.

Community management and ownership of assets represents a local response to economic and social change in rural locations. In recent years, many communities have responded to challenges such as public service withdrawal, limited economic development due to low profitability, and social and community change due to demographic and geographical factors, by assuming greater responsibility for the provision of services and amenities (Moore and McKee, 2014). This trend also reflects continued political support for such initiatives. For example, within the UK, the Conservative Party's 2015 election manifesto set out a vision for a 'more engaged nation, one in which we take more responsibility for ourselves and our neighbours; communities working together' (Conservative Party, 2015: 45), reaffirming the principles introduced in the Localism Act 2011 that gave communities greater power and control over local planning processes.

While practices of community engagement and participation in rural development have a long history (Edwards, 1998), the political era of localism and trends in state and market provision have led to a new wave of rural social enterprise and asset-holding organisations such as development trusts, and community land trusts (Moore, 2015; Steiner and Teasdale, in press). Community asset ownership has spread to a range of areas, including pubs (Cabras, 2011), energy schemes (van Veelen and Haggett, 2017), and community retail shops (Calderwood and Davies, 2013). Housing is a particular area where the role of communities has significantly grown, with the formation and development of community land trusts responding to issues of growing disparities between rural incomes and house prices, high second home ownership in some communities, and limited supply (Satsangi et al., 2010). This chapter uses community land trusts as a case study for the increasing involvement of communities in the planning, management and ownership of rural assets, tracing the sector's development, their purpose and function, and opportunities and challenges associated with voluntary-led, community-based housing provision.

The growth of community land trusts in England

Community land trusts (CLTs) are organisations formed and managed by residents of a local community seeking to facilitate the delivery of affordable housing and other community facilities. CLTs are non-profit organisations and place an emphasis on community leadership through democratic governance structures, usually involving a management group composed of local volunteers and a broader community membership base. Those who lead CLTs are rarely, if ever, the beneficiaries of the housing they provide, distinguishing CLTs from forms of resident-owned housing such as cooperatives. Therefore, they perform a dual role as affordable housing provider and as a platform for community influence of the planning, delivery and management of housing and other local assets.

While the community land trust concept was popularised in the United States, growth in England emerged in rural communities in response to rural housing affordability and supply crises, as house prices outstrip incomes and supply fails to keep pace with demand (DEFRA, 2017). This contributes to declines in the provision of rural services and sustainability of communities. It is in this context that the CLT model emerged, initially supported by demonstration programmes between 2006–2008 and 2008–2010 that supported volunteers with CLT development. This work led to the creation of a national membership organisation – the National CLT Network – that provides support and assistance to CLTs across the country, while lobbying Government to create favourable conditions for their development. The result of this has been the creation of a Government-backed 'Community Housing Fund', providing ring-fenced funding for community-led development until 2020. The consequence of this is that there are now approximately 225 CLTs around the country in both rural and urban communities. It is also important to note that while the concept of CLTs has taken root in other parts of the UK, its use in England is largely linked to the provision of affordable housing, and occasionally other local amenities, whereas in places such as Scotland CLTs may have a broader focus on taking land from previously feudal ownership into the ownership of communities.

Local homes for local people

The dual and interrelated issues of rural housing affordability and a lack of supply to meet local housing needs provide a major impetus for the formation of rural CLTs. In rural areas, the cheapest homes are estimated to cost 8.3 times lower incomes (DEFRA, 2017), pricing many households out of communities. Therefore, there is a recognised need to increase affordable housing supply due to high demand (particularly in some areas from second home buyers and holiday lets) and in some areas, such as National Parks, tight planning constraints on new house building and difficulties in accessing land (Rural Services Network, 2015). An additional barrier to rural housing development is that, rather than facilitating development, communities are often resistant to new house building due to concerns over its impact on the countryside and stigma attached to occupants of social and affordable housing (Sturzaker, 2011).

CLTs attempt to overcome these obstacles through provision of affordable housing that is guaranteed to benefit households with local connections. CLTs provide housing for sale or for affordable rents, usually at 80 per cent of market rents. When housing is sold, CLTs typically retain the freehold of land and sell leaseholds or equity stakes at affordable levels, meaning the CLT retains some control and influence over the future use of the housing. Resale restrictions are usually put in place to retain affordability and restrict house price inflation, thus ensuring a stock of housing that remains affordable for future generations. CLT housing in rural areas is also usually strictly allocated in a way that prioritises local connections, taking into account both a potential resident's housing need and financial circumstance and their connection to the local community in question.

Local connection in CLT allocation policies can be defined in a number of ways, but often prioritises applicants from the village or parish in question, before cascading out to adjacent communities as necessary. The dual emphasis placed on affordability and meeting local needs, as defined by the CLT, has helped CLTs overcome some of the traditional barriers to rural housing development. There are instances of CLTs accessing parcels of land from landowners that would ordinarily be reluctant to sell land for housing, but have been reassured of the community and locally specific benefits of releasing it into CLT ownership (Moore, 2015). Similar examples are evident in urban areas, where local authorities have facilitated land release for nominal values, though there remain tensions between the social value of land and the pressure to obtain full market rates (Field and Layard, 2017).

In addition to land release, the prioritisation of local connection in allocation policies, over and above more general needs usually addressed by housing associations, has helped to build community support and mitigate potential opposition to development. CLT commitments to meeting the needs of those with local connections – such as family links, past residency, or employment contributions – help to justify the need for new housing to residents that may often oppose development (Moore, 2015).

That there is a preference for local allocation of affordable housing in rural areas reflects previous research in this area. That communities are now able to exercise this preference through a legally incorporated housing provider, formed and managed by local residents, offers further insights into the growing importance and nature of community empowerment in rural housing planning and management (Sturzaker, 2011). However, previous studies have also indicated that more prominent roles for community organisations can sometimes give voice and power to those seeking to exclude particular groups from rural spaces (Yarwood, 2002; Sturzaker, 2011). While research on CLTs show they adopt a proactive approach to development, broader support for this is often contingent on housing availability for local people, suggesting a need to be alive to the potential consequences of such initiatives for those that may wish to live in rural areas, but lack the requisite local connections or financial capital to do so.

Community participation and engagement

As noted, the role of communities in rural planning processes is longstanding. While historically this has been articulated through traditional forms of representative democracy, such as Parish Councils (Connelly et al., 2006), CLTs can be considered as part of a new wave of initiatives that combine elements of representative and participatory democracy in order to facilitate community influence (Sturzaker and Gordon, 2017 also cite neighbourhood planning processes introduced in the Localism Act 2011 as another relatively new example).

The local residents that form and manage CLTs on a voluntary basis are usually, though not exclusively, retired individuals from professional backgrounds. They are often people of long residency in communities, who possess strong attachment to place and have been resident long enough to observe social trends and changes, including the sale of former social housing through the Right to Buy and growing housing difficulties faced by local young people and key workers (Moore, 2015). Volunteers have often been involved with other forms of voluntary and community activity in their communities, though choose to participate in CLTs due to the organisation's specific focus on housing issues (Moore, 2014). They are therefore reflective of what Mohan (2012) refers to as the 'civic core': groups of people who are more likely to engage in voluntary activity, and share characteristics of higher educational qualifications and long residency in communities.

An argument made in favour of CLTs is that they offer greater community-based legitimacy for the planning and management of local housing developments, due to their local accountability and specific focus on housing (Moore, 2018). This manifests in two ways. Firstly, through the day-to-day management board composed of local volunteers, who are often motivated by specific housing issues that have negatively affected their communities (Paterson and Dayson, 2011). Secondly, evidence suggests CLTs undertake extensive community planning exercises, which seek to reach beyond both their management board and membership base in order to fully involve other local residents (Moore, 2015). Practically, this may involve surveys, consultation events, interactive techniques of collaborative planning, and regular public meetings and feedback sessions, which not only inform the decision-making of CLTs but help to strengthen the legitimacy, understanding and acceptance of local development. While it is true that other housing providers and local authorities would typically undertake community consultation exercises, distrust of authorities has often created a perceived 'democratic deficit' among rural communities, where their influence is perceived as limited and constrained (Sturzaker, 2011). As such, CLTs may represent an effort to embed community influence into the planning, design, delivery and management of local housing from the outset, offering opportunities for influence to people that would ordinarily oppose development and providing assurance to residents as to the CLT's commitment to meeting local needs (Moore, 2018). This offers credence to the views that levels of local opposition are often contingent on the extent to which citizens that are likely to be affected by the outcome of development, are afforded opportunities to shape and influence local outcomes (Armour, 1991). Indeed, research on Neighbourhood Planning processes has highlighted that many of the emerging Neighbourhood Plans demonstrate evident demand and need for forms of community-led housing (Field and Layard, 2017).

While the emergence of CLTs has it advantages, including the development of housing in areas perceived as neglected by state and market provision, there are a number of challenges related to their reliance on voluntary initiative. Most CLT boards are staffed by multiple people, but are usually reliant on one or two figureheads who lead CLT activity, creating the risk of knowledge concentration and overreliance on a small number of CLT volunteers (Moore, 2014). This issue is reflective of the challenges CLTs face in recruiting board members: rural areas are, by the nature of their small populations, limited with respect to human resources and research has shown that CLTs sometimes struggle to access necessary skills and resources and have concerns over the future sustainability of their boards (Paterson and Dunn, 2009; Moore, 2015). This is particularly crucial, given that there are a number of technical issues and requirements that CLTs require navigating. As legal entities, there may be regulations that govern their activities and the way in which funds and surpluses are distributed, restrictions on the activities they may undertake, particularly if registered as a Charity, and obligations to maintain open democratic governance. Furthermore, while the 'ideal type' of CLT governance is often presented as a tripartite structure, involving co-management of trusts between CLT leaders and wider community members, CLT residents, and local authorities (Davis, 2010), most CLTs – particularly in rural areas – are in reality solely led by local residents (Moore, 2015).

Building CLT capacity through partnerships

One way in which CLTs attempt to overcome issues of capacity is by accessing support and assistance from partners, including dedicated CLT support bodies and housing associations. CLT support bodies – sometimes referred to as umbrellas or intermediaries – work to advise on issues of formation, development and delivery of CLT schemes. They usually operate at regional or sub-regional scales and aim to help CLTs overcome barriers, including explanation and promotion of the concept to partners and authorities, access to specialist skills and knowledge required for organisational formation and housing development, and additional networking knowledge and skills that can help to expedite resource acquisition and housing delivery (Moore, 2015). The role of these bodies has been acknowledged for many years. Research conducted for the Countryside Agency in 2005 highlighted that, for the CLT sector to grow, there would need to be a network of specialist support that could 'disseminate good practice and promote public understanding and acceptance of this mutual approach to ownership of land and property' (Countryside Agency, 2005: 53). Research has shown that the provision of technical support and promotion of CLTs within local, regional and national policy networks has contributed to CLT expansion in areas where support bodies operate (Moore and Mullins, 2013). Their important role offers important lessons for understanding the operation of CLTs, as access to specialist support seems crucial for CLTs to thrive, and reduces risks related to lack of human resources and capacity within communities.

An additional way in which CLTs have overcome reliance on volunteers is through partnerships with housing associations, where associations provide technical support and expertise to help expedite the development of CLT housing (Paterson and Dayson, 2011). This is a growing trend in the CLT sector, partly motivated by the creation of state funding streams made available to CLTs, but with encouragement to partner with housing associations to share expertise in development and defray investment risk in small, voluntary-led organisations (Moore, 2018). While this financial imperative provided an initial stimulus for partnerships, the view that CLTs can enhance community engagement and influence in the planning of housing offers benefits for housing associations that often face significant opposition to new developments. However, many partnerships are often structured on financial terms that are less advantageous to communities, with housing associations often assuming management responsibilities for homes but retaining all or the majority of rents paid. Some proponents of CLT schemes have argued that this dilutes the potential of CLTs to generate future surpluses that could be regenerated into the local community and used for local economic development, as well as undermining their community-led nature, though others balance this against the reduced risk for, and reliance on, volunteers (Moore, 2018).

It is clear that partners such as CLT support bodies and housing associations play an important role in helping to popularise and spread the CLT model, as their provision of specialist support and skills supports volunteers, overcomes human resource limitations evident in some rural communities, and helps to expedite the planning and development of housing (Moore and Mullins, 2013). It is therefore clear that community-led development is not only reliant on the role of active citizens, but on the involvement of a broader range of stakeholders and involving varied degrees of community participation and partnership. Czischke (2018) argues that forms of community housing progress through 'multi-stakeholder collaboration', involving varied patterns of co-production between state, market and civil society actors. As such, CLTs and other schemes are understood not only in terms of self-organisation and citizen initiative, but also through the different forms of partnership and support provided by a range of actors.

Conclusion

Forms of community asset ownership have a long history in England (Wyler, 2009), but recent political support for such initiatives through localism – combined with the economic climate of austerity that has affected state and market provision – has led to a rise in community asset organisations. Community land trusts are one such example of this, and their formation to provide affordable housing has helped to meet rural housing needs that may otherwise have not been met.

It is evident that CLTs have managed to overcome many of the barriers associated with rural housing planning and development, using narratives of community, local benefit, and local need to mitigate opposition and build support. This has been further supported by access to partners able to offer specialist support, which helps reduce risks associated with their reliance on voluntary initiative. There is an emerging structure to the CLT sector, as partnership models with CLT support bodies and housing associations begin to emerge (Moore, 2018), which has been given a further boost by the Government's decision to create a Community Housing Fund providing revenue and capital funding until at least 2020. This gives contemporary CLTs clearer access to funds than their predecessors and should result in a scaling-up of the sector. Yet, this also creates dilemmas, as scaling-up activity – whether it be the size of individual schemes, the size of the sector as a whole, or a scaling-up of the services provided by CLT support bodies – may risk distancing CLTs from the very local, community-based roots by which they identify and through which they overcome many of the obstacles faced by conventional housing providers.

Studies of CLTs have so far largely focused on 'successful' schemes; that is CLTs that have progressed through planning and developed CLTs. As the sector develops and forms of community-led housing emerge through Neighbourhood Plans, an interesting line of inquiry may be to track and understand schemes which struggle or fail to get off the ground. As Field and Layard (2017) note, even in collective ownership housing does still not lend itself to being shared in the way that other forms of collective resources might, such as community centres or parks, and this may create challenges and difficulties in some communities where land use is contested. The development of CLTs, as asset-owning organisations, offers another insight into the ways in which community empowerment in rural planning can help facilitate rural housing provision (Sturzaker, 2011). Further research and thinking can help inform the extent to which the CLT model can be replicated to provide affordable housing and other community amenities that challenge dominant modes of provision.

References

Armour, A. (1991) The siting of locally unwanted land uses: towards a cooperative approach. *Progress in Planning*, 35(1): 1–74.

Cabras, I. (2011) Industrial and provident societies and village pubs: exploring community cohesion in rural Britain. *Environment & Planning A*, 43(10): 2419–2434.

Calderwood, E. and Davies, K. (2013) Localism and the community shop. *Local Economy*, 28(3): 339–349.

Connelly, S., Richardson, T. and Miles, T. (2006) Situated legitimacy: deliberative arenas and the new rural governance. *Journal of Rural Studies*, 22(3): 267–277.

Conservative Party. (2015) *Strong Leadership, a Clear Economic Plan, a Brighter More Secure Future: The Conservative Party Manifesto*. London: Conservative Party.

Countryside Agency. (2005) *Capturing Value for Rural Communities: Community Land Trusts and Sustainable Rural Communities*. Cheltenham: Countryside Agency.

Czischke, D. (2018) Collaborative housing and housing providers: Towards an analytical framework of multi-stakeholder collaboration in housing co-production. *International Journal of Housing Policy*, 18(1): 55–81.

Davis, J. E. (2010) *The Community Land Trust Reader.* Cambridge, MA: Lincoln Institute of Land Policy.

DEFRA. (2017) Statistical digest of rural England: May 2017. Retrieved from www.gov.uk/government/uploads/system/uploads/attachment_data/file/615855/Statistical_Digest_of_Rural_England_2017_May_edition.pdf

Edwards, B. (1998) Charting the discourse of community action: perspectives from practice in rural Wales. *Journal of Rural Studies,* 14(1): 63–77.

Field, M. and Layard, A. (2017) Locating community-led housing within neighborhood plans as a response to England's housing needs. *Public Money & Management,* 37(2): 105–112.

Mohan, J. (2012). Geographical foundations of the Big Society. *Environment and Planning A,* 44(5): 1121–1127.

Moore, T. (2014) Affordable homes for local communities: the effects and prospects of community land trusts in England. Retrieved from www.academia.edu/24005961/Affordable_homes_for_local_communities_The_effects_and_prospects_of_community_land_trusts_in_England

Moore, T. (2015) The motivations and aspirations of community land trust volunteers in Somerset, Dorset and Devon. Retrieved from www.academia.edu/18500007/The_motivations_and_aspirations_of_community_land_trust_volunteers_in_Somerset_Dorset_and_Devon

Moore, T. (2018) Replication through partnership: the evolution of partnerships between community land trusts and housing associations in England. *International Journal of Housing Policy,* 18(1): 82–102.

Moore, T. and McKee, K. (2014) The ownership of assets by place-based community organisations: political rationales, geographies of social impact and future research agendas. *Social Policy & Society,* 13(4): 521–533.

Moore, T. and Mullins, D. (2013). Scaling-up or going viral? Comparing self-help housing and community land trust facilitation. *Voluntary Sector Review,* 4(3): 333–353.

Paterson, B. and Dayson, K. (2011) Proof of concept – community land trusts. Retrieved from http://usir.salford.ac.uk/19312/2/Proof_of_Concept_Final.pdf

Paterson, E. and Dunn, M. (2009) Perspectives on utilising community land trusts as a vehicle for affordable housing provision. *Local Environment,* 14(8): 749–764.

Rural Services Network. (2015) Affordable housing: policy briefing note. Retrieved from www.rsnonline.org.uk/images/files/policy-briefing-affordable-housing-august2015.pdf

Satsangi, M., Gallent, N. and Bevan, M. (2010) *The Rural Housing Question: Community and Planning in Britain's Countryside.* Bristol: Policy Press.

Steiner, A. and Teasdale, S. (in press) Unlocking the potential of rural social enterprise. *Journal of Rural Studies.*

Sturzaker, J. (2011) Can community empowerment reduce opposition to housing? Evidence from rural England. *Planning Practice & Research,* 26(5): 555–570.

Sturzaker, J. and Gordon, M. (2017) Democratic tensions in decentralised planning–Rhetoric, legislation and reality in England. *Environment and Planning C: Politics and Space,* 35(7): 1324–1339.

Van Veelen, B. and Haggett, C. (2017) Uncommon ground: the role of different place attachments in explaining community renewable energy projects. *Sociologia Ruralis,* 57(S1): 533–554.

Wyler, S. (2009) A history of community asset ownership. Retrieved from http://locality.org.uk/wp-content/uploads/A-History-of-Community-Asset-Ownership_small.pdf

Yarwood, R. (2002) Parish councils, partnership and governance: the development of 'exception' housing in the Malvern Hills District, England. *Journal of Rural Studies,* 18(3): 275–291.

13

The dark side of community

Clientelism, corruption and legitimacy in rural planning

Linda Fox-Rogers

Introduction

Over the last 30 years, the principles of community participation have become deeply embedded in planning theory and practice where formal statutory and community processes of public participation are considered pivotal to the delivery of positive planning outcomes (see Healey, 1997). However, it remains clear that even in instances where communities have actively engaged in the planning process, the end result arrived at can all too often reflect dominant interests at the expense of the 'common good'. Within the literature, the tendency towards regressive planning outcomes is frequently analysed through a neoliberal lens where the reorientation of the planning system towards market interests is used to explain the disjoint between the normative ideals of planning on the one hand, and the realities of planning practice on the other (Harvey, 1985; Brindley et al., 1989). While this theoretical framework is rich in analytical value, much of what we see in relation to the failure of participatory approaches to deliver just outcomes (and not simply just processes) essentially boils down to the uneven power dynamics that exist between key stakeholders in the planning process. Indeed, many critics argue that the so-called 'communicative turn' (Healey, 1992) in planning fails to address the distorting effects of power in the planning system (Tewdwr-Jones and Allmendinger, 1998; Neuman, 2000). Specifically, they highlight how the uneven distribution of resources (e.g. money, knowledge, skills) means that stakeholders do not come to the negotiating table on a level playing field and this becomes manifested in the development outcomes ultimately arrived at (Hibbard and Lurie, 2000). Moreover, these power disparities can foster the emergence of clientelist exchanges, and in more extreme cases corruption, which largely operate outside of the planning system's formal participatory processes, as well as its broader institutional framework (Jiménez et al., 2012; Fox-Rogers and Murphy, 2014).

Strangely, however, the distorting effects that clientelist or corrupt exchanges can have on the democratic nature of planning practice has largely been neglected by academics despite there being a general appreciation of the potential for planning corruption among the media, the general public and practitioners themselves (Dodson et al., 2006). Even less consideration has been given to such issues within the context of rural planning where the power dynamics at play in rural communities may reflect subtle, yet important, differences to those which permeate urban centres where there is a general tendency towards a greater concentration of political and economic power.

This chapter seeks to address this gap by examining the 'dark side' of community planning within a rural setting. It begins by examining the conditions which have been identified in the literature as being central to understanding the planning system's vulnerability to clientelism and corruption. It then seeks to explore how clientelist exchanges arise and operate within the context of rural communities before analysing how they play out in terms of specific social, economic and environmental outcomes. In doing so, lessons will be drawn from the Republic of Ireland, a country which has become synonymous with unethical planning practice in recent years (see Government of Ireland, 2012). The chapter concludes with a critical reflection on the broader implications of clientelist relations for planning as a discipline and sets out a research agenda aimed at stimulating more empirical enquiries in this field.

Clientelism and corruption in planning

The terms 'patronage', 'clientelism' and 'corruption' are generally understood as 'deviant and disqualified practices in the democratization processes, which are reproduced in the informal spaces of political life' (Robles-Egea and Aceituno-Montes, 2011: 3). While these phenomena are undoubtedly interrelated and often used interchangeably within the literature, they do have distinct characteristics. For instance, the term 'patronage' or 'clientelism' generally refers to personal favours or discretionary decisions which may be morally questionable, but not illegal *per se*. For example, clientelism can be used to describe situations in which the politician uses their influence to obtain certain state benefits for the constituent in return for their electoral support (Komito, 1985). These types of acts tend to 'straddle the border' of the legal and illegal (Robles-Egea and Aceituno-Montes, 2011: 4). However, once they cross the border, they fall under the rubric of corruption. For scholars such as Jiménez et al. (2012: 365), corruption is simply defined as 'the use of any public power for private gain'. In essence, these unsavoury practices work to undermine the democratic principles underpinning key institutions, not least the planning system, that seek legitimacy through carefully designed legal and administrative frameworks intended to promote transparent and accountable decision-making that serves public, over personalised, interests.

As already mentioned, there exists a dearth of empirical enquiry into issues of corruption and clientelism from a planning perspective. The peculiar lack of research activity in this area is likely to stem from methodological issues rather than a lack of acceptance among planning scholars that the normative ideals of planning practice can be undermined by unsavoury relationships between key actors in the planning system. For instance, Dodson et al. (2006) highlight the problematic nature of relying on secondary sources of information, such as media reports, as allegations of corruption may be ill-founded or may go unreported altogether. Furthermore, primary methods of data collection, such as interviews or surveys, also present particular challenges as the sensitive nature of the subject can preclude key actors from responding in a frank and honest manner and might even deter them from participating in the first instance.

Nevertheless, a number studies exist which explicitly explore issues of clientelism and corruption in planning (see Chiodelli and Moroni, 2015; Jiménez, 2009; Jiménez et al., 2012). Much of the literature in this area has centred on analysing the conditions which promote corrupt planning practice, with a particular focus on Mediterranean countries such as Spain where corruption scandals in the land development process have been well documented. Among the various factors identified, central is the level of discretion that exists within planning policy coupled with 'the economic repercussions of the decisions taken on this matter' (Jiménez, 2009: 255). With regard to the former, Jiménez et al. (2012) warn that the assumed benefits of devolution and subsidiarity which underpin the decentralisation strategies being implemented across Europe can be quickly undermined where there are strong patronage networks. Specifically, they argue that giving local actors greater discretion in terms of decision-making power can result in them becoming subservient to the needs of local elites.

The other key feature of the planning system which encourages land speculation and corruption is the windfall gains which can be derived from zoning decisions which deem land as being suitable for development. This has been a long recognised as the planning system's 'Achilles heel' in terms of its vulnerability to corruption. For some, the solution lies in the introduction of value-capture mechanisms to disincentivise corruption as windfall gains on the unearned increases in land values are returned to the community (Day, 1995; Chiodelli and Moroni, 2015). However, there is evidence to suggest that the instrument of planning gain has been largely co-opted by landowners and development interests as an effective lobbying tool in seeking planning permission (see Fox-Rogers and Murphy, 2015).

In addition to the formal institutional context that makes the planning system particularly vulnerable to corruption, authors such as Jiménez et al. (2012) highlight that informal institutional arrangements also play an important role in this regard. For example, they highlight how the Local Integrity System in Spain (which sets out various rules to promote transparent and accountable local governance) has been weakened by the encroachment of informational institutions like patronage networks of local politicians. Moreover, they highlight how social context can help explain breaches of planning policy in Spanish local government, where the smaller populations in villages and towns means that mayors (who have responsibility for planning enforcement) know most of their residents on a personal level, making it harder for them to apply sanctions to their prospective voters. This is a particularly prominent issue where mayors have not enforced the strict prohibition to build on rural land if it affects their voters and actually use this lenient approach as an effective strategy to gain political support.

While Jiménez et al.'s (2012) work sets out a valuable framework for understanding how corruption flourishes within particular contexts, their high level analysis does not offer insights into how corrupt relationships become established between individuals in the first place or on what basis they are formed. Similarly, there is a need to consider the mechanics of how clientelist relations operate in practice and what impacts they have in terms of the planning outcomes they breed. Moreover, while various authors have suggested that clientelist and corrupt relations can be more acute in more peripheral settings (see Komito, 1985; Jiménez et al., 2012; Robles-Egea and Aceituno-Montes, 2011), there has been a failure to examine this phenomenon through a rural lens specifically. The remainder of this chapter is dedicated to redressing these issues and in doing so, now turns its attention to the Republic of Ireland.

Planning, power and politics in rural Ireland

Using the Republic of Ireland as a case study to explore issues of clientelism and corruption from a rural planning perspective makes sense for a number of reasons. First, while the country's urban population is growing at a faster rate than rural areas, Ireland nevertheless remains one of the most rural countries in the European Union with rural dwellers accounting for 37.3 per cent of the population (CSO, 2017) compared with an EU average of 27 per cent and 12 per cent in Britain (McMahon, 2016). Second, Ireland has become well known for its sprawling pattern of low rise, low density development which has emerged due to the gradual encroachment of development on the countryside (see Scott and Murray, 2009; Murphy and Scott, 2013; EEA, 2006). This became particularly pronounced during the Celtic Tiger period when the intense development pressures and rising land values experienced in Dublin led to 'overspill' development which took place up to 100 km from the metropolitan area in rural counties such as Meath, Kildare, Wicklow and Westmeath (Williams and Shiels, 2000). Such trends have been exacerbated further by the proliferation of 'one-off' housing, dispersed throughout the Irish open countryside (EEA, 2006; Murphy and Scott, 2013). The leapfrogging of development activity to the rural fringes and the scale of one-off rural housing development that has occurred in recent years has been associated with weak planning regulation and the reckless land-zoning decisions of councillors where vast swathes of agricultural land have been routinely rezoned for development, most notably residential. Indeed, a report compiled by the European Environment Agency highlights that Ireland's planning regime imposes far 'few constraints on the conversion of agricultural areas to low-density housing areas' (EEA, 2006: 22).

Third, there is a long history of clientelist and corrupt practices in Irish political life at national and local level. Indeed, several empirical enquiries have examined this phenomenon from an Irish perspective such as Komito's (1985) anthropological study of clientelist exchanges in Dublin. While the social, economic and political context in Ireland has changed dramatically since the time of Komito's work, it is clear that the problem of cosy relationships between the political and economic sphere persist in Irish society. For instance, the Irish planning system in particular has been subjected to intense scrutiny in recent years as evidence of misconduct and corruption in the planning process served as a catalyst to a 15-year-long Tribunal of Enquiry into Certain Planning Matters and Payments (more commonly referred to as 'the Mahon Tribunal'). The Tribunal was established to investigate the facts behind a number of dubious planning decisions that took place in the Greater Dublin Area from the mid-1980s to 1997 (Grist 2012), the central findings of which stated that corruption in Ireland 'was both endemic and systemic' and its 'existence was widely known and tolerated' (Government of Ireland, 2012: 1). In a similar vein, An Taisce's (2012: 17) (Ireland's National Trust) independent review of Ireland's planning system highlighted that 'endemic parochialism, clientelism, cronyism and low-level corruption' was inherent in Ireland during the Celtic Tiger era. More recently, widespread public outrage was generated after an undercover documentary was aired by the state broadcaster in 2015 which strongly alleged a host of wrongdoing from politicians flouting rules about non-disclosure of business interests, as well as councillors seeking money from developers and business interests in return for 'assistance' through the planning process (see RTÉ, 2016). The unethical practices that have been laid bare through the aforementioned channels have been closely implicated in terms of driving urban sprawl into the countryside and fuelling Ireland's property bubble which collapsed in dramatic fashion in 2008 with devastating social and economic consequences (see Fraser et al., 2013; Kitchin et al., 2012; Murphy and Scott, 2013).

The drivers, mechanics and effects of clientelism in rural planning

Political power in rural planning

In seeking to understand the distorting effects of clientelism and corruption in rural communities, it is important to establish what gives rise to the emergence of these relationships in the first instance and how the dynamics at play might differ from what we see in urban centres. The basic premise underpinning these interactions is that politicians have the power to deliver certain benefits to an individual or group in return for some form of reward. In Ireland, the power of local elected representatives to secure benefits in the land development process is significant.

Specifically, the Irish planning system is a plan-led system where the main instrument for the regulation and control of development is the development plan which must be updated every six years. The adoption or making of a development plan is the 'reserved function' of the local elected representatives who have ultimate decision making-powers in terms of crucial land zoning decisions and policy formulation and are under 'no obligation to take the recommendations of experts [planning officials] into account' (Kitchin et al., 2012: 1315). The fact that politicians have a direct role in deciding specific, individual policy decisions of high value to wealthy businesses interests makes them particularly vulnerable to pressures from landowners and development interests (Collins and O'Shea, 2003; Kitchin et al., 2012; Fox-Rogers and Murphy, 2015).

Within rural areas, these pressures can be particularly pronounced on the basis of the dramatic rise in land value that is derived from a decision to zone a piece of farmland for development. For instance, while the average price of agricultural land in Ireland in 2017 stands at around €9,550 per acre (Finnerty, 2017), once zoned it can realistically achieve a multiple of that figure and such decisions routinely turn fortunate landowners into 'paper millionaires overnight' (Fox-Rogers, 2014: 298). Although the same phenomenon undoubtedly exists in large urban centres, the pressure on councillors to zone land for development can be more acute in peripheral locations given the widespread availability of agricultural land which presents lucrative opportunities for landowners with otherwise limited options to make a high return on their landholding. Another distinctive feature worth noting is that greenfield land banks in urban centres are scarce and tend to be tightly controlled (and at times hoarded) by a smaller pool of speculative landowning/development interests – an issue which has become critical in the Dublin region (see Murphy, 2017). By way of contrast, land in rural areas is generally acquired through family inheritance rather than speculative endeavours. As such, landownership is somewhat more dispersed resulting in councillors facing regular pressures from a broader range of landowning interests ranging from large-scale farmers to more modest landowners, all of whom are seeking to achieve the windfall gains that can arise on the foot of a favourable zoning decision.

It is important to note that while the development management process (the consent procedure for authorising development) is an 'executive function' of the chief executive (or delegated officials) of a local authority, councillors nevertheless also come under pressure from individuals regarding individual applications in the hope of obtaining planning permission – a phenomenon that is particularly prevalent in relation to applications for one-off rural houses (Fox-Rogers and Murphy, 2014). This peculiarity stems from the fact that councillors have historically had powers to direct the chief executive on how to perform any their executive duties and could thus override specific planning decisions by passing a 'Section 140' motion. Indeed, Grist (2012) and Komito (1984) highlight that these motions were widely abused by councillors for many years to ensure that certain applications were approved. While this mechanism has been removed in relation to planning matters, there is evidence to suggest that councillors still seek to influence individual planning applications albeit through more informal channels (see Fox-Rogers and Murphy, 2014). For instance, elected members may make repeated calls to planners about specific applications and can be caught 'mooching around' planners' desks in a bid to influence their recommendations (ibid.). This has led to fraught relationships between planners and councillors and various steps have been taken to alleviate this phenomenon including the redesign of local authority buildings in a bid to reduce councillors' access to planning officials (Fox-Rogers, 2014).

Community power in rural planning

Although councillors may seek to resist the pressures mentioned above, it is important to note that there are various inducements which can motivate councillors to exert their influence (through formal or informal channels) in ways that reflect individual, over public, interests. At the more general level, councillors may act on behalf of individual constituents in the hope that the 'incurred debt' will be repaid on election day (Komito, 1984: 180). This can mean that within rural communities, where the desire to build one-off housing is particularly robust, councillors may find themselves at considerable risk of losing their seats if they take a strong stance against this practice which has become normalised throughout rural Ireland (see Gkartzios and Scott, 2009; Scott and Murray, 2009).

Moreover, well-resourced individuals in the community (i.e. local capital interests) can exert considerable influence over councillors by employing a diverse range of lobbying techniques which require economic resources. These can range from traditional approaches such as political donations to help fund electoral campaigns, to more extreme cases of corrupt payments being made directly to councillors for their personal benefit. In recent years, it apparent that there has been a shift away from these more explicit forms of lobbying towards more subtle approaches that are less likely to invite criticism of wrongdoing (see Fox-Rogers and Murphy, 2015). More specifically, recent corruption scandals have forced landowners and developers to devise more innovative ways to reward local politicians for their assistance in the planning process that are less transparent and negate the need for declaring political donations. Examples include offers of planning gain (e.g. provision of additional local sports facilities) as part of a development proposal to generate wider political support among the electorate for the politician/s who lend their support to the proposals (ibid.). This phenomenon spans both rural and urban communities and in many respects can be more a more effective strategy in more peripheral local authorities where the dearth of public investment since the onset of the economic crash in 2008 has exacerbated the inherent reliance on the private sector for the development of new facilities even further (ibid.).

While the uneven distribution of economic resources is central to understanding the basis of clientelist (or corrupt) exchanges, it is important to note that less structural factors also need to be considered. Specifically, personal contacts and friendships can play an important role in terms of determining one's ability to exercise their stake in the planning process. For instance, the ability of councillors to resist the temptation to lend their support for a zoning submissions or planning application can be reduced if the proposal comes from a person whom they know and like on a personal level based on human nature. While these connections can (and do) develop along structural lines (e.g. business relationships etc.), they can also emerge organically on the basis of living in close proximity to one another, drinking in the same pub, attending the same school, or having children playing on the same football team etc. (Fox-Rogers and Murphy, 2014). In more rural areas, this phenomenon can be particularly pronounced as local elected representatives have a tendency to have a closer connection to their constituents and know them on a first name basis relative to urban areas (see Jiménez et al., 2012). These connections can be powerful in terms of determining who exercises power in the planning process as 'the resource of 'personal contacts' is unevenly distributed' (Komito, 1984: 187).

Manifestations of rural clientelism

The above suggests that slightly different power dynamics are at play within rural communities. While economic resources (including landownership) remain critically important in terms of understanding the uneven power relations that exist in the planning process, personal contacts and friendships can also play a central role in terms of exercising influence over local decision-makers. In this regard, the centre of power in rural areas is somewhat more dispersed than in urban centres as access to central decision-makers is easier based on personal friendships and there is less of a monopoly in terms of landownership. Although any mention of more dispersed power might signal the delivery of more democratic planning outcomes, it is clear that the relationships between rural landowners and local elected representatives outlined above can have serious implications in terms of sustainable planning decision-making. Central to this is the fact that elected representatives' desire to remain in power means that the professional recommendations and advice of professional planners in local authorities are often overlooked, with decisions being reached which reflect the interests of political 'clients' rather than any notion of the common good. In Ireland, the rejection of planners' advice by councillors has been associated with a whole host of planning issues, most notably the excessive zoning of land. This has precluded a more gradual and incremental pattern of growth from taking place, with development increasingly occurring on the outskirts of villages and towns with little connections to existing built up areas. Moreover, the informal influence that councillors can exert in the development management process to secure planning permission on behalf of their constituents (particularly one-off rural houses) has exacerbated the problem further. These trends have serious social, economic and environmental consequences. For instance, dispersed development patterns not only encroach on the countryside, they also foster greater levels of car dependency, longer commutes and create immense difficulties in terms of environmental protection (e.g. groundwater contamination from private waste water treatment systems etc.) and the provision of high quality physical and social infrastructure nationally (e.g. high speed broadband, transport infrastructure, post offices etc.; see EEA, 2006; Williams and Shiels, 2000; Williams et al., 2010; Murphy, 2012). Moreover, political interference in the planning process has also been implicated in terms of driving the country's property bubble, the collapse of which has created enormous problems not just in terms of economic austerity but has also left physical scars on the environment in the form of unfinished developments or 'ghost estates', many of which are located in more rural areas where there was never sufficient demand to sustain the level of speculative development taking place in these locations (see O'Callaghan et al., 2014).

Aside from the social, economic and environmental manifestations of rural clientelism, there are also broader implications for planning as an institution. Clientelist exchanges not only preserve the status quo by helping local politicians retain their power, they also deliver planning outcomes that reflect private interests at the expense of others. In this regard, it is clear that such relationships undermine the democratic nature of planning and discredit the rhetoric surrounding collaborative or participatory approaches that are so often advocated within the context of community planning. In essence clientelism and corruption undermine the functioning of a democratic planning system in the same way that it endangers the functioning of democracy more generally (see Della Porta and Meny, 1997). As Robles-Egea and Aceituno-Montes (2011: 9) put it: 'a quality democracy . . . is incompatible with . . . patronage-clientelism and corrupt practices'.

Conclusions

Combatting issues of clientelism and corruption remains a fundamental challenge for planning. In Ireland, recent attempts at tackling this issue have been geared towards putting more checks and balances in place to curb the problem of 'excessive zoning and permissions being granted' (Kitchin, 2012: 1314). Like many other countries that have experienced similar problems, the Irish response has largely centred on legal reforms and more monitoring mechanisms (see Grist, 2012). For instance, the concept of 'core strategies' has been introduced to the development plan making process to ensure that local policies (including zonings) are consistent with regional planning policy. The power of councillors in the development management process has also been curtailed significantly and a web-based lobbying register (www.lobbying.ie) has been established to identify those communicating with designated public officials on specific policy, legislative matters or prospective decisions. While these interventions have largely arisen based on the recommendations arising from the Mahon Tribunal, other key recommendations such as the establishment of a Planning Regulator have yet to be implemented 6 years after the publication of the final report.

While the effectiveness of these measures remains to be seen, these types of reforms that focus on the formal institutional sphere have yielded poor results elsewhere (Jiménez, 2009). This can be largely attributed to the 'local and informal institutional features which affect the formal rules and regulations' and the problem of collective action as there is a 'shared expectation that people do not respect the law' (Jiménez et al., 2012: 366). Moreover, it is apparent that influence can still be exercised in more opaque processes that exist within the shadows of the planning system's formal structures (see Fox-Rogers and Murphy, 2014). In many respects, the tightening up of existing checks and balances has had the unintended consequence of reducing accountability and transparency in the planning system as clientelist networks develop more innovative channels of influence that are more difficult to detect (Fox-Rogers and Murphy, 2015). What is clear from the above discussion is that the existence of a shadow planning system is not a distinctly urban phenomenon and more attention needs to be given to how rural stakeholders gain access to, and utilise these informal channels of influence to serve their own ends.

This chapter attempts to shed light on some of these issues but more needs to be done to redress the dearth of empirical enquiry surrounding the convergence of political and landowning/development interests in rural planning. In doing so, attention needs to focus on how power relations develop and play out within rural communities in ways which distort the democratic nature of planning practice. While Dodson et al.'s (2006) call for more empirical investigation into the impacts of corruption on the built environment is certainly warranted, care needs to be given to ensure that sufficient attention is also afforded to the social and economic implications. Further research is also needed to explore the impacts that clientelism and corruption can have on public perceptions of planning with specific regard to levels of public trust and confidence in the system and profession as a whole. In doing so, in-depth qualitative research which allows for a much deeper penetration into this sensitive subject area is needed rather than relying solely on more quantitative-based surveys as suggested by Dodson et al. (ibid.). It is hoped that this chapter stimulates more empirical enquiries in this field and can help enhance our understanding of how clientelism and corruption can be contested through interventions which take account of both the formal and informal institutions that constitute the planning system.

References

An Taisce (2012) *State of the Nation: A Review of Ireland's Planning System 2000–2011*. Dublin: An Taisce.

Brindley, T., Rydin, Y. and Stoker, G. (1989) *Remaking Planning: The Politics of Urban Change in the Thatcher Years*. London: Unwin Hyman.

Chiodelli, F. and Moroni, S. (2015) 'Corruption in Land-Use Issues: A Crucial Challenge for Planning Theory and Practice'. *Town Planning Review*, 86(4), 437–455.

Collins, N. and O'Shea, M. (2003) Political Corruption in Ireland. In M. J. Bull and J. L. Newell (eds), *Corruption in Contemporary Politics*, 164–177. Basingstoke: Palgrave Macmillan.

CSO (2017) 'Census 2016 Summary Results – Part 1'. Retrieved from www.cso.ie/en/media/csoie/newsevents/documents/census2016summaryresultspart1/Census2016SummaryPart1.pdf (accessed 15 December 2018).

Day, P. (1995) *Land: The Elusive Quest for Social Justice, Taxation Reform and a Sustainable Planetary Environment*. Brisbane: Australian Academic Press.

Della Porta, D. and Meny, Y. (eds) (1997) *Democracy and Corruption in Europe*. London: Pinter.

Dodson, J., Coiacetto, E. and Ellway, C. (2006) 'Corruption in the Australian Land Development Process: Identifying a Research Agenda'. In P. Troy (ed.), *Refereed Proceedings of the 2nd Bi-Annual National Conference on The State of Australian Cities*, 1–25. Brisbane: Griffith University. Retrieved from https://research-repository.griffith.edu.au/bitstream/handle/10072/11501/Dodson2006Corruption_LandDevt_SOAC.pdf?sequence=1&isAllowed=y.

EEA (2006) *Urban Sprawl in Europe: The Ignored Challenge*. Copenhagen: European Environment Agency.

Finnerty, C. (2017) 'What was the average cost of agricultural land in the first half of 2017?'. Retrieved from www.agriland.ie/farming-news/what-was-the-average-cost-of-agricultural-land-in-the-first-half-of-2017 (accessed 22 January 2018).

Fox-Rogers, L. (2014) 'Power and the Planning System: A Political Economy Perspective'. Unpublished PhD thesis. Dublin: University College Dublin.

Fox-Rogers, L. and Murphy, E. (2014) 'Informal Strategies of Power in the Local Planning System'. *Planning Theory*, 14(3), 244–268.

Fox-Rogers, L. and Murphy, E. (2015) 'From Brown Envelopes to Community Benefits: The Co-option of Planning Gain Agreements under Deepening Neoliberalism'. *Geoforum*, 67, 41–50.

Fraser, A, Murphy, E. and Kelly, S. (2013) 'Deepening Neoliberalism via Austerity and "Reform": The Case of Ireland'. *Human Geography*, 6, 38–53.

Gkartzios, M. and Scott, M. (2009) 'Planning for Rural Housing in the Republic of Ireland: From National Spatial Strategies to Development Plans'. *European Planning Studies*, 17(12), 1751–1780.

Government of Ireland (2012) *The Tribunal of Enquiry into Certain Planning Matters and Payments*. Dublin: Dublin Stationary Office.

Grist, B. (2012) *An Introduction to Irish Planning Law* (2nd edition). Dublin: Institute of Public Administration.

Harvey, D. (1985) *The Urbanisation of Capital*. Oxford: Basil Blackwell.

Healey, P. (1992) 'Planning Through Debate: The Communicative Turn in Planning Theory'. *Town Planning Review*, 63(2), 143–162.

Healey, P. (1997) *Collaborative Planning: Shaping Places in Fragmented Societies*. Basingstoke: Macmillian.

Hibbard, M. and Lurie, S. (2000). 'Saving Land but Losing Ground: Challenges to Community Planning in the Era of Participation'. *Journal of Planning Education and Research*, 20(2), 187–195.

Jiménez, F. (2009) 'Building Boom and Political Corruption in Spain'. *South European Society and Politics*, 14(5), 255–272.

Jiménez, F., Villoria, M. and Quesada, M. G. (2012) 'Badly Designed Institutions, Informal Rules and Perverse Incentives: Local Government Corruption in Spain'. *Journal of Local Self-Government*, 10(4), 363–381.

Kitchin, R., O'Callaghan, C., Gleeson, J., Keaveney, K. and Boyle, M. (2012) 'Placing Neoliberalism: The Rise and Fall of Ireland's Celtic Tiger'. *Environment and Planning A*, 44(6), 1302–1326.

Komito, L. (1984) 'Irish Clientelism: A Reappraisal'. *The Economic and Social Review*, 15(3), 173–194.

Komito, L. (1985) 'Politics and Clientelism in Urban Ireland: Information, Reputation and Brokerage'. Unpublished PhD thesis. University of Pennsylvania. Retrieved from www.ucd.ie/lkomito/thesis.htm (accessed 11 November 2017).

McMahon, A. (2016) 'Ireland's Population One of Most Rural in European Union'. *The Irish Times*. Retrieved from www.irishtimes.com/news/health/ireland-s-population-one-of-most-rural-in-european-union-1.2667855 (accessed 15 December 2017).

Murphy, D. (2017) 'Clampdown on Land Hoarders Part of Housing Crisis Solution'. Retrieved from www.rte.ie/news/business-analysis/2017/0715/890325-clamp-down-on-land-hoarders-part-of-housing-solution (accessed 28 November 2017).

Murphy, E. and Scott, M. (2013) 'Mortgage-Related Issues in a Crisis Economy: Evidence from Rural Households in Ireland'. *Geoforum*, 46, 34–44.

Neuman, M. (2000) 'Communicate This! Does Consensus Lead to Advocacy and Pluralism?'. *Journal of Planning Education and Research*, 19, 343–349.

O'Callaghan, C. Boyle, M. and Kitchin, R. (2014) 'Post-Politics, Crisis, and Ireland's "Ghost Estates"'. *Political Geography*, 42, 121–133.

Robles-Egea, A. and Aceituno-Montes, J.M. (2011) 'An Impossible Democracy: Political Clientelism and Corruption in Andalusia'. Paper presented to at the ECPR General Conference, Reykjavik, 24–27 August 2011.

RTÉ (2016) 'RTÉ Investigates – Standards in Public Office'. Retrieved from www.rte.ie/news/investigations-unit/2015/1207/751833-rte-investigates (accessed 15 January 2018).

Scott, M. and Murray, M. (2009) Housing rural communities: Connecting rural dwellings to rural development in Ireland. *Housing Studies*, 24(6), 755–774.

Tewdwr-Jones, M. and Allmendinger, P. (1998) 'Deconstructing Communicative Rationality: A Critique of Habermasian Collaborative Planning'. *Environment and Planning A*, 30, 1975–1989.

Williams, B. and Shiels, P. (2000) 'Acceleration into Sprawl: Causes and Potential Policy Responses'. *Quarterly Economic Commentary*, June, 37–73.

Williams, B., Walsh, C. and Boyle, I. (2010) 'The Functional Urban Region of Dublin: Implications for Regional Development Markets and Planning'. *Journal of Irish Urban Studies*, 7–9, 5–30.

Part III
Planning for the rural economy

The contributions to this section explore theories and practices of rural development in diverse settings. They focus on approaches to development, its potential drivers, urban–rural linkages and the ways in which different rural assets can be mobilised in support of economic security and wellbeing.

The first chapter in Part III, from Gkartzios and Lowe, provides a broad view of the rural development literature, revisiting key thinking on – and critiques of – exogenous and endogenous development 'approaches' before re-examining neo-endogenous development as an attempt to 'rationalise . . . efforts on the ground'. The neo-endogenous 'model' sets local actions within a broader governance frame, comprising elements of external support and control. The chapter concludes with the view that ideas guiding development practice have to be grounded in a deep and nuanced understanding of that practice and the contexts in which it occurs. Bock's reference to a 'nexogenous' approach to development is used to highlight how innovative practice connects to place and how disconnected 'self-help' is too crude a pathway to development in marginal rural areas. Greater *reflexivity* is presented, finally, as a route to more effective, varied and locally innovative development practice. That reflexivity needs to centre on the co-production of knowledge between, for example, academia and practice and reject any artificial boundary between internal and external actors or between approaches, embracing instead a full range of groups and ideas.

Focusing on the 'rural' in regional development, Tomaney and colleagues return the discussion to a broader focus. Chapter 15's concern is with the city-region as a concept and the rural as 'hinterland'. Regional development frameworks adopt a relational perspective on rural–urban places and this chapter seeks to unpack the changing nature of those relationships, showing how, despite the endurance of food security concerns, the broader focus of rural economies has shifted away from agricultural production. Small towns play an increasingly important role in regional development as anchor points and development hubs. The chapter explores the broadening regional development focus in five contexts – France, the Netherlands, Poland, Australia and the Nordic countries: the latter represented by Finland Norway and Sweden. France is presented as exemplifying an integrated approach to regional planning and rural development – achieved through plans that position regions 'as lead actors in the field of planning and sustainable development'. Implementation by, and cooperation between, France's large number of rural communes remains a challenge however, as does the use of consultants to design and develop strategies, which often do not connect with local knowledge. A key challenge in the Netherlands is the management of flood waters, which requires significant infrastructure investments. A move is afoot to integrate existing regulations into a single Environment and Planning Act that will aim to ensure oversight of potential planning and development conflicts which, it is claimed, have been accentuated by past separation of responsibilities and distribution of regulations between different pieces of legislation. Poland has a nested hierarchy of spatial development strategies. At the lowest level, spatial plans are concerned with implementation of polices. But there is fragmented plan coverage at that level – which works against cooperation and integrated planning, and is seen as being responsible for a high rate of uncoordinated conversion of arable land to urban uses in peri-urban areas. In Australia, the State of Victoria has developed regional plans that aim to integrate 'land use and infrastructure planning at the sub-regional scale'. Moreover, 'regional growth plans' seek to manage the distribution of population in peri-urban areas and protect high-amenity landscapes. Finally, in the Nordic countries, regional planning must address the needs of remoter areas with often very low population densities. Particular issues here include balancing the rights of indigenous populations with significant infrastructure investments needed to develop new sectors and realise development. Those sectors, including forestry, impact on the environment in ways that sometimes conflict with the tourism sector. Big trade-offs need to be addressed through spatial planning at a regional scale. More broadly, the chapter concludes by emphasising what can be achieved in rural areas at the regional level – a coordination of plans and actions, including infrastructure investments, which help build / sustain connections and foster the wellbeing of communities.

Rural areas are, according to Papadopoulos, 'part of an interconnected world driven by market forces, mobilities and globalisation'. His contribution to this section is concerned with this connectivity and with the impacts of the 2007/8 global financial crisis on rural areas, owing to interconnectedness and because of the austerity that arrived in the wake of the crisis. Continuing in the same vein as Tomaney and colleagues, Chapter 16 explores the uneven geographies of regional development and the particular vulnerabilities of rural areas in Europe. It starts with an overview of the financial crisis, borrowing from Varoufakis's analysis of the causes of the crisis and its impacts on the Eurozone. High ratios of government debt to GDP triggered swingeing cuts in public spending, which disproportionately impacted on more vulnerable, remoter regions. Housing markets in southern Europe and Ireland faced a particularly stern test, with prices contracting and household indebtedness rising. This corresponded with escalating unemployment and an exodus from many rural areas, again especially from southern Europe. The conclusion reached is that space/place matters: the financial crisis accentuated existing patterns of spatial disadvantage, magnifying vulnerabilities and impacting most on the poorest and remotest parts of Europe. The concept of resilience – 'a loose antonym to vulnerability' – is used to frame Papadopoulos's analysis of rural responses to crisis. Sweeping through the broad literature on this concept, it is accepted that resilience refers to the adaptive capacity of regions. Examples of rural resilience are drawn from research in Greece and Spain. Different territorial dynamics, including infrastructural / institutional capacity and economic composition, can support different adaptive capacities and hence resilience to shocks. These can be exploited by coordinated interventions.

Innovation, and the role of small businesses in driving growth and productivity, is a theme carried forward by Cowie and colleagues in Chapter 17. A broad overview is provided of innovation and entrepreneurship, which is often associated with opportunities for clustering and agglomeration, more often found in urban than in rural areas. However, the view of rural areas as 'static and traditional' is now outdated: despite fewer patents being registered in these areas, they nevertheless contain 'a diverse range of engines for economic growth'. Many rural areas are also on an upward track in terms of new product development, spurred on by the barriers to growth that often characterise such locations. The chapter focuses on the nature of these barriers, presenting them as inhibitors in the first instance but ultimately as incubators of necessary innovation. Face to face contact often seeds innovation practice and cannot be wholly substituted with ICT as a platform for knowledge sharing activities. In response to this barrier, 'enterprise hubs' have been established in some rural areas, notably in the north east of England. The authors examine the role that such hubs have played in centring the flow of knowledge in rural areas and thereby fostering innovation through the co-location of businesses, concluding that the hubs are an innovative translation of an urban idea into a rural setting, which have achieved some success in facilitating the expansion of businesses into new products and markets. They have provided essential incubators for start-ups in their early years of development, often being promoted by the state – which plays a key role in establishing the essential infrastructure for growth and prosperity, through investment and effective planning.

Herslund (Chapter 18) returns the focus to rural business start-ups, innovation and the role of the 'creative class' (as defined by R. Florida) in rural locations. She examines what it means to 'be creative' in the countryside and how that creativity allows individuals to overcome many of the regular barriers to business success identified by Cowie and colleagues. The motivations of creative individuals arriving in rural areas are examined alongside the challenges that policy-makers face in retaining and nurturing 'creative talent'. The chapter draws on interviews with in-migrants starting micro-businesses in Denmark, some undertaken a decade ago and some more recent second interviews with the same subjects – to gain a view of their unfolding experiences. Interviews addressed the challenges of adapting business plans to rural contexts and how adaptation evolved over time, with the setting up of support groups (micro-networks) that provided opportunities to share experiences and learn from the mistakes/successes of others. These are found to be hugely important in incubating business success but also difficult to establish and operate as funding is seldom available. Rather, they rely on the enthusiasm and commitment of volunteers, although very occasionally municipalities will step in and provide limited support. Herslund ends her contribution by asking whether the 'creative class' is a valid label for newcomer-led business start-ups in rural areas. She argues that creativity, however, is not confined to any one sector/place but rather references the way in which micro-businesses display adaptive capabilities, often achieved through effective networking and knowledge sharing.

In Chapter 19, Juntti shifts the focus away from broader questions of development to another core concept – ecosystem services (ES). More specifically, the chapter looks at payment for ecosystem services (PES) as acknowledgement of 'the value of the natural capital on which our economies depend'. It is concerned with the way in which sustainable economic activities in rural areas are encouraged and incentivised: specifically by establishing the 'market signals' (costs) that will 'incentivise . . . sustainable use' of resources during the conduct of economic activities (farming, tourism development etc.). Payments can help diversify local economies towards conservation activities that are assigned monetary value and which constitute defined sustainable management solutions. However, Juntti identifies the challenges of establishing such a system and the difficulties of assigning value to different ES. Values have been catalogued for a range of purposes – not only as a basis for payments but also for estimating the cost of environmental change or to calculate the value of investment in rural land assets. It is payments to farmers, however, that would appear to have the clearest role in growing incomes and supporting economic diversification. Farmers benefit directly and other businesses, dependent on the quality of ES (the example given is bottled water producers), are unaffected by activities that might otherwise have negative environmental impacts and threaten business income. But not all PES arrangements involve cash payments: some, especially in developing countries, involve in-kind payments (including 'free' labour) to ensure that forest managers, for example, can undertake the kinds of local projects that support sustained or enhanced water access and quality. Elsewhere in the global south, access to communal land for purposes of tourism or hunting may be granted in exchange for direct payments to a community: hence 'recreational ES' is monetarised. This can be problematic if foreigners (for example) engage in illegal big-game hunting and threaten already-declining species. But more broadly, 'environmental values and assets . . . can be engaged to ease the dependency of rural economies on the volatile and declining returns from agriculture'. PES potentially add to the adaptive capacity of rural regions, by providing a source of diversification and also safeguarding natural capital. Moreover, they incentivise the sorts of behaviours that seem crucial to tackling 'planetary challenges' such as climate change – hence extending well beyond the economic development domain and providing rural planning and policy with a means of delivery many of its core objectives, if it is prepared to embrace the 'neo-liberal shift in environmental governance approaches'.

The final chapter in Part III, from Scott, examines the 'role of spatial planning in promoting economic development and sustainable livelihoods in rural places'. It begins by arguing that 'planning' is too often pre-occupied with protection and preservation, and sometimes poorly connected with issues of economic development. There are opportunities, however, for planning to 'coordinate the spatial and territorial dimensions of rural development'. Diversification of rural economies is presented as a key public policy goal, which spatial planning might have a more central role in delivering. Examining how its role might evolve is the ultimate aim of the chapter, though the initial focus is on defining the challenge of rural economic diversification – what obstacles stand in its way, and what opportunities might be grasped. A picture of significant spatial variability is painted, with different rural areas enjoying/enduring different locational and resource opportunities/constraints. Garrod and colleagues' conceptualisation of rural resources as constituent 'capitals' is picked out as a useful frame for thinking about the interventions and assets that might be mobilised through planning. But the planning system has hitherto been marginal in enabling rural economic growth, focusing narrowly on protecting farming interests, halting development and framing sustainability as something only deliverable through urban-style concentration. Emerging from this critique, Scott tries to draw together key principles for supporting rural economies. Numerous ways in which planning might enable economic diversification and vitality are catalogued, stretching from rural enterprise zones to other place-based interventions delivering greater flexibility for development. Bridging the gap between environmental and job-creation priorities, perhaps through support for the eco-economy, is another area in which planning might contribute more. Agreeing with Tomaney and colleagues, Scott argues that taking an integrative / relational approach to rural development through regional plans and planning also seems sensible, as 'meaningful scales of action' are often only possible at a regional/strategic level. Similarly, harnessing the economic potential of heritage assets is another area in which planning could do more, by coordinating traditional management activities with strategies to valorise such assets. The right infrastructural investment – which is well coordinated – will also be instrumental in diversifying rural economies, with ICT and the potential for seeding 'smart rural economies' flagged here as particularly important. Finally, planning's role in delivering an open and inclusive countryside seems critical, with affordable and accessible housing being key to the future strength of rural economies. Taking a cue from other contributions to this section, and other chapters in this volume, Scott argues that rural areas must be open to new ideas and new contributions, often from migrants, if they are to prosper. Planning, it is contended, must be an enabler and a mobiliser of rural development. At the same time, it must mediate rural change in such a way as to not stifle innovation or development opportunities. Movements in this direction will represent a paradigmatic change for some planning systems in some countries, which have displayed a tendency to resist rather than embrace change, often at the behest of vested conservatism.

The contributions to Part III track multiple dimensions of rural economic development, starting with explanatory models and ending with planning principles. Their coverage extends from broad economic development strategy through business innovation, creativity and entrepreneurialism, to the rural resource dimension of economic development, represented here by a focus on payments for ecosystem services. While this introduction seeks to tie these themes together for the reader, it is Scott's closing contribution that gathers the different threads together into a coherent overview of the context for economic development and spatial planning in rural areas.

14

Revisiting neo-endogenous rural development

Menelaos Gkartzios and Philip Lowe

Introduction

The quest for a comprehensive rural development theory has been long. Various efforts have emerged focusing primarily on the governance mechanisms of rural development as well as the priorities of rural policy remits, across a mosaic of production and consumption interests, and the characteristics of rural areas. What all attempts towards a theory have in common is the acceptance that it is vital to understand how things work in practice, suggesting a more discursive approach to the production of rural development knowledge. As van der Ploeg and colleagues point out:

> The hard core of what constitutes the essence of rural development will emerge as the strength, scope and impact of current *rural development practices* become clear. Much will depend on the capacity of scholars to develop an empirically grounded theory.
>
> *(Van der Ploeg et al., 2000: 391)*

In this context, this chapter reviews attempts to engineer rural development theory from the early modernisation project of exogenous rural development to more 'bottom-up' models encapsulated by the concept of endogenous development, and eventually towards a neo-endogenous approach. The chapter offers critical insights for developing such approaches based on more recent contributions (for example Bock's suggestion of nexogenous rural development), and less desired practices observed in rural planning policy (for example Gkartzios and Scott's observations of pseudo-endogenous development in a housing context). It is argued that neo-endogenous thinking can move forward through an explicitly reflexive approach in the production of knowledge on rural development, facilitated by practices such as networks across polity and the academy, interdisciplinary methodologies and international comparative research.

Exogenous model

In the post-Second World War period in Europe, a modernisation model of rural development emerged, usually termed 'exogenous' rural studies (i.e., 'derived from outside'), with its key characteristics described in Table 14.1. In this model, rural areas were treated as dependent (technically, culturally and economically) on urban centres, while the main function of rural areas was to provide food for the ever expanding urban populations (Lowe et al., 1998). Lowe et al. (1995: 89) criticise various assumptions in the exogenous model, which operated within a narrow productivist policy frame: 'The spatial category of rural was often viewed as a residual category and became equated with the sectoral category of agriculture'.

As a consequence, a discourse around marginality and peripherality was often used to address and seek solutions to rural development problems. This 'rural pathology' is often witnessed in European policy of the time, particularly in the first phase of this model which refers to the consolidation of farm structures (such as land reforms), land improvement schemes (such as drainage and irrigation) and the development of farm-oriented infrastructure. The second phase acknowledges a new focus on attracting new types of employment to rural areas, for example through tourism, supporting firms to relocate in rural settings and by investing in transportation and communication links between urban and rural areas (Lowe et al., 1995).

Woods (2005) argues that exogenous rural development had its successes, such as increased employment rates in rural areas, improvements in technology, communication and infrastructure as well as combating prolonged rural depopulation in certain cases. However, he criticises this model on the grounds that, first, exogenous development is dependent on external investment (and consequently the profits of the development are often exported and not diffused locally) and, secondly, that the non-participatory nature of the model can create a democratic deficit. Similar concerns have been expressed by many social scientists (see for example Mitchell and Madden, 2014; van der Ploeg et al., 2000). In summary, exogenous approaches to development have been heavily criticised for promoting the following (Lowe et al., 1998):

- *dependent* development, reliant on continued subsidies and the policy decision of distant agencies or boardrooms;
- *distorted* development, which boosts single sectors, selected settlements and certain types of business, but leaves others behind and neglects non-economic aspects of rural life;
- *destructive* development, as it erases the cultural and environmental differences of rural areas; and,
- *dictated* development, as it is devised by external experts and planners.

Table 14.1 Exogenous model of rural development.

Key principle	Economies of scale and concentration
Dynamic force	Urban growth poles; the main forces of development conceived as emanating from outside rural areas
Function of rural areas	Food and other primary production for the expanding urban economy
Major rural area problems	Low productivity and peripherality
Focus of rural development	– Agricultural industrialisation and specialisation – Encouragement of labour and capital mobility

Source: Lowe et al. (1998: 7)

Endogenous model

These criticisms eventually found expression in rural development policy which sought to address not only the productivist myopia of rural policies, but also the top-down governance in which they were framed. In particular, a fundamental shift has characterised rural policy, from sectoral supports (predominantly about agriculture) to territorial development and spatial approaches (Moseley, 1997, 2000; Shortall and Shucksmith, 2001; OECD, 2006). While the position of agriculture in rural development is still pre-eminent for many countries (van der Ploeg and Renting, 2000), this has been in relative decline (Woods, 2005), thus eroding the distinctiveness of rural economies. Indeed, in some contexts, such as the UK, urban and rural economies are broadly similar in composition, with differentiation *across* rural areas being of greater significance (OECD, 2011). Parallel to these changes, the emergent policy discourse advocates that territorial approaches should integrate the delivery of separate sectoral dimensions of public policy (agriculture, housing, employment creation, transport, etc.) and offer a holistic approach to balancing the economic, social and environmental processes that shape rural areas (these themes are also discussed by Lapping and Scott in Chapter 3 of this Companion).

Examples of such multi-sectoral and multi-scalar initiatives can be seen in many countries (for example, in Japan: OECD, 2016; in Chile: OECD, 2014). In Europe, since the 1990s, much of the focus in rural development practice has been targeted on local action and endogenous ('emerging from within') development initiatives, exemplified by the European Union's Liaisons Entre Actions de Developpement de l'Economie Rurale (LEADER) programme (EC, 2006). In practice, LEADER put at the heart of the development process autonomous local action groups (LAGs) 'working in partnership' across public, private and voluntary sectors (Edwards et al., 2001). The essential elements of this approach to rural development are identified by researchers such as Moseley (1997) and Ray (2000) and include: a territorial and integrated focus; the use of local resources; and local contextualisation through active public participation. As Picchi (1994: 195) argues endogenous development is to be understood as 'local development produced mainly by local impulses and grounded largely on local resources' (cited in Lowe et al., 1995).

In this context of policy transformation, social scientists were faced with the challenge to provide useful models of social science theory to capture these trends. As Lowe et al. (1995: 91) observe, 'the switch from a concern with exogenous to endogenous development strategies has been driven by practical realities and not by theory' (see also van der Ploeg et al., 2000). Neither was there a simple switch from one to the other: endogenous approaches (such as LEADER) sat beside exogenous approaches (such as most of the Common Agricultural Policy).

Endogenous development draws attention to the distinction between local and external actors having *control* of the development process (Lowe et al., 1995) with the endogenous model favouring a 'mosaic' of local action (Ray, 2006). According to Ray (1997) an endogenous approach to rural development has the three following main characteristics:

- It sets development activity within a territorial rather than sectoral framework, with the scale of territory being smaller than the nation.
- Economic and other development activity is restructured in ways so as to maximise the retention of benefits within the local territory by valorising and exploiting local resources – both physical and human.
- Development is contextualised by focusing on the needs, capacities and perspectives of local people.

Similarly, Lowe et al. (1998) summarise the characteristics of the endogenous model of rural development as shown in Table 14.2.

The discourse of endogenous/exogenous development has been criticised, not least because it creates a 'development dichotomy' (Lowe et al., 1998), but also because it fails to appreciate wider links and power struggles throughout the development process (drawing on Whatmore, 1994). Indeed, the endogenous model in principle implies a very different style of policy, dependent on a significant transfer of power away from centrally defined top-down policy, towards participative, community-led action. However, in practice, while endogenous approaches became the norm in rural development policy prescription across Europe (Ray, 2000), the LEADER experience has typically demonstrated problems of participation, elitism and the limitations of local action and control (e.g. Barke and Newton, 1997; Storey, 1999; Bosworth et al., 2016). Shucksmith, for example, argues that 'there is a tendency for endogenous development initiatives to favour those who are already powerful and articulate, and who already enjoy a greater capacity to act and to engage with the initiative' (Shucksmith, 2000: 215). Similar concerns have been expressed by many social scientists in various European contexts (Kovach, 2000; Osti, 2000; Shortall, 2008). The result is often mixed outcomes. For example, Navarro et al. (2016) recently identified limited participation of marginal groups such as unemployed people and young people in LEADER programmes, but increased engagement of women who have been traditionally underrepresented in rural development fora. Certain policy areas such as agricultural production policy, taxation and transport infrastructure policy remain strongly exogenous in their outlook. This is also the case with housing. For example, a property-led regeneration policy through tax incentives in the Republic of Ireland, aimed to repopulate a marginal rural region though sponsoring private housing construction. The policy was developed outside mainstream rural development channels and was top-down in its conception, monitoring and implementation despite the evidence of bottom-up rural policy approaches in the Republic of Ireland (Gkartzios and Norris, 2011).

Even where policy delivery embraces an endogenous rhetoric, it can have adverse effects on rural localities. In a rural planning context for example, in the Republic of Ireland again, a policy that supported local people's housing needs included specific measures devised to support the construction of new private houses in the countryside on the basis of local need criteria such as bloodline, residency, language requirements, etc. in keeping with notions of housing development associated with healthy, diverse and growing rural communities (Gkartzios and Scott, 2009, 2014; Scott, 2012). However, the particular policy in practice has been associated with a clientelist system of local governance and corruption, and has been criticised for contributing to unsustainable rural settlement patterns with associated environmental, social and economic costs, as well as for supporting inequality of access to rural housing. For this reason, Gkartzios

Table 14.2 Endogenous model of rural development.

Key principle	The specific resources of an area (natural, human and cultural) hold the key to its sustainable development
Dynamic force	Local initiative and enterprise
Function of rural areas	Diverse service economies
Major rural area problems	The limited capacity of areas and social groups to participate in economic and development activity
Focus of rural development	– Capacity building (skills, institutions and infrastructure) – Overcoming social exclusion

Source: Lowe et al. (1998: 11)

and Scott (2014) present such policies as pseudo-endogenous development. Similar adverse effects have been described in Britain, when rural planning policies ostensibly aim to prioritise the needs of local people, but which usually result in favouring only certain, and more powerful, social groups (Satsangi et al., 2010; Shucksmith, 1981).

From *models* to *approaches*: the neo-endogenous thinking

The increased influence of external pressures and actors on rural areas has been recognised by Brunori and Rossi (2007), who highlight the role of capital, consumers and regulatory bodies in shaping rural localities through processes of economic globalisation. Similarly, Woods (2007), developing the ideas of a 'globalised countryside', explores how local and global forces reconstitute rural spaces within an extremely differentiated geography, producing hybrid relations. Within this context of a globalised rural economy, it has been argued that endogenous development is not a realistic paradigm. For Ward et al., for example:

> The notion of local rural areas pursuing socio-economic development autonomously of outside influences (whether globalization, external trade or governmental or EU action) may be an ideal but is not a practical proposition in contemporary Europe.
>
> *(Ward et al., 2005: 5)*

It was consequently suggested that there is a need for a hybrid model that goes 'beyond endogenous and exogenous modes', by focusing on the dynamic interactions between local areas and their wider political and other institutional, trading and natural environments' (Ray, 2001: 3–4). Drawing on network analysis (e.g. Dicken and Thrift, 1992; Cooke and Morgan, 1993) and local/non local hybrids (e.g. Amin, 1993), original conceptualisations were made by Lowe et al. to consider hybrid local development narratives too:

> We should, however, recognise and indeed celebrate interconnections between areas and between networks. Then the pressing task becomes studying associations and links as sets of power relations. The object of this analysis should be to ascertain where inequalities and asymmetrics within networks lead to a weakening of already weak actors in peripheral or declining areas.
>
> *(Lowe et al., 1995: 104)*

Ray (2001) eventually proposed the term neo-endogenous development to describe an approach to rural development that is locally rooted, but outward-looking and characterised by dynamic interactions between local areas and their wider environments. Ray argues that neo-endogenous development:

> . . . requires us to recognize that development based on local resources and local participation can, in fact, be animated from three possible directions, separately or together. First, it can be animated by actors *within* the local area. Second, it can be animated *from above* as national governments and/or the EU respond to the logic of contemporary political administrative ideology. Third it can be animated from *the intermediate level*, particularly by non-governmental organizations which see in endogenous development the means by which to pursue their particular agendas. The manifestation of neo-endogenous development in any territory will be the result of various combinations of the *from the above* and *intermediate level* sources interacting with the local level.
>
> *(Ray, 2001: 8–9)*

Neo-endogenous development was thus a perspective on the governance of rural development. It was not a policy prescription devised by social scientists at the time on how development should work in practice. In fact social scientists were late in recognising the potential of rural areas to steer or inflect development pathways to their own advantage:

> The so called population turn-around and the urban–rural employment shift surprised policy makers as much as academics. While social scientists have struggled to come to terms with the meaning and significance of these empirical trends, development agencies have realised that rural areas and regions may possess a growth potential of their own just waiting to be unlocked.
>
> *(Lowe et al. 1995: 92)*

Various works have illustrated the networked nature of knowledge production, contrary to a hierarchical and unidirectional transfer from the academy to policy circles, acknowledging that knowledge is not the exclusive reserve of academics (Lowe and Philipson, 2006; Shortall, 2012). In this context, neo-endogenous development was an effort to rationalise what was actually happening on the ground, a way of thinking about how things work in practice, accepting that rural development knowledge is produced by various agents.

Neo-endogenous thinking embraces the previous endogenous model, in the way for example that rural development is multi-sectoral, 'territorial' and moves this forward by focusing on networks, realising that the development potential requires the merging of both internal and external networks (see also Bosworth et al., 2016). From a planning perspective, this advocates institutional integration (local, regional, national and European) and brokering connections between town and country and new urban–rural and local–global relationships (Scott and Murray, 2009), ideas well embedded in spatial planning practice (Albrechts et al., 2010; Zonneveld and Stead; Davoudi and Stead, 2002). A neo-endogenous approach suggested by Gkartzios and Scott (2014) (drawing on the earlier models described by Lowe et al., 1998), is presented in Table 14.3. The authors here drew particular attention to rural development research, calling for more international comparative perspectives in rural studies, as most research on endogenous and exogenous models has been primarily conducted on a single country and case study basis. More critically, the authors observe that where internationally comparative projects have emerged, these have primarily focused on North American and European experiences.

Various theoretical and policy applications have been described within this hybrid neo-endogenous context, trying to understand local and extra-local agency in rural governance and development processes. This has ranged from the role of universities in rural development through the creation and support of a research-practice rural network (Atterton and Thompson, 2010), to the role of architects in designing affordable and vernacular houses in the countryside (Donovan and Gkartzios, 2014). Special attention has been given to in-migrants in rural areas as extra-local agents of neo-endogenous development given their contributions to local identity, knowledge and skills, even though in terms of employment they might generate few full-time jobs for locals (see contributions by: Steel and Mitchell, 2017; Mitchell and Madden, 2014; Atterton et al., 2011; Bosworth and Atterton, 2012). In the context of innovative service network technologies that support planning practice, Tolón-Becerra et al. (2010) propose a knowledge exchange model for the co-construction of local development plans following neo-endogenous principles. More recent work has also looked at art practice through a neo-endogenous lens, in the ways that art practice supports the merging of local and extra local networks facilitating an understanding of community-nature relations and/or even fractures (Crawshaw and Gkartzios, 2016). Bock (2016), drawing on the concept of social innovation,

Table 14.3 Neo-endogenous model of rural development.

Key principle	Socio-spatial justice and balancing local needs while competing for extra-local people, resources, skills and capital
Dynamic force	Fostering a new urban–rural and local–global relationship through inclusive, multi-scalar and multi-sectoral governance arrangements
Function of rural areas	– Sustaining rural livelihoods, while maintaining natural capital – A mosaic of re-emerging productivist functions and consumerist uses (including housing, services)
Major rural area problems	– Exclusive countrysides – Neoliberal deregulation versus policy apathy and lack of regulation – Climate change challenges – Economic crisis
Focus of rural development	– Place-making and community wellbeing – Building resilient rural places – Coping with the new politics of austerity – Coping with emerging geographies of exclusion and (im)mobility triggered by economic crises – Realising and valorising alternatives to development (especially non neoliberal) in times of crisis
Focus of rural development research	International comparative analysis, dialogues and shared lessons (inclusive of, but not exclusive to the USA/EU contexts)

Source: Gkartzios and Scott (2014: 11)

describes the opportunity it offers in understanding the shifting nature of local and non-local development agents, who in turn produce more nuanced understandings of place, beyond the local and the rural that have occupied development thinking thus far. She makes the case for a new approach to rural development in the particular context of marginal rural areas (see also Chapter 9 of this Companion), introduced as a *nexogenous approach*, that:

> . . . departs from the importance of reconnecting and binding together forces across space. It borrows from the Latin noun '*nexus*' for bond and the Latin verb '*nectere*' for binding together. It underlines the importance of reconnection and re-established socio-political connectivity of especially marginal rural areas. The linkage and collaboration across space give access to exogenous resources, which allow for vitalisation if matched with endogenous forces. The development of marginal areas is seriously hampered if social innovation is understood simply as self-help and an indication that marginal rural areas have to rescue themselves. Then, social innovation reconfirms their material, symbolic and political disconnection.
>
> *(Bock 2016: 570)*

What these contributions demonstrate, *inter alia*, is that the demand for a single, all-inclusive model or theory of rural development is no longer a requirement or a realistic expectation. Diverse approaches will emerge given the unique cultural and knowledge linkages emerging in different spatial contexts. In all cases, however, there is an explicit focus on the creation, valorisation and continuation of both local and extra-local networks that facilitate knowledge exchange that create opportunities for the benefit of rural areas.

Conclusions: knowledge and the quest for reflexivity

Exogenous and endogenous 'ideal' models were somewhat, but not equally, naïve in trying to theorise the rural development process. What both models lacked was (self) reflexivity concerning the actors involved in local development as well as their agency, although the endogenous model and associated rural development practice (i.e. the LEADER programme) were relatively pioneering in transferring – with observed limitations – power from central experts to local communities. The contribution of the neo-endogenous thesis was not to present a *model* of development but rather a *way* of thinking about rural development and understanding how things work on the ground. Research on neo-endogenous development approaches has thus drawn attention to power struggles centred on the interactions of local and non-local actors in steering the development potential of rural areas. An example of understanding the terrain of power in rural development processes is provided by notion of the 'differentiated countryside' which, drawing on England, offered a new rural 'non-typology' of understanding the development trajectories of rural areas and the associated positionality of various actors (farmers, landowners, in-migrants, commuters, the state, etc.; Murdoch et al., 2003).

Implicitly at least, neo-endogenous thinking constitutes a reflexive practice which requires and embraces the involvement of multiple actors (including academics too) and seeks to understand their positionality on the development of rural areas. Central to this practice of neo-endogenous thinking is the ability to (self) reflect on one's own disciplinary perspectives and agency in the production of knowledge in support of rural policy-making. Despite numerous contributions to neo-endogenous practices in Europe and elsewhere, what remains absent in these debates is reflexivity on the role of the academy and in particular of social scientists in the production of rural development policy discourse. Power struggles regarding competing forms of knowledge is not a new proposition in the rural domain (see Shortall, 2012). The model of knowledge transfer where social scientists produce theories that are consequently applied in practice by policy makers is long gone (Lowe and Philipson, 2006); however, what is less evidenced is an appetite to produce reflexive discourses, across multiple rural development stakeholders who contribute to rural development debates.

To understand reflexivity we draw attention here to the pioneering works of Schön, who views reflection as 'a dialogue of thinking and doing through which I become more skilled' (Schön, 1983: 31). Schön's reflective model involves: the construction of personal meaning from knowledge and experience; the quest, participation and analysis of feedback; the evaluation of one's skills, attitudes and knowledge; and the identification and exploration of new possibilities for professional action (Schön, 1983, 1987). The quest for a model for reflection has been further developed by other social scientists particularly in pedagogy, due to the discipline's need to continue developing from practical experience (see also Kolb, 1984; Brookfield, 1995). However, these contributions of experiential learning offer original insights into understanding the continuous co-construction of knowledge in the rural development process. In line with neo-endogenous thinking, these contributions highlight that reflection does not only occur within the self, or within particular actors, but also requires actors to continuously seek feedback, to understand other actors' experiences, and to question attitudes and accepted behaviours. It is in fact a combination of such experiences that both constitute and produce reflexivity. Central to such a reflexive approach is social scientists' understanding of their own agency in producing narratives and knowledges about rural development. This requires an open discussion about the nature of reflexivity, how it is an essential element in the co-construction of knowledge and expertise, and how it can be engaged in development policy.

Ways to promote reflexivity within neo-endogenous thinking have been described in the literature, implicitly or explicitly perhaps, and include at least: first, research-practice rural development networks (such as the experiences of the Rural Learning Network of Central and Western New York co-ordinated by Cornell University in the US; the Northern Rural Network co-ordinated by Newcastle University in the UK; see also Atterton and Thompson, 2010); secondly, interdisciplinary methodological synergies between different social scientists, and also between natural and social scientists working in rural development contexts (see Lowe and Philipson, 2006, 2009); and, thirdly, international comparative research (Gkartzios and Scott, 2014; Lowe, 2012). Comparative research is particularly important not only for the transfer of knowledge or lesson sharing as it is commonly argued, but, more critically, because it mirrors and challenges one's own assumptions about the development process (Gkartzios and Shucksmith, 2015). All these channels of production of knowledge are essential elements of neo-endogenous thinking as they promote a culture of reflexivity challenging normative thinking, own agency and power dynamics. They all are networked approaches too in the way they require collaboration with multiple actors who may hold different forms of knowledge. The continuous co-production of discourse on *what* is and *how* to do rural development, as well as the transfer of lessons across different cultural and/or rural contexts rests within these reflexive platforms.

References

Albrechts, L., Healey, P. and Kunzmann, K.R. (2010) Strategic spatial planning and regional governance in Europe. *Journal of the American Planning Association*, 69, 113–129.

Amin, A. (1993) The globalization of the economy: an erosion of regional networks. In G. Grabher (ed.), *The Embedded Firm: On the Socioeconomics of Industrial Networks*, 278–295. London: Routledge.

Atterton, J. and Thompson, N. (2010) University engagement in rural development: A case study of the northern rural network. *Journal of Rural and Community Development*, 5(3), 123–132.

Atterton, J., Newbery, R., Bosworth, G. and Affleck, A. (2011) Rural enterprise and neo-endogenous development. In G.A. Alsos, S. Carter, E. Ljunggren, and F. Welter (eds), *The Handbook of Research on Entrepreneurship in Agriculture and Rural Development*, 256–280. Cheltenham: Edward Elgar.

Barke, M. and Newton, M. (1997) The EU LEADER initiative and endogenous rural development: the application of the programme in two rural areas of Andalusia, southern Spain. *Journal of Rural Studies*, 13(3), 319–341.

Bock, B.B. (2016) Rural marginalisation and the role of social innovation: a turn towards nexogenous development and rural reconnection. *Sociologia Ruralis*, 56, 5, 552–573.

Bosworth, G. and Atterton, J. (2012) Entrepreneurial in-migration and neo-endogenous rural development. *Rural Sociology*, 77(2), 254–279.

Bosworth, G., Annibal, I., Carroll, T., Price, L., Sellick, J. and Shepherd, J. (2016) Empowering local action through neo-endogenous development: the case of LEADER in England. *Sociologia Ruralis*, 56(3), 427–449.

Brookfield, S. (1995) *Becoming a Critically Reflective Teacher*. San Francisco, CA: Jossey-Bass.

Brunori, G. and Rossi, A. (2007) Differentiating countryside: social representations and governance patterns in rural areas with high social density: the case of Chianti, Italy. *Journal of Rural Studies*, 23(2), 183–205

Cooke, P. and Morgan, K. (1993) The network paradigm: new departures in corporate and regional development. *Environment and Planning D: Society and Space*, 11, 543–564.

Crawshaw, J. and Gkartzios, M. (2016) Getting to know the island: artistic experiments in rural community development. *Journal of Rural Studies*, 43, 134–144.

Davoudi, S. and Stead, D. (2002) Urban–rural relationships: an introduction and a brief history. *Built Environment*, 28(4), 269–277.

Dicken, P. and Thrift, N. (1992) The organisation of production and the production of organisation: why business enterprises matter in the study of geographical industrialisation. *Transactions of the Institute of British Geographers*, 11, 279–291.

Donovan, K. and Gkartzios, M. (2014) Architecture and rural planning: 'Claiming the vernacular'. *Land Use Policy*, 41, 334–343.

EC (2006) *The Leader Approach: A Basic Guide – Fact Sheet*. Brussels: European Commission.

Edwards, B., Goodwin, M., Pemberton, S. and Woods, M. (2001) Partnerships, power, and scale in rural governance. *Environment and Planning C*, 19(2), 289–310.

Gkartzios, M. and Norris, M. (2011) 'If you build it, they will come': governing property-led rural regeneration in Ireland. *Land Use Policy*, 28(3), 486–494.

Gkartzios, M. and Scott, M. (2009) Planning for rural housing in the Republic of Ireland: from national spatial strategies to development plans. *European Planning Studies*, 17(12), 1751–1780.

Gkartzios, M. and Scott, M. (2014) Placing housing in rural development: exogenous, endogenous and neo-endogenous approaches. *Sociologia Ruralis*, 54(3), 241–265.

Gkartzios, M. and Shucksmith, M. (2015) 'Spatial anarchy' versus 'spatial apartheid': rural housing ironies in Ireland and England. *Town Planning Review*, 86(1), 53–72.

Kolb, D. (1984) *Experiential Learning: Experience as the Source of Learning and Development*. Englewood Cliffs, NJ: Prentice Hall.

Kovach, I. (2000) LEADER, a new social order, and the central and east European countries. *Sociologia Ruralis*, 40(2), 181–189.

Lowe, P. (2012) The agency of rural research in comparative context. In M. Shucksmith, D.L. Brown, S. Shortall et al. (eds), *Rural Transformations and Rural Policies in the US and UK*, 18–38. New York: Routledge.

Lowe, P. and Philipson, J. (2006) Reflexive interdisciplinary research: the making of a research programme on the rural economy and land use. *Journal of Agricultural Economics*, 57(2), 165–184.

Lowe, P. and Philipson, J. (2009) Barriers to research collaboration across disciplines: scientific paradigms and institutional practices. *Environmental and Planning A*, 41, 1171–1184.

Lowe, P., Murdoch, J. and Ward, N. (1995) Networks in rural development: beyond exogenous and endogenous models. In J.D. Van der Ploeg and G. Van Dijk (eds), *Beyond Modernisation: The Impact of Endogenous Rural Development*, 87–106. Assen: Van Gorcum.

Lowe, P., Ray, C., Ward, N. et al. (1998) *Participation in Rural Development: A Review of European Experience*. Newcastle: Centre for Rural Economy, University of Newcastle.

Mitchell, C.J.A. and Madden, M. (2014) Re-thinking commercial counterurbanisation: evidence from rural Nova Scotia, Canada. *Journal of Rural Studies*, 36, 137–148.

Moseley, M. (1997) New directions in rural community development. *Built Environment*, 23, 201–209.

Moseley, M. (2000) Innovation and rural development: some lessons from Britain and western Europe. *Planning Practice and Research*, 15, 95–115.

Murdoch, J., Lowe, P., Ward, N. and Marsden, T. (2003) *The Differentiated Countryside*. London: Routledge.

Navarro, F.A., Woods, M. and Cejudo, E. (2016) The decline of the bottom-up approach in Rural Development Programmes: the cases of Wales and Andalusia. *Sociologia Ruralis*, 56(2), 270–288.

OECD (2006) *The New Rural Paradigm: Policies and Governance*. Paris: OECD Publishing.

OECD (2011) *OECD Rural Policy Reviews: England, United Kingdom 2011*. Paris: OECD Publishing.

OECD (2014) *OECD Rural Policy Reviews: Chile 2014*. Paris: OECD Publishing.

OECD (2016) *OECD Territorial Reviews: Japan 2016*. Paris: OECD Publishing.

Osti, G. (2000) LEADER and partnerships: the case of Italy. *Sociologia Ruralis*, 40(2), 172–180.

Picchi, A. (1994) The relations between central and local powers as context for endogenous development. In J. van der Ploeg and A. Long (eds), *Born from Within: Practice and Perspectives of Endogenous Rural Development*, 195–203. Assen: Van Gorcum.

Ray, C. (1997) Towards a theory of the dialectic of local rural development within the European Union. *Sociologia Ruralis*, 37(3), 345–362.

Ray, C. (2000) The EU LEADER programme: rural development laboratory. *Sociologia Ruralis*, 40(2), 163–171.

Ray, C. (2001) *Culture Economies*. Newcastle: Centre for Rural Economy. Newcastle University. Retrieved from www.ncl.ac.uk/media/wwwnclacuk/centreforruraleconomy/files/culture-economy. pdf (accessed 13 December 2017).

Ray, C. (2006) Neo-endogenous rural development in the EU. In P.J. Cloke, T. Marsden and P. Mooney (eds), *Handbook of Rural Studies*, 278–291. London: Sage.

Satsangi, M., Gallent, N. and Bevan, M. (2010) *The Rural Housing Question: Communities and Planning in Britain's Countrysides*. Bristol: Policy Press.

Schön, D. (1983) *The Reflective Practitioner: How Professionals Think in Action*. New York: Basic Books.

Schön, D. (1987) *Educating the Reflective Practitioner.* San Francisco, CA: Jossey-Bass.

Scott, M. (2012) Housing conflicts in the Irish countryside: uses and abuses of postcolonial narratives. *Landscape Research,* 37(1), 91–114.

Scott, M. and Murray, M. (2009) Housing rural communities: connecting rural dwellings to rural development in Ireland. *Housing Studies,* 24(6), 755–774.

Shortall, S. (2008) Are rural development programmes socially inclusive? Social inclusion, civic engagement, participation, and social capital: exploring the differences. *Journal of Rural Studies,* 24(4), 450–457.

Shortall, S. (2012) The role of subjectivity and knowledge power struggles in the formation of public policy. *Sociology,* 47(6), 1088–1103.

Shortall, S. and Shucksmith, M. (2001) Rural development in practice: issues arising in Scotland and Northern Ireland. *Community Development Journal,* 36(2), 122–134.

Shucksmith, M. (1981) *No Homes for Locals?* Farnham: Gower.

Shucksmith, M. (2000) Endogenous development, social capital and social inclusion: perspectives from LEADER in the UK. *Sociologia Ruralis,* 40(2), 208–218.

Steel, C.E. and Mitchell, C.J.A. (2017) Economic transition in the Canadian north: is migrant-induced, neo-endogenous development playing a role? *Journal of Rural and Community Development,* 12(1), 55–74.

Storey, D. (1999) Issues of integration, participation and empowerment in rural development: the case of LEADER in the Republic of Ireland. *Journal of Rural Studies,* 15(3), 307–315.

Tolón-Becerra, A., Lastra-Bravo, X. and Galdeano-Gomez, E. (2010) Planning and neo-endogenous model for sustainable development in Spanish rural areas. *International Journal of Sustainable Society,* 2(2), 156–176.

Van der Ploeg, J.D. and Renting, H. (2000) Impact and potential: a comparative review of European rural development practices. *Sociologia Ruralis,* 40(4), 529–543.

Van den Ploeg, J.D., Renting, H., Brunori, G., Knickel, K., Mannion, J., Marsden, T., de Roest, K., Sevilla-Guzman, E. and Ventura, F. (2000) Rural development: from practices and policies to theory. *Sociologia Ruralis,* 40(4), 391–407.

Ward, N., Atterton, J., Kim, T.Y., et al. (2005) *Universities, the Knowledge Economy and 'Neo-endogenous' Rural Development.* Newcastle: Centre for Rural Economy, Newcastle University. Retrieved from www.ncl.ac.uk/media/wwwnclacuk/centreforruraleconomy/files/discussion-paper-01.pdf.

Whatmore, S. (1994). Betterment revisited: issues in contemporary land use planning: an introduction. *Land Use Policy,* 11(3), 163–167.

Woods, M. (2005) *Rural Geography.* Sage: London.

Woods, M. (2007) Engaging the global countryside: globalization, hybridity and the reconstitution of rural place. *Progress in Human Geography,* 31(4), 485–507.

Zonneveld, W. and Stead, D. (2007) European territorial cooperation and the concept of urban–rural relationships. *Planning Practice & Research,* 22, 3, 439–453.

15

Regional planning and rural development
Evidence from the OECD

John Tomaney, Tamara Krawchenko and Chris McDonald

Introduction

Regional planning provides a critical framework for rural development. Yet, the 'rural' and the 'regional' typically have been treated as separate analytical categories, implying different developmental challenges and requiring different policy responses. This cleavage is accentuated in a context where rates of urbanisation are accelerating and large, densely developed cities, drawing on agglomeration economies act as the drivers of productivity growth and economic development. In fact, the relationships between the rural and urban are important, multiple and complex. Moreover, rural areas are diverse in terms of their development problems and potential. Frameworks of spatial planning must accommodate, respond to and shape diverse capacities in pursuit of sustainable development.

Land-use planning systems are highly variable between countries reflecting different national governance frameworks, economic structures, tenure structures and cultural preferences – all of which reflect the outcomes of material and discursive conflicts (e.g. see Chapter 3 of this Companion). The design and effectiveness of regional planning systems play a large part in shaping the conditions for rural development. In some countries planning systems have legal frameworks that treat rural and urban places separately while others encourage integrated approaches. Ineffective or misguided land-use planning systems may provide perverse incentives that promote inefficient and unsustainable forms of peri-urbanisation. A lack of appropriate governance arrangements may limit the capacity of smaller rural areas to manage land use and pursue inter-municipal strategies. Poorly designed or absent land-use plans may promote the loss of vital agricultural land.

While differences between and within countries are important, shared challenges mean that developing a broad understanding of the relationship between rural development and regional planning is critical. More effective regional planning suggests the need for better data and observation of land-use change that allows rural communities to plan better and deal with challenges such as climate adaptation, waste management and address land-use issues in an integrated way across functional areas and in relation to other places. Better planning needs to be based on an understanding of social, economic and environmental drivers of spatial change in the context of multi-level governance systems.

In this chapter we identify the origins and trace the connections between regional land-use policies, spatial planning and rural development. We examine how a concern with the rural is incorporated into regional planning and we look in depth at different national examples, focusing on France, the Netherlands, Poland, the State of Victoria in Australia and the sparsely populated Nordic regions. The chapter concludes by emphasising the importance of strategic spatial planning as a mechanism for promoting rural development.

Conceptualising the regional and the rural

The region represents a fundamental category for understanding the division of space (Tomaney, 2017). The region is a foundational concept in geography (Dickinson, 1969). Pioneers of the modern discipline, such as Alfred Hettner in Germany, drawing on the earlier work of Kant and von Humboldt, aimed to define the character of places through a deep understanding of their human and physical characteristics and the internal relations that gave them their character (Hartshorne, 1958). Early geographical enquiry in France, notably in the work of Élisée Reclus and Paul Vidal de la Blache, was concerned also with defining regions in terms of their chorological characteristics. (Similarly, early French sociology, influenced by the work of Frédéric Le Play, focused on the relationship between place, work and family.) In England, Halford Mackinder promoted the idea of regional geography on similar lines. Although later criticised and lapsing out of fashion, the importance of these early contributions lay in their emphasis on the human and physical relationships that distinguish one region from another.

The ideas of early geographers influenced the theories of planning offered by Patrick Geddes and, later, Patrick Abercrombie. For Geddes, the region was the natural unit for planning. His famous image of 'The Valley Section' linked climate, land and ways of life as the focus for planning. In an early essay he argued, 'In short, then, it takes the whole region to make the city' (Geddes, 1905: 166). Influential early initiatives such as Ebenezer Howard's 'Garden Cities' concept or Lewis Mumford's 'Regional City', together with emergence of 'Town and Country Planning' in the United Kingdom (Abercrombie, 1933) and later the concept of the city-region (Dickinson, 1947) indicate that a concern with urban–rural relationships were present at the origins of regional planning. Abercrombie's richly detailed 1944 Greater London Plan, presented a vision for London situated within a much larger region stretching across the south east of England. After the Second World War regional planning in developed countries, mainly in the hands of central governments, typically sought to accelerate the development of rural areas and to limit the growth of urban areas through the designation of green belts and urban growth boundaries.

Recent analysis and policy has given great attention to the role of the city as the driver of productivity growth with size and density as the key components of urban development (Glaeser, 2012). Such a focus risks neglecting the contribution to development and wellbeing of places outside major urban centres, including rural areas and overlooks how a wide variety of places can contribute to growth. In the developing world, rural–urban linkages embody flows of agricultural and other commodities to urban markets and manufactured and other goods from urban centres to rural areas, shorter- and longer population movements and financial flows, such as remittances. As Tacoli (2003: 4), notes: while 'these flows and linkages exist between all rural and urban areas, their scale and strength are determined by the nature of economic, social and cultural transformations'.

The diversity of rural–urban linkages means that local and regional governments are best placed to identify local needs and develop appropriate plans for their management. Decentralisation of government and planning systems is a global trend with diverse outcomes (OECD, 2012; Pike et al., 2017). In principle, decentralisation allows a more tailored focus on local problems and greater democratic accountability of regional planning, but also raises challenges where local governments struggle to provide services due to their inability to raise sufficient revenue. These problems can be mitigated by inter-municipal cooperation, collaboration in multi-level governance systems, and 'borrowing scale' by improving the access of rural places to services provided in urban areas (Tacoli, 2003).

Within developed countries, rural regions are home to 25 per cent of the OECD population and contain most of the land, water and other natural resources which are crucial for urban development. Rural areas are heterogenous and develop in diverse ways, but low densities of development, small workforces, high reliance on extractive industries and processing activities that are sensitive to transport costs and vulnerable to competition, together with low levels of innovation, can make rural regions less resilient than large, diverse urban economies. In some cases, though, rural regions are making novel use of their traditional resource base. Food security concerns and the desire to reduce food miles raise new development possibilities for some rural and peri-urban regions. New or expanding activities create value, including rural tourism, wildlife conservation and the stewardship of landscapes. The shift to renewable energy production has boosted the development of rural areas and there is a new recognition of the ecosystem services, such as carbon capture and water management, which are provided by rural areas. Rural areas can be places of opportunity and some rural regions are growing at a fast rate, but this throws up planning challenges (OECD, 2016a).

The OECD identifies three categories of rural regions: (i) rural regions within a functional urban area (FUA) – a city plus its commuting zone; (ii) rural regions close to an FUA; and (iii) remote rural regions. These regions have different regional potentials, challenges and policy needs. Rural regions within or near to FUAs generally demonstrate the greatest potential for economic development. Proximity allows urban residents easy access to rural areas, while rural residents have easier access to advanced public and private services. Within OECD countries a large majority of rural residents and rural economic activity are found near urban areas. Urban and rural regions become increasingly connected and interdependent but remain distinct in terms of their economic functions, settlement patterns and ways of life. Moreover, the pattern of these links varies significantly between countries.

An evolution of regional planning for rural development can be charted from a centralised, sector-focused, top-down approach to one shaped by multi-level governance and integrated policies, which recognises the diversity of rural places and their relationships to urban areas (Table 15.1).

Regional and rural planning in OECD countries

The OECD has undertaken a programme of work on the governance of land use which provides an overview of regional and rural planning (OECD, 2017d). Among all types of spatial and land-use plans in the OECD countries surveyed, it is found that regions are responsible for roughly a third of all plans (32% in 2016) (OECD, 2017c: 14). The majority of regional plans contain general policy guidelines (70%) and elements of strategic plans (75%); around a third contain some kind of boundary plans – but only to a limited degree. Around three-quarters of all regional plans are approved by regional governments while the remaining quarter are approved by national governments. Over a third (34%) of the regional plans in the OECD are binding

Table 15.1 Rural policy 3.0.

	Old paradigm	New Rural Paradigm (2006)	Rural Policy 3.0 (2017)
Objectives	Equalisation	Competitiveness	Well being considering multiple dimensions of: i) the economy; ii) society and; iii) the environment
Policy focus	Support for a single dominant resource sector	Support for multiple sectors based on their competitiveness	Low-density economies differentiated by type of rural area
Key actors and stakeholders	Farm organisations and national government	All levels of government and all relevant departments plus local stakeholders	Involvement of: i) public sector – multi-level governance ii) private sector – for-profit firms and social enterprise; and iii) third sector – non-governmental organisations and civil society
Policy approach	Uniformly applied top down policy	Bottom-up policy, local strategies	Integrated approach with multiple policy domains
Rural definitions	Not urban	Rural as a variety of distinct types of place	Three types of rural: i) within a functional urban area, ii) close to a functional urban area, and iii) far from a functional urban area

Source: OECD (2016b)

with either no or rare exceptions; 24 per cent are binding with frequent exceptions and 28% are not binding. The remainder are either not specified (11%) or not binding, but incentivised (3%) (OECD, 2017c). While the majority of countries surveyed have precise and up-to-date cadastre maps in digital format that are often available online, in 13 countries experts judged the quality of cadastre as insufficient; often, this concerns predominantly rural areas.

Adopting an integrated approach

Regional planning today is connected to broader agendas such as the transition to a low-carbon economy, reducing social-spatial inequality, and creating opportunities for economic growth and prosperity. It is thus linked to policy ambitions at multiple scales, extending across sectoral issues and involving an ever wider array of actors in structures of governance. This has fundamentally challenged planning systems to adapt, both in terms of the formal institutional rules, but also informal roles and ways of working. These discussions have often focused on 'joining up' or 'holistic' government, at a sub-national level of government – regions, local authorities, towns and villages. The objective here is to connect different governance initiatives focused in the arena of the planning system, through the system's concern with the use and development of land, with spatial organisation and the qualities of places.

Regional plans are increasingly cross-sectoral, covering a number of thematic areas and provide strategic direction and general policy guidance. For example, in a survey of 32 OECD countries it is found that the majority of regional plans (69%) cover three or more thematic areas such as transport, the environment, housing, industry, commerce or agriculture (OECD, 2017c). The most common thematic areas covered by regional plans are the environment and transport. For rural areas, participation in these integrated planning frameworks poses a challenge. As regional level planning becomes more complex, it can be harder for rural communities to make sense of and engage in these processes – there are inherent power asymmetries to overcome.

Aligning policy incentives for coordinated spatial development

One of the main arguments forwarded in the OECD's work on the governance of land use is that a wide range of policies beyond spatial planning impact upon how land is used. Since almost any economic and social development eventually affects land use, any public policy that affects social and economic factors ultimately impacts on land as well. Among the most important of these are tax policies and fiscal systems; they influence the incentives for using land in different locations by affecting the financial costs and benefits of doing so.

One of the most obvious examples how tax policies affect land use is the tax deductibility of commuting expenses. Commuting expenses are tax deductible in 12 of 26 analysed OECD countries (Harding, 2014). Given the marginal tax rates in OECD countries, this preferential tax treatment reduces the costs of commuting by up to 50 per cent, which provides incentives to people to live further away from their place of work than they would otherwise. These policies facilitate peri-urbanisation. Another important example of a fiscal policy affecting land use is agricultural subsidies. Without these subsidies, agricultural activity would likely become unprofitable in some parts of the OECD and agricultural land use would decline (Renwick et al., 2013). Since agriculture uses between 15 and 50 per cent of the total land area in most OECD regions the potential impact of agricultural subsidies on land use is obvious. Tourism promotion is another example of a policy with significant consequences for land use. If successful, it is likely to encourage development in attractive and especially sensitive areas, such as along coastlines or in mountain areas which can contradict policy objectives concerning land use in those areas and may result in development that conflicts with existing plans.

Although none of the policies mentioned above pursue explicit land-use outcomes, they all have significant consequences for land use that may work in favour or against the objectives of planners. By using them actively to provide incentives that are aligned with land-use objectives, land-use planning can become more effective, while being less restrictive and more flexible at the same time.

Regional planning for rural development in practice

France – integrated regional plans and the capacity of small communes

Regional planning in France exemplifies an integrated approach. The Planning, Sustainable Development and Territorial Equality Regional Plan (Schéma Régional d'Aménagement, de Développement Durable et d'Egalité du Territoire) places regions as lead actors in the field of planning and sustainable development. Unlike the previous planning process (which was not compulsory), the new law requires regions to develop an integrated spatial plan. The new law replaces the essential elements of the three sectoral plans (on transportation, ecological networks and climate, air and energy) and adds a requirement for the region to develop a specific plan on the prevention and management of waste. Beyond the reorientation of plans, regional council presidents are obliged by law to host regular conferences to promote the concerted use of planning competences by every level of local authority in order to support an integrated and cross-disciplinary planning process, instead of a sector-specific one.

While most OECD countries place land-use planning at a local government level, the particular challenge in France is that the capacity of communes – particularly rural communes – to carry out and implement effective planning is highly variable. The majority of the 36,000 communes in France have populations of less than 1,000 inhabitants. While responsibility for specific land-use plans continues to remain at the level of the commune because it is at this level of government that the granularity of actual decisions about particular parcels of land can best be managed, the ability of small communes to develop plans that are in compliance with the plethora of laws, regulations and agreements, that have to be respected is in doubt. Places that lack the resources to have in-house professional services have to contract this function out to consultants. A consequence is that a plan may meet the requirements or constraints that are imposed on the commune, but it may not adequately address local interests in land use, because the outside consultant has no real knowledge of local concerns, and elected officials with this knowledge cannot adequately participate in defining objectives for the planning process.

The resulting planning framework is also often incapable of addressing common spatial issues. Planning could be much more integrated and cross-disciplinary if more planning and land-use decisions were made at a higher level of government. This is the whole challenge of encouraging the greater use of multi-commune plans (of which France has several types, including for rural communes) where there are important spill-over effects on neighbouring jurisdictions. Presently, there is a combination of 'carrot' and 'stick' processes to achieve such collaboration. In some cases, the national government compels joint action by communes because it requires action and cannot trust that the competing interests of individual places will be co-operatively resolved. This is most common in instances where protection of the natural environment cuts across multiple communes. In other instances, modest incentives to co-operate can provide the impetus for communes to see that they all can benefit from joint action and once this process is started it can continue. In an assessment of the French system of spatial planning, Geppert remarks that: 'although there is co-ordination between different levels of government, it results in joint investments rather than in shared spatial visions and/or common objectives' (Geppert, 2015: 109).

The Netherlands – from a regulatory to an active approach to land management

The Netherlands is a small country which is notoriously challenged by the need to manage flood waters through its polder system. Rural areas in the Netherlands are not remote – and yet, given the size of the country, it has been imperative to protect this resource for both its high amenity value and agricultural output (two uses which are not always compatible). Provinces in the Netherlands prepare rural development plans for rural areas in their territory. Rural Development Plans are zoning plans for rural areas, where no major change in use is foreseen, but major changes in local infrastructure or land readjustments are required.

A major legislative reform is taking place in 2018 which will likely reshape rural planning practices. Under the Environment and Planning Act (*Omgevingswet*) all national environmental legislation will be consolidated under one framework. The new Act will integrate, modernise, harmonise and simplify current rules on the wide array of activities. These affect for example the environment, land-use planning, urban and rural development, water management, construction of buildings, protection of cultural heritage, and the development of major public and private works. This marks an important shift from the old environmental law dispersed across 26 separate acts into one consolidated piece of legislation and is expected to take effect in 2019.

The law specifies 'environmental values' (*omgevingswaarden*) which describe the desired state or quality of the physical environment and acceptable levels of environmental degradation or nuisance. These can be set as thresholds or goals to be achieved and apply across different geographies and locations. These 'environmental values' must, however, conform to upper level ones set by the national or provincial government, EU directives or other international obligations. This mechanism shifts the regulatory approach towards implementation of key objectives or goals, and monitoring and evaluation of their effectiveness. Each administrative body that establishes an 'environmental value' must then develop a programme outlining how this will be met, including a package of policy tools and regulations. It requires that a monitoring system be established on the state or quality of the physical environment for each 'value'. This scheme of monitoring environmental values is derived from EU quality standards.

This cyclical approach has been described by the national government as no less than a paradigm shift – from preservation and protection towards an active approach that continuously monitors and assesses the quality of the physical environment. The role of the government within this context is to link the initiatives of various actors who are developing projects that impact upon the physical environment, and to monitor their outcomes. Where there are undesirable outcomes, the relevant administrative bodies will need to establish a programme to remedy this. It is in this way that compliance with environmental objectives is ensured. For rural areas, this approach will help to more effectively monitor physical and environmental change. It may also, over time, lead to a shift in the 'rural idyll'. A multiplicity of visions about rural landscapes and functions has long been a source of conflict in the Netherlands (Frouws, 1998) but has intensified with the decentralisation of land-use policies, a focus of regional policy on promoting economic development and a move away from restricting land use. This new approach will require ongoing monitoring, negotiation and adjustment – it ushers in a new governance of land use for rural and urban areas alike (OECD, 2017e).

Poland – planning in the face of rapid peri-urbanisation and a lack of valid plans

Like most OECD countries, Poland has a nested hierarchy of spatial development strategies wherein strategies of higher levels of governments are meant to inform those of the government below. The national government sets the overarching legal framework that regulates land use and building law in the country and has also developed a national spatial strategy that provides an assessment of key challenges and puts forward a vision for the country's medium and long term spatial development. In turn, there are regional plans which describe general development conditions and demarcate the regional settlement system. However, it is at the level of local government where the most detailed decisions about how land is used are taken through spatial studies, local spatial development plans and planning (or development) decisions.

One of greatest obstacles to co-ordinated spatial planning in Poland has been the very low coverage of spatial plans across the country and recourse to planning decisions instead (one-off building permissions that are not linked to land-use plans). Only 30 per cent of the national territory falls within an applicable local spatial management plan, and in seven *voivodeships* this share of territory is below 20 per cent. This lack of local plan coverage together with the widespread use of the planning decision mechanism is one of the greatest challenges to coherent spatial development in Poland. Substantively, it has facilitated rapid peri-urbanisation and unco-ordinated developments. Poland has one of the highest rates of agricultural land conversion in Europe; between 2004 and 2011, Polish municipalities converted 545,000 hectares of agri-cultural land to non-agricultural uses (Kowalewski et al., 2013). This is not a universal trend; regions such as Podlaskie have a strong agricultural sector and the share of agricultural land has increased. This conversion of agricultural land is problematic when high-quality agricultural land is converted to other uses or where land uses are incompatible (e.g. animal husbandry next to residential zones) or otherwise inefficient in their allocation. The scope of these issues are dif-ficult to gauge, in part because there is poor co-ordination between the regional and local levels in terms of monitoring land-use change. There are, for instance, no data on planning appeals and the regional government does not assess the impact of land-use planning in communities.

The national government has long recognised that additional reforms to the framework of spatial planning in the country are needed and some incremental changes have been made; how-ever, more remains to be done. More effective legal regulations are required to protect high-quality soil from being used for non-agricultural purposes and monitoring should be enhanced. Moreover, Poland needs to adopt a comprehensive approach to land-use planning that can meet the challenges facing different types of rural communities. There is an appetite by regions to play a greater role in integrated and functional planning, but they do not have the tools to undertake such a role (e.g. the statutory authority or incentives for municipalities to adopt functional or integrated planning). For example, the region of Zachodniopomorskie aims to establish func-tional areas where smaller communities cooperate with each other and are supported by both the regional and national governments in a range of strategic areas (e.g. attracting investment, developing transport, enhancing vocational education). Małopolskie is interested in encourag-ing villages to develop a town centre in order to more efficiently deliver services to residents; however, they too have no tools with which to implement such an approach. This lack of tools to link up sectoral investments in a spatially co-ordinated manner is a missed opportunity. As a final note, Poland's regional governments have recently established regional territorial observa-tories in order to monitor the impact of policies with a territorial impact. These entities collect and analyse data in order to evaluate the impact of development policies locally. But rather than just reporting up, these observatories could serve to enhance the analytical capacities of local communities themselves, in order to promote, for example, urban–rural partnerships. This is particularly important for rural communities that have limited internal capacity to undertake such functions.

Australia – integrating metropolitan, regional and rural planning

Over the past decade, the State of Victoria, Australia has introduced a number of reforms to bet-ter integrate metropolitan, regional and rural planning. An important aspect of these reforms has been collaboration between local municipalities and state agencies to develop regional develop-ment plans that establish economic, social and environmental goals and priorities for sub-regions within the State. These provided a platform for integrating land use and infrastructure planning at a sub-regional scale. Prior to these reforms local municipalities had the primary role in land use and local infrastructure planning while state agencies undertook long-term infrastructure and service planning on a sectoral basis. These regional plans are also supported by a number of mechanisms to support implementation.

In 2010, the Victorian Government released its regional development policy *Ready for Tomorrow: A Blueprint for Regional and Rural Victoria* which included strategies to invest in skills, export development, economic and social infrastructure, and regional planning. The Blueprint was developed under the auspice of a Ministerial Taskforce and supporting inter-departmental committee. The Blueprint established a long term state-wide framework for regional development to support regions in identifying their priorities, and was organised around a triple bottom line framework (economic, social and environmental goals). This approach was based on the principle that regional communities are best placed to identify and plan for their priorities, future challenges and opportunities (McDonald, 2014). This was realised through each region developing a Regional Strategic Plan (RSP). RSPs are framed around this triple bottom line approach and priorities for action over a medium term horizon. The broad scope of the plans vary, however, all include at a minimum:

- An analysis of current regional economic, social and environmental performance and future challenges and opportunities.
- A strategic vision highlighting long term objectives for the region around the broad goals of prosperity, liveability and sustainability.
- Strategies and actions for the short and medium term.

Each region was then supported to develop a Regional Land-Use Plan – called a Regional Growth Plan (RGP). RGPs provide broad direction for land use and development for each region. This includes a framework for future settlement planning, balancing different needs in peri-urban areas (related to housing, extractive industries, water, and agricultural activities), protecting productive and high amenity landscapes, and infrastructure corridors. Each RGP was developed through a partnership between local government and state agencies and authorities through consultation with the community and key stakeholders. They have been included in the Victorian Planning Scheme and, as an ongoing part of the State of Victoria's planning system, the regional growth plans will be reviewed every four to six years.

A good example of this initiative is the G21 region, which is adjacent to metropolitan Melbourne. The main urban centre in the G21 region is Geelong which has a population of 238,603 (Australian Bureau of Statistics, 2018). The population of the city grew by 18.4 per cent in the period 2006–2016. The RGP for the Geelong region provides a framework for managing this growth and forecast land-use pressures to 2050. It incorporates the strategic land use and growth planning already done at a local municipal level across the region and builds on this to identify where future residential and employment growth will occur. The overall strategy of the region is to contain growth within, and in proximity to, existing urban settlements to reduce pressures on high amenity landscapes, areas of biodiversity, and agricultural land. It also identifies the key infrastructure required to manage and support this growth.

Importantly, the RGP for G21 is supported by a number of implementation mechanisms. The RGP identifies four key actions to be investigated, developed and packaged into an Implementation Plan:

- An Infrastructure Plan that will identify the key regional level infrastructure projects critical to supporting the housing and employment growth directions of the Growth Plan.
- A Residential and Industrial Land Supply Report and a Land Supply Monitoring and Reporting Tool that will enable more accurate reporting and monitoring of land supply across the region.
- A Housing Strategy Incentives Report that will outline actions to support strategic housing objectives, including the take up of land for higher density living at identified key nodes and for the development of housing markets in centres such as Winchelsea and Colac.

- An analysis of the identified Further Investigation Areas to identify constraints and opportunities, key infrastructure requirements and planning outcomes and set out a timing and land planning/development process (G21 Regional Alliance, 2013).

Finland, Norway and Sweden – planning in remote and sparsely populated areas

Planning in very remote and low density places brings its own unique challenges. The northern regions of Finland, Norway and Sweden characterise such conditions. Collectively referred to at Northern Sparsely Populated Areas (NSPA) – these remote regions have just 5 persons every square kilometre, for a total of about 2.6 million people over an area of 532,000 square kilometres.[1] This is comparable to the population of Rome inhabiting the entire area of Spain. The NSPA regions are linked by a set of common territorial characteristics which are absent in other European regions that include the presence of long distances from major markets; limited connectivity to large urban places including national capitals; a large number of small isolated settlements that are too far apart to allow significant interaction; and a narrow resource base, which while valuable limits economic opportunities.

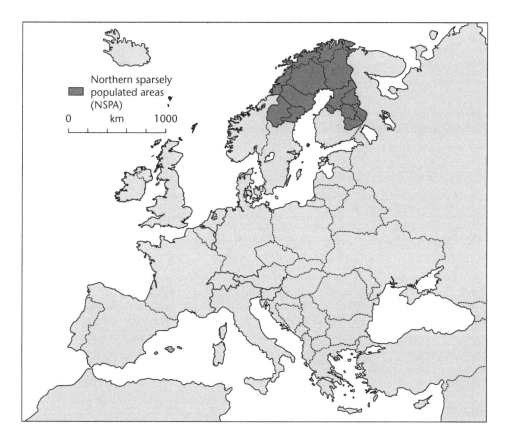

Figure 15.1   Northern and sparsely populated areas in Finland, Norway and Sweden.

Source: OECD (2017b)

179

One of the unique aspects of land management in these regions is the relationship with Indigenous peoples. The Sami are the indigenous people of Finland, Norway, Sweden and the Kola Peninsula in the north-western part of Russia. Although defining themselves as one people, the Sami encompass distinct communities with different identities, culture, social structures, traditions, and livelihoods. Sami language and culture is closely tied to the use of land through traditional reindeer herding, hunting and fishing. There are numerous examples of large scale infrastructure and mining developments leading to conflicts with the Sami community. These land-use conflicts are dealt with through national mining and environmental permitting, and associated judicial processes. One way to address these challenges in a strategic way is to support the inclusion of Sami in land governance. This has occurred in the northern Norway region of Finnmark. The vast majority of land in the region is publicly owned under the auspice of the Finnmark Estate Agency. The governing board of the agency includes representatives from the Sami Parliament who have an important stake in land management through cultural heritage and reindeer herding. Sweden has also taken steps in this direction through the inclusion of Sami representatives in the governing board for World Heritage Laponia, which is an area of protected nature reserves. This enables the incorporation of Sami reindeer herding interests and perspectives in decisions about natural resource management.

The second issue is related and is associated with land-use conflicts between different industry sectors. Economic growth and change generates new land use and infrastructure requirements in rural areas. In the NSPA traditional industries have particular land-use requirements. The forestry industry requires large tracts of land, and infrastructure that enables heavy haulage of wood to nearby towns and cities for processing and export. Mining is a more intensive land use and can impact surrounding land users in terms of noise, dust and pollution. A sector of increasing importance across the NSPA is nature-based tourism. Fostering the growth of the tourism industry may require new transport and communications linkages, and the protection of environmental assets, landscapes, and amenities. These can conflict with traditional industries such as forestry and emerging ones like renewable energy and aquaculture. As nature-based tourism increases the interaction between people and nature also becomes more intense and can lead to concerns about impacts on ecosystems.

One way to better manage these negative externalities is to strengthen strategic spatial planning at the regional scale. Across the NSPA decisions about land use tend to be made at the local or national level with the regional level having a weak role (OECD, 2017a). As a result there is no mechanism to have a strategic dialogue about desired future landscapes at the regional level, and integrating this with the region's vision and priorities for future development. Although this will not resolve conflicts between land users and proponents, it could provide the opportunity to discuss trade-offs, and where future industrial land use should occur. It would also enable stronger coordination between municipalities about land use at the scale of functional labour markets and regional supply chains. In the case of Sweden, the NSPA review recommended a way forward for the national government was to allocate a competency for spatial planning to the body responsible for regional development in the region, and ensuring these regional spatial plans are integrated with planning for regional transport and communications infrastructure.

Conclusion

How land is used has a wide-ranging impact on economic development and quality of life in rural communities. The types of land-use issues faced by rural communities depend a great deal on the nature of their local economy, their natural endowments, and proximity to urban centres and major roads. Rural communities may, for instance, need to balance the demands of industry in proximity to agricultural and forestry activities; the demand for natural amenities and tourism facilities; protection of biodiversity; the often conflicting mix of these activities with residential uses; and the growing need for climate-change mitigation and adaptation (e.g. flood water management). Managing these diverse uses is not just a technical endeavour, it requires community buy-in and sensitivity to historical and culturally embedded ideas about rural landscapes and their functions by residents, investors, regional policy makers etc. Rural communities – being land-rich with lower population densities – also need to consider how best to provide infrastructure and services to citizens in a cost-effective way while maintaining accessibility. This includes providing connections and access to waste, sewage and water systems, which in urban locales are often more established. Good land-use planning is therefore critical – it brings spatial order to individualised decisions about where to live, work, grow food and manufacture products and helps manage environmental risks.

Integrated spatial planning has arisen as a new orthodoxy to address the complexity of these issues. This stems from the recognition that effective spatial management is connected to a broader range of considerations such as economic and social development and wellbeing and that sectoral policies have spatial dimensions that need to be co-ordinated (e.g. the location of services and transportation infrastructure). And yet, many countries lack the structures to achieve the required co-ordination on spatial development between levels of government, let alone sectors. Rural communities – as inherently small administrations – have more limited capacity to undertake the elaboration of both spatial and land-use plans let alone the types of integrated plans that connect sectoral issues. As such, regional governments have a critical role of play in supporting rural planning efforts. There is a need for more effective tools for rural areas to monitor and analyse land-use changes for both their own communities and that of surrounding ones and to provide the relevant information in an accessible format for both communities and residents to make use of. This is pressing in places that are seeing major land-use changes either due to peri-urbanisation or major industrial developments and where there is conflict over use. But it is equally important in rural regions facing population decline due to a combination of out-migration and population aging. In these places, rural planning linked to broader regional development strategies can help to manage fixed assets so that those who remain can still deliver high quality of life to their citizens.

Finally, it is important to take a broad view of policies that impact how land is used. As the OECD's work on the governance of land use reveals, spatial planning and land-use policies are just one aspect of this. In rural areas, fiscal incentives and a wide range of sectoral policies may be of equal or greater importance. It is hence critical that governments adopt a spatial lens on these polices to gauge their effects.

Note

1 The Northern Sparsely Populated Areas referenced here encompass: the four northernmost counties of Sweden (Norrbotten, Västerbotten, Jämtland, and Västernorrland), the seven northernmost and eastern regions of Finland (Lapland, Northern Ostrobothnia, Central Ostrobothnia, Kainuu, North Karelia, Pohjois-Savo and South-Savo) and North Norway (Finnmark, Troms and Nordland).

References

Abercrombie, P. (1933) *Town and Country Planning*. Oxford: Oxford University Press.

Australian Bureau of Statistics (2018) Regional population growth, Australia (3218.0). Retrieved on 4 March 2018 from www.abs.gov.au/ausstats/abs@.nsf/mf/3218.0.

Dickinson, R. (1947) *City Region and Regionalism: A Geographical Contribution to Human Ecology*. London: Routledge & Kegan Paul.

Dickinson, R. (1969) *Makers of Modern Geography*. London: Routledge.

Frouws, J. (1998) The contested redefinition of the countryside: An analysis of rural discourses in the Netherlands. *Sociologia Ruralis*, 38(1), 54–68.

G21 Regional Alliance (2013) G21 Regional growth plan – implementation plan background report. Retrieved on 4 March 2018 from www.g21.com.au/sites/default/files/resources/g21-rgp-ip_imple mentation_plan_background_report_-_decnov_2013.pdf.

Geddes, P. (1905) Civics as applied sociology, part 1. In V.V. Bradford (ed.), *Sociological Papers*, 58–119. London: Macmillan.

Geppert, A. (2015) Planning without a spatial development perspective? The French case. In G.-J. Knaap, Z. Nedovic-Budic and A. Carbonell (eds), *Planning for States and Nation-States in the U.S. and Europe*. Cambridge, MA: Lincoln Institute of Land and Policy.

Glaeser, E. (2012) *Triumph of the City*. London: Pan.

Harding, M. (2014) *Personal Tax Treatment of Company Cars and Commuting Expenses: Estimating the Fiscal and Environmental Costs*. OECD Taxation Working Papers, no. 20. Paris: OECD Publishing.

Hartshorne, R. (1958) The concept of geography as a science of space, from Kant and Humboldt to Hettner. *Annals of the Association of American Geographers*, 48(2), 97–108.

Kowalewski, A., Mordasiewicz, J., Osiatyński, J., Regulski, J., Stępień, J. and Śleszyński, P. (2013) *Raport o ekonomicznych stratach i społecznych kosztach niekontrolowanej suburbanizacji w Polsce*. Warsaw: Fundacja Rozwoju Demokracji Lokalnej, IGiPZ PAN.

McDonald, C. (2014) Developing information to support the implementation of place-based economic development strategies: a case study of regional and rural development policy in the State of Victoria, Australia. *Local Economy*, 29(4–5), 309–322.

OECD (2012) *Promoting Growth in all Regions*. Paris: OECD Publishing.

OECD (2016a) *Governance of Land Use in Poland: The Case of Lodz*. Paris: OECD Publishing.

OECD (2016b) *Regional Outlook 2016. Productive Regions for Inclusive Societies*. Paris: OECD Publishing.

OECD (2017a) *OECD Territorial Reviews: Northern Sparsely Populated Areas, Case Study of Finnmark*. Paris: OECD Publishing.

OECD (2017b) *OECD Territorial Reviews: Northern Sparsely Populated Areas*. Paris: OECD Publishing.

OECD (2017c) *Land-use Planning Systems in the OECD: Country Fact Sheets*. Paris: OECD Publishing.

OECD (2017d) *The Governance of Land Use in OECD Countries: Policy Analysis and Recommendations*. Paris: OECD Publishing.

OECD (2017e) *The Governance of Land Use in the Netherlands: The Case of Amsterdam*. Paris: OECD Publishing.

Pike, A., Rodríguez-Pose, A. and Tomaney, J. (2017) *Local and Regional Development*, 2nd edition. London: Routledge.

Renwick, A., Jansson, T., Verburg, P.H., Revoredo-Giha, C., Britz, W., Gocht, A. and McCracken, D. (2013), Policy reform and agricultural land abandonment in the EU. *Land Use Policy*, 30(1), 446–457.

Tacoli, C. (2003) The links between urban and rural development. *Environment and Urbanization*, 15(1), 3–12.

Tomaney, J. (2017) Region. In R. Kitchin and N. Thrift (eds), *International Encyclopedia of Human Geography*, volume 9, pp. 136–150. Oxford: Elsevier.

16

Rural planning and the financial crisis

Apostolos G. Papadopoulos

Introduction

Rural areas are part of an interconnected world driven by market forces, mobilities and globalisation. Rural regions and localities are reconstituted by means of these wider processes, while there is interplay of local, regional, national and global actors to produce new forms and interactions. Studying rural areas through a relational understanding of space (Massey, 2005) enables us to see that globalisation shapes rural places through politics of domination and sub-ordination, while at the same time there are local/regional politics of negotiation, hybridisation and adjustment (Woods, 2007). The recent external shocks, triggered by the globalised financial market, though posing economic constrains and challenges, do not imply a predetermined future for rural places.

The long-forgotten conditions of uneven regional development are important for conceiving the impact of globalisation upon rural areas. For Massey (1979: 234), geographical inequality is a 'historically relative phenomenon', which is the result of two processes: firstly, changes in the geographical distribution of the requirements of production (spatial, locational, surface) and more particularly in the distribution of population or of resources, or changes in relative distances caused by developments in transport and communication; and secondly, changes in the requirements of production process itself: that is, because of changes in the locational demand of profitable economic activity. The concept of spatial inequality is pivotal for considering the processes of uneven regional development, which is an important component of the recent financial crisis of capitalism, both globally and more particularly in the EU and the Eurozone (Harvey, 2010; Hadjimichalis, 2011; Hadjimichalis and Hudson, 2014).

In this context, recent research has revealed that to varying degrees, rural areas in Europe are becoming more vulnerable to different sets of environmental, social, economic, global and state-led processes. This is related to the changing and expanding public expectations about the function(s) and purpose(s) of rural areas and responding to highly volatile shifts in the demand for rural goods and services (Marsden, 2009; Freshwater, 2015). Especially in rural areas of poorer countries within the EU, there is significant disadvantage shown by objective indicators of welfare and quality of life (Shucksmith et al., 2009). It may be argued that processes of mobility and vulnerability, operating at multiple scales, tend to question the overall sustainability of rural areas. The rural domain, which is an important container for natural resources and socio-cultural relations for urban dwellers and rural communities, has been an arena for contestation and development (Marsden, 2009).

Despite the wider challenges for rural places in the context of globalisation, the emergence of the global financial crisis of 2007–2008 has caused to rural regions and localities a number of shocks and disruptions which initially affected the labour and housing markets, but later on escalated to major cut-backs in the delivery of rural public services (Scott, 2013). The severity, the length and depth of the financial crisis created a feeling of certainty among social scientists and human geographers of a paradigm shift in the conceptualisation of development and planning, with particular attention to rural areas. The current post-crisis, prolonged recessionary and deep austerity era within the EU has offered a new impetus to re-theorising and reflecting upon the new rural realities.

The main aim of this chapter is to explore the diverse impacts of global financial crisis on rural places and to examine more closely how the latter have responded and/or adapted in the relation to the rapidly changing socio-economic environment deeply affected by the hegemony of neoliberal austerity policies. The remainder of the chapter is structured as follows: the first section offers a very brief account of the main stages and characteristics of the financial crisis, placing emphasis on the changing situation and developments at regional level. The main focus here is to illustrate the various impacts in rural regions and localities. The second section reports on the different practices, strategies and resilient adaptation of rural regions and localities in response to the challenges posed by the financial crisis as a major external shock. Finally, some concluding remarks and insights are recorded, building on the preceding discussion.

The financial crisis and its impact

This section will address financial crisis mainly as a creator of a series of external shocks on countries, regions and particularly rural areas. Looking across the literature, there is limited exposition of the impact of the financial crisis on rural areas, but it can be inferred from the regional analysis or spatial categories (e.g. remoteness, proximity to cities) applied in the analysis. Certainly, we have tried to respond to the major challenge of connecting quantitative higher spatial scale analyses with qualitative low/local scale interpretations with particular focus on rural areas.

The turbulent story of the recent financial crisis is the final stage of a long chain of events that occurred in the course of the last few decades, while the global capitalist system emerged as the result of a financial system built on trade deficits (Harvey, 2010; Varoufakis, 2011) and an unsustainable growth model (Kitson et al., 2011).

For Harvey (2011), the financial crisis originated in the actions taken to resolve the crisis of stagflation and the political threat to capitalist class power in 1970s in advanced capitalist economies. He refers to five prerequisites of the crisis: the hegemony of neoliberal fiscal policies in relation to the decline of organised labour institutions; intensification of global competition and increased capital mobility; deregulation and empowering of the most fluid and highly mobile form of capital; the gap between declining wages and increasing effective demand giving rise to expansion of debt; and the creation of bubbles in the asset market which compensated for the lack of investment opportunities. In other words, capital deregulation together with the decline of organised labour and the take over of fiscal policies by capital, led to an uncontrolled financialisation of capital at a global scale. In a nutshell, the financial crisis was a demonstration of the impasse created by the overdevelopment of global financial capital, which, nevertheless, had a geographical focus in the EU and, more particularly, the Eurozone:

> In 2008, following Wall Street's collapse under the weight of its hubris and of the mountains of risk that financialization had amassed, America could no longer provide the European Union with the demand for its exports that had, until then, stabilized it. . . . Europe would soon discover that its private banks were replete with Wall Street–sourced toxic debt and that countries like Greece had insolvent states. The death embrace, or doom loop, between insolvent banks and insolvent European states had begun. The rest is history. The Eurozone's architecture was incapable of sustaining the shockwaves of the 2008 earthquake. Since then it has been in a deep crisis reinforced largely by the European Union's denial that there was anything the matter with its currency's rules, as opposed to their enforcement.
>
> *(Varoufakis, 2016: 270–271)*

Looking more closely to the characteristics of the financial crisis, we may argue that since 2007–2008 there has been a sequence of crises. In 2007 the US subprime crisis, caused by the collapse of the housing bubble based on a housing boom fuelled by low interest rates and toxic mortgages, led to a banking crisis in the country (Aalbers, 2009; Engelen and Faulconbridge, 2009). By 2008 there was a spill over of the banking crisis to Europe, resulting in a global financial crisis, which developed into a deep recession. By 2009, the Great Recession emerged at the global level, but also in Europe, and it was connected to a stock market crash, drop in the GDP and increase in unemployment (Kitson et al., 2011). The next step was a series of asymmetric shocks, which affected the economically vulnerable EU member states, and, therefore, a Euro crisis emerged in 2010. The PIIG countries (i.e. Portugal, Ireland, Italy, Greece and Spain) suffered the most from these asymmetric external shocks. More particularly, Ireland and Spain were hit by a housing crisis in conjunction with a banking crisis. In Greece and Italy, and to a lesser extent in Portugal and Spain, there was an increasing public debt which became unsustainable. The crisis in the Eurozone was identified with widening public debts, macro-imbalances and the need for a new European Monetary Union (EMU) economic governance. The policy design of the EMU was not crisis-proven, but in the midst of the Euro crisis a number of measures were taken: fiscal policy coordination was strengthened, new rescue mechanisms were introduced and the banking sector was re-organised (Breuss, 2016).

The main lesson to be learnt is that international financial institutions proved unable to prevent the US mortgage crisis from transforming into a global recession, while the EU failed to take timely measures to avoid the escalation of the crisis and it was unable to predict a series of sovereign debt crises, which led to increased interventions to member states, generating in the medium term contestations to the so-called European project (Bermeo and Pontusson, 2012).

As mentioned already, the financial crisis has exerted a very significant impact on GDP and employment. Among the more affected were the three Baltic States, Ireland, Greece and Spain. The Baltic States and Ireland started recuperation in 2010 or 2011. Spain started growing in 2011, but its GDP contracted in 2012. Two thirds of the EU regions suffered a GDP contraction of up to −6 per cent a year between 2007 and 2010 (European Commission, 2013). In the period 2010–2015, Greece lost 26 per cent from its real GDP, while unemployment rate peaked at 27 per cent in 2013 (OECD, 2016). For Greece, it was estimated that partial recovery would take place by 2017, but full recovery would take much longer. The country's portrait is that of hysteresis (Martin, 2012), which implies a negative growth path with significant implications for future development. Indeed, the countries with the deepest recessions have experienced the largest long-term damage (Ball, 2014).

Moreover, the financial crisis has resulted in significant increases in total development debt due to bank recapitalisation, increased social spending (e.g. unemployment benefits) and funding to boost demand. A national increase in debt was higher in Ireland, Portugal, Greece and Spain. In the period 2011–2013, many member states cut expenditure aiming at fiscal consolidation, which meant that expenditure linked to medium-term growth was eliminated. Nevertheless, some member states recorded high and relatively high ratios of government debt to GDP: Greece, Italy, Portugal, Ireland and Belgium (European Commission, 2013), with obvious consequences to the funding of public services in all regions and especially to remote regions.

Housing market bubbles are a major indication of the main macro-economic imbalances linked to the financial crisis. Housing indebtedness is closely related to housing market developments. The mechanism which links the two ends of this equation is the following: firstly, significant growth of credit to households, followed by increases in house prices that gave boost to high residential investment and, finally, higher indebtedness of the private sector. The tendency to increase the indebtedness of individuals, households and companies has been notable in weaker and/or deficit countries of the south and it has been a consistent strategy pursued by financial institutions in the EU (Varoufakis, 2016).

There is significant variation in the length and speed of increases in housing prices across countries, but it should be mentioned that by 2006–2007 half of the EU member states recorded price increases above 6 per cent annually, which is a threshold alert of internal imbalances. Moreover, in the period 2007–2012, house prices contracted considerably in Ireland (–49.5%), Latvia (–35.7%), Estonia (–30.2%), Spain (–28.0%) and Romania (–26.1%) (European Commission, 2013). Ireland has been an illustrative case of how the financialisation relating to mortgage markets and home ownership has led to high household vulnerability in general, and more importantly in rural areas (Murphy and Scott, 2014).

In the wake of the financial crisis, the population at risk of poverty or social exclusion increased considerably. In the period 2008–2014, the highest increase is recorded in Greece (+7.9%), Spain (+4.7%), Malta (+3.7%), Hungary (+2.9%) and Italy (+2.8%). The crisis has resulted to greater flexibility in the labour market and to an expansion of precarious jobs (i.e. temporary contracts, part-time jobs) often associated with low income (European Parliament, 2014).

Most of the regions affected by rising unemployment and also those who have the highest unemployment scores are mostly found in the Southern European countries (Hadjimichalis, 2018). High scores of youth unemployment, especially in Southern European, East European and Baltic countries, has been a major reason for emigrating to the prosperous regions in Europe (European Commission, 2013). Rural areas, and more particularly remote areas, have been depopulated further due to the increased outflows of youth and especially of women. Moreover, migrants who has become a well-examined key driver of population dynamism in many economically developed urban and rural areas before the crisis (Deas and Hincks, 2014; Woods, 2016), but since the outbreak of the financial crisis, migrant flows have declined – emigration from poorer regions has increased and effectively rural areas have been affected (European Commission, 2013; Ballas et al., 2017).

To sum up, despite that the financial system has been reinventing itself in the midst of the crisis, geography still matters since space has been a constituent factor in the formation and unfolding of the crisis. Spatial interconnectivity appears to be a major aspect for the (re)production or expansion of the crisis. Moreover, there has been a map of financial flows, of differential wealth effects, of areas hardest hit and of crises of various types (French et al., 2009; Engelen and Faulconbridge, 2009). The financial crisis triggered national recessions of varying severity; the hardest-hit economies include those in the periphery of the Eurozone, which experienced severe banking and debt crises (Ball, 2014). However, recent comparative regional analysis has shown that since the financial crisis, urban and rural remote regions performed below the growth rate of intermediate and rural regions with high proximity to cities (Dijkstra et al., 2015). Finally, it is important to highlight that the actual differences in the quality of life and types of challenges faced by European populations are not entirely found across national borders, but between regions within countries, between rural localities and urban centres and/or between rich and poor neighbourhoods in cities/towns (Ballas et al., 2017).

Rural responses to the crisis

The starting point in this section has to do with the capacity of rural regions to respond to the challenges posed by the financial crisis. I will resist from reproducing the neoclassical discussion on the convergence of regions, the new regionalism and other theorisations on the capacity of regions to improve their position in the context of financial crisis (see, for example, Hadjimichalis, 2011; Hadjimichalis and Hudson, 2014; Dijkstra et al., 2015). I will focus my discussion instead on the notion of resilience, which has become the new buzzword in regional/rural development (see also Chapter 5 of this Companion).

The rapidly expanding literature has dealt with the issue(s) of community, rural and regional resilience to the financial crisis, which is considered as a means of illustrating the adaptability of specific spatial units to externally induced shocks (Hassink, 2010; Hudson, 2010; Martin, 2012; Scott, 2013; Wilson, 2013; Martin and Sunley, 2015; Martin et al., 2016; Evenhuis, 2017). Social resilience has become a fashionable concept over the last decade or so in the social sciences and human geography literature in particular, and due to its origins in the natural and behavioural sciences, it remains under constant scrutiny. One of the main questions is whether resilience increases the capacity of communities or regions to cope with external disruptions and hence it can be a loose antonym to vulnerability. Secondly, the relationship between resilience and sustainability remains a critical question. Thirdly, this concept was launched to capture the ability of systems to cope and/or respond to changes that come from outside to challenge their systemic coherence and sustainability (Adger, 2000).

The usefulness of resilience has been questioned or interrogated by some writers (Hassink, 2010; Hudson, 2010; Davoudi, 2012; Welsh, 2014), while others argue in favour of the concept on the basis that it offers an opportunity to re-frame development and practice (Shaw, 2012; Scott, 2013; Evenhuis, 2017). Moreover, the ambiguity of the concept and its contested character has been exposed (Christopherson et al., 2010; Porter and Davoudi, 2012), its characteristics are thoroughly analysed and its various facets are interrogated against the social reality of the financial crisis (Martin, 2012; Martin and Sunley, 2015). Also, the complexity of resilience is well documented (Welsh, 2014; Evenhuis, 2017).

In this context, four major questions seem relevant for the critical conceptualisation of the concept: resilience of what, to what, by what means and with what outcome (Martin and Sunley, 2015: 12)? According to the rapidly expanding literature, the reclaiming of the concept and its distancing from the hegemonic neoliberal policy frame (Peck, 2013) requires the integration of human agency, the recognition of the operation of power structures and the re-politicisation of its components so that it will be possible to pursue a political economy of regional/rural/community resilience (Hudson, 2010; Scott, 2013; Bristow and Healy, 2014). Finally, resilience needs to gain a normative content and become more connected to the notion of socio-spatial justice (Storper, 2011; Fainstein, 2014; Hadjimichalis, 2018), although it is not completely clear how this can be achieved.

As Scott (2013) suggests, resilience may well become a useful conceptual lens to rural development and practice; however, to provide a useful framework, resilience approaches should shed its evolutionary connotations, its depoliticised frame and its identification with either competitiveness or sustainability. In most of the literature, resilience is identified with the adaptive capacity of regions in times of crisis (Martin and Sunley, 2015; Evenhuis, 2017), while only a few writers argue that resilience offers nothing less than a 'paradigm shift' that leads to fundamental questioning of the central tenets of contemporary approaches to planning, particularly in relation to linear assumptions (Hudson, 2010; Shaw, 2012).

There is a firm belief among some analysts that rural areas cannot be resilient or that rural areas were less affected by the recent economic crisis. It is argued that the real estate bubble and the stronger connection of urban regions to international markets are the main reasons for their underperformance (Dijkstra et al., 2015). However, as Artelaris (2017: 80) argues: 'Although the ramifications of the crisis have been proved anything but spatially uniform, both at national and EU levels . . . the literature of the geographies of crisis remains sporadic, incomplete and inconclusive'.

To examine the geographies of the crisis, and specifically the rural dimension, the next section examines examples of rural resilience to the financial crisis based on research evidence in Greece and Spain.

Firstly, evidence from Greece shows that the less urbanised and developed regions, which tend to focus economic activities around more labour intensive sectors, such as tourism and agriculture, performed better during the years of the crisis. The predominantly rural regions lack a number of typical advantages, such as highly skilled labour, large market size, openness, good transportation connections, etc. However the lack of these parameters has operated as a 'protection buffer', due to the fact that these regions are less exposed to the national and international environment. Moreover, the significant degree of self-consumption and self-sufficiency as well as the dependence of these regions on agriculture and subsidies has created a 'safety net' against the current economic recession (Artelaris, 2017).

Moreover, there is evidence that following the financial crisis, the role of agriculture in certain rural areas grew due to the fact that numerous people followed a 'back to the land/roots' movement, enhancing rural and individual resilience. This led to a revitalisation of rural areas through the strengthening of the farming sector (Kasimis and Papadopoulos, 2013), sometimes in search of a 'farming' or 'rural idyll' (Anthopoulou et al., 2017). The significant proportion of ownership-occupancy in rural areas acts as a 'safety net' for those already living in rural areas, but also for those who have inherited a small farm and a house in the countryside (Zografakis and Karanikolas, 2012). Further evidence from regional towns illustrate patterns of counterurbanisation referring to ex-urbanites who seek a better quality of life and/or security in the midst of the financial crisis (Gkartzios, 2013; Gkartzios et al., 2017).

Secondly, evidence from Spain, along with other countries (Fieldsend, 2013), has shown that there are successful territorial dynamics which improve the resilience of certain rural regions in times of crisis. More specifically, a number of factors such as economic diversification, agricultural employment, existence of infrastructure and facilities, retention of immigrant population, availability of natural resources and institutional capacity and governance play a positive role for the resilience of rural regions. For those factors to act together, it is important to design and implement rural public policies that are comprehensive, complementary and effectively coordinated (Sanchez-Zamora et al., 2014).

Moreover, it is important to highlight that because of the financial crisis, the locational preferences of immigrants have changed, while the number of new residents in Spain rapidly decreased. The role of international immigrants has adjusted following the overall development pattern in the country, eliminating reasons for conflict or discontent in rural areas (Alama-Sabater et al., 2017). Finally, local studies have offered strong indications of changing household strategies in order to be more resilient to economic recession. As an example, rural households started combining residencies and stays between village and city to address the changing needs of their family in view of cutbacks in public services (Camarero and Oliva, 2016), while intergenerational support and greater personal mobility have become mechanisms for rural resilience in times of recession (Camarero et al., 2016).

To sum up, the responses of rural areas to the financial crisis have not been clearly identified in the relevant literature. Despite the ambiguity and its contested character, the notion of resilience has been useful as a conceptual lens to look more closely at how rural areas can cope in times of crisis. There is only slim evidence on rural resilience to the financial crisis, which is rather challenged by dominant urban regional discourses.

Conclusion

The so-called Great Recession of 2007–2008 has been a major example of the way global capitalist processes under the hegemony of neoliberal policies have reshaped the uneven geographies of Europe. It has been evident that the impact of the crisis has varied significantly across countries and regions. The greatest impact of the crisis has been on the periphery of the Eurozone with particular emphasis on Southern European countries. Greece looms large due to the exceptionally fierce impact of the crisis and the prolonged austerity that followed. The paradigmatic treatment of the Greek case on part of the EU and its institutional mechanisms has had devastating effects for the well-being of the general population and also for its rural component.

The relatively new concept of rural resilience is considered appropriate for analysing rural responses to the financial crisis on the condition that it can be re-claimed so that it is inclusive of human agency, it takes into account the power structures and becomes less neutral. More attention should be paid to rural areas and their adaptive capacity in view of the crisis effects, while sustainability may be strengthened as a component of resilience. Rural actors devise resilient practices to ensure their sustainability. Finally, public policies, which are formulated under the hegemony of the neoliberal agenda, should follow existing rural practices linked to regional/community resilience.

References

Aalbers, M. (2009), Geographies of the Financial Crisis, *Area* 41(1), 34–42.

Adger, W.N. (2000), Social and Ecological Resilience: Are They Related? *Progress in Human Geography* 24(3), 347–364.

Alama-Sabater, L., Alguacil, M. and Serafi Bernat-Marti, J. (2017), New Patterns in the Locational Choice of Immigrants in Spain, *European Planning Studies* 55,1–22.

Anthopoulou, T., Kaberis, N. and Petrou, M. (2017), Aspects and Experiences of Crisis in Rural Greece. Narratives of Rural Resilience, *Journal of Rural Studies* 52, 1–11.

Artelaris, P. (2017), Geographies of Crisis in Greece: A Social Well-Being Approach, *Geoforum* 84, 59–69.

Ball, (2014), Long-term Damage from the Great Recession in OECD Countries, *European Journal of Economics and Economic Policies: Intervention* 11(2), 149–160.

Ballas, D., Dorling, D. and Hennig, B. (2017), Analysing the Regional Geography of Poverty, Austerity and Inequality in Europe: A Human Cartographic Perspective, *Regional Studies* 51(1), 174–185.

Bermeo, N. and Pontusson, J. (2012), Coping with Crisis: An Introduction, in N. Bermeo and J. Pontusson (eds), *Coping with Crisis: Government Reactions to the Great Recession*, New York: Russell Sage Foundation, 1–31.

Breuss, F. (2016), The Crisis in Retrospect: Causes, Effects and Policy Responses, in H. Badinger and V. Nitsch (eds), *Routledge Handbook of the Economics of European Integration*, New York: Routledge, 331–350.

Bristow, G. and Healy, A. (2014), Regional Resilience: An Agency Perspective, *Regional Studies* 48(5), 923–935.

Camarero, L., Cruz, F. and Oliva, J. (2016), Rural sustainability, intergenerational support and mobility, *European Urban and Regional Studies* 23(4), 734–749.

Camarero, L. and Oliva, J. (2016), Mobility and Household Forms as Adaptive Strategies of Rural Populations, *Portuguese Journal of Social Science* 15(3), 349–366.

Christopherson, S., Michie, J. and Tyler, P. (2010), Regional Resilience: Theoretical and Empirical Perspectives, *Cambridge Journal of Regions, Economy and Society* 3, 3–10.

Davoudi, S. (2012), Resilience: A Bridging Concept or a Dead End? *Planning Theory & Practice* 13(2), 299–307.

Deas, I. and Hincks, S. (2014), Migration, Mobility and the Role of European Cities and Regions in Redistributing Population, *European Planning Studies* 22(2), 2561–2583.

Dijkstra, L., Garcilazo, E. and McCann, P. (2015), The Effects of the Global Financial Crisis on European Regions and Cities, *Journal of Economic Geography* 15, 935–949.

Engelen, E. and Faulconbridge, J. (2009), Introduction: Financial Geographies – The Credit Crisis as an Opportunity to Catch Economic Geography's Next Boat, *Journal of Economic Geography* 9, 587–595.

European Commission (2013), *The urban and regional dimension of the crisis*, Eighth progress report on economic, social and territorial cohesion, Luxembourg: Publications Office of the European Union.

European Parliament (2014), *Poverty in the European Union: The Crisis and its Aftermath*, Luxembourg: Publications Office of the European Union.

Evenhuis, E. (2017), New Directions in Researching Regional Economic Resilience and Adaptation, *Geography Compass* 66, 1–15.

Fainstein, S. (2014), Resilience and Justice, *International Journal of Urban and Regional Research* 66, 157–167.

Fieldsend, A.F. (2013), Rural Renaissance: An Integral Component of Regional Economic Resilience, *Studies in Agricultural Economics* 115, 85–91.

French, S., Leyshon, A. and Thrift, N. (2009), A Very Geographical Crisis: The Making and Breaking of the 2007–2008 Financial Crises, *Cambridge Journal of Regions, Economy and Society* 2, 287–302.

Freshwater, D. (2015), Vulnerability and Resilience: Two Dimensions of Rurality, *Sociologia Ruralis* 55(4), 497–515.

Gkartzios, M. (2013), 'Leaving Athens': Narratives of Counter-urbanisation in Times of Crisis, *Journal of Rural Studies* 32, 158–167.

Gkartzios, M., Remoundou, K. and Garrod, G. (2017), Emerging Geographies of Mobility: The Role of Regional Towns in Greece's 'Counterurbanisation Story', *Journal of Rural Studies* 55, 22–32.

Hadjimichalis, C. (2011), Uneven Geographical Development and Socio-Spatial Justice and Solidarity: European Regions after the 2009 Financial Crisis, *European Urban and Regional Studies* 18(3), 254–274.

Hadjimichalis, C. (2018), *Crisis Spaces: Structures, Struggles and Solidarity in Southern Europe*, London: Routledge.

Hadjimichalis, C. and Hudson, R. (2014), Contemporary Crisis Across Europe and the Crisis of Regional Development Theories, *Regional Studies* 48(1), 208–218.

Harvey, D. (2010), *The Enigma of Capital*, London: Profile.

Harvey, D. (2011), Crises, Geographic Disruptions and Uneven Development of Political Responses, *Economic Geography* 87(1), 1–22.

Hassink, R. (2010), Regional Resilience: A Promising Concept to Explain Differences in Regional Economic Sustainability? *Cambridge Journal of Regions, Economy and Society* 3, 45–58.

Hudson, R. (2010), Resilient Regions in an Uncertain World: Wishful Thinking or a Practical Reality? *Cambridge Journal of Regions, Economy and Society* 3, 11–25.

Kasimis, C. and Papadopoulos, A.G. (2013), Rural Transformations and Family Farming in Contemporary Greece, in A. Moragues Faus, D. Ortiz-Miranda and E. Arnalte Alegre (eds), *Agriculture in Mediterranean Europe: Between Old and New Paradigms*, Bingley: Emerald Publications, 263–293.

Kitson, M., Martin, R. and Tyler, P. (2011), The Geographies of Austerity, *Cambridge Journal of Regions, Economy and Society* 4, 289–302.

Marsden, T. (2009), Mobilities, Vulnerabilities and Sustainabilities: Exploring Pathways from Denial to Sustainable Rural Development, *Sociologia Ruralis* 49(2), 113–131.

Martin, R. (2012), Regional Economic Resilience, Hysteresis and Recessionary Shocks, *Journal of Economic Geography* 12, 1–32.

Martin, R. and Sunley, P. (2015), On the Notion of Regional Economic Resilience: Conceptualization and Explanation, *Journal of Economic Geography* 15, 1–42.

Martin, R., Sunley, P., Gardiner, B. and Tyler, P. (2016), How Regions Reach to Recessions: Resilience and the Role of Economic Structure, *Regional Studies* 50(4), 561–585.

Massey, D. (1979), In What Sense a Regional Problem? *Regional Studies* 13, 233–243.

Massey, D. (2005), *For Space*, London: Sage.

Murphy, E. and Scott, M. (2014), Household Vulnerability in Rural Areas: Results of an Index Applied During a Housing Crash, Economic Crisis and Under Austerity Conditions, *Journal of Rural Studies* 51, 75–86.

OECD (2016), *Economic Surveys: Greece*, Paris: OECD.

Peck, J. (2013), Explaining (with) Neoliberalism, *Territory, Politics, Governance* 1(2), 132–157.

Porter, L. and Davoudi, S. (2012), The Politics of Resilience for Planning: A Cautionary Note, *Planning Theory & Practice* 13(2), 329–333.

Sanchez-Zamora, P., Gallardo-Cobos, R. and Cena-Delgado, F. (2014), Rural Areas Face the Economic Crisis: Analyzing the Determinants of Successful Territorial Dynamics, *Journal of Rural Studies* 35, 11–25.

Scott, M. (2013), Resilience: A Conceptual Lens for Rural Studies? *Geography Compass* 7(9), 597–610.

Shaw, K. (2012), 'Reframing' Resilience: Challenges for Planning Theory and Practice, *Planning Theory & Practice* 13(2), 308–312.

Shucksmith, M., Cameron, S., Merridew, T. and Pichler, F. (2009), Urban–Rural Differences in Quality of Life across the European Union, *Regional Studies* 43(10), 1275–1289.

Storper, M. (2011), Justice, Efficiency and Economic Geography: Should Places Help one Another to Develop? *European Urban and Regional Studies* 18(1), 3–21.

Varoufakis, Y. (2011), *The Global Minotaur: America, the True Origins of the Financial Crisis and the Future of the Global Economy*, London: Zed Books.

Varoufakis, Y. (2016), *And the Weak Suffer What They Must? Europe's Crisis and America's Economic Future*, New York: Nation Books.

Welsh, M. (2014), Resilience and Responsibility: Governing Uncertainty in a Complex World, *Geographical Journal* 180(1), 15–26.

Wilson, G.A. (2013), Community Resilience, Policy Corridors and the Policy Challenge, *Land Use Policy* 31, 298–310.

Woods, M. (2007), Engaging the Global Countryside: Globalization, Hybridity and the Reconstitution of Rural Place, *Progress in Human Geography* 31(4), 485–507.

Woods, M. (2016), International Migration, Agency and Regional Development in Rural Europe, *Documents d'Analisi Geografica* 62(3), 569–593.

Zografakis, S. and Karanikolas, P. (2012), Tracing the Consequences of Economic Crisis in Rural Areas: Evidence from Greece, in R. Solagberu Adisa (ed.), *Rural Development: Contemporary Issues and Practices*, Shanghai: InTech, 311–336.

17

Rural innovation and small business development

Paul Cowie, Pattanapong Tiwasing, Jeremy Phillipson and Matthew Gorton

Introduction

Rural economies are some of the most dynamic places in terms of growth rates and SME innovation (Deller and Conroy, 2017; Frontier Economics, 2014). However, how is rural innovation understood in policy terms, what hinders it, and how might barriers and obstacles be overcome? To explore these issues in more detail the chapter uses a case study of the rural enterprise hubs initiative in England, established to provide the vital networking and infrastructure needed to support rural innovation and small business development.

Innovation is the process of creating something novel, a product, ways of working or service (the invention), and introducing that novelty to the real world (OECD, 2005). This process is important as it is a mechanism by which value can be created and productivity increased. Schumpeter (1961) argues this occurs through 'creative destruction' – old, inefficient ways of producing products or providing services are superseded by new, more efficient methods. In this way productivity can be increased and value created for the economy. Closely linked to the concept of innovation is the concept of entrepreneurship. Entrepreneurs are often identified as the key actors in the process of innovation. For some, entrepreneurship is connected to new business formation (Stewart et al., 1999). For others, just forming a business is insufficient; the process also involves a degree of novelty and improvement to qualify as being entrepreneurial (Schumpeter, 1961).

Typically, innovation and entrepreneurial processes are most often associated with urban rather than rural areas (Carlino et al., 2007). This is in part due to the importance placed on flows of knowledge within innovation and entrepreneurial processes. With agglomeration of firms and higher density of knowledge networks in urban areas, new ideas can flow more readily between innovators and entrepreneurs (Cooke, 2002; Caniëls and Romijn, 2005; OECD, 2014). There are also intangible benefits from this clustering of actors and high density of networks (Bathelt et al., 2004). Individuals circulate throughout this environment freely, both through formal processes (i.e. higher turnover of staff, business networking events or training and learning events) and more informal social networks, and it is through these interactions that both codified knowledge and tacit knowledge are transmitted. Studies demonstrate (Saxenian,

1994; Schoenherr et al., 2014) that to create an innovative entrepreneurial environment, the ability to access both forms of knowledge are important.

In contrast to the dynamic image of urban economies, the narrative for rural areas tends to be that they are static and traditional, dominated by primary sectors, and comprising low growth firms and 'lifestyle' entrepreneurs who choose to start a business to achieve a work/life balance and are motivated by social and cultural aspirations rather than economic imperatives (Kalantaridis, 2004). However, a review conducted by OECD (2014) details how these framings of rural innovation fail to capture the contemporary nature of rural areas, and that sustained, high levels of regional performance are no longer necessarily dependent on high population densities. Rural regions contain a diverse range of engines for economic growth, driven by enterprise, innovation and new technologies, across a host of new and old industries. However, rural areas tend to underperform in terms of the generation of patents (Carlino et al., 2007) the metric most often used to measure place-based innovation. The reach of formal, patent focused linear research-based innovation systems, reliant on scientific or technological discovery in universities, national governments and corporate R&D facilities, remains weak.

However, an overreliance on patent-based measures has led to the 'mistaken assumption that rural areas are not innovative' (OECD, 2014: 10) and that to allow the potential for growth in all regions a place-based approach is required along with a 'better understanding of how innovation can emerge in a rural setting and how governments can support and encourage it' (ibid.: 9). Contrary to this assumption, prior research identifies rural businesses to be more innovative than urban businesses in terms of new product development (Anderson et al., 2005; Phillipson et al., 2017). Moreover, the characteristics of rural economies, their limited local markets, lower business densities and distance to major markets can act as a spur to innovation as entrepreneurs innovate to overcome barriers to growth (Hubbard and Atterton, 2014). Alternative framings of innovation associated with open innovation models and networked and collaborative approaches may be more relevant in rural areas (Chesbrough, 2006). But what key barriers must be overcome to realise the full potential of rural innovation?

Barriers to rural innovation and small business development

Successful innovation requires a critical mass of human, financial and social capital (Drabenstott and Henderson, 2006). However, in rural areas several barriers may hinder the accumulation and successful application of such capital. For instance, rural businesses may suffer from small pools of local labour, limited business diversification, poor availability of external funding, and low-quality information communication technology (ICT) infrastructure (Huggins and Hindle, 2010; Kotey and Sorensen, 2014). Some of these problems may be more acute in rural areas where labour markets are typically shallower, with a smaller pool of skilled workers. Distance from major markets may also hinder innovation, particularly where ICT infrastructure is weak (Henderson, 2007).

Recent EU wide research (Tisawang et al., 2017), from the ongoing INNOGROW Interreg project, examines the adoption of innovative technologies by small and medium sized enterprises (SMEs). Data, from owners or directors of SMEs (in this case from farming and food related businesses), were collected from 8 countries, namely Bulgaria, Czech Republic, Greece, Hungary, Latvia, Italy, Slovenia and the UK. Survey respondents reported information on the type of technology adopted, their motives for doing so, barriers encountered on adoption of the new technology, as well as impacts on the business in terms of job generation, ability to access new markets, and firm profitability. SMEs encounter multiple barriers during the integration/

adoption of new technologies, with the most common relating to a lack of funding/finance, integration costs, lack of expertise/skilled labour, inability to hire new employees with relevant skills/expertise, competition in the industry, and regulation / limited support by local agencies. Other barriers include a lack of appropriate external advice/technological skills, difficulties in establishing effective collaboration with supply chain partners, and lack of consumer demand or stakeholders' interest.

The same research explores the link between the impacts of new technology adoption on the business and the barriers encountered in adopting the technology. For example, whether SMEs benefit or not from job generation following adoption was not found to be associated with any particular barriers. However, rural SMEs that have not experienced improved access to new markets were more likely to have faced lack of funding/financial resources, lack of consumer demand or stakeholders' interest as their main barriers. SMEs with improved access to new markets were more likely to highlight competition in the industry as a barrier. Firms that did not witness improved profitability were more likely to report regulation/limited support by local agencies, lack of funding/financial resources, lack of consumer demand or stakeholders' interest, difficulties in establishing effective collaboration with supply chain partners as barriers to the adoption of innovative technologies. Overall, the study highlights that financing difficulties, unfavourable regulatory conditions and market risk are significant obstacles to the adoption of innovative technologies by rural SMEs.

Improving flows of knowledge is a vital component of the innovation process and in overcoming hurdles to rural innovation. For example, the barriers to knowledge impact collaboration within supply chains and more generally the ability of businesses to obtain information about the innovation process (i.e. finding funding for innovation and advice from those with specialised knowledge). Two characteristics of rural businesses combine to create this knowledge barrier. Firstly, rural businesses tend to be smaller, often having no employees other than the business owner (Cosh and Hughes, 1998); there is often little spare capacity to engage in knowledge exchange and networking activity. Secondly, a higher proportion of rural businesses have no independent business premises, operating instead from the home (Taylor, 2008), which can act as a barrier to meeting other businesses and other commercial contacts.

The need to stimulate knowledge transfer (NESTA, 2007) both within rural areas and between rural and urban areas has long been recognised. While developments in ICT provide a platform for many rural areas to participate in knowledge sharing activities irrespective of distance, physical agglomeration and location of the business (Cairncross, 1997), not all areas have benefitted and, given the quality of ICT connectivity, reach of ICT has been uneven. Moreover, the benefits of personal face-to-face connections remain a key component of innovation practice to facilitate the transmission of tacit as well as codified knowledge (Storper and Venables, 2004). This can be seen in an urban context with the proliferation of co-working spaces particularly in the digital and creative sectors (see for example Dunlop, 2017). Co-working and shared office spaces are one way of fostering inter-personal contact between entrepreneurs in a way that fosters collaboration and innovation. This is also being encouraged in several rural areas, with the next section presenting a case study of this in greater detail.

Case study: rural enterprise hubs in the northeast of England

This section outlines a case study of a policy initiative to establish a series of 'enterprise hubs' in the rural northeast of England to support small business development and innovation. In this project, an enterprise hub is the central point in a business network. This relates to both a *physical* network, a geographical place where individuals meet, and also to a *node* in the flow of knowledge, a place which transmits and circulates knowledge. In 2012, the northeast of England was named as one of five Rural Growth Network (RGN) pilot projects. Its aim was to improve rural growth and productivity. As part of the RGN programme a work-package was to focus on the provision and development of rural business premises. It has been known for some time that the provision of suitable rural business premises is a significant barrier to growth for rural businesses. The NE RGN wanted to test a number of possible models for rural business premises: live/work units; 'smart working' and incubator space.

There was also a more fundamental issue of what a rural enterprise hub is and what it should do. To answer this question a review of rural business premises was conducted. The review found 18 enterprise hubs in the pilot area. The majority of these hubs were owned by the private sector (50%) with the not-for-profit sector owning a further 33 per cent and the public sector owning the remainder. Interviews with hub managers revealed most offered very little additional services or support beyond renting the physical space. There was also little flexibility in either the tenure offered to occupiers or the space they could rent (Cowie et al., 2013).

Based on the initial review of rural enterprise hubs within the pilot area, a working definition of enterprise hubs was developed. From the initial interview data it appeared there are two factors which differentiate an enterprise hub from any other business premises. The first relates to the physical characteristics of the enterprise hub. Enterprise hubs tend to offer additional facilities and services which are not offered within a general business premises. For example Bergek and Norrman (2008: 21) argue that there are four features common to all business enterprise hubs:

- shared office space, which is rented under more or less favourable conditions;
- a pool of shared support services to reduce overhead costs;
- professional business support or advice; and
- network provision, internal and/or external.

In addition, there are also other benefits which relate to the flexibility in the terms of letting the premises. Flexibility in terms of length of tenure and the ability to move between larger and smaller spaces over time as the business grows or shrinks is extremely beneficial to growth businesses.

The second, less tangible factor which differentiates an enterprise hub from other business premises is the opportunity to share and exchange knowledge. Having space in an enterprise hub provides opportunities to the businesses to share knowledge both with other businesses within the hub and externally across the wider economy. This knowledge brokering is a key source of additionality which enterprise hubs can provide. It adds value over and above the physical bricks and mortar of the building. Enterprise hubs become key nodes in the transmission and use of knowledge within the rural and regional economy.

An initial analysis of the rural hubs with the Rural Growth area also highlighted an interesting distinction between hub types. These were characterised as either 'Hive Hubs' or 'Honey Pot Hubs'. Hive Hubs contained mainly business to business enterprises without necessarily having customers attend the premises. In contrast Honey Pot Hubs contained mainly business to customer enterprises often in the craft and arts sector. These hubs relied on getting customers to visit the hub and often contained ancillary attractions such as a café or were built around an existing tourist attraction (Cowie et al., 2013).

The first phase of the project also highlighted the needs of the hubs themselves. Many of the hubs were quite vulnerable. They needed business and innovation support in the same way as their tenants. This was recognised in the second phase of the project which instigated a number of projects to support hubs and allow them to be more innovative. One example of such an innovation was the use of pre-fabricated office spaces, known in the project as 'office pods'. Two of the enterprise hubs experimented with these office pods as a cost-effective way of creating more flexible office space. The pods were fully serviced and self-contained offices. The hubs were able to commission a variety of pod sizes, which allowed them to provide flexibility of office space to their tenants. One of the hubs that commissioned the pods via the RGN found them so successful they bought a further three themselves. They were then able to accommodate businesses as they grew. One business moved from a home location to a two-person pod. As they grew, they moved again to a four-person pod and ultimately graduated from the hub to their own office space elsewhere. In terms of innovation, the hubs themselves have innovated to create new business space. Research from the pilot project also found two innovations – the live/work units and 'smart workspace' – did not succeed. These types of hubs were not seen as being attractive and needed in the pilot project: the live/work unit idea was abandoned before any hubs were created, while the smart workspace hubs failed to attract sufficient interest and so were converted to more traditional work spaces.

Interestingly, these experiments – the office pods, live/work units and 'smart workspace' – were tested by the not-for-profit and public-sector hubs. This highlights the need to understand that innovation can be instigated by the public as well as the private sector. Indeed Mazzucato (2015) argues that the state has an integral part to play in the innovation process through fostering an entrepreneurial environment.

In phase two of the project, the hub occupiers became the focus of the research. A survey of rural enterprise hubs in the pilot area found that on the whole businesses occupying enterprise hubs were younger but employed more people than the general NE rural business population. In terms of their markets, hub businesses had fewer very local customers but more regional customers than the general NE rural business population. Hub based businesses were also more

Figure 17.1 Office pods at the Northumberland National Park's enterprise hub.

likely to serve other businesses – this is particularly striking given the sample included businesses in honey pot hubs which target private customers.

In terms of the role enterprise hubs play in the rural economy, one of the most interesting findings of the project was the fact that 58 per cent of hub occupiers had moved to the hub from home. In additional a further 10 per cent had started their business in the hub. This indicated that enterprise hubs were playing an important role in incubating new and early-stage businesses. The exact nature of this role was explored further with businesses asked what their primary motivation was for moving to the hub. The majority of respondents cited that rent and flexibility of tenure was the biggest influence in their decision. More intangible benefits such as the opportunity to network or collaborate with other businesses and gain access to business support was much less important. This can be contrasted with what businesses perceive to be the barriers to growth. In both the INNOGROW and Enterprise Hubs research, collaboration and better business support were cited as significant barriers to growth. The finding suggests that hub occupiers do not see the move to an enterprise hub as a way of achieving better access to knowledge – at least not initially.

The nature of hub occupiers may be key to understanding this issue. As mentioned above, two-thirds of hub occupiers have either started their business in the hub or are moving on from a home-based business. In both instances taking on commercial business premises will be a significant commitment. Having flexible terms will therefore play a significant role in the decision-making process. Once a business has secured suitable space on favourable terms, it can start to develop the networks needed to overcome the barriers to growth. The need to overcome isolation and gain access to better business support start off as latent needs but then crystallise once the functional problem of finding suitable staff has been resolved. The business owner was aware of the potential to collaborate and gain access to business support but it played a secondary role in the decision making process. This has implications for the way rural enterprise hubs are promoted and how business support and networking activity is delivered through the hubs.

A post-hoc evaluation to the RGN programme highlighted the rural enterprise hubs as one of the more successful elements in the overall project (SQW, 2016). The project originally planned to create 6 hubs but this was expanded during the project to 11 as they proved to be very successful. One aspect of the rural enterprise hubs that was not as successful as expected was the provision of hot-desk co-working spaces. It was anticipated that the dispersed, isolated nature of home-based rural businesses would mean that there would be a demand for space to work outside the home. This was not the case and many of the hot-desking spaces have been removed. Notwithstanding the failure of hot-desking in a rural context, the remaining programme highlighted how it is possible to create environments which mirror what might be considered a more 'urban' way of working. The high occupancy rates found in the hubs and the anecdotal evidence (from author's personal communication with hub managers in the summer of 2017) of businesses growing and collaborating with other hub occupiers highlights the opportunities for rural entrepreneurship once some of the barriers are removed. There has also been individual examples of innovation by the businesses in the hubs directly as a result of their presence in a hub. One business was able to adapt and transfer their service to a new market following discussion with other hub members and the hub manager. Another hub occupier is attempting to introduce a new product to the market in the UK and is using many support services from other members of their hub. Both these examples would not have happened were it not for the connections made through the hub and the ability to gain both tacit and codified knowledge through the hub network.

The hubs initiative is itself an example of rural innovation. An essentially urban concept has been translated into a rural context. As outlined above, it has not been a complete success.

Innovations such as smart working and live/work units have not succeeded in a rural context. The hubs have also created a network to support their activity, foster collaboration between hubs and to share knowledge between them. These activities start to overcome some of the barriers to innovation often found in rural economies. What is not as clear is the degree to which the hubs have stimulated innovation within the rural economies in which they are located. Further research on this issue is needed.

Conclusions

Innovation is a complex process which requires a consideration of the spatial dimension. There are substantive differences between the way innovation occurs in rural and urban areas. The density of knowledge networks, limited local labour force and poor access to markets and finance in rural areas do act as barriers to innovation. That said, innovation can and does occur in rural areas. The opportunities afforded by developments in ICT, while acknowledging that this is not fully deployed in many rural areas, offers opportunities for rural businesses to engage in more open forms of innovation.

To foster greater innovation requires proactive initiatives to overcome some of the barriers to rural innovation. The EU INNOGROW project quantifies many of these barriers and highlighted their interdependencies (Tisawang et al., 2017). Enterprise hubs offer a potential, place-based, solution to overcoming these barriers. The flow of knowledge is at the heart of the innovation process and often an inhibited flow of knowledge within rural economies is a major barrier to innovation. This can be as a result of the physical properties of rural economies. It can also be through more intangible barriers which occur as a result of certain narratives attaching to rural economies, that they are not innovative and not a location in which innovation can gain a foothold.

The rural enterprise hubs programme has shown a form of innovation: the translation of an idea from one context to another. An urban innovation often associated with high tech industry has been translated to a rural context. The case study highlights the process of innovation is not a smooth linear one. Certain avenues of innovation can be dead-ends. In the case of rural enterprise hubs, it was the smart workspace and live/work innovations that failed to succeed. The case study also highlights the role that the state can play in innovation. In the case of the rural enterprise hubs, it was able to bring together the key actors in this field: the public, private and third sector hub owners. The state, in collaboration with others, was able to mobilise resources, not just financial but also academic resources, to tackle the issues at hand.

References

Anderson, D., Tyler, P. and McCallion, T. (2005) Developing the rural dimension of business-support policy, *Environment and Planning C: Government and Policy*, 23(4), 519–536.

Bathelt, H., Malmberg, A. and Maskell, P. (2004) Clusters and knowledge: local buzz, global pipelines and the process of knowledge creation, *Progress in Human Geography*, 28(1), 31–56.

Bergek, A., and Norrman, C. (2008) Incubator best practice: A framework. *Technovation*, 28(1–2), 20–28.

Cairncross, F. (1997) *The Death of Distance: How the Communications Revolution will Change Our Lives*, Cambridge, MA: Harvard Business Press.

Caniëls, M. C. and Romijn, H. A. (2005) What drives innovativeness in industrial clusters? Transcending the debate. *Cambridge Journal of Economics*, 29(4), 497–515.

Carlino, G. A., Chatterjee, S. and Hunt, R. M. (2007) Urban density and the rate of invention, *Journal of Urban Economics*, 61(3), 389–419.

Chesbrough, H. (2006) *Open Innovation: The New Imperative for Creating and Profiting from Technology*, Cambridge, MA: Harvard Business Press.

Cooke, P. (2002) *Knowledge Economies: Clusters, Learning and Cooperative Advantage*. London: Routledge.

Cosh, A. D. and Hughes, A. (1998) *Enterprise Britain: Growth, Innovation and Public Policy in the Small and Medium Sized Enterprise Sector 1994–1997*, Cambridge: Centre for Business Research, University of Cambridge, retrieved from www.cbr.cam.ac.uk/fileadmin/user_upload/centre-for-business-research/downloads/other-publications/survey-enterprisebritain-summary.pdf.

Cowie, P., Thompson, N. and Rowe, F. (2013) *Honey Pots and Hives: Maximising the Potential of Rural Enterprise Hubs*, Newcastle-upon-Tyne: Centre for Rural Economy.

Deller, S. C. and Conroy, T. (2017) Business survival rates across the urban–rural divide, *Community Development*, 48(1), 67–85.

Drabenstott, M. and Henderson, J. (2006) A new rural economy: A new role for public policy, *Main Street Economist*, 1(4), 1–6.

Dunlop, T. (2017) Co-working spaces are the future of work but that could be a good thing, *Guardian*, 24 April, retrieved from www.theguardian.com/sustainable-business/2017/apr/24/co-working-spaces-are-the-future-of-work-but-that-could-be-a-good-thing (accessed 22 September 2017).

Frontier Economics. (2014) Drivers of rural business growth, decline and stability, retrieved from http://sciencesearch.defra.gov.uk/Default.aspx?Menu=Menu&Module=More&Location=None&Completed=0&ProjectID=18782

Henderson, J. (2007) The power of technological innovation in rural America, *The Main Street Economist: Regional and Rural Analysis*, 2(4), 1–5.

Hubbard, C. and Atterton, J. (2014) Unlocking rural innovation in the north east of England: the role of innovation connectors, in OECD (eds), *Innovation and Modernising the Rural Economy*, 95–112, Paris: OECD.

Huggins, R. and Hindle, R. (2010) *Opportunities and Barriers to Business Innovation in Rural Areas: A Report Produced for Department for Environment, Food and Rural Affairs (Defra)*, London: GHK.

Kalantaridis, C. (2004) Entrepreneurial behaviour in rural contexts, in L. Labrianidis (ed.), *The Future of Europe's Rural Peripheries*, 62–85. Aldershot: Ashgate.

Kotey, B. and Sorensen, A. (2014) Barriers to small business innovation in rural Australia, *Australasian Journal of Regional Studies*, 20(3), 405–422.

Mazzucato, M. (2015) *Entrepreneurial State: Debunking Public vs Private Sector Myths*, London: Anthem Press.

NESTA (2007) *Rural Innovation*, London: NESTA.

OECD (2005) *Oslo Manual: Guidelines for Collecting and Interpreting Innovation Data*, 3rd edition, Paris: Eurostat.

OECD (2014) *Innovation and Modernising the Rural Economy*, Paris: OECD Publishing.

Phillipson, J., Gorton, M., Maioli, S., Newbery, R., Tiwasing, P. and Turner, R. (2017) *Small Rural Firms in English Regions: Analysis and Key Findings from the UK Longitudinal Small Business Survey, 2015*, Newcastle upon Tyne: Centre for Rural Economy and Newcastle University Business School.

Saxenian, A. (1994) *Regional Advantage: Culture and Competition in Silicon Valley and Route 128*, Cambridge, MA: Harvard University Press.

Schoenherr, T., Griffith, D. A. and Chandra, A. (2014) Knowledge management in supply chains: the role of explicit and tacit knowledge, *Journal of Business Logistics*, 35(2), 121–135.

Schumpeter, J. A. (1961) *The Theory of Economic Development: An Inquiry into Profits, Capital, Credit, Interest, and the Business Cycle*, trans. R. Opie, Cambridge, MA: Harvard Business Press.

SQW (2016) *Final Evaluation of the Rural Growth Network Pilot Initiative: Final Report to the Department for Environment, Food and Rural Affairs*, Cambridge: SQW.

Stewart, W. H., Watson, W. E., Carland, J. C. and Carland, J. W. (1999) A proclivity for entrepreneurship: a comparison of entrepreneurs, small business owners, and corporate managers, *Journal of Business Venturing*, 14(2), 189–214.

Storper, M., and Venables, A. J. (2004) Buzz: face-to-face contact and the urban economy, *Journal of Economic Geography*, 4(4), 351–370.

Taylor, M. (2008) *Living Working Countryside: The Taylor Review of Rural Economy and Rural Housing*, London: Department for Communities and Local Government.

Tisawang, P., Gorton, M. and Phillipson, J. (2017) *Impact of Main New Technologies for Rural Economy SMEs*, Newcastle-upon-Tyne: Centre for Rural Economy, Newcastle University.

18

The creative class doing business in the countryside

Networking to overcome the rural

Lise Herslund

Introduction

Over the years in Western countries, migration to cities has been taking place at the same time as counter-urbanisation movements involving urban dwellers who, in many cases, display the outward characteristics of Florida's so-called creative class. These have moved from the city to further away towns and rural areas. These urban newcomers have been described as middle-class with a service class identity (Urry, 1995), mobile professionals bringing know-how and expertise (Thrift, 1987), and as more 'growth-oriented' than their local counterparts (Bosworth, 2010). Research into why creative individuals are attracted to rural areas point to lifestyle preferences such as quality of life and finding creative inspiration away from high rents and the homogeneity of urban areas (Duxbury and Campbell, 2011; Roberts and Townsend, 2016; Benson and O'Reilly, 2009).

Following Florida's (2002) argument, attracting the creative class should also be a route to higher economic growth. Traditional rural businesses – including retail or activities located in workshops – operate in a context of scattered population, lower levels of education, distance from customers, suppliers and advisory services and limited and slow internet access (Salemink et al., 2017; Smallbone et al., 2002). The question then is how the arrival of the creative class impacts on this patterning of economic activity. As established in the literature, the in-migration of people to rural areas affects both ideas about the quality of life and the nature of business development. For example, in the North-East of England, half of rural micro-businesses are run by in-migrants (Atterton and Affleck, 2010). Murdoch et al. (2003) also described shifts in rural economies as a 'counter urbanisation-led' diversification. Stockdale (2006) portrays newcomers as promoting economic regeneration. Bosworth (2010) applies the term 'commercial counter-urbanisation' to the growth of rural economies stimulated by the arrival of professionals from the city. According to Roberts and Townsend (2016), incoming creative individuals can produce a more culturally inspiring countryside but rural areas may struggle to retain this 'creative talent' owing to deficiencies in infrastructure and a lack of policies supporting creativity and/or infrastructure investment and upgrading (Roberts and Townsend, 2016).

Bosworth and Willett (2011) question the impact that this group, and their various enterprises have on rural areas: if they are 'turning away' from the pressures of modern urban life, searching instead for the rural idyll, they are perhaps unlikely to bring dynamism to rural communities. There will only be an impact if the businesses build local networks and become 'embedded', referring to the extent to which an enterprise becomes part of the local economy, rather than merely located in it (Bosworth and Willett, 2011). Embedded businesses can contribute to sustainable rural development by extending the extra local networks.

Murdoch (2000) stressed the importance of building 'horizontal' networks that link rural actors and spaces into more general and non-agricultural processes of economic change to complement the mainly 'vertical networks' of the modern agricultural sector to national and EU bodies as well as global markets (ibid.). Also the ideas of neo-endogenous – and the more recent nexogenous rural development (see Chapter 14 of this Companion) – point to the need for rural areas to 're-connect' to urban development and to a broader set of public and private actors in order to break the isolation and solve the challenges they face they (Ray, 2006; Bock, 2016). This network building is regarded by Bock as essential social innovation (ibid.).

This chapter looks at urban creative class newcomers starting micro-business in rural areas: why they do it, whether they can stay creative and what keeps them in the rural setting. It also discusses the impact they have on the rural area with a particular focus on the networks they create. The chapter draws on interviews with 30 Danish rural in-migrants who established micro-business in rural locations almost a decade ago. That research is detailed in Herslund (2012). Now, almost ten years on, four of the respondents have been re-interviewed and asked to reflect on their business activities and the networks they have initiated or participated in.

Methods

The selection of the initially interviewed respondents took place through a network of micro-businesses (micro-net) started in the former Storstrøm county in Denmark. Today, counties have been merged to larger regions and the former county now comprises six municipalities. Respondents were contacted through the network's homepage and newsletter. The selection criterion was that the businesses should be run by a newcomer from the metropolitan area of Copenhagen, who had relocated to a rural area. A 'newcomer' was defined as an individual who had moved to the area within the last six years and started a business. All respondents meeting these requirements ($n = 30$) were interviewed (Herslund, 2012).

Three original respondents were re-interviewed in 2017; a 'copy-writer', a 'networking consultant' and a 'dietician' alongside a new respondent whose business started in 2011; a 'management consultant'. In this chapter, the main results from the initial study (Herslund, 2012) are presented and then reflected on by the four respondents interviewed.

Results

Pushed to business by lifestyle

Respondents moved to the rural area because they found a more attractive place to live which meant a bigger and cheaper home in a more attractive location. An attractive location meant a scenic spot away from immediate neighbours and also an opportunity to become part of a local community. The respondents could be divided into two groups according to their age and life cycle stage. More than three-quarters of the respondents (24/30) were over 50 years of age, with either no children or grown-up children. The remainder were young families with children. The young families tended to be seeking a 'sense of community' whereas the middle-aged respondents were drawn by the quiet scenic locations. More than 90 per cent of the respondents were women, although the selection process did not use gender as a criterion.

All respondents had expected their everyday life to become less stressful when moving to the countryside because of their new quiet dwelling and, for the young families, the hope that they would thrive in their new community and more natural surroundings. Members of the middle-aged group were all employed in creative and knowledge-jobs in the city, for example within the media, communication and business services sectors. Two thirds also did consultancy and freelance work on the side, especially those working in media and communications. These included journalists, and people in marketing and advertising. Others, working in biotech and business consulting, were more traditionally employed in larger private companies. The women with young families were mainly engaged in public sector occupations, such as teachers, health workers, social workers and in administration, IT and communication. They expected to find jobs locally.

The move to the rural area did not turn out as expected for many respondents. The impact of a busy working life in the city undermined the wish for time at the new attractive dwelling in the countryside, leaving little time for enjoying the new home and/or scenic location. For the young family respondents, finding employment close to their new home proved difficult and several ended up having to commute considerable distances to find jobs that matched their competences.

After a few years in the new residence, the respondents had registered as self-employed. Self-employment was a strategy that enabled them to combine a country home with a continued career and achieve greater flexibility in everyday life. The wish was to keep on doing what they liked professionally while becoming more flexible. They were, in a way, being pushed into self-employment by their lifestyle aspirations. The copy-writer explains:

> By moving to the countryside I was not so tied up financially so I felt I now should live my business dream in copy-writing instead of just being stressed. Now I had the chance to do what I probably would never have dared to do if I had stayed in the city.

The starting point was that self-employment in advertising and copy-writing was possible with a good internet connection. The networking consultant adds:

> I took leave from my job in the national radio to see if I could get a business going in web marketing and the use of social media and got an ISDN connection installed because at that time the internet was even more impossible than today.

In another example, the dietician says:

> I never had any intention of starting a business but I simply could not take becoming re-trained so I started as a privately practising dietician without knowing how.

In the start-up phase, there were two groups of self-employed businesses. One started as a continuation of a 'city working life'. The middle-aged knowledge workers became self-employed consultants in a field similar to their former job and mainly depended on market, business and collaborators in the city – 'city businesses'. Those working in media and communication focused on their freelance activities in copywriting, graphic design, web marketing etc., which in most cases had been a freelance activity running alongside their regular employment. The business consultant, such as the respondent employed in a biotech company, established herself as a 'tech transfer consultant'; and the engineer working in a big engineering company became a self-employed 'consulting engineer'.

The young family respondents established private services focused towards the local area ('local businesses'), often using skills developed in their former occupations.. For example, a teacher and an IT worker established web businesses selling, respectively, garden accessories and children's clothes. Another teacher started a language school. A communication worker became a life coach and the dietician set up a private practice. They often combined social or maternity benefits with self-employed activities in the first years of start-up. Like the business services, they started businesses that were not common to the rural area.

Creative continuous adaptation

Becoming self-employed was a way to reduce commuting, but running a knowledge business with customers and business partners mainly in the city also generated its own need to commute. Thus, the first years in business, especially for the middle-aged group, were spent finding ways to reduce the need for commuting and face-to face contact. This meant finding customers and collaborators in the regional area. For the women in more locally oriented businesses, it was also about finding more customers in the regional or city context as the local market turned out to be too limited.

For example, the management consultant looked out for business services closer by:

> My accountant, my IT help and my web designer. They were all so far away but when I actually searched I could find some in my neighbourhood. They were just not visible. Not even a sign outside the house because who just drops by here in the countryside?

The copy-writer had a similar experience, starting an informal network of other people in advertising mainly comprising newcomers: 'In advertising you need to meet face-to face with all the other people around a product, like those doing graphic design, those doing web etc. for the product to turn out as you want it. Finding those closer by rather than in the city was a way to cut down on the commuting'. The networking consultant also started a network locally:

> I started meeting up with handful of also newly started self-employed because I missed being with like-minded people and also wanted to share experiences from the life of being self-employed. We have, over the years, also written some books together on local foods and arts.

The strategy adopted by these entrepreneurs was to find ways to combine a few customers in the urban area with services or product oriented towards the regional market. They continued to undertake work for their most well-established city or national customers while also offering more all-round business services for regional clients like organising courses, talks, project management etc. This tension between pursuing urban-based, regional or local markets is illustrated in the following examples and interview extracts. The networking consultant outlines that:

> After all these years I still go to the city around twice a month just to show my face, hold courses and meet with publishers and working partners. I just pay for the hours I need to use an office or a meeting room in an office partnership.

The copy-writer makes this comment on the nature of the regional market:

> The companies keep more in-house and you get smaller jobs than you would in the city. But they have gotten used to us and we have gotten bigger jobs over the years.

The young family respondents who had started businesses oriented to private individuals were now looking to broaden their customer base and seeking opportunities among other businesses, municipalities and institutions. They were also trying out new activities such as writing blogs and articles or giving talks reaching other customers than private people. Finally, the dietician comments on her changing activities:

> When I finally got an assignment to make a diet plan for children with ADHD, I realised that it should not just be the kitchen personnel I talked to but all the staff I should give courses and talks on how diet influences learning and behaviour. Inspired by other micro-businesses in the micro-net I started promoting giving talks, writing articles and now have a blog for municipal staff, kindergartens, parents and all.

The dietician felt that she used her skills and professional experience more than she ever thought she would following her move to the countryside.

While young family respondents reported greater satisfaction with their businesses now than in the early years, several had found it more difficult to live in the rural area with children than they had originally anticipated.. The dietician says:

> You are very dependent on your local area with children. Country living is more than just forest, beach and a large home. It is schools, kindergartens, activities for children and a local social life.

If the school closes and families leave, everyday life becomes more difficult and several of the respondents moved from their village to the nearest market town. In contrast, the middle-aged respondents felt that they benefited from greater freedom than they had experienced in their former metropolitan lives. They describe the rural area as a place where they can find time to relax and be creative, also in their 'free time'.

The copy writer reflects on the social life:

> Moving to a cheaper dwelling meant I dared to start living out my dream professionally, but after some years I also felt I could live it out socially and push for a more creative social life. Now I feel there is so much going on in music, arts, workshops and networks for women and it is us newcomers that started it.

Overcoming the rural by networks

As outlined above, the respondents were found through a network of micro-businesses. The network was started in 2005 by the networking consultant, inspired by her smaller network of 'like-minded' people that she started some years earlier. With funding from the EU, alongside State grants, she organised a three-day conference inviting self-employed people from the whole of Storstrøms county, but micro-businesses from the whole country also attended: 'It is becoming normal to be self-employed, so people from all over needed to come together and share experiences'. Today, the micro-network holds regular events, workshops and has sub-groups on various themes such as lecturing, web media, writing etc. There is also a 'market' platform where members offer each other services of a more commercial nature and a blog where the members share experiences of their self-employment. Facebook is today the most important media for information and coordination.

Yearly survey returns show that most of the 200 active micro-businesses present in the area were operating full-time in 2017. More women than men run these businesses and three-quarters of the owners/operators are university educated. Furthermore, three-quarters are in the 41–60 age bracket. The founder of the micro-net further adds:

> Most members are newcomers or return migrants and they join [the network] to reflect on life as self-employed, to get to know people and see it as a way into the local area. The network can create a feeling of belonging.

The founder concludes that the micro-net attracts well-educated adults which she finds particularly interesting for a remote and rural area.

The dietician compares the network with business services:

> This network inspired me to give talks and write. The traditional business services just said my kind of business could not survive here and they were right. Even after so many years I still join micro-net events because they feel almost like family but I also take part in other networks like one for 'storytellers' also started by a newcomer.

The copywriter adds:

> The business services could not help people like me. The micro-net was important during the first years of adapting. Today I am still part of it mainly for the social dimension and for meeting people.

The management consultant adds the following concerning business services:

> Still today they focus on the 20 big companies and you still need to create jobs to attract much interest. This is why I applied for funding to start a network in my home area, a business house and then also found various funding to repeat the success in another local area.

The networks started by the management consultant are more locally based networks between small businesses in two local areas and include workshops, (web) shops, therapists and knowledge and creative businesses. The networks' activities centre on a Facebook page and monthly meetings. In one of the local areas a 'development house' has also been established in rooms above the local library. Here, businesses can share office space and there are rooms for local associations. According to the management consultant the problem is now to keep the networks going when the funding is running out:

> I am also a business and I cannot run too many networks without any pay. I started the networks because I was concerned about my local area but it is naïve to think that all networks can run by themselves after the initial start-up.

According to her, the success and reason why the micro-net still runs after 12 years is that it is run by an 'activist' and a person who writes books and consults on networking so she can also use the micro-net as a learning case in her business.

According to the networking consultant it is demanding to run networks. Since the set-up grants ran out, she and two others have run the network on a voluntary basis:

> When you are an unconventional group it is difficult to seek funding. We have had to create a formal society but the administrative burdens are exhausting. We are not geared in the same manner as more established societies, businesses and training centres. So now we do not even try to seek funding anymore as it is too time consuming.

Inspired by the micro-net, in 2013 two municipalities started a project to attract more newcomers. Part of this was to make micro-businesses in the two municipalities more visible. When the project finished, the business society in one of the municipalities took over and today still coordinates monthly meetings and social events such as Christmas lunches for the micro-businesses in that municipality.

Any impact? Giving but also receiving

Becoming self-employed is a way to combine continuing a career with enjoyment of a rural lifestyle. Respondents report that they do not want to return to conventional jobs because they want to stay free and flexible. But what do rural areas gain from the presence of newcomers and their business start-ups? Evidence suggests that they gain growth in a skilled and qualified workforce, a population that values the qualities of a rural location, and new knowledge that is freely exchanged through networks. But, from the point of view of respondents, authorities and established business networks do not always acknowledge the value of incoming entrepreneurs, focusing their support on conventional businesses that deliver direct employment growth.

Yet, respondents see great value – to rural places and their economies – flowing from new patterns of working and business start-ups. As the dietician commented:

> We can encourage more people to become self-employed as the self-employed way of living is also a way to combine a career with senior life in the rural area. We are getting older but can still contribute after retirement age.

The copywriter adds:

> We bring the qualified workforce they always say we lack in the region. Freelance living and self-employed consultants have exploded and is now not just an urban phenomenon. Even if we are just a handful of copywriters in the area, I have been able to live well for over the 15 years I have been in business.

According to the management consultant, it is the energy and knowledge they can bring into the networks:

> We can motivate and coordinate, keep them focused and take the learning out of monthly meetings and Facebook discussions and make a difference for the local small businesses as they are often overlooked by the advisory services.

The networking consultant further adds:

> We are an agile workforce and are not expensive to run. But what we need is some recognition for all the knowledge we bring into play and the efforts for sharing it.

The networking consultant refers to a survey undertaken as part of the project by the two municipalities, which showed that 40 per cent of businesses in the municipalities are micro-businesses (Mikronet, 2013). The network consultant is happy that one municipality now coordinates a kind of micro-network: 'They do not try to take over but let the businesses decide what input they need'. She feels that she has put much effort into trying to get interest and recognition from municipalities and business societies without much luck other than the promise of free membership for a year in business societies, start-up courses and new start-up accommodation by the waterfront. However, as the network consultant suggests:

> But we are not start-ups, we are way beyond business advice and we do not need higher expenses for fancy houses or expensive memberships. What we need is visibility and recognition from all around, a free place to meet and some small continuous funds to cover practical help and some of our time spent.

Discussion

Bosworth (2010) described newcomer businesses as growth-oriented. In this study, the businesses are generally in growth sectors and are adaptation-oriented. They are in sectors that are new to the rural area and they also represent new ways of running a business in a rural area. It is characteristic that the respondents have worked to diversify their services and project management, securing funding and running networks is also part of this portfolio.

They are very flexible and go far to adapt mainly due to lifestyle and everyday needs, and not strictly because of growth concerns. They can be badged 'lifestyle businesses' even though they are not confined to lifestyle sectors and do not comprise hobbyists. The newcomers fit the creative class label, settling in their rural dwelling and local area mainly for lifestyle reasons – but rather than being attracted to the creative urban buzz, they are instead searching for the rural idyll.

When 'creative class' people move to a rural area, becoming self-employed provides a means of balancing lifestyle and business aspirations. The rural idyll is, however, different for different newcomers and for people at different life-stages. Middle-aged newcomers, without young/dependent children, perhaps adapt more easily than younger families with their wider set of service needs. Older people look for support networks and cultural events and often contribute to setting these up.

Bottom-up networks for knowledge businesses, local businesses, self-employed in advertising, story-tellers etc. serve as fora for sharing experiences on writing, social media and self-employment but also have a lifestyle dimension as they serve as meeting places and a way into the regional and local area.

So are these new businesses locally embedded? In short, yes: they bring together locals and newcomers. They are trying to become even more locally embedded by getting business services and municipalities to recognise and engage with these kinds of start-ups. They provide examples of social innovation, coalescing into knowledge networks and bringing new activities to rural areas, which offer new service opportunities. But are the networks that support business innovation sustainable once initial funding draws to an end? They rely on the energy of the creative class and are themselves a sign of the entrepreneurialism that has been drawn to the countryside. But given the clear value of those networks to start-ups and the perennial problems of funding, it is questionable whether they should perhaps become more traditional business networks with membership fees, taken over by municipal/advisory service staff or stay as flexible as they are with some running costs covered from municipal budgets.

Starting a business can be a way to stay or even become part of the creative class and less vulnerable to some rural obstacles such as limited markets, lack of advisory services and limited opportunities for re-training. The locally oriented businesses started by respondents in this study needed to be creative in diversifying their products and customer bases. The inspiration to adapt was found through networks, which also extended the local and horizontal linkages for local businesses. The use of the 'creative class' badge for rural newcomers starting micro-businesses is questionable. Certainly, creativity is not confined to any particular sector – the 'creative arts' or media. Rather, it refers to the adaptability of businesses to rural settings and their thinner markets – their ability to break through barriers to succeed and thrive in the countryside. Entrepreneurial business networks play a critical role in their success.

References

Atterton, J. and Affleck, A. (2010) *Rural Businesses in the North East of England: Final Survey Results (2009)*. Newcastle: Centre for Rural Economy Research Report.

Benson, M. and O'Reilly, K. (2009) Migration and the search for a better way of life: a critical exploration of lifestyle migration. *The Sociological Review* 57, 608–625.

Bock, B. (2016) Rural marginalisation and the role of social innovation: a turn towards nexogenous development and rural reconnection. *Sociologia Ruralis* 56, 552–573.

Bosworth, G. (2010) Commercial counterurbanisation: an emerging force in rural economic development. *Environment and Planning* 42, 966–981.

Bosworth, G. and Willett, J. (2011) Embeddedness or escapism? Rural perceptions and economic development in Cornwall and Northumberland. *Sociologia Ruralis* 51, 195–214.

Duxbury, N. and Campbell, H. (2011) Developing and revitalizing rural communities through arts and culture. *Small Cities Imprint* 3, 111–122.

Florida, R. (2002) *The Rise of the Creative Class*. Cleveland: Perseus

Herslund, L. (2012) The rural creative class: counterurbanisation and entrepreneurship in the Danish countryside. *Sociologia Ruralis* 52, 235–255.

Mikronet (2013) 25 bud til politikerne ved Kommunalvalg 2013. Retrieved from http://mikronet.dk/25-bud-til-politikerne-ved-kommunalvalg-2013 (accessed 12 December 2017).

Murdoch, J. (2000) Networks – a new paradigm of rural development? *Journal of Rural Studies* 16, 407–419.

Murdoch, J., Lowe, P., Ward, N. and T. Marsden (2003) *The Differentiated Countryside, Routledge Studies in Human Geography*. London: Routledge.

Ray, C. (2006) Neo-endogenous development in the EU. In P. Cloke, T. Marsden and P. Mooney (eds), *Handbook of Rural Studies*, 278–310. London: Sage.

Roberts, E. and Townsend, L. (2016) The contribution of the creative economy to the resilience of rural communities: exploring cultural and digital capital. *Sociologia Ruralis* 56, 197–219.

Salemink, K., Strijker, D. and Bosworth, G. (2017) Rural development in the digital age: a systematic literature review on unequal ICT availability, adoption, and use in rural areas. *Journal of Rural Studies* 54, 360–371.

Smallbone, D., North, D., Baldock, R. and I. Ekanem (2002) *Encouraging and Supporting Enterprise in Rural Areas*. Hendon: Centre for Enterprise and Economic Development Research, Middlesex University Business School.

Stockdale, A. (2006) Migration: pre-requisite for rural economic regeneration? *Journal of Rural Studies* 22, 354–366.

Thrift N. (1987) Manufacturing rural geography? *Journal of Rural Studies* 3, 77–81.

Urry, J. (1995) A middleclass countryside? In T. Butler and M. Savage (eds), *Social Change and the Middle Classes*, 205–219. London: UCL Press.

19

Payments for ecosystem services and the rural economy

Meri Juntti

Introduction

The concept of ecosystems services (ES) frames the relationship between society and nature in terms of the functions of ecological systems that directly or indirectly benefit humans (MEA, 2005; Chapter 11 in this Companion). This departs in many respects from the prevalent focus on the notions of scientific value and rarity that inform the conservation status of species, habitats and landscapes. ES based approaches highlight the central role that ecosystem components and processes such as aesthetically pleasing landscapes and nutrient cycling, which are not extractable resources, play in supporting human wellbeing and indeed most economic processes (Norgaard, 2010; NCC, 2015). Conventionally, ES are divided into four categories (MEA, 2005):

1   *supporting ES* such as soil formation and Nitrate and Phosphate cycling;
2   *regulating ES* such as climate regulation, flood regulation and water purification;
3   *provisioning ES* such as food production and habitat provision; and
4   *cultural ES* such as the aesthetic and recreational enjoyment and associated physiological benefits that humans gain from interacting with nature.

In contrast to the protected species and areas approach that reduces conservation objectives to static target figures with debatable value, the notion of flowing services such as habitat provision, water purification and flood regulation captures the dynamic character of natural systems and their functions. This fluidity is better poised to evoke an integrated and flexible approach to conservation efforts. The emphasis on services from nature to societies can be seen as an articulation of the utility value of the elements and processes of nature, even where they do not constitute extractable goods, and is aimed at enabling and incentivising sustainable economic activities and decisions (NCC, 2015). Payments for ecosystem services (PES) are increasingly used to mediate the relationship between resource managers such as farmers or forest authorities and ES recipients such as water companies (and their customers) or national governments (and their citizens) who have signed up to conservation and emission mitigation targets. A PES scheme is an economic contract aiming to secure the provision of specific ES and the goods and benefits derived from them. Contracting to a PES scheme will likely bind the resource manager to specific management practices and ES output targets against which payments will be made.

The rest of this chapter provides a closer examination of the rationale underpinning PES and some of the possible formats that PES can be delivered through. Prevalent debates regarding legitimacy and effectiveness of PES are referenced throughout and the concluding section collates some recommendations for 'good PES'.

The governance rationale underpinning payments for ecosystem services

The need for economic incentives

PES can be seen as a manifestation of the so called governance shift which has seen the initial top-down regulatory approach increasingly complemented by new economic policy instruments that speak to broader pressure for deregulation and the scaling down of the role of the state (Wurzel et al., 2013). The UN Convention for Biological Diversity (UNCBD) outlines 12 principles of ecosystem management based on this integrated approach highlighting the need to maintain ecosystem functionality (and thereby services) and to consider ES in their economic context (Secretariat to the UNCBD, 2007). Specifically, this latter point is taken to mean:

1   reducing those market distortions that adversely affect biological diversity;
2   aligning incentives to promote biodiversity conservation and sustainable use;
3   internalising costs and benefits in the given ecosystem to the extent feasible.

This approach recognises that since many ES and the goods and benefits derived from them constitute so called public goods or at best, common pool resources, governance intervention is required to establish the kind of market signals that incentivise provision and sustainable use (Keohane and Olmstead, 2016). Characteristic to such non-market resources, the exclusion of ES users is difficult and ES are non-rivalrous (whatever one individual consumes does not directly limit the consumption by others) (Ostrom, 2008). Therefore, there is little to incentivise their production especially if other lucrative uses of the means of production are available. Where resource managers are in a position to secure and enhance the provision of ES by, for example, undertaking farming practices that enhance water infiltration and pollutant removal capacity of soil, the managers need to be rewarded for their positive contributions to incentivise this. Environmental economists argue that economic incentives are more efficient at mediating public good provision than public regulation (Keohane and Olmstead, 2016).

Indeed, in 2002 Ferraro and Kiss published a seminal paper in the *Science Journal* entitled 'Direct payments to conserve biodiversity', where they set out a compelling argument for providing direct economic compensation for the provision of biodiversity through land management practices. Their argument was based on evidence demonstrating that long standing efforts aimed at achieving conservation targets by other means had produced very little conservation value for money. Their suggestion rested mainly on the reported inability of community-led conservation efforts, aspiring to reconcile community level development needs with conservation targets, to deliver actual conservation benefits (Roe et al., 2001). The assumption was that communities are first and foremost opportunistic in their resource use decisions: they are unlikely to take responsibility for sustainable management practices unless there is a clear return and no significant opportunity costs. Therefore, according to Ferraro and Kiss, direct payments in return for producing the desired conservation outcomes would offer the most cost-efficient route to success. They would also avoid the so called 'marginalisation by conservation' problem (e.g. Robbins, 2012) where livelihood strategies of communities are displaced or even criminalised through conservation initiatives. This logic underpins the development of most PES schemes today. It is worth noting that, direct payments are not the only form of PES, and as Ferraro and Kiss (2002) point out, there are several more indirect formats through which economic compensation for the provision and maintenance of ES can be accrued (see e.g. Dunn, 2011).

However, the PES approach is not without its critics and on closer examination the advantage that PES hold over community-led efforts is not as convincing as it might first appear. Swart (2003) and Muradian et al. (2013), point out that it is often the underlying and broad institutional weaknesses and associated power imbalances that compromise the ability of community based conservation efforts to yield results – particularly in a developing world context – and PES are unlikely to address these. An example of this would be insecure access and land rights that, in the face of a potential future conservation designation, may fuel speculative clearing of forest land for farming in order to diminish its conservation value (Roe et al., 2001). Indeed, many argue that direct payments in return for conservation outputs may also produce unintended outcomes due to speculative behaviour by stakeholders, such as land grabbing by those who are in a powerful position and wish to secure the PES income from the land (Green and Adams, 2015). Findings from the ESPA programme (ESPA, 2017) suggest that due to unequal distribution of PES income among local households, a stronger incentive for conservation could be found in framing conservation outputs in terms of community goals such as food security – continuing access to a crop or game species. Moreover, a significant problem is posed by the complexity of most ES and subsequent scientific uncertainty over how exactly they are produced and how their presence should be detected – what biophysical processes are key, what management practices exactly influence ES and how (Robertson, 2006).

Anticipating these complexities, the CBD highlights the importance of 'involv[ing] the necessary expertise and stakeholders at the local, national, regional and international level, as appropriate' as well as 'involv[ing] all stakeholders and balanc[ing] local interests with the wider public interest' through management practices that are devolved to the lowest possible grassroots level (Secretariat to the UNCBD, 2007). Proponents of PES argue that they can do just this: engage local resource managers and scientific experts, render ecosystem functions meaningful to people through articulating the services that they provide and linking those whose activities impact directly on ES supply with those who benefit from the ES in question (e.g. Dunn, 2011). Paying for management practices that enhance and conserve attractive landscapes that people want to visit or protect and improve water quality at drinking water source presents a solution that not only improves environmental quality but also diversifies local economies and stimulates the creation of new sustainable management solutions. However, to address the thorny issues of legitimacy and scientific uncertainty, the principle of broad involvement in negotiations regarding ES value and actually accrued benefits is crucial. Ideally, value negotiations involve direct and indirect resource users and managers, ES beneficiaries, scientific experts and public sector actors that can preside over procedures and oversee contract arrangements that will ensure that the ES in question are actually delivered against the payment (Bullock et al., 2017). Corbera et al. (2007) refer to procedural justice of PES decision-making (equal representation of all whose interests are concerned) and to the distribution of benefits (including payments) and finally, equity in 'access', which is particularly pertinent in a developing country context of smallholders where levels of knowledge about sustainable management practices and information about available compensation mechanisms can be limited. As with any policy instrument, decision-making procedures as well as distributive outcomes are central to fairness and legitimacy (Paavola, 2007) but remain elusive in existing PES (Corbera et al., 2007).

A cautionary note on economic valuation of ecosystem services

While scientific uncertainty may compromise the effectiveness of PES, even more debate has been generated by issues surrounding the economic valuation of ES, which is seen to pose a further significant challenge to legitimacy (De Groot et al., 2012). Value transfer, where an ES valuation made in one context is used to support the monetary valuation of the same ES in another context is a reasonably quick method increasingly used by policy-makers (ibid.). The Ecosystem Services Value Database (ESVD; supported by the Ecosystem Services Partnership, www.es-partnership.org) collates data on original ES values calculated on empirical case study basis and supports the transferability of the recoded values by disclosing relevant information about the used valuation methodology and the specific case study context (biome, sub-biome or ecosystem type, geographical location and year of valuation). The values in the database have been converted into to international dollars using appropriate purchasing power parity (PPP) conversion factors. So for example, the food provisioning service provided by one hectare of marine biome is estimated to be worth $93 per annum, whereas the same ES provided by a hectare of grassland biome is worth $1119 per annum. Similarly, the climate regulation value of one hectare of tropical forest is $2044 per annum whereas the same ES provided by one hectare of woodland biome over a year is only worth $7. This makes for compelling comparisons but it must be kept in mind that value transfer integrates uncertainties and remains highly problematic (De Groot et al., 2012). The purpose of ES valuation may vary and is not necessarily tantamount to commercialising ES or establishing PES. Bullock et al. (2017) list a range of purposes such as environmental liability assessment, estimating the cost of environmental change, and assessing the value of investment. De Groot et al. (2012) argue that the monetary values listed in the ESVD should be treated as indicative rather than absolute and only used as a supporting element in weighing and contrasting opportunity costs of resource use. They also highlight the uneven availability of data for different biomes and ES – some tend to be over studied whereas others have been neglected, not due to their insignificance but because of a lack of data and complexity. Though the standardisation methods render the value or ES transferrable, cultural and socio-economic context, and availability of and access to substituting sources of similar services matter hugely.

Payments for ecosystem services in practice

PES are often described as market-based policy instruments where markets for scarce public goods are created to incentivise their optimal provision. Most PES are not strictly speaking market-based however, not least because the provision of these ES is not strictly conditional on the reimbursement, and the number of involved producers and consumers tends to be limited (Muradian et al., 2013). In fact, it is useful to distinguish between location specific PES, for example compensation from water users to land managers whose actions impact on the quality of the water source in question; and payments for ES such as carbon sequestration where ES providers are not tied to a specific location. In this latter case a closer resemblance to a market situation can be achieved, where a number of ES providers may compete for contracts, and efficiency of provision may factor in securing one (Milder et al., 2010). A competitive bidding system may disadvantage farmers who do not have access to specialist conservation knowledge or have higher per-unit costs. This would be counterproductive if the PES in question were intended to augment farm incomes among poor small farmers (Milder et al., 2010; Corbera et al., 2007). Nevertheless, PES very rarely if ever establish ideal market like conditions where the law of supply and demand would guarantee that ES provision is maintained at an optimal level. Instead, PES are best described as negotiated contracts between stakeholders that establish a reimbursement aimed at securing the provision of specific ES. The sections below set out a range of existing PES formats and examples.

Direct payments to farmers

An often cited example of voluntary *direct payments by a private company* is the PES scheme set up by the French mineral water company Vittel, which pays cattle farmers that manage land in the catchment area of the aquifer where Vittel sources its spring water (e.g. Dunn, 2011). The payments are in return for the adoption of less intensive farming methods such as outdoor grazing. This, it is hoped, will ensure that the spring water that Vittel bottles will not exceed drinking-water standards for nitrates. Outdoor grazing reduces the need for feed cultivation and other practices that are associated with high nitrate application to soil and subsequent risk of leaching and infiltration. The payments also compensate for the planting of trees to improve soil conditions and to promote soil-based filtration services. According to Perrot-Maitre (quoted in Dunn, 2011), Vittel has managed to engage the majority of farmers in the catchment into PES contracts that run for up to 30 years.

While the Vittel PES scheme is an example that speaks to a private business rationale, the EU agri-environmental schemes are a long standing example of *public subsidy payments to farmers*. These PES schemes address one of the core priority areas of EU Rural Development Policy (Regulation 1305/2013 of the European Parliament and of the Council) that recognises the potential role of agriculture in restoring, preserving and enhancing a range of ecosystems. All EU Member States are directed to establish regional Programmes for 'Agri-Environment-Climate Payments' that cover environmental commitments going beyond the regulatory standards. While national Agri-Environmental Programmes respond to Member State specific environmental circumstances, values and issues, the payments are calculated in a uniform way throughout the region. Article 28.6 of the regulation states:

> . . . payments shall be granted annually and shall compensate beneficiaries for all or part of the additional costs and income foregone resulting from the commitments made. Where necessary, they may also cover transaction costs up to a value of 20% of the premium paid for the agri-environment-climate commitments.

Contracts to these PES schemes are offered for a minimum of five years. Testament to the complexity of measuring ES provision, the Commission's mid-term evaluation of the 2007–2013 Rural Development Policy laments the difficulty of establishing causal links between detected changes in biodiversity and the measure associated with the schemes (Schuh et al., 2012).

Attention is drawn to the inadequacy of existing indicators, and instead, impacts are projected based on observed adoption rates of conservation measures such as declining nitrate inputs, buffer zones and lighter tillage methods. Nevertheless, both Schuh et al. (2012) and Poláková et al. (2011: xxi) argue that there is good evidence to say that 'as a whole, the biodiversity status of agricultural habitats subject to agri-environmental measures is significantly better than would have been the case if the policy had not been in place'. Moreover, the schemes that are co-funded from the central EU budget and by Member State governments have indirectly contributed to rural economies and particularly the economic viability of less productive but high nature value farming. Some Member States such as the UK have introduced competitive bidding for the higher tier schemes, with the aim of ensuring that contracted bidders are those that offer and deliver greatest conservation value for money.

Reciprocal benefits in watershed management

Meyer and Lundy (2017) describe a PES scheme in Chon Ak-Suu river basin in Kyrgyzstan, an area where access to improved drinking water sources remains a challenge. This is a *PES in kind*, where ES recipients – local water users such as residents and agricultural producers – pay the local forest managers in terms of days of work, undertaking forestry and pasture maintenance tasks such as tree planting and bridge building within the river catchment. The forest managers provide the tools and trees, with the labour provided helping to maintain and restore forest habitat and associated pasture lands. In return for these in-kind payments, the ES recipients receive water with a lower sediment load (leading to fewer numbers of blockages within irrigation systems) and reduced levels of contamination by manure, enhancing drinking water quality and reducing associated illnesses. This kind of a PES in-kind arrangement may be particularly suitable in a context of weak or absent institutions that could oversee and execute economic contracts – a frequent challenge for PES in developing countries (Milder et al., 2010). This PES arrangement was supported by the Regional Environmental Centre for Central Asia (CAREC), in partnership with the Norwegian Government (Meyer and Lundy, 2017). As with the Indian example cited below, facilitation by this external project-based intervention played a significant role in enabling the PES.

Kovacs et al. (2016) describe a *reciprocal water access agreement* (RWA) between the municipal council of the town of Palampur in the Himalayan foothills of India, and the inhabitants of upstream villages who forage and consume the forest resources in the catchment of the Bohal spring, the most reliable supply of drinking water to the town. The RWA involves an annual monetary compensation to a democratically elected community body (the Village Forest Development Society, VFDS) that oversees the implementation of a sustainable forest management plan sanctioning resource use and management practices in the catchment of the Bohal spring. Kovacs et al. (2016) emphasise the centrality of trust and legitimacy to the functioning of the RWA that has been in place since 2010. A key tenet is the consensus among villagers that some kind of a sustainable management plan was needed and should be implemented in a transparent, accountable and just way. The VFDS was established for this specific purpose and replaced previous community groups that had failed to maintain the trust of the communities involved. While the RWA was instigated by an external project run by the German Development Agency, the setting up of the terms of the Forest Management Plan as well as the financial compensation form the water users to the forest managers involved intense negotiations between representatives of three different villages and the municipal administration of the user community. An external expert report was commissioned to ascertain that the proposed forest management practices really would have a beneficial impact on water quality in the spring. The RWA therefore managed to gain high trust in the relationship between the upstream forest resource users and the downstream water users and met several procedural justice criteria (Corbera et al., 2007). Kovacs et al. (2016) emphasise the centrality of local institutions to the success of any such scheme and describe ongoing issues of trust within the upstream communities that may yet pose a risk to the functioning of the PES scheme.

Indirect payments for ecosystem services from wildlife tourism

The CAMPFIRE programme in Zimbabwe focusses on the ability of communities to derive *direct and indirect income from local ES* (Frost and Bond, 2008; http://campfirezimbabwe.org/index.php/projects-t/12-community-based-tourism) through sanctioning eco-tourism and access rights for hunting on communal lands. This is hoped to provide new livelihood opportunities and encourage sustainable management practices and conservation of wildlife. The CAMPFIRE Association is a voluntary not for profit organisation and works with the regional administration to support rural communities in setting up and enforcing hunting quotas, sanctioning access rights and establishing agreements with tourism operators so that locals gain revenue from tourism on their lands. Essentially therefore, CAMPFIRE enables the monetisation of recreational ES while trying to ensure that local communities have an incentive for their sustainable management. Controversially, the highest revenue recreational ES has for long been trophy hunting (distinct from illegal poaching; Frost and Bond, 2008). This is far more lucrative then eco-tourism or safari tourism and therefore favoured by locals. This risks sustainable management especially where the game species are in decline. However, Lindsey et al. (2006) point out that hunting is not necessarily detrimental if sustainable quotas are set and observed. While CAMPFIRE trains and deploys 'Resource Monitors' and 'Game Scouts' to support sustainable practice, quota setting is a highly politicised exercise and frequently ineffective. Nevertheless, PES such as payments for hunting access and locally embedded tourism (e.g. home tourism or locally run lodges) valorise the local ES of remote rural communities and help achieve community led wildlife management. However, their ultimate social and environmental sustainability hinges on design and implementation that is based on a fine tuned understanding of the local context – institutions, conflicts and power relations included (Green and Adams, 2015).

Rural economies, livelihoods and payments for ecosystem services

As discussed above, the PES approach sits in the context of a broader neo-liberal shift in environmental governance approaches. The flipside of PES is commodification and neo-liberalisation associated with market-led conservation and 'green economies' where the end goal remains the ultimately unsustainable imperative of perpetual economic growth (e.g. Green and Adams, 2015). A similar tendency can be detected in the rural development narratives that emphasise the economic potential in landscape and heritage assets (NCC, 2015). While commodification of ES is unlikely to deliver sustainability targets unless framed by robust environmental standards and regulations, environmental values and assets painted as ES can be engaged to ease the dependency of rural economies on the volatile and declining returns from agriculture and to diversify farm economies and rural economies more broadly. While many forms of tourism can be seen as indirect forms of PES, Schomers and Matzdorf (2013) suggest that the majority of direct PES schemes in existence are public subsidies such as the EU agri-environmental schemes. PES schemes hold significant potential for increasing livelihood opportunities of poor rural communities in a developing country context (Milder et al., 2010), but the majority of large scale PES programmes are operating in developed countries. However, this may be changing as publically funded PES are spreading fast for example in Latin America, where the ESPA (2017) report describes a PES format that covers more than half a million acres of water-producing forests in 50 Bolivian municipalities. Nevertheless, in a developing country context PES often remain piecemeal project-based initiatives. This may be because underdevelopment often goes hand-in-hand with features that hinder the establishment of well-functioning and legitimate PES. The lack of robust institutions that could take initiative and oversee contracts, and the lack of agency among rural communities and resource managers are hallmarks of poor rural economies. Roe et al. (2001) argue that what matters for successful conservation efforts are the four 'R's of rights, responsibilities, returns and relationships. Existing and traditional use rights are important as they are founded on accepted, and thus legitimate, hierarchies and traditions and often informed by a long standing experiential way of knowing the local resource base. However, there may not be robust arrangements to uphold these in the face of conservation efforts such as PES that involve external, often state-led interventions and may overrule access rights and management responsibilities, and introduce what to locals may appear to be a questionable knowledge base. The watershed PES described above demonstrate the importance of context sensitive efforts to work with existing institutions and where these are inadequate, establish new ones in negotiation with local stakeholders. Again, procedural justice is crucial but may be challenged by tacit local alliances, or relations with external businesses who promise investments in return for access to resources. This is a risk factor with the payments for wildlife ES facilitated by CAMPFIRE. While such pacts may justifiably be seen as essential development efforts that warrant resource extraction, they often have demonstrable detrimental impact on livelihoods and conservation in the long run (Roe et al., 2001).

Nevertheless, the valorisation of the components and processes of ecosystems that can be termed public goods does have the potential to significantly contribute to rural economies, frequently framed as 'thin markets', unable to diversify and attract investment (OECD, 2014). Where rural economies depend on the pull of urban centres, ES can be an important articulation of the contingency of urban quality of life on rural assets as demonstrated by the watershed PES and the direct payments to farmers, above. The role of rural areas in delivering water and food security and in tackling major planetary challenges such as climate change is increasingly recognised and ES and PES can be central in consolidating this role. Environmental and climate change policy could and should act as a significant push factor in the articulation of these reverse dependencies to the benefit of rural economies and ES. Sustainable cities need not only the provisioning ES but also flood regulation and water quality benefits that rural resource managers are in a position to provide. This is increasingly recognised by planners, water providers and environmental managers (see for example www.upstreamthinking.org).

Concusions: do payments for ecosystem services support sustainable rural development?

ES and PES are a part of a broader effort to recognise the economic value inherent in natural capital and especially the components that are not readily priced though market mechanisms. This is not just a manifestation of unfettered commodification. Many argue that these approaches constitute a broader recognition and an important articulation of the dependency of humans and economic processes on healthy ecosystems, and respond to the failure of nature conservation via conventional means such as designated areas. At their best, ES and PES signify an emerging economic sustainability thinking that takes as it basis economic diversity and the sustainable embeddedness of economic activities in local socio-environmental contexts. While embedded economic engagements with local ES can be seen to support and valorise them, they may also open doors to 'green grabbing' (Green and Adams, 2015) – the exploitation of weak land-ownership and access rules. Well established and monitored land and access rights and economic accountability mechanisms are necessary for PES to function to the intended purpose. The ability of local stakeholders to represent their values and needs in any processes where income is gained from the valorisation of local ES is integral to the legitimacy of any single PES scheme. The distribution of monetary or in-kind gains from PES among households and communities is also central to just and legitimate PES, and must be acceptable to the providing communities.

References

Bullock C., Joyce D. and Collier M.J. (2017) *Guidance on the application of the Ecosystem Services Approach for Local Authorities (ESLA).* Dublin: University College Dublin.

Corbera E., Cosoy N. and Martinez Tuna M. (2007) Equity implications of marketing ecosystem services in protected areas and rural communities: Case studies from Meso-America. *Global Environmental Change* 17: 365–380.

De Groot R., Brander L., Van Der Ploeg S., Costanza R., Bernard F., Braat L., Christie M., Crossman N., Ghermandi A., Hein L. and Hussain S. (2012) Global estimates of the value of ecosystems and their services in monetary units. *Ecosystem Services* 1(1): 50–61.

Dunn H. (2011) *Payments for Ecosystem Services, DEFRA Evidence and Analysis Series.* London: Department for Environment, Food and Rural Affairs.

ESPA. (2017) *Annual Report 2016–2017.* Swindon: ESPA. Retrieved from www.espa.ac.uk/files/espa/ESPA_16-17_Annual_Report_ONLINE_1.PDF.

Ferraro P.J. and Kiss A. (2002) Direct payments to conserve biodiversity. *Science* 298(5599): 1718–1719.

Frost B.G.H. and Bond I.B. (2008) The CAMPFIRE programme in Zimbabwe: Payments for wildlife services. *Ecological Economics* 65: 776–787.

Green K.E. and Adams W.M. (2015) Green grabbing and the dynamics of local-level engagement with neoliberalization in Tanzania's wildlife management areas. *The Journal of Peasant Studies* 42(1): 97–117.

Keohane N.O. and Olmstead S.M. (2016) *Markets and the Environment.* Washington, DC: Island Press.

Kovacs E.K., Kumar, C., Agarwal C., Adams W.M., Hope R.A. and Vira B. (2016). The politics of negotiation and implementation: a reciprocal water access agreement in the Himalayan foothills, India. *Ecology and Society* 21(2): 37.

Lindsey, P.A., Alexander, R., Frank, L.G., Mathieson, A. and Romanach, S.S. (2006). Potential of trophy hunting to create incentives for wildlife conservation in Africa where alternative wildlife-based land uses may not be viable. *Animal Conservation* 9: 283–291.

MEA. (2005). *Ecosystems and Human Well-being: Synthesis Report.* Washington, DC: Island Press.

Meyer B.C. and Lundy L. (2017) *The Introduction of an Ecosystem Services Concept in Central Asia: Towards a Framework for the Sustainable Management of Nature and Land Use.* Almaty: CAREC.

Milder J.C., Scherr S.J. and Bracer C. (2010) Trends and future potential of payment for ecosystem services to alleviate rural poverty in developing countries. *Ecology and Society* 15(2): 4.

Muradian R., Arsel M., Pellegrini L., Adaman F., Aguilar B., Agarwal B., Corbera E., Ezzine de Blas D., Farley J., Froger G. and Garcia-Frapolli E. (2013) Payments for ecosystem services and the fatal attraction of win-win solutions. *Conservation Letters* 6(4): 274–279.

NCC. (2015) *The State of Natural Capital: Protecting and Improving Natural Capital for Prosperity and Wellbeing.* Third Report to the Economic Affairs Committee. London: NCC.

Norgaard R.B. (2010) Ecosystem services: from eye-opening metaphor to complexity blinder. *Ecological Economics* 69: 1219–1227.

OECD. (2014) Innovation and modernising the rural economy. OECD Rural Policy Reviews. Retrieved from www.keepeek.com/Digital-Asset-Management/oecd/urban-rural-and-regional-development/innovation-and-modernising-the-rural-economy_9789264205390-en#page1.

Ostrom E. (2008) Tragedy of the commons. In S.N. Durlauf and L.E. Blume (eds), *The New Palgrave Dictionary of Economics Online*, 2nd Edition. Palgrave Macmillan: London. Retrieved from www.dictionaryofeconomics.com/article?id=pde2008_T000193.

Paavola J. (2007) Institutions and environmental governance: a reconceptualization. *Ecological Economics* 63: 93–103.

Poláková J., Tucker G., Hart K., Dwyer J. and Rayment M. (2011) *Addressing Biodiversity and Habitat Preservation through Measures Applied under the Common Agricultural Policy.* Report Prepared for DG Agriculture and Rural Development, Contract No. 30-CE-0388497/00-44. London: Institute for European Environmental Policy.

Robbins P. (2012) *Political Ecology: A Critical Introduction.* Malden, MA: Wiley-Blackwell.

Robertson M.M. (2006) The nature that capital can see: science, state, and market in the commodification of ecosystem services. *Environment and Planning D: Society and Space* 24: 367–387.

Roe D., Mayers J., Grieg-Gran M., Kothari A., Fabricius C. and Hugheset R. (2001) *Evaluating Eden: Exploring the Myths and Realities of Community-Based Wildlife Management.* London: International Institute for Environment and Development.

Schomers S. and Matzdorf B. (2013) Payments for ecosystem services: A review and comparison of developing and industrialized countries. *Ecosystem Services* 6: 16–30.

Schuh B., Beiglböck S., Novak S., Panwinkler T., Tordy J., Fischer M., Zondag M.J., Dwyer J., Bański J. and Saraceno E. (2012) *Synthesis of Mid-Term Evaluations of Rural Development Programmes 2007–2013: Final Report.* Vienna: Österreichisches Institut für Raumplanung.

Secretariat to the Convention on Biological Diversity. (2007) Ecosystem approach principles. United Nations Environment Programme. Retrieved from www.cbd.int/ecosystem/principles.shtml (accessed 19 December 2017).

Swart J.A.A. (2003) Will direct payments help biodiversity? Response to Ferraro and Kiss. *Science* 299: 1981.

Wurzel R.K.W., Zito A.R. and Jordan A.J. (2013) *Environmental Governance in Europe: A Comparative Analysis of the Use of New Environmental Policy Instruments Europe.* Edward Elgar: Cheltenham.

20

Spatial planning and the rural economy

Mark Scott

Introduction

Previous chapters in this section have explored various dimensions of the rural economy, from models of economic development (e.g. endogenous, neo-endogenous and regional) to potential development pathways, including rural enterprise innovation, the potential of payments for ecosystem services for the rural economy, and the role of the so-called creative class as a new driver of rural growth. This chapter aims to further contribute to this work by specifically exploring the role of spatial planning in growing and developing rural economies and supporting sustainable rural livelihoods. As discussed extensively in the literature, recent decades have witnessed considerable change in the rural economy (see, for example, van der Ploeg et al., 2000; Murdoch et al., 2003; Wilson, 2008), particularly the widely discussed declining dominance of traditional productivist models of agricultural modernisation and sectoral support policies (Shucksmith, 2000). Within this context, diversification of the wider rural economy represents a key public policy goal within advanced economies with increased emphasis placed on enhancing competitiveness of rural areas, valorisation of local assets, a focus on local specificities as a means of generating new competitive advantages, such as amenities (environmental or cultural) or local products, and a multi-sectoral approach – exemplified by the EU's LEADER Programme (e.g. Ray, 2000, 2006; Scott, 2002) and the OECD's New Rural Paradigm (OECD, 2006). This policy direction represents a fundamental shift away from sectoral support policy for agriculture and top-down policy interventions towards a spatial, territorial and integrated approach to rural development (Shucksmith, 2000), which Woods (2011) characterises as a shift from a modernisation paradigm to new rural development paradigm, illustrated in Table 20.1.

Within this context of territorial rural development, this chapter will explore the role of spatial plans in supporting rural enterprise and diversification strategies. At its most basic level, the planning system is concerned with regulating land-use through a statutory system of spatial plans and development control. On this level therefore, planning, particularly at a local authority level, plays a key role in controlling the development and location of economic activities in rural areas through its regulatory functions. Moreover, the planning system potentially

Table 20.1 Features of the modernisation paradigm and new rural development paradigm.

Modernisation paradigm	New rural development paradigm
Inward investment	Endogenous development
Top-down planning	Bottom-up innovation
Sectoral modernisation	Territorially based integrated development
Financial capital	Social capital
Exploitation and control of nature	Sustainable development
Transport infrastructure	Information infrastructure
Production	Consumption
Industrialisation	Small-scale niche industries
Social modernisation	Valorisation of tradition
Convergence	Local embeddedness

Source: Woods (2011)

performs a much wider remit within economic diversification through its place-making functions and its role as a coordinator of public policies within a spatial framework at local, regional and national levels.

However, spatial policies and planning also represent a challenge for many rural areas. As Moseley and Owen (2008) observe, spatial planning is often dominated by sustainable development discourses, which are often translated into spatial planning practice as a justification for concentrating development into key settlements and larger urban centres on the grounds of reducing the need to travel by private motor car and thus reduce CO_2 emissions. Such a restrictive policy of spatial planning and narrow approach to sustainable development can have negative impacts on smaller and remote rural communities including eroding local services and preventing development that may provide employment for local people (Gallent and Tewdwr-Jones, 2007; Gallent et al., 2015; Moseley and Owen, 2008). This chapter explores these tensions through examining the role of planning policy and practice within rural economic development, by critiquing rural planning orthodoxies and then developing a more positive framework for spatial planning in supporting rural economies through creating mutually reinforcing relationships between the economy, environment and community. Firstly, the chapter will examine the challenge of rural diversification and economic performance in rural localities.

Rural economic diversification

Contemporary rural economies are characterised by diversity and variable outcomes, reflective of rural area types, place-specific assets, local political capacity to realise rural potential and the availability of social and human capital, which combine and interact to produce differentiated patterns of change across rural regions. *Structurally*, agriculture is no longer the backbone of many rural economies in the global north, either in terms of its contribution to GDP or in relation to employment in rural areas. Although agriculture retains an important role in shaping rural landscapes in many advanced capitalist societies, its weight in rural economies is often low and declining. For example, a 2006 OECD study highlighted that in the 25 EU countries (pre Bulgaria, Romania and Croatia joining the EU), 96 per cent of rural land use is agricultural (including forestry), but only 13 percent of rural employment is in agriculture, producing only 6 per cent of gross value added in rural regions (OECD, 2006). However, comparative analyses of rural regions in advanced economies from the 1980s onwards illustrate that the decline in agriculture is often paralleled by a growing diversity of employment in the manufacturing and

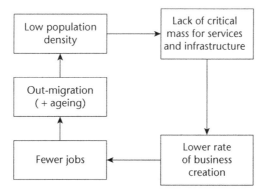

Figure 20.1 Predominantly rural regions and a cycle of decline.
Source: OECD (2006)

services sectors (Terluin, 2003). Indeed, a number of studies have found more rapid employment growth in firms in rural areas than in urban areas (see for example, Patterson and Anderson, 2003), although patterns have been uneven with the emergence of rural regions as 'winners, in between and losers' (Terluin, 2003) within these restructuring processes. *Spatially*, differences in rural economies and performance are pronounced. On aggregate, predominantly rural regions face problems of decline with out-migration, an ageing population, a lower skills base and lower average labour productivity, that then reduces the critical mass needed for effective public services, infrastructure and business development and innovation, thereby creating a vicious circle (OECD, 2006), represented in Figure 20.1. These trends were further consolidated following the 2007–2008 financial crisis and the impact of austerity and contraction of the wider economy (see Chapter 16).

However, 'rural' is not always synonymous with decline and many rural regions have seized the opportunities to build on their existing assets to forge successful new development pathways. Indeed, rural regions in the OECD area recorded an average annual rate of growth of GDP per capita of around 1.7 per cent over the period 1995–2011 – higher than the average growth rates of urban and intermediate regions growing at 1.5 and 1.4 per cent, respectively (OECD, 2016). In this context, Phillipson et al. (2011) argue that while policies for economic development, innovation and growth typically assume an urban focus, rural areas have the potential to make significant contributions to national economic growth as both incubators and catalysts for development, illustrated by the example of rural England outlined in Box 20.1.

Box 20.1   The contribution of the rural economy in England

The rural economy in England:

- Contains 86 per cent of England's land area and 19 per cent of its population.
- Contributes £200 billion or 19 per cent of national Gross Value Added (GVA).
- Is home to 4.6 million employees, with a further 0.8 million working in urban workspaces and contributing directly to the productivity of these areas.
- Contributes to UK's 72 per cent self-sufficiency in indigenous food products.

(continued)

(continued)

Rural localities in England:

- Have more businesses and more start-ups per head of population than urban areas, apart from London, with rural firms having a higher propensity to export their goods and services.
- Display higher levels of self-employment and entrepreneurial activity.
- Have higher growth rates in knowledge intensive businesses, including business and financial services which now account for a quarter of rural economic output.
- Contain a greater proportion of employees in manufacturing.
- Have higher employment rates.

Source: Phillipson et al. (2011)

This growing diversification and growth of the rural economy has been underpinned by three trends:

1. there has been a 'ruralisation' of business activity from conurbations to smaller settlements and rural areas (Phillipson, et al., 2011);
2. the almost universal provision of infrastructure means that the rural economy has become a continuation of the urban economy but at lower densities (Patterson and Anderson, 2003); and
3. the emergence of the 'new rural economy' based on utilising place-fixed assets linking economic growth more closely with enhancing personal, social and environmental well-being (Argent and Measham, 2014).

Thus, the rural economy and rural enterprises can be considered as comprising the traditional land-based economy, business activities located in rural areas as a continuation of the urban economy, and the new rural economy as actors seek to develop new opportunities through reappraising the rural resource base, illustrated in Table 20.2.

Within the context of the restructuring countryside, the differential performance of rural regions and localities has been the focus of much debate over the last decade or more. As recorded by Bryden and Munro (2001), differences in economic development success between rural localities may be explained by the interplay of global and local factors. The external environment of rural regions, for example, is affected by current globalisation processes and by macro-economic conditions. These relate to international conditions – including global economic growth, exchange rates, interest rates, global commodity prices (energy costs etc.), and domestic supply side issues – international competitiveness, wages and broader input costs, and regulatory burdens (Riordan, 2005). However, increasingly both academic literature and policy-makers have focused on territorial dynamics to denote a set of specific regional and local factors that influence relative economic performance (Terluin, 2003). These factors include developing both tangible and intangible aspects of local development and enhancing 'non-mobile' and 'less mobile' assets in the form of exploiting economic, social, cultural and environmental capital which are specific to individual rural localities (Bryden and Munro, 2001). For example, the OECD in its publication, *The New Rural Paradigm: Policies and Governance* (OECD, 2006) highlights the need for a shift towards rural development strategies that focus on local specificities as a means of generating new competitive advantages within the context of spatial and territorial development.

Table 20.2 A diverse rural economy.

Traditional land-based rural economy	*Food* – agriculture and the wider agri-food sector. Models of agricultural development vary considerably, from rural areas where farming is marginal and in decline to areas characterised by super-productivism and globalised agri-food networks
	Fibre
	Fuel extraction – carbon intensive practices, oil and gas extraction, including fracking and biofuels
	Minerals – extraction fuelled by recent boom in minerals market
Activities located in rural areas	Largely service sector based employment located in rural areas e.g. retail, financial services.
	Relocation of business activities from urban to rural due to perceived quality of life factors in rural areas and amenity factors
	Home-working
The new rural economy	Linking economic development to the revalorisation of place-fixed rural resources. *Examples include*:
	• Tourism based on rural heritage and cultural assets
	• Payments for ecosystem services and environmental goods
	• Renewables and low carbon economy
	• The 'circular' rural economy
	• The 'eco-economy': viable businesses and economic activities that utilise the varied and differentiated forms of environmental resources in rural areas in sustainable ways (Kitchen and Marsden, 2006)
	• Multifunctional agriculture characterised by on-farm diversification, on-site added value and landscape management

This theme has been further developed by Courtney and Moseley (2008) in a study of local economic performance in rural England. These authors examine economic performance through a framework of unevenly distributed capitals or inherited resources, referring to endowments which can influence a locality's capacity for economic success based around 'five capitals': economic, social, human, cultural and environmental, outlined in Table 20.3. This framework suggests the importance of a broad range of factors including both tangible (traditional) and less tangible (or softer) factors in explaining territorial variations in development performance. This framework is also echoed by Garrod et al. (2006) who argue for a re-conceptualisation of rural resources as 'countryside capital' to recast the rural resource base as a kind of capital to be invested in and from which a stream of benefits may be drawn. Although broad economic performance is influenced by national and global conditions, local action also provides an important arena to influence job creation and to secure sustainable livelihoods in rural areas. In particular, local action can address specific local challenges, tailor action towards local needs, and effectively engage with local people to identify priorities through a partnership of local stakeholders.

Spatial planning and the rural economy

Given these shifts in the rural economy and the importance of place-based and territorial approaches to developing new pathways for sustainable economic growth in rural areas, spatial planning has the *potential* to move centre-stage in rural diversification processes. This includes assessing the endogenous potential of rural areas (Kitchen and Marsden, 2009), recognising the

Table 20.3 Framework for local economic development on rural areas.

Groups of resources	Main elements
Economic capital	Transport and communications infrastructure, workspace, local economic linkages, past private investment by firms and households, range of businesses in existence
Human capital	Education, skills, health, attitudes, confidence, entrepreneurship and capacity for risk taking of the local population
Social capital	Networks and partnerships linking the public, private and voluntary sectors, the quality of local institutions and governance, trust and the shared norms that facilitate cooperation
Cultural capital	Political consensus, civic engagement, local history, customs and heritage, 'place identity' and people's sense of place, valorisation of culture and culture as a collective resource
Environmental capital	Natural and man-made assets that are valorised by local residents, investors and visitors; location, embracing peripherality, perceptions of peripherality and proximity to other places

Source: Courtney and Moseley (2008)

diversity of rural places (Ray, 1999), holistically addressing economic, social and environmental concerns (Scott, 2008), enhancing urban–rural relations within a regional spatial planning context (Hudson, 2010), and harnessing territorial identity for effective governance (Ray, 2006).

However, in practice, planning has often been marginal in enabling rural economic growth and diversification. For example, in a review of rural planning policy in post-war England, Curry and Owen (2009) are highly critical of the embedded land-use planning practices that undermine sustainable economic development in rural localities, including:

- *The enduring primacy of agriculture*: Curry and Owen chart the entrenched primacy of agricultural production in rural areas – 'at all costs' – within planning policy narratives of the rural, even as the economic, social, environmental and demographic realities of rural areas continue to change suggesting that the protection of land resources has been at the expense of economic and social welfare.
- *The creation of a 'no development ethic' in countryside planning*: whereby development in the countryside is framed as an environmental detractor, particularly the impact of physical development and new buildings on the rural landscape. Reinforcing this perspective, some rural interests groups have mobilised around the 'environment' to campaign against and prevent development in rural areas (Murdoch et al., 2003).
- *Framing sustainable development as urban development*: suggesting that planning has been characterised by an 'urbanising of sustainable development' with the widespread institutionalising of compact cities, brownfield development and raising residential densities. While these objectives represent policies to enhance urban sustainable development, Curry and Owen (2009: 584) argue that a focus on urban measures residualises the 'countryside as a theatre for sustainable development', condemning some rural localities to a downwards cycle of decline through restricting development.

Similarly, within an Irish context, Scott (2008) and Scott and Murray (2009) argue that rural policy remains fragmented between environmental and economic objectives, with economic actors casting planning and the environment as a barrier to economic growth, representing a largely discredited and outdated economic argument whereby the environment is perceived as a key obstacle to development (see Kitchen and Marsden, 2009). In contrast, local planning policy prioritises landscape protection and urban focused development, with limited assessment of rural resources, needs and potentialities of rural economies. Therefore, rural sustainable development is often translated within planning practice into narrow environmental concerns, with limited understanding of the inter-relationships between economic, social and environmental processes within rural localities. This suggests the need to consider the economic and social health of rural communities as important elements of sustainability alongside environmental aspects (Owen, 1996) and for spatial plans to create mutually reinforcing relationships between environment and economy to bridge this limiting divide (Kitchen and Marsden, 2006).

Developing planning principles for supporting the rural economy

To address this disconnect between regulatory planning systems and wider policies for promoting rural enterprise, this section will attempt to identify a series of policy principles to re-orientate planning towards a more positive supporting role for economic development in spatial planning. These principles, developed from reviewing policy, practice and research, are considered under the following headings:

- An enabling and mobilising agenda for supporting place-based rural diversification.
- Bridging the gap between environment and job creation activities.
- Embedding the rural economy in regional spatial plans.
- Linking heritage, rural economic development and local well-being.
- Addressing rural connectivity.
- Open and inclusive countryside.

An enabling and mobilising agenda for supporting place-based rural diversification

As outlined above, rural planning policy has often focused on landscape protection and a narrow view of sustainability, which often neglects the social and economic dimensions of rural well-being. To balance this perspective, spatial planning should also be underpinned by a positive engagement with rural economic development delivered through community-led planning (Scott, 2008). For example, the UK government-commissioned Taylor Review on the rural economy and affordable housing emphasises that:

> Greater recognition of the ways that economic growth can improve sustainability, especially by providing opportunities for people to work near where they live, needs to be central to planning decisions to underpin rural economic regeneration. Taking into consideration local circumstances and conditions, development of all types of business and enterprise should be considered.
>
> *(Taylor, 2008: 120)*

This suggests the need to consider the endogenous economic potential of rural areas through place-based approaches. For Phillipson et al. (2011), this requires a sea-change in how planners view rural areas to recognise both the existing significance and multidimensional nature of rural economies and to enable rural localities to realise their economic potential as a means of sustaining rural communities and managing natural capital. A place-based approach recognises that rural regions are not homogenous spaces with a single set of shared experiences. Depending on economic contexts, geographic location and local institutional capacities, rural areas face a diversity of challenges and opportunities, whereby a 'one size fits all' approach towards enterprise creation and planning often fails to address the diversity of rural areas.

An enabling role for the planning system in rural economic development refers to the need for planning practice to more fully engage with the needs and aspirations of the rural economy and the need to support economic growth rather than a presumption against development in rural places. At a very basic level, this suggests the need for planning policy to develop responsive, flexible and proactive policies relating to rural workspace and business premises requirements, often viewed as a key constraint on rural enterprise development, particularly as local businesses grow and expand. In this regard, innovative or experimental practices are emerging, for example:

- Phillipson et al. (2011) suggest the case for experimenting with establishing Rural Enterprise Zones in suitable, sparse rural locations to achieve business growth objectives. These zones may include: business rate discounts, rate retention and refunds, tax holidays, simplified planning regulations for physical development, and additional support for rollout of superfast broadband.
- Noting that the rural economy is often characterised by new business start-ups and microbusinesses, Taylor (2008) outlines a series of recommendations to ensure a good supply of suitable business premises through the planning system. These include support for incubator units as enterprises grow from home businesses to more ambitious enterprises; the flexible adaptive reuse of farm buildings for suitable enterprises; and better support for home-based businesses and home-working. For Taylor, home-working is a critical issue, noting (in an English rural context) that this takes place in many different forms, is crucial for new business start-ups and is significantly more important in rural areas than urban areas. A flexible planning system may include positive support for home extensions to facilitate home-based enterprises, or facilitate the building of dedicated multiple 'live/work' units (with homes and business premises on the same site), combining the benefits of home-working with working in a business community or hub.
- Local authorities in rural Ireland are currently experimenting with new planning approaches to develop rural business hubs, including fast-track/streamlined planning in designated zones. For example, Mayo County Council is currently applying a Strategic Development Zone (SDZ) approach to the further development of Knock Airport in the rural west of Ireland (Mayo County Council, 2012). SDZs have generally been applied in urban areas to deliver large housing and mixed-use development projects. The objective is to designate the airport and its environs as a rural business hub, with the planned expansion of the airport itself serving to improve connectivity to this remote rural region (and improve access to tourism and business markets), while developing the area around the airport as hub for rural businesses in the region. Once an SDZ with a masterplan has been approved, planning consent is not required for individual development, enabling a speedier development process. Other planning authorities in the west of Ireland have also been experimenting with more flexible approaches including enterprise zones and developing 'electronic courtyards' to generate incremental business activity by offering the entrepreneur a turn-key residential and business solution in a high amenity location (implemented in Roscommon).

In addition to *enabling* rural economic development through local policy and appropriate development management, spatial planning has the potential to perform a *mobilising* role in supporting local economies. Local spatial plans, for example, provide a means to mobilise communities to work towards developing a shared narrative of rural challenges and opportunities and to collectively (re)imagine their places (Phillipson et al., 2011) or to explore 'new story-lines' (Healey, 1999) through collaborative action to produce consensus-driven development strategies or to mediate between conflicting conservation and development goals. As Murray (2010) suggests, for planners this can also be viewed as an opportunity to engage in a more interactive style of plan-making partnerships with rural communities, linked to interest group mediation, the building of trust-relations, and re-orientating policy towards local needs, capacities and perspectives of local people.

Bridging the gap between environment and job creation activities

While a number of commentators have highlighted the tensions between environmental and economic objectives for the countryside, good practice in rural planning suggests the urgent need to realise the value of interdependencies between economic growth, social well-being and environmental quality in developing sustainable and resilient rural places (Phillipson et al., 2011). In other words, environmental protection should not be viewed as a barrier to economic and social well-being, but rather the environment adds value to the rural economy. There have been a number of overlapping conceptualisation of this integration, including:

- *The eco-economy* (Kitchen and Marsden, 2006: 5): defined as the effective management of environmental resources in ways designed to mesh with and enhance the local and national ecosystem rather than disrupting and destroying it. That is, the eco-economy consists of viable businesses and economic activities that utilise the varied and differentiated forms of environmental resources of rural localities in sustainable ways that do not result in a net depletion of resources but provide net benefits and add value to the environment, intertwining to create both economic and environmental added value.

Table 20.4 Planning principles for supporting rural economies in the UK and Ireland.

The Rural Coalition	Government of Ireland
<ul><li>To seize the potential presented by the shift to a low carbon economy</li><li>To promote businesses in all rural communities and the wider countryside (subject to conditions)</li><li>To protect and maintain a good supply of appropriate sites and premises for all kinds of businesses</li><li>To unlock the potential of re-using historic farm buildings</li><li>To encourage home-based businesses</li><li>To invest in extending high quality broadband throughout rural communities</li></ul>	<ul><li>Enhancing our unique rural settings and communities who live there</li><li>Planning for the future growth and development of rural areas, including addressing decline, with a special focus on activating the potential for the renewal and development of small towns and villages</li><li>Putting in place planning and investment policies to support job creation in the rural economy</li><li>Addressing connectivity gaps</li><li>Better coordination of existing investment programmes dealing with social inclusion, rural development and town and village renewal</li></ul>

Sources: Rural Coalition (2010); Government of Ireland (2017)

- *The green economy* as a framework for rural investment: defined by the OECD (2011) as a means fostering economic growth and development, while ensuring that natural assets continue to provide the resources and environmental services on which our well-being relies. Spatial plans have the potential to contribute to 'short circuit' rural economies through enabling complementary businesses in rural localities such as circular bio-economy activities relating to the re-use of farm waste products. Natural capital and ecosystem services are both valued and commoditised by rural enterprises, with planning policy performing a role in ensuring rural economic activities enhance environmental resources through effective development management practices.
- *The low carbon economy*: rural planning frameworks should attempt to capture the potential afforded by the shift to a low carbon economy while managing how rural communities emerge from the ageing carbonist model of spatial development (Kitchen and Marsden, 2011). While much attention has been directed at low carbon urban transitions, rural planning offers the potential to capture the benefits of low carbon economies. As outlined by the Rural Coalition (2010), this includes examining the renewable energy potential of rural places and the role of the planning system in mediating conflicts over the siting of renewable energy infrastructure and capturing benefits of sustainable energy production for local rural communities through community or planning gain. Assessing the potential for local community ownership models of renewable energy schemes can be explored in local place-making strategies while also sensitively enabling facilities that extract energy from agricultural and forestry by-products.

Embedding the rural economy in regional spatial plans

Spatial planning traditions vary significantly within advanced economies in relation to varying degrees of emphasis placed on the regional dimension of planning, from mature traditions of regional planning and development (e.g. Canada, Netherlands), to new experiments with national and regional plans (e.g. Ireland), countries with a start-stop approach to regional issues (e.g. the UK, particularly England) and countries without any significant tradition of regional planning (e.g. the USA). However, embedding rural economies into wider regional planning frameworks offers clear potential for promoting rural economic development, more fully explored by Tomaney et al. in Chapter 15 of this Companion. For rural economies, two advantages can be identified:

- A regional approach enables networks of smaller towns and villages to develop complementarities or a collaborative network to develop a critical mass needed to access or support networks and to develop local supply chains, therefore developing greater local economic closure through, for example, the (re)localisation of food supply chains, enhancing local transactions or creating business clusters (Hudson, 2010; Scott, 2013). This approach may also facilitate the use of a regional or territorial identity in promoting goods and services through place-marketing or branding of products. Developing regional clusters of rural settlements is of particularly relevance in predominantly rural regions, whereby rural towns can develop into important local nodes (Planning Institute of Australia, 2010).
- Regional planning also has potential for strengthening urban–rural partnerships to enable the integration of rural and urban economies, especially for rural localities within urban functional areas. In this context, a regional approach often represents a more meaningful scale of action in terms of labour and housing markets (Healey, 2002), and can promote

relations between rural and urban enterprises, urban markets and rural suppliers, rural areas as consumption areas for urban dwellers, and rural areas as suppliers of natural capital for urban areas (Bengs and Zonneveld, 2002). This approach therefore facilitates rural economies to mobilise extra-local resources suggesting a more outward looking approach across spatial boundaries. For instance, urban–rural partnerships and a regionalisation of food supply chains may open new market opportunities and provide alternative pathways than dependency on large supermarkets for food products (Morgan and Sonnino, 2007).

Linking heritage, rural economic development and local well-being

Rural regions are critical repositories of cultural heritage. This includes tangible heritage assets such as natural heritage, cultural landscapes, antiquities and historical sites, historic settlements and traditional settlement patterns containing a mosaic of small towns, villages, and scattered settlement, and built heritage, which in turn are inextricably linked to local identity and culture through, for example, minority languages, customs, music, arts, artisanal food and agricultural products and practices. Planning policy and practice has a key role to play in managing and protecting tangible heritage assets. A key challenge, however, is to move beyond protection towards harnessing its potential for regeneration while at the same time preserving the values embedded in inherited rural landscapes. As noted in the report *Getting Cultural Heritage to Work for Europe* (EC, 2015), a positive experience of cultural heritage is not yet universal and is not completely understood in the richness of its diversities. Firstly, in many places, rich cultural assets have not been recognised for the potential they hold to regenerate and renew. Secondly, many municipalities and civil society groups often lack the resources, critical mass, capacities and knowledge to move beyond a simple conservation, restoration and physical rehabilitation of a site towards harnessing heritage potential as a powerful economic, social and environmental catalyst for regeneration, sustainable development, economic growth and improvement of people's well-being and living environments. Therefore, a key challenge for spatial planning policy is to rethink the management of rural heritage from being thought of as a potential cost or barrier to rural regeneration towards harnessing rural cultural heritage to become a driver for improving rural quality of life. Rather than preservation framed as a cost or restraint on rural communities, how can cultural heritage become regarded a central innovative stimulant for enhancing rural well-being? This suggests the need for spatial plans to identify, valorise and 'exploit' heritage for rural regeneration to unlock the potential of rural regions. Local place-making strategies, in this context, can identify heritage assets and landscapes in need of protecting, but also positive support policies for the adaptive reuse of historic buildings (e.g. underused historic farm buildings), effective linkages between local heritage assets (e.g. linking historic settlements and cultural landscapes through greenways, cycle routes), and exploring place-based narratives as a means of utilising territorial identity for collective place-branding.

Box 20.2 Examples of heritage-led rural regeneration

Built heritage and place-branding: Røros (Norway) is an historic mining town (established in the seventeenth century) on a highland plateau located in a predominantly rural area. Its townscape comprises historic industrial sites and an old wooden town of around 2000 wooden

(continued)

(continued)

one- and two-storey houses, surrounded by a mountainous landscape (see Figure 20.2), which is a UNESCO World Heritage Site. Following decades of decline following the cessation of mining in the 1970s, Røros has reinvented itself as a tourism hub following the adaptive reuse of its industrial heritage, which in turn has led to the creation of local craft-businesses, high quality tweed products and eco-friendly food products. The town's heritage has been used to diversify the local economy beyond tourism as a means of quality place-branding for local products, creating local webs of innovation.

Linking heritage to tourism and community well-being: the Cooley and Carlingford Peninsula (Ireland) is located close to the border with Northern Ireland and comprises a mix of tangible natural and built heritage and intangible cultural assets, including Carlingford Lough, Cooley Uplands and historic settlements e.g. Carlingford (a walled town dating from the twelfth century) and Greenore nineteenth-century industrial village, and *An Táin* trail (based on ancient Irish literature). Natural and built heritage has been central to local development and community-led strategies (exploiting mythology and landscape alongside walking tourism, food festivals), and increasingly these initiatives are being linked with projects promoting local well-being and health through adapting abandoned railways for community greenways that further link to regional tourist walking trails (e.g. the Ancient East trail, the Great Eastern Greenway).

Figure 20.2   Example of built heritage in Røros utilised as a driver of local economic innovation.

Source: photo supplied by Dag Kittang and reproduced with permission

Addressing rural connectivity

While extending high-quality broadband into rural localities may be beyond spatial planning policy, increasingly local development strategies frame rural connectivity as *essential* infrastructure and a prerequisite for growing the local economy. The roll-out of high quality broadband has the potential to further diversify rural economies, enabling small scale entrepreneurs in rural localities to connect to larger regional and global markets through web-based services and e-commerce. As indigenous businesses grow or as urban entrepreneurs potentially relocate to rural places, policy should ensure a supportive approach to encourage local business growth including provision for rural business workspace (e.g. flexible business units) or to support home-based businesses as they grow. Rural connectivity also offers the potential to not only assist individual businesses and entrepreneurs but also to enable rural areas to collectively create opportunities for creating new rural value chains. In this context, the European Union (EU) has recently launched its Smart Villages strategy as a means to capture the potential of new digital technologies for rural economic and social well-being:

> In Smart Villages traditional and new networks and services are enhanced by means of digital, telecommunication technologies, innovations and the better use of knowledge, for the benefit of inhabitants and businesses. Digital technologies and innovations may support quality of life, higher standard of living, public services for citizens, better use of resources, less impact on the environment, and new opportunities for rural value chains in terms of products and improved processes.
>
> *(EC, 2016: 3)*

A smart village or rural locality may be one whose economy is increasingly driven by technically inspired innovation, creativity and entrepreneurship, enacted by smart people. This suggests that key interventions in human capital and technological infrastructures and programmes will attract businesses and jobs to the rural economy, create efficiencies and savings and raise the productivity and competitiveness of local government and businesses.

While rural connectivity offers significant potential for local economies, the paradox, as noted by Salemink et al. (2017), is that rural communities most in need of improved digital connectivity to compensate for remoteness, are the least connected and included in broadband and digital network provision. This suggests the need for customised policy and hybrid delivery of digital services and the potential for community-led broadband delivery (Roberts et al., 2017); however, Ashmore et al. (2017) also highlight that tailored approaches may further consolidate uneven rural development. While offering much potential, the increased penetration of ICT into rural areas also increases possibilities for individual local consumers to 'shop around' for services, particularly undermining local retail functions (Moseley and Owen, 2008).

Box 20.3   Smart rural economies: some illustrative examples

- Utilising local and ICT networks for organising and adding value to supply chains.
- Enhancing existing rural assets with ICT e.g. digitisation of cultural heritage for tourism.
- Real-time management of the circular economy through more efficient connecting of supply and demand.

(continued)

(continued)

- Creative economy clustering.
- Creating a rural 'market place' of ideas for the knowledge economy.
- Attracting new 'creative economy' in-migrants and mobile capital attracted to the rural living environment, enabled by ICT infrastructure.
- Alternative finance models to support community enterprises e.g. crowdfunding through blockchain technologies or crypto currencies for payments within the local community ('community coins').

Open and inclusive countryside

The performance of local rural economies is dependent not only on land-based resources and fixed assets (such as cultural heritage), but also on human capital and an entrepreneurial culture within rural communities. In this context, sustainable local economies are also interlinked to housing provision in rural areas and the ability of rural localities to attract or retain human capital. For example, a lack of affordable housing in some rural areas constrains the ability of rural businesses to recruit and retain staff (Phillipson et al., 2011). Moreover, research has also examined the relationship between in-migration processes and rural economic development in more remote rural contexts. Stockdale et al. (2000) highlight the potential positive impacts of in-migration into rural areas, particularly when this is associated with in-migration of middle class residents into areas previously experiencing depopulation – these include the maintenance of local services, employment creation and the prospects for enhanced rural expenditure. Similarly, Kilpatrick et al. (2011) examine middle-class rural in-migration in remote rural localities, drawing on research from Australia and Canada. In these predominantly rural areas, an influx of newcomers represents an opportunity as new residents can add new skills, entrepreneurial capacity and political capital, adding much needed capital and skills to underpin rural regeneration initiatives. Bosworth (2010) terms the influence of in-migrants on the rural economy as 'commercial counter-urbanisation', linking demographic counter-urbanisation with new rural in-migrants stimulating local economies through new business start-ups, new entrepreneurial activity and embedding new enterprises into local business networks. However, as noted by Mitchell and Madden (2014), Bosworth's conceptualisation neglects a more inclusive model of rural in-migration which may involve newcomers' origins including other rural places as common in Canada, Scotland and Ireland among others.

This relationship between in-migration and economic development challenges existing assumptions that dominate planning policy. As examined by Gallent and Scott (Chapter 22 of this Companion), planning policy is often characterised by a no house-building ethic (as in the English case) – thereby restricting the ability of new residents to move into rural localities – or policies that favour accommodating 'locals only' for new residential development – which in turn can create 'closed' places without the ability to attract new skills into the locality. Placing housing within a rural development framework suggests the need to balance local needs, while also competing for extra-local resources, particularly human resources, skills and capital (social, political and financial) that provide key resources for rural development (Gkartzios and Scott, 2014). Rural housing policies are often shaped by wider political landscapes; however, placing housing within a rural development framework attempts to move beyond short-term politics towards developing an appropriate model of housing supply in rural areas. This would comprise an approach whereby top-down programmes meet bottom-up approaches. A rural development framework approach to housing should move beyond housing policy that sets and constructs locals against non-locals, recognising that outward-looking rural communities should not only facilitate the needs of the local population, but also require open communities to develop entrepreneurial and risk-taking capacity critical for local economic development (Courtney and Moseley, 2008). Policies favouring 'locals only' often fails to acknowledge opportunities for rural economies resulting from extra-local populations and capital, as in the case of counter-urbanisation (see for example Stockdale et al., 2000; Bosworth and Atterton, 2012).

Conclusion

While planning is often portrayed as a barrier to economic development in the countryside (Scott and Murray, 2009), this chapter has argued that a positive planning approach can provide a supportive policy environment for promoting rural economies. The chapter has identified three key roles for spatial plans and practices in diversifying the rural economy. Firstly, planning should be positioned as an *enabler* (rather than barrier) of rural economic development based on engaging and understanding the changing nature of the economy across different types of rural areas. This moves beyond simplistic assumptions around the primacy of agriculture and reactionary development control in rural areas to integrating the needs of the local economy and rural livelihoods alongside protecting landscapes and the environment. Secondly, planning can provide a key role as a *mobiliser* of local economic development through creating shared discourses of rural futures through place-making and community-led planning approaches. This would enhance the role of planning as a spatial coordinator of rural development, while providing a means for balancing community engagement and local needs along with strategic and regional priorities. Thirdly, planning plays an important role as a *mediator* of rural change, particularly in reconciling and seeking mutually reinforcing relationships between environmental, economic and social objectives, and bridging the traditional gap between environment and economy.

Finally, learning the lessons of the financial crisis of 2007–2008, planning has a role in building the *resilience* of rural economies and enabling rural communities to identify sustainable pathways to promote local economies. Rural economies are firmly integrated into global relationships and deeply affected by external factors, such as deepening neo-liberalisation and austerity. This suggests that planning for local economic development must be forward looking and anticipate, or at least debate, opportunities and challenges arising from wider global factors, such as climate change, post carbon transitions or new disruptive technologies and practices. For example, the negative impacts of online retail and banking have already undermined traditional retail roles of rural towns and it remains to be seen how scenarios relating to the growth in so-called sharing economies, greater automation and artificial intelligence will play out in reshaping local economies, labour markets and service delivery. Maximising the benefits of these wider changes at the local scale will be an enduring challenge for rural stakeholders.

References

Argent, N. and Measham, T. (2014) New rural economies: introduction to the special themed issue. *Journal of Rural Studies*, *36*, 328–329.

Ashmore, F.H., Farrington, J.H. and Skerratt, S. (2017) Community-led broadband in rural digital infrastructure development: Implications for resilience. *Journal of Rural Studies*, *54*, 408–425.

Bengs, C. and Zonneveld, W. (2002) The European discourse on urban–rural relationships: a new policy and research agenda. *Built Environment*, *28(4)*, 278–289.

Bosworth, G. (2010) Commercial counterurbanisation: an emerging force in rural economic development. *Environment and Planning A*, *42*(4), 966–981.

Bosworth, G. and Atterton, J. (2012). Entrepreneurial in-migration and neoendogenous rural development. *Rural Sociology*, 77(2), 254–279.

Bryden, J. and Munro, G. (2001) New approaches to economic development in peripheral rural regions. *Scottish Geographical Journal*, 116(2), 111–124.

Courtney, P. and Moseley, M. (2008) Determinants of local economic performance: experience from rural England. *Local Economy* 23, 305–318.

Curry, N. and Owen, S. (2009) Rural planning in England: a critique of current policy. *Town Planning Review*, *80*(6), 575–596.

EC (2015) *Getting Cultural Heritage to Work, Report of the H2020 Expert Group on Cultural Heritage* (Brussels: CEC).

EC (2016) *EU Action for Smart Villages* (Brussels: CEC).

Gallent, N. and Tewdwr-Jones, M. (2007) *Decent Homes for All* (London: Routledge).

Gallent, N., Hamiduddin, I., Juntti, M., Kidd, S. and Shaw, D. (2015) *Introduction to Rural Planning: Economies, Communities and Landscapes* (London: Routledge).

Garrod, B., Wornell, R. and Youell, R. (2006) Re-conceptualising rural resources as countryside capital: the case of rural tourism. *Journal of Rural Studies*, *22*(1), 117–128.

Gkartzios, M. and Scott, M. (2014) Placing housing in rural development: exogenous, endogenous and neo-endogenous approaches. *Sociologia Ruralis*, *54*(3), 241–265.

Government of Ireland (2017) *Ireland 2040: Draft National Planning Framework* (Dublin: Government of Ireland).

Healey, P. (1999) Institutionalist analysis, communicative planning and shaping places. *Journal of Planning Education and Research*, *19*, 111–121.

Healey, P. (2002) Urban–rural relationships, spatial strategies and territorial development. *Built Environment* 1978, 331–339.

Hudson, R. (2010) Resilient regions in an uncertain world: wishful thinking or a practical reality? *Cambridge Journal of Regions, Economy and Society*, *3*, 11–25.

Kilpatrick, S., Johns, S., Vitartas, P. and Homisan, M. (2011) In-migration as opportunity for rural development. *Planning Theory & Practice*, *12*(4), 625–630.

Kitchen, L. and Marsden, T. (2006) *Assessing the Eco-economy of Rural Wales*, Research Paper No 11 (Cardiff: Wales Rural Observatory).

Kitchen, L. and Marsden, T. (2009) Creating sustainable rural development through stimulating the eco-economy: beyond the eco-economic paradox? *Sociologia Ruralis*, *49*(3), 273–294.

Kitchen, L. and Marsden, T. (2011) Constructing sustainable communities: a theoretical exploration of the bio-economy and eco-economy paradigms. *Local Environment*, *16*(8), 753–769.

Mayo County Council (2012) *Knock Airport Local Area Plan* (Ballina: Mayo County Council).

Mitchell, C.J. and Madden, M. (2014) Re-thinking commercial counterurbanisation: evidence from rural Nova Scotia, Canada. *Journal of Rural Studies*, *36*, 137–148.

Morgan, K. and Sonnino, R. (2007) Empowering consumers: the creative procurement of school meals in Italy and the UK. *International Journal of Consumer Studies*, *31*, 19–25.

Moseley, M.J. and Owen, S. (2008) The future of services in rural England: the drivers of change and a scenario for 2015. *Progress in Planning*, *69*(3), 93–130.

Murdoch, J., Lowe, P., Ward, N. and Marsden, T. (2003) *The Differentiated Countryside* (London: Routledge).

Murray, M., 2010. *Participatory Rural Planning: Exploring Evidence from Ireland* (Oxford: Ashgate).

OECD (2006) *The New Rural Paradigm – Policies and Governance* (Paris: OECD).

OECD (2011) *New Rural Policy: Linking Up for Growth* (Paris: OECD).

OECD (2016) *OECD Regional Outlook 2016: Productive Regions for Inclusive Society* (Paris: OECD).

Owen, S. (1996) Sustainability and rural settlement planning. *Planning. Practice and Research*, *11*, 37–47.

Patterson, H. and Anderson, D. (2003) What is really different about rural and urban firms? Some evidence from Northern Ireland. *Journal of Rural Studies*, *19*(4), 477–490.

Phillipson, J., Shucksmith, M., Turner, R., Garrod, G., Lowe, P., Harvey, D., Talbot, H., Scott, K., Carroll, T., Gkartzios, M., Hubbard, C., Ruto E. and Woods, A. (2011) *Rural Economies: Incubators and Catalysts for Sustainable Growth* (Newcastle: Centre for Rural Economy, Newcastle University).

Planning Institute of Australia (2010) *Rural and Regional Development – A Planning Response* (Kingston, ACT: PIA).

Ray, C. (1999) Endogenous development in the era of reflexive modernity. *Journal of Rural Studies*, *15*, 257–267.

Ray, C. (2000) The EU LEADER programme: rural development laboratory. *Sociologia Ruralis*, *40*(2), 163–171.

Ray, C. (2006) Neo-endogenous rural development in the EU. In P. Cloke, T. Marsden and P. Mooney (eds), *Handbook of Rural Studies* (London: Sage), 29–36.

Riordan, R. (2005) Economic conditions in the baseline 2025 scenario. In *Rural Ireland 2025 Foresight Perspectives* (Dublin: NUIM/UCD/Teagasc), 29–36.

Roberts, E., Anderson, B.A., Skerratt, S. and Farrington, J. (2017) A review of the rural-digital policy agenda from a community resilience perspective. *Journal of Rural Studies*, *54*, 372–385.

Rural Coalition (2010) *The Rural Challenge: Achieving Sustainable Rural Communities for the 21st Century* (London: Town and Country Planning Association).

Salemink, K., Strijker, D. and Bosworth, G. (2017) Rural development in the digital age: A systematic literature review on unequal ICT availability, adoption, and use in rural areas. *Journal of Rural Studies*, *54*, 360–371.

Scott, M. (2002) Delivering integrated rural development: Insights from Northern Ireland. *European Planning Studies*, *10*(8), 1013–1025.

Scott, M. (2004) Building institutional capacity in rural Northern Ireland: the role of partnership governance in the LEADER II programme, *Journal of Rural Studies 20*, 49–59.

Scott, M. (2008) Managing rural change and competing rationalities: insights from conflicting rural storylines and local policy making in Ireland. *Planning Theory & Practice*, *9*(1), 9–32.

Scott, M. (2013) Resilience: a conceptual lens for rural studies? *Geography Compass*, 7(9), 597–610.

Scott, M. and Murray, M. (2009) Housing rural communities: connecting rural dwellings to rural development in Ireland. *Housing Studies*, *24*(6), 755–774.

Shucksmith, M. (2000) Endogenous development, social capital and social inclusion: Perspectives from LEADER in the UK. *Sociologia Ruralis*, *40*(2), 208–218.

Stockdale, A., Findlay, A. and Short, D. (2000) The repopulation of rural Scotland: opportunity and threat. *Journal of Rural Studies*, *16*(2), 243–257.

Taylor, M. (2008) *Living and Working Countryside – the Taylor Review of Rural Economy and Affordable Housing* (London: Department for Communities and Local Government).

Terluin, I.J. (2003) Differences in economic development in rural regions of advanced countries: an overview and critical analysis of theories. *Journal of Rural Studies*, *19*(3), 327–344.

Van der Ploeg, J.D., Renting, H., Brunori, G., Knickel, K., Mannion, J., Marsden, T., de Roest, K., Sevilla-Guzman, E. and Ventura, F. (2000) Rural Development: From Practices and Policies towards Theory. *Sociologia Ruralis*, *40*, 391–408.

Wilson, G.A. (2008) From 'weak' to 'strong' multifunctionality: Conceptualising farm-level multifunctional transitional pathways. *Journal of Rural Studies*, *24*(3), 367–383.

Woods, M. (2011) *Rural* (London: Routledge).

Part IV
Social change and planning

Part IV deals with planning issues in the context of a continuously changing rural society. Social change and rural restructuring and social change have been discussed from the 1970s, with new social agents, sometimes the middle classes, rediscovering the countryside and bringing new meanings and expectations about what it is to be rural. These changes have had profound effects in rural communities and have created new contested arenas of planning policy and conflict in terms of infrastructure and services. The aim of this section is to discuss these planning issues in the context of continuous social change.

The part starts with a chapter by Smith, Phillips, Kinton and Culora on emerging rural geographies. The authors briefly discuss the influential concept of the 'differentiated countryside' in order to introduce their study of population dynamics in the cases of England and Wales in the UK. Georeferenced data from the censuses of 2001 and 2011 are analysed and mapped to produce a nuanced understanding of uneven contemporary rural geographies. The hypothesis of an increasingly differentiated countryside is confirmed through particular explorations regarding the distribution of rural populations (by employment, class, age and family formation types) and what this might mean for services and planning policy. Key processes highlighted in the chapter are counterurban-led rural gentrification as wells as the restructuring of agriculture and, wider, rural economies.

The following chapter presents undoubtedly the most debated issue as a result of shifting preferences for rural residential environments and associated social change in the countryside, housing. This is discussed by Gallent and Scott who, in the second chapter of this section, provide an overview of the demand and supply factors resulting in well observed housing inequalities. Rural housing research is complicated because the policy challenges faced by the planning systems are very different internationally. This is evidenced for example through the insightful comparison provided here between imposing occupancy conditions in a case study in England (a context characterised by limited housing provision in rural areas, housing market distortions and issues of exclusion) and the case of 'locals only' planning restrictions in the Republic of Ireland (a context characterised by a period of housing oversupply and a pro-development rural planning ethos). Both cases demonstrate the multiple ways that new consumer preferences and social change in rural areas, have resulted in spatially selective inequalities.

Continuing on the topic of housing, Chapter 23 by Paris focuses on holiday homes in countryside areas, including coastal localities. Paris rejects the notion of 'second homes' as it does not capture the complexities of holiday homes (for example some social groups might own more than one home), but, more importantly, because it does not recognise the hybridity of residential and recreational uses of housing, as well as the translational and mobility aspects of such housing preferences. Paris reviews the characteristics of holiday homes in the context of the post-productivist countryside and highlights areas of concern for planning intervention, including issues that require further research (for example the implications of short-term letting of holiday homes through airbnb and other internet platforms). He also provides a comparative perspective of the evolution of preferences and demand for holiday homes internationally, by looking at very different planning contexts across Britain, Ireland and Australia.

Moving on from housing, the following chapter by Kilpatrick, Auckland and Woodroffe looks at health as a key issue for rural planning in the context of diversified rural communities. The focus here is on Australian rural communities and the authors, drawing on three case studies, offer examples about how health needs can be addressed. The authors first conceptualise rural health across different models, biomedical, economic and community-strength approaches, favouring the latter as they produce a more nuanced understanding of the heterogeneity of rural communities. The authors identify the main characteristics of community health planning, before they offer three examples of how strength-based approaches can support planning. The cases draw on the attributes identified earlier by the authors, including, *inter alia*, the role of both strength-based and placed-based approaches that can produce positive outcomes on health and wellbeing, community participation and consultation to gather evidence of need, and partnerships across local government and health providers.

Chapter 25, by Oliva and Camarero, positions debates of mobility, accessibility and social justice at the core of rural planning research and practice. Access to services and mobility options frame key concerns for rural service provision, and thus are critical for rural sustainability and social inclusion. The chapter reviews the evolution of mobility approaches in rural contexts towards more complex theorisations that move beyond sectoral (transport) policies. It also critiques reliance on private automobility, in the way that it introduces new geographies of exclusion. The chapter finally discusses what particular emerging rural geographies, such as demographic ageing, mean for rural mobility and policy interventions.

In the final chapter of this part, by Crawshaw, we change slightly the focus to look at a particular social agent in the context of rural planning policy and research, and that is artists. While questions of creativity and the application of the 'creative class' thesis is explored in Chapter 18 by Herslund, in this chapter, Crawshaw looks in detail how artists (and 'artist-led practice') can support rural planning inquiries. Based on the pragmatist philosophy of John Dewey and his ideas on the 'experience of art' reviewed in this chapter, Crawshaw provides two empirical examples that demonstrate how artists and their practice can support 'transactional' understandings of rural communities in the context of wider rural restructuring processes. Instead of doing research *of* artists or of particular social agents and communities, Crawshaw suggests doing research *with* them. Drawing on ethnographic accounts in an English and Swedish island, Crawshaw explores interdisciplinary collaborations with artists for understanding the rural condition. Rather than putting a social agent under the research microscope, Crawshaw offers insightful ideas on how to work with a particular social group.

The contributions in this section aim to provide an understanding of the demographic and social change that has occurred in rural areas and what issues might arise for planning design and intervention, particularly in the context of housing and services. The last chapter offers an insightful account of exploring such issues with new epistemologies including artistic practice for understanding rural social change and wider challenges within planning practice.

21

Rural population geographies in the changing differentiated countryside

Darren P. Smith, Martin Phillips, Chloe Kinton and Andreas Culora

Introduction

A prominent hallmark of contemporary rural places is the changing demographics and make-up of local populations (Shucksmith and Brown, 2016), with identities and differentials of age and lifecourse, social class, gender, ethnicity, and family formations all being in states of flux and diversification (Smith, 2007; Phillips, 2010; Shortall, 2015). Of course, this is not a new phenomenon, as stressed by Lowe et al. (2003: 32): 'For most of its history the British countryside has been characterised by considerable diversity'.

During the last 50 years, numerous subdisciplinary-defining studies of rural social change (e.g. Pahl, 1965; Newby, 1980; Cloke and Thrift, 1987; Murdoch and Marsden, 1994) have identified how and why rural populations are reshaped, often in profound ways, by shifting political, economic, social and cultural conditions. This body of work has increasingly shown that long-standing, traditional representations of rural social relations, often founded on the powerful (paternalistic) landed gentry and rural working classes (i.e. deferential workers), have faded in many rural contexts, as local agricultural-based economies and communities have been restructured under globalisation and new phases of capitalism (Cloke and Goodwin, 1992; Nelson and Nelson, 2010; Woods, 2010).

Although historical rural relations may persist in some contexts (Sutherland, 2012), most have been superseded by very different sets of rural social relations (Woods, 2004). To capture this changing context, a plethora of new terms have been introduced into the rural studies lexicon to describe the incipient strata of new rural populations, such as rural gentrifiers (Phillips, 1993), 'greentrifiers' (Smith and Phillips, 2001), the 'rural squirearchy' (Heley, 2010), 'crisis-counterurbanisers' (Gkartzios, 2013), and the 'rural precariat' (Kasimis et al., 2015; McKee et al., 2017; Lever and Milbourne, 2017).

Within Britain, population and demographic processes continue to be fuelled by counter-urban impulses and predilections within large segments of British society and the enduring lure of the 'rural idyll' (Halfacree, 2014, 2018; Scott et al., 2017), but are forging new and multiple political, socio-economic, and cultural ruralities. Geographical perspectives continue to reveal the unevenness of rural population outcomes (e.g. Milbourne and Kitchen, 2014), and show the differential ways through which rural populations are reconstituted by the cross-cutting factors of demographic processes (birth/ageing/death rates) and different intra, in- and out-migration flows, both sub-nationally and internationally.

In tandem, key general transformative forces are widely recognised within rural studies scholarship, such as: the global restructuring of rural economies and labour markets (Bosworth and Venhorst, 2017); changing regimes of agriculture and forestry (post-production) and work (Robinson, 2014); increasing pressures on greenbelts and planning for new-build housing and rural living (Gkartzios and Scott, 2010); legislative protection of specific rural environments (Holloway and Hubbard, 2014), and; advances in information technology and media allowing more flexible work practices (Roberts and Townsend, 2016), and changing patterns of commuting from rural places to metropolitan centres (Clark, 2018).

Such factors are integral to Lowe et al.'s. (2003) framework of the *differentiated countryside*. Evidenced by dedicated case studies in England of the so-called 'preserved', 'contested' and 'paternalistic' countryside, it is shown 'how economic, social and political networks act to differentiate rural regions and how the "new" networks of middle-class counterurbanisers interact with more traditional rural socio-economic formations' (ibid.: 75). Although the global application of this context-specific types have been challenged (i.e. Gkartzios and Remoundou, 2018), these interactions are viewed as being pivotal to the growing diversity of rural geographies, and, crucially, 'decisively determines the shape of the differentiated countryside' (Lowe et al., 2003: 75).

Within the context of the diversifying countryside, the key aim of this chapter is to deepen understandings of the unevenness of rural population geographies within the particular cases of England and Wales. Using UK census data, the focus of the discussion is centred on six dimensions of contemporary rural population change that are prevalent themes within recent scholarship within rural studies. As a precursor, the next section briefly outlines the methods and data.

Methods

The following sections present analyses of 2001 and 2011 UK census data that was extracted and manipulated for an on-going ESRC-ORA project, investigating processes and forms of international rural gentrification (iRGENT).[1] Undertaking temporal and spatial comparisons at geographic resolutions that can effectively identify the distinct scale of rural change (Smith et al., 2018) is a key component of this cross-national project, and a widely employed method of comparison is to employ standardised measures or metrics, albeit not the only method (Phillips and Smith, 2018a). This chapter therefore uses Census variables to explore how rural places and populations have changed between 2001 and 2011, making use of high-resolution areal comparable units. To analyse change between 2001 and 2011, 2011 output areas (OAs)[2] (181,408 in total) were matched to 2001 OAs (175,434 in total). Of the additional 5,974 OAs, we found that 641 had been merged between 2001–2011, and 6,615 had been split. As a result, our analysis identified 182,049 comparable areal units based around OAs. We then used the Rural–Urban Classification (RUC), established by Bibby and Brindley (2013), to identify categories of rural areas (Village; Village in a Sparse Setting; Hamlets and Isolated Dwellings, and; Hamlets and Isolated Dwellings in a Sparse Setting) at the level of OAs. In this chapter we have excluded the RUC rural categories of 'Town and Fringe', and 'Town and Fringe in a Sparse Setting' from our analysis, due to the semi-rural/urban character of many of these locations. This resulted in 17,666 rural units of analysis, of which 54.8% were Village (9,992); 5.9% were Village in a Sparse Setting (1,045); 34.2% were Hamlets and Isolated Dwellings (6,035), and 5.1% were Hamlets and Isolated Dwellings in a Sparse Setting (894). We focus on these rural areas in this chapter, although we map urban and town and fringe areas within the following figures (shaded in light grey).

Of course, the spatial intensity of rurality varies at a regional level in England and Wales, and this needs to be borne in mind when investigating concentrations of sub-populations. As a percentage of the total rural OAs, most are within the South West (18.5%), South East (16.9%), East of England (16.4%), and, to a lesser extent, East Midlands (10.7%), Wales (9.6%) West Midlands (9.6%), Yorkshire and Humber (8.0%), and North West (7.3%). Fewest rural OAs at a regional level are evident within the North East (2.9%) and, not surprisingly, London (0.2%).

In the following sections we present spatial analyses of key population–related variables, highlighting the relatively high (shaded in black) and low (shaded in dot pattern) extremes within the decile range: very high (90th decile), high (80th decile), low (20th decile) and very low (10th decile) for the mapping of the 2011 census data. We term deciles (30th–70th) between these extremes as 'middling rural' (in white) (see Figures 21.1–21.9). Our decile calculations are based on rural OAs only, as the inclusion of urban OAs would skew the data. We also undertook analyses of percentage changes between 2001 and 2011, with a focus on very high change (90th decile) and very low change (10th decile). Finally, we also present analyses for the ten regions of England and Wales by category of rural area for the 90th decile (see Tables 21.1–21.6) of variables.

Processes of rural population change: some key dimensions

Dominant representations of rural population change often hinge on the replacement of populations geared towards agricultural (or forestry) production by affluent middle-class households, often perceived to be commuting to nearby metropolitan centres for employment (Smith and Phillips, 2001; Phillips, 2005). Of course, this over-simplistic binary reading of rural population change is readily disrupted when the diversity of rural geographies is woven into analyses (see Stockdale, 2016), not least by the gamut of different physical (and perceptual) distances between rural and urban places, and associated meanings of proximity, distance, attachment and separation between the rural and urban. Also of significance are social differentiations, with work highlighting, for instance, variability in the location of groups within and beyond the middle-class (e.g. Hoggart, 1997, 2007; Phillips, 2005), as well as the presence of populations established through lines of difference such as age, gender, household structure, and sexuality (e.g. Lowe and Speakman, 2006; Philip et al., 2013; Phillips, 1998; Smith and Holt, 2005; Stockdale, 2014). In the following sections we will explore some of these spatial and social lines of differentiation drawing on comparative analyses of the 2001 and 2011 Censuses.

Agricultural workers

As a starting point to explore the unevenness of rural population changes, Figure 21.1 shows the marked distribution of 'agricultural workers' (NS-SEC[3] 4.8.2 and 4.9.2) in 2011 across England and Wales. It can be seen that OAs in the 90th decile (8.6–32.8%) are highly concentrated in North and Mid Wales (excluding Anglesey), Pembrokeshire, East Cornwall and Devon, North Pennines and Northumberland, Cumbria and Lake District, Yorkshire Dales and North York Moors, and, to a slightly less concentrated extent, the Peak District. As shown in Table 21.1 these areas have proportions of OAs classified as sparse. These are the sparsely-populated upland areas of England and Wales, where there are distinct pastoral agricultural economies and might be viewed as areas of wilderness (see Smith et al., 2018). At the same time, agricultural workers as a percentage of total usual residents is generally not lowered in these areas by the presence of other occupational traits, given it is difficult in these more remote rural locations to commute to nearby metropolitan centres, when compared to the occupational profiles of residents in rural locations on the fringes of metropolitan centres.

Indeed, it is interesting to note the distribution of OAs in the 10th and 20th agricultural deciles (0–0.1%) within the rural areas relatively close to large metropolitan centres, but beyond the Town and Fringe. Figure 21.1 may thus serve to demonstrate the differences between the rural economies of sparsely-populated upland areas and those of rural areas in relative proximity to metropolitan centres. In this sense, longstanding notions of the countryside as spaces for agricultural workers appear to be a socio-cultural construct that may be largely applicable only to sparsely-populated upland regions in the English and Welsh context. It is also important to note

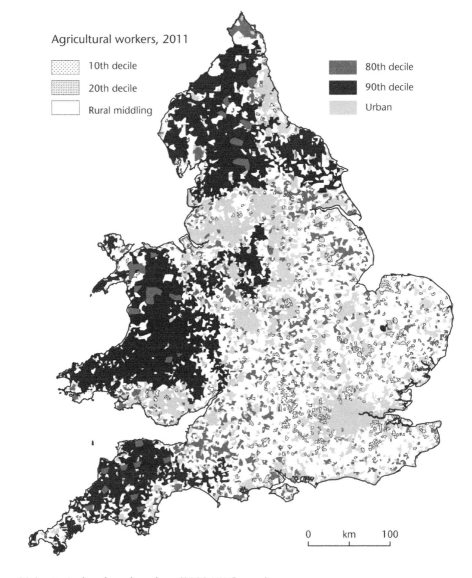

Figure 21.1 Agricultural rural workers (2011 UK Census).

Table 21.1 Regional distribution of agricultural workers by 90th decile (2011 UK Census).

	Village	Village in a Sparse Setting	Hamlets and Isolated Dwellings	Hamlets and Isolated Dwellings in a Sparse Setting	Total
East Midlands	18	0	62	0	80 (4.5%)
East of England	1	1	24	0	26 (0%)
London	0	0	0	0	0 (0%)
North East	6	5	33	43	87 (4.9%)
North West	19	26	121	68	234 (13.3%)
South East	2		18		20 (1.1%)
South West	29	19	284	123	455 (25.8%)
Wales	14	71	87	263	435 (24.6%)
West Midlands	18	4	147	52	221 (12.5%)
Yorkshire and The Humber	28	14	118	48	208 (11.8%)
Total	135	140	894	597	1,766

that there are some areas of upland agricultural land use that have low proportions of agricultural workers, which may point to the presence of relatively high proportions of non-agricultural residents in these areas (e.g. the Trans-Pennine region of Lancashire and Yorkshire).

Table 21.1 shows a regional breakdown of agricultural workers by RUC category of rural location, and reveals distinct differences between south and central England, and areas to the north and west. The highest concentrations of agricultural workers (90th decile) appear in OAs within the South West (25.8%) and Wales (24.6%), and, to a lesser extent, North West (13.3%), West Midlands (12.5%) and Yorkshire and Humberside (11.8%). Agricultural workers in the 90th decile are relatively absent in East Midlands (4.5%), North East (4.9%), South East (1.1%) and East of England (0%). Although the latter may appear surprising, it may be due to the dominance of large commercial organisations and the implementation of technological advances in agricultural production systems in the East of England (and/or perhaps the use of seasonal labour that does not get fully recorded in the UK census) (Woods, 2010). It is also striking in Table 21.1 that the relatively high number of OAs in the 90th decile are in the RUC category Hamlets and Isolated Dwellings in a Sparse Setting (597). Indeed, of the total number of OAs defined as Hamlets and Isolated Dwellings in a Sparse Setting (894), it can be seen that the majority (66.8%) are in the 90th decile for agricultural workers, emphasising the link between agricultural workers and rural locations in sparse settings.

Non-agricultural petite-bourgeoisie

The 2011 census data allows the identification of a non-agricultural petite-bourgeoisie (NS-SEC 4.8.2 and 4.9.2) social class (i.e. self-employed workers such as artists, craft workers, Bed and Breakfast and retail owners catering for local tourism). Interestingly, when compared to Figure 21.1, Figure 21.2 reveals that OAs in the 90th decile (19.4–41.3%) point to some overlaps between the concentrations of agricultural- and petite-bourgeoisie workers in North Devon (Exmoor), Pembrokeshire, North (Snowdonia), Lake District and Cumbria, and North Yorkshire Moors. The upland agricultural economies and landscapes that are conserved and preserved by the National Park Authorities in these areas clearly seem to foster distinctive local rural tourism economies, which may account for the presence of non-agricultural petite-bourgeoisie workers/populations (see Chapter 38 for a wider discussion of national parks and planning). Moreover, Figure 21.2 also expresses the relative concentration of OAs in the 90th decile within coastal North/South Cornwall, North Devon, South Devon AONB, Dartmoor, Dorset AONB and Dorset, Cotswolds, Isle of Wight, South Downs, M11 corridor, coastal Norfolk, Southwold coastal area of Suffolk, Shropshire, and, some parts of Lincolnshire. It is possible that these latter places may point to the effects of the restructuring of some local rural economies from agriculture/fishing to more tourism-dependent economies. This may also be connected to the unfolding of processes of rural gentrification, which is discussed below.

Table 21.2 shows that the regional pattern of petite-bourgeoisie non-agricultural workers by 90th decile tends to be most concentrated in OAs within the South West (31.6%), South

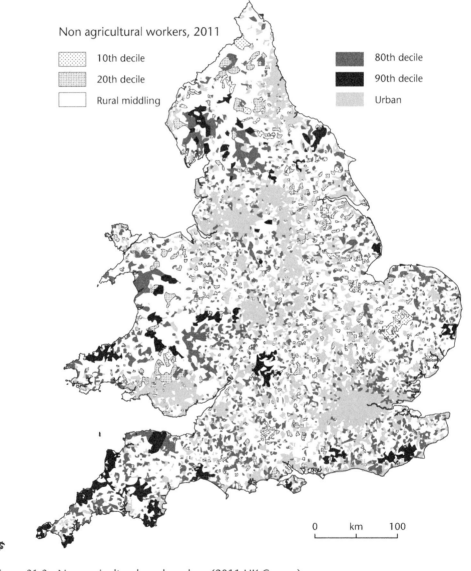

Figure 21.2 Non-agricultural rural workers (2011 UK Census).

Table 21.2 Regional distribution of petite-bourgeoisie non-agricultural workers by 90th decile (2011 UK Census).

	Village	Village in a Sparse Setting	Hamlets and Isolated Dwellings	Hamlets and Isolated Dwellings in a Sparse Setting	Total
East Midlands	41	8	45	4	98 (5.5%)
East of England	111	16	130	7	264 (14.9%)
London	3				3 (0.2%)
North East	2	3	3	2	10 (0.6%)
North West	17	20	54	14	105 (5.9%)
South East	140		169		309 (17.5%)
South West	199	60	233	66	558 (31.6%)
Wales	7	52	12	62	133 (7.5%)
West Midlands	37	5	101	24	167 (9.5%)
Yorkshire and The Humber	32	27	42	18	119 (6.7%)
Total	589	191	789	197	1766

East (17.5%) and East of England (14.9%). In these regions petite-bourgeoisie non-agricultural workers tend to be clustered along the rural coastline, and, hence, again appears closely tied to the prevalence of rural coastal tourism industries, and perhaps this is linked to the growing rise of a rural creative class (Bell and Jayne, 2010; Anderson et al., 2016).

Middle-class individuals

Figure 21.3 presents the distribution of middle-class individuals, calculated based on aggregations of NS-SEC 1, 2 and 4. Particularly notable are relatively high concentrations of OAs in the 90th decile (70.4–84.8%) across the counties surrounding London, followed by South Wales, South Devon, South of Manchester, North Lancashire and across to Cumbria, and in some parts of Northumberland. It is interesting to note that within this specific set of rural geographies there is a prevalence of Areas of Outstanding Natural Beauty (AONB), such as: Surrey Hills, High Weald, Chilterns, Cotswolds, Shropshire, Quantock Hills/Blackdown Hills, Mendips, Malvern Hills, Nidderdale, Howardian Hills AONBs. There are also some concentrations within the National Park areas of: South Downs, Brecon Beacons, Dartmoor, (High) Peak District, Yorkshire Dales, North York Moors, Lake District, and Northumberland. What this may point to is the distribution of the rural middle-class populations in highly prized rural locations, which are protected by dedicated state legislation, regulations and management, and where there is available and suitable rural housing stock to gentrify in conspicuous total numbers. Thus, the general lower availability of housing stock in Exmoor and Snowdonia may be a significant factor in the seemingly lower intensity of middle-class populations in these National Park areas, as represented by our mapping. What this may emphasise are the differential ways that rural gentrification processes shape National Park areas in the UK, given the inherent diversity of local labour and housing markets in National Park areas.

It is plausible that many of these rural locations are being dramatically changed by the exclusionary processes of gentrification, and are displacing or marginalising lower income groups to less desirable rural places. At the same time, the wide reporting of these processes in the national

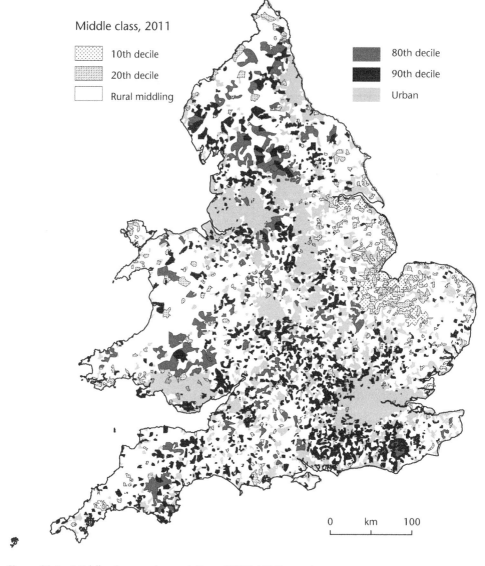

Middle class, 2011

10th decile
20th decile
Rural middling

80th decile
90th decile
Urban

0   km   100

Figure 21.3   Middle-class rural populations (2011 UK Census).

media may be an exaggeration of the scale and magnitude of gentrification: for example, according to the national media during the last two decades, 70 per cent of the British countryside is gentrified by the middle class (*The Times*, 30 August 2015), while 'gentrification is sweeping through the countryside, riding shotgun on the collapse of Britain's farming industry and in many places pricing locals out of the market' (*Guardian*, 18 April 2006: 28). In this context, recent studies have importantly drawn attention to the diversity of rural gentrification processes (Stockdale, 2010; Smith et al., 2018), and also to the different phases and magnitudes of gentrification processes in rural places with high middle-class populations (e.g. Phillips, 2005).

Table 21.3 Regional distribution of middle-class individuals by 90th decile (2011 UK Census).

	Village	Village in a Sparse Setting	Hamlets and Isolated Dwellings	Hamlets and Isolated Dwellings in a Sparse Setting	Total
East Midlands	116	0	44	0	160 (9.1%)
East	80	0	65	1	146 (8.3%)
London	1	0	0	0	1 (0.1%)
North East	21	1	26	7	55 (3.1%)
North West	62	1	102	14	179 (10.1%)
South East	288	0	286	0	574 (32.5%)
South West	124	5	144	17	290 (16.4%)
Wales	34	7	26	18	85 (4.8%)
West Mids	54	0	81	13	148 (8.4%)
Yorkshire & The Humber	51	3	65	9	128 (7.2%)
Total	831	17	839	79	1,766

Figure 21.3 also reveals the relative absence (i.e. dot pattern, 10th decile [10.3–42.2%]) of middle-class populations in the fenland areas of Lincolnshire, Cambridgeshire and Norfolk, coastal and north Lincolnshire, the Solway coast and Arnside in Cumbria, Silverdale in Lancashire, the South Wales Valleys, and Goss Moor Cornwall. Perhaps conditions and connotations of entrenched socio-economic decline, coupled with perceptions of less appealing rural landscapes and cultures, might be important factors for explaining the low proportions of middle-class populations.

Working-class individuals

Reflecting the findings above, Figure 21.4 shows that OAs in the 90th decile (41.4–69.8%) of working-class populations (NS-SEC 5, 6 and 7) are also concentrated in Goss Moor area in Cornwall (with declining china clay industries), South Wales Valleys, the Solway coast and Silverdale in Cumbria, Arnside in Lancashire, coastal, the fenland areas of Lincolnshire, Norfolk and Cambridgeshire, and coastal and north Lincolnshire. This points to a seemingly dichotomous relationship between the presence of middle-class and absence of working-class populations, and vice versa. It is possible that these social class divisions in rural England and Wales may be widening, as middle-class-dominated rural places become more exclusive, via rising property prices and the increasing lack of affordable housing, and working-class populations become relatively absent (i.e. the dot pattern areas on the map showing the 10th decile [3.8–15.9% and 20th decile [15.9–18.8%]).

Table 21.4 shows that at a regional level, the 90th decile for working-class individuals is most concentrated in Wales (18.4%), East of England (18.4%), East Midlands (15.4%) and South West (14.7%), in part, exemplifying the sub-regional clusters of working-class populations, outlined above. What is particularly striking in Table 21.4 is the predominance of working-class individuals in villages (68.9%), as opposed to more sparsely-populated rural areas. This may be due to more sparsely-populated areas being either gentrified (see below) or dominated by agricultural workers (see above).

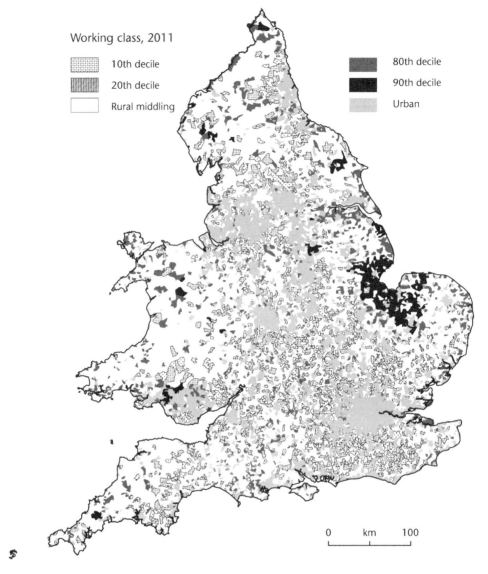

Working class, 2011

▦	10th decile	
▥	20th decile	
☐	Rural middling	
▓	80th decile	
■	90th decile	
░	Urban	

0   km   100

Figure 21.4   Working-class rural populations (2011 UK Census).

Ties between different rural populations and property prices

The uneven rural social class geographies expressed by Figures 21.3 and 21.4, and Tables 21.3 and 21.4, are clearly reflected in Figure 21.5, which shows median house price change between 2001 and 2014 using land registry data (Middle Super Output Area mapped at OA). As can be seen, lowest price changes are mostly evident in the rural areas with highest proportions of working-class populations, outlined above. The most marked house price changes tend to be evident in rural areas with the high middle-class populations. Interestingly, proximity to metropolitan centres does not appear to be an *a priori* marker of higher property prices, perhaps

Table 21.4 Regional distribution of working-class individuals by 90th decile (2011 UK Census).

	Village	Village in a Sparse Setting	Hamlets and Isolated Dwellings	Hamlets and Isolated Dwellings in a Sparse Setting	Total
East Midlands	188	13	62	9	272 (15.4)
East of England	251	19	47	7	324 (18.4)
London	1	0	0	0	1 (0.1)
North East	74	13	16	7	110 (6.2)
North West	59	24	14	5	102 (5.8)
South East	114		18		132 (7.5)
South West	193	25	39	2	259 (14.7)
Wales	149	129	33	14	325 (18.4)
West Midlands	94	5	18	1	118 (6.7)
Yorkshire and The Humber	92	8	20	2	122 (6.9)
Total	1215	236	267	47	1765

pointing to the importance of other possible factors such as rural aesthetics, village morphology, and rural service provisions on property price differentials (see also the following chapter by Gallent and Scott on rural housing).

Some coastal locations appear in the 90th decile in Figure 21.5, such as Dorset, south Devon, north Cornwall, and north Norfolk. This may well reflect the continuing gentrification of these prized coastal locations by relatively affluent, in-migrant mature populations at pre- and post-retirement phases of their lifecourse, these areas having long been identified as sites for such migration (e.g. Clout, 1972; Parsons, 1980; Phillips and Williams, 1984; Phillips, 2005). Indeed, there is growing scholarship within both policy and academic discourses (e.g. Lowe and Speakman, 2006; Davies and James, 2016; Maclaren, 2018) stressing issues of ageing within in rural locations in the UK. It is widely noted that this has serious implications for the supply and demand of public (e.g. health and social care) and private (e.g. retail and leisure) services as discussed in the following chapters in this section, with wider knock-on consequences for levels of poverty and deprivation, housing affordability, local labour markets, sustainability of rural schools, and expressions of social capital/volunteering and community cohesion (Hardill, 2009). In the next section we will explore the distribution of this population in more detail.

Populations aged 65 or more

Figure 21.6 shows the distribution of households with one of more individuals aged 65 or over, as a percentage of total households. Clearly, there are concentrations of OAs in the 90th decile (35.1–90.8%) within Snowdonia, Ceredigion, Quantock Hills, South Devon AONB, Dorset AONB, north Devon and Somerset, rural environs of Bognor Regis, Romney, Southwold, the north Norfolk coast, and coastal Lincolnshire. Away from the coastline, there are also relatively high concentrations of OAs in the 90th decile in parts of the Lake District, Yorkshire Dales, North York Moors and south Warwickshire.

Given the character of many of these rural locations, it is likely that in-migration may be influential here in the reproduction of these rural places as locations for retirement. It is important to recognise that the relative strength of a retired population in an area might be reflective of the out-migration of younger age groups, and it also likely that there will be some important social class differentials tied to particular places. For example, Lincolnshire has seen a

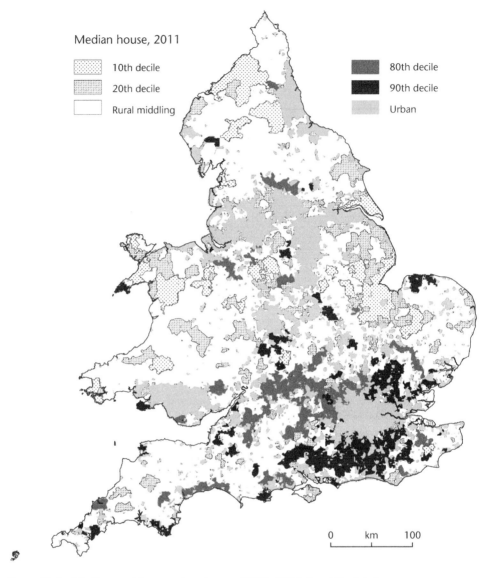

Median house, 2011

- :::::: 10th decile
- ▓▓▓ 20th decile
- ☐ Rural middling

- ▓▓ 80th decile
- ■ 90th decile
- ☐ Urban

0   km   100

Figure 21.5   Median rural house price change (2001–2014, Land Registry).

proliferation of 'retirement parks' (Bevan, 2010; Beatty et al., 2011), in contrast, for example, to the development of high-cost, new-build, retirement apartments in parts of Cornwall and Dorset (Leger et al., 2016).

What Figure 21.6 would appear to express is that representations of rural households with one or more member aged 65 or over are less salient in the rural areas (shaded in dot pattern) of the 'home counties' (surrounding London) of Buckinghamshire, Essex, Hertfordshire, Surrey, Sussex and, to a lesser extent, Kent, and other counties in relative proximity to London, such as Berkshire and Cambridgeshire, plus Cheshire, the Trans-Pennine corridor; Wiltshire, Porth Cawl, and inland Cornwall.

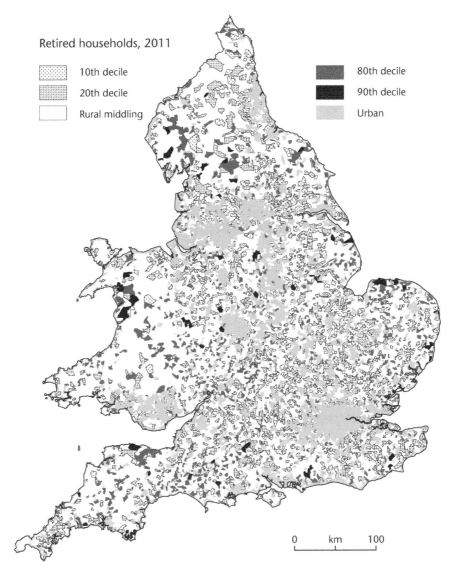

Figure 21.6 Households with one or more aged 65 or over (2011 UK Census).

At a regional level, further analysis (Table 21.5) supports this characterisation, revealing that there are highest concentrations of households (90th decile) with one more adults aged 65 or over in OAs in the South West, South East and East of England (55% of the total OAs). It also suggests that there is a preponderance of such OAs within the Villages and Hamlets and Isolated Dwellings categories (89.5%), and a limited presence in more sparse rural locations (10.5%). It is unsurprising that these age groups and households tend to be relatively absent from the most sparse rural locations where access to health, social care and other services is often less accessible. Although there is evidence that elderly in such areas may be resistant to move despite the lack of such services (e.g. Burholt, 2006), Champion and Shepherd (2006) have argued that there

Table 21.5 Regional distribution of households with one or more aged 65 or over by 90th decile (2011 UK Census).

	Village	Village in a Sparse Setting	Hamlets and Isolated Dwellings	Hamlets and Isolated Dwellings in a Sparse Setting	Total
East Midlands	129	2	31	4	166 (9.4%)
East of England	148	3	95	1	247 (14.0%)
London	1				1 (0%)
North East	38	11	30	7	86 (4.9%)
North West	61	14	56	13	144 (8.1%)
South East	186		125		311(17.6%)
South West	210	24	145	34	413 (23.4%)
Wales	46	31	17	23	117 (6.6%)
West Midlands	86		58	8	152 (8.6%)
Yorkshire and The Humber	73	7	45	4	129 (7.3%)
Total	978	92	602	94	1,766

is also net out-migration from village and dispersed settlement areas amongst people over 75, particularly in those categorised as sparse (see also Chapter 30 of this Companion).

Lone parents

The next population group we wish to focus upon are lone parents with dependent children. As Hughes (2017) has remarked, the presence of lone parents in rural areas has been widely overlooked, in stark contrast to the attention given to this social group within urban space (e.g. Hawkes, 2017). This is important given the overall mean for lone parents with dependent children is higher for rural OAs (8.8%), when compared to the urban OAs (7.0%).

In the rural context, it would appear that rural OAs in the 90th decile (10.7–34.5%) are concentrated in Pembrokeshire, the South Wales Valleys, and North Wales, a distribution that might connect to the low levels of house price inflation in these areas (see Figure 21.5). Conversely, the OAs in the 10th decile in relation to lone parent with dependent children (0–2.6%) may be seen to lie in areas of high housing price increases such as the counties surrounding London, and in areas of strong middle-class presence such as the Forest of Bowland. Of course, there may be a melee of cross-cutting factors underpinning these different rural geographies of lone parenting that perhaps span different expressions of social class, economic capital and income, and cultural lifestyles, emphasising the need to transcend the stereotypes of lone parenting and social deprivation. For instance, rural residence for some middle-class, and, equally working-class, lone parents may possibly have socio-cultural meaning as 'escape areas', enabling a 'fresh start' after the breakdown and/or dissolution of a marriage or cohabiting union (Smith and Culora, 2017).

The distinct sub-regional geographies of lone parents are, in part, masked in the wider regional breakdowns of lone parents 90th decile (see Table 21.6). It can be seen that the regions with highest levels of lone parents are South West (18.0%), South East (17.1%), East of England (16.1%), West Midlands (12.6%) and North West (10.0%). This may point to the presence of lone parents across rural locations, and raise questions about linking lone parenthood with socio-economic differentiation, notwithstanding the appearance of some sub-regional 'hot-spots'[4] within Figure 21.7.

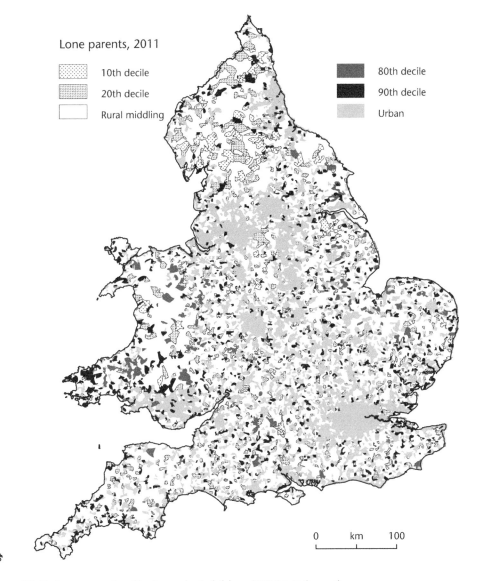

Figure 21.7 Lone parents with dependent children (2011 UK Census).

Couples with children

Figure 21.8 shows the geographical distribution of couples (married/cohabiting) with depend-ent or non-dependent children, as a percentage of total households. This is an important vari-able given idyllic representations of rurality often reinforce rural places as a bucolic spaces for family forming and the rearing of children, often coupled to the presence of high-performing rural schools (Smith and Higley, 2012), and the possibility of a healthy and safer family life that is closer to nature.

It can be seen that there is a distinctive ring of OAs in the 90th decile (36.7–85.5%) around London, which extends northwards along the routes of the M1 (Buckinghamshire/

Table 21.6 Regional distribution of lone parents with dependent children by 90th decile (2011 UK Census).

	Village	Village in a Sparse Setting	Hamlets and Isolated Dwellings	Hamlets and Isolated Dwellings in a Sparse Setting	Total
East Midlands	110	5	39	1	155 (8.8%)
East of England	185	6	93	1	285 (16.1%)
London	4		1	0	5 (0.3%)
North East	8	2	3	0	13 (0.7%)
North West	85	11	69	11	176 (10.0%)
South East	187		115		302 (17.1%)
South West	172	13	121	11	317 (18.0%)
Wales	52	53	39	27	171 (9.7%)
West Midlands	107	6	92	18	223 (12.6%)
Yorkshire and The Humber	63	5	45	6	119 (6.7%)
Total	973	101	617	75	1,766

Bedfordshire), A1 (Hertfordshire) and M11 (Essex/Cambridgeshire) road networks. Three other notable concentrations of OAs in the 90th decile can be identified in: Cheshire/Shropshire; Forest of Bowland; Vale of York/North Yorkshire, rural environs of Cardiff, and the West/East Midlands.

It would therefore appear that there is a relationship between large metropolitan urban centres and relatively high concentrations of couples with children on the rural fringes. In rural locations that are more remote from the metropolitan centres, such as Cornwall, Dorset, Ceredigion, Snowdonia, Lake District/Cumbria, Northumberland, Lincolnshire, Norfolk, South-East coastline, there are relatively low concentrations of couples with children in rural areas. This distribution raises a series of questions. First, are couples with children gravitating to rural areas in striking distance of urban metropolitan centres for employment reasons or to access cultural infrastructures in urban centres? Second, are other factors, perhaps working in combination with the above, important in attracting couples to certain rural locations, such as the provision of high performing schools (see Smith and Higley, 2012)? Or, third, is the distribution of couples with children a reflection of the presence of other groups in particular areas, such as elderly households in some retirement locations. To address such issues, it is worthwhile considering the distribution of couples without children.

Couples without children

Figure 21.9 shows the geographies of childless couples (married/cohabiting) as a percentage of total households. What can be seen from Figure 21.9 is that OAs in the 90th decile (30.8–54.4%) tend to be generally more prevalent in the north and east of England, with the exception of the south-east of England. By contrast, OAs in the 10th decile (1.5–17.2%) are concentrated in North Wales, Pembrokeshire, North Cornwall, and the rural environs of London. Perhaps one of the factors for the low proportion of childless couples in these rural locations is the specific local and regional labour markets of North Wales, Pembrokeshire, and North Cornwall. It is possible that (young) childless couples do not tend to gravitate towards or remain in rural locations that are in relative distance from metropolitan labour markets.

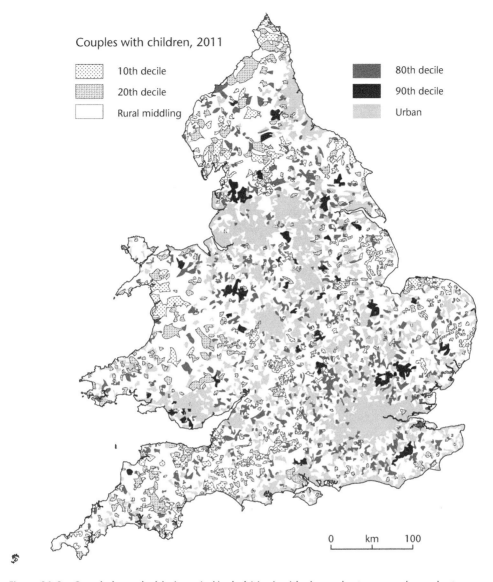

Figure 21.8   Couple households (married/cohabiting) with dependent or non-dependent children (2011 UK Census).

Conversely, the relative absence of childless couples in the close rural environs of London may reflect the high cost of housing in these rural locations, with childless couples at pre-family forming phases of their lifecourse as yet not having amassed appropriate levels of economic capital or earning power to compete in these housing markets and/or to take on the economic burdens of raising a family, or may have a predilection for more of an urban-based way of living/lifestyle. Overall, Figures 21.7 to 21.9 disrupt simplistic representations of rural places based on dichotomous relationship that hinges on the presence or absence of children in rural places.

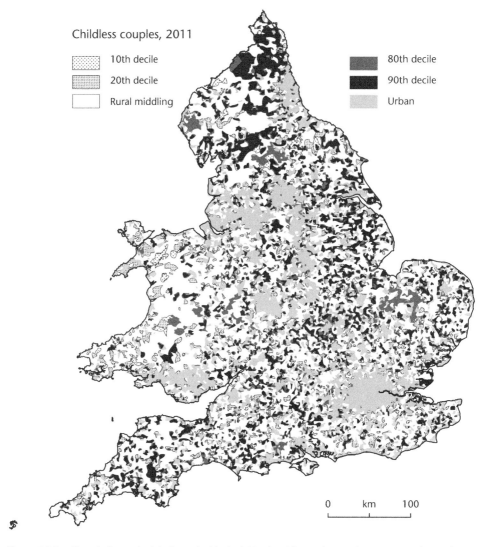

Figure 21.9 Couple households (married/cohabiting) without dependent or non-dependent children (2011 UK Census).

Discussion and conclusion

Our main aim in this chapter was to explore the unevenness of rural populations with a focus on some key rural population dimensions using census data. Our analysis has pointed to an increasingly differentiated countryside, whereby relatively affluent and less-affluent rural households increasingly live apart both socially and spatially, and/or may undertake disconnected, 'parallel lives' in distinct rural spaces across England and Wales.

It would appear that these sociospatial divisions are further complicated by lifecourse-related factors and different family formations, and possibly the predilection for different living arrangements. Crucially, we have crafted a representation of rural populations based on 2001 and 2011 Census data, which perhaps illustrates the exclusionary processes of rural change that unfolded during the 2000s, including the prominence of counterurban-led rural gentrification (see Phillips and Smith, 2018b), alongside the restructuring of agriculture under conditions of capitalist restructuring and globalisation.

In this context, what is particularly intriguing to contend is that it is possible that the pace and rate of rural population change may have slowed down during the 2000s, and that the dynamics of rural population change may not be as pronounced as the previous decade. Key here is the influence of migration on geographies of rural population change, including in-, out- and intra-migration within rural locations. For example, *The State of the Countryside Report* stressed: 'the slowing of in-migration to rural areas noted in 2008 has continued' (Commission for Rural Communities, 2010: 1). Likewise, there is a rapidly growing academic scholarship that provides empirical evidence to affirm the reduction of internal migration flows *per se* (e.g. Smith and Sage, 2014; Shuttleworth and Champion, 2017). What this may mean for rural populations is that local changes in the future may be more tied to the natural ageing processes of rural populations, as opposed to being most significantly shaped by the effects of rural in-, out- and intra-migration.

However, our own analysis of 2001 and 2011 census data found that the overall mean proportion of in-migrants within rural OAs was relatively static between 2000–2001 (12.9%) and 2010–2011 (12.2%) (using the one-year measure of migration). Of course, this may represent pre- and post-recession flows of rural in-migration, and may not capture the slow-down of in-migration during the economic recession years of 2007–2009. Clearly, further research is required to pin down how rural places are being reconfigured by contemporary migration processes, and how this compares to population movements into, out of and between urban counterparts.

To conclude, it is clear from our analysis that processes of population change have yielded a more diverse set of rural geographies, the implications of which are explored in more detail in the following chapters. The British countryside is an insightful case to study rural population dynamics because it is increasingly being sub-divided, both socio-culturally and spatially, through the exclusionary workings of rural (and urban) labour, housing and education markets (Hemming and Roberts, 2018). Furthermore, the ways in which both general and contingent geographic conditions stimulate commonalities and diversities within and between rural locations warrants the urgent attention of social scientists, particularly as the outcomes will have serious planning implications for provision of health, housing, education, social care, policing, and other public and private rural services.

Notes

1 See www.i-rgent.com for more details of the project. The UK element of this project was supported by the Economic and Social Research Council (grant number ES/L016702/1).
2 Output areas (OAs) are the lowest level of geography produced across all census topics, providing insight into local patterns of change. OAs contain approximately equal numbers of usual residents, and are intended to provide geographies that allow reporting of statistics across time on a consistent geographical base
3 National Statistics Socio-economic Classification (NS-SEC) is used by official UK statistics for socio-economic classifications. Categories 4.8.2 and 4.9.2 refer to employers in small organisations (agriculture) and own account workers (agriculture), respectively.
4 One issue to bear in mind when interpreting OA-based maps is that apparent hotspots might simply reflect the presence of larger scale OAs in spatial extent.

References

Anderson, A.R., Wallace, C. and Townsend, L. 2016. Great expectations or small country living? Enabling small rural creative businesses with ICT. *Sociologia Ruralis*, 56(3), 450–468.

Beatty, C., Fothergill, S., Powell, R. and Scott, S. 2011. *The Caravan Communities of the Lincolnshire Coast*. Sheffield: CRESR, Sheffield Hallam University.

Bell, D. and Jayne, M. 2010. The creative countryside: policy and practice in the UK rural cultural economy. *Journal of Rural Studies*, *26*(3), 209–218.

Bevan, M. 2010. Retirement lifestyles in a niche housing market: park-home living in England. *Ageing & Society*, *30*(6), 965–985.

Bibby, P. and Brindley, P. 2013. *The 2011 rural-urban classification for small area geographies: a user guide and frequently asked questions (v1. 0)*. London: UK Government Statistical Service.

Bosworth, G. and Venhorst, V. 2017. Economic linkages between urban and rural regions: what's in it for the rural? *Regional Studies*, 1–12.

Burholt, V. 2006. 'Adref': theoretical contexts of attachment to place for mature and older people in rural North Wales. *Environment and Planning A*, 38, 1095–1114.

Champion, T. and Shepherd, J. 2006. Demographic change in rural England. In P. Lowe and L. Speakman (eds), *The Ageing Countryside: the Growing Older Population of Rural England*. London: Age Concern, 29–50.

Champion, T. and Shuttleworth, I. 2017. Are people changing address less? An analysis of migration within England and Wales, 1971–2011, by distance of move. *Population, Space and Place*, *23*(3).

Clark, M.A. 2018. *Teleworking in the Countryside: Home-Based Working in the Information Society*. London: Routledge.

Cloke, P. and Goodwin, M. 1992. Conceptualizing countryside change: from post-Fordism to rural structured coherence. *Transactions, Institute of British Geographers*, 17, 321–336.

Cloke, P. and Thrift, N. 1987. Intra-class conflict in rural areas. *Journal of Rural Studies*, *3*(4), 321–333.

Clout, H. 1972. *Rural Geography: An Introductory Survey*. Pergamon, Oxford.

Commission for Rural Communities. 2010. *The State of the Countryside Report (2010)*. Retrieved from http://webarchive.nationalarchives.gov.uk/20110303181543/http://ruralcommunities.gov.uk/files/sotc/sotc2010.pdf.

Davies, A. and James, A. 2016. *Geographies of Ageing: Social Processes and the Spatial Unevenness of Population Ageing*. London: Routledge.

Gkartzios, M. 2013. 'Leaving Athens': narratives of counterurbanisation in times of crisis. *Journal of Rural Studies*, *32*, 158–167.

Gkartzios, M. and Remoundou, K. 2018. Language struggles: representations of the countryside and the city in an era of mobilities. *Geoforum*, *93*, 1–10.

Gkartzios, M. and Scott, M. 2010. Residential mobilities and house building in rural Ireland: evidence from three case studies. *Sociologia Ruralis*, *50*(1), 64–84.

Goodwin-Hawkins, B. 2015. Mobilities and the English village: moving beyond fixity in rural West Yorkshire. *Sociologia ruralis*, *55*(2), 167–181.

Halfacree, K. 2014. Jumping up from the armchair: beyond the idyll in counterurbanisation. In M. Benson and N. Osbaldiston (eds), *Understanding Lifestyle Migration*. Basingstoke: Palgrave Macmillan, 92–115.

Halfacree, K. 2018. From Ambridge to the world? Class returns to rural population geographies. *Dialogues in Human Geography*, *8*(1), 26–30.

Hardill, I. 2009. Introduction: geographies of aging. *The Professional Geographer*, *61*, 1–3.

Hawkes, D. 2017. Child health and lone motherhood: evidence from the UK Millennium Cohort Study. In F. Portier-Le Cocq (ed.), *Fertility, Health and Lone Parenting*. London: Routledge, 88–100.

Heley, J. 2010. The new squirearchy and emergent cultures of the new middle classes in rural areas. *Journal of Rural Studies*, *26*(4), 321–331.

Hemming, P.J. and Roberts, C. 2018. Church schools, educational markets and the rural idyll. *British Journal of Sociology of Education*, *39*(4), 501–517.

Hoggart, K. 1997. The middle classes in rural England, 1971–1991. *Journal of Rural Studies*, *13*, 253–273.

Hoggart, K. 2007. The diluted working classes of rural England and Wales. *Journal of Rural Studies*, *23*, 305–317.

Holloway, L. and Hubbard, P. 2014. *People and Place: The Extraordinary Geographies of Everyday Life*. London: Routledge.

Hughes, A. 2017. Geographies of invisibility: the 'hidden' lives of rural lone parents. In M. Kneafsey (ed.), *Geographies of Rural Cultures and Societies*. London: Routledge, 126–144.

Kasimis, C., Papadopoulos, A.G. and Zografakis, S. 2015. The precarious status of migrant labour in Greece: evidence from rural areas. In D. della Porta, S. Hänninen, M. Siisiäinen and T. Silvasti (eds), *The New Social Division*. Palgrave Studies in European Political Sociology. London: Palgrave Macmillan, 101–120.

Leger, C., Balch, C. and Essex, S. 2016. Understanding the planning challenges of brownfield development in coastal urban areas of England. *Planning Practice & Research*, *31*(2), 119–131.

Lever, J. and Milbourne, P. (2017) The structural invisibility of outsiders: the role of migrant labour in the meat-processing industry. *Sociology*, *51*(2) pp. 306–322.

Lowe, P., Marsden, T., Murdoch, J. and Ward, N. 2003. *The Differentiated Countryside*. London: Routledge.

Lowe, P. and Speakman, L. (eds.) 2006. *The Ageing Countryside: the Growing Older Population of Rural England*. London: Age Concern.

Maclaren, A.S. 2018. Affective lives of rural ageing. *Sociologia Ruralis*, *58*(1), 213–234.

McKee, K. Hoolachan, J.E. and Moore, T. 2017. The precarity of young people's housing experiences in a rural context. *Scottish Geographical Journal*, *133*(2), 115–129.

Milbourne, P. and Kitchen, L. 2014. Rural mobilities: Connecting movement and fixity in rural places. *Journal of Rural Studies*, *34*, 326–336.

Murdoch, J. and Marsden, T. 1994. *Reconstituting Rurality: Class, Power and Community in the Development Process*. London: Routledge.

Nelson, L. and Nelson, P.B. 2010. The global rural: gentrification and linked migration in the rural USA. *Progress in Human Geography*, *35*, 441–459.

Newby, H. 1980. *Green and Pleasant land? Social Change in Rural England*. London: Penguin.

Pahl, R.E. 1965. *Urbs in rure*. Geographical Paper 2. London: London School of Economics.

Parsons, D. 1980. *Rural Gentrification: The Influence of Rural Settlement Planning Policies*. Brighton: Department of Geography, University of Sussex.

Philip, L., Macleod, M. and Stockdale, A. 2013. Retirement transition, migration and remote rural communities: evidence from the Isle of Bute. *Scottish Geographical Journal*, *129*, 122–136.

Phillips, D. and Williams, A. 1984. *Rural Britain: A Social Geography*. Oxford: Basil Blackwell.

Phillips, M. 1993. Rural gentrification and the processes of class colonisation. *Journal of Rural Studies*, *9*(2), 123–140.

Phillips, M. 1998. Investigations of the British rural middle classes, part 1: from legislation to interpretation. *Journal of Rural Studies*, *14*(4), 411–425.

Phillips, M. 1999. Gender relations and identities in the colonisation of rural 'Middle England'. In P. Boyle and K. Halfacree (eds), *Gender and Migration in Britain*. London: Routledge, 196–214.

Phillips, M. 2005. Differential productions of rural gentrification: illustrations from North and South Norfolk. *Geoforum*, *36*(4), 477–494.

Phillips, M. 2010. Counterurbanisation and rural gentrification: an exploration of the terms. *Population, Space and Place*, *16*(6), 539–558.

Phillips, M. 2016. Assets and affect in the study of social capital in rural communities. *Sociologia Ruralis*, *56*(2), 220–247.

Phillips, M. and Smith, D.P. 2018a. Comparative approaches to gentrification: lessons from the rural. *Dialogues in Human Geography*, *8*(1), 3–25.

Phillips, M. and Smith, D.P. 2018b. Comparative ruralism and 'opening new windows' on gentrification. *Dialogues in Human Geography*, *8*(1), 51–58.

Roberts, E. and Townsend, L. 2016. The contribution of the creative economy to the resilience of rural communities: exploring cultural and digital capital. *Sociologia Ruralis*, *56*(2), 197–219.

Robinson, G. 2014. *Geographies of Agriculture: Globalisation, Restructuring and Sustainability*. London: Routledge.

Scott, M., Murphy, E. and Gkartzios, M. 2017. Placing 'home' and 'family' in rural residential mobilities. *Sociologia Ruralis*, *57*(S1), 598–621.

Shortall, S. 2015. Gender mainstreaming and the common agricultural policy. *Gender, Place & Culture*, *22*(5), 717–730.

Shucksmith, M. and Brown, D.L. (eds). 2016. *Routledge International Handbook of Rural Studies*. London: Routledge.

Smith, D.P. 2007. The changing faces of rural populations: '(re)fixing the gaze' or 'eyes wide shut'? *Journal of Rural Studies*, *23*(3): 275–282.

Smith, D.P. and Culora, A. 2017. Uneven family geographies in England and Wales: (non)traditionality and change between 2001 and 2011. In J. Stillwell (ed.), *A Handbook for Census Resources, Methods and Applications: Unlocking the UK 2011 Census*. Aldershot: Ashgate, ch. 18.

Smith, D.P. and Higley, R. 2012. Circuits of education, rural gentrification, and family migration from the global city. *Journal of Rural Studies*, *28*(1), 49–55.

Smith, D.P. and Holt, L. 2005. 'Lesbian migrants in the gentrified valley' and 'other' geographies of rural gentrification. *Journal of Rural Studies*, *21*(3), 313–322.

Smith, D.P. and Phillips, D.A. 2001. Socio-cultural representations of greentrified Pennine rurality. *Journal of Rural Studies*, *17*(4): 457–469.

Smith, D.P. and Sage, J. 2014. The regional migration of young adults in England and Wales (2002–2008): a 'conveyor-belt' of population redistribution? *Children's Geographies*, *12*(1): 102–117.

Smith, D.P., Phillips, M. and Kinton, C. 2018. Wilderness gentrification: 'moving off the beaten rural track'. In L. Lees and M. Phillips (eds), *International Handbook of Gentrification*. Cheltenham: Edward Elgar, 363.

Stockdale, A. 2010. The diverse geographies of rural gentrification in Scotland. *Journal of Rural Studies*, *26*(1), 31–40.

Stockdale, A. 2014. Unravelling the migration decision-making process: English early retirees moving to rural mid-Wales. *Journal of Rural Studies*, *34*, 161–171.

Stockdale, A. 2016. Contemporary and 'messy' rural in-migration processes: comparing counterurban and lateral rural migration. *Population, Space and Place*, *22*(6), 599–616.

Sutherland, L.A. 2012. Return of the gentleman farmer? Conceptualising gentrification in UK agriculture. *Journal of Rural Studies*, *28*(4), 568–576.

Woods, M. 2004. *Rural Geography: Processes, Responses and Experiences in Rural Restructuring*. London: Sage.

Woods, M. 2010. *Rural*. London: Routledge.

22

Housing and sustainable rural communities

Nick Gallent and Mark Scott

Introduction

Housing is a principal concern of rural policy and planning in many Western countries (see Gallent et al., 2003) and beyond. When urban housing conditions began, very slowly, to improve during the early years of the twentieth century, rural conditions remained persistently poor (Satsangi et al., 2010). The plight of agricultural workers, and the condition of the housing they occupied, reflected the deep economic restructuring of the previous century. Rural areas had become economic backwaters, seldom sharing in the benefits of industrial advancement which, during the interwar period, disproportionately benefited a growing urban middle class. But this changed in the second half of the twentieth century. City populations had swelled. Economic prosperity and urban living came with social costs and as new road and rail infrastructure penetrated into the surrounding countryside, a process of counter-urbanisation began. That process brought new wealth to rural areas alongside rising land and housing costs, and growing inequality between many sections of the established rural population and newcomers. This is the familiar story of rural house change through the process of industrialisation, rural depopulation and rural return. It is sometimes punctuated by the changing role of land-use planning; its attempts to promote development in some instances, but more usually its fixation with rural landscape preservation, despite widening economic and social inequalities (Gallent et al., 2017).

In this chapter, the familiar narrative of rural housing challenges – and the shift from concerns over housing conditions to questions of inequality – foreground a focus on the centrality of housing to the creation of more sustainable rural communities. Building on Smith et al.'s analysis of shifting social geographies in the previous chapter, the aim here is to examine the implications of social restructuring for housing markets and planning regulation. Counter-urbanisation and the free market for housing has generated inequalities that threaten the well-being of rural economies and communities. Rural housing in many advanced economies is one of the first ports of call for investment in fixed assets. Amenity-led migration, second home and retirement buying all began in earnest in the 1960s, and in countries with stricter planning regimes and less permissive approaches to residential new-build in the countryside, housing soon gained increased 'scarcity value'. Today, those restrictions give housing an obvious investment value, which is often defended by middle-class households through local and community-level planning (Coelho et al., 2017). Rural housing poverty is a mirror to new wealth in the countryside, reinforced by inequalities centred on property and the shifting function of housing in the wider economy. This chapter shines a light on housing inequalities and the challenge around sustainable rural communities.

The chapter has two parts. The first looks at rural housing dynamics – the forces and the pressures that generate housing tensions in rural areas. The second then examines the sorts of interventions and planning approaches adopted to engineer alternative housing outcomes, often aimed at sustaining (and making sustainable) rural communities, but which frequently bring unintended consequences.

Rural housing dynamics

Across much of the globe, urbanisation is thought to be at the heart of housing stress. The pace of that urbanisation means that governments, and their various partners, struggle to deliver the infrastructure, the services and the new homes needed to support concentrated population growth. Therefore, the housing question is largely an urban one – as Friedrich Engels observed in Germany almost 150 years ago (Engels, 1872). But urbanisation is as much a trigger for rural housing tensions as it is for crises in major cities. This was true during the western industrial revolution and remains true today, as many cities in the Global South and in Southeast Asia experience growth rates far in excess of those recorded in European countries from the mid- to late nineteenth century. Urbanisation sets in train a range of immediate and longer-term processes that impact on near urban and remoter rural areas. In the first instance, cities expand beyond their current boundaries, consuming vast swathes of previously undeveloped land and overwhelming near-urban agriculture. At the same time, it sucks people from the countryside, depriving many rural areas of their youngest and most able residents – the so-called 'rural exodus'. In its early phases, urbanisation will drive rural depopulation. This happened in Europe nearly two centuries ago and has been happening in much of the Global South for the last 50 years. But when sections of the new urban population (or more likely their successors) begin to prosper and governments attempt to 'contain' urban growth within fixed boundaries – causing pressures on urban resources, especially on housing and schools – those sections (now a prospering middle class) start to abandon the city. Improved transport for commuting facilitates this process. Initial rural depopulation becomes eventual repopulation, as witnessed in the United States. The economic crisis of 1929 to 1932 hit the remaining population of rural America particularly badly, triggering sudden depopulation in many agricultural areas and the disintegration of communities, leaving abandonment and dereliction in its wake. This situation persisted for the next three decades, but the arrival of post-war prosperity and the building of America's freeways precipitated an urban flight and gentrification of the newly accessible countryside.

As revealed by Smith et al. in the previous chapter, greater accessibility combined with post-war prosperity to produce a sharp rise in demand for rural living. In response, planning was tightened in various countries in order to protect the openness of rural areas from urban encroachment (Hall, 1974). However, not all planning systems sought to defend the countryside. After decades or even centuries of abandonment, the response in many places was to welcome new investment as an indicator of rural vitality (Scott, 2012). The historic practice of keeping a second home was revived in many parts of Europe, especially in the south and in Scandinavia, and seemed to offer an economic life-line to some areas and a respite from their isolation (Gallent, 2014, 2015). In Canada and the US, rural resort development was actively encouraged with second home and 'vacation cottage' ownership becoming a feature of local economic strategies. However, the combined effect of key demographic, social and economic changes was to alter radically the nature of rural housing markets. Demand, triggered by absolute population increase, outstripped housing supply; social change brought new people – new 'housing classes' – to the countryside; and economic change deprived local people of any capacity to compete against 'adventitious' homebuyers from nearby cities (Robertson, 1977). Property that was previously worth very little suddenly became highly sought after as constrained rural housing markets began to over-heat.

A market overheats when there are key impediments to supply despite rising demand. During the early years of counter-urbanisation, rural housing markets were weak or stagnant. Demand for private purchase was low, with most existing residents in privately rented or tied accommodation. There was also a surplus of housing, often in a derelict or dilapidated state, in many places across Europe and North America. This situation persisted in France, for example, into the 1990s (Buller and Hoggart, 1994). When demand rises, 'the market' (a composite of transactions focused on the most desirable parts of the available supply) absorbs 'prime' property: whatever is most sought after by those adventitious buyers seeking to exploit the urban-rural rent gap. The first target might be the scattered former farmsteads, which have some land attached and can be upgraded to prestigious residences. The next (or simultaneous) target may be prime village property: those homes that, for many, epitomise the rural idyll. And the next target may be the bulk of village property, either smaller workers' cottages or more substantial homes. When these become a focus for adventitious purchases, tied workers or private renters may be displaced to public housing alternatives, if available, or are forced to move away. Thereafter, once there has been full absorption of the current stock, pressure grows behind the weight of demand to open up new sites for development. Some executive housing is permitted and it may look likely that an imminent supply response (to rising prices) will restore market equilibrium. However, the restoration of that equilibrium is never an easy or simple task.

In urban areas, the supply of land will quickly be depleted and there may be few opportunities to respond to strong demand. Consequently, prices in the most sought-after urban locations – parts of New York City's lower Manhattan or key streets in London's Kensington and Chelsea – become truly astronomical. But in the countryside, it is often political rather than physical constraint that keeps prices high. On seeing the first batch of executive homes built, newcomers to the countryside – residing in their prime village houses or farmsteads – may experience anxiety over the likely effect of new building on both the value of their own homes and the character of the village they live in or near. After all, they have invested not only in a home but in a way of life and in villages that should be 'picturesque, ancient and unchanging' (Newby, 1979: 167). At that point, the debates that take place in community councils seldom lead to the opening up of farmland for development, in response to demand (and the needs of displaced local households), but rather to the rejection of further development for the sake of 'protecting the countryside': a laudable aim, with sometimes dire social consequences.

This imagined narrative, however, does not play out in the same way everywhere. Two countries that stand in stark contrast to one another in terms of attitudes to rural development (and therefore in the nature of their respective housing challenges) are the Republic of Ireland and England (see also Gkartzios and Shucksmith, 2015 for a detailed Irish–English comparison). Dispersed rural settlement has been a characteristic feature of many rural areas in Ireland (McGrath, 1998), comprising small towns, villages and, more commonly, houses in the open countryside often developed on a self-build basis. Traditionally, within rural Ireland the planning system has been facilitative of new development, reflecting historically high levels of outmigration, lack of development pressures, and the protection of private property rights rooted in the struggle for land and past colonial experiences. Until recently, the overriding historical pattern of population change in Ireland has been one of sustained emigration, resulting in rural areas characterised by higher rates of economically dependent population groups, gender imbalances, and a loss in ability to create new employment opportunities leading to weakened rural communities. Indeed, the scar of past rural depopulation has ingrained into a local political culture an ethos of celebrating physical development in the countryside as a visible indicator of community wellbeing. This standpoint has led to the operation of a liberal planning system in rural Ireland, described as one of the more lax rural planning regimes in Europe (Duffy, 2000) leading to relaxed attitudes towards building new single, dispersed dwellings in the open countryside. While there is a strong cultural predisposition to living in the countryside, recent years have been marked by an increasing pace of rural housing growth related to the demographic recovery of many rural areas and the relatively lower costs associated with developing a 'one-off' house on a self-build basis. Critics have drawn attention to the environmental cost of this development 'free-for-all', but for many rural stakeholders additional housing is seen as essential to maintaining viable rural communities and the debate is often expressed in terms of a right to live in rural areas and freedom to build (Scott, 2008). In England, although there have been episodes of displacement from the land, particularly as a result of the historical land enclosures, there has never been the same level of animosity towards landed interests – certainly no revolution or lurch towards republicanism. Attitudes towards development, therefore, are rooted in nostalgia rather than in militant struggle. Ireland gets its sporadic development; England its blanket curfew on all sorts of rural change.

Land-use planning can be a key impediment to supply, but only in instances where social and political discourse leads to a restrictive brand of planning. This can happen where powerful interests monopolise that discourse, where the need for additional development is unacknowledged or ignored, or where other considerations regularly outweigh the socio-economic case for new housing.

Demand pressures on rural housing

Counter-urbanisation has created a multitude of competing 'property classes' in rural areas (Weber, 1968). These property classes are distinguished by their life-stage (and hence earning potential or accumulation of wealth), by current and past tenure (and hence by stored equity in property), by wealth accumulated through other means (for example, inheritance), by employment prospects and borrowing potential, and by family support. In composite, these characteristics create relative advantage or disadvantage. The young person, with no past history of home ownership, no accumulated wealth, few skills and no family support will, quite clearly, suffer access disadvantage relative to the forty-something buyer about to withdraw accumulated equity from a house in Milan, who has substantial savings from a professional city job, and whose family already have a portfolio of urban and rural property. Saunders (1984) has argued that past history of ownership has become a key determinant of housing advantage. This is perhaps more true today than it was 30 years ago, as prices (especially in the UK) have soared, and the deposit required to purchase a home is now far too large for any first-time-buyer to find without substantial family support, or a very high salary. Current owner-occupiers form a privileged 'domestic property class' and can easily outgun those in rented accommodation who are trying to make the jump to home-ownership. This is essentially what happened in England in the 1960s when external demand for rural property began to take off, and rural renters (who were in a majority at the time) with 'low income and low wealth' lost out to 'more prosperous groups' (Shucksmith, 1990a) as their landlords sold their homes to incoming buyers. The same divisions persist today; in a constrained private market where prices are high and rising, wealth and income will determine the general pattern of housing access.

Historically, the rural poor (with low income and wealth) were tenant farmers or labourers. This has changed today and it is often unskilled workers serving the tourist economy, renting their homes and earning only a minimum wage, who struggle to access decent housing in locations close to their work. They may also find themselves in competition with migrant workers for jobs and accommodation. In the southern US, Mexican migrants – legal and illegal – compete with US citizens for unskilled farming and service jobs. In western Europe, migrants from the eastern accession states have arrived in large numbers. The majority have headed to the major cities, but many have taken up rural jobs in farming and in construction (see Commission for Rural Communities, 2007).

In relation to the operation of housing markets, it is useful to distinguish between those groups who express 'effective' demand (because of their relative economic power) and those that have clear *need*, but are unable to compete in a more aggressive market. The former now comprise second home buyers, investors (usually in holiday lets), lifestyle downshifters/changers (fleeing from cities and urban lifestyles), salaried migrants looking for more space, and retired households. The latter are a much smaller group, with particular wealth, income and life-stage characteristics. Although certainly a generalisation, it is young first-time-buyers or renters who often experience the greatest housing access difficulties in many rural areas. This group includes young singles and couples, occasionally with children. They are either Pahl's (1975) *rural poor* or his successors to the agricultural workers once housed in tied accommodation. But if these groups are considered in the context of urbanisation and counter-urbanisation, the obvious question is this: why have these groups remained in the countryside? Ford and colleagues (1997) have categorised young people in rural areas as 'committed stayers', 'committed leavers', 'reluctant leavers' and 'reluctant stayers'. Genuine housing stress occurs when young people who are committed to remaining in a rural area cannot meet their needs in either the private rented sector or through home-purchase, and where state alternatives are unavailable. The difficulties faced by this group are expressed through a range of inadequate *in situ* housing solutions: living with friends or relatives for prolonged periods, making use of short-term seasonal lets (and then facing periodic episodes of homelessness), or finding themselves in temporary forms of accommodation (including caravans). Similarly genuine stress is also evident when young people leave rural areas, not because they wish to but because they cannot match their housing and employment needs (Rugg and Jones, 1999).

Both the 'committed stayers' and the 'reluctant leavers' presumably had some reason to remain in their village; including family connections and support networks, employment opportunities and a desire to stay put for a range of social reasons. Arguably, a *housing need* exists where it is housing access *alone* that causes individuals to endure the hardship of staying or of leaving. But without socio-economic anchors in a place, the existence of that need may be doubted. Without job opportunities, in particular, the case for lower income groups having a genuine housing need may appear weak; and consequently, so too is the case for intervening in market processes to deliver different forms of non-market state housing. Such intervention could leave the 'rural poor' stranded in gentrified villages without job prospects or the support of social networks. Some commentators have argued that intervention is necessary in order to maintain the right of poorer households to live in the villages of their birth: to protect an 'ancestral right' (Gallent and Robinson, 2012). But the 'rights' argument seems unnecessary. The need to create opportunities for lower-paid service sector workers to remain in rural areas is not merely an individual need, but a broader socio-economic need. Without a cross-section of people living in the countryside, rural economies cannot function and key services cannot be provided. This has been observed on the Norfolk coast in England. This area has seen widespread gentrification since the 1960s, followed by retirement, second-home buying and *in situ* ageing. Many villages are entirely devoid of young people; but yet there is thriving demand for services, particularly care and social services for the elderly residents. Those providing these services need to travel considerable distances from more affordable market towns, adding to their living costs and making it difficult to provide those care and social services when and where they are needed (Gallent et al., 2002). Without young people, the service economy will falter, and farms and tourism businesses will be starved of their labour supply – or accessing their labour will be more difficult, costlier, and lead to unsustainable patterns of car dependency and rural commuting.

The need for accessible housing in the gentrified countryside is not a need encountered solely by individuals; it is a wider socio-economic need. This is a crucial point. The countryside benefits from having a mixed profile of housing types and opportunities; but it should also be acknowledged that delivering *universal access to market housing* in particular, specific hotspots is impossible. Bramley and Watkins (2009) draw a distinction between 'within area', 'outward' and 'inward' affordability in rural areas. Affordability is often expressed as a ratio between lower quartile earnings and lower quartile rents or house prices, inferring the likelihood of those on the lowest incomes being able to meet rent costs or service a mortgage: that is, 'enter' the housing market via one of these two routes. Bramley and Watkins define 'within area' affordability as the percentage of households in a village/neighbourhood able to enter the housing market in that village/neighbourhood; 'outward' affordability as the percentage of the same households able to enter the market in the wider area; and 'inward' affordability as the percentage of households in that wider area able to enter the market in that village/neighbourhood.

These distinctions are important as they express the essential problem of housing access and affordability in a gentrified countryside, in which key villages have been the targets of adventitious purchasers. A typical situation, confronting planning and public policy, is for a hot-spot village to have a misleadingly high level of within area affordability (as the village is populated by wealthy households) and therefore high outward affordability (the same wealthy households could enter the market anywhere), but *low* inward affordability: a high proportion of people not already in that village would not be able to buy or rent there.

Two issues arise from this 'village /neighbourhood' perspective on affordability. First, the concentration of wealth in gentrified villages masks *in situ* housing poverty (experienced by a small number of committed stayers). And second, it is far easier to address that poverty in the wider area than in the hot-spot villages where rents, property prices and land values are likely to be high and all forms of development will be contested.

Rural housing supply pressures

The social transformation of many countrysides has altered dramatically the context for a supply response to rural housing problems. Concentrated wealth leads to a questioning of the case for additional development; it is often suspected that gentrifiers are seeking to protect property values through their opposition to additional housing. But in many instances, it is obvious that such housing would simply attract further inward investment from adventitious buyers and do little or nothing to calm prices or help lower income groups. Housing is therefore rejected on amenity grounds, because prized views (across open fields) or the character of a village might be threatened. The rejection of new housing is only problematic in instances where that housing would have otherwise met an identified need: perhaps because starter homes were being proposed, some with restrictive occupancy covenants, or the intention was to build some form of public housing.

Although additional market housing in heavily gentrified and extremely desirable locations is unlikely to assist with housing access, more specific interventions – building particular types of housing to help the 'rural poor' – has become a key battleground in rural planning. Over thirty years ago, David Clark (1984) set out five possible responses to rural housing access challenges. The first was a *general market response*: open up more land for development and try to restore equilibrium between supply and demand. The second was a *density market response*: whenever permissions for new housing are given, use the planning system to encourage building more housing on less land at higher densities. The third, fourth and fifth were all *non-market, targeted responses*: give priority, through planning, to meeting the needs of the 'rural poor'; build housing for that group on agricultural land (see Gallent and Bell, 2000); and give priority to third-sector and community groups within the planning process.

The two market responses were acknowledged to be blunt instruments. Under circumstances of strong external demand, which was at the root of access difficulties, building more open market housing would merely attract additional external investment. The density response might well have the same effect, and would also alter the pattern of rural settlement. The idea that demand should simply be absorbed through further development clearly ignores the highly specific nature of rural housing need. In some instances, a supply response may be appropriate – for example, in addressing *in situ* growth in larger settlements or market towns, or in-migration triggered by economic growth – but in small village locations, which have been bearing the full force of gentrification, a much more targeted, non-market response would seem appropriate. Therefore, in countries affected by the meta-narrative of counter-urbanisation and restrictive planning, the answer to the rural housing question is often a highly targeted one, focused on addressing the specific needs of lower income groups. This might be achieved in a variety of ways – as illustrated in the second part of this chapter – and is rationalised as a proportionate response set against the need to protect the wider countryside from development pressure.

However, in rural areas affected by continuing depopulation, there is often a very different supply response. Attempts to diversify and strengthen economic activity are regularly led by national governments or supra-national organisations. There is often a focus on tourism and the (infra) structural development of peripheral rural areas. If these attempts have any effect on local economies, they may be followed by private speculation on (potential) rising land values and the development of housing for adventitious purchasers. This all adds up to 'property-led rural regeneration' (Gkartzios and Norris, 2011; Norris et al., 2013) aimed at promoting development and re-population.

There are broadly two situations in which land-use planning is called upon, by the state or by communities, to tackle rural housing challenges. The first is where the forces of counter-urbanisation make a general supply response undesirable, as it would risk environmental degradation without solving underlying market tensions or access difficulties. This is the general situation in many near-urban areas in lowland England, especially in the south. The second is where economic stagnation and depopulation suggest an approach to rural regeneration (with a view to assisting the current population) that may include an element of general house building. There are certainly poorer rural areas in England where such a response might be appropriate; though across England, Scotland and Wales *both* situations are regularly encountered. The broader issue of rural regeneration is addressed in Part III of this Companion. The remainder of this chapter focuses mainly on the first situation, in which targeted responses to market pressures are developed under the watchful eye of sometimes highly conservative rural communities.

Housing's role in sustainable rural communities

It was suggested above that rural areas benefit from having a mixed profile of housing types and opportunities. This is because the sustainability of those areas depends on their 'life-cycle offer' to a mix of households (or property classes), distinguished by age, wealth, income and occupation. Many gentrified villages in England have developed extremely skewed age and wealth profiles. This has resulted in concerns over the long-term sustainability of communities which lack young people and families with children, and are instead dominated by retired households and seasonal residents. Concerns are often prosaic ones: who will run local services as the population ages? What will happen to schools if there are no children to fill classrooms? What will give communities their vitality if large proportions of homes are empty? Such concerns are not ubiquitous, but many rural communities in relatively easy reach of large towns and cities or those with particular character and amenity value, face the pressure of rapid socio-economic change and its housing consequences. Housing access is thought to play a pivotal role in sustaining the life-cycle of communities and, for that reason, great effort is expended on developing interventions that seek to deliver wider access. Two examples illustrate this point. The first concerns attempts to prioritise full-time residency through housing occupancy restrictions in St Ives, Cornwall. The second relies on 'locals only' restrictions in the Republic of Ireland.

Broadening housing access through planning – St Ives, Cornwall

Forty years ago, a dominant concern for rural planning in England was the revitalisation of rural communities following decades of decline. Second homes were seen, in some quarters, as a lifeline for rural communities. Grants were available to prospective purchasers wishing to renovate derelict rural property (Gallent, 1997). But in some areas, the pressures were such that occupancy restrictions were designed and introduced, with the intention of prioritising the needs of 'local' and full-time residents over demand from non-local and seasonal buyers.

In May 2016, the residents of St Ives in Cornwall voted to adopt a new neighbourhood development plan containing a 'full time principal residence requirement'. High demand for second homes in and around the town was generally regarded as a cause of rising house prices and declining affordability. Second homes had also hollowed out parts of the town, with some streets largely empty for much of the year. In that context, the Town Council won support for the imposition of a planning condition on all permissioned development: no new homes will be permitted unless they are to be occupied for a minimum of 270 days each year. Similar approaches have been used in the past to give priority to 'local' people in otherwise open housing markets. During the 1970s, a planning condition in the Lake District (in the northwest of England) sought to 'restrict completely all new development to that which can be shown to satisfy a local need' (LDSPB, 1977). This condition was written into the Cumbria and Lake District Joint Structure Plan. The condition was eventually deleted by national government, which agreed with inspectors that it was 'unreasonable' to use 'planning powers to attempt to ensure that houses should only be occupied by persons who are already living in the locality', adding that planning should be 'concerned with the manner of the use of land, not the identity or merits of the occupier' (DoE, 1981). Attitudes have changed since the 1970s and it is now generally accepted that planning should negotiate with developers for the inclusion of affordable housing in private schemes. But that usually means that a proportion of homes are available for social renting, discounted purchase or some form of intermediate tenure. The intention is never to restrict the onward sale of all housing produced by private enterprise, which is viewed by many as unwarranted intrusion in the operation of the housing market.

It is too early to judge the impacts of the St Ives policy, but the restriction in the Lake District remained in place for a full seven years between 1977 and 1984, prior to its deletion. During those years, the impacts of the policy were studied by Shucksmith (1990b). He made a number of observations. First, households seeking second homes in the Lake District were largely undeterred by the restriction; they had always preferred second hand property anyway (older village homes rather than new build) but now demand from them became entirely concentrated in that segment of the market. Second, although new build housing was now targeted at 'local need' – at full time residents – the supply of that housing began to dip. It was observed that 'builders ceased speculative residential development, partly because of the uncertainties raised by the new policy, but principally because of the greater difficulty of acquiring suitable land with planning permission' (ibid.: 122). Third, the aggregate impact across the entire housing market – comprising second hand housing and a declining quantity of new build – was a slightly faster rate of house price inflation. This, combined with the restriction on non-local purchase in the market for new build, choked off some of the external demand. Some aspiring second home buyers found the Lake District suddenly too expensive and shifted their attention elsewhere, outside the area of restriction.

But overall, price adjustments for second hand and new build property were largely balanced out across the market. Excess external demand shifted entirely to second hand property, benefiting existing homeowners. That same demand was removed from the new build segment, but prices there were largely unaffected owing to changes in land values, development activity and therefore reduced supply. Indeed, 'local people who could afford to buy new housing will have found prices roughly the same as before, once the shifts in the demand and supply schedules had worked through' (Shucksmith, 1990b: 123). Those unable to access the market before were not assisted greatly by the policy.

Back in St Ives, the Town Council is responding to the full range of housing pressures described at the beginning of this chapter. It seems to have taken the view that the town needs more full time residents and fewer empty homes if the community is to be 'sustainable'. Sustainability also means securing job opportunities for local people, and tourist income – from occasional and seasonal visitors – is important in this respect. But second home buying has now reached a level that has brought significant disruption to the local housing market. Prices have risen apparently beyond the reach of some local buyers, including younger people and those with families. Local action in this context seems reasonable. But second homes are merely symptomatic of the investment and consumption pressures that many areas, rural and urban, now face. Such local action – achieved through planning – is likely to either push the problem elsewhere or generate unintended consequences. Second homes and other forms of seemingly inessential consumption evidence the commodification of housing and a long transition away from thinking about housing as home to housing as asset. Arguably, the tensions in St Ives speak to a much deeper and more fundamental housing crisis that planning cannot resolve.

Prioritising 'local needs' in rural Ireland

As outlined earlier in this chapter, dispersed rural housing has been a longstanding feature of the Irish countryside. However, during the so-called Celtic Tiger house-building boom, the scale and pace of rural housing development became an increasingly controversial issue with estimates suggesting that one in three new houses built in Ireland during the property boom (approximately 1996–2006) were single houses in the open countryside (see Gkartzios and Scott, 2009). At a local political level, facilitating new housing in rural areas is deeply engrained, echoed by a longstanding state commitment to maintain viable rural communities as a public policy goal. In this context, national planning policies have attempted to balance socio-economic and environmental goals through distinguishing between the needs of rural communities, which should be identified in county development plans, and encouraging more restrictive policies in near urban contexts to manage pressure for 'overspill' development from urban areas, suggesting a belated recognition of differentiated rural contexts. Moreover, since the publication of the *National Spatial Strategy* in 2002 (DOELG, 2002) and related publication of central government planning guidelines for *Sustainable Rural Housing* (DEHLG, 2005), planning policy distinguishes and prioritises 'rural-generated housing' – housing needed in rural areas within the established rural community or by those who are an intrinsic part of the rural community by way of background or employment – from 'urban-generated' rural housing – development driven by urban centres, i.e. housing sought in rural areas by people living and working in urban areas, including second homes.

Therefore, in managing local housing supply, planning policies prioritise and accommodate people perceived to be a part of the rural community – in other words, in assessing planning consent for a single house in the countryside, planners are asked to examine not only the physical development but also the *applicant* for planning permission as well. Local authorities, in turn, include local needs criteria within county development plans, and while these vary across local government areas (Gkartzios and Scott, 2009), these often refer to:

1 existing residency or former residency in the county;
2 bloodline links to the locality (i.e. to have relatives among the local residents or family links to the area);
3 local employment;
4 those engaged in agricultural activities would be eligible for planning permission; and
5 in relation to *Gaeltacht* or Irish language speaking areas, linguistic ability requirement (i.e. where permission to build a residence is dependent on Irish language proficiency).

In this context, local need is conceived as an individual's need to live in the locality rather than a wider assessment of sustainable rural communities e.g. understanding the level of local development that would sustain local services or enable new human capital resources to contribute to a vibrant community. Local needs rules are, however, proving increasingly controversial in light of the European Commission's goal of promoting and protecting the free movement of capital. By 2017, a number of Irish local planning authorities had begun to review local needs assessments (with their focus on individuals rather than contexts) following a Court of Justice of the European Union ruling (*Libert and Others versus Flemish Government and Others*) concerning the use of restrictive housing policies in Belgium, which were similar in intent to Ireland's local need restrictions. It is thought that the days of focusing on applicants' local links and 'bloodline' when determining planning permission may well be numbered.

In parallel to these local needs assessment, another approach to managing rural housing in Ireland has been through the use of rural housing design guidelines aimed at mitigating the visual impact of new housing by putting tighter restrictions on the design and siting of houses. More than 80 per cent of local authorities have published design guidelines for rural housing. Given the lack of political will to control the quantity of new houses in Ireland – which is in sharp contrast to the situation in England and St Ives – planning authorities have been tightening design controls in the hope preventing, or at least ameliorating, the impact of housing on rural environments and landscapes.

Conclusions

Shucksmith observed, more than 30 years ago, that 'the essence of the housing problem in rural areas is that those who work there tend to receive low incomes, and are thus unable to compete with more affluent 'adventitious' purchasers from elsewhere in a market where supply is restricted' (Shucksmith, 1981: 11). This statement retains its currency, at least in countries that have experienced counter-urbanisation, amenity-driven migration and some degree of planning restriction aimed at protecting rural landscapes and productive land. People often move to the countryside from towns and cities because of their income and wealth advantage and the desire to live somewhere less 'spoilt' by intrusive development. They frequently enjoy competitive advantage in the housing market and introduce more conservative attitudes to development. It is noteworthy that in the case of St Ives, 'affluent newcomers' were more supportive of the full-time residency restriction than many 'local workers' (see Gallent et al., 2016). There was clear sky between those wishing to slow the pace of development and those worried about the local economy and jobs. In the Republic of Ireland, policies promoting the needs of 'local households' recognise the disadvantage suffered by people on local and low incomes relative to incomers seeking homes in the countryside, for leisure or investment. The movement of people alongside income differentials and development restriction are key determinants of rural housing outcomes, bringing gentrification to some rural areas. However, this picture of housing stress and community change does not hold true for all areas. Poverty and peripherality remain more important drivers of housing and social outcomes in some parts of the world. The relative inaccessibility of peripheral areas – genuine rural 'backwaters' – means that there has been little external interest or investment. Job markets are very weak and housing conditions remain poor. There is another 'rural' that does not fit Shucksmith's useful generalisation, but which is examined elsewhere in this Companion.

Acknowledgement

Parts of this chapter previously appeared in N. Gallent, I. Hamiduddin, M. Juntti, S. Kidd and D. Shaw, *Introduction to Rural Planning*, 2nd edition (Routledge, 2015).

References

Bramley, G. and Watkins, D. (2009) Affordability and supply: the rural dimension, *Planning Practice and Research*, 24, 2, 185–210.

Buller, H. and Hoggart, K. (1994) *International Counter-Urbanisation: British Migrants in Rural France*, Avebury, Aldershot.

Clark, D. (1984) Rural housing and countryside planning, in M. Blacksell and I. Bowler (eds), *Contemporary Issues in Rural Planning*, SW Papers in Geography, Plymouth Polytechnic, Plymouth, 93–104.

Coelho, M., Dellepiane-Avellaneda, S. and Ratnoo, V. (2017) The political economy of housing in England, *New Political Economy*, 22(1), 31–60.

Commission for Rural Communities (2007) *A8 Migrant Workers in Rural Areas*, Briefing Paper, CRC, Cheltenham.

DEHLG (2005) *Planning Guidelines for Sustainable Rural Housing: Consultation*, Stationery Office, Dublin.

DoE (1981) *Proposed Modifications to the Proposed Cumbria and Lake District Joint Structure Plan*, Department of the Environment, London.

DOELG (2002) *The National Spatial Strategy 2002–2020: People, Places and Potential*, Stationery Office, Dublin.

Duffy, P. (2000) Trends in nineteenth and twentieth century settlement, in T. Barry (ed.) *A History of Settlement in Ireland*, Routledge, London.

Engels, F. (1872) *The Housing Question*, Progress Publishers, Moscow (published in 1954 from Engels's articles for the Leipzig Volksstaat).

Ford, J., Quilgars, D. and Burrows, R. with Pleace, N. (1997) *Young People and Housing*, Rural Research Report No. 31, The Rural Development Commission, Salisbury.

Gallent, N. (1997). Practice forum improvement grants, second homes and planning control in England and Wales: a policy review, *Planning Practice & Research*, 12(4), 401–410.

Gallent, N. (2014) The social value of second homes in rural communities, *Housing, Theory and Society*, 31(2), 174–191.

Gallent, N. (2015). Bridging social capital and the resource potential of second homes: the case of Stintino, Sardinia, *Journal of Rural Studies*, 38, 99–108.

Gallent, N., and Bell, P. (2000). Planning exceptions in rural England: past, present and future, *Planning Practice and Research*, 15(4), 375–384.

Gallent, N. and Robinson, S. (2012) Community perspectives on localness and priority housing policies in rural England, *Housing Studies*, 27(3), 360–380.

Gallent, N., Mace, A. and Tewdwr-Jones, M. (2002) *Second Homes in Rural Areas of England*, Countryside Agency, Cheltenham.

Gallent, N., Shucksmith, M. and Tewdwr-Jones, M. (eds) (2003) *Housing in the European Countryside*, Routledge, London.

Gallent, N., Kelsey, J. and Hamiduddin, I. (2016) Swimming against a Cornish tide? *Town and Country Planning*, 85(6), 237–240.

Gallent, N., Tewdwr-Jones, M. and Hamiduddin, I. (2017) A century of rural planning in England, *Planum*, 35(2), 93–106.

Gkartzios, M. and Norris, M. (2011) 'If you build it, they will come': governing property-led rural regeneration in Ireland, *Land Use Policy*, 28(3), 486–494.

Gkartzios, M. and Scott, M. (2009) Planning for rural housing in the Republic of Ireland: from national spatial strategies to development plans, *European Planning Studies*, 17(12), 1751–1780.

Gkartzios, M. and Shucksmith, M. (2015) 'Spatial anarchy' versus 'spatial apartheid': rural housing ironies in Ireland and England, *Town Planning Review*, 86(1), 53–72.

Hall, P. (1974) The containment of rural England, *The Geographical Journal*, 140(3), 386–408.

LDSPB (1977) *Draft National Park Plan*, Lake District Special Planning Board, Kendal.

McGrath, B. (1998) Environmental sustainability and rural settlement growth in Ireland, *Town Planning Review*, 3, 227–290.

Newby, H. (1979) *Green and Pleasant Land? Social Change in Rural England*, Hutchinson, London.

Norris, M., Gkartzios, M. and Coates, D. (2013) Property-led urban, town and rural regeneration in Ireland: positive and perverse outcomes in different spatial and socio-economic contexts, *European Planning Studies*, 22(9), 1841–1861.

Pahl, R.E. (1975) *Whose City? And Further Essays on Urban Society*, Penguin, Harmondsworth.

Robertson, R.W. (1977) Second home decisions: the Australian context, in J.T. Coppock (ed.), *Second Homes: Curse or Blessing?* Pergamon Press, London, 119–138.

Rugg, J. and Jones, A. (1999) *Getting a Job, Finding a Home: Rural Youth Transitions*, Policy Press, Bristol.

Satsangi, M., Gallent, N. and Bevan, M. (2010) *The Rural Housing Question: Communities and Planning in Britain's Countrysides*, Policy Press, Bristol.

Saunders, P. (1984) Beyond housing classes: the sociological significance of private property rights in the means of consumption, *International Journal of Urban and Regional Research*, 8, 202–227.

Scott, M. (2008) Managing rural change and competing rationalities: Insights from conflicting rural storylines and local policy making in Ireland, *Planning Theory and Practice*, 9(1), 9–32.

Scott, M. (2012) Housing conflicts in the Irish countryside: uses and abuses of post-colonial narratives, *Landscape Research*, 37(1), 91–114.

Shucksmith, M. (1981) *No Homes for Locals?* Gower Publishing, Farnham.

Shucksmith, M. (1990a) A theoretical perspective on rural housing: housing classes in rural Britain, *Sociologia Ruralis*, 30(2), 210–229.

Shucksmith, M. (1990b). *Housebuilding in Britain's Countryside*, Routledge, London.

Weber, M. (1968) *Economy and Society*, University of California Press, Berkeley, CA.

23

Second homes, housing consumption and planning responses

Chris Paris

Introduction

This chapter provides a critical review of concepts and issues relating to planning and rural second homes. It introduces the concept of post-productivist dwelling in countryside areas and explores the evolution of leisure-oriented dwelling as a form of tourism, with increasingly blurred and over-lapping relationships between 'housing' and 'tourism'. These issues are developed further, in the contrasting cases of Britain, Ireland and Australia.

'Planning', 'second homes' and 'holiday homes'

Planning is defined here as a generic form of future-oriented human activity, undertaken to various degrees by individuals and corporate entities, including governments. Specific government land-use planning organisations are *political* agencies operating ecologically over time in interaction with other socio-economic and political relations. The notion that such planning comprises professional technical skills and activities separate from socio-economic and political relations is absurd and unworthy of further examination, though planning regimes and processes are important *constitutive* elements of cities, regions and the countryside.

The Australian term 'holiday home' is preferred to 'second home' to describe dwellings owned (or occasionally rented) for leisure use in addition to 'primary residences', partly because many affluent households own more than one 'second' home (Paris, 2011). The term 'holiday home' emphasises the *hybrid* nature of the ownership and use of multiple dwellings for leisure and investment purposes as it makes no artificial distinction between housing and leisure 'markets'. It also places tourism and mobility (Hall, 2005, 2017) at the core of our understanding of 'rural second homes' (Müller, 2011; Paris, 2014).

Marsden (1969: 57–58) noted ambiguities in the term 'holiday home' over 40 years ago and proposed a typology of their use; he was writing about Australia, but the logic applies in all cases:

1   Private holiday homes: used at weekends and holidays by owners, family and friends; diverse dwelling types usually located within 'generally acceptable recreational commuting times'.
2   Intermittently commercial holiday homes: mainly for private use, but occasionally let to others especially during high seasons.

3   Intermittently private: often purchased for possible future retirement, but mainly used as commercial lettings with some private use.
4   Commercial: investment properties generally let and maintained by agents.

Such diversity of use of holiday homes (or 'second homes') is reflected in the difficulty involved in trying to count how many there *are*. Official data sources and series are quite good for England, through the English Housing Survey (EHS), but we must rely on indicative census data, industry estimates and expert analysis for other UK countries, Ireland and Australia.

Rural areas or post-productivist countrysides?

The *Oxford English Dictionary* (2017) defines rural as 'in of, relating to the countryside rather than the town' and the term is used typically to refer to areas of low population density, where economies are typically dominated by primary industries especially farming, forestry, and fishing. Simple commonsense usage is fine, and is the only way in which the term 'rural' is used here, but it becomes problematic if used as an *analytical* category to contrast 'rural' and 'urban' phenomena, ways of life or social relations. Sociologists and geographers comprehensively demolished simplistic urban–rural distinctions by the 1970s and the UN Department of Economic and Social Affairs (2017) argues that distinctions between urban and rural areas are 'blurred' in industrialised countries, with rapid urbanisation doing the same in developing countries (see also discussion by Gallent and Gkartzios in Chapter 2 of this Companion).

Practical public policy analysis, however, requires evidence of spatial variations within and between countries, using comparable objective statistical criteria in spatial hierarchies based on variations in population density (UN Department of Economic and Social Affairs, 2017). Such categories have changed over time and vary between countries, with most contemporary spatial hierarchies descending from major metropolitan areas through smaller urban centres and so on. The US Census Bureau (2017) describes 'rural' as a residual category to refer to 'all population, housing, and territory not included within an urban area'; the Australian Bureau of Statistics (2017) simply notes that 'any population not contained in an urban centre or locality is considered to be rural balance'.

British and Irish statistical spatial hierarchies include a 'rural' category but use different definitions. The British Rural Urban Classification defines areas as rural if they fall outside of settlements with more than 10,000 resident population (DEFRA, 2013); the Irish Central Statistics Office uses the statistical concept of Aggregate Rural Area for the population living in all areas outside clusters of 1,500 or more inhabitants.

The term 'countryside' is preferred to 'rural areas' here simply to refer to areas of relatively low population density, including 'coastal' areas where over half British, most Irish and almost *all* Australian holiday homes are located, and excluding wilderness areas. The countrysides of industrial societies have become almost entirely the sites of 'post-productivist' dwellings, with the word 'dwellings' used both as noun and as verb. Most countrysides remain sites of production, but that is separated from most dwellings within countryside settings, except in areas of very low population density such as country Australia.

The industrialisation of farming, consolidation of agricultural holdings and growth of transnational agribusinesses have resulted in widespread depopulation of countrysides across Europe, with big variations between countries in timing, extent of depopulation, and degree of gentrification, and with huge differences in wealth. Settler capitalist countries *never* had pre-capitalist agrarian geographies, but they have undergone many of the same processes of agricultural mechanisation and farm consolidation since 1945.

No countryside is 'natural', despite advertising hype of some contemporary examples of commercial holiday home developments in artfully designed areas in the English Cotswolds or gated Tasmanian trout fishing enclaves. All are historical artifacts, made through the interaction of human agency and natural processes, and still changing in many ways, as explored in Ireland (Aalen et al., 2011) and Britain (Rackham, 1986). Post productivist countrysides are valued as high amenity locations for holiday homes, with attractive views, distinctive terrains, access to water and valued countryside and coastal leisure pursuits: that is, diverse residential and touristic consumption. There are many variations in terms of scale and the specific amenities available in post productivist countrysides, and in the costs associated with accessing such locations. The analysis of holiday homes and their countryside settings in the following sections can only provide an outline of their national and transnational histories of urbanisation, industrialisation and agricultural transformation.

Land-use planning regimes, housing and tourism in countrysides

The term 'planning regime' refers here to all public policies and practices relating to land-use regulation and development, affecting physical uses of land and dwellings, and/or dimensions socio-economic of dwellings. Most statutory land-use planning systems only came into being after 1945 and barely exist in some countries. Planning regimes have varied enormously between and within countries, especially federal nations where planning powers are the responsibility of sub-national levels of government (e.g. USA and Australia), as discussed also in Chapter 2 of this Companion by Gallent and Gkartzios.

The role of land-use planning is debated widely in relation to holiday homes and many commentators want planning 'solutions' to 'problems' relating to holiday homes, as if somehow planning was separate from socio-economic and political relations. In many cases, however, the operation of planning regimes, has been a *constitutive* element of the development of contemporary countrysides and holiday homes, exacerbating the 'problems' of scarcity and house price inflation that commentators ascribe to holiday homes.

There are very few ways that land-use planning can meaningfully differentiate between holiday homes and dwellings used as main residences. Four analytically separable sets of issues can be identified regarding the scope and limits of land-use planning and holiday homes:

- Environmental concerns: visual and physical pollution, potential damage to sensitive ecosystems, unsafe or risky locations (e.g. risk of flooding or erosion).
- Protection of specified areas of high amenity or heritage values (especially National Parks).
- Strategic planning for housing development, redevelopment or replacement.
- Local land-use zoning and regulation.

There is no meaningful difference between dwelling use as holiday home and as permanent residence in terms of issues relating to environmentally sensitive development: both are dwelling *uses*, so the issue is how many dwellings of what kind may be permitted. The only way for planning to stop new dwellings being used as holiday homes in high amenity or heritage areas is to stop *any* housing development, but this leads to gentrification and deflection of development to *other nearby* areas.

If governments or citizens wish to stop dwellings from being used as holiday homes, the role of planning is just as problematic. Many have proposed that holiday homes should be designated as a land-use class separate from other residential development. It is relatively easy to specify that certain dwellings be used *only* as holiday accommodation, and planning permission for holiday accommodation often requires that dwellings are not occupied for specific periods. But 'primary residences' also are often unoccupied for various parts of the year and people could occupy 'holiday' dwellings as their main residence for much of the year and take holidays elsewhere while their 'holiday' home is closed!

Strategic planning for housing has great difficulty in dealing with holiday homes. There are no technical problems in including estimated demand for holiday homes in calculations of additional land required for housing. In practice, however, well-orchestrated anti-development lobbies routinely oppose any additional housing construction in Britain and even in countries without powerful anti-housing lobbies, such as Australia, regional planning strategies struggle to deal with the implications of increasingly mobile populations and growing ownership of multiple residences.

Most land-use planning systems and associated building regulations have assumed that dwellings are used *either* as residences or commercially, but this distinction collapses in highly mobile contemporary societies. The concept of 'hybridity' highlights the overlapping dimensions of dwellings, within *both* 'housing' and 'leisure/tourism' markets. Dwellings are used for residential *and* recreational purposes, for private consumption *and* commercial leisure investments, with diverse changing temporal dimensions. The expansion of 'Airbnb' and other internet-based lettings platforms highlight the planning issues for communities and governments, resulting from how dwellings are *actually* used, rather than how they have been classified bureaucratically for planning purposes. (These developments, moreover, force us to reconsider the very nature of housing tenures, as the categories 'private renting', 'home ownership' and 'holiday accommodation' become blurred and much more fluid than conventional classifications would allow.)

The empirical focus below on national case studies simply uses official area classifications of 'rural' and 'regional' areas of relatively low population density to compare and contrast the stories of holiday homes in Great Britain and Ireland, and to contrast those cases with Australia. Variations in planning regimes, polices and practices have been significant constitutive elements of the differences in holiday homes in these three countries, not some external force with potential to resolve 'problems'.

There is no clear distinction between holiday homes used by their owners and holiday homes available for short-term rental letting in any of these cases. Many holiday rental agencies advertise houses, cottages and other types of accommodation on the internet and in tourist brochures, but it is impossible to quantify how many of these are also used at least part time by their owners.

Holiday homes in British and Irish countrysides

Holiday home ownership grew rapidly both in the UK and Ireland from the early 1990s to the global financial crisis (GFC) in 2007 (Norris et al., 2010; Paris, 2011). The main drivers of growing second home ownership in both countries were growing affluence and mobility, combined with a strong belief in the utility of investment in property as a core element of households' investment and consumption strategies. After 2007, however, the stories of holiday home ownership in Britain and Ireland went in very different directions.

Holiday homes in British countrysides

Holiday home ownership in Britain took off slowly in the 1960s, but grew rapidly during the 1990s. Any abandoned former farming dwellings in Britain were quickly snapped up as growing demand for holiday homes fuelled gentrification of existing dwellings in high amenity areas, and spilled into adjacent areas. Holiday home owners purchased dwellings from owner occupiers or landlords seeking to capitalise on increasing property values. The number of English households owning holiday homes doubled between 1997 and 2007 with strongest growth *outside* the UK (Paris, 2011; Savills, 2007).

British holiday home ownership differs from most other EU countries in two distinctive ways:

- The depopulation of the English countryside areas occurred earlier and more extensively than in continental Europe, through agrarian reforms, enclosures of common land, and development of large country estates. The 'Highland Clearances' in Scotland also removed tenant farmers to make space for sheep and sporting estates. Thus post-1945 Britain contained far fewer former farming dwellings for occupation by new holiday homeowners than most other EU countries.
- The restrictive post-1945 British planning regime was introduced earlier and more comprehensively than in other EU countries, officially separating 'urban' and 'rural' areas, constraining housing development generally, and stopping most building in the countryside (Gkartzios and Shucksmith, 2015). 'Green belts' were introduced supposedly to constrain metropolitan expansion but in practice turned into iconic sacred spaces to be savagely protected by anti-development lobbies. 'New towns' were supposed to take 'overspill' population and jobs across the green belts from the major cities but just became nodes within complex multi-functional metropolitan regions. Other development 'leap-frogged' green belts, former working villages were gentrified by commuters and retirees, and medieval industrial towns were re-branded as 'villages' to be protected at all costs from the metropolitan poor.

Planning restrictions are tightest in National Parks or Areas of Outstanding Natural Beauty (AONBs), which together cover over 30 per cent of the land area in England and Wales and are all included within 'rural' areas in the English rural–urban classification of local authority districts shown in Figure 23.1. Figure 23.2 shows the distribution of holiday homes in England and Wales in the 2011 Census (ONS, 2012). The data describe the rate, per 1,000 usual residents, of people with 'holiday' second addresses. The great majority are in 'rural' areas with very high concentrations in or adjacent to National Parks, AONBs, and other high amenity areas, especially in Devon, Cornwall, Norfolk and the Lake District.

There has been widespread diversification of developments as well as growth of developments aimed at holiday home purchasers just outside zones of severe development constraint. Many recent up-market developments have obtained planning permission from local authorities eager for prestigious new development, based on existing use rights or replacement of like for like (e.g. replacing old hotels with new holiday apartments). Mobile and static prefabricated dwellings in 'parks' are widespread, often carefully packaged and marketed for different market segments, with complex overlaps between housing and leisure markets (Paris, 2011, 2013).

Planning has influenced the location of some developments through a combination of initially *unrelated* planning policies. Post-war policies permitting sand and gravel extraction in river valleys resulted in sites which were subsequently allocated 'brownfield' status as former 'industrial' areas, and developers could obtain planning permission for holiday homes that may not be occupied as permanent residences. Some of these developments cater exclusively for upmarket holiday home ownership in private landscaped areas of mixed woodland and lakes around former gravel pits in the Cotswolds Water Park area, some 150 kilometres from central London. The Cotswolds is one of the most expensive housing areas of England, with extremely high house price to income ratios. Unlike older villages in the countryside, however,

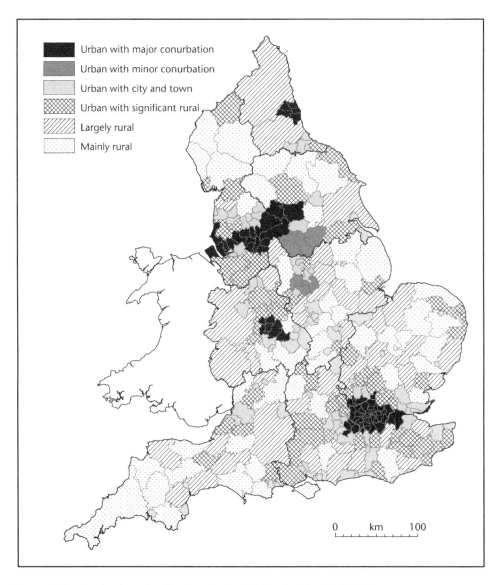

Figure 23.1 Rural–urban classification for English local authority districts.

the development of up-market holiday homes estates on former mining sites did not directly involve gentrification or replace locals with incomers.

Luxurious holiday home developments in gated estates include Lower Mill Estate, where you can 'discover a naturally beautiful family home in a 550-acre secure estate and find inspiration for life' (http://lowermillestate.com) and The Lakes, with luxury holiday homes providing a range of ownership options and short-term luxury holiday rental options (www.thelakesbyyoo.com). These are a distinctively British type of holiday home development, crucially influenced by the planning system, depending on high levels of affluence and mobility of households with incomes generated far away: delightful gated leisure enclaves for hyper-consumption within a post-productivist countryside marketed through commodified concepts of 'nature', 'community' and 'the Cotswolds'.

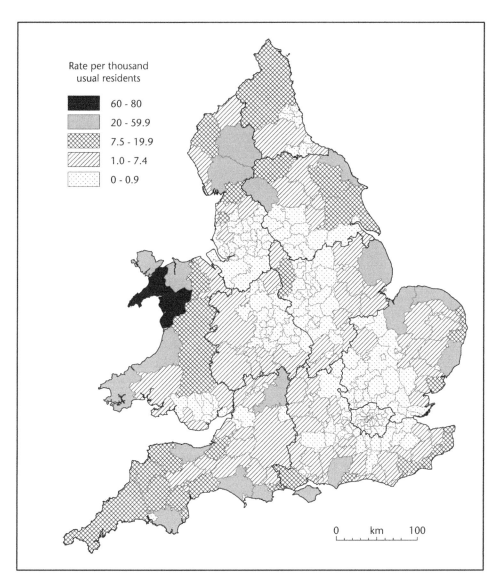

Figure 23.2 Location of holiday homes relative to usual residents, England and Wales 2011.

The GFC had negligible impact of the growth of holiday home ownership by English house-holds (see Table 23.1); rather, the story of British second home ownership since the GFC has been one of increasing socio-economic polarisation. The higher incomes and greater assets of households in London and South East England ensure that this region provides the bulk of continuing demand for additional luxury holiday homes. Some commentators had suggested that there could be a shift back towards a preference for owning second homes within Britain (Direct Line Insurance, 2005) and Table 23.1 indicates that this *has* been happening. The English Housing Survey showed the number of English-owned holiday homes *within* England increasing by *more* than the total increase of 80,000 between 2008 and 2014 (DCLG, 2017). Industry commentators reported strong demand for new holiday homes in England, especially *luxury* homes in Devon and Cornwall, as the weaker pound made investment 'at home' better value than buying overseas (Paris, 2016). Social housing organisations reported growing holiday home ownership fuelling house price increases in 'new second homes hotspots' in English countryside areas, as the number of holiday homes increased by over 50 per cent between 2004 and 2009 (Shelter, 2011). The overlap between housing and leisure markets is a key element in the demand for English holiday homes purchased for letting to holiday makers and occasional private use. The strong demand for up-market second homes and continuing gentrification of the post-productivist countryside so evident since the GFC, however, have stood in stark contrast to falling construction more generally, sluggish sales of primary residences and static house prices in many areas.

Irish holiday homes

The history of Irish countrysides stands in stark contrast to Britain. Ireland was Britain's first overseas colony with a series of land grabs and 'plantation' of ruling Protestant elites from the sixteenth century, minimal agrarian reform, and an impoverished peasantry paying rent to absentee landlords. Devastating famine in the nineteenth century was followed by mass starvation and emigration mainly to Britain and the USA. Continued emigration after the 1850s resulted in long-term population decline in stark contrast to the population in nineteenth and twentieth century Britain. Ireland was one country within the United Kingdom up to 1921 when 'partition' resulted in the establishment of an independent state in the south and Northern Ireland as a constituent jurisdiction of the UK. Both parts of Ireland were economic backwaters in the 1970s and 1980s, and Northern Ireland suffered dreadfully in the 1970s and 1980s through 'The Troubles' – the euphemism for intense conflicts over sovereignty and national orientation. Many thousands of abandoned dwellings were scattered in Irish countryside areas readily available if there ever should be demand for use as holiday homes.

Rapid growth of holiday home ownership came relatively late to the Republic of Ireland, during the 1990s 'Celtic Tiger' economic boom, as overall housing production increased enormously: from below 20,000 a year in 1990 to 93,000 in 2006. The laissez faire planning regime permitted rapid growth of new construction, including thousands of so-called 'one-off' homes in the countryside, in stark contrast to static levels of new housing construction and savage restrictions on new building in the countryside across the Irish Sea in Britain.

The Irish housing boom was further stimulated by generous tax breaks for home owners and property investors, including area-based tax incentives (the 'seaside resort scheme' and the

Table 23.1 The location of English-owned holiday homes, 2008 to 2014.

	England	Other UK	Europe	Other foreign	Total
	thousands of homes				
2008/09	244	36	280	100	660
2013/14	342	40	245	117	744
	percentages				
2008/09	37	5	42	15	100
2013/14	46	5	33	16	100

Source: DCLG English Housing Survey, Table FT2611

'rural renewal scheme'), resulting in explosive growth of *newly* built holiday homes and holiday villages in attractive coastal and countryside areas, while leaving most abandoned dwellings untouched (Norris and Winston, 2010; Norris et al., 2010; Gkartzios and Norris, 2011). The high proportion of vacant dwellings recorded in the Irish 2006 census was a sign of increased holiday home ownership, as well as many recently constructed dwellings as yet unoccupied, after 15 years of continuously rising housing production and anticipation of ever-continuing growth.

The holiday home boom was also noted in Northern Ireland, especially in hot spot areas including the Causeway Coast, Newcastle and the Mournes, and the Fermanagh Lakelands. Although Northern Ireland is part of the UK, its planning regime is similar to the Republic and most holiday homes were newly built in country and coastal areas, with very little gentrification. In Northern Ireland, as in the Republic, there were signs by 2007 of significant over-production of new dwellings for sale as holiday homes and concern that house prices had doubled overall in less than three years with no substantive demographic or economic drivers (Norris et al., 2010).

The speculative housing, land and investment booms in both Irish jurisdictions came to a sudden end in 2007. The Irish economy in the Republic of Ireland was decimated by the GFC, exacerbated by wild bank lending practices, leading to extensive job losses and deteriorating public finances. Housing completions fell dramatically with high vacancy rates marking over-optimistic speculative building projects, with thousands of unoccupied and unfinished new dwellings including hundreds of 'ghost estates' (Drudy and Collins, 2011; Kitchen et al., 2010). Thousands of new homes remained unsold in 2017, especially in coastal areas with high proportions of holiday homes. Negative equity was widespread and house prices had fallen by 50 per cent from their peak of 2007. Northern Ireland shared in the collapse of new construction, falling overall house prices, and the ending of the holiday homes boom (Paris, 2011).

Australia: holiday homes in coastal countrysides

This section provides a comparative historical analysis of the evolution of holiday homes in Australia and explores overlapping dimensions of Australian holiday homes as items of private leisure consumption *and* as sites of capital investment in contemporary land, housing and commercial holiday markets.

There were no pre-capitalist 'rural' landscapes in Australia, as there had not been any 'urban' ones with which to contrast, though indigenously used and managed landscapes were destroyed by invasion, genocides and 'settlement'. The story of Australian holiday homes differs sharply from European experiences and is like other affluent 'new world' countries with most being newly constructed rather than representing changing use of older dwellings (Marsden, 1969; Murphy, 1977). The Australian case, however, differs from the USA or Canada, especially in timing and geographical context. Mass holiday home ownership and developer-driven holiday homes projects emerged some 20 years later in Australia than in the USA and Canada. Colonial settlement and subsequent development in Australia was highly concentrated in coastal centres, with most holiday developments occurring within relatively easy access to main population concentrations.

Australia was relatively prosperous compared to most European countries from the 1880s gold rushes to the 1920s. The economy suffered during the 1930s depression but quickly regained prosperity during the post-war boom. Australia had one of the highest home ownership rates of the world by the 1960s, due to strongly supportive government policies and the prosperous and expanding economy (Paris et al., 1993).

Early holiday homes developments from the 1930s to the early 1960s were typically informal structures in unregulated environments, often involving self-construction of modest cottages or 'shacks' on Crown land. Many 'holiday homes communities' were developed on previously undeveloped places, typically by the coast or inland rivers and lakes within easy weekend commuting distance from metropolitan centres, including hundreds of 'shacks' in the Central Highlands of Tasmania built by trout anglers and their families. They were places of escape from a rapidly changing world, albeit only possible thanks to enhanced affluence and mobility in that new world (Marsden, 1969; Murphy, 1977).

Holiday home ownership grew substantially during the 1950s and 1960s for family recreational use, increasingly concentrated in coastal areas with more formally built structures regulated by new systems of planning and building. Some were near to or soon accompanied by other tourist accommodation, especially hotels; thus some holiday home places subsequently became much bigger, more mixed use places The growth of holiday home ownership was driven by growing affluence, increased paid leisure time and rapidly expanding car ownership as 'the weekender was being turned into the holiday home' (White, 2005: 136). Subsequent developments of all housing, including construction designed to be sold as holiday homes, has taken place within more comprehensive statutory land-use planning systems. Land-use planning systems vary somewhat between states and territories, generally resembling a British approach to the regulation of development, albeit with a much more permissive and pro-development orientation.

The growth of holiday homes ownership slowed in the late 1960s, but Robertson (1977) predicted that the proportion of Australian households with second homes would increase from around 5 per cent in 1971 to 20 per cent by the end of the century. But actual growth has been *much* lower and holiday home ownership has changed and diversified within a changing context. The long post-war boom ended in the mid-1970s and the growth of second home ownership was arrested by the same processes affecting Australian society and housing: economic restructuring, demographic change, the shift to neo-liberalism across Australian politics and society, and growing socio-economic polarisation.

As in many other wealthy countries, there have been major transformations in Australian housing policies and markets since the 1970s, with a switch away from the 'social project' of mass homeownership to commodified systems of housing production and ownership (Forrest and Hirayama, 2014). Home ownership fell from 71 per cent of households in 1971 to 67 per cent in 2011 as the public policies that had promoted growth of home ownership were replaced with supply-side subsidies (Beer et al., 2011). The small social housing sector was in decline, with growth in private renting fuelled by tax breaks for wealthy investors. Booming capital city house prices were further inflamed by extensive overseas purchase of Australian residential property (Paris, 2016). These changes in the housing system have also affected leisure-related investment in dwellings.

The wider environmental context of holiday homes has also changed as metropolitan areas have expanded massively, mainly through low-density developments catering for a rapidly growing population fuelled by high levels of in-migration. There has been extensive development in coastal 'sea change' and peri-urban areas, reflecting wider socio-economic and demographic changes (Burnley and Murphy, 2004). Many holiday homes places developed in the 1960s have been overtaken by suburbia as cities expanded outwards; others became parts of new large communities, such as the Sunshine Coast to the north of Brisbane. State governments have removed or are removing many shack settlements from Crown land and National Parks and have introduced stricter environmental planning regulations and building controls. Some other early post-war holiday homes developments have endured but shacks are increasingly replaced with larger and more substantial dwellings.

One Australian fly fishing writer lamented the decline of shacks and development of 'lake houses':

> Once, shacks were built with whatever materials you could scrounge, 'borrow' or steal, and constructed in the quickest way possible, the object being to secure shelter from the elements without having to sacrifice too much fishing time either while physically building the shack or by working at-out in your day job to pay it off. Today, lake houses everywhere are becoming trophy items where status and investment, or sometimes art and design take precedence over living and affordability.
>
> *(French, 2008: 164)*

There have also been upmarket developments of private trout fishing estates with guaranteed water supply and ample supplies of fish, most recently in 'Highland Waters Estate'. The estate's glossy advertising has parallels with designer holiday homes developments in the Cotswolds: gated security with a resident ranger, large allotments for the construction of waterfront or hillside holiday homes, and shared ownership of an expansive private fishery in a 'small community of like-minded fly fishing neighbours' (www.highlandwaters.com.au).

Greater affluence for some Australians and enhanced mobility for most has resulted in complex patterns of seasonal movement and migration for leisure purposes (Bell and Brown, 2006; Charles-Edwards et al., 2008; Hugo et al., 2013). Holiday homes are a vital element of tourist accommodation in many areas, and sometimes the *only* form of holiday rental accommodation, as in the Yorke Peninsula (Paris et al., 2014). Demand for seasonal rental holiday homes in Australia has distinctive peaks compared to northern hemisphere tourism, due to the coincidence of summer holidays and the Christmas-New Year period. The peak effect is most concentrated in southern states (SA, Victoria and Tasmania) where cool and dismal winters contrast with the pleasant winter weather in northern NSW, Queensland and WA.

Our study of the holiday homes and retirement migration in South Australia included four case studies of second homes 'hotspots' with rapid land and house price inflation in particular localities, the imposition of high service provision costs on local governments, as well as numerous conflicts between 'permanent' and 'temporary' residents (Paris et al., 2014). The study concluded that there was no meaningful difference between 'second homes' as sites of family leisure, and commercial 'holiday rentals': holiday homes were sites of leisure and amenity *and* business premises used for commercial letting. The emergence of this hybrid type of dwelling use was not recognised in Australian State and Territory planning systems as the basis for distinguishing between holiday accommodation and permanent residences had become muddled. Holiday homes were a vital form of tourist accommodation, the *main* form in some local government areas, but this was often ignored in local planning and tourism strategies.

The study identified the development of distinctive local housing markets, with growing proportions of large and expensive holiday homes left empty for much of the year and available for holiday letting at high costs, but little or no other rental housing (Paris et al., 2014). The distinctive local housing markets in holiday home hot spots reflected the mismatch between the incomes and wealth of permanent residents and second homeowners. This in turn affects attitudes to holiday homes and aspirations for the future of local areas. Positively, the growth of holiday home ownership has been a vital element in sustaining and growing local businesses, with more substantial restaurants and shops than would be viable based on permanent population alone. The study found different attitudes to aspects of development: tourism is vital to the local economy, but holiday home owners oppose the provision of facilities for tourists near 'their' beaches and seek to privatise public spaces. Holiday homeowners and permanent residents, however, came together to oppose proposed developments such as wind farms or mining; growing proportions of retirees often fall into both camps.

The wider implications of the growth of holiday homes and other forms of short-term letting, including Airbnb and other internet platforms, are being explored in ongoing research on Australian local government and housing for the twenty-first century (https://localgovernmen tandhousing.com) and is expected to lead to a fuller understanding of the impacts of changing inter-relations between housing and tourism, private housing use and investment in housing for touristic uses.

Conclusions

Britain and Ireland

Holiday home ownership in the British countryside has been massively affected by the restrictive planning regime affecting all new building, leading to widespread gentrification as well as diverse and innovative forms of new developments. The restrictive planning regime also helped to maximise the value of existing dwellings and encouraged the search for cheaper holiday homes overseas especially between 1997 and 2005 when the pound was strong against the euro.

The growth of holiday home ownership in the Irish countryside mainly involved the purchase of new dwellings, with no significant gentrification, during a period of unprecedented prosperity after 1990. Strong growth in incomes, housing production and house values growing came to a sharp halt in 2007 as the onset of the GFC highlighted systemic failure of financial institutions. The impact on Irish holiday home ownership is hard to assess, though over-production was clearly highest in holiday homes areas.

The recent story of holiday home ownership in Britain is sharply different from Ireland, though it is one of growing socio-economic polarisation and deepening regional divisions, not all-round prosperity. New housing construction fell to levels not seen since the 1920s. House prices initially fell across the whole UK after 2007 but subsequently moved in different directions in different places, and were stagnant in Northern Ireland which had been caught up in the contagion of the Irish property bubble.

The strong demand from affluent English, primarily *southern* English households for holiday homes shows no signs of abating. There has been a shift back to more domestic-based holiday home ownership with wider spread of gentrification and redevelopment of existing leisure use sites, such as caravan parks, rather than additional green field construction. Holiday homes are also used extensively as short-term rental accommodation for tourists and holidaymakers, so any distinction between 'second homes' for family use and 'holiday lets' is completely blurred.

The combined effects of the British planning and tax systems appear to be crucial drivers of some of these recent trends. Planning restricts new development, boosting the price of existing dwellings and preserving the British countryside for royalty, the aristocracy and rich households, including a growing stream of overseas-based country landowners.

Australia

Australian holiday homes have changed from low cost vernacular structures into commoditised forms with high levels of investor activity. Self-built shacks and cottages have given way to commercially driven upmarket housing sold as holiday homes, with growing numbers of commercial developments of large expensive houses, often with marinas or other leisure facilities.

These changes have been ecological, rather than planned, though planning policies have been one factor affecting these changes, especially the demise of 'shacks'. The growth of holiday home ownership in Australia since the 1970s has been much lower than contemporary writers had predicted, and evolved into a hybrid form of dwelling ownership by affluent households for leisure use *and* investment. Like second homes worldwide, they are inter-changeable with other dwellings in terms of use as permanent or non-permanent homes, and they will continue to relate ambiguously to housing and leisure markets.

References

Aalen, F., Whelan, K. and Stout, K (2011) *Atlas of the Irish Rural Landscape*, 2nd edition. Cork: Cork University Press.

Australian Bureau of Statistics (2017) Australian statistical geography standard. Retrieved from www.abs.gov.au/websitedbs/D3310114.nsf/home/Australian+Statistical+Geography+Standard+(ASGS).

Beer, A. and Faulkner, D. with Paris, C. and Clower, T. (2011) *Housing Transitions Through the Life Course: Aspirations, Needs and Policy*. Bristol: Policy Press.

Bell, M. and Brown, D. (2006) Who are the visitors? Characteristics of temporary movers in Australia. *Population, Space and Place*, 12(2), 77–92.

Burnley, I. and Murphy, P. (2004) *Sea Change: Movement from Metropolitan to Arcadian Australia*. Sydney: University of New South Wales Press.

Charles-Edwards, E., Bell, M. and Brown, D. (2008) Where people move and when: temporary population mobility in Australia. *People and Place*, 16(1), 21.

DCLG (2017) English housing survey. Retrieved on 10 November 2017 from www.gov.uk/government/statistical-data-sets/owner-occupiers-recent-first-time-buyers-and-second-homes.

DEFRA (2013) *2011 Rural Urban Classification*. London: DEFRA. Retrieved on 20 November from www.gov.uk/government/statistics/2011-rural-urban-classification

Direct Line Insurance (2005) *Direct Line Second Homes in the UK*. London: Direct Line.

Drudy, P. and Collins, M. (2011) Ireland: from boom to austerity. *Cambridge Journal of Regions, Economy and Society*, 4(3): 339–354.

Forrest, R. and Hirayama, Y. (2014) The financialisation of the social project: Embedded liberalism, neo-liberalism and home ownership. *Urban Studies*, 52(2): 233–244.

French, G. (2008) *Artificial*. Sydney: New Holland Publishers.

Gkartzios, M. and Norris, M. (2011) 'If you build it, they will come': governing property-led rural regeneration in Ireland. *Land Use Policy*, 28(3): 486–494.

Gkartzios, M. and Shucksmith, M. (2015) 'Spatial anarchy' versus 'spatial apartheid': rural housing ironies in Ireland and England. *Town Planning Review*, 86(1): 53–72.

Hall, C. M. (2005) *Tourism: Rethinking the Social Science of Mobility*. Edinburgh: Pearson.

Hall, C. M. (2017) Resilience in tourism: development, theory, and application. In J. Cheer and A. Lew (eds), *Tourism, Resilience and Sustainability*, 18–33. London: Routledge.

Hugo, G., Feist, H. and Tan, G. (2013) *Mobility and Migration: the Australian Experience*. Farnham: Ashgate.

Kitchen R., Gleeson, J., Keaveney, K. and O'Callaghan, C. (2010) *A Haunted Landscape: Housing and Ghost estates in Post-Celtic Tiger Ireland*. NISRA Working Paper 59. Maynooth: NUI Maynooth.

Marsden, B. S. (1969) Holiday homescapes of Queensland. *Australian Geographical Studies*, 7(1), 57–73.

Müller, D. (2011) Second homes in rural areas: reflections on a troubled history. *Norwegian Journal of Geography*, 65(3): 137–143.

Murphy, P. A. (1977) Second homes in New South Wales. *Australian Geographer*, 13(5), 310–316.

Norris, M., Paris, C. and Winston, N. (2010) Second homes within Irish housing booms and busts: North-South comparisons, contrasts and debates. *Environment and Planning C, Government and Policy*, 28: 666–680.

Norris, M. and Winston, N. 2010. Rising second home numbers in Ireland: distribution, drivers and implications. *European Planning Studies*, 17(9), 1303–1322.

ONS (2012) 2011 Census: number of people with second addresses in local authorities in England and Wales, March 2011. *ONS Statistical Bulletin*, 22 October. Retrieved on 10 November from www.ons.gov.uk/peoplepopulationandcommunity/housing/bulletins/2011censusnumberofpeoplewithsecondaddressesinlocalauthoritiesinenglandandwales/2012-10-22#second-addresses-not-in-the-uk.

Oxford English Dictionary (2017) Rural. Retrieved from https://en.oxforddictionaries.com/definition/rural.

Paris, C. (2011) *Affluence, Mobility and Second Home Ownership*. London: Routledge.

Paris, C. (2014) Critical commentary: second homes. *Annals of Leisure Research, 17*(1): 4–9.

Paris, C. (2016). Second home ownership since the global financial crisis in the United Kingdom and Ireland. In Z. Roca (ed.), *Second Home Tourism in Europe: Lifestyle Issues and Policy Responses*, 3–32. London: Routledge.

Paris, C., Beer, A. and Sanders, W. (1993). *Housing Australia*. London: Macmillan Education.

Paris, C. and Thredgold, C. with Jorgensen, B. and Martin, J. (2014) Second homes and changing populations, impacts and implications for local government in South Australia, South Australia LGA. Retrieved from www.lga.sa.gov.au/page.aspx?u=522&t=uList&ulistId=0&c=37472.

Rackham, O (1986) *History of the Countryside*. London: Phoenix Press.

Savills (2007) Second homes abroad 2007. Retrieved on 25 July 2008 from www.savills.co.uk/research/Report.aspx?nodelD=8522.

Shelter (2011) Taking Stock. Retrieved on 20 July 2011 from http://england.shelter.org.uk/__data/assets/pdf_file/0008/346796/Shelter_Policy_Briefing_-_Taking_Stock.pdf.

UN Department of Economic and Social Affairs (2017) Population density and urbanization. Retrieved on 20 November 2017 from https://unstats.un.org/unsd/demographic/sconcerns/densurb/densurb methods.htm#B.

US Census Bureau (2017) Census urban area FAQs. Retrieved on 10 November 2017 from www.census.gov/geo/reference/ua/uafaq.html.

White, R. (2005) *On Holidays*. Melbourne: Pluto Press.

24

Community health planning

Rural responses to change

Sue Kilpatrick, Stuart Auckland and Jessica Woodroffe

Introduction

The challenges facing rural communities have been highlighted increasingly in past decades as rural communities seek to adapt to significant social and economic change (Flora et al., 2015). This chapter introduces community health as a key theme for rural planning. It argues that effective community health planning is a process that should be fundamentally underpinned by consultation and involvement of community members in order to balance private verses public interests in the delivery and allocation of health resources. We explore a number of key elements which underpin how rural communities including individuals, groups and key industries can act as drivers of change in enabling strategic and innovative local responses to improving the 'availability, accessibility and quality of services as a means to improving health status' (Steen, 2008: 1). Focused case studies of health planning by rural Australian communities and partner organisations showcase how a strengths-based approach to community health can assist in planning and driving a variety of health and wellbeing support programmes and strategies.

This chapter discusses health planning in the context of Australian rural communities, which differ from those in many other countries, and discusses key factors informing how rural health needs can be addressed. The National Strategic Framework for Rural and Remote Health uses the term 'rural and remote' to encompass all areas outside Australia's major cities (Department of Health, 2012). The Framework is a response to the 'unique characteristics, needs, strengths and challenges experienced in rural and remote parts of the country' (Australian Health Ministers' Advisory Council Rural Health Standing Committee, 2011: 6).

Background

Rural areas in Australia and around the world have long been known to have poorer health outcomes than metropolitan areas (Beard et al., 2009). Australians living in rural and remote areas tend to die younger, suffer higher levels of injury and disease, and have poorer access to health services (Australian Institute of Health and Welfare, 2017).

An understanding of 'what is health' is useful before discussing rural health planning. Society's understanding of what constitutes good health has evolved over recent decades.

In 1948, the World Health Organization (WHO) defined health as a 'state of complete physical, mental and social wellbeing and not merely the absence of disease or infirmity' (World Health Organization, 1948). In recent years the perceptions of what health is have evolved beyond viewing health as a fixed state, to health and wellbeing as influenced by systems and resources that support an individual's capacity to function in wider society including an ability to adapt to change (Giacaman et al., 2009). Additionally, key WHO directives such as the Ottawa Charter on Health Promotion (1986) have stressed the importance of communities having the opportunity and ability to actively participate in planning for community health and wellbeing which has led to increasing capacity building and engagement of communities around rural health planning (Kilpatrick, 2009; Hawe et al., 1997).

Conceptualising rural health requires an understanding of how the rural setting influences an individual's health outcomes. The value of community to health and wellbeing in rural areas, while difficult to quantify, cannot be underestimated, particularly in communities with high levels of connectivity and social cohesion. It is widely recognised that there are a number of social determinants of health, including socioeconomic status, race and ethnicity, gender, sociocultural and psychosocial factors (Dixon and Welch, 2000). Individual and neighbourhood socioeconomic disadvantage, sociocultural factors, and physical and other environmental factors interact to produce health outcomes, for example low population density, often high proportions of Indigenous populations and geographic isolation, are related to limited health services and limited availability of fresh and healthy food choices. On the other hand, fresh air and natural environmental amenity and close knit communities can promote wellbeing; while limited education and outmigration of young people reduce community income and resilience (Beard et al., 2009). Outmigration, in particular, can be seen to have two major negative effects on health: a reduction in health status as the remaining population is proportionally older, and a reduction in health services as populations fall below the critical mass deemed as needed to support them (Rickards, 2011).

Pressures on current health care services in Australia have increased the demands on policy makers to implement effective health planning. However, planning is often underpinned by the adoption of both biomedical approaches and economic rationalist models of health care provision which have embedded a deficit way of thinking (Bourke et al., 2010). Such models focus on problems that need 'fixing' from an individual health perspective or considers services from an economic rationalist perspective where community members are seen as consumers not citizens. Often a deficit model of rurality is used to justify an economic rationalist approach to rural health policy and planning, most often to the exclusion of rural communities themselves.

This contributes to the stereotyping of rural areas as problematic environments in which to live, work and for which to plan and deliver health services, and tends to downplay provision of preventative health and wellbeing programs and services. Furthermore, the outcome of biomedical and economic rationalist approaches is a tendency to see all rural areas as the same, and to disregard the local contexts and environments of rural settings which can impact significantly on health determinants and the effectiveness of programmes.

Focusing on key strengths is an alternative way to move rural planning in a time of change from 'problem-describing' to 'problem-solving' paradigms. Such an approach starts from a position of local knowledge and values and engages the community in shaping goals and pathways (Rickards, 2011). Many traditional approaches to community health planning start with a needs analysis or some other way of focusing on the community's needs (Kretzmann and McKnight, 1993; Mathie and Cunningham, 2003).

A strengths-based approach tends to build relationships between community members, while deficit-based approaches tended to focus more on the relationship between the organisation and community members.

The next section explores approaches to community health planning and capacity building for health including how rural communities can provide a basis to improve what works, promote positive aspects, challenge stereotypes, attract staff, encourage local responsibility for health and develop further strategies to ensure optimal health and wellbeing (Bourke et al., 2010; White, 2013).

Community health planning: key issues and considerations

Community health planning has been acknowledged as a major field of practice and policy in nations across the world. The ultimate goal of any health planning process is to improve or optimise the health and wellbeing of a community. The World Health Organization (2018: 8) defines community health planning as 'the orderly process of defining community health problems, identifying unmet needs and surveying the resources to meet them' which also involves establishing priority goals that are realistic, feasible and actionable particularly within the context of rural areas where they may be increasing demands for health services with increasingly limited resources.

While health planning can vary across countries and settings, within developed nations, it is argued that 'the challenge' which effective community health planning addresses is making any healthcare system 'more accountable to the average residents in their communities' (Steen, 2008: 1) and thus better balancing commercial interests in healthcare delivery with public interests. Traditionally, health planning has tended to be driven and controlled by government bodies with the aim only to identify priority issues. However, the ways in which information has been gathered through health needs assessments and used by governments has been criticised for not only the quality and validity of community consultation, but the ways in which has been used to align with bureaucratic pressures and priorities rather than those of a particular community or area (Jordan et al., 1998; Hancock and Minkler, 1997).

Health service planning in Australia in recent years has seen a significant policy shift in the way in which rural health services are designed and funded. This shift has been characterised by the emergence of Primary Health Networks and support for strategic and less linear based funding approaches. Primary Health Networks are national, government funded policy implementation and service purchasing agencies. Such approaches challenge the traditional methods of service provision, based on identifying a need, funding a service and expecting a casual outcome. This policy shift is driven by the five outcomes and objectives defined within the National Strategic Framework for Rural and Remote Health (Australian Health Ministers' Advisory Council Rural Health Standing Committee, 2011):

- improved access to appropriate and comprehensive health care for people living in rural and remote Australia;
- effective, appropriate and sustainable health care in rural and remote settings;
- health service design that better meets local consumer and community needs;
- collaborative health service planning and policy development in rural and remote Australia; and
- strong leadership, governance.

While the importance of community health planning is argued internationally (Manaf and Juni, 2017), there is a paucity of information on best practice in health planning for rural areas, particularly in the context of social change, economic and new technologies for health service delivery. Recent developments have seen the emergence of community development approaches that work at the grassroots, engage and empower local residents encourage participation in processes that build on strengths and address weaknesses (Johns et al., 2007).

White (2013) examined key issues and best practice for rural health population planning and delivery within published literature and found that there were six key elements that should be considered by those responsible for the development of policies and strategies in rural communities. Many of these areas are concerned with identifying and understanding the local contexts of health including defining rural communities, identifying local assets and challenges, multiple levels of community support and understanding social determinants of health which considers a 'full range of factors that influence and contribute' (ibid.: 34) to a community's health and wellbeing. These findings are consistent with our own research and practice which has identified that viewing communities themselves as authentic partners is essential in any process of community health planning (Auckland et al., 2007; Johns et al., 2007; Kilpatrick and Auckland, 2009; Whelan et al., 2009). We have identified the following five attributes as key components of evidence based community health planning:

- an integrated and holistic approach to community health based on social determinants of health;
- a strengths-based approach which builds capacity within communities to address community priorities;
- partnerships and coalitions based on engagement and shared aims and visions;
- a strong commitment to community consultation and engagement, through strategic approaches such as community health needs assessments or health improvement projects; and
- an understanding of research and evaluation, and the collection of valid information from a variety of areas which best provide informed views on community priorities.

Planning in practice: case studies from rural Australia

This section provides three focused case studies developed by rural Australian communities and partner organisations which showcase how a strengths-based approach to community health can assist in planning and driving a variety of health and wellbeing support programmes and strategies.

Case study 1: planning for the health and wellbeing of fishing families

The Australian wild-catch fishing industry, through its producer-led state-based associations, identified a need to address industry economic losses incurred through poor health and wellbeing of fishers' who are dispersed along rural coastal areas of Australia (King et al., 2015). This economic motivation contrasts with traditional social, often equity focused, motivation for planning Australian rural health services. Three industry funded research projects saw researchers working with fishing communities and associations to develop programmes and tools to improve fisher health (Fisheries Research and Development Corporation, 2017; Kilpatrick et al., 2012; King et al., 2015).

Alongside the disease and health risk factors experienced by Australian rural populations, fishers are at particular risk of certain kinds of illnesses (for example, skin and diet-related), as well as injury (fatality rates are more than double those in the agricultural sector). While both women and men are at risk, 87 per cent of fishers are male, a factor placing them at greater risk of suicide. Insecurity of fishing quotas and licences drives chronic livelihood insecurity, resulting in reports of stress, depression and suicide (King et al., 2015).

The first project worked with both farming and fishing communities. It found farming industry associations, unlike their fishing counterparts, were proactive in working with local and national health services to develop health and wellbeing programmes that matched farmers' needs. Using credibility of industry events as 'soft entry points' to health checks, mental wellbeing and other programmes was key to farmer uptake (Kilpatrick et al., 2012). In the second project, three fishing communities worked with the researchers and local health services to develop health and wellbeing plans and health toolkits of useful information and contacts for each of the fishing communities, and prioritise soft entry points (King et al., 2015). This project identified women as key players in the fishing industry (Kilpatrick et al., 2015). Their knowledge of, and credibility within, fishing businesses makes them valuable sources of information about health issues facing the industry and effective strategies to address them. A key finding was women's expertise should be applied in conjunction with industry associations and health providers to achieve better health outcomes (Kilpatrick et al., 2015). The final project adapted the Sustainable Farm Families programme for the fishing industry and collected detailed self-reported health data from 872 commercial fishers (over 20% response rate), to be used in planning health services.

The fishing industry example highlights the power of planning for rural health using a geo-occupational lens to engage community and better target not only evidence-based needs of various groups within rural populations, but also take account of gender and behaviour norms and preferences for service access. Embedding responsibility for health in industry organisations fosters sustainability.

Case study 2: planning for healthy and resilient communities (HaRC)

The Centre for Rural Health in Tasmania has collaborated with Rural Alive and Well Inc. (RAW) to identify community based service interventions that enhance community resilience and build the capacity of rural Tasmanian communities to react to challenging life experiences as individuals, families and whole communities. Rural Alive and Well Inc. is a not-for-profit organisation that has a primary goal of helping individuals, families and the community through mental health issues with a focus on suicide prevention. The underpinning rationale for RAW's grassroots approach is the acknowledgement that the relationship between people, place and health is fundamental to the success of health interventions within rural and remote communities.

Rural communities are often characterised, as being resilient in the way in which they draw on their underlying social capital to deal with life challenging events, however, in areas that experience greater social isolation, the social capital may not be adequate to cope with the adversity. The design and delivery of the HaRC programme focuses on the concepts of community strength, preparedness and resilience within the context of enhancing both individual and collective wellbeing within those communities. The programme is underpinned by an understanding that rural people construct 'community' both as people relating to each other in a shared locality, and the extent to which people relate to each other in cooperative, sharing, supportive, caring, and trusting ways.

HaRC programme staff focused on firstly understanding the community's level of readiness in the design of engagement processes and interventions. Through the adoption of different measures across multiple dimensions including a sense of community, desire for the programme, capacity to manage and develop the programme, commitment to programme goals and outcomes as well as a commitment to collaborate with programme staff to achieve programme goals, the programme staff were able to match community interventions to the community's level of readiness.

Pivotal to this approach has been the programme's ability to recruit 'field workers' considered to have broad skillsets described as a 'specialist-generalist'. Programme staff are flexible and responsive to community needs, which included them being a single point of contact, providing care coordination, assisting the community to become self-reliant, while clearly outlining their role with clients and maintaining boundaries.

The goal of the HaRC programme is the development of twenty 'community owned' suicide prevention strategies over a three-year period commencing 2016. The evaluation identified pre and post programme measures to help assess the programme's performance using indicators of the effective implementation of rural community health programmes. Findings from the evaluation have been used to refine programming and planning, including the assessment of new sites and delivery of services.

Case study 3: planning for whole of municipality/community health and wellbeing

Within Australia and elsewhere, health planning including strategic responses to whole of municipality (local government area) health and wellbeing issues, including service provision, infrastructure and resource allocation, has increasingly moved from being the remit of *only* federal or state governments and their administration to other levels of government such as municipal local governments or regional authorities.

Increasingly health and wellbeing plans for regions, municipalities and communities underpinned by community consultation have become accepted practice in Australia - particularly in rural areas. This case study focuses on one municipal area within the state of Tasmania to demonstrate how effective health and wellbeing planning was undertaken in partnership with the University of Tasmania and other stakeholders.

The West Tamar municipality has a population of approximately twenty thousand people spread across 690 square kilometres and includes both rural and regional townships and communities, some with significant socio-economic disadvantage. Over 18 months, the municipal government worked with the local university to conduct a whole of community health needs assessment which would inform a number of strategic areas for the authority including sports and recreation planning, youth and positive ageing strategies, infrastructure allocation and cross sectoral collaborations and partnerships.

Led by a steering group of municipal staff, university researchers, community representatives and other local stakeholders including industry, business and health professionals, the West Tamar Community Health and Wellbeing Project consulted with more than five thousand residents of the municipality to identify, understand and prioritise the health and wellbeing issues facing the whole municipality at that time and for the next three to five years. This was achieved using a variety of information collecting methods including community forums, focus groups, interviews with stakeholders within and outside the community, a whole of community survey, youth surveys and analysis of existing policies, documents and other relevant information which could inform planning for the future and for expected changes in population such as ageing and new housing development.

Informed by a social determinants of health model, the project took an integrated approach to sustainable planning for the health and wellbeing of the West Tamar. It considered six key areas through which health and wellbeing planning for the municipality would be guided and prioritised including (a) Services and Resources (b) Partnerships and Collaborations (c) Information,

Facilitation and Support (d) Inclusion, Engagement and Participation and (e) Infrastructure, Environment and Safety.

The project also incorporated a strong capacity building element to how it was conducted, in that university researchers worked with community members and municipal staff to develop understanding, skills and knowledge in the areas of research and collection of information. Community members were immediately involved where determined to be effective; for example, in the distribution of surveys to other members of the community and in the validation of findings from other data collection methods such as community forums. Community participation was based on the project team's strong view that in any community only those who live and inhabit community spaces can truly understand the experiences, priorities, traditions, needs, issues, barriers and enablers that exist there (Whelan et al., 2009; Langwell, 2009).

Key messages from case studies

The three case studies illustrate how a strengths-based approach works with community and health services to identify assets and community characteristics which can positively contribute to health and wellbeing outcomes. Place is a key resource contributing to community strength and resilience, even in the context of a rural industry community such as in the fishing case study.

Both the West Tamar and HaRC case studies drew heavily on the importance of adopting an integrated and holistic approach to community health based on social determinants of health, particularly the value of social connectivity and support systems in HaRC. Community engagement and evidence gathering was evident in the fishing case study where the industry associations facilitated the engagement of a significant proportion of wild catch fishers nationally in a detailed self-reporting of health status. In the HaRC study, data was collected about community readiness though engaging with the community. In the West Tamar, more than six months was dedicated to community consultation and the gathering of evidence to inform the community health plan including the allocation of resources into the future.

A partnership between industry and researchers over three projects built evidence to refine health planning for fishers. Significantly, the HaRC project participants identified the need for stronger collaborations and partnerships both between service providers and between community and service providers. In West Tamar, the municipal local government understood that they alone could not assess capacity to meet or successfully address or the health and wellbeing priorities and needs of their municipality, and that intersectoral and interagency partnerships, including those with health providers and industry were essential to the sustainability of the plan.

Understanding community health and wellbeing needs, assets, challenges and priorities was evident in the fishing and HaRC case studies, but was at the heart of the West Tamar case study. Regardless of context, the community should be a vested partner in the process of health planning and if done effectively can not only reflect public interest and experience but can direct agencies to areas where rural health and wellbeing issues can best be addressed, improved or approached. Understanding rural place and health needs facilitates alignment between health programmes and community, and helps incorporate community resources into health care (Kilpatrick, 2009). Rural communities, health services and other community organisations need skills in working together to develop effective partnerships that transfer some power from health systems.

Monitoring and evaluation must be embedded in planning. In the HaRC programme, research indicated that there needed to be more structuring, monitoring and evaluation and as part of programme planning. Evaluation and review was also built into the West Tamar project, and is an integral part of any cycle of health planning or strategic response.

Implications for rural health planning

This chapter moves our understanding of rural health planning from a desktop and external activity towards an activity which includes the characteristics which underpin how healthy communities can be created, strengthened and understood. We outline a number of elements which underpin our work with rural communities, particularly taking a strengths-based community development approach that embraces the importance of the knowledge of communities themselves and builds partnerships to achieve an integrated and holistic approach to community health based on social determinants of health.

Strength–based, community health development approaches to rural health planning build relationships between community members and between community and health services. This planning approach provides a basis to improve service delivery for the local context, promote positive aspects, challenge stereotypes, help attract and retain staff, encourage local responsibility for health and develop further strategies to ensure optimal health and wellbeing. A strengths-based approach is effective because it takes account of local rural context, including geo-spatial, occupational and other demographic contexts. The five key attributes listed above are essential for successful and effective health and wellbeing service planning, and crucial for planning for healthy communities.

References

Auckland, S., Whelan, J., Barrett, A. and Skellern, K. (2007) Mapping community health needs and priorities: Reflections on community engagement from the Tasmanian University Department of Rural Health and the Meander Valley Community, *Australian Journal of University Community Engagement* 2(1), 237–245.

Australian Health Ministers' Advisory Council Rural Health Standing Committee (2011) *National Strategic Framework for Rural and Remote Health*, Canberra: Commonwealth of Australia, retrieved 5 December 2017 from www.health.gov.au/internet/main/publishing.nsf/Content/national-strategic-framework-rural-remote-health.

Australian Institute of Health and Welfare (2017) *Rural and Remote Health*, retrieved 11 September 2017 from www.aihw.gov.au/reports/rural-health/rural-remote-health/contents/rural-health.

Beard, J., Tomaska, N., Earnest, A., Summerhayes, R. and Morgan, G. (2009) Influence of socioeconomic and cultural factors on rural health, *Australian Journal of Rural Health* 17(1), 10–15.

Bourke, L., Humphreys, J., Wakerman, J. and Taylor, J. (2010) From 'problem-describing' to 'problem-solving': challenging the 'deficit' view of remote and rural health, *Australian Journal of Rural Health* 18, 205–209.

Department of Health (2012) *National Strategic Framework for Rural and Remote Health*. Canberra: Commonwealth of Australia.

Dixon, J. and Welch, N. (2000) Researching the rural–metropolitan health differential using the 'social determinants of health', *Australian Journal of Rural Health* 8(5), 254–260.

Fisheries Research and Development Corporation (2017) Social sciences economics research coordination program: sustainable fishing families: developing industry human capital through health, wellbeing, safety and resilience, retrieved 5 December 2017 from http://frdc.com.au/project?id=2960

Flora, C. Flora, J. and Gasteyer, S. (2015) *Rural communities: Legacy+ change*, fifth edition, Boulder, CO: Westview Press.

Giacaman, R., Khatib, R., Shabaneh, L., Ramlawi, S., Sabri, B., Sabatinelli, G., Khawaja, M. and Laurance, T. (2009) What is health? The ability to adapt, *The Lancet* 373(9666), 781–866.

Hancock, T. and Minkler, M. (1997) Community health needs assessment of healthy community assessment? Whose community? Whose health? Whose assessment? In M. Minkler (ed.), *Community Organising and Community Building for Health*. New York: Rutgers University Press, 148–152.

Hawe, P., Noort, M., King, I. and Jordens, C. (1997) Multiplying health gains: The critical role of capacity building within health promotion programs, *Health Policy* 339, 29–42.

Johns, S., Kilpatrick, S. and Whelan, J. (2007) Our health in our hands: Building effective community partnerships for rural health service provision, *Rural Society* 17(1), 50–65.

Jordan, J., Dowswell, T., Harrisson, S., Lilford, R. and Mort, M. (1998) Health needs assessment: Whose priorities? Listening to users and the public, *BMJ* 316, 1668–1670.

Kilpatrick, S. (2009) Multi-level rural community engagement in health, *Australian Journal of Rural Health* 17(1), 39–44.

Kilpatrick, S. and Auckland, S. (2009) Community participation in a socially fragmented region: understanding social capital to support a primary health care approach. In G. Gregory (ed.) *Proceedings of the 10th National Rural Health Conference*, Cairns, Qld, 17–20 May, Canberra: National Rural Health Alliance, retrieved 9 January 2018 from http://10thnrhc.ruralhealth.org.au/papers/docs/Kilpatrick_Sue_D2.pdf.

Kilpatrick, S., King, T. and Willis, K. (2015) Not just a fisherman's wife: Women's contribution to health and wellbeing in commercial fishing, *Australian Journal of Rural Health* 23, 62–66.

Kilpatrick, S., Willis, K., Johns, S. and Peek, K. (2012) Supporting farmer and fisher health and wellbeing in 'difficult times': communities of place and industry associations, *Rural Society* 22(1), 31–44.

King, T., Kilpatrick, S., Willis, K. and Speldewinde, C. (2015) A different kettle of fish: Mental health strategies for Australian fishers, and farmers, *Marine Policy* 10(60), 134–140.

Kretzmann, J. and McKnight, J. (1993) *Building Communities from the Inside Out. A Path toward Finding and Mobilizing a Community's Assets*, Evanston, IL: Center for Urban Affairs and Policy Research.

Langwell, K. (2009) *Conducting Community Health Needs Assessments: Objectives and Methodologies and Uses*, Washington, DC: United States Department of Health and Human Services.

Manaf, R. and Juni, M. (2017) Analysis of health planning theories- a systematic approach, *International Journal of Public Health and Clinical Sciences* 4(3), 14–22.

Mathie, A., and Cunningham, G. (2003) From clients to citizens: Asset-based community development as a strategy for community-driven development, *Development in Practice*, 13(5), 474–486.

Rickards, L. (2011) Rural health: Problems, prevention and positive outcomes, retrieved 9 January from www.futureleaders.com.au/book_chapters/Health/Lauren_Rickards.php.

Steen, J. (2008) Community health planning, retrieved 9 January 2018 from www.ahpanet.org/Community_Health_Planning_09.pdf.

Whelan, J., Spencer, J. and Dalton, L. (2009) Building rural health care teams through interprofessional simulation-based education. In G. Gregory (ed.), *Proceedings of the 10th National Rural Health Conference, Cairns Queensland, 17–20 May* 2009, 1–10. Canberra: National Rural Health Alliance.

White, D. (2013) Development of a rural health framework: implications for program service planning and delivery, *Healthcare Policy* 8 (3), 27–41a.

World Health Organization (1948) What is the WHO definition of health? Retrieved 9 January 2018 from www.who.int/suggestions/faq/en.

World Health Organization (2018) Health Systems Strengthening Glossary. Retrieved 9 January 2018 from www.who.int/healthsystems/hss_glossary/en/index6.html

25

Mobilities, accessibility and social justice

Jesús Oliva and Luís Camarero

Introduction

There is a long tradition in rural planning of examining the transport and mobility challenges arising for rural populations due to infrastructure sparsity. Accessibility to opportunities and services is one of the fundamental premises of modern societies, but rural areas usually show severe disadvantages, and this very often becomes a defining trait. Daily commuting to work, school, health facilities or shopping is a way to rebalance these inequalities, and is one of the challenges that planning has tried to respond to first.

Mobility is directly linked to rural welfare. The trend of concentration of services, the difficulties in developing an efficient public transport system, and the differentiation of modern lifestyles have made the private automobile an essential resource for rural residents. This raises issues with regard to social inclusion. Mobility deprivation, resulting from the lack of these resources or skills, is a factor of rural poverty that particularly affects the most vulnerable groups, such as the elderly, children and women. Moreover, the policies that combat these inequalities also have to face a further disadvantage resulting from rural socio-demographic imbalances, such as ageing and population loss experienced in many rural regions.

In this chapter we first analyse the debates on inequality in rural accessibility and mobility, and the evolution of approaches towards a more complex theorisation of the problem beyond the traditional focus on transport policies. Then, we explore the importance of accessibility in relation to social justice and exclusion. The third section explores the paradoxical impact of private automobility as solution to rural transport deficiencies, because its prevalence involves new exclusions and dependencies. Finally, three main processes of rural change that represent key challenges and opportunities for rural planning in the mid-term are evaluated: population ageing, new communication and information technologies, and the transition towards a new mobility paradigm.

Accessibility and mobility deprivation: more than a transport problem

Inequality in rural areas, in terms of accessibility to services and opportunities, is the result of an accumulation of factors and processes. In the first instance, due to geographical constraints, disadvantage is concentrated in remote regions; for example, mountainous areas and areas disconnected from the urban nuclei where resources and facilities are focused. Moreover, rural areas usually have more dispersed and less dense populations, which means a limited number of employment opportunities, while public and private services are also more limited and have witnessed a decline in recent decades. As such, from the premise of a right to equal social conditions, the very terrain produces disadvantages of accessibility suggesting that mobility becomes essential to overcome these territorial disadvantages.

In urban areas this discriminatory geography (Soja, 2010) is eliminated by means of infrastructure and by collective transport networks that support the right to mobility. However, in rural areas mobility becomes a more complex problem. Two forms of mobility may be defined. Firstly, direct mobility whereby people travel to obtain resources, and secondly, inverse mobility, when the merchandise and services move towards the consumers. In urban planning the use of the right to mobility criterion tends to focus on a provision of mobility with reduced costs at fairly equal distances in accessing public transport networks. The key to this is a public transport network, which guarantees social cohesion. Nevertheless, in rural areas not only is there a greater demand for mobility due to the greater distances between places, but also because there is a significant lack of public transport. Thirdly, rural regions possess much weaker demographic structures that are characterised by more intense ageing than in urban areas. Thus, rural areas can be isolated and vulnerable social environments, but also desperately in need of services such as healthcare. Finally, we are dealing with a world supported to a large extent by private automobility (Urry, 2004). This dependency on the car allows people to confront and redress some of the aforementioned disadvantages; for example, accessing outside training and employment, but which also bring about new inequalities.

A first generation of policies of rural modernisation had as their objective to incorporate rural regions into national economies by creating critical new infrastructures for mobility (i.e. roads, trains and bus routes). We might say that they form part of the first modernisation, or 'hardware' for development as defined by Bauman (2000). However, since the middle of the last century, investment in rural infrastructure provision has declined as a political priority in the face of, migration from the country to the city, the progressive erosion of rural services, and the transfer of responsibilities from the state to the voluntary and community sector. In some countries, such as the United Kingdom, the problem of accessibility was theorised and integrated by researchers into a spatial planning agenda. As Clout (1972) suggested, we are dealing with a problem linked to the progressive spread of the automobile, the decline of public transport and their impact on disadvantaged groups. The benchmark definition has been conceptualised by Moseley (1979) after an extensive review of the different dimensions of the 'rural accessibility problem' and the policies aimed at their resolution.

The subsequent criticism of the deregulation of public services carried out since the mid 1980s under the neoliberal model raises further challenges relating to inequality of accessibility. On one hand, and as stated by the European Commission (2008, 2011), a lack of access and mobility are determining factors in 'absolute poverty'. Rural areas better connected to urban centres have experienced significant growth, due to counterurbanisation, relocation of production and holiday homes. In contrast, those less accessible regions show clear disadvantages 'in terms of easiness to access to all those services and activities which represents common facilities for people living in urban centres (such as schools, hospitals, sports and cultural facilities)' (European Commission, 2008: 5). However, this problem is also considered as a determining factor in regional development: as the OECD proposes, 'the physical accessibility of rural areas is a key objective for both equity and efficiency reasons' (OECD, 2006: 63).

The formulae employed within planning to address rural accessibility have been based on different perspectives. Pacione (1995), for example, distinguishes between those that deal with locational accessibility, and those that understand the issue in terms of personal accessibility. The former orientation defines accessibility in terms of the distance from the services to potential population that may or may not reap their benefits. From this perspective, the issue of accessibility, and implicitly its relationship with social inclusion or inequality, is linked to a problem of transport and physical accessibility. Solutions are, therefore, sought to enhance transport for the profile of an average citizen. This makes other dimensions of exclusion and sociological processes invisible (Yago, 1983). In the latter perspective, 'the challenge for rural accessibility planning is the first to determine the dimensions of the action-space of significant population subgroups' (Pacione, 1995: 286).

Planning interventions to enhance mobility have often estimated the population as being homogeneous, without considering the characteristics of the different groups or the responsibilities assigned to each of them by their different social roles. By applying the principle of neutrality as a concept of social justice, the complex inequalities that accessibility entails are hidden (Cullinane and Stokes, 1998) as is the case with mobility (Camarero et al., 2016). For example, those deprived of mobility and suffering poverty of access are repeatedly identified in rural studies as the poor, the elderly, housewives, children and young people (Clout, 1972; Moseley, 1979; Gray et al., 2006; Velaga et al., 2012; Shergold and Parkhurst, 2012; Ahern and Hine, 2012).

Spatial inequality in some rural regions further compounds other social, economic and demographic inequalities, for example, ageing, poverty, unskilled groups and resources for automobility. In this way, together with the aforementioned, we can also consider other contributions from locality-based studies – for example, early studies from a rural restructuring perspective identify the key characteristics of rural accessibility (Cloke, 1984; Lowe et al., 1986; Bell and Cloke, 1991; Cloke and Little, 1990). In this context, other experiences and local initiatives have proposed alternatives for gaining mobility and accessibility of specific rural groups. For example, in the context of deregulation of transport services in England, Nutley (1988) identifies a key role for unconventional public and private provision of transport, which are based on flexible routeing and timing, multipurpose operations and non-profit operation by community groups (social car, community bus, shared taxi, flexibus, hospital car).

The necessity to more fully consider all these relationships that combine in accessibility and mobility led some scholars to explore them from other perspectives. Kaufmann et al. (2004) conceptualise the idea of 'motility' as a combination of resources, personal skills and potential for mobility; it analyses them as a social capital that can increase or exchange itself for other types of capitals such as economic or political capital. As summarised by Canzler et al. (2008: 3), motility is 'the capacity of an actor to move socially and spatially'. Further to this, the relationships between exclusion, social networks and mobility have been further studied by Urry (2002) and Cass et al. (2005), examining how the flexibility that the car provides becomes a coercive power that normalises a general standard which only private automobility provides, and leaves the excluded with more vulnerable and precarious accessibility.

Mobility deprivation, social justice and geographies of exclusion

As a result of this debate, policies for addressing rural accessibility have shifted towards more holistic perspectives, whereby a focus on social inclusion is a prioritised (Farrington and Farrington, 2005). This treatment of inequality and accessibility represents an evolution. It changes from considering the mobility-deprived as being focused exclusively on transport to conceiving multiple deprivation, which certain groups experience, while also considering issues such as the role of social networks and social participation.

Mobility is not the only source of accessibility, exclusion or social inequality, but it is very strategically related to all those that requires special attention. Moreover, as Shergold and Parkhurst (2012: 413) point out:

> the relationship between mobility and exclusion is not static; accessibility is a dynamic phenomenon. As society becomes more mobile, and those that provide the 'opportunities' respond to this, so there is a greater risk of exclusion for those who are less mobile.
>
> *(Shergold and Parkhurst, 2012: 413)*

For example, Urry (2002) suggests that the greater the commute to places of work, residence and services, then the greater the continual mobility needed to reach them; this progressively erodes the local community links, and thus there is greater dependence on the private car. However, Gray et al. (2006) argue, in remote rural regions, this local relational capital is vital for groups that do not have cars and only attain accessibility through drivers in their networks of family members, neighbours and friends.

This focus on social cohesion incorporates the right to mobility as a fundamental pillar. The spatial movement highlights the effect that the exclusion has on the reproduction of social inequalities, and there is greater agreement that mobility belongs to universal rights and freedoms (Cresswell, 2006). This includes both the capacity of choice of places and ability to travel, and the very possibility of renouncing the movement (Sager, 2006). Canzler et al. (2008) further propose that mobility can be considered a principle of modernity, similar to equality or individuality. Moreover, only in modern society do we find a connotation between mobility and social change: 'this idea of using spatial movement as a "vehicle" or an instrument for the transformation of social situations and in the end to realise certain projects and plans' (ibid.: 6).

The right to mobility has been forged, from an urban perspective, as freedom to travel through the city and to access places and opportunities (Verlinghieri and Venturini, 2017). This right maintains an evident relationship with the right to the city theorised by Lefebvre (1967), which is defined as a place of civic production through the formulation of public spaces. Lefebvre claims that the city is produced through the attainment of public spaces subtracted from the logic of private accumulation (ibid.). In this commitment to the right to city life, mobility is an absolutely necessary condition; it is an instrument for the production of democracy.

In rural areas, the spaces of citizens, as well as the rules for their exclusion, are constructed and function differently. Bell and Osti (2010) emphasise that they are characterised by the persistence of place, by the importance that the community and local orientation has to the consumption, cultural and economic activities, such as conditions for attaining welfare. In rural areas there is a tacit recognition of belonging, accessibility and of conforming with spatial identities. However, at the same time, the geography feeds and reproduces unequal regional development. In this sense, the forms of citizens' exclusion become defined not by the habits of access to local space but by external forms of socio-territorial segregation.

From a global perspective, rural accessibility must be considered in the context of its power to configure, together with other processes (i.e. demographic patterns, new geographies of inclusion and exclusion). The effect of the erosion of the public services and the ideal of universal infrastructures, which Graham and Marvin (2001) explored in the city, may be extended to the analysis of territorial policies in general, and particularly to those that favour a kind of self-sufficient rurality. Considering their joint effects, all these processes function as true socio-territorial engineering that produce regions with privileged connections, remote peripheries and interstitial spaces. Moreover, without policies orientated at combating the exclusion and promoting regional development, an unprecedented rural exodus will be brought about (McDonagh, 2006: 33). It is necessary to bear in mind the potential impact that these dynamics, in order to understand the enormous social awareness that accessibility awakens and the resistance rural populations demonstrate when it is restricted. As Farmer et al. (2011: 136) explain, when local residents protest because they lose a nurse, 'they are protesting about losing a complex range of assets: an accessible health-preserving service, human, social and economic resources and a symbol of community vitality'.

Automobility, gender, age and rural sustainability

The sustainability of rural areas is heavily dependent on the private automobile. Daily journeys define many of the deep-rooted strategies regarding work, upbringing and caring, as well as general access to services (Oliva, 2010). This generalisation has allowed for the maintenance and even the repopulation of certain areas surrounding the cities with new residents. This, however, also has become the principal factor in the erosion of public transport, while car dependence creates new forms of exclusion and social inequality. The progressive diffusion of the automobile, from its inception as a family resource to its current individualised use (Gartman, 2004), clearly indicates its impact on social contemporary life. To understand its importance in rural areas we must also consider the progressive contemporary deroutinisation of times and personal spaces characterised by the postmodern society (Urry, 2004).

Studies in various national contexts have detected that the ratio of cars per household is greater in small localities than in bigger ones (Cullinane and Stokes, 1998). As Nutley (1996) explains, in countries where universal car use exists, such as the United States, rural transport is not even considered to be a problem. Nutley and Thomas describe the role of automobility in rural areas, which is illustrated by the fact that even those groups that do not have that resource, because they lack the skills or aptitudes to drive or do not own a vehicle, also organise their mobility and accessibility strategies 'accommodated not by the bus service but by lifts in other people's cars' (Nutley and Thomas: 1995: 35). Thus, social networks substitute public services.

Nevertheless, the personal car exerts contradictory effects. On one hand, the distinction between car ownership and car availability has been proposed in the context of unequal accessibility for certain groups, for example, housewives. But, on the other hand, access to the automobile may reinforce gender roles at the same time as increasing women's mobility and accessibility (Noack, 2010; Camarero et al., 2016). As Noack (2010: 90) summarises, 'private cars might just make it easier to reach shopping facilities and result in more chauffeuring of children' and mobility takes the form of a succession of tasks more orientated towards caring for others.

With regard to the effects that spatial inequities have exerted, we must also incorporate the demographic composition of rural areas within accessibility debates. Each socio-demographic group has specific mobility requirements and unequal positions of access to them. The socio-demographic structure of places is additional to the territorial determinants in accessibility. The urban versus rural differences with respect to mobility are determining factors in accessibility, depending on the socio-demographic condition. Urban public transport minimises the differences among the old, young and those dependent on it, and makes mobility universal. Conversely, private automobility, which dominates rural mobility, increases and amplifies those differences.

In the Spanish case, Camarero et al. (2016) explore the role of the 'support generation', the cohort between thirty and fifty years of age for rural sustainability. They describe this generation as:

a group responsible not only for the maintenance of productive activities, but also for the provision of care and assistance to dependent groups (children and the elderly), especially in those geographic areas undergoing demographic decline and with problems of accessibility.

(Camarero et al., 2016: 734)

This is a process that emphasises again the importance of social networks in maintaining welfare and quality of rural lives, as well as promoting inclusion and social participation. In some rural areas, therefore, long-distance commuters guarantee a locality's social sustainability, but at a personal cost. As an example, Milbourne and Kitchen (2014), in their case study of Wales, came up with the hypermobility concept to describe the continued, forced and almost strenuous character of movements that characterise the daily life of some rural inhabitants.

In these social environments, and with the lack of public transport, resorting to family members and the community is vital. These limitations condition and oblige the mobility of other generational groups, transforming the composition of households and the relationship among generations into determining factors in the possibilities of mobility. Certain groups demonstrate demands and determinants of a distinctive mobility. This is the case for young people living in sparsely populated rural areas, and who often require a high degree of mobility between leisure venues. In this context, proposals for shared transport have been developed in accordance with demand. But sudden immobility produces pockets of marginality for vulnerable groups, such as elders, children, immigrants and those dependent on the mobility of others. Furthermore, and this should be stressed, it impedes and slows down the arrival of newcomers. Neglecting mobility policies contributes to depopulation and to rural decline.

Conclusion: demographic shift, mobility transition and rural planning

Because of the shifting rural population, technological innovations and means of transport, the rural accessibility system is always changing. As Moseley stated, it 'is in a state of flux' (Moseley, 1979: 28). Nowadays, planning for accessibility must confront the challenges and opportunities that are derived from the convergence of three great processes of transformation: demographic change, the communication and information technology revolution and the transition towards another mobility. With respect to the first of the three, new technologies are changing the very configuration of accessibility to services and resources. In the same way that accessibility does not reduce transport, not all of these solutions necessarily lie in mobility. In this sense, the technological innovations may bring about a reformulation of times and spaces of rural accessibility for its residents. In fact, together with the car, they have shattered the patterns and routines as much for modern societies as they have for rural residents, giving rise to some lifestyles organised around the continuous planning of meeting and activities.

The characteristics of rural territory, such as remoteness, low density and habitat dispersion, necessitate the provision of particular home services. This type of inverse mobility, in which merchandise and services travel to people, can play an essential function in rural sustainability. Thus, an important role in the condition of a rural citizen is produced by the capacity of reception of services; for example, mail, medical visits and travelling vendors. New technological applications in fields such as education, administrative procedures, medical diagnoses, 3D printing, and delivery using drones, herald huge possibilities to avoid unnecessary travelling.

Secondly, the transition towards another mobility, such as autonomy, shared, low-demand and post-carbon, is one of the processes that has received most attention; this is as much for the different levels of public administration as for the social agents, and may acquire multiple functionality for the rural environment. These opportunities should be adequately governed, incorporating the right to mobility as a pillar of social justice and adapting them to avoid other forms of exclusion and inequality with regard to accessibility. For example, groups of immigrants tend to have more limited and more dispersed social support networks, as well as greater requirements for public attention. For them, access to mobility is crucial, but the elevated cost of its acquisition presents barriers to social inclusion.

Finally, the main medium term challenge for rural areas may be rooted in the third of these processes related to demographic change. In Europe, the rural accessibility issue is deeply affected by the increasingly ageing population, and this process is expected to intensify. As the European Commission explains, rural poverty in many areas is related to ageing, since these social groups usually have lower incomes and lower levels of education, as well as specific physical conditions and problems of accessibility:

> for 'rural transport poor', a group which includes many elderly who cannot drive, do not have access to their own transport or who are badly served by limited public transport facilities, access to basic healthcare services can be extremely difficult. This is exacerbated by the decline of private services (e.g. shops, banks and pharmacies) as well as the centralisation of services.
>
> *(European Commission, 2008: 32)*

The extreme ageing that characterises many rural areas produces a social space in terms of mobility that is very different to that of urban areas. In the first case, the elderly population has more restricted automobility, as they own fewer vehicles and hold a lower percentage of driving licences. Often older people are limited to shorter journeys to familiar places, often during daylight hours and when there are good weather conditions. However, ageing also affects demand-side requirements and itineraries of mobility due to the relocation of services, such as health centres and chemist shops; this imposes obligatory mobility.

Demographic ageing impacts generally upon sustainability and the social organisation of the rural environment in a process of accumulative coincidences. Firstly, it significantly affects the economic and social system, given their importance as a highly immobilised group, and at the same time they require health and welfare services. Secondly, the impact of this tendency is immediate since the rationalisation of services to a large extent in transferred to their own families, specifically women, and the 'support generation'. Public policies, social innovation and solidarity projects, often emerging from voluntary organisations offer the potential for mobility services for others or promote cooperative residential projects that share resources, will play a decisive role in this transformation.

References

Ahern, A. and Hine, J. (2012) Rural transport: valuing the mobility of older people. *Research in Transportation Economics*, 34: 27–34.

Bauman, Z. (2000) Time and space reunited. *Time & Society*, 9(2–3): 171–185.

Bell, P. and Cloke, P. (1991) Deregulation and rural bus services: a study in rural Wales. *Environment and Planning A*, 23(1): 107–126.

Bell. M. and Osti, G. (2010) Mobilities and ruralities: an introduction. *Sociologia Ruralis*, 50(3): 199–204.

Camarero, L., Cruz, F. and Oliva, J. (2016) Rural sustainability, inter-generational support and mobility. *European Urban and Regional Studies*, 23(4): 734–749.

Canzler, W., Kaufmann, V. and Kesselring, S. (eds) (2008) *Tracing Mobilities: Towards a Cosmopolitan Perspective*. London: Routledge.

Cass, N., Shove, E. and Urry, J., (2005) Social exclusion, mobility and access. *The Sociological Review*, 53: 539–555.

Cloke, P. (1984) *Wheels within Wales: Rural Transport and Accessibility Issues in the Principality*. Lampeter: Centre for Rural Transport.

Cloke, P. and Little, J. (1990) *The Rural State? Limits to Planning in Rural Society*. Oxford: Oxford University Press.

Clout, H. (1972) *Rural Geography: An Introduction Survey*. Oxford: Pergamon.

Cresswell, T. (2006) The right to mobility: the production of mobility in the courtroom. *Antipode*, 38(4): 735–754.

Cullinane, S. and Stokes, G. (1998) *Rural Transport Policy*. Oxford: Pergamon Press.

European Commission (2008) *Poverty and Social Exclusion in Rural Areas. Final Study Report*. Brussels: European Commission.

European Commission (2011) *Poverty in Rural Areas of the EU*. EU Agricultural Briefs, 1 May. Brussels: European Commission.

Farmer, J., Nimeggeer, A., Farrington, J. and Rodger, G. (2011) Rural citizens' rights to accessible health services: an exploration. *Sociologia Ruralis*, 52(1): 134–144.

Farrington, J. and Farrington, C. (2005) Rural accessibility, social inclusion and social justice: towards conceptualization. *Journal of Transport Geography*, 13: 1–12.

Gartman, D. (2004) Three ages of the automobile: the cultural logic of the car. *Culture & Society*, 21(4–5): 169–195.

Graham, S. and Marvin, S. (2001) *Splintering Urbanism*. London: Routledge.

Gray, D., Shaw, J. and Farrington, J. (2006) Community transport, social capital and social exclusion in rural areas. *Area*, 38(1): 86–98.

Kaufmann, V., Bergman, M. and Joy, D. (2004) Motility: mobility as capital. *International Journal of Urban and Regional Research*, 28(4): 745–756.

Lefebvre, H. (1967) Le droit à la ville. *L'Homme et la société*, 6(1): 29–35.

Lowe, P., Bradley, T. and Wright, S. (eds) (1986) *Deprivation and Welfare in Rural Areas*. Norwich: Geobooks.

McDonagh, J. (2006) Transport policy instruments and transport-related social exclusion in rural Republic of Ireland. *Journal of Transport Geography*, 14: 355–366.

Milbourne, P. and Kitchen, L. (2014) Rural mobilities: connecting movement and fixity in rural places. *Journal of Rural Studies*, 34: 326–336.

Moseley, M. (1979) *Accessibility: the Rural Challenge*. Methuen, London

Noack, E. (2010) Are rural women mobility deprived? A case study from Scotland. *Sociologia Ruralis*, 51(1): 79–97.

Nutley, S. (1988) Unconventional modes of transport in rural Britain: progress to 1985. *Journal of Rural Studies*, 4: 73–86.

Nutley, S. (1996) Rural transport problems and non-car populations in the USA. *Journal of Transport Geography*, 4(2): 93–106.

Nutley, S. and Thomas, C. (1995) Spatial mobility and social change: the mobile and the immobile. *Sociologia Ruralis*, 35(1): 24–39.

OECD (2006) *The New Rural Paradigm: Policies and Governance*. Paris: OECD.

Oliva, J. (2010) Rural melting-pots, mobilities and fragilities: reflections on the Spanish case. *Sociologia Ruralis*, 50(3): 277–295.

Pacione, M. (1995): *Rural Geography*. London: Paul Chapman.

Sager, T (2006) Freedom as mobility: implications of the distinction between actual and potential travelling. *Mobilities*, 1(3): 465–488.

Shergold, I. and Parkhurst, G. (2012) Transport-related social exclusion amongst people in rural Southwest England and Wales. *Journal of Rural Studies*, 28: 412–421.

Soja, E. W. (2010) *Seeking Spatial Justice*. Minneapolis, MN: University of Minnesota Press.

Urry, J. (2002) Mobility and proximity. *Sociology*, 36: 255–274.

Urry, J. (2004) The 'system' of automobility theory. *Culture & Society*, 21(4–5): 25-39.

Velaga, N., Beecroft, M., Nelson, J., Corsar, D. and Edwards, P. (2012) Transport poverty meets the digital divide: accessibility and connectivity in rural communities. *Journal of Transport Geography*, 21: 102–112.

Verlinghieri, E. and Venturini, F. (2017) Exploring the right to mobility through the 2013 mobilizations in Rio de Janeiro. *Journal of Transport Geography*, 67: 126–136

Yago G. (1983) The sociology of transport. *Annual Reviews of Sociology*, 9(1): 171–190.

26

Art as rural planning inquiry

Julie Crawshaw

Introduction

As forged by the narratives of the 'creative city', the role of art in planning is primarily understood as a mode of delivery in service of economic and social development ambitions (see also Chapter 18 of this Companion). By drawing on ethnographic-artistic fieldwork, this chapter offers an alternative view of art in rural planning as a mode of *inquiry*. To develop this alternative position, this chapter introduces two empirical projects where art practice was included as part of the inquiry. Theoretically the chapter draws on classical pragmatism to suggest that the work of the artist acts as a *way of knowing* human and non-human relations (Bryson et al., 2009: 173). From here, as a way of knowing transactional space, I suggest artistic practice has much to offer deliberative rural planning. In support of future practice, I suggest artist in residency programmes can be shaped as opportunities for planners and planning academics to design research *with* artists.

Pragmatism

As paying attention to the 'situated-ness' of practice, pragmatism has a long history with planning (Healey, 2009: 287). As promoting a relational ontology of dynamic continuities, in recent years the tropes of classical pragmatism have enjoyed a resurgence across the social sciences, science studies, political science and philosophy (Bridge, 2013). Taking direction from understanding the world as in the making (James, 1910), a foundational concept of this philosophy is that of 'transactional space' – where objects and humans are understood to be *within* each other (Bridge, 2013: 307). As departing from Ingold's call for understanding art as a way of reawakening our senses 'to allow knowledge to grow from the inside in the unfolding of life' (Ingold, 2013: 8), this chapter draws on the heritage of pragmatism to explore the potential of art's inquiry in support of rural planning explorations.

To keep alert to the continuity of transactions I draw on John Dewey (1859–1952), whose particular strand of phenomenological pragmatism is rooted in commitment to the body acting in the world; where the epidermis is regarded not as a rigid border but a crossing between things 'outside' the skin and 'within' it (Sullivan, 2001: 24); and who regards the art experience as one of transacting these border crossings by suggesting the art experience alters 'outer' physical materials (such as paint, plaster and plastic) as well as our 'inner' emotional selves (Dewey, 1934: 15). Readers will be familiar with the notion that making art makes changes in materials such as marble, plaster and paint; but not so readily attuned perhaps with adjustments made to our memories and emotions through art making. From a Deweyan perspective we are drawn to understand art (across the disciplines) as transactional – not simply in terms of the relations between objects, and objects and people; but being *within* each other as produced through an exchange of inner and outer materials. By drawing on artistic-ethnographic fieldwork, I will argue that it is this transactional experience that underscores art's ability for knowing transactional space (or in other words awakening ourselves to it); and from here, I will suggest that art's transactional quality has methodological value for planning as a way of knowing.

With this way of knowing in mind, in current planning studies it is relevant to note that the pragmatist concern of transactional space is perhaps particularly evidenced by the use of actor–network theory (e.g. Murdoch, 1998; Abram, 2011; Rydin and Tate, 2017) as a 'useful way of thinking about how spatial relations come to be wrapped up into complex networks' (Murdoch, 1998: 357). Central to actor–network theory (ANT) is the epistemology and ontology of networks as a chaining of associations between human and non-human actors. ANT scholars are critical of studies that are purely concerned with human relations, and support the view that any methodology should allow actors to 'build their own space' (Latour, 2005: 20). Inspired by the popular pragmatist method of ANT, I will return to the foundational tropes of pragmatism with particular reference to Deweyan philosophy to explore the potential for art in the rural planning domain.

Artistic-ethnographic fieldwork

I will draw on two artistic-ethnographic projects. The first project is undertaken in the Holy Island of Lindisfarne (2013) in the northeast of England, as part of a larger interdisciplinary knowledge exchange project exploring the role of art in rural development. For the purpose of engaging island residents in local planning and decision-making processes, the fieldwork was developed in partnership with the island's strategic planning agency (the Holy Island Partnership). I designed the research with residents, including a series of artistic workshops developed by five artists (from visual, performance and sonic disciplines). Supplementing the Holy Island fieldwork, the second study draws on fieldwork undertaken in the island of Öland in Sweden, as part of a curatorial and ethnographic project exploring artist-led practice. When writing with this fieldwork, I specifically draw on my one-week residency with Kultivator (an art-farming collective).

In Holy Island I worked as an ethnographer in tandem with planners and community members, who demonstrated interest in artistic workshops – resulting in artists being commissioned (Crawshaw, 2018a). In Öland I worked as an ethnographer with the Kultivator artists, who were interested in how to better articulate their working practice – resulting in the development of a collaborative mode of performance writing (which is included in this chapter). The Holy Island study was undertaken at the invitation of island planners with the view of influencing future planning and enhancing community participation; and as such has direct application in the rural planning field. I include the Öland study to both expand on my argument and introduce a mode of artistic practice that is less usual in planning scholarship.

The term 'artistic workshops' refers to sessions that artists design to enable participants to engage with some kind of artistic process which in this case included: drawing and painting, dance, photography, sonic mapping and theatre performance. Perhaps a less familiar term, 'artist-led practice' is generally traced from the alternative space movement (Anderson, 2015); and in this particular research is used to denote a 'stretched' mode of art practice where artists develop their own space of work (www.stretchedartistled.com). As following collaborative anthropology (Rabinow and Marcus, 2008), both projects are designed as ethnographic studies *with* art practice. The Holy Island fieldwork was developed via monthly one-week residential research trips. The Öland fieldwork was undertaken while in residence for one week with the art-farming collective Kultivator.

Holy Island, England

A tidal island with a resident population of 150, Holy Island lies off the Northumberland coast, a rural region in the northeast of England. The island's significance is reflected by the number of strategic agencies with governance responsibilities locally and the approximately half a million visitors who check the timetable to cross the twice-daily tide every year (Tristram, 2009: 144). In 2009, the Holy Island Partnership (HIP) was developed to facilitate discussions between the community and agencies with a purpose to develop a coordinated response to island issues. The coordinator of the HIP was keen to explore the role of art as way of engaging island residents in the planning process. As combining a professional background in cultural management with arts scholarship, my invitation was to design a project that explored the role of art in the island planning context, with particular focus on community participation.

The fieldwork began by attending an island community meeting, where residents outlined how the HIP coordinator and other officials lived on the mainland; meaning they arrived and left each day with the tide. From here, the group suggested I get to know the island by 'crossing the tide'. As shaped by this meeting, I designed the research to stay on the island for one week a month. I followed the residents' interest to develop a series of artistic workshops to explore the island from the island's perspective (rather than developing activities for tourists). Twenty 3-hour photography, sonic mapping, dance, theatre-performance and drawing and painting workshops took place over July and August 2013. Of 150 island residents, twenty-five participated in the workshops. Apart from the two dance sessions, I participated in all of the artistic workshops.

Getting to know the island

As an ethnographer guided by pragmatism, when staying on the island I write with my experience of the relational landscape by drawing on differing collisions: when I drove in to water (when crossing the tide) for example; and when I got sun burnt. As aligned with the principles of ANT, my diary entries record the island not through human relationships as distinct; but as associations between humans and materials, objects and elements:

> When walking the island, I notice the pathways made by my boots in the ground . . . My line is made by my stepping rhythm and body weight pushing in sand, soil and grasses. The line has a depth to it . . . As I curve round the headland, up banks and down dunes, I come to know the interplay of living: deer, cows, sheep; orchids; grasses; a bucket on the beach; the wind whipping the sand.
>
> *(Crawshaw, 2018b)*

However, when writing *with* the artistic experience, there is a step-change in the narrative (Crawshaw and Gkartzios, 2016; Crawshaw, 2018a), as conveying a deeper or more sensory sense of transactions: 'a sense of connecting with the land in a way I hadn't before [as] my full body is engaged in a process of knowing actively as part of the working landscape' (Crawshaw, 2018a: 218). As such, I suggest the workshops aid knowing the island in a pragmatist sense.

Pragmatism of course is anti-Cartesian in that it does not divide human organisms into 'mind' and 'body', but rather sees them as a continuum in response to the environment: as 'body-minds' as Dewey called it. Bodying as a continuous process produces a continuum of knowledge. ANT views relations as networks of human and non-human associations. As inferring ongoing co-constitution, John Dewey called his version of relationality 'transaction'. He called embodied knowledge 'had' knowledge and cognitive or reflective knowledge 'known' knowledge. This knowledge is not hierarchical but interactive or what Dewey called involving 'transaction'; as including 'ongoing speculation and experimentation about the nature of the environment and what it does and does not yield to ongoing action' (Bridge, 2014: 1648). I suggest the workshops aid pragmatist knowing via their transactive quality; they are experimental and speculative.

As writing with my bodying experience, I have suggested that 'art has a diagnostic capacity to induce processes of reflexivity, revealing community relationships [as] nature-human relationships' (Crawshaw and Gkartzios, 2016: 142). As a 'diagnostic reading' of human and non-human relations, my proposition (as developed with Gkartzios) might be understood to be one of aligning artistic inquiry with ANT. As 'reading' human and non-human relations we are certainly suggesting that the workshop experience follows ANT's sensibility as rejecting the discreet human. As 'had', however, I also underscore that the workshops go beyond the flat ontology of ANT (Bridge, 2013: 311); that they stretch towards a transactional way of knowing. Through participating in the workshops I am introduced to a more Deweyan way of getting to know the landscape – as 'had' with my body; via dance and performance improvisation, and drawing, photography and sonic mapping. As such the workshops work in support of coming to know the multiverse of transactional space; as taking place between bodies and landscape to take us beyond a surface (or spectator) view. The invitation was to mediate participation by residents in planning processes. This took place via mediating knowing of the space of planning practice.

Öland, Sweden

Approximately two years after the Holy Island fieldwork, Mathieu picked me up from Kalmar's train station on the Baltic Sea. From here, together we drove across the Kalmar Strait via a concrete bridge to Öland. After twenty minutes the tall trees faded to a sharp fork. Taking the left turn led us to park next to what looked like a stable. A woman dressed in beige overalls walked towards us. 'This is Malin'. My right hand stretched to hold her left before saying 'Hello' and thanking them both for putting me up for the week. As I notice the assorted constructions strewn in the paddocks and pastures around us, Mathieu smiles in response while saying: 'I used to make sculpture and now I build.' I wonder, build what?

Malin and Mathieu live on this farm in Dyestad with their two children. There are structures of various kinds across the landscape. As artists, and partners, they work under the name of Kultivator – a project they began with two organic farmers in 2005. Kultivator's 'activities' include exhibitions and events, a residency programme and a dairy farm. Informed by the fieldwork in Holy Island, I set out to explore the work of Kultivator in tandem with them.

As we focused on developing a new story to articulate the work they do, Malin and Matheiu and myself decided to start our research together by writing a short text, with the purpose of orienting the ongoing research. This resulting text was first prepared for presentation at a conference at Lund University called Co-laborations: Sharing Authorship and Space in Architectural

Figure 26.1 Some of the structures of Kultivator.

Source: Crawshaw (2015)

and Urban Research (11–12 February 2016) and then published as a small booklet for inclusion in an exhibition at Kultivator called New Horse Cultures. The text that follows is taken from the booklet, which is called *The Stable (A Collaborative Story)*. Shaped as performance ethnography (Denzin, 2003), this section acknowledges that developments in planning practice may require alternative approaches to research which might be more creative (Umemoto in Ryan et al., 2017: 297). The co-produced text specifically embraces Latour's (1988) call to storytelling as promoted by Baum (in Ryan et al., 2017) in the planning field. For practical reasons when performing at the conference the script to follow was written for Julie and Malin to dual role. As such they play Mathieu and Aslambek, who are also in the story, as well as themselves.

Now I build

Malin: When you arrived Mathieu showed you our house and around the farm a little, and then to the other house, the *vita huset* where people stay.

Julie: Yes, I noticed immediately, of course, the many structures – houses, farm buildings, and other shelters. I remember, Mathieu said, 'I used to make sculpture and now I *build*'.

Malin: Yes, we build. We re-built the *vita huset* and our own house. And we build different structures for exhibitions like our root cellar, which is also an archive, under an open-air classroom. Or our chicken house, which was originally was set up as a community art piece in Stockholm.

Julie:	You asked me for dinner, which was nice.
Malin:	Yes, we thought it would be good for you to meet Henric (the farmer) as well, my sister (who lives close by) and her son came also. It was a shame that our kids weren't there (ours and Henric's). They were on a bus coming back from Stockholm – their first adventure without their parents!
Julie:	I arrived just before your sister. I wasn't sure how to 'get started', so I began by asking you and Mathieu some basic questions about what you do. I asked if it was OK to tape our conversations and also to take photographs from a camera around my neck. I think I explained that it was set to take pictures every minute?
Malin:	Yes, later in the week, when we looked at the pictures together, I remarked that the images look like they were taken by a small girl. Like someone else was with us. In the kitchen, I told you about how Mathieu and I met at art school.
Julie:	After Henric and your sister left we sat by the stove. We talked about what artists 'do' and how research that describes the 'reality' of practice is much needed. Sitting by the stove, you said:
Malin:	Everyone thinks they know what farmers do. They are walking around with hay coming out of their mouth. And the artist is standing in his *atelier* with his hat like this. This is a strong image that everyone knows. Today's reality is very different. Not many people know what an artist does now, or a farmer. This is not the way it is. We need a different story.
Julie:	*The Stable (A Collaborative Story).*

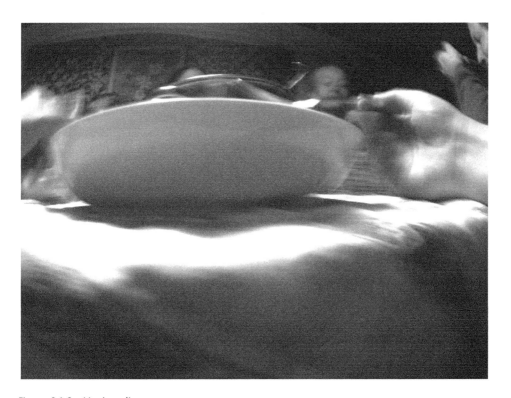

Figure 26.2   Having dinner.

Source: Crawshaw (2015)

Malin:	To set the scene: Mathieu explains that during the week he will mostly be at the school, teaching film studies and the refugee integration programme. There will be a music evening on Thursday for the refugee families and other students. This is an unusual week for me, as I will be mostly at the farm, apart from Thursday when I also need to be at the school for meetings.
	During the week, we decide that Julie should stay on the farm a bit, but also go to the school and interview a few people we work with: another artist-led project called Yellow Box, the director of the Kalmar art museum, our regional art coordinator and the headmaster of the school.
I ask Julie:	What would you like to do tomorrow?
Julie:	When people come here, what do *they* do? Do they get involved in the farm?
Malin:	Mathieu responded, 'Yes, we encourage them to have the farm experience. And they can continue if they want to'.
Julie:	I think tomorrow I will stay at the farm. I will join Henric for the milking.
Malin:	OK, milking starts at 7 a.m. and continues until around 10.30. What size are your feet? It is better that you wear capped boots. The cows can kick. Come here first. I will give you the boots, we can take hay to the horses, and then I will walk you to the milking shed.
Julie:	On the way back to my house I notice a man in a red jacket.
Malin (as Aslambek):	Hej.
Julie:	Hej.
Malin:	The man in the red jacket is Aslambek. Last year he was a student of the integration course. He has experience of farming and building, so we asked him to work with us to help build a stable. He is from Chechnya. He speaks Russian and is learning Swedish. He speaks to Julie in Swedish. She doesn't understand. He probably just asked if she had seen us, or maybe he asked if she was lost. 'Svenska?', he asks.
Julie:	Engelska? My name is Julie [pointing to the white house]. The next day I go to the main house to pick up some logs for the fire. Malin invites me for a coffee. Aslambek comes in with Mathieu. 'Svenska. Lerra svenska' Mathieu says. Then looking at me, 'she wants to learn Swedish'. I say, 'I have learned that I am in the *vita huset* (the white house)'.
Malin:	Mathieu says, 'The idea is that he helps to build *and* learns Swedish but he is always by himself'. Looking at Julie, Aslambek asks 'Ingelska?'. She nods. He tells of his time in Chechnya. I try to translate parts for Julie. Sometimes I realise that I am translating things he *hasn't* said, he is 'speaking' physically.
Julie:	When you are translating the spoken exchange, something you, or Mathieu say, I remember is, 'when you have electric shock all the water disappears from your mouth'.
Malin:	After coffee, Julie goes back to the *vita huset* and Mathieu and I help Aslambek. We need to move the stable door so it can be painted. It is heavy, so I ask Julie if she will help.
Julie:	I say, 'Of course'. And help carry it, to rest on the stone wall in front of the horses. It looks like a sculpture.
Malin:	Yes, it is a very nice collaboration.

Building deliberative space

In this performance, Malin calls for a new story; for which of course *this* story is written in response. The story asks us to overturn any essentialist notion of the artist – as sitting in their

Figure 26.3 The making of the stable.

Source: Crawshaw (2015)

atelier (making sculptures and paintings), in favour of building something else. But what is this 'new story'? What is building?

As an object of study the social-ness of art is the foundation of the institutional theory of the 'art world' where art is socially produced, via a network of relations and conventions as exchanged between art personnel (such as artists, curators and museum directors) (Becker, 1982). When reading *The Stable* story, however, we can perhaps take the socialness of Kultivator further. As introduced to Aslambek we might position the work of Kultivator within the art historical canon of socially-engaged or participatory art practice (e.g. Bishop, 2012) as broadly denoting work that involves other people as well as the artist or artists. The workshops on Holy Island can also be aligned with this socially-engaged or participatory category. Indeed, a study *of* the Holy Island workshops might have discussed the merits of the sessions in relation to these terms. As a study *with* the workshops, however, we are rather introduced to a transactional understanding of Holy Island itself. So what if we avoid conducting a study *of* Kultivator in favour of studying *with* Kultivator? What is the work *doing*? What is building?

As with the first reading of the workshops, in the way of ANT we might suggest that the collaborative story introduces us to the farm as a network of human and non-human associations; or as Murdoch (1998: 359) suggests, a 'stable set of relations . . . by which the world is built and stratified'. Indeed, we are introduced to the workings of the farm beyond human associations which is akin to Harbers's (2010) account where 'farm life formed a miniature society – a network in which people, animals, plants and things co-existed, a hybrid collective in which the diverse constituent elements relationally defined and determined one another'. Following ANT scholarship *of* the day-to-day work we might trace associations between Aslambek, screws, wood and other people, objects and materials (Crawshaw, 2015). Instead, by trying to position myself as working *with* Kultivator, I want to suggest that their work enables a more classical Deweyan sensibility.

On arrival Mathieu points to the structures strewn around the farm area and says 'I used to make sculptures and now I build'. What is being built? Malin refers to the structures as an end point. I suggest, however, that what is being built through the day-to-day of the farm is deliberative space; and what Malin and Mathieu are doing as Kultivator is inviting people to be part of this deliberative activity: through being in residence and taking part in activities.

The thinking elements of a pragmatist enquiry are treated as 'embodied and purposeful and involving social coordination and deliberation over current problems and future courses of action' (Bridge, 2013: 304). The deliberative element has informed a well-established strand of research in what is termed both deliberative and critical pragmatist urban planning (ibid.: 15); where a 'fully pragmatic planning might push the boundaries of deliberation back beyond the focus on participatory regimes to include how issues actually become matters of concern, and how people are affected or learn to be affected by them' (ibid.: 311). From here the work of Kultivator goes beyond the participatory regime as fully pragmatic.

I propose that the day-to-day work of Kultivator goes beyond encouraging Aslambek to engage in new work as a new arrival but creates circumstance for coming to know the transactions of our every day. And as such the Kulivator space makes us switched on to matters of concern – and in this case in the rural context of the island of Öland. For Dewey, 'Objects are "taken" rather than merely perceived and their "objectness" is an ongoing outcome of networks of human purposes or in the way they disrupt those purposes' (Bridge, 2013: 307). In this way I suggest the making of these objects (the stable and other structures) are disruptive; and as such, they support *having* deliberative thought as made available via the transactions of the workings of the space. We might say that the work of Kultivator is building space for working out the transactions of space.

Art as planning research

Although there is scant mention of artistic research in planning studies, the notion of *art as research* has taken hold within and beyond artistic disciplines. The visual, literary and performing arts are increasingly framed as modes of inquiry; and this turn to research in the arts has nurtured interest from across social science disciplines including anthropology (e.g. Ingold, 2013) and geography (Hawkins, 2013). So how might positioning art *as inquiry* benefit a rural planning and development project?

Although interest in 'relationality' is evidenced in planning scholarship via the use of ANT, the transactional qualities of art are rarely accounted for in rural studies (Crawshaw and Gkartzios, 2016). Instead, as following the urban story, after arguing for art as an instrument for economic development, rural research usually positions art as a prescriptive way to solve community problems and improve social sustainability (Balfour et al., 2018; Markusen, 2006; Anwar Mc Henry, 2011). Instead of focusing on delivering against economic and social strategies, I call for the potential for the transactional qualities of art and how engaging with art as enquiry might contribute to rural research.

So how can planners and planning academics work *with* art practice? Like in the examples discussed here, one way is to make a step change to conduct rural planning research – not *of*, but *with* artists. Drawing on the Holy Island case study, I have proposed that art can be understood as a way of knowing transactional space; and supplemented this with fieldwork from Öland, where

I expand from this proposal to suggest that the work of Kultivator can be understood as building deliberative space in support of fully pragmatic planning. Furthermore, in support of conducting research *with* artists, I propose that artist-in-residence programmes can be utilised in support of this ambition. One example is the model of Newcastle University's Centre for Rural Economy (CRE), where CRE co-hosts a six-month annual artist in residence with Berwick Visual Arts (BVA) in the North of England. Researchers are given the opportunity to develop research projects *in tandem with* artists (CRE, 2017), which have also given rise to engagement with planners themselves (Gkartzios and Crawshaw, 2017). Such collaborations between the academy and art professionals are very important, because although artist in residency programmes are common place in urban and increasingly rural settings, academic partnerships are fairly unique.

Acknowledgements

This chapter draws on fieldwork as part of 'Northumbrian Exchanges' (NX), funded by the UK Arts and Humanities Research Council and 'Stretched: Expanding Notions of Artistic Practice through Artist-Led Cultures', funded by the Swedish Research Council. The artistic workshops on Holy Island were supported by the Newcastle Institute for Social Renewal and developed by artists: Claire Pencak; Tess Denman-Cleaver; Emma Rothera; James Wyness and Jenny Moffitt. *The Stable (A Collaborative Story)* was supported by the Research Board at Valand Academy, University of Gothenburg and designed by Kjell Caminha. I would like to thank NX colleagues and Stretched colleagues and Kultivator (www.kultivator.org) for supporting this project.

References

Abram, S. (2011) *Culture and Planning*. Farnham: Ashgate.
Anderson F. (2015) Preserving and Politicising the Alternative Space. *Oxford Art Journal* 38(3), 448–451.
Anwar McHenry, J. (2011). Rural Empowerment through the Arts: The Role of the Arts and Social Participation in the Mid West Region of Western Australia. *Journal of Rural Studies* 11, 245–253.
Balfour, B., Fortunato, M.W.P and Alter, T.R. (2018) The Creative Fire: An Interactional Framework for Rural Arts-Based Development. *Journal of Rural Studies* 63, 229–239.
Becker, H. (1982) *Art Worlds*. Berkeley, CA: University of California Press.
Bishop, C. (2012) *Artificial Hells: Participatory Art and Politics of Spectatorship*. London: Verso.
Bridge, G. (2013) A Transactional Perspective on Space. *International Planning Studies* 18(3–4), 304–320.
Bridge, G. (2014) On Marxism, Pragmatism and Critical Urban Studies. *International Journal of Urban and Regional Research* 38(5), 1644–1659.
Bryson, J. M., Crosby, B. C. and Bryson, J.K. (2009) Strategic Planning as a Way of Knowing. *International Public Management Journal* 12(2), 172–207.
Crawshaw, J. (2015) Working Together: Tracing the Making of Art as Part of Regeneration Practice. *Anthropological Journal of European Cultures* 24(2), 76–96.
Crawshaw, J. (2018a) Island Making: Planning Artistic Collaboration. *Landscape Research* 43(2), 211–221.
Crawshaw, J. (2018b) Drawn Together: Stories of Holy Island. In Y. Holt, W. Jones and D.M. Jones (eds), *Imagining Islands: Visual Culture and the Northern British Archipelago*, 153–168. New York: Routledge.
Crawshaw, J. and Gkartzios, M (2016) Getting to Know the Island: Experiments in Rural Community Development. *Journal of Rural Studies* 43, 134–144.
CRE (2017) CRE Celebrate Artistic Partnerships. Retrieved from www.ncl.ac.uk/cre/news/item/crecelebratesartisticpartnerships.html.
Denzin, N. K. (2003) *Performance Ethnography: Critical Pedagogy and the Performance of Culture*. London: Sage.
Dewey, J. (1934) *Art as Experience*. New York: Perigree.
Gkartzios, M. and Crawshaw, J. (2017) Making Homes, Making Rural: With an Artist-in-Residence. Paper presented at XXVII European Society for Rural Sociology Congress, Krakow, Poland.
Harbers, H. (2010). Animal Farm Love Stories: About Care and Economy. In A. Mol, I. Moser and J. Pols (eds), *On Tinkering in Clinics, Homes and Farms*, 141–170. Piscataway, NJ: Transaction Publishers.
Hawkins, H. (2013) *For Creative Geographies*. London: Routledge.

Healey P. (2009) The Pragmatic Tradition in Planning Thought. *Journal of Planning Education and Research* 28, 277–292.

Ingold, T. (2013) *Making: Anthropology Archaeology, Art and Architecture*. New York: Routledge.

James, W. (1910) *Pragmatism*. Seaside: Watchmaker Publishing.

Latour, B. (1988) The Politics of Explanation: An Alternative. In S. Woolgar (ed.), *Knowledge and Reflexivity: New Frontiers in the Sociology of Knowledge*, 155–176. London: Sage.

Latour, B. (2005) *Reassembling the Social*. Oxford: Oxford University Press.

Markusen, A. (2006) An arts-based state rural development policy. *Journal of Regional Analysis and Policy* 36, 47–49.

Murdoch, J. (1998) The Spaces of Actor-Network Theory. *Geoforum* 29(4), 357–374.

Rabinow, P. and Marcus, G. E. (2008) *Design for an Anthropology of the Contemporary*. Durham, NC: Duke University Press.

Ryan, M. et al. (2017) Confronting the Challenge of humanist Planning/Towards a Humanist Planning/a Humanist Perspective on Knowledge for Planning: Implications for Theory, Research, and Practice/ To Learn to Plan, Write Stories/Three Practices of Humanism and Critical Pragmatism/Humanism or Beyond? *Planning Theory & Practice* 18(2), 291–319.

Rydin, Y. and Tate, L. (2017) *Actor Networks of Planning: Exploring the Influence of Actor Network Theory*. London: Routledge.

Sullivan, S. (2001) *Living Across and Through Skins: Transactional Bodies, Pragmatism and Feminism*. Bloomington, IN: Indiana University Press.

Tristram, K. (2009) *The Story of Holy Island*. London: Canterbury Press.

Part V
Planning the inclusive countryside

This part focuses on issues of inclusion and diversity and looks at some specific literatures around ethnicity, gender, sexuality and age. Issues of class and social justice have already been explored in Part I (see for example Chapter 8), but here we aim to draw attention to specific planning issues and literatures around sectional inequalities. In the following four chapters these issues are discussed through particular cases.

Chapter 27, by Satsangi and Gkartzios, starts with an overview of sectional inequalities and positioning ideas of diversity and equality within ongoing debates about social inclusion. The authors recognise a diversity of approaches to social inclusion, but highlight Amin's four registers of solidarity and their application to the rural context by Shucksmith as a way to approach social inclusion. The authors present the key theories of ethnicity displacement, spatial assimilation and spatial segregation, and then proceed to briefly describe other sectional inequalities in rural planning debates drawing specifically on gender, sexuality, disability and age. Disproportionally perhaps, the authors cover most of the literature on ethnicity, as other literatures of inclusion are explored through detailed case studies in the following chapters (with the exception of disability issues – a relatively emerging and critical arena in rural planning debates). As the authors contend in their chapter, the focus of this section is to inspire more research on issues of inclusion, diversity and equality by drawing on literatures that might have been considered fugitive within the planning research domain. In many cases, the reluctance to look at certain issues around inclusion is not only observed in policy, but within planning academia too.

The role of gender in planning and development debates is considered in more detail by Shortall in Chapter 28, which presents a case of gender bias through the (non) planning of the farmyard. Shortall discusses the farmyard is a hybrid space, whose boundaries between the family and work are blurred. She argues that the farmyard is a critical micro-space that has not been given adequate attention in planning research, despite being central to an occupation with the highest rates of accidents and fatalities. In the tradition of resisting planning interventions in agriculture observed in British and Irish planning systems, Shortall further argues how the exclusion of the farmyard from planning scrutiny demonstrates gendered constructions regarding agricultural work and, wider, rural spaces, despite increasingly changing gender roles. Thus, as more women enter farming, more effort is needed to consider its politics: not only in terms of the gendered spatiality of the farmyard, but also in terms of health safety and risks it poses to women farmers, family members and agricultural workers. Indeed her fieldwork in Scotland evidences how women in agriculture take risks to 'prove' themselves as 'authentic' farmers. The problem here of course is not that these entrants are women – the problem lies specifically with the planlessness of the farmyard.

Continuing the discussion of sectional exclusions, Doan and Hubbard use queer theory to critique the heteronormative and metronormative qualities of planning. The authors draw a parallel between the binaries of urban-rural and gay-straight and by queerying both these taxonomies, they make a case for bias. The obvious bias here is to think of urban areas as places for LGBTQ persons, but the authors review a series of queer geographies beyond the city and discuss how some non-normative groups have used rural spaces and rural imaginery on their own terms. However, although rurality and queerness are not contradictory terms, community services and infrastructure that targets the needs of LGBTQ community (e.g. gay cafes, guesthouses, gay pride parades, etc.) are often lacking in rural areas. An area that the authors draw attention is the HIV/AIDS experience in rural areas, with rural areas being more limited in treatment and prevention services. Bias is also observed within the planning academy and profession with limited publications on the intersection of rurality and queerness. As a way of moving forward the authors highlight a few research approaches to promote queer perspectives in rural planning debates, such as understanding the impacts of LGBTQ counterurbanisation and gentrification experience, as well as the needs of the LGBTQ community in terms of social services, including health and education.

The chapter by Bevan concludes this section with an examination of the issues of rural ageing. The population in western industrial societies is ageing, but rural areas in many cases age faster. The chapter explores the social construction of ageing, and argues that ageing refers to wider shifting experiences in social roles, activities and health as people grow older, than merely chronological ageing. Bavan calls for better understanding of the challenges that older people face in planning and wider public policy debates, avoiding models that view older people as a problem and realising their capabilities and opportunities. In this context, Bevan, reviews two approaches emerging from georontological research in articulating the diversity of experiences in later life. The first refers to the notion of healthy ageing, and the actions that can be incorporated in planning, both physical and social, so that older people will avoid disease and disability while remaining socially engaged and active. The second approach refers to understanding the impact of reduced mobility for older people, which is a key concern for people living in rural and remote areas. Finally, Bevan explores the emerging concept of 'age friendly communities' and how such ideas can be used in planning to support people in their later life.

The contributions in this section aim to highlight some of the key emerging literatures in the areas of inclusion. All contributions draw on academic literature and research to highlight cases of bias and discrimination against minority ethnic groups, women, non-normative LGBTQ and older people. The list is not exhaustive, and special attention is given by Satsangi and Gkartzios to the intersectionality of inequalities, including those to do with rural identity. Rather than arguing that these are the most important issues, the collection aims to inspire a future generation of planners and planning researchers who will question their own understandings of inclusion and diversity and contribute to the research in exiting new ways.

27

Social inclusion, identities and planning practice

Madhu Satsangi and Menelaos Gkartzios

Introduction

Across the EU, there has been debate over some two decades on the meaning and measurement of social inclusion, how it relates to notions of social justice and what this might mean for rural development and planning (e.g. Shucksmith and Chapman, 1998; Commins, 2004; Shortall, 2008). The spatial, and indeed rural, connotations of inclusion have most recently been considered by Shucksmith's reflections on the possible moral underpinnings and constitution of a 'good countryside' (Shucksmith, 2016). Shucksmith's essay develops Amin's (2006) practical urban utopianism, based on four registers of solidarity: repair, relatedness, rights and re-enchantment, and enquires as to their application in rural areas. It concludes with 'a call . . . to encourage and assist deliberation and debate on rural futures . . . engaging and reflecting critically on these processes' (Shucksmith, 2016: 9). In this introductory chapter, we specifically consider how issues of ethnicity, gender, sexuality, and age might feature in that debate as none of these characteristics was explicitly discussed in Shucksmith's paper. We begin by summarising Shucksmith's rural application of the four registers of solidarity and then examine some of the ways in which the sectional exclusions have figured in the rural studies literature. We offer here a more focused account on ethnicity and the spatial assimilation and spatial segregation literature, as detailed accounts and examples of rural planning implications across gender, sexuality and age follow in this section. We also explore the way in which inequalities may be multiple and intersectional. The chapter concludes by suggesting a role for collaborative planning practice in addressing these inequalities, while acknowledging that current evidence bases may limit progress in practice.

Registers of solidarity

Shucksmith (2016) seeks to apply Amin's (2006) four registers of solidarity, developed as a way of thinking how a 'good city' might be framed, to rural areas. Both authors are centrally concerned with moral arguments, as these are intrinsic to the notion of 'good'. What is meant by 'good' is clearly open to debate, with perspectives varying over time and space and reflecting different ethical and ideological standpoints as well as varying with individual experiences and socio-economic characteristics. The first register, *repair*, is concerned with 'maintaining the invisible systems which enable the complexity of everyday life . . . to proceed' (Shucksmith, 2016: 4), and Shucksmith notes that in rural areas, this should be concerned with maintaining social as well as natural capital. How capital is maintained raises questions of rights and responsibilities, the role of the state and the individual. Secondly, *relatedness* is a matter of 'social justice, an ethic of care to insider and outsider . . . welcoming difference and diversity', which in rural areas may present a paradox: 'rural areas are proclaimed as inclusive and neighbourly, and yet these can only be protected from corrosive urban values through being exclusive and drawing tight bounds' (ibid: 5). This paradox is very important in thinking of the social constitution of rural areas and how it changes, or how change is minimised. The attributes attested to rural areas are also intrinsic to the conception of the rural idyll (Satsangi et al., 2010). Third, *rights* 'include the right to participate in the public realm, to shape and enjoy . . . life and respect others' rights', which needs to contend with 'rural disadvantage, exclusion, participation, citizenship, governance and power' (ibid.). Power and status and their maintenance are important dimensions of whether disadvantage and exclusion are recognised. Finally, *re-enchantment* is having the 'hopes and rewards of association and sociality' or, for some in the countryside of 'solitude and being an elite apart' (ibid.: 6). It is in the definition of relatedness and rights that Shucksmith comes closest to explicitly identifying equality as a component of 'the good countryside' and within this inclusivity in terms of social class, ethnicity, gender, sexuality or physical ability. The title of 'a good countryside' would not be merited by one characterised by exclusion or segregation, on any criterion alone or in combination with others.

Ethnicity, spatial assimilation and spatial segregation

How rurality and ethnicity are related can be approached by looking at the prevailing models in the literature on the residential location of minority ethnic groups and if and how this changes over time. These can be described as falling into two broad camps: spatial assimilation and spatial segregation. The (modified) assimilation thesis would suggest that as incomes and resources rise, migrants increasingly move from neighbourhoods of greater to lesser minority ethnic concentration – and in many cases rural areas are viewed as such. For at least the past 50 years, a significant shift in the UK population as a whole, and particularly with increasing affluence, has been a movement out of the larger cities to towns and other settlements of different but smaller scale. This is noted in both the growth of the suburbs and of rural towns, particularly those readily accessible to major urban centres. It is characteristic that the smaller settlements concentrate the more affluent populations, with rural housing demand the most critical driver of this well observed social change in the UK (Gkartzios and Shucksmith, 2015). Champion (2001) has noted that there has been debate as to whether the movements reflect a pattern of counterurbanisation and whether this is distinct from suburbanisation and that the trends appears to have abated a little more recently (Champion, 2016). The question that emerges here is the extent to which the trend exhibited among the majority ethnic population is also shown by minority ethnic populations. And, if there are differences observed, are these explained by individual characteristics other than affluence?

The spatial segregation thesis would suggest that households from minority ethnic backgrounds largely maintain residence in areas of minority concentration as a result of choices made by majority households to select areas with no or few minority households perhaps accompanied by discriminatory practices, and/or minority preference to stay in those areas. The balance of majority push or minority pull factors and how these operate over space and time is the emergent empirical question. Peach (2009) also notes that neither process – assimilation nor segregation – is inevitable or unidirectional.

These models are also associated with debate on the abatement or persistence of disadvantage. Thus, the assimilation thesis contends that the disadvantage experienced by first-generation migrants abates over time and particularly for second and subsequent generations. The segregation thesis counters that there is limited or no abatement of disadvantage over time. Ballarino and Panichella (2015) compare experiences of older and newer male immigrants in six European countries. They found evidence of assimilation for the UK and Sweden and of persistent disadvantage in France and Germany while no conclusive patterns emerged for Spain and Italy. Obucina's (2014) analysis confirms this in Sweden, modelling poverty entry, exit and duration for immigrants and native Swedes. In the Netherlands, Boschman et al. (2017) report that ethnic minorities and low-income households are less likely than other households to realise their wish to leave their neighbourhoods. Looking at the USA, Jargowsky's (2009) analysis supports the assimilation hypothesis except for Mexican immigrants who do not move out of poorer neighbourhoods as quickly as other minority ethnic groups.

Analysis of Census data has shown that Britain is generally characterised by increased mixing of majority and minority ethnic groups (save for a relatively small population of Chinese background (Peach, 2009; Simpson, 2012, 2014; Farley and Blackman, 2014; Catney, 2016). At the same time, there is some evidence of self-segregation of minority ethnic groups in social housing stocks, which is seemingly exacerbated where there is a greater degree of user choice and a lesser degree of 'objective' data that match applicants to vacancies (van Ham and Manley, 2009). People from minority ethnic backgrounds are more likely to live in deprived neighbourhoods than the white majority and typically face greater employment-related disadvantage (Jivraj and Khan, 2013). It also appears that there is significant differentiation of experience according to the specific minority ethnic background a person has: an Indian background tends to be less disadvantageous than Bangladeshi or Black Caribbean (Platt, 2003, 2007). The need to consider ethnic backgrounds separately is well-recognised in the international research literature (Jargowsky, 2009; Kaplan and Douzet, 2011; Kazemipour and Halli, 2000; McConnell, 2012; Walks and Bourne, 2006).

The most extensive literature on segregation, its path, causes and consequences, comes from the USA. It is rich and powerful, though it is not suggested here that it can adequately model the experiences of other countries and places. Kaplan and Douzet (2011) review a wide national and international literature on the many individual, neighbourhood and societal consequences of race/ethnicity- based segregation. Madden (2014) recognises substantial racial desegregation in large metropolitan areas in the period 1970–2009, while poverty segregation increased from 1970 to 1990, then fell to 2000 and remained stable to 2009. Using US Census tract data for large metropolitan areas, she fits a fixed-effect model to test relationships between poverty segregation and racial segregation. She finds that poverty segregation showed no change with decreasing racial segregation largely as racial integration had taken place among people of the same poverty status. There had also been some sorting of poor non-African Americans into areas that predominantly housed African American households. Massey and Fischer (2000) further find that during 1970–1990, the same structural factors (income inequality and class segregation) are responsible for high and increasing concentrations of poverty for groups experiencing high and increasing racial segregation as well as the opposite directions in both outcomes.

While most analysis and debate has focused on metropolitan areas, there is recognition of how patterns and processes repeat or differ in suburban and rural areas. Thus, Walker (2018) looks at how diversity decreases with distance from a metropolitan core. Jensen's (1994) focus is on rural racial and ethnic minorities in the USA. Access to employment and job quality (together making for employment hardship) are particularly strongly associated with poverty for rural poor minorities and he argues that processes of stratification that occur over the life-course need to be analysed to understand point in time disadvantage.

More recent work by Lichter et al. (2012) looks above the small neighbourhood scale. Their analysis shows that in this century, there has been increasing sorting of poor people into poor, economically declining communities; they emphasise that poor places include both inner cities and pockets of poverty across rural America (see also Saenz, 2012). Indeed, Slack and Jensen (2002) conclude that underemployment is more acute among non-metropolitan minorities than city-based minorities with underemployment being a significant component of economic disadvantage.

Analysis of internal migration data from the British Census tends to suggest that ethnic differentiation in migration patterns is associated with socio-demographic composition and urban location (Finney and Simpson, 2008; Finney et al., 2015). Thus, households from minority ethnic groups have predominantly tended to retain residence in the city, moving short distances therein. Looking at migration in the year before the 2001 Census, Stillwell and Hussain (2010) show that the major non-white ethnic groups show evidence of net movement out of inner London, but not for other British cities. While Phillips (1998) had recognised that clustering may be the wish of ethnic minority households, suggesting that minority households may have limited desire to leave big city locations for residence in small town and rural Britain, she later (Phillips, 2006) questioned this and Stillwell and Hussain's (2010) analysis supports the latter position. The general tenor of much of the literature is that greater integration is indicative of a fairer society and thus a better countryside. Merry (2013, for example) has argued, however, that where separation is the outcome of freely expressed choice, it is benign and, perhaps, more desirable an objective than forced integration. Away from Britain, Ibraimovic and Masiero (2014) report survey results in the Swiss city of Lugano that showed a preference towards ethnic co-location but that this played a relatively limited role in locational choice. Skifter Andersen's (2016) analysis of a large database of linked population and housing characteristics in Denmark from 1985–2008 points to ethnic retention (minority groups choosing to remain co-located) lacking any empirical substantiation, with segregation resulting from avoidance of concentrated minority ethnic neighbourhoods and minority groups' attraction to those areas. Tammaru and Kontuly (2011) use longitudinal elements of Censuses in Estonia to look at minority ethnic migration from their cities of arrival. In line with the assimilation thesis, their analysis shows this vindicated by minorities' movement, though diluted by the higher out-migration from the same origin cities by majority groups.

In accordance with the spatial segregation thesis, some analyses have suggested that households from minority ethnic backgrounds have been excluded from rural Britain and that part of the contemporary construction of the rural idyll aligns with a construction of white British identity (Neal and Agyeman, 2006; Phillips, 1998; Reeve and Robinson, 2007). However, with respect to more recent migration from central and Eastern Europe, Reeve and Robinson (2007) note that a significant proportion have had small town and rural locations being their first in Britain, rather than the larger urban centres characterising earlier waves of migration from India, Pakistan, Bangladesh, China and Afro-Caribbean countries (also Robinson, 2010). Reeve and Robinson (2007) note that more recent arrival and a different location did not prevent European migrants facing similar inequalities in housing to earlier immigrants. The literature thus indicates that a number of countries in the advanced industrial world have countrysides marked by limited presence of people from minority ethnicities and, where minorities are present, with inequality in treatment on housing and other dimensions.

Gender

In relation to rural studies, Jo Little and Ruth Panelli (2003: 281) comment that 'gender analyses of rural societies have an increasingly flourishing existence'. Barbara Pini (2003: 249), however, argues that 'rural sociology is . . . a field of scholarship in which feminist theory remains largely un-noticed and unacknowledged'. With both of these reviews being written at approximately the same time and with access to the same range of literature, there would seem to be only one way in which these opposing conclusions can be reconciled. That is by referring to the basis of judgement: with Little and Panelli (2003) seeming to comment on the (then-) recent growth of gendered analyses of rural society and Pini (2003) seeming to reflect on the point that rural sociology had reached in comparison to other fields of sociological enquiry.

Central to rural planning issues, a significant amount of research attests to gendered difference in who gets what housing and land, when, how and why and, wider, the gendered construction of rural spaces (see also the literature review by Shortall in Chapter 28 of this Companion). The analysis of evidence indicates both direct and indirect gender bias in housing and labour market processes (see Satsangi, 2011, 2013 for further discussion). It also shows housing market processes' embedding in cultural systems of value, conferred via the gendered meaning of home, family and community (discussed by Satsangi, 2011, 2013). Housing in rural areas, particularly where labour market and family roles retain 'traditional' highly gendered divisions, seems likely to be prone to gendered differences in outcomes.

Disability

The ways in which rurality and disability might intersect are summarised by Pini et al. (2017). They suggest that dimensions include: 'the limited availability or accessibility of provisions which can potentially assist a disabled person in treating their impairment (health and rehabilitation facilities), managing it (social and welfare services) or contributing in other ways; whether and how differential resources possessed by different rural dwellers complicate limitations in the locally available or accessible service landscape; the conflation of disability with physical access and, concomitantly, the concentration of research energies on physical disability rather than the full spectrum of mind-body differences; how rural areas may in some ways be productive of disability, or at least of specific impairments, in that something about rural lifestyles may carry the potential to be disabling' (Pini et al., 2017: 225–226) and, how disability is represented in 'the rural'.

If a legitimate function of the rural in advanced, industrial societies is for leisure and recreation, the broad conclusion from extant research seems to be that that is not equally open to all. Thus, Kitchin's (1998) view for the UK that 'disabled people are made to feel out of place in outdoor space' was no less applicable a decade later (Burns et al., 2009), and there is little reason to believe that the current situation is different. Burns et al. (ibid.: 403) argue that relatively little attention has been placed on disabled people's access to the countryside in research and policy terms compared to access to the built environment. Bruton (2006) is more sanguine about the role that access initiatives have played, while acknowledging that access irrespective of (dis-)ability is far from being achieved.

Referring to the experience of blind and visually-impaired people, Macpherson (2009) summarises previous work showing that the sight of landscape may depend on our age, race, class, gender, and socioeconomic positioning as well as the embodied practices we are engaged in. Her work concludes that 'landscape becomes present as a tangible, tactile relationship between sighted guide and walker; as a historical relationship with one's own embodied past; and as a sociocultural relationship with what is popularly understood to constitute "the" landscape' (ibid.: 1052).

Sexuality

The representation of the rural through an idealisation of particular characteristics has been recognised to have an exclusionary impact on people who do not fit the 'norms', and research highlights the difficulties for example for young people who resist a 'laddish culture' or gendered expectations about their future (Shucksmith, 2004). Following the seminal paper by Bell and Valentine (1995), many works have demonstrated exclusionary processes faced by lesbian, gay, bisexual, and transgender (LGBT) people in rural contexts (see a review by Kazyak, 2011), although discourses on sexuality and dealing with heterosexism hardly find their way to formal rural development policy objectives nor academic accounts on social exclusion. Watkins and Jacoby (2007), for example, reflect on this with respect to the lived experiences of single people and LGBT people in the English countryside. Experiences of homophobia and heterosexism have been reported even in rural settlements hosting significant events for LGBT communities (Gorman-Murray et al., 2008). However, the rural does not only hold negative associations for LGBT people compared to the promised sexual freedom and equality of the metropolis (McGlynn, 2018). In France and the USA, research by Annes and Redlin (2012) challenges the presumably unidirectional mobility patterns of rural gay men (i.e. from the countryside to the city) and highlights the importance of 'the rural' in constructing their gay male identity. As an example of resisting the 'othering' of LGBT households from academic accounts on rural in-migration, Smith and Holt (2005) examine lesbian households as actors of rural gentrification in England.

The LGBT movement has made significant steps in recent years, with marriage equality gaining momentum across the western world. Similarly, the application of queer theory in planning practice is growing, although still primarily focusing on urban debates (see a review by Doan and Hubbard in Chapter 29 of this Companion; see also Doan, 2011). Still, issues of heterosexism, homophobia and the heteronormative construction of community and family are well embedded in planning practice in many advanced, industrial societies (Forsyth, 2011; Frisch, 2002). Research has shown how these are also particularly evident in cases of crisis and disaster (Dominey-Howes et al., 2014) and even more so in rural and remote contexts, as for example in the case of the 2011 Tohoku earthquake in Japan where LGBT individuals in the affected rural communities were faced with bullying and discrimination in the recovery process (Yamashita et al., 2017). Criticising planning practice from a queer lens has focused on planners' actions to promote diversity that they recognise, i.e. which is visible to *them* (Muller Myrdahl, 2011) and the need of planners to be reflexive on their own assumptions about the communities they work with (Doan, 2011).

Age

Age concerns in rural planning practice refer both to the needs of an ageing countryside as well as the conditions of young people living in rural areas in terms of access to employment, opportunities and housing. The countryside is ageing across industrial societies, and this creates implications for housing and service provision (see also Chapter 30 of this Companion). Ageing primarily occurs as young people continue to move out (Brown and Glasgow, 2008), while older migrants, usually pre-retirement groups, come to stay in rural settings particularly in high amenity areas (Stockdale, 2006). The latter tend to be more affluent, so they occupy better quality housing creating problems of affordability particularly in contexts such as the UK with rural housing shortages (Satsangi et al., 2010). The profile of older rural residents creates of series of inequalities that cause concern for rural planning practice (Gkartzios and Ziebarth, 2016). First, it draws attention to the provision of age-specific housing and community service needs for an increasing number of elders living in rural areas (Bevan, 2009). Secondly, although older homeowners have property assets, incomes may be quite low pointing to processes of social exclusion (ibid.).

At the same time, there is a need to move on from an ageist view of the countryside that tends to problematise older people (Satsangi et al., 2010). Part of this lies in recognising that in many cases older people are integral to the development of rural communities as they offer skills and time for volunteering activities (Glasgow and Brown, 2012) as well as bringing new capital that can stimulate entrepreneurship.

The needs of younger people in the countryside as well as provisions for engagement in formal policy development are common concerns in the literature (Shucksmith, 2004), although the responses to such issues need not only to think about the support of young people to stay in rural areas, but also support to leave (Jentsch, 2006). With respect to young people in the rural housing market, just as more widely, leaving the parental home is a significant part of the transition into adulthood. Motivations include relationships, to become more self-sufficient, education and employment but can be problematic for young people due to employment, access to finance, types of accommodation available and the access to social rental accommodation compared to those classed as being more in need. The particularly rural features of these issues are that first, employment opportunities are often more limited and salaries are typically lower than in cities. Second, young people looking to continue with further or higher education are likely to need to migrate. Thirdly, access to the housing market is constrained by limited choices resultant from supply-side factors (Satsangi and Campbell, submitted; Satsangi et al., 2010).

Conclusions

This introductory chapter has outlined some of the ways in which sectional exclusions on the criteria of ethnicity, gender, sexuality, disability and age operate in rural society and thus inform rural planning. The following chapters attempt to approach some of these issues in greater depth. Our aim was to refer to these selective exclusions in an effort to inspire further research that specifically addresses inequalities in the rural context, rather than to imply that knowledge and debate on social exclusion is framed only by these, explored here, issues. On the contrary, rather than researching single or 'competitive' categories of exclusion, we need research on their intersection, accepting that inequalities and their associated identities are multiple, fluid and overlapping (drawing on McCall, 2005; Nash, 2011). Questions of citizenship, governance and spatial justice are of course entangled with such explorations. Amin's registers of solidarity adapted for the rural context by Shucksmith offer a useful lens, particularly through the register of relatedness and rights explored earlier.

In all cases research is required on how certain social groups are engaged or not in the planning process, and what steps planning institutions take to demonstrate their commitment to emerging equality and diversity agendas specifically to the rural context. We recognise that our evidence base is very far from complete: it is partial, deriving from telling but nonetheless small-scale case study work in usually specific contexts (the dominance of US–UK perspectives is undeniable in any review of the literature), and limited in scope and the extent to which we have good international data. Aiming to reach a consensus regarding experiences of exclusion globally is neither feasible nor desirable. What we aim for is research to demonstrate the diversity and intersectionality of such experiences, and how they might be embedded within planning practice across differentiated cultural and socio-economic contexts. We further observe that in some cases exclusions remain not only under the radar of policy makers, but of academics too. It rests with future planning researchers not only to debate how exclusions are played out (or even ignored) in the rural planning context, but also within academia and the research domain.

References

Amin, A. (2006) The good city, *Urban Studies*, 43 (5/6): 1009–1023.

Annes, A. and Redlin, M. (2012) Coming out and coming back: rural gay migration and the city, *Journal of Rural Studies*, 28 (1): 56–68.

Ballarino, G. and Panichella, N. (2015) The occupational integration of male migrants in western European countries: assimilation or persistent disadvantage? *International Migration*, 53 (2): 338–352.

Bell, D. and Valentine, G. (1995) Queer country: rural lesbian and gay lives, *Journal of Rural Studies*, 11 (2): 113–122.

Bevan, M. (2009) Planning for an ageing population in rural England: the place of housing design, *Planning, Practice and Research*, 24 (2): 233–249.

Boschman, S., Kleinhans, R. and van Ham, M. (2017) Ethnic differences in realizing desires to leave urban neighbourhoods, *Journal of Housing and the Built Environment*, 32(3), 495–512.

Brown, D. L. and Glasgow, N. (2008). *Rural Retirement Migration* (Vol. 21). Springer Science & Business Media, New York.

Bruton, M. (2006) Equality and diversity in the countryside: a disability perspective, *Countryside Recreation*, 14 (3/4): 6–11.

Burns, N., Paterson, K. and Watson, N. (2009) An inclusive outdoors? Disabled people's experiences of countryside leisure services, *Leisure Studies*, 28 (4): 403–417.

Catney, G. (2016) Exploring a decade of small area ethnic (de-)segregation in England and Wales, *Urban Studies*, 53 (8): 1691–1709.

Champion, A.G. (2001) The continuing urban–rural population movement in Britain: trends, patterns, significance, *Espace, Populations, Sociétés*, 1–2: 37–51.

Champion, A.G. (2016) Internal migration and the spatial distribution of population, in A.G. Champion and J. Falkingham (eds), *Population Change of the United Kingdom*, 125–142. London: Rowman and Littlefield.

Commins, P. (2004) Poverty and social exclusion in rural areas: characteristics, processes and research issues, *Sociologia Ruralis*, 44 (1): 60–75.

Doan, P.L (2011) *Queering Planning: Challenging Heteronormative Assumptions and Reframing Planning Practice*, Farnham: Ashgate.

Dominey-Howes, D., Gorman-Murray, A. and McKinnon, S. (2014) Queering disasters: on the need to account for LGBTI experiences in natural disaster contexts, *Gender, Place & Culture*, 21 (7): 905–918.

Farley, K. and Blackman, T. (2014) Ethnic residential segregation stability in England, 1991–2001, *Policy and Politics*, 42 (1): 39–54.

Finney, N. and Simpson, L. (2008) Internal migration and ethnic groups: evidence for Britain from the 2001 Census, *Population, Space and Place*, 14 (1): 63–83.

Finney, N., Catney, G. and Phillips, D. (2015) Ethnicity and internal migration, in D.P. Smith, N. Finney, K. Halfacree and N. Walford (eds), *Internal Migration: Geographical Perspectives and Processes*, 31–46. Farnham: Ashgate.

Forsyth, A. (2011) Queerying planning practice: understanding non-conformist populations, in P. Doan (ed.), *Queerying Planning: Challenging Heteronormative Assumptions and Reframing Planning Practice*, 21–52, Farnham: Ashgate.

Frisch, M. (2002) Planning as a heterosexist project, *Journal of Planning Education and Research*, 21 (3): 254–266.

Gkartzios, M. and Shucksmith, M. (2015) 'Spatial anarchy' versus 'spatial apartheid': rural housing ironies in Ireland and England, *Town Planning Review*, 86 (1): 53–72.

Gkartzios, M. and Ziebarth, A. (2016) Housing: a lens to rural inequalities, in M. Shucksmith and D. Brown (eds), *International Handbook of Rural Studies*, 495–508, New York: Routledge.

Glasgow, N. and Brown D.L. (2012) Rural ageing in the United States: trends and contexts, *Journal of Rural Studies*, 28 (4): 422–431.

Ibraimovic, T. and Masiero, L. (2014) Do birds of a feather flock together? The impact of ethnic segregation preferences on neighbourhood choice, *Urban Studies*, 51 (4): 693–711.

Jargowsky, P.A. (2009) Immigrants and neighbourhoods of concentrated poverty: assimilation or stagnation? *Journal of Ethnic and Migration Studies*, 35 (7): 1129–1151.

Jensen, L. (1994) Employment hardship and rural minorities: theory, research, and policy, *Review of Black Political Economy*, 22 (4): 125–144.

Jentsch, B. (2006) Youth migration from rural areas: moral principles to support youth and rural communities in policy debates, *Sociologia Ruralis*, 46 (3): 229–240.

Jivraj, S. and Khan, O. (2013) Ethnicity and deprivation in England: How likely are ethnic minorities to live in deprived neighbourhoods? Manchester: University of Manchester Centre on Dynamics of Ethnicity. Retrieved on 14 April 2016 from www.ethnicity.ac.uk/medialibrary/briefingsupdated/ethnicity-and-deprivation-in-england-how-likely-are-ethnic-minorities-to-live-in-deprived-neighbourhoods%20(1).pdf.

Kaplan, D.H. and Douzet, D. (2011) Research in ethnic segregation (iii): segregation outcomes, *Urban Geography*, 32 (4): 589–605.

Kazemipour, A. and Halli, S. (2000) The colour of poverty: a study of the poverty of ethnic and immigrant groups in Canada, *International Migration*, 38 (1): 69–88.

Kazyak, E. (2011) Disrupting cultural selves: constructing gay and lesbian identities in rural locales, *Qualitative Sociology*, 34 (4): 561–581.

Kitchin, R. (1998). 'Out of place', 'knowing one's place': space, power and the exclusion of disabled people, *Disability and Society*, 13 (3): 343–356.

Lichter, D.T., Parisi, D. and Taquino, M.C. (2012) The geography of exclusion: race, segregation, and concentrated poverty, *Social Problems*, 59 (3): 364–388.

Little, J. and Panelli, R. (2003) Gender research in rural geography, *Gender, Place and Culture: A Journal of Feminist Geography*, 10 (3): 281–289.

Macpherson, H. (2009) The intercorporeal emergence of landscape: negotiating sight, blindness, and ideas of landscape in the British countryside, *Environment and Planning A*, 41: 1042–1054.

Madden, J.F. (2014) Changing racial and poverty segregation in large US metropolitan areas, 1970–2009, *International Regional Science Review*, 37 (1): 9–35.

Massey, D.S. and Fischer, M.J. (2000) How segregation concentrates poverty, *Ethnic and Racial Studies*, 23 (4): 670–691.

McCall, L. (2005) The complexity of intersectionality, *Signs* 30 (3): 1771–1800.

McConnell, E.D. (2012) House poor in Los Angeles: examining patterns of housing-induced poverty by race, nativity, and legal status, *Housing Policy Debate*, 22 (4): 605–631.

McGlynn, N. (2018) Slippery geographies of the urban and the rural: public sector LGBT equalities work in the shadow of the 'Gay Capital', *Journal of Rural Studies*, 57: 65–77.

Merry, M.S. (2013) *Equality, Citizenship, and Segregation: A Defense of Separation*, New York: Palgrave.

Muller Myrdahl, T. (2011) Queerying creative cities, in P. Doan (ed.), *Queerying Planning: Challenging Heteronormative Assumptions and Reframing Planning Practice*, 157–167, Farnham: Ashgate.

Nash, J.C. (2011) 'Home truths' on intersectionality, *Yale Journal of Law and Feminism*, 23 (2): 445–470.

Neal, S. and Agyeman, J. (eds) (2006) *The New Countryside? Ethnicity, Nation and Exclusion in Contemporary Rural Britain*, Bristol: Policy Press.

Obucina, O. (2014) Paths into and out of poverty among immigrants in Sweden, *Acta Sociologica*, 57 (1): 5–23.

Peach, C. (2009) Slippery segregation: discovering or manufacturing ghettoes? *Journal of Ethnic and Migration Studies*, 35 (9): 1381–1395.

Phillips, D. (1998) Black minority ethnic concentration, segregation and dispersal in Britain, *Urban Studies*, 35 (10): 1681–1702.

Phillips, D. (2006) Parallel lives? Challenging discourses of British Muslim self-segregation, *Environment and Planning D: Society and Space*, 24 (1): 25–40.

Pini, B. (2003) Feminist methodology and rural research: reflections on a study of an Australian agricultural organisation, *Sociologia Ruralis*, 43(4), pp. 418–433.

Pini, B., Philo, C. and Chouinard, V. (2017) On making disability in rural places more visible: challenges and opportunities [Introduction to a special issue], *Journal of Rural Studies*, 51: 223–229.

Platt, L. (2003) Ethnicity and inequality: British children's experience of means-tested benefits, *Journal of Comparative Family Studies*, 34 (3): 357–377.

Platt, L. (2007) Child poverty, employment and ethnicity in the UK: the role and limitations of policy, *European Societies*, 9 (2): 175–199.

Reeve, K. and Robinson, D. (2007) Beyond the multi-ethnic metropolis: minority ethnic housing experience in small town England, *Housing Studies*, 22 (4): 547–571.

Robinson, D. (2010) The neighbourhood effects of new immigration, *Environment and Planning A*, 42: 2451–2466.

Saenz, R. (2012) Rural race and ethnicity, in L.J. Kulcsár and K.J. Curtis (eds), *International Handbook of Rural Demography, International Handbooks of Population, Volume* 3, 207–223. New York: Springer.

Satsangi, M. (2011) Feminist epistemologies and the social relations of housing provision, *Housing, Theory and Society*, 28 (4): 398–409.

Satsangi, M. (2013) Synthesizing feminist and critical realist approaches to housing studies, *Housing, Theory and Society*, 30 (2): 193–207.

Satsangi, M. and Campbell, R. (Submitted) Young people and constrained housing markets: the potential of pathways and institutionalist approaches, under submission at *Housing, Theory and Society*.

Satsangi, M., Gallent, N. and Bevan, M. (2010) *The Rural Housing Question: Communities and Planning in Britain's Countrysides*, Bristol: Policy Press.

Shortall, S. (2008) Are rural development programmes socially inclusive? Social inclusion, civic engagement, participation, and social capital: exploring the differences, *Journal of Rural Studies*, 24 (4): 450–457.

Shucksmith, M. (2004) Young people and social exclusion in rural areas, *Sociologia Ruralis*, 44 (1): 43–59.

Shucksmith, M. (2016) Re-imagining the rural: from rural idyll to good countryside, *Journal of Rural Studies*, 59: 163–172.

Shucksmith, M. and Chapman, P. (1998) Rural development and social exclusion, *Sociologia Ruralis*, 38 (2): 225–242.

Simpson, L. (2012) More segregation or more mixing? Dynamics of Diversity: Evidence from the 2011 Census, Manchester: University of Manchester Centre on the Dynamics of Ethnicity. Retrieved on 18 April 2016 from www.ethnicity.ac.uk/medialibrary/briefingsupdated/more-segregation-or-more-mixing.pdf.

Simpson, L. (2014) How has ethnic diversity changed in Scotland?, Dynamics of Diversity: Evidence from the 2011 Census, Manchester: University of Manchester Centre on the Dynamics of Ethnicity. Retrieved on 18 April 2016 from www.ethnicity.ac.uk/medialibrary/briefings/dynamicsofdiversity/code-census-briefing-scotland_v2.pdf.

Skifter Andersen, H. (2016) Selective moving behavior in ethnic neighbourhoods, white flight, white avoidance, ethnic attraction or ethnic retention? *Housing Studies*, 32 (3): 296–318.

Slack, T. and Jensen, L. (2002) Race, ethnicity, and underemployment in nonmetropolitan America: a 30-year profile, *Rural Sociology*, 67 (2): 208–233.

Smith, D.P. and Holt, L. (2005) 'Lesbian migrants in the gentrified valley' and 'other' geographies of rural gentrification, *Journal of Rural Studies*, 21 (3): 313–322.

Stillwell, J. and Hussain, S. (2010) Exploring the ethnic dimension of internal migration in Great Britain using migration effectiveness and spatial connectivity, *Journal of Ethnic and Migration Studies*, 36 (9): 1381–1403.

Stockdale, A. (2006) The role of a 'retirement transition' in the repopulation of rural areas, *Population, Space and Place*, 12 (1): 1–13

Tammaru, T. and Kontuly, T. (2011) Selectivity and destinations of ethnic minorities leaving the main gateway cities of Estonia, *Population, Space and Place*, 17: 674–688.

Van Ham, M. and Manley, D. (2009) Social housing allocation, choice and neighbourhood ethnic mix in England, *Journal of Housing and the Built Environment*, 24 (4): 407–422.

Walker, K. (2018) Locating neighbourhood diversity in the American metropolis, *Urban Studies*, 55 (1): 116–132.

Walks, R.A. and Bourne, L.S. (2006) Ghettoes in Canada's cities? Racial segregation, ethnic enclaves and poverty concentration in Canadian urban areas, *Canadian Geographer-Geographe Canadien*, 50 (3): 273–297.

Watkins, F. and Jacoby, A. (2007) Is the rural idyll bad for your health? Stigma and exclusion in the English countryside, *Health and Place*, 13: 851–864.

Yamashita, A., Gomez, C. and Dombroski, K. (2017) Segregation, exclusion and LGBT people in disaster impacted areas: experiences from the Higashinihon Dai-Shinsai (Great East-Japan Disaster), *Gender, Place & Culture*, 24 (1): 64–71.

28

Planning the farmyard
Gender implications

Sally Shortall

Introduction

Ruth: *It got us thinking that maybe it's time for a bit of a refurb.*
Pip: *Of the milking parlour?*
Ruth: *Actually a full redesign.*
Pip: *That would be amazing Mum. It's such a struggle in there. It would have to be flexible.*
Ruth: *I agree.*

Later . . .

Ruth: *You've worked so hard to get the finances back on track, looked after the farm really well when we were away.*
Pip: *Wow you give me a pay rise Mum!*
Ruth: *I wish. No what we were thinking. . .only if you want to, is that you lead on the redesign of the milking parlour.*
Pip: *You're asking me to do the refurb? Oh Mum that's amazing.*
Ruth: *Watch your coffee!*
Pip: *Sorry, sorry but that's amazing! I hate the parlour. It's so out of date.*
Ruth: *Now you can bring it up to scratch.*
Pip: *I'm so excited!*

(Extract from The Archers, BBC Radio 4, 24 November 2017)

Many countries have popular radio or television shows that are used to communicate agricultural messages through drama. *The Archers* is the popular UK radio drama that has run for over 60 years. The quote above is from a conversation between Pip and Ruth Archer. Pip in the Archers is unusual. She is the eldest child, and despite having two younger brothers, it is she who farms full-time with her parents. Redesigning the milking parlour is a rare and expensive event on any farm, and it is again the exception that a woman would be given oversight of this task. What many people will be unaware of, is that Pip will not need planning permission to redesign the milking parlour. She can erect, extend or alter a building under permitted farm development.

The farmyard is central to the farm. It is the functional space where family labour meets with farm production. It is purposeful, and more recently the aesthetics of the farmyard have been considered (Tassinari et al., 2013). The farmyard is also recreational, and many farm children grow up playing in farmyards. They are also dangerous places, with persistent levels of farm accidents occurring over the years (Shortall, 2017) due to unsafe practices and structures in the farmyard. Farmyards are not just a collection of buildings. It is where machinery is stored, equipment is repaired, and grain and straw is kept. It is remarkable that they are subject to such light planning regulation.

This chapter has two central arguments. First, it is argued that men's farming identity is linked to being an independent autonomous farmer, who is responsible for the family farm. Planning legislation (in a UK context) has accommodated this identity by only recently having any farm developments subject to legislation, and even now there are many caveats that allow exemptions. For a long time, farmers could build what they wanted on their farms. Second, it is argued that the farmyard, the physical space where the family meets the farm, has not been subject to planning scrutiny, in both research and policy terms. This has gender implications because farms have evolved and are premised on the idea that farming relies on brawn and a masculine physique. There is little research on farmyards, but what there is focuses on the preservation of farmyard buildings and maintaining historical buildings of cultural heritage interest. Yet, as women increasingly enter farming, there is a need to redesign the yard to ensure it does not continue to presume the physique of a male farmer. The farmyard is a site of unconscious gender bias.

The chapter is structured as follows. First a brief overview of planning legislation related to farming is given for the British Isles. How this intertwines with farming individuality and autonomy is considered. Next some of the literature on gender, masculinity and farming is reviewed. A farmer is typically seen as a hard man dealing with dangerous substances and undertaking work that requires strength. Then a Scottish study (Shortall et al., 2017) is described which considered the role of women in Scottish agriculture. Through this research it became clear that women are taking risks with farm safety to 'prove' their farming identity. It is concluded that planning a farmyard is central to the safety of women. Further it is argued that the current lax approach of planning legislation is reminiscent of a pre-modern view of farming and farmers' rights to do what they wish on their property. The farmyard is a work place, but exemptions allow farmers incredible freedom to construct this workplace as they wish.

Planning regulation and farming

In Ireland and the UK, agriculture-related land-use changes have been largely exempted from planning. It is characteristic that both in Ireland and the UK the post-war introduction of planning frameworks (in 1947 in the UK and much later, in 1964, in Ireland) either ignored the rural as a space for planning intervention (i.e. in the case of Ireland), or over-emphasised the conservation of rural landscapes (i.e. in the case of the UK) (Gkartzios and Scott, 2009). In both cases the hegemony of agriculture and agriculture-related activities is supported with minimum planning intervention, despite significant reforms across different contexts (England, Scotland, Wales, Ireland, Northern Ireland) and the introduction of planning regulations for new structures (see for example Regan, 2002). Revised legislation in the 2000s for example was tied to the expectation that farms were about to undertake modernisation and farm diversification activities that would require new structures (ibid.). Planning regulations relate to new activities, and previous buildings are not subject to planning legislation. It is of significance that in each context there are many more concessions for farming than for any other occupation.

The planning regulations are confusing, but many farming developments are classified as 'permitted development'. Permitted developments do not need planning permission. In Pip Archer's case, she is altering a building. She does not need planning permission. Farmers are also allowed to excavate and carry out engineering operations needed for agricultural purposes. The allowances made for agriculture are interesting. Farming is one of the few family occupations that persists, and it is the occupation with the highest rate of heredity – in general, sons take over their fathers' farms. It means farming retains some pre-industrial features, and the planning regulation seems to accommodate this expectation that farmers have a right to do what they wish on their private agricultural land. It is similar to the exceptions made for agricultural practices with regard to health and safety on farms (Shortall, 2017). While young people cannot drive cars until they are seventeen years old, exceptions are made for farm children who are allowed to drive heavy machinery at thirteen and fourteen years of age. In both cases planning regulation and health and safety regulation are treating farming as a special case, evidencing the hegemony of agriculture in rural space and its legitimacy within planning and public policies. The farming expectation that they should retain their independent identity to act on their private property is clear in the UK National Farmers' Union (NFU) response to the Rural Planning Review. They argue 'it is unfortunate that to the majority Britain's countryside is regarded as simply an asset to be protected rather than used productively, as it has been for millennia' (NFU, 2016: 2). The preservation of the English countryside as opposed to the productive countryside is a long running English debate (Murdoch and Lowe, 2003), particularly in relation to rural housing (Gkartzios and Shucksmith, 2015). In their response to the Rural Planning Review, the NFU makes a strong case for farm families to be able to build retirement homes on their farms and they suggest not being able to do so threatens the future of farming. They also argue that it is necessary for the size of buildings which can be built on farms without planning permission to be raised as farm machinery has increased in size. In concluding, the NFU outline that they are:

> frequently asked why local planning authorities are so negative towards and don't understand farming . . . this lack of integration and understanding is often seen as a barrier by our members and the NFU . . . will encourage the consideration of farming priorities and do not deter them because of, for example, restrictive landscape designations and sustainable transport policies.
>
> *(NFU, 2016: 18)*

The Farmers' Union displays the expectation that farmers should have autonomy over their actions on their own farms. It is part of the farming identity of autonomous, self-employed and self-sufficient workers. The interesting point is that planning regulations, along with health and safety regulations, make considerable exceptions for farming. They facilitate this sense of autonomy and self-sufficiency. The farmyard, a work place, is very lightly regulated.

Gender, masculinity and the farmyard

Space is gendered (Shortall, 2016). This of course has a cultural and temporal context; some cultures and religions have more gendered space than others, and in general, the rigidities of gendered space have dissipated. This is true for rural areas as well as urban areas. Nonetheless, rural space, and in particular, farming space, continues to be used to signify and maintain distinctive gender identities. A considerable amount of work has focused on how the farm shapes the farm family, gender roles, and the identity of family members. All of these are bound up together. It is only possible to touch on some of that literature here, and it is done so for illustrative purposes. The farmer is typically understood to be male. It defines his role identity, his group identity, and his gender identity. The patrilineal line of inheritance also means that it is deeply embedded in culture and traditional practices that agriculture is male (Shortall, 2014). The outdoors are coded as masculine, while indoor activities are coded as feminine (Pini et al., 2014; Brandth and Haugen, 2010; Brandth et al., 2014; Little, 2014; Little and Panelli, 2003; Campbell and Bell, 2000; Campbell et al., 2006). In the outdoors, men undertake hard, physical, and sometimes dangerous work such as handling heavy machinery, being foresters and dealing with chemicals. The tractor, for example, is argued to have become a symbol of male power and spatial domination by men of the outdoors. Men have appropriated agricultural technology to underline their identity as farmers (Brandth and Haugen, 1998; Pini, 2005; Saugeres, 2002). Specific activities such as hunting and mining are seen as outdoor male activities (Campbell and Bell, 2000). Women who breach this male space and work with heavy machinery often seek other ways to reconfirm their feminine identity (Brandth, 1995). Women's indoor work is predominantly seen as domestic or as socially reproductive work. It is seen as sustaining the household (Whatmore, 1991; Trauger et al., 2010) This significantly contributes to the invisibility of components of women's farm work such as management of accounts, and decision making, because the indoor nature of this work means it is not seen as authentic farm work and thus reduces women's identity as farm workers (Sachs, 1983; Alston, 1995; Bock and Shortall, 2006).

Further gender segregated space is evident in the provision of agricultural training. Most agricultural training is structured in a vocational way for those that will enter the occupation, so in many ways it is not surprising that most agricultural programmes have a majority of male students. However, the socially constructed identities of women as home makers and farmers' wives mean that they do not obtain a knowledge transfer appropriate to their farming roles. Women farmers are underserved in agricultural education and technical assistance (Trauger et al 2008; Shortall, 1996; Alston, 1998; Liepins and Schick, 1998). Women often view training groups and programmes as being for men and feel unwelcome and conspicuous in this space. Agriculture extension workers do not always see women as 'authentic' farmers, because they do not occupy outdoor space and hence do not invite them to training initiatives or address programmes to their work (Barbercheck et al., 2009; Trauger et al., 2010; Teather, 1994). Here we see women's self-verification of not being the farmer being institutionally reinforced by agricultural extension workers. It is remarkable that this gender divided space persists. It is problematic, because increasingly off-farm employment to support the farm is decided between the couple, and educational levels and life cycle issues determine who will work on the farm and who will work off the farm (El-Osta et al., 2008; Benjamin and Kimhi, 2006). Seeing men as the authentic farmer means the relevant person on the farm may not receive appropriate training. The literature on women's role in agriculture is characterised by themes of continuity and change. On the one hand, the hegemonic discourse of the masculine prevails in farming despite changing gender roles (Brandth and Haugen, 2016). At the same time, change occurs. In the study which will be described in the next section, it is clear that women are taking a more central role in farming. When they do so, they often seem to take risks to 'prove' their farming identity, to prove that they can undertake the same dangerous and physically arduous work of men. This will make them 'authentic' farmers. The unconscious gender bias in how farmyards are planned becomes of significance here.

The research: women and the farmyard

In 2016 the Scottish Government's Rural and Environmental Science and Analytical Services Division commissioned research on 'Women and Farming in the Agricultural Sector'.

The overall purpose of this research was to establish a baseline position on women in farming and the agricultural sector in order to inform policies to enhance the role of women in these sectors. The specific aim of this research project was to investigate the role of women in farming and the agriculture sector in Scotland under five headings specified by the Scottish Government: daily life, aspirations, career paths, leadership and comparative analysis with women in other family businesses. During the research, inheritance, training and farm safety also emerged as important issues.

The research questions were explored through interviews and focus group discussions with women and men involved in farming, crofting and the agricultural sector across Scotland. Specifically, the interviews and focus groups included women who are new entrants to farming, as well as those who are involved in agricultural industry leadership, estates and large-scale farms, crofting, and farm diversification. A sample of men involved in farming, crofting and the agricultural sector were also included as interviewees and focus group participants. Interviews and focus groups were also arranged with women who work in non-farming family businesses, in order to provide a comparison to farming businesses.

The research comprised a literature review, 10 focus groups, 34 interviews and two online surveys: in total, over 1300 women and 17 men from across Scotland participated. The research was undertaken from June 2016 to March 2017 all over Scotland. The qualitative and quantitative components ran side by side because of the specified time constraints set by the Scottish Government. This chapter only uses the qualitative research. Full details on the methodology can be found in Shortall et al. (2017).

Findings: the gendered farmyard and occupational entry

The research brief did not ask for this study to consider farm safety. However, it came up repeatedly and, soon into the research, a number of questions about safety were asked.

Both men and women were of the view that women are more safety conscious than men. This did not seem to be the case in our study, and it has not been possible to locate any other study that specifically spoke to women farmers. Both men and women take risks on farms. As already discussed, taking risks is seen as part of the process of establishing the masculinity of the occupation. Women sometimes recounted taking risks to 'prove' that they could farm as well as men, that they are 'authentic' farmers. Consider the dialogue that occurred within this focus group with women active in agriculture (emphasis added):

> I suppose in a way for me there's . . . especially with the background of my father him wanting a son and being . . . I love farming and it is what I want to do but there is *that little kind of devil on my shoulder that says you need to prove them wrong. You're a girl and I'm just as . . .* and I am very . . . when there's a guy on the farm and they're lifting heavy . . . *they say do you want a hand with that? I'm like no I can do it!*
>
> *It's amazing what you can lift when they offer you help isn't it?* [Laughter]
>
> *Yeah I'm the same whenever somebody offers me like do you want me to hitch that trailer up for you or whatever, or do you want me to do this for you, do you want me to do that? I'm like no I will do it!* Yeah there is almost like a point –
>
> No but with . . . *within the industry and things like that yeah it's . . . yeah there's definitely something to prove isn't there?*
>
> *(Women in agricultural industry focus group)*

In this dialogue, women are sharing their experiences of taking risks to prove themselves as authentic farmers. They talk about needing to prove farming men 'wrong' and also the wider farm 'industry'. Farming is the occupation that has persisted as a masculine one more than any other. Women seek to demonstrate their ability to lift objects typically too heavy for a woman's frame. They see the farming industry as one where they need to 'prove' they are as able. The issue is that in the process of proving themselves to the industry and displaying this prowess associated with the masculine performance of farming, they are taking risks with their own well-being and safety. This is not unconscious – one woman above talked about 'that little devil on my shoulder'. The devil on your shoulder pushes you to undertake activities that you know you should not.

Farming is the occupation with the highest rates of accidents and fatalities (Shortall, 2017). One difficulty is that farms tend to be family farms, with all family members in and out of the farmyard. Men report being in the zone, or not thinking about what they are doing when something dangerous happens. In the quote below, this woman nearly suffered a fatality. However, again, she was consciously trying to prove her authenticity as a farmer:

> *My big accident I nearly had last year was . . . and it was part of my own stubbornness and not asking for help*, I was carting the grain in and out on the combine and we were putting the winter barley into the bins and my partner was on the combine and I was loading the pits which then . . . and I was having to check the bins to make sure they weren't over filling and when one bin fills you need to move the shoot that comes out of the top. *You have to physically lift it up and this is overhanging a big empty sixty ton bin on your left hand side and you have to move it along and put the shoot to the next bit. Open the hole and this is quite heavy and its quite an awkward . . .* you're hanging out over a big empty bin and its very awkward because it breaks in half as well if you don't hold it properly and I got it off and I was shifting backwards and it came in half and I . . . *and it went into the bin and I – Luckily I didn't go into the bin but I gave myself a heck of a fright . . . but that was a learning curve. I would never do that again.*
> *(Women in agricultural industry focus group)*

This woman knew that she was undertaking dangerous farm activities and that her own stubbornness nearly resulted in her fatality. The work is presented as hard physical labour, and she knew that she was pushing herself to a dangerous limit. She told the group that she had learned her lesson and would never do that again. However, this could easily have been a fatality, and research on farm accidents suggests that the farming community does not actually change its behaviour after near misses (Shortall, 2017).

This dangerous environment for women is linked to the unconscious gender bias inherent in how the farmyard is planned. This became evident with some of the new entrant women interviewed. New entrants are understood in our report as people who did not inherit their farm, but tend to come into the sector through renting land. The new entrants interviewed were very dynamic and cutting edge. Many of the women had realised they would not inherit the farm and had trained in agriculture and taken agriculture related employment. Typically they and their partner took on a rented farm. Their partners also tended to be highly skilled, perhaps because renting a farm requires a considerable amount of capital. In most cases both worked off the farm and ploughed their resources back into the farm, and women tended to reduce their off-farm work when they had children and increase their full-time farming hours. Consider this quote from a new entrant woman below:

> Because I just think well . . . there's not someone else about that if something happened, *it's not a very safe . . . environment sometimes.* So as well as like technically I could do all that, but sometimes I actually just want another pair of hands. *I think that's definitely more so when*

you're a female. Like he wouldn't bat an eyelid going out himself and just doing it. Yeah I think so, yeah I think women probably think about it, definitely there's a lot of that and *I always go on to him because . . . if we just had the right set up we wouldn't . . . because there are times, situations that . . . I think well actually if this gating was all adjusted a little bit I could run that cow from there to there, lock it in the yoke, I don't have to carry a six foot hurdle myself and pin it up, and move dung to get it in the right place.*

(New entrant woman, Orkney)

This new entrant woman recognises that the farmyard is not a safe environment and particularly so for women. This particular woman, as told to us during the interview, was five foot, two inches. She recognises that the issue is not her size or strength, but rather how the farmyard is designed. She 'goes on' at her husband about the need for them to have the 'right set up'. This was repeated time and again during interviews with new entrants. Women talked about the need to have the 'right kit' and gates on wheels, which eliminate the need for strength to move heavy gates. The woman above discusses the need to adjust gating to allow her to move cattle from here to there and to be able to secure the cattle without having to carry a six-foot hurdle.

Below is another example of a young woman new entrant describing how her work would be easier if the farmyard was planned in a way that was gender friendly:

And I'll often say right I'll do the PDA [i.e. Personal Digital Assistant] stuff and I've said to him do you know what I could do that quite happily but if you're there and you're going to wrestle with them crack on and do that! And I'll . . . I don't know just operate the pens. *And we're having a bit of a conversation at the moment because we need to upgrade, or we're going to try and put in an application for gates and some handling equipment and I would love what's called a Combi Clamp System so the sheep run up into the Combi Clamp. You can clamp them with your weight and dose them.*

(Focus group new entrants)

Here again, we see a new entrant woman who recognises how the gendered planning of the farmyard has implications for women. She negotiates with her husband that she will undertake the less strenuous work. They have to upgrade their farmyard so they are discussing equipment that means she can undertake more strenuous work safely. The new system will allow her to close the clamp with her weight, and she will then be able to dose the sheep. This woman recognises that the problem is not that she is a woman, the problem is how the farmyard is (not) planned.

Conclusions

Farmyards are subject to remarkably light planning legislation and, despite their importance, have remained invisible spaces in planning policy and research. This chapter has focused on the British Isles, but a cursory look at international legislation suggests this pattern is universal, but this also needs more careful examination. Why is this the case? On one hand it follows arguments that agriculture has been traditionally seen as the legitimate activity for rural areas and therefore, in parallel with the powerful farming lobby, it should be accommodated as far as possible. Furthermore, the heredity nature of farming and the involvement of the family from a young age suggests that it is seen as a special, pre-industrial occupation to which rules do not apply as elsewhere. The state protects our right to own private property and perhaps exceptions are made because farms are privately owned business properties. Yet other privately owned businesses do not enjoy the same exemptions that farms are afforded. The farmyard is the merging of a social and productive space, it is where the family meets the farm. It is a workplace, and the most dangerous occupation with the persistently highest rates of farm fatalities and life changing accidents every year. Hard questions need to be asked about why this hazardous occupation has so many exemptions from planning regulations (and health and safety regulations).

In some land tenure systems, there seem to be an increasing number of women entering farming. This is the case for Scotland and Scottish research presented in this chapter shows how the unconscious gender bias in the construction of the farmyard is putting women in danger. Farmyards are constructed for men; they presume strength and brawn. The relevant literature shows that undertaking dangerous activities and working with dangerous substances is one of the ways men demonstrate their masculinity and authenticity as farmers. Women undertake dangerous activities in farmyards to 'prove' they are authentic farmers. Interestingly, dynamic new entrant women recognise that the problem is not that they are women, but rather the gendered design of the farmyard. Yet, these issues hardly appear in planning research, or in policy discourses. It is high time for planning to consider the gendered construction of such spaces and their implications. The invisibility of farmyards in planning literature continues well embedded notions about farming and rural identities that are not only outdated but pose concerns for health and safety.

Acknowledgments

This research was funded by the Scottish Government (tender ref. CR/2015/21). Sally Shortall was the Principal Investigator, and Lee-Ann Sutherland and Annie McKee were co-investigators. The 'Gender, Masculinity and the Farmyard' section borrows heavily from Shortall (2016). I am grateful to Menelaos Gkartzios for critical comments and Anne Liddon for transcribing the Archers quote.

References

Alston, M. 1995. *Women on the land: the hidden heart of Australia.* UNSW Press, Australia.
Alston, M. 1998. Farm women and their work: why is it not recognised? *Journal of Sociology* 34(1): 23–34.
Barbercheck, M., Brasier, K., Kiernan, N., Sachs, C., Trauger, J., Fiendeis, J. and Moist, L. 2009. Meeting the extension needs of women farmers. *Journal of Extension* 47(3): 1–11.
Benjamin, C. and Kimhi, A. 2006. Farm work, off-farm work, and hired farm labour: estimating a discrete-choice model of French farm couples' labour decisions. *European Review of Agricultural Economics* 33(2): 149–171.
Bock, B. and Shortall, S. (eds). 2006. *Rural gender relations: issues and case studies.* London: CAB International.
Brandth, B. 1995. Rural masculinity in transition: gender images in tractor advertisements. *Journal of Rural Studies* 11(2): 123–133.
Brandth, B. and Haugen, M. S. 1998. Breaking into a masculine discourse: women and farm forestry. *Sociologia Ruralis* 38(3): 427–442.
Brandth, B. and Haugen, M. 2010. Doing farm tourism: the intertwining practices of gender and work. *Signs* 35(2): 425–446.
Brandth, B. and Haugen, M. 2016. Rural masculinity. In M. Shucksmith and D. Brown (eds), *The Routledge International Handbook of Rural Studies*, 412–424. New York: Routledge/Falmer.
Brandth, B., Follo, G. and Haugen, M. 2014. Paradoxes of a women's organisation in the forestry industry. In B. Pini, B. Brandth and L. Little (eds), *Feminisms and Ruralities*, 57–69. London: Lexington Books.
Campbell, H., and Bell, M. 2000. The question of masculinities. *Rural Sociology* 65(4): 532–546.
Campbell, H., Bell, M.M. and Finney, M. (eds). (2006). *Country Boys: Masculinity and Rural Life.* Philadelphia, PA: Pennsylvania State University Press.

El-Osta, H., Mishra, A. and Morehart, M. 2008. Off-farm labor participation decisions of married farm couples and the role of government payments. *Applied Economic Perspectives and Policy* 30(2): 311–332.

Gkartzios, M. and Scott, M. 2009. Planning for rural housing in the Republic of Ireland: from national spatial strategies to development plans. *European Planning Studies* 17(12): 1751–1780.

Gkartzios M. and Shucksmith, M. 2015. 'Spatial anarchy' versus 'spatial apartheid': rural housing ironies in Ireland and England. *Town Planning Review* 86(1): 53–72.

Liepins, R. and Schick, R. 1998. Gender and education: towards a framework for a critical analysis of agricultural training. *Sociologia Ruralis* 38(3): 286–302.

Little, J. and Panelli, R. 2003. Gender research in rural geography. *Gender, Place and Culture: A Journal of Feminist Geography* 10(3): 281–289.

Murdoch, J. and Lowe, P. 2003. *The Differentiated Countryside*. London: Routledge.

NFU. 2016. NFU: Consultation response: rural planning review. Retrieved from www.nfuonline.com/assets/62032.

Pini, B. 2005. Farm women: driving tractors and negotiating gender. *International Journal of Sociology and Food* 13(1): 1–12.

Pini, B., Brandth, B. and Little, J. 2014. *Feminisms and Ruralities*. London: Lexington Books.

Regan, S. 2002. *Planning Requirement for Farm Building Development*. Teagasc: Athenry.

Sachs, C.E. 1983. *The Invisible Farmers: Women in Agricultural Production*. Totowa, NJ: Rowman and Allanheld.

Saugeres, L. 2002. Of tractors and men: masculinity, technology and power in a French farming community. *Sociologia Ruralis* 42(2) 143–159.

Shortall, S. 1996. Training to be farmers or wives? Agricultural training for women in Northern Ireland. *Sociologia Ruralis* 36(3): 269–286.

Shortall, S. 2014. Farming, identity and well-being: managing changing gender roles within Western European farm families. *Anthropological Notebooks* 20(3): 67–81.

Shortall, S. 2016. Gender and identity formation. In D. Brown and M. Shucksmith (eds), *The Handbook of Rural Studies*, 349–356. New York: Routledge/Falmer.

Shortall, S. 2017. People and the land: understanding the family farm. Inaugural lecture, 16 November, Newcastle University, Newcastle Upon Tyne.

Shortall, S., Sutherland, L.A., McKee, A. and Hopkins, J. 2017. *Women in Farming and the Agriculture Sector*. Edinburgh: Scottish Government.

Tassinari, P.,Torreggiani, D., Benni, S. and Dall'Ara, E. 2013. Landscape quality in farmyard design: an approach for Italian wine farms. *Landscape Research* 38(6): 729–749.

Teather, E. 1994. Contesting rurality: country women's social and political networks. In S. Whatmore, T. Marsden and P. Lowe (eds), *Gender and Rurality*, 31–50. London: David Fulton Publishers.

Trauger, A., Sachs, C., Barbercheck, M., Kiernan, N., Brasier, K. and Findeis, J. 2008. Agricultural education: gender identity and knowledge exchange. *Journal of Rural Studies* 24(4): 432–439.

Trauger, A., Sachs, C., Barbercheck, M., Kiernan, N., Brasier, K. and Schwartzberg, A. 2010. The object of extension: agricultural education and authenic farmers in Pennsylvania. *Sociologia Ruralis* 50(2): 85–103.

Whatmore, S. 1991. *Farming Women: Gender, Work and Family Enterprise*. London: Macmillan.

29

Queerying rural planning

Petra L. Doan and Daniel P. Hubbard

Introduction

The goal of this chapter is to use a queer theory lens to explore the ways in which planning and sexuality intersect in rural areas. Planning literature on rural spaces in both the United States and the United Kingdom does not often address issues linked to the LGBTQ population (Lapping, 1989; Daniels and Lapping, 1996; Gallent et al., 2008), although other disciplines such as geography and rural sociology have significantly more coverage of issues linked to this vulnerable population in rural areas. It is important to note that the authors are academic and professional planners whose primary experience is in the United States, and this context clearly colours the text. In the United States context for many people the very idea of rural planning is a virtual oxymoron since planning has been for most of the past 100 years a largely urban phenomenon. For example, Popper (1988: 292) argues that rural communities traditionally resist land use planning because these places 'espoused the American ideal of rugged individualism; many of their residents thought zoning verged on socialism'. Furthermore, in some rural areas residents view planning as an attempt by the government to impose regulations which unduly restrict property rights (Teitz, 2002). In contrast, planning in the United Kingdom and elsewhere has been influenced by a long tradition of town and country planning that reflects a more inclusive view of rural regions, though questions of how to define 'rural' remain a sticky question (Green, 2005).

The queer theory framework used in this chapter provides insights that help to deconstruct critical concepts about the location of LGBTQ individuals and explores the ways that LGBTQ subjectivities intersect with planning processes in rural areas. Knopp and Brown (2003: 409) suggest that queer theory

> applies the analytical and political tools of poststructuralism (including deconstruction and decentering) to the categories commonly associated with lesbian and gay studies and experiences (lesbian and gay identities, gender, sexuality, closeting, etc.) . . . The aim is to reveal not only the indeterminacy, contingency, malleability, and often oppressive nature of taken-for-granted concepts (such as 'the state'), but also the subtle ways in which the hidden power relations behind these categories depend crucially on the ways in which they are gendered and sexualised.

One example of the use of this technique is to complicate the ways that planners divide complex phenomena into binary categories. This dichotomous reduction enables the use of tools such as the two-by-two contingency table that streamline analysis, permit cross-tabulations, and enable the use of chi square statistics. For example, the simplification of a region into urban and rural might be combined with the division of identities into gay and straight to produce an easily understood contingency table represented below as Table 29.1.

A common misconception of using such a table is the assumption that all gays are urban and rural areas are straight. Such generalisations mask the underlying complexity which queer theory seeks to unpack.

De-constructing the urban–rural spectrum

The idea of a rural–urban dichotomy is deeply rooted in much LGBTQ culture with an implicit assumption that queer people are an urban phenomenon. For instance, Kath Weston's (1995) advice to young queer people was to move to the city to escape the narrow and traditional biases of rural areas. LGBTQ migrants were assumed to be attracted to the lure of a more anonymous life in the city with its greater freedom to explore sexuality and identity as well as find community. However, deconstructing this binary spatial frame allows a much more nuanced understanding of the ways in which queer people actually experience rural and urban spaces. Gorman-Murray (2007: 106) suggests that queer migration may be more usefully understood as a journey of self-discovery that

> shifts explanatory power from the rural-to-urban scale of movement per se to the embodied motivations of individual migrants; from the fixed, symbolic contrast between the city and the country to the actual movement of the queer body through space. And since self-discovery and personal identity-formation is ongoing and fluid, never complete and fixed, a focus on embodied displacement also encourages us to question the teleology of rural-to-urban relocations, and enables us to contemplate peripatetic migrations by queers as progressive quests for sexual identity.

Furthermore, a closer analysis of actual migration patterns suggests that the city does not always function as a final destination for queer rural migrants, but as a stopover in movements that follow a more circular pattern such that migration 'between and within these two spaces more accurately define informants' life itineraries' (Annes and Redlin, 2012: 67).

The deconstruction of urban and rural as opposites is critical to redefining rurality in ways that may enable planners to more accurately understand the nature of the communities for which they are planning. In general, urban areas are characterised as places where people live, develop land, and offer amenities in close proximity to one another. The opposite of urban is generally termed rural and practicing planners often rely on the US Census Bureau for guidance.

Table 29.1 Simplified contingency table.

| | | Identity | |
		Gay	Straight
Location	Urban		
	Rural		

In addition, the notion of rural is often embedded in comprehensive plans and land development regulations, as well as plans or regulations governing 'community character'.

A significant limitation of the US Census Bureau is its failure to count the entire spectrum of the LGBTQ community (Black et al., 2000). An atlas of same sex partners constructed by Gates and Ost (2004) shows clearly that same sex partners exist in nearly every county in the United States, no matter how rural. However, these numbers do not distinguish single gays, lesbians, and bisexuals. Furthermore, they say nothing about the complex identities included in the transgender community (Doan, 2016). Researchers wishing to make more accurate counts must rely instead on population-based sample surveys to estimate the LGBTQ presence at national, state, and local levels (for example, see Gates, 2014). In contrast, the United Kingdom has recently made a concerted effort toward counting 'sexual identities' by region, sex, age, marital status, ethnicity and socio-economic classification (Office for National Statistics, 2016) which holds some promise for more nuanced rural and urban planning policies.

Another element of the rural–urban distinction that must be examined is the way that planners differentiate spaces and places on the urban–rural spectrum through future land use designations and/or land development regulations. Frequently these land use tools employ urban and rural definitions to establish densities, intensities, and other development standards. Because these tools were based on heteronormative and patriarchal conceptualisations of families and appropriate geographical locations for them (Ritzdorf, 2000: 175), they are of interest to queer and feminist planners examining the relationship between land use planning and social equity (Forsyth, 2001).

A more useful approach would be to use 'rural' simply as a descriptive element of the character of a community. Defining the urban–rural spectrum in this way enables practitioners and researchers to describe attributes of a geographic area qualitatively, generating more accurate representations of a space or place. A prime example of this comes from Herring (2010) who acknowledged that connected metropolises of the northeast and areas such as Dothan, Alabama or Macon, Georgia are classified similarly using Census Bureau guidance (i.e. urbanised, metropolitan), but are qualitatively different in terms of culture and amenities. Similarly, Kirkey and Forsyth (2001) explored life for queers living on the suburban–rural fringe (i.e. exurban or semi-rural). These areas have reasonable access to amenities, but are typically closer to truly rural areas and considered by some to feel more rural. The variation in these rural designations complicates the efforts of researchers seeking insight into the lives of rural queers (Boso, 2012).

Unpacking bias: metronormativity

The notion of an urban–rural dichotomy, rather than a spectrum, has too often shaped colloquial and academic attitudes about these spaces. As they relate to sexuality, urban areas have more often been associated with exploration, expression, and openness, whereas rural areas have been portrayed as repressive and backwards (Halberstam, 2005). Halberstam termed this approach 'metronormativity', described as follows:

> The metronormative narrative maps a story of migration onto the coming-out narrative . . . the metronormative story of migration from 'country' to 'town' is a spatial narrative within which the subject moves to a place of tolerance after enduring life in a place of suspicion, persecution and secrecy.
>
> *(Halberstam, 2005: 42)*

Annes and Redlin (2013) found that urban environments are often considered the only places that permit the exploration of one's sexual identity. Their findings also suggest that metropolitan spaces allow for visibility of the LGBTQ community and the opportunity for LGBTQ identifying individuals to disappear among the crowd. Rural queer dwellers, on the other hand, might struggle with heightened individual visibility due to a largely invisible queer community in these areas. Most importantly, Annes and Redlin (2013: 57) also found that gay 'men build their sense of self and their subjectivities through back-and-forth movements from the country to the city'.

Jerke (2011: 263) argued that '[r]uralism is best cast as a form of *structural* discrimination' in which others view rural residents as overly traditional, such that that rural queer dwellers suffered doubly from queer ruralism. That is, queers occupying rural spaces are disadvantaged by queer metronormativity because they are seen as backwards and can remain invisible within their community. Others have acknowledged that appraising rural groups as unenlightened juxtaposes the group to urbanites both temporally and spatially (Weston, 1995; Halberstam, 2005; Herring, 2010). This juxtaposition incorporates the assumption that queer persons from non-metropolitan areas lack refinement and class, while urban queers are endowed with greater tastes. This essentialising belief is held by those with an urban bias. These individuals may then contribute to discursive queer spaces within urban centers by creating a 'pressure to follow a hegemonic gay identity' (Annes and Redlin, 2013: 65). Perhaps this discourse could partially explain findings from Wienke and Hill (2013) that indicate rural settings may provide some queers with a greater sense of well-being than do larger cities.

Queerying the lives of rural dwellers

The Williams Institute (Boso, 2012) estimated that in the United States there were 64,000 LGBTQ people living in rural areas in 2010, a number that the consortium considers conservative since it was formulated using census data of cohabiting couples. This estimation contributes to a growing body of work disrupting the commonly held belief that rural areas are categorically unwelcoming to non-normative individuals. Previous research has also challenged this by documenting stories from queer culture as well as directly from experiences of rural queer dwellers. These complex subjectivities are reinforced by descriptions of 'bachelor farmers' in North Dakota (Kramer, 1995) who may struggle to define precisely their identities, but are clear in their attractions. Similarly Bell's depiction of farm boys and wild men (Bell, 2000) and cowboy love (Bell, 2007) suggest a much more complicated subjectivity that defies simplistic categorisation. For example, Annie Proulx's short story 'Brokeback Mountain' and the subsequent film directed by Ang Lee describe the ways that two ultra-masculine cowboys explore their attraction for each other, but are unable to fully express themselves as gay men. Although these characters exhibit characteristics of the MSM (men who have sex with men) label used by the Centers for Disease Control, even this label does not fit precisely (Manalanson, 2007).

Furthermore, Bell and Valentine (1995: 114) noted that some prominent queer authors spoke of the 'secret seclusion' in countryside locations where 'men [could] escape the law's prying eyes'. There are a few discernable implications of this: (1) public centers were unsafe for gay men, (2) rural settings could be safer, and (3) rural settings were, by some, romanticised. This idealisation of rural areas likely influenced a portion of the queer groups who sought respite from city life. In addition, in some cases this rural imaginary resulted in a revisioning of queer lifestyles. Browne (2008) highlighted the need to move beyond the idea of a rural idyll to a more explicit consideration of more utopian projects, such as the womyn's land movement of the 1970s as well as the Michigan Womyn's Festival. Although utopian ideas may not be achievable, especially when exclusive 'womyn-born-womyn entry policies' undermine those utopian visions, planners should recognize that sometimes temporary constructions of idealised and empowering 'queer space' can be a vital element in re-imagining more equitable rural spaces.

Some lesbian women responded to the perceived ills of city living by moving to the country (Valentine, 1997). Many of these women appear to have left the city for the purposes of both rural scenery and to escape from aggressive heterosexual pressures from certain men. Smith and Holt (2005) explored the motivations of lesbian migrants to the Hebden Valley of West Yorkshire, finding that while they might be considered rural gentrifiers, they 'often visit metropolitan locations to 'dip into' and consume commercial lesbian venues and scenes' (ibid.: 319). Similarly, in the United States, Lewis (2015) suggested that those gay men who moved away from the District of Colombia towards the rural fringe still return to gay bars and other community institutions in the urban core for weekend entertainment and for community events.

The topic of queer migration from town to country has implications for rural gentrification. Queer movement to the fringe does contribute to increased property values and changes to socioeconomic structures (Smith and Holt, 2005), undermining rural gentrification research that has traditionally disregarded impacts by queers on rural areas. Recognising these impacts is important to understanding ways in which planning practitioners might reduce unfavorable consequences of urban to rural queer migration. Furthermore, generating research to inform academics on how queer rural dwellers are pioneers in lower income rural areas may lead to new insights on gentrification processes in general.

Deconstructing rurality and rural communities

Part of urban bias is the idea that only urban areas can provide visibility and opportunity for LGBTQ identifying persons (Gray, 2009). The logical opposite to urban visibility and opportunity is therefore rural invisibility and inopportunity. However, it is important to deconstruct rurality since rural areas exhibit large variations in the degree of isolation, including distance from urban areas and overall levels of population density.

One disruption to the notion of rural invisibility was highlighted by Detamore (2013) in his auto-ethnographic account of experiences while attending a festival in one of Kentucky's small rural towns. Detamore shared how he was able to both hold a conversation on sexuality with a heterosexual man, and openly display same-sex affection in public. Furthermore, he described his experience in bar-hopping with a group of queer friends in this rural space and noted that rather than being threatened, they were incorporated into the social fabric, adding an element of queerness to those spaces. These observations of rural Kentucky highlight a weakness in conflating rurality, invisibility, and inopportunity, showing that queerness and rurality coalesce within some non-urban heteronormative social environments.

However, many rural areas have rather limited 'gay community infrastructure' that is essential for 'social and sexual networking opportunities' (Rosser and Horvath, 2008: 2) creating challenges for rural males who have sex with other males. Kramer (1995) suggested that prior to the internet the lack of gay-friendly institutions was a significant constraint to identity formation for both gay men and lesbians in North Dakota. Furthermore, D'Augelli and Hart (1987) suggest that while rural isolation is linked to higher rates of depression in the wider community, among sexual minorities these effects are even further exacerbated. Similarly, in Wyoming, Leedy and Connolly (2007) found that there are important differences across counties. Residents of the most rural (least densely populated counties) experience higher rates of suicide, lower earnings, and less health insurance than other counties.

> It is those residents in these counties [most rural] which report the highest levels of discrimination at both the community/personal level and at the institutional level. Further, these respondents report the fewest opportunities for services and programs for GLB persons.
>
> *(Leedy and Connolly, 2007: 31)*

In short, deconstructing rurality suggests that there are significant variations across rural areas in terms of their LGBTQ residents' ability to make connections and find supportive services. Rural is not simply the opposite of urban, but is a complex concept with significant variations. Planners seeking to ensure equitable social service deliveries across rural areas must be cognizant of the diversity of communities and services that are open to and inclusive of LGBTQ rural residents.

Rurality, queerness, and HIV/AIDS

The lack of community institutions and reliable information sources certainly exacerbated the HIV/AIDS experience in rural areas. In urban areas AIDS devastated gay enclaves prior to the development of adequate treatments since these neighbourhoods attracted dense concentrations of gay men who practised unprotected sex. In particular, urban area bathhouses were criticised for their contribution to spreading the virus (Gross, 1987) in places such as the Castro District and Greenwich Village. Infection rates in these highly populated areas increased substantially from 1982 to 1984 (Lam and Liu, 1994), and this likely contributed to the false assumption that rural persons were not at-risk.

However, Kramer (1995) suggests that in rural areas highway rest stops were frequent venues for significant amounts of unprotected sex between men, as were other remote areas such as viaducts, public parks, and the fringes of the North Dakota State Fair. For example, a participant in a study of rural MSM activities indicated he did not use protection because he was located in a rural setting (Williams et al., 2005: 52). In a sense, the belief that rural MSM populations are not at-risk for sexually transmitted diseases is a rural bias unsupported by facts, which suggest that from 1988–1990 many counties with lower populations tended to report higher rate increases of HIV/AIDS than did highly populated counties. Moreover, areas such as the rural south are reporting higher rates than other rural areas (CDC, 2016). In short HIV/AIDS is a scourge that knows no specific geography, and the need for services from a population stricken by HIV/AIDS is likely to be found in nearly every community.

In the United States national expenditures for HIV/AIDS funding in fiscal year 2018 is requested to be approximately $32.9 billion (Kaiser Family Foundation, 2017). This is roughly $840 million less than what was allocated in 2017. Many of the major drops in funding were for global programmes and research, but notable programmes such as the Ryan White Program and the Department of Housing and Urban Development's Housing for Persons with AIDS, both of which are discretionary allocations, may experience a loss of $84.8 million in funding. These programmes aid both urban and rural residences, but with competitive application processes, rural areas may be significantly limited in the care, support services, and prevention they may be able to provide (Albritton et al., 2017). The lack of supportive funding for rural areas has also been found to reduce the capacity of various entities to fund ancillary programmes such as housing and transportation assistance. Furthermore, programmes such as Ryan White HIV/AIDS Program are also at risk (Health Resources and Services Administration, 2018).

In contrast, the United Kingdom's Parliament passed the Equality Act of 2010, which expanded protected statuses to be inclusive of sexual minorities – specifically sexual orientation and gender assignment (Browne and McGlynn, 2013). This legislation could have been useful for protecting the interests of sexual minorities in rural settings; however, the potential for change in these areas was limited by a provision that required 'evidence' of a need for the public sector to meet. As Browne and McGlynn stated, 'The challenge then was to prove LGBTQ people existed, used services, and had a particular need' (ibid.: 42) for which they could request care.

As noted above, gay community infrastructure (e.g. gay cafes, guesthouses, bookstores, gay parades, gay-affirming churches) is often lacking in rural areas, but it has been linked with successful HIV prevention implementation in places where it exists and is robust (Rosser and Horvath, 2008). Differences between population characteristics (e.g. population density, ethnic-racial composition) and successful examples of HIV prevention were not found, suggesting stereotypical rural demographics generally were not barriers to successful implementation. Still, impediments to prevention and treatment of HIV/AIDS in rural areas exist and sometimes relate to the vastness of these areas and the lack of agglomerated services (Heckman et al., 1998). Issues such as lengthy commutes to find adequate facilities and too few transportation options are planning problems that can be ameliorated. In addition to gay community infrastructure, funding was also found by Rosser and Horvath (2008) to be important in HIV prevention efforts, especially when these funds target MSM populations, which continue to have the highest prevalence rates (CDC, 2016).

Recommendations and conclusions

Planning literature has begun to explore the needs of the LGBTQ community (Forsyth, 2001; Doan, 2011, 2015), but there are few publications that acknowledge the intersection of the LGBTQ population in rural areas. Further expansion beyond this chapter is clearly warranted to examine the problems unique to rurality and queerness as well as to provide resources for practicing planners to understand how to conceptualise and ameliorate issues facing this population.

Frisch (2015) noted that planning has traditionally not been inclusive of sexual minorities in theory or in practice. Normative thinking on what constitutes the 'public interest' and the 'core of planning' have prevailed in years past; however, more recently, theory and practice have begun to take a more inclusive perspective especially pertaining to marginalised populations including LGBTQ populations, but these vulnerable individuals are often overlooked or ignored in rural areas. A concerted effort should be made to more carefully enumerate the LGBTQ community wherever they may be located.

Planners should seek to incorporate diverse perspectives into strategic and long range planning efforts and encourage broader expression of what rural space might become, including the occasional utopian perspective. At the same time planners should develop a more realistic understanding of the benefits and costs of LGBTQ urban to rural migration that might contribute to revitalising communities, as well as the factors that may exacerbate further rural gentrification.

The rural LGBTQ population is likely to be at significant risk due to inadequate access to basic social services since some rural facilities may not be supportive of non-normative individuals. Accordingly, rural planners should consider the utility of multi-function rural centres as loci for a dispersed queer community to increase the accessibility of critical health care, mental health, and community-based services. Furthermore, a special focus on funding for HIV/AIDS care and education in rural areas seems warranted. These services should also include related housing, transportation, and meals on wheels to those who are most severely stricken by this disease.

References

Albritton, T., Martinez, I., Gibson, C, Angley M., and Grandelski, V.R. (2017) What about us? Economic and policy changes affecting rural HIV/AIDS services and care. *Social Work in Public Health*, 32, 4, 273–289.

Annes, A and Redlin, M. (2012) Coming out and coming back: rural gay migration and the city, *Journal of Rural Studies*, 28, 56–68.

Bell, D. (2000) Farm boys and wild men: rurality, masculinity, and homosexuality. *Rural Sociology*, 65, 547–561.

Bell, D. (2007) Cowboy love. In H. Campbell, M. Bell, and M. Finney (eds), *Country Boys: Masculinity and Rural Life*. Pennsylvania State Press, University Park, 163–182.

Bell, D. and Valentine, G. (1995) Queer country: rural gay and lesbian lives. *Journal of Rural Studies*, 15, 113–122.

Black, D., Gates, G., Sanders, S., and Taylor, L. (2000) Demographics of the gay and lesbian population in the United States: evidence from available systematic data sources. *Demography*, 37, 2, 139–154.

Boso, L. (2012) Urban bias, rural sexual minorities, and the courts. *UCLA Law Review*, 60, 562.

Browne, K. (2008) Imagining cities: living the other: between the gay urban idyll and rural lesbian lives. *Open Geography Journal*, 1, 25–32.

Browne, K. and McGlynn, N. (2013) Rural lesbian, gay, bisexual, and trans equalities: English legislative equalities in an era of austerity. In A. Gorman-Murray, B. Pini, and L. Bryant (eds), *Sexuality, Rurality, and Geography*. Plymouth: Lexington Books, 35–50.

CDC. (2016) *HIV Surveillance Report, 2016: vol. 28*. Retrieved on 25 October 2018 from www.cdc.gov/hiv/library/reports/hiv-surveillance.html.

Daniels, T.L. and Lapping, M.B. (1996) The two rural Americas need more, not less planning. *Journal of the American Planning Association*, 62, 285–288.

D'Augelli, A.R. and Hart, M. (1987) Gay women, men, and families in rural settings: towards the development of helping communities. *American Journal of Community Psychology*, 15, 70–93.

Detamore, M. (2013) Queering the hollow: space, place, and rural queerness. In A. Gorman-Murray, B. Pini, and L. Bryant (eds), *Sexuality, Rurality, and Geography*. Plymouth: Lexington Books, 35–50.

Doan, P.L. (ed.) (2011) *Queerying Planning: Challenging Heteronormative Assumptions and Reframing Planning Practice*. Farnham: Ashgate.

Doan, P.L. (ed.) (2015) *Planning and LGBTQ Communities: The Need for Inclusive Queer Space*. London: Routledge.

Doan, P. (2016) To count or not to count: queering measurement and the transgender community. *Women's Studies Quarterly*, 44, 89–110.

Forsyth, A. (2001) Sexuality and space: nonconformist populations and planning practice. *Journal of Planning Literature*, 15, 339–358.

Frisch, M. (2015) Finding transformative planning practice in the spaces of intersectionality. In P. Doan (ed.), *Planning and LGBTQ Communities: the Need for Inclusive Queer Space*. London: Routledge, 129–146.

Gallent, N., Juntti, M., Kidd, S., and Shaw, S. (2008) *Introduction to Rural Planning*. London: Routledge.

Gates, G.J. (2014) LGBT *Demographics: Comparisons among Population-Based Surveys*. The Williams Institute. Retrieved from http://williamsinstitute.law.ucla.edu/wp-content/uploads/lgbt-demogs-sep–2014.pdf.

Gates, G.J. and Ost, J. (2004) *The Gay and Lesbian Atlas*. Washington, DC: Urban Institute Press.

Gorman-Murray A. (2007) Rethinking queer migration through the body. *Social and Cultural Geography*, 8, 1, 105–121.

Gray, M. (2009) *Out in the Country: Youth, Media, and Queer Visibility in Rural America*. New York: New York University Press.

Green, R. (2005) Redefining rurality. *Town and Country Planning*, 74, 6, 202–205.

Gross, J. (1987) AIDS threat brings new turmoil for gay teenagers. *The New York Times* (Metropolitan News), 21 October.

Halberstam, J. (2005) *In a Queer Time and Place: Transgender Bodies, Subcultural Lives*. New York: New York University Press.

Health Resources and Services Administration. (2018) HIV/AIDS in rural America. Retrieved from https://hab.hrsa.gov/livinghistory/issues/rural_1.htm.

Heckman, T.G., Somlai, A.M., Peters J., Walker J., Otto-Salaj, L., Galdabini, C.A., and Kelly, J.A. (1998) Barriers to care among persons living with HIV/AIDS in urban and rural areas. *AIDS Care*, 10, 3, 365–375.

Herring, S. (2010) *Another Country: Queer Anti-Urbanism*. New York: New York University Press.

Jerke, B.W. (2011) Queer ruralism. *Harvard Journal of Law and Gender*, 34, 259–312.

Kaiser Family Foundation. (2017) US federal funding for HIV/AIDS: trends over time. Retrieved from http://files.kff.org/attachment/Fact-Sheet-US-Federal-Funding-for-HIVAIDS-Trends-Over-Time.

Kirkey, K. and Forsyth, A. (2001) Men in the valley: gay male life on the suburban–rural fringe. *Journal of Rural Studies*, 17, 421–441.

Knopp, L. and Brown, M. (2003) Queer diffusions. *Environment and Planning D: Society and Space*, 21, 409–424.

Kramer, J.L. (1995) Bachelor farmers and spinsters: gay and lesbian identities and communities in rural North Dakota. In D. Bell and G. Valentine (eds), *Mapping Desire: Geographies of Sexualities*. London: Routledge, 200–213.

Lapping, M.B. (1989) *Rural Planning and Development in the United States*. New York: Guilford Press.

Leedy, G. and Connolly, C. (2007) Out in the Cowboy State. *Journal of Gay and Lesbian Social Services*, 19, 1, 17–34.

Lewis, N. (2015) Fractures and fissures in 'Post-Mo' Washington, DC: the limits of gayborhood diffusion and transition. In P. Doan (ed.), *Planning and LGBTQ Communities: the Need for Inclusive Queer Spaces*. London: Routledge, 56–75.

Manalanson, M.F. (2007) Colonizing time and space: race and romance in Brokeback Mountain. *GLQ: Gay and Lesbian Quarterly*, 13, 1, 97–100.

Office for National Statistics. (2016) Sexual identity, UK: 2016. Retrieved from www.ons.gov.uk/peoplepopulationandcommunity/culturalidentity/sexuality/bulletins/sexualidentityuk/2016.

Popper, F.J. (1988) Understanding American land use regulation since 1970: a revisionist interpretation. *Journal of the American Planning Association*, 54, 3, 291–301.

Ritzdorf, M. (2000) Sex, lies, and urban life: how municipal planning marginalizes African-American women and their families. In K.B. Miranne and A.H. Young (eds), *Gendering the City: Women, Boundaries, and Visions of Urban Life*. Lanham, MD: Rowman and Littlefield, 169–181.

Rosser, B.R.S. and Horvath, K.J. (2008) Predictors of success in implementing HIV prevention in rural America: a state-level structural factor analysis of HIV prevention targeting men who have sex with men. *AIDS and Behavior*, 12, 2, 159–168.

Smith, D. and Holt, L. (2005) Lesbian migrants in the gentrified valley and 'other' geographies of rural gentrification. *Journal of Rural Studies*, 21, 313–321.

Teitz, M. (2002) Meanwhile, out on the edge. *Town and Country Planning*, 71, 2, 57.

Valentine, G. (1997) Making space: lesbian separatist communities in the United States. In P. Cloke and J. Little (eds), *Contested Countryside Cultures: Otherness, Marginalisation, and Rurality*. London: Routledge, 109–122.

Weston, K. (1995) Get thee to a big city: sexual imaginary and the great gay migration. *GLQ: Gay and Lesbian Quarterly*, 3, 253–277.

Wienke, C. and Hill, G. J. (2013) Does place of residence matter? Rural–urban differences and the wellbeing of gay men and lesbians. *Journal of Homosexuality*, 60, 9, 1256–1279.

Williams, M.L., Bowen, A.M., and Horvath, K.J. (2005) The social/sexual environment of gay men residing in a rural frontier state: implications for the development of HIV prevention programs. *Journal of Rural Health*, 21, 1, 48–55.

30

Planning for an ageing countryside

Mark Bevan

Introduction

Population ageing has been described as one of the most significant trends of the 21st Century, and the Madrid Declaration and International Plan of Action 2002 highlights how planning in the widest sense has responded globally to this trend. In 2012 one in nine persons in the world were reported as aged 60 or over and this figure was projected to increase to one in five by 2050 (United Nations Population Fund, 2012). Rapid ageing of rural populations has been highlighted across Asia, Latin America, and to a lesser extent in Sub-Saharan Africa (Heide-Ottosen, 2014). The focus for this chapter is on the Global North, with particular reference to the policy context in England. Population estimates for 2013 for this latter country showed that people aged 65 or over comprised about 23 per cent of the rural population in England, compared with 16 per cent of the urban population (Wilson, 2017). Projections suggest that the most rural local authority areas will see the highest rate of population increase in the people aged 80 and over.

The United Nations Population Fund (2012) differentiates between population ageing, which can be defined as the process whereby older individuals become a proportionately larger share of the total population, and individual ageing, which is the process of individuals growing older. Although planning for broad demographic shifts encapsulated by population ageing in rural areas has become an established focus for policy, recent research has highlighted the variable extent to which local authorities in England were undertaking needs mapping and developing tailored responses (Connors et al., 2013).

Social constructions of ageing

In defining the process of individual ageing, the United Nations Population Fund (2012) have also highlighted that the notion of 'growing old' goes beyond thinking about chronological age, and can be defined instead as a social construct often associated with the changes that people may experience in social roles, activities or health. Changes and transitions that may take place across the lifecourse have become an important focus for policy and planning, identifying key moments that can be targeted for support and services. Various reviews have explored transitions in later life such as retirement, bereavement and health status and how policy might best respond to these diverse experiences among older people (Robertson, 2014; Centre for Ageing Better, 2016).

Transitions in later life

The experience of transitions in later life has been considered in greater depth by Grenier (2015) who has critiqued expectations of growing old that are organised according to chronological age and based on fixed stages of development across a linear lifecourse, which also includes the way that transitions in later life have been traditionally framed by policy. By drawing on the notion of liminality, Grenier (2012) suggests our understandings of continuity and change over the lifecourse need to be reconsidered to embrace a greater subjectivity and fluidity in the way that individuals live with, and experience, change. This analysis resonates with the perspectives offered by Warburton et al. (2017: 476) in relation to the experiences of exclusion as people get older in rural areas. They highlight how policy and practice tends to view individuals as either excluded or included based on whether they are in 'at-risk' groups, rather than more complex constructions of exclusion that acknowledge that people can move in and out of exclusion as they age, or that they can experience inclusion in one area of life while simultaneously experiencing exclusion in other areas.

Framing later life within policy

As highlighted above by the United Nations Population Fund, definitions of old age are socially inscribed processes that risk ascribing characteristics to individuals based on numerical age (Boyle et al., 2015). Other studies have reflected specifically on the way that older people may be described and defined by policy in rural contexts (Burholt and Dobbs, 2012; Hennessy et al., 2014). A challenge for planning is encapsulating the diversity of experience in rural areas across later life, including frailty and vulnerability, while avoiding the use of pejorative deficit models which either frame people as a problem because of their age, or take a limited and narrow view of how ageing may be experienced (Murakami et al., 2008). An example within the field of housing are strategies that define the housing needs of older people primarily in terms of congregate housing settings or housing with care models, even though the vast majority of people in later life live in mainstream housing. Grenier (2012) suggested that a distinction must be made between using dominant policy and academic frameworks of ageing as a way of helping to understand older people's lives, as opposed to their use as fixed frameworks that set expectations of the way that individuals should or could experience later life, especially in relation to notions of healthy or successful ageing. For instance, Wild et al. (2013) caution that the concept of resilience might be used as a way of diminishing and downplaying the diversity of experiences in later life that do not conform or live up to the notion of the successful or productive individual, and might also underplay disadvantage and precarity.

Gerontological perspectives on later life in rural areas

Furthermore, as Glasgow and Brown (2012: 428) have observed in relation to population ageing in rural areas, 'demography is not destiny', and population changes are mediated by a community's social structure, by macro-structural processes external to localities, and by the policy environment in which places are embedded. Critical perspectives within gerontological studies across a range of disciplines have highlighted how patterns and processes that occur at the micro, meso and macro scales affect how ageing is experienced (Boyle et al., 2015). Various commentators have put forward frameworks that not only attempt to capture the tensions and convergences between structure and agency, but also to ensure that the diversity of experiences in later life can be articulated and represented within policy and practice. This can be illustrated with two examples in the way that wider gerontological perspectives have examined the policy concerns of firstly, healthy ageing and secondly, mobility, and how these can be related to the rural context.

Healthy ageing and resilience in rural contexts

The notion of healthy ageing is an important policy driver, and provides a significant body of evidence that planning can draw on in guiding the design of built environments, as well as supporting social and community engagement. Healthy ageing can be described as health promotion interventions that are aimed at how older people can avoid disease and disability, maintain high mental and physical functioning, and remain socially engaged. Older people are responsible for engaging in exercise, diet and social engagement to produce good health. However, research cautions against interpretations of these initiatives that impose neo-liberal discourses of healthy or successful ageing in a way that homogenises experiences and individualises risk and responsibility, with little or no attention to the wider socio-economic contexts in rural areas in which people live (Warburton et al., 2017).

Instead, Stephens et al. (2015) offer an alternative framework for analysing individual needs and aspirations from a health perspective, with a focus on the aspects of life valued by older people themselves, whatever their circumstances. They argue that rather than interventions and policy focused on an individual responsibility to produce health as if it was a kind of commodity, they suggest a focus on the things that people actually value, and how planning can build out from these perspectives to facilitate environments that enable these values to be pursued. Mansvelt (2014: 407) drew upon her own research to illustrate this point further. Her work with older people who were mostly confined to their homes highlighted the very diverse and relational ways in which people enabled mobility, and the ways in which environments and technologies caused frictions. She argued for a shift in emphasis within policy from expectations of very frail individuals doing things 'normally' to enabling 'the normality of doing things differently'.

This argument comes back to how far planning in rural areas can draw upon a user centred approach that can draw meaningfully on the very diverse experiences in later life, and capture the nuances that this diversity expresses. A difficulty may be the very limited nature of research evidence that is available for planners to draw on in rural contexts to assist planning specific areas of service delivery (Bacsu et al., 2014). Stephens et al. (2015) noted that their analysis could not be applied without empirical work to identify the locality specific contexts in which older people lived. To take one example of planning for end of life care in rural areas, a review of the literature on this subject by Rainsford et al. (2017) emphasised the paucity of research in rural contexts that takes account of the views of individuals at end of life and their family caregivers.

The salience of social connectedness (Hennessy et al., 2014) links with potential future directions in public health investment in rural areas. There is a growing recognition within the health sector in England that facilitating the things that really matter to people – social relationships – may have positive outcomes in relation to health, with the potential for reducing costs for primary and acute care. One mechanism for taking forward this approach is social prescribing, which can be defined as enabling healthcare professionals to refer patients to a link worker, so that people with social, emotional or practical needs are empowered to find solutions which will improve their health and wellbeing, often using services provided by the voluntary, community and social enterprise sector. Although the evidence base is mixed, the potential for social prescribing to help achieve better health outcomes for older people in rural areas has been highlighted in recent policy discussions (All-Party Parliamentary Group for Ageing and Older People, 2017), and has been put forward in relation to addressing loneliness and social isolation among older people in the countryside (De Koning et al., 2017).

The notion of resilience has also been much discussed in relation to individual responses to transitions over the lifecourse, and resilience has also been framed in the way that wider community and societal contexts are likely to support people in later life (Bennett, 2015). A significant rural body of work in this regard has been brought together to discuss ageing not only in relation to individuals, but the resilience of settlements that are themselves ageing in terms of their original economic rationale (Skinner and Hanlon, 2016). These include rural areas that have traditionally relied on primary resources, especially declining or redundant extractive industries, but which are now reconfiguring to meet the needs and aspirations of their ageing populations.

Stephens et al. (2015) draw on the notion of resilience to situate individual experiences within a wider context, using a capabilities approach. Here, capabilities were framed not just in relation to the resilience of individuals, but how the capabilities of the wider social and physical environment enabled or constrained the ability of people to pursue values that mattered to them. Using the concept of wellness, Winterton et al. (2016) also draw upon the notion of capabilities to analyse how far rural environments can create opportunities and capabilities for older adults to achieve desired potentials as well contributing to health and wellbeing. Winterton et al. (ibid.) go on to conclude that service planning and provision would benefit from future research that explores how different resources can interact to compensate for limitations in the way that specific communities support older people, and crucially, which resources are critical in sustaining people in rural communities. This latter point goes to the heart of the age friendly communities approach discussed later on in the chapter.

Rural mobilities

Wild et al. (2013) utilise the notion of 'mobility resilience' to shift the focus from the resilient individual and situate the individual within overlapping and inter-related scales of household, family, community and neighbourhood as well as societal resilience in enabling or constraining mobility. This area of work around health and mobility in later life has resonances with key issues in rural environments around connectedness and travel. To take just one example, an important transition in later life can be driving cessation (Musselwhite and Shergold, 2013), and especially the implications of stopping or reducing driving on mobility and access to services and facilities in rural areas. Ormerod et al. (2015) highlighted that the lack of, or loss in, mobility disproportionately affects people living in rural and remote areas, which can have knock on effects on individual health outcomes as a consequence of a reduced use of preventative services, primary care and hospital care due to the geographical inaccessibility of the services and the costs and inconveniences of longer journeys. At the same time, other research has also emphasised the wider motivations of older people for travel in rural areas, especially the importance of journeys for social reasons, which may include community activities and volunteering (Shergold et al., 2012). The latter research emphasised the localised nature of most journeys that older people made in the rural areas they studied and suggested that more emphasis should be placed in rural transport policy on facilitating these short-range journeys for social purposes. A key conclusion was that an over-reliance on cars as a policy solution risks not only undermining community engagement and activity, but also the alternative travel options (including walking, cycling and mobility scooters), especially for people without access to cars.

These examples illustrate how wider critical gerontological perspectives could provide useful frameworks for taking forwards a better understanding of the characteristics of rural areas that support, or detract from, ageing well in rural areas, leading to more nuanced planning approaches in rural contexts. The concern with diverse perspectives of older people, and situating these within wider community and societal contexts resonates with a strong and developing body of research that has focused specifically on ageing in rural areas (Keating, 2008). A key element in these latter approaches has been the use of the ecological model that makes explicit the links between environment and the individual, focused on the notion of age friendly rural communities. This approach aims to provide a better understanding of whether rural settings are good places to grow old.

Age-friendly communities in rural areas

There are diverse definitions of age friendly communities (Menec et al., 2013), and a variety of approaches as to how communities might be assessed for 'age friendliness' (Fitzgerald and Caro, 2014). Age-friendliness can be broadly described as a measure of the goodness of fit between communities and their older residents. Menec et al. (2013: 205) note that the main premises underlying the concept are that older adults are valued participants in society and that they may require a wide range of supports and services in order to remain independent and healthy and enjoy a high quality of life in old age. Globally there has been considerable interest in the development of age friendly communities, illustrated by the World Health Organization (WHO) guidelines that include the encouragement of active ageing by optimising opportunities for health, participation, and security in order to enhance people's quality of life as they age. The guidelines encapsulate a range of dimensions including service provision, characteristics of the built environment as well as social aspects (such as civic and social participation). As Lui et al. (2009) highlight, an assumption widely shared by policy-makers and planners is that an enabling social environment is just as important as material conditions in determining well-being in later life.

A key aspect of the WHO's age friendly agenda has been how to move from generic principles to locality specific approaches that reflect the articulated needs and aspirations of local populations. While the focus of attention has been on age friendly cities, Canada developed a specific set of guidance to assist communities in the assessment and development of age friendly communities in rural and remote areas (Federal, Provincial, Territorial Ministers Responsible for Seniors, 2007). Menec et al.'s (2013) subsequent study of what makes rural areas age friendly in Canada identified particular aspects of communities that may be easier to support or more amenable to change, including the social environment, opportunities for participation, and communication and information. They found that other factors were more intractable such as the physical environment, housing, health care services and transportation. In this respect, they suggest that policy makers need a wide-ranging approach that can strengthen the resources that rural areas have in order to meet the needs of diverse groups of older people.

The study by Menec et al. (2013) also noted the complexity of experiences in different rural contexts. One expression of this geographical dimension was apparent problems experienced in small, remote communities. Certainly in England, research has highlighted that older people in rural areas are more likely to move (Wu et al., 2015) and especially from small, remoter communities to larger communities (Champion and Shepherd, 2006). A key issue is the extent to which these moves are an expression of choice or planning by individuals, or how far they reflect 'forced' moves as a result of a crisis, or a view that their home or wider environment can no longer sustain them.

A recent body of work set out to refine how the age friendly communities framework might be applied in rural areas in order to develop a more nuanced approach that can more readily assess what community resources and needs might be most compatible with which groups of older residents. This research also aimed to provide a platform from which to potentially test the efficacy of age-friendly interventions and to inform policy and practice across a wide range of communities (Keating et al., 2013). These authors suggested that an understanding of what makes rural areas age friendly requires an appreciation of the diversity among both older persons and communities; the complexities of the connections between older people and their communities, and change over time in people and in place. The latter temporal element was also emphasised by Walsh et al. (2012) who highlighted the dynamic nature of rural communities as constantly shifting and changing, which shapes (and is shaped by) people in later life. Recent discussions on the capacity to progress rural age friendly communities have also been set in the wider context of structural economic changes and the impact of financial austerity (Walsh, 2015), as well as ongoing limited state support for programmes to support rural ageing in diverse contexts such as the United States (Nash et al., 2011). A difficulty for planning, for example, is that housing might be built close to existing services and facilities, but can be left high and dry as services subsequently erode around them through closures and rationalisation.

The concept of 'friendly communities' as a generic framework

The notion of age friendly communities is part of a much wider discourse, in which the term 'friendly community' has become an important generic organising framework for assessing how diverse locations meet the needs and aspirations of distinct groups across society, as well as a focus for fostering or sustaining the conditions that make communities work for these various discrete groups. Other 'friendly community' frameworks may include older people as part of the wider constituency being considered, such as 'carer friendly' communities (Carers UK, 2016). Other 'friendly community' frameworks are distinct from people in later life, such as 'child friendly communities', which is part of a global programme of action headed by UNICEF. A further expression of this approach can be seen in the tight focus on very specific groups predominantly in later life, such as dementia friendly communities. While guidance for taking forwards how dementia friendly communities might be designed and fostered by planners has been set out in England (Lewis, 2017), Wiersma and Denton (2016) illustrate how dementia friendly communities might be taken forwards in rural communities in Canada.

Balancing needs and prioritising between groups

The 'friendly communities' approach enables a focus on the distinct issues facing specific groups, as well as a way of raising awareness of their needs among policymakers, service providers and the wider public. However, it also illustrates a challenge for broader planning in terms of balancing the needs of these diverse groups and assessing the extent to which there is convergence between the measures that feature across the various 'friendly community' agendas. This issue goes to the heart of universal design principles of working towards meeting the needs of as wide a proportion of the population as possible. Recent discussions have posited an overlap between factors that are conducive to supporting age friendly communities and wider agendas such as the promotion of liveable communities for all ages, especially in relation to the built environment and health (Fitzgerald and Caro, 2014; Jackson et al., 2013). This overlap suggests that the development of age friendly features may have shared benefits for other groups within the wider population, in addition to older people (Fitzgerald and Caro, 2014). The evidence base on the wider benefits of age friendly design for other groups within the general population remains limited, however (Scharlach and Lehning, 2013). While there certainly may be overlaps between the needs of different groups this should not be taken for granted. Research has highlighted that design features intended to promote mobility for people with specific conditions or impairments may lead to inadvertent barriers for others (Goodman and Watson, 2010), as well as the extent to which design solutions may be contested between different groups (Risser et al., 2010). The extent to which the needs of diverse groups in later life coincide or diverge in relation to design features in the rural built environment highlights the difficulty of reconciling competing needs, and is an issue that requires nuanced policy and practice responses. However, in an era of austerity and ongoing retrenchment, this highlights the dilemmas that planning faces with respect to prioritising between diverse groups.

Conclusion

What comes through strongly from a wide-ranging body of research is that what really matters for people is the social and emotional facets of their lives, and Boyle et al. (2015) argue that what makes for successful ageing in place is the potential for places to support meaningful relationships. Critical gerontological perspectives across a range of disciplines offer frameworks for planning to assess how rural areas can support and enable people in later life to fulfil their potential, though while acknowledging the increasingly practical constraints in which planning has to operate. Nevertheless, the experience of transitions in later life within rural contexts is currently underexplored, and would add a new perspective for helping to understand how resources in rural communities are supportive, or not, of individuals within the context of their social networks.

References

All-Party Parliamentary Group for Ageing and Older People (2017) *Ageing in Rural Areas*, London: House of Commons.

Bacsu, J., Jeffery, B., Abonyi, S., Johnson, S., Novik, N., Martz, D. and Oosman, S. (2014) Healthy aging in place: perceptions of rural older adults, *Educational Gerontology*, 40 (5), 327–337.

Bennett, K. (2015) *Emotional and Personal Resilience through Life: Future of an Ageing Population: Evidence Review*, London: Government Office for Science.

Boyle, A., Wiles, J. L. and Kearns, R. A. (2015) Rethinking ageing in place: the 'people' and 'place' nexus, *Progress in Geography*, 34 (12), 1495–1511.

Burholt, V. and Dobbs, C. (2012) Research on rural ageing: where have we got to and where are we going in Europe? *Journal of Rural Studies*, 28, 432–446.

Carers UK (2016) *Building Carer Friendly Communities*, London: Carers UK.

Centre for Ageing Better (2016) *Managing Life Changes Roundtable – Summary*, London: Centre for Ageing Better.

Champion, T. and Shepherd, J. (2006) Demographic change in rural England, in P. Lowe and L. Speakman (eds), *The Ageing Countryside: The Growing Older Population of Rural England*, London: Age Concern, 29–50.

Connors, C., Kenrick, M. and Bloch, A. (2013) *Impact of an Ageing Population on Service Design and Delivery in Rural Areas*, London: DEFRA.

De Koning, J., Stathi, A. and Richards, S. (2017) Predictors of loneliness and different types of social isolation of rural-living older adults in the United Kingdom, *Ageing & Society*, 37, 2012–2043.

Federal, Provincial, Territorial Ministers Responsible for Seniors (2007) *Age-Friendly Rural and Remote Communities: A Guide*, Ottawa: Minister of Industry.

Fitzgerald, K. and Caro, F. (2014) An overview of age-friendly cities and communities around the world, *Journal of Aging and Social Policy*, 26 (1–2), 1–18.

Glasgow, N. and Brown, D. (2012) Rural ageing in the United States: trends and contexts, *Journal of Rural Studies*, 28, 422–431.

Goodman, C. and Watson, L. (2010) *Design Guidance for People with Dementia and for People with Sight Loss: A Comparative Review*, http://pocklington-trust.org.uk/wp-content/uploads/2016/02/Dementia-Sight-Loss.pdf.

Grenier, A. (2012) *Transitions and the Lifecourse: Challenging the Constructions of 'Growing Old'*, Bristol: Policy Press.

Grenier, A. (2015) Transitions, time and later life, in J. Twigg and W. Martin (eds), *Routledge Handbook of Cultural Gerontology*, 404–411.

Heide-Ottosen, S. (2014) *The Ageing of Rural Populations: Evidence on Older Farmers in Low and Middle-Income Countries*, London: HelpAge International.

Hennessy, C., Means, R. and Burholt, V. (2014) *Countryside Connections: Older People, Community and Place in Rural Britain*, Bristol: The Policy Press.

Jackson, R. J., Dannenberg, A. L. and Frumkin, H. (2013) Health and the built environment: 10 years after, *American Journal of Public Health*, 103 (9), 1542–1544.

Keating, N. (2008) *Rural ageing: A good place to grow old?* Bristol: Policy Press.

Keating, N., Eales, J. and Phillips, J. (2013) Age-friendly rural communities: conceptualizing 'best-fit', *Canadian Journal on Aging / La Revue canadienne du vieillissement*, 32 (4), 319–332.

Lewis, S. (2017) *Dementia and Town Planning: Creating Better Environments for People Living with Dementia*, London: Royal Town Planning Institute.

Lui, C., Everingham, J., Warburton, J., Cuthill, M. and Bartlett, H. (2009) What makes a community age-friendly: a review of the international literature, *Australasian Journal on Aging*, 28, 116–121.

Mansvelt, J. (2014) Elders, in P. Adey, D. Bissell, K. Hannam, P. Merriman and M. Sheller (eds), *The Routledge Handbook of Mobilities*, London: Routledge, 398–408.

Menec, V., Hutton, L., Newall, N., Nowicki, S., Spina, J. and Veselyuk, D. (2013) How 'age-friendly' are rural communities and what community characteristics are related to age-friendliness? The case of rural Manitoba, Canada, *Ageing & Society*, 35, 203–223.

Murakami, K., Atterton, J. and Gilroy, R. (2008) *Planning for the Ageing Countryside in Britain and Japan: City-Regions and the Mobility of Older People*. Newcastle: Centre for Rural Economy, Newcastle University, Research Report 49.

Musselwhite, C. and Shergold, I. (2013) Examining the process of driving cessation in later life, *European Journal of Ageing*, 10, 89–100.

Nash, B., Folts, W., Muir, K., Peacock, J. and Jones, K. (2011) Elderly populations and the rural housing continuum, in D. Marcouiller, M. Lapping and O. Furuseth (eds), *Rural Housing, Exurbanisation, and Amenity-Driven Development*, Farnham: Ashgate, 75–94.

Ormerod, M., Newton, R., Phillips, J., Musselwhite, C., with McGee, S. and Russell, R. (2015) *How Can Transport Provision and Associated Built Environment Infrastructure be Enhanced and Developed to Support the Mobility Needs of Individuals as They Age? Future of an Ageing Population: Evidence Review*, London: Foresight, Government Office for Science.

Rainsford, S., MacLeod, R., Glasgow, N. Phillips, C., Wiles, R. and Wilson, D. (2017) Rural end-of-life care from the experiences and perspectives of patients and family caregivers: a systematic literature review, *Palliative Medicine*, 31 (10), 1–18.

Risser, R., Haindl, G. and Stahl, A. (2010) Barriers to senior citizens' outdoor mobility in Europe, *European Journal of Ageing*, 7 (2), 69–80.

Robertson, G. (2014) *Transitions in Later Life: Scoping Research*, London: Calouste Gulbenkian Foundation.

Scharlach, A. and Lehning, A. (2013) Ageing-friendly communities and social inclusion in the United States of America, *Ageing and Society*, 33 (1), 110–136.

Shergold, I., Parkhurst, G. and Musselwhite, C. (2012) Rural car dependence: an emerging barrier to community activity for older people, *Transportation Planning and Technology*, 35 (1), 69–85.

Skinner, M. and Hanlon, H. (2016) *Ageing Resource Communities: New frontiers of rural population change, community development and voluntarism*, London: Routledge.

Stephens, C., Breheny, M. and Mansvelt, J. (2015) Healthy ageing from the perspective of older people: A capability approach to resilience, *Psychology & Health*, 30 (6), 715–731.

United Nations Population Fund (2012) *Ageing in the Twenty-First Century: A Celebration and a Challenge*, New York: United Nations Population Fund (UNFPA).

Walsh, K. (2015) Interrogating the 'age-friendly community' in austerity: myths, realities and the influence of place context, in K. Walsh, G. Carney and Á. Ni Léime (eds), *Ageing through Austerity: Critical Perspectives from Ireland*, Bristol: Policy Press, 80–96.

Walsh, K., O'Shea, E., Scharf, T. and Murray, M. (2012) Ageing in changing community contexts: cross-border perspectives from rural Ireland and Northern Ireland, *Journal of Rural Studies*, 28, 347–357.

Warburton, J., Scharf, T. and Walsh, K. (2017) Flying under the radar? Risks of social exclusion for older people in rural communities in Australia, Ireland and Northern Ireland, *Sociologia Ruralis*, 57 (4), 459–480.

Wiersma, E. and Denton, A. (2016) From social network to safety net: dementia-friendly communities in rural northern Ontario, *Dementia*, 15 (1), 51–68.

Wild, K., Wiles, J. and Allen, R. (2013) Resilience: thoughts on the value of the concept for critical gerontology, *Ageing and Society*, 33 (1), 137–158.

Wilson, B. (2017) *State of Rural Services, 2016*. Tavistock: Rural England. Retrieved from https://ruraleng land.org/wp-content/uploads/2017/01/SORS-2016-full-report.pdf.

Winterton, R., Warburton, J., Keating, N., Petersen, M., Berg, T. and Wilson, J. (2016) Understanding the influence of community characteristics on wellness for rural older adults: A meta-synthesis, *Journal of Rural Studies*, 45, 320–327.

Wu, Y. T., Prina, A. M., Barnes, L. E., Matthews, F. E. and Brayne, C. (2015) Relocation at older age: results from the Cognitive Function and Ageing Study, *Journal of Public Health*, 37 (3), 480–487.

Part VI

Rural settlement, planning and design

Part VI casts a broad net over the planning and 'built environment' challenges of rural areas. Its concern is with rural settlements and services, with infrastructure and with the design and planning of 'villages' as focal points in rural economies and within the social fabric of remoter or peripheral regions that have been subject to significant structural change over recent decades. The initial focus is on England, both in terms of service challenges and settlement planning. That focus then broadens to encompass examples of countries grappling with urban encroachment in rural hinterlands and rural decline in more peripheral areas.

The part opens with a general account from Gallent (Chapter 31) of the service challenge arising in areas of distributed and sparser population. How can private enterprise, public investment and voluntary action come together to provide the essential 'rural infrastructures' needed to give viability and vibrancy to communities? Rural infrastructures are said to possess 'community' and 'social' components, with those labels denoting the hardware (of material things) and software (of immaterial social supports) that are essential for the functioning of rural communities. It is argued that in conventional perspectives on services, 'community action is . . . one of a triad of inputs potentially supporting the delivery of essential hardware. The others are private enterprise and pubic investment'. But the infrastructures perspective set out in Chapter 31 'shifts the rural services debate in two significant ways: firstly, key hardwares are brought together and it is accepted that these will need to be delivered through a mix of means: through private and public investments, through voluntary action and through the protection and promotion of essential services through land-use planning and other policy mechanisms'. Secondly, and perhaps more significantly, 'social infrastructure is elevated to the status of software, not merely part of the delivery triad but an essential resource in its own right'. This perspective borrows heavily from previous work dealing with 'entrepreneurial social infrastructure' and invokes a broad literature on social capital and active citizenship. It shines a light on the importance of 'social software' in delivering against the 'services challenge'. Evidence suggests that such software has universal importance, shaping outcomes not only in poorer or depopulating rural areas but in outwardly affluent areas, where social infrastructure still plays a vital role in delivering rural liveability.

In the next chapter, Sturzaker contrasts more traditional planning approaches to service viability – through concentration of rural assets in key settlements – with softer community support strategies. He starts by outlining the rationale for service/development and asset concentration in higher tier key settlements. The objective is to create viable service markets by putting everything in one place. The corollary of this is that lower tier settlements – small hamlets and villages that are not designated as key settlements – are starved of development and residents of those communities become car-dependent and need to travel daily to access many essential services. Key settlement policy (KSP) is motivated by the pursuit of sustainability but it often generates unsustainable patterns of travel and threatens the wellbeing of small village locations. Sturzaker begins by re-tracking the history of KSP in Great Britain, with a particular focus on Durham County Council's development of this approach to settlement planning in the early 1950s. As the economic vitality of villages drained away, there was an expectation that population would / should 'regroup' in larger centres. Planning, however, had no power to force such regrouping and despite a loss of development and services, counter-urbanisation from the 1970s onwards brought the repopulation of hitherto declining villages. In response, KSP became part of a broader urban containment strategy, only being rationalised by the pursuit of 'sustainable' development in the 1980s and 1990s. The latter goal was used to justify restraint policies, though there was seldom any consideration of their implications on rural economies and communities. Other countries have devised their own key settlement / concentration policies, sometimes backed up with an insistence that populations should regroup (achieving this by force or coercion). However, there has been opposition to KSP in different places, with that opposition often flagging the many reasons for adopting less restrictive approaches to settlement planning. In the face of counter-urbanisation, restraint policies tend to accentuate gentrification – making it more difficult for 'local' buyers to access housing at a reasonable price in small village locations. Moreover, declining housing affordability sits alongside a squeeze on services and rural employment as a triad of impacts brought by KSP. Sturzaker highlights the growing body of research pointing to the problems of KSP, its narrow interpretation of what constitutes 'sustainable development' (preventing car journeys), and its many negative impacts on the social and economic life of rural areas. Broadly, KSP is said to offer an interpretation of sustainability that ignores issues of equity and social justice – and this is being formally and informally challenged in many places via community-based planning practice. In England, neighbourhood planning provides an avenue through which communities can legally challenge restrictive local plan policies and develop more permissive approaches. Alternatively, groups may embark upon illegal action, directly challenging formal planning restriction by taking forward 'low impact development' that contravenes planning regulation. This tends to provoke a swift enforcement response by authorities, but it evidences a degree of frustration with planning's sometimes narrow definition of what constitutes sustainable and acceptable development in the countryside. Sturzaker concludes by noting that despite there now being some experiments with more permissive approaches to rural settlement planning, occasionally allowing one-off developments in open countryside, the mantra of 'urban good, rural bad' retains its potency, as does a conveniently narrow definition of sustainability that has proven so effective at gentrifying many rural areas and limiting economic diversification.

Sticking with the same theme, but broadening it to encompass the local processes of rural settlement planning, Murray (Chapter 33) examines the way in which planning practice must embrace economic, social and cultural considerations – thus avoiding the narrow regulatory focus highlighted by Sturzaker in the previous chapter. Of particular concern in Chapter 33 is the role of 'local stakeholders' – notably professional planners and village residents – in shaping 'more complete village futures' than those envisaged by land-use planning alone. There is inherent complexity in resident input as residents are not a homogeneous group. Rather, they display great variety in terms of place attachment and their particular ambitions for village planning and development. Murray contrasts different approaches to the production of village plans, starting with those that are planner-led. A 'technocratic approach' to village plans in the UK was supported by the KSP already analysed by Sturzaker. It was often highly controversial, prompted dissent, and gave the impression of a planning system unsympathetic to 'complete village futures'. In Northern Ireland, housing land allocations have been recently made according to a settlement hierarchy and matched to infrastructure capacity – giving continuity to past KSP. The 'regulatory planning arena' has, however, made space for community input in recent years – with Murray noting the past role of design statements in England and the more recent arrival of neighbourhood development plans. The latter have been popular in villages and market towns under pressure from housing development and Murray concludes that the focus of such plans has been on developing 'locally acceptable regulatory frameworks for housing development' rather than harnessing 'the potential that village communities possess to shape their own destinies'. For that reason, he steps away from this regulatory interface to look at broader 'participatory and strategic village planning'. Under that heading, the ways that communities might mobilise and develop inclusive strategies, which aim to spark a conversation about future development, are discussed. Such strategies aim to catalyse a 'process of discovery to produce a transformative trajectory that is much more appropriate for local community and institutional realities'. Murray views that wider process as crucial for broadening the focus of village planning beyond restriction / control towards engagement with the aforementioned economic, social and cultural considerations.

Parkinson (Chapter 34) adds a further dimension to this debate through focusing on design and built environment form within small settlements. Parkinson begins by noting the erosion of local distinctiveness in the appearance of rural village reflecting wider trends of growing standardisation and homogeneity in both rural and urban places more generally. The chapter sets out to explore notions of place distinctiveness and the different interpretations of place qualities among professional and lay voices. This brings into play varying narratives concerning quality, place identity, an design traditions and how these relate to contemporary development trends related to retail and house-building. Parkinson then examines how these concepts and narratives relate to local planning processes and the management of the built environment. Specifically, the chapter examines the practice of village design statements, emerging initially in England, but explored here as the practice was transferred to the Irish context. Parkinson uses the experience of preparing village design statements to explore how local distinctiveness is represented within place-making strategies. The chapter concludes by emphasising the importance of harnessing local knowledge and community values in deliberating local distinctiveness, creatively building bridges between design professionals and the public in accommodating new development in villages while protecting, preserving and enhancing the distinctive character of rural settlements.

A further contribution from Parkinson, co-authored with Pendlebury (Chapter 35), pushes the narrative further through exploring rural heritage and conservation and its importance to place-making in rural contexts. While initially focused on built heritage, including historic buildings such as wealthy ancestral homes and historic assemblages of buildings within villages, the chapter also relates these debates to natural heritage and the importance of landscape as a form of heritage – this links with chapters in Part VII that focus on landscape management and preservation. In this chapter, Parkinson and Pendlebury examine how heritage values are culturally constructed and vary over time and across spatial contexts. In turn, these representations of heritage shape policy priorities and frames how competing demands on rural places relate to rural and place identity. The authors examine heritage through two contrasting case studies of England and Ireland, exploring 'rural heritage imaginaries' and how these frame contrasting public attitudes and heritage conservation policies within the two countries. The chapter argues that in England, the rural landscape and built heritage is central in debates on national identity, are highly prized and generally subject to strict planning controls. In an Irish context, much less emphasis has traditionally been placed on heritage protection, with built heritage in particular often associated with postcolonial memories. In contrast to England, regulatory protection of heritage has been relatively recent in its introduction and national landscape policies are poorly developed and piecemeal.

The last two chapters in this part look at the built environment of rural areas in very different settings. Kan's focus in Chapter 36 is on mainland China and the ways in which urban expansion has impacted on peri-urban and remoter rural locations. The chapter begins with a broad perspective on change drivers, identifying the commodification of land as central to urban and economic growth – under China's free-market reforms, land that could hitherto be freely used for farming and industry, became a commodity that the state could lease for a fee. This precipitated the creation of 'growth coalitions' between local states and property developers, accelerating urban expansion into rural locations. The broader impacts of urban growth have included a transformation of near-urban areas and an exodus of population from remoter rural areas, resulting in abandonment or hollowing out of some villages. Kan provides a substantial focus on the transformation of near-urban farmland to urban land, through a variety of channels including expropriation by urban authorities. This has led to the creation of 'urban villages' that have come to represent a 'distinct spatial form that traverses rural-urban boundaries'. These villages are a feature of many Chinese cities, with land often remaining in collective ownership. Interestingly, this prompts the comment that 'the state [therefore] lacks effective means of spatial intervention'. State control of land is viewed as a prerequisite for effective intervention; where control is more limited, 'blight and informality' gain a foothold. On the other hand, collective ownership can be used as a means of sucking community groups into the development process and into the 'pro-growth development agenda'. Communities, incentivised by generous compensation packages, have become instrumental in driving forward urban growth – and many have benefited personally. But away from this expansion, villages in remoter countrysides have frequently experienced neglect and abandonment. De-collectivisation of land during the early period of reform resulted in increased pressure on household farm units to sustain and grow income. This proved difficult and became another driver of the rural exodus, fuelled more generally by the pull of urban growth and promise of an easier/better life in the cities. By the mid-2000s, the exodus had triggered considerable official concern and a programme of infrastructure/housing investment and modernisation was seen as a means of stemming the outflow and addressing rural 'backwardness'. The programme sat alongside a strategy of relocating villagers to concentrated 'new rural villages' in more accessible locations – something akin to the key settlement policy described in previous chapters, only with more muscle. Relocation has attracted a mixed response: in some quarters it is viewed as problematic, generating difficulties as households struggle to meet the higher costs of semi-urban living; but it is also viewed positively by many relocated households, who see themselves as achieving a more modern and urban lifestyle. Kan's account of the transformations affecting China's rural areas points to a literal and conceptual blurring of 'the rural' and 'the urban', with characteristics of the latter often seen as an antidote to the problems of the former. There are of course parallels with earlier episodes of rural development in western countries, characterised by modernisation and exogenous development approaches.

The final contribution to this part, from Tietjen and Jørgensen, shifts the focus to one of rural decline in Denmark. Local projects aiming to address the cyclical effect of population loss, reduced economic activity and service failure are analysed by the authors, who illustrate how planning can help achieve new development potentials. They note, early on, that there has been a recent shift in policy focus – away from tax redistribution and subsidy in support of peripheral areas to a more place-based, developmental emphasis. Following local government reform in 2007, two large peripheral rural authorities – Thisted and Bornhold (island) – emerged with new responsibilities for rural planning and development. The challenges of population decline were to be addressed through a range of physical development projects which aimed to improve infrastructure, anchor population and grow the tourism sector. Tietjen and Jorgensen detail five development projects: one in Thisted that delivered new infrastructure with the aim of increasing its attractiveness to surfers (and hence grow the tourism sector) and four on the island of Bornholm that exploited industrial heritage assets (former fishing harbours and granite quarrying) to create new focal points for tourism and community use. They show, using actor–network theory, how physical assets (sea conditions off Thisted and former fishing harbours and granite quarries on Bornholm) formed part of the networks that were instrumental to delivering strategic visions and underpinning projects. They also show how these deliver tangible population and economic gains but also outlived the projects themselves and continue to provide networks for further actions and interventions – achieving strategic longevity that is central to further physical and social development in peripheral rural areas. The experiences in the two case studies are used by the authors to reveal the nuances and challenges of strategic spatial planning, linking those experiences to dominant theories and particular ideas around the way strategic processes, drawing together professional and lay actors, incubate in 'soft spaces'. But ultimately, they come back to the importance of sustained networks in addressing the population, economic and service challenges arising from spatial peripherality.

The contributions to this part begin and end with a broad focus on the importance of a mix of actions and actors in bringing new activity to rural areas. They deal with contexts of poverty, wealth, growth and decline. But they all share a sense that planning is an instrumental force in shaping rural places, and that planning is at its most effective and positive when making strong connections to rural communities and business groups.

31

Rural infrastructures

Nick Gallent

Introduction

This chapter is concerned with rural service provision, framed as rural infrastructures. The service challenge in rural areas is commonly presented as one of increased per capita costs of servicing the needs of a more spatially dispersed population. For the public sector (and its private subcontractors), it costs more to run (or subsidise) a bus if there are more stops, fewer passengers and longer journeys. Likewise, higher costs of collecting household waste are incurred if homes are distributed over a wide area. Primary schools will be more expensive to operate, per capita, if there are fewer children enrolled. Private enterprise will fare no better. Shops will see less walk-in trade and those that are able to operate will stock a much smaller selection of goods than their urban counterparts. Choice for the rural consumer is more limited. People will have to journey by private car to access that choice, often because bus services are irregular. There is a certain inevitability around the service challenge faced by rural places. Markets are thinner and service levels – either determined directly by that market or by public investment – are commensurately lower. In the past, responses to this challenge have involved smarter public intervention (including relaxations of planning rules), supported enterprise and a large dose of voluntary action. While the standard concerns of rural service provision are addressed in this chapter, a broader view is also offered. Rural infrastructures include the full-range of community infrastructure (public and private, from health care, through affordable housing, to local transport) while social infrastructures are those that provide communities with soft support (community groups and local networks). Rather than dividing services between public and private types, infrastructures comprise the items and supports that communities cannot do without. They are provided through combined inputs: different sources of funding and finance, from within and outside the community, and through the social exchanges that deliver community capacity (i.e. the social capital dimension of community development).

This chapter is divided into two parts. The first takes a general look at service and accessibility challenges in rural areas through a more conventional lens. The second then looks across the 'software' and 'hardware' of rural infrastructures and the general strategies adopted now, or that might be taken up in the future, to deliver broader access to those essential infrastructures on which rural liveability depends.

The rural service challenge

Questions of rural service provision are usually combined with issues of 'get-at-ability' and 'accessibility' (Moseley, 1979). A sparser population distribution in all rural areas, but especially those further from large towns and other key service centres, creates a difficult environment for public services and a thinner, less attractive, market for private enterprise. In the UK, availability and access to rural services is viewed as a critical challenge for rural development policy and planning. This challenge, however, has altered considerably over the last 50 years. Counter-urbanisation has brought new people to the countryside, and with them have come new needs and capabilities. Rural communities are more mixed, socially and economically, than they once were and therefore express a diversity of needs and are able to meet those needs (or not) in different ways. New affluence means more cars; and while more cars means greater mobility for some households, it also reduces the demand for public transport and leaves those unable to afford a car (or unable to drive) potentially more isolated. The Department for Transport in England (referenced in DEFRA, 2014) has developed measures of 'reasonable' accessibility using different travel modes, assuming that it is reasonable to walk, cycle, or drive to services depending on the distance of those services from home. Eight services are deemed to be key: places of employment, primary schools, secondary schools, further (adult) education, primary (basic) healthcare, secondary (specialist) healthcare, and town centres (offering a mix of retail and banking) and shops selling groceries. Not surprisingly, levels of accessibility (measured as the percentage of local residents enjoying 'reasonable' access) fall in rural areas and especially in those areas furthest from service centres. This is judged to be an 'access gap' and is equivalent to the 'get-at-ability' challenge that Moseley wrote about almost 40 years ago. The challenge is a simple one: live centrally in a town and enjoy access to most things within walking distance; live remotely in a small hamlet and expect to drive to most basic and all specialist services. If driving is not an option, then expect a long and difficult journey often punctuated with many inter-changes.

This rural planning challenge confronts both the consumers and producers of services. 'Traditional' public services – education and healthcare – are costlier to fund where the population is more dispersed. The challenge for the public sector is how to stretch budgets and provide equivalent service levels. In 2011, 100 per cent of the urban population in England lived within 4km of a doctors' surgery. The figure of 69 per cent for villages and hamlets masks the longer distances that some rural people need to travel to access basic healthcare. A further challenge is how to deliver more flexible services that counter the inequalities that distance produces. Mobile services or secure virtual consultations with doctors have been trialled in some places. Elsewhere, trained volunteers are asked to monitor the wellbeing of older people, alerting health care professionals to arising needs (Institute of Rural Health, 2012). But such initiatives are unable to bridge the gap in emergency care or replicate the range and quality of services enjoyed in population centres. The experience of healthcare is bound to be different in the countryside.

This is also true in relation to school provision. Ribchester and Edwards (1999) have shone a light on the problems facing small rural schools and their pupils. There are educational arguments for and against the retention of small schools, but these seem evenly balanced: their curricula may be more limited, but can be enhanced through the sharing of specialist teaching across a network of similar schools; the teaching environment may be too 'sheltered' but that same environment can nurture a stronger sense of community; and they may lack many of the facilities of bigger urban schools, but these again can be pooled and shared. The weight of economic argument seems, however, to favour rationalisation. Per pupil costs in rural areas (where there is a proliferation of smaller schools) are 77 per cent higher than in urban areas (where schools are generally larger and may have more than a single-form annual intake) (Hindle et al., 2004).

Yet the 'community' function of such schools generates considerable support for their retention, pushing 'planners and educationalists towards devising alternative reorganisation and support strategies' (Ribchester and Edwards, 1999: 58). As with healthcare, the response to higher costs is a search for innovation. Pooling staff and facilities is a common strategy, with teachers moving between sites in order to deliver curriculum diversity. After 2010, a big policy shift in England allowed parents (and other community members) to establish their own 'free schools'. This can allow the setting up of schools, managed by 'founding' volunteers, in vacant buildings. But core funding may still be insufficient to overcome those same barriers – that is, higher per pupil costs – facing established schools.

Ideas around healthcare and education – traditional public services – hint at the preferred solutions for all kinds of service provision: costs must be reduced, innovation is essential, and volunteers need to be engaged. Private rural services, like public ones, need all of these things too. Many villages in England have been losing their dedicated grocery shops (Woods, 2005: 97). Such businesses are already marginal, but final closure often reflects changing shopping habits: commuting households doing their weekly shops in supermarkets or purchasing groceries through on-line box schemes. A decade ago, the Commission for Rural Communities (2006) warned that general stores do not have the purchasing power needed to drive down supplier costs, and therefore carry a smaller selection of pricier goods, which in turn drives customers to use supermarkets or the internet. Although such stores play a vital social function – hosting post-office counters and providing a meeting place for residents – they are ultimately micro-businesses that must be viable to survive. That viability often comes from an input of volunteer time. Stores at risk of closure can, and sometimes are, turned into not-for-profit cooperatives. They become community enterprises: the number of such enterprises in England rose from 23 in 1993 to 303 in 2012 – with more than 200 more planned (Plunkett Foundation, 2013). This is possible where communities can draw on the good will and hard work of volunteers, where the ownership of assets can be secured through a long-lease, and where tax relief for such enterprises further enhances their viability. Much depends, of course, on the capacity of communities to organise and deliver their own solutions: on their store of 'social capital' and on the leadership of 'active citizens'.

Small stores are often seen as one component of a triad of private services available in viable communities. Those viable communities have a base level of core infrastructure. The other two components are post offices and pubs. Arguably this is a very Anglo-centric view of what communities need to thrive and very particular and peculiar to the archetypal English village. But within this triad are a set of basic services that are central to the socio-economic life, and liveability, of rural areas: basic provisions, for when these run out; simple banking services; and a social focus. Something similar to the pub could substitute elsewhere: the small bar or *caffetteria* in rural Italy, for example.

In 2000, there were just over 9,000 post offices in rural England. Two-thirds of these had closed within just over a decade (Rural Services Network, 2013). That contraction is likely to have been a result of the switch away from traditional services: pensions and other benefits being credited directly into bank accounts or car tax, for example, being renewed online. The opportunities of internet-based services have resulted in a sharp decline in face-to-face counter provision. This is perhaps inevitable but it leaves those without access, or ability to access, e-services hugely disadvantaged and potentially more isolated. For that reason, some communities are keen to maintain their post-office counters and will run these from village stores (where these survive), community centres (if there is one) or even pubs. Again, innovation, reduced costs and volunteer time are all important requirements if some level of key service provision is to be maintained.

One might imagine that pubs fare a little better than other depleted rural services: it might be thought that they are a little further up the community pecking order given their likely role in the social life of villages. But pubs have faced very particular challenges. First, changing patterns of 'sociability' have deprived many of them of regular customers. And second, the desirability of 'a home in the country' means that it makes economic sense to transform 'public' houses into 'private' ones. This can be achieved through the de-licencing of a pub and permission to convert it to solely residential use. Across the UK, around 1,300 pubs become private houses each year. Concern expressed by nearby residents has become a political concern with planning authorities in England now instructed to 'promote the retention and development of local services and community facilities in villages such as . . . public houses' (DCLG, 2012: para. 28). Planning is more broadly called upon to enhance the 'sustainability of communities' by affording protection to 'valued services and facilities' and thereby enhancing communities' capacities to meet 'day-to-day needs' (ibid: para. 70). Pursuant to this, rural parish councils in England are able to designate pubs and other facilities as 'assets of community value' and, if they can raise sufficient funds, purchase these facilities should they come onto the market. At that point, the new community owners will face the same challenges as the previous private ones and will only be able to sustain services if they can generate new business and off-set running costs with voluntary contributions.

But pubs – and other community 'hubs' – are different from the other two parts of the triad. Post office and basic retail services can be delivered on-line, although for this to work, the average household will need to carefully plan out the likelihood of ever running out of milk or needing an extra loaf of bread. But the hub-function is difficult to replicate on-line, unless one imagines the re-creation of a virtual village, populated with virtual villagers – young and old – sitting at home and interacting via Facebook, Twitter and other social media platforms rather than wanting to meet face-to-face. Whether this would constitute a loss or gain of community is unclear, but it is unlikely to be the rural community that most people imagine or buy into. That said, the 'judicious introduction of ICT' was viewed, nearly twenty years ago, as part of the answer to the rural service conundrum (Moseley, 2000: 415). It has some potential to plug education and healthcare gaps and deliver against regular retail needs. Research into digital service provision tends to view it as generally a good thing, while drawing attention firstly to a 'digital divide' (between areas that have good wired and wireless connection and those which do not) and secondly to 'digital differentiation' (between groups who tend to make use of the technology and those who are less comfortable with it) (Riddlesden and Singleton, 2014). Longley and colleagues (2008) have shown that knowledge and use of ICT is spatially differentiated: there is a greater concentration of the 'e-unengaged' in rural areas while the 'e-experts' seem to congregate in cities. This may be the result of infrastructure shortcomings. However, in the US, LaRose and colleagues (2007) have shown that 'digital differentiation' remains even when there is infrastructure equivalence: demography, relative poverty and educational background are therefore important factors in determining whether people go on-line.

'Digital differentiation' is not thought to be as big an issue in the UK as it is in the US where some rural areas are characterised by concentrated poverty. Rather, the challenge for UK rural areas is one of broadband access. Providing this access is the business of private companies, which have a vested interest in servicing those markets where there is both high demand and where costs of provision are less. It has been estimated that the cost of connecting the final 10 per cent of potential customers – those living in the least accessible locations – will be three times greater than it was to connect the first two-thirds (Townsend et al., 2013). Poor ICT infrastructure continues to compound the accessibility challenges faced by communities at the widest point in the digital divide. Some rural households, especially older and more vulnerable people, may find themselves deprived of information, paying more for utilities because of their reliance on paper billing, or unable to access those virtual networks that often prelude real social opportunities (Wellman, 2001).

The conventional service challenges confronting people and communities in rural areas have generally been presented, over the last 50 years of rural socio-economic change, as material ones delivered by traditional providers: usually the public or private sectors. Increasingly, community life – expressed through voluntary action and sometimes conceived as a store of social capital (see Falk and Kilpatrick, 2000 for a particular rural take on this broad concept) – has been viewed as a resource for communities, enabling them to offset costs and do more for themselves. But that community life, and everything it generates and sustains, should be seen as an important part of the infrastructure of rural communities. The service challenge has generally been regarded as one of 'hard' infrastructure and its non- or limited availability in rural places. Soft infrastructure is viewed as a compensation or substitute. And yet, soft infrastructure is more than this. It is a resource in its own right that shapes the experience of living and working in the countryside.

Rural infrastructures

The focus on 'rural services' can seem a little old fashioned, limited and perhaps just a bit quaint. The outcomes of analyses of service provision depend on which items are deemed significant. Rural pubs are closing at an alarming rate and this seems, to some observers, to threaten and undermine the social life of communities. On the other hand, volunteers are now running more community shops than ever before, and this points to a sustained vitality that is evident in many smaller villages. Likewise, post offices are closing but the need for them is less than it was – and many general stores, where they exist, continue to provide transaction services (for the payment of bills etc.). Past concern for rural services has generally tried to capture two things: changes in the availability of provisioning infrastructure (goods and services) and changes in the availability of physical places for interaction. Provisioning infrastructure can be sourced publicly or privately. Interaction generates a social infrastructure that either becomes a further source of provisioning (running a community facility or bus) or offers social support.

This infrastructure perspective on 'services' captures more of what rural communities need, divided into the requisite operational 'hardware' and 'software'. Although the labels 'community' and 'social' infrastructure are often used inter-changeably, the former includes those hard items supplied by the market or by tax-payers and the latter comprises those soft things that people do for themselves. The idea of community infrastructure is seldom extended to market goods such as retail, public houses or broadband. Shops and pubs arise from private enterprise and broadband may be viewed as part of a 'growth infrastructure', installed alongside other utilities to facilitate development. But in many rural areas, these are often marginal activities, installed or operated with considerable difficulty. They are provisioning infrastructure, delivering against community goals, which require either capital or labour subsidy. In thinner markets, the line between private and public provisioning is blurred. Therefore it seems reasonable to place supported private services in a general community infrastructure group.

The hardware of community infrastructure includes the services discussed in the last section plus other items. Transport infrastructure delivering accessibility is crucial, as are primary and secondary health care services. Schools and nurseries are equally important alongside public community facilities – such as libraries – that are likely to be concentrated in key service centres. Less obvious community infrastructure include emergency (or 'blue light') services, crematoriums and utilities – water supply and treatment, power, waste disposal and of course broadband, as previously mentioned. Other, even less obvious infrastructure in this group can include flood defences and affordable housing. The inclusion of housing as a community infrastructure is significant. Elsewhere in this Companion, it is shown that market processes in many rural areas generate acute housing injustice and that housing should not be viewed merely as a market exchange good but as basic need that underwrites the sustainability of communities.

The real benefit, however, of an infrastructures perspective on rural services is that it elevates soft support to the status of social infrastructure – essential in its own right and not purely as an alternative to private enterprise or public investment, that could be jettisoned if only the market or the state would step in. Social infrastructure centres on people, voluntary action and local networks (Flora and Flora, 1993). Social infrastructures include community facilities run by volunteers (shops, village halls, buses, car-sharing schemes): in some cases these bring together the hardware paid for by others (e.g. buses) with the labour / software provided by residents. They also include local networks that coalesce into support groups and organisations (e.g. residents' associations, mother and baby groups, support for the elderly or for lonely and isolated neighbours) which can, eventually, develop into local projects and perhaps secure private, public or charitable funding. Networks may have a 'serious' focus – on personal learning, mental health, or the development of skills – or a comparatively more 'trivial' focus, on hobbies and pastimes. Such networks bring people together, give them common purpose, and underpin more formal community structures such as village or parish councils (Gallent and Robinson, 2012). One of the essential infrastructures that flows from social interaction is *leadership*: the emergence of individuals who are able to generate broader interest in local service challenges, who go on to lead a community land trust which acquires buildings and establishes a village hall, or who provide a bridge to private enterprise or public bodies and win support for local projects from multiple sources. These have variably been described as 'everyday fixers' (Hendriks and Tops, 2005; van der Pennen and Schreuders, 2014) or 'active citizens' (Vilà, 2014). They are part of the social infrastructure of communities and play a key role in growing that infrastructure (by expanding interest and assembling volunteers) and directing it to essential projects and initiatives.

Around the world, there are many examples of (rural) social infrastructure in action. Kilpatrick and colleagues (2014) outline its role in providing mental health support in farming and fishing communities in Australia. Its input into bus services in England was noted by Moseley (2000) and more recently, car sharing has been viewed as an important expression of social capital in rural areas (Currie and Stanley, 2008). In Northern Ireland, evidence suggests that elderly people in rural areas are less likely to require admission to care homes than those living in urban areas, partly as a result of family support but also because of a greater likelihood of 'informal care' (McCann et al., 2014). The capacities of community land trusts in rural areas, to deliver affordable homes and other essential hard infrastructures are also well-documented (see Satsangi, 2014 for Scottish examples). More generally, social networks are important in sustaining marginal economic activities, providing an infrastructure to farmers and others, and helping them reduce costs and increase profitability (Magnani and Struffi, 2009). The last example strays beyond provisioning infrastructure and social support into the realm of rural economic development. But here too, social structures – and social capital – have an important part to play in progressing development (Woods, 2011) and are seen as a crucial driver of endogenous growth. Across multiple domains, the social infrastructure of communities plays a critical role in the functioning of rural places. For that reason, it is more than a substitute for external investment but rather a basic and fundamental resource for rural people.

Conclusions

This chapter has sought to shift the focus of 'rural service' debates. Conventional perspectives have drawn attention to a shortlist of key hardwares that communities would struggle to do without – usually a mix of healthcare, education and retail functions. Regular bus services are also viewed as important, delivering the accessibility to services that dispersed populations frequently need. In this view of rural services, community action is seen as one of a triad of inputs potentially supporting the delivery of essential hardware. The others are private enterprise and pubic investment. An infrastructures perspective shifts the rural services debate in two significant ways: firstly, key hardwares are brought together and it is accepted that these will need to be delivered through a mix of means: through private and public investments, through voluntary action and through the protection and promotion of essential services through land-use planning and other policy mechanisms. Second, social infrastructure is elevated to the status of software, not merely part of the delivery triad but an essential resource in its own right. Clearly, this perspective tries to bring together different strands of existing debates. Flora and Flora (1992) have flagged the importance of entrepreneurial social infrastructure to rural development and many others have analysed the role of community action in delivering services and initiatives that would not otherwise have been funded by the public sector (Edwards, 1998). The contribution of this short chapter to these debates is an integrative one: rural infrastructures are multiscale and cross-domain. They are diverse and potentially delivered in a variety of ways. But two ideas bind them together: first, the essential operating system of rural areas necessarily comprises a hardware of material things and a software of social processes; second, it is this combination of material and immaterial things – *rural infrastructures* – that make places liveable and sustain rural wellbeing, for many groups and across many fronts.

References

Commission for Rural Communities (2006) *Rural Disadvantage: Reviewing the Evidence Base*, Cheltenham: CRC.

Currie, G. and Stanley, J. (2008) Investigating links between social capital and public transport, *Transport Reviews*, 28(4), 529–547.

DCLG (2012) *National Planning Policy Framework*, London: Department for Communities and Local Government.

DEFRA (2014) *Statistical Digest of Rural England*, London: Department for Environment, Food and Rural Affairs.

Edwards, B. (1998) Charting the discourse of community action: perspectives from practice in rural Wales, *Journal of Rural Studies*, 14(1), 63–77.

Falk, I. and Kilpatrick, S. (2000) What is social capital? A study of interaction in a rural community. *Sociologia Ruralis*, 40(1), 87–110.

Flora, C. B. and Flora, J. L. (1993) Entrepreneurial social infrastructure: a necessary ingredient, *The Annals of the American Academy of Political and Social Science*, 529(1), 48–58.

Gallent, N. and Robinson, S. (2012) *Neighbourhood Planning: Communities, Networks and Governance*, Bristol: Policy Press.

Hendriks, F. and Tops, P. (2005) Everyday fixers as local heroes: a case study of vital interaction in urban governance, *Local Government Studies*, 31, 4, 475–490.

Hindle, T., Spollen, M., and Dixon, P. (2004) *A Review of the Evidence on the Additional Costs of Delivering Services to Rural Communities*, London: SECTA.

Institute of Rural Health (2012) Rural proofing toolkit, retrieved on 17 July 2014 from www.ruralproofingforhealth.org.uk.

Kilpatrick, S., Willis, K. and Lewis, S. (2014) Community action in Australian farming and fishing communities, in N. Gallent and D. Ciaffi (eds), *Community Action and Planning*, 79–96, Bristol: Policy Press.

LaRose, R., Gregg, J., Strover, S., Straubhaar, J. and Carpenter, S. (2007) Closing the rural broadband gap: promoting adoption of the internet in rural America, *Telecommunications Policy*, 31, 359–373.

Longley, P., Webber, R. and Li, C. (2008) The UK geography of the e-society: a national classification, *Environment and Planning A*, 40, 362–382.

Magnani, N., and Struffi, L. (2009) Translation sociology and social capital in rural development initiatives: a case study from the Italian Alps, *Journal of Rural Studies*, 25(2), 231–238.

McCann, M., Grundy, E., and O'Reilly, D. (2014) Urban and rural differences in risk of admission to a care home: a census-based follow-up study, *Health & Place*, 30, 171–176.

Moseley, M. (1979) *Accessibility: the Rural Challenge*, London: Methuen.

Moseley, M. (2000) England's Village Services in the 1990s: entrepreneurialism, community involvement and the state, *Town Planning Review*, 74, 1, 415–433.

Plunkett Foundation (2013) *A Better Form of Business 2013: Community-Owned Village Shops*, Woodstock: Plunkett Foundation.

Ribchester, C. and Edwards, W.J. (1999) The centre and the local: policy and practice in rural education provision, *Journal of Rural Studies*, 15, 1, 49–63.

Riddlesden, D. and Singleton, A.D. (2014) Broadband speed equity: A new digital divide? *Applied Geography*, 52, 25–33.

Rural Services Network (2013) *State of Rural Service Provision 2013*, Tavistock: Rural Services Network.

Satsangi, M. (2014) Communities, Land-ownership, housing and planning: reflections from the Scottish experience, in N. Gallent and D. Ciaffi (eds), *Community Action and Planning*, 117–130, Bristol: Policy Press.

Townsend, L., Sathiaseelan, A., Fairhurst, G. and Wallace, C. (2013) Enhanced broadband access as a solution to the social and economic problems of the rural digital divide, *Local Economy*, 28, 6, 580–595.

Van der Pennen, T., and Schreuders, H. (2014) The Fourth Way of active citizenship: case studies from the Netherlands, in N. Gallent and D. Ciaffi (eds), *Community Action and Planning*, 135–156, Bristol: Policy Press.

Vilà, G. (2014) From residents to citizens: the emergence of neighbourhood movements in Spain, in N. Gallent and D. Ciaffi (eds), *Community Action and Planning*, 59–78, Bristol: Policy Press.

Wellman, B. (2001) Physical place and cyber-place: the rise of personalized networking, *International Journal of Urban and Regional Research*, 25, 2, 227–252.

Woods, M. (2005) *Rural Geography: Process, Responses and Experiences in Rural Restructuring*, London: Sage.

Woods, M. (2011) *Rural*, London: Routledge.

32

Settlement, strategy and planning

John Sturzaker

Introduction: key settlement policy

One answer to the question of viability of services considered in the previous chapter has been concentration of new housing and other development in so-called 'key settlements', or 'key service centres'; as opposed to an alternative more dispersed pattern of development. These key settlements may already be centres of employment, service provision, etc., or the approach may be intended to ensure they develop in such a way through targeted investment. The corollary of this, of course, is that settlements which are not designated as 'key' see less development and investment – with the long term result being likely decline. Indeed, this is the explicit intention in some cases.

Perhaps the first, and certainly the most well-known example of the key settlement policy (KSP), situated in a framework of strategic 'structure planning', is found in Great Britain. Great Britain's experience therefore provides the focus of this chapter. Firstly the history of that experience is discussed, from its origins in the 1940s in a context of strong centralised planning, through the use of KSPs in response to trends of counterurbanisation, to the more recent linking of the policy to the notion of 'sustainability'. Examples of similar policies from other international contexts, including China and several European cases, are then introduced. The chapter then returns to the contested concept of sustainability, highlighting the impacts of its application to rural settlements and illustrating alternative approaches that communities have tried to adopt to promote a different interpretation of what a sustainable rural community might be.

Key settlement policy in Great Britain

It has been argued that 'the key settlement policy may be regarded as the principal agent of planned change in post-war rural Britain' (Cloke, 1979: vii). KPSs have indeed been used across Great Britain, but are used more extensively, and more rigorously, in England than in Wales or Scotland, as with other policies which aim at achieving 'urban containment; (Hall et al., 1973), including green belts (Sturzaker and Mell, 2016). Much analysis of KSPs and similar policies focuses, therefore, on England, and this section does likewise.

The early days – efficiency of service provision?

Cloke (1979), in his landmark study of KSPs, traces their theory and practice in Great Britain back to the first half of the twentieth century, but they were formalised through interventions such as the wartime *Report of the Committee on Land Utilisation in Rural Areas* (Scott, 1942). At this time, and perhaps still today, the rural and the urban were dichotomised in Great Britain (see Chapter 2 of this Companion): rural land was to be preserved for agriculture and urban land was provide the location for housing and industry.

As urban areas were thus to be the focus of new development, the concern in rural areas was for the economic provision of services and infrastructure. The Ministry of Town and Country Planning in 1948 advised local planning authorities in 1950 that 'in certain extreme economic circumstances the only course open to planners was to demolish and clear the village and resettle the inhabitants in new centres' (Cloke, 1979: 56). This advice was quickly acted upon in what has come to be seen as the zenith/nadir of the KSP, the infamous 'Category D' policy adopted in 1951 by Durham County Council in the north east of England, in response to a decline in mining activity. The policy is worth repeating in full here, as it includes the evidently socially reforming spirit behind it:

> Many of the rows of houses which grew up around the pitheads have outlived their useful-ness. As the uneconomic pits close and coal working is concentrated in more economic workings, a gradual regrouping of population should take place. Indeed the very reason for the existence of some of these small and isolated places will disappear completely and new development and redevelopment in some of the better placed settlements will not only be better adjusted to the future pattern of employment opportunity but will also offer better living conditions than ever before to many of the inhabitants.
>
> *(Quoted in Pattison, 2004: 316)*

All villages in the county were categorised as A, B, C or D. The former were considered likely to grow in population and thus receive considerable investment, while the latter were assumed likely to see population decline and thus receive no investment, with the people who lived there expected to move to other villages or towns. Very few villages were actually demolished, but the lack of investment, in a time when most infrastructure and housing was supplied by the state, meant that many people left the Category D villages between 1951 and 1977, when the policy was officially abandoned. The unsurprising opposition to the policy was one factor in its eventual abandonment, as was the lack of power held by Durham County Council to acquire privately owned houses, but Pattison (2004: 327) argued that most impor-tant was 'the changing geography and circumstances of the villages themselves', as suburbani-sation and counterurbanisation meant that villages comparatively close to urban centres such as Durham City and Newcastle became more popular places to live. This change, aligned to factors such as hypermobility, was reflected all over Great Britain and subsequently led to a reconfiguration of the stated purpose of KSPs.

Later years – a response to counterurbanisation

Cloke (1979) observed that as more residential development began to occur in rural areas in the 1960s, planning authorities in closer proximity to urban areas became more likely to adopt KSPs. This was done for reasons including 'to permit the economic provision of public services and ensure that the selected communities are conveniently situated with regard to employment . . .

[and] *local* environmental conditions and services' (Cloke, 1979: 70–71; emphasis added). This approach quickly became the norm in more peri-urban rural authorities in the same way as it was near universal in more remote rural authorities, with justifications including those highlighted by Cloke, along with a broader 'urban containment' philosophy – by 1983, Cloke and Shaw (1983: 350) found that 'pressured rural areas appear to be increasingly dominated by urban and regional planning objectives, rather than more localised rural needs'.

This approach, with KSPs justified in peri-urban areas on a similar basis to green belts, continued through the 1980s. In the late 1980s and early 1990s a new purpose for KSPs, and indeed planning policy more broadly, began to be conceptualised – the pursuit of 'sustainable development'.

Popularised by the 1987 Brundtland Report (World Commission on Environment and Development, 1987), the concept of sustainable development swiftly became 'a principal, if not *the* principal, consideration in planning' (Owen, 1996: 38) in Great Britain and elsewhere. As was equally swiftly noted, a problem with 'sustainable development' is that it is not well defined – and without such a definition, there is a risk that 'some local planning authorities might use the term to justify restraint policies without fully examining the likely implications' (ibid.: 40). This is indeed what happened – sustainable development has become the dominant justification for KSPs in national and local policy. Sturzaker and Shucksmith (2011) tracked the discursive construction of sustainability through successive iterations of national policy in the 2000s, regional policy (now abolished), local policy and decision-making on planning applications.

The current national policy in England (current at the time of writing) continues to refer to sustainable development as regards rural housing: 'To promote sustainable development in rural areas, housing should be located where it will enhance or maintain the vitality of rural communities' (DCLG, 2012: 14). While the NPPF allows the possibility that 'development in one village may support services in a village nearby' (ibid.), in practice there is little evidence that there has been significant change in policy at the local authority level, with key settlement policies part of a continued overall policy of restraint in rural areas.

Key settlement policies in other places

The north and south of Ireland are different nations, but they share a connection with Great Britain (constitutional on the one hand and post-colonial on the other) yet a different attitude to rural life, wherein 'a dispersed settlement pattern comprising single dwellings in the open countryside is a longstanding feature of rural areas' (Murray, 2010: 55). Despite this history, in both Northern Ireland and the Republic of Ireland, attempts have been made to limit housing development to urban areas, for similar reasons as in Great Britain (Scott and Murray, 2009). Some have resisted the influence of this 'urban containment' orthodoxy on the grounds of 'colonialism' (Scott, 2012), but it can also be seen as reflective of broader trends at the European scale (Scott, 2006).

Countries which have adopted approaches which can be seen as being essentially those of a KSP include Romania, which in the 1980s adopted a similar policy to that of County Durham in England, whereby investment was channelled into 'key villages with the highest potential for growth' (Turnock, 1991: 251). Turnock notes that there was some logic for this, with many villages in Romania lacking services. As with County Durham, the effects of the policy in practice were not as radical as it was on paper, but a critical distinction is that under Ceausescu, Romania was a dictatorship, so the policy could be backed up with coercion. Some villages were destroyed in the late 1980s, with Turnock identifying some deaths among those opposing the policy.

A different approach, but one also involving strong top-down policy direction, is that found in China, perhaps the most extreme example of urbanisation at this point in the twenty-first century. Justification for attempts to focus activity in key settlements is most closely aligned to the early years of KSPs in Great Britain, though the nature of the policy differs. As China has urbanised, with consequent rural depopulation, some villages have been abandoned and others have seen significant depopulation and accompanying sparsity of active population. In response the Chinese state has begun experimenting with various forms of state-led rural settlement consolidation, in which 'rural collectives are encouraged or pushed (and sometimes even coerced) to relocate to newly built villages of a higher density at another location' (Guo and Zhong, 2016: 20). In some cases these experiments have led to opposition and resistance from disenfranchised local communities (Sturzaker and Verdini, 2017).

In many other contexts which may have urbanised earlier than China, in both developing (Geyer and Geyer, 2017) and developed (Stockdale et al., 2000) countries, some degree of counterurbanisation has been found, with planners consequently developing policies to control it, and/or to manage the impact of it on rural areas. In the Netherlands, the preservation of the countryside for agriculture remains a prominent discourse, with housing supply tightly controlled 'in the open country as well as in villages' (van Dam et al., 2002: 473) – instead of a KSP approach, the Netherlands is exploring 'creating rural residential environment in urban or suburban areas (ibid.: 468) to meet the demands of aspirant counter-urbanisers. National policy in Finland is to move towards a key settlement policy, as opposed to the existing pattern of dispersal, for similar reasons to those espoused in Great Britain and elsewhere – the Finnish Ministry of the Environment claiming 'Climate change will mainly be curbed by reducing the volume of traffic, which is the aim of creating a more coherent urban structure' (cited in Sireni, 2016: 193). However, unlike in Great Britain, this approach is being strongly resisted by local politicians so has yet to enter local policy discourse or exert a strong role in decision-making. But what are the reasons for resisting and questioning the key settlement approach?

Resistance and opposition to key settlements

In most cases, beyond the extremes of Romania or China, KSPs and similar policies do not actively seek to relocate the residents of non-key settlements, an approach which tends to cause an unsurprising degree of resistance. Yet even in non-coercive examples of KSPs, there remains a backlash against them, both because the outcomes of the policies can be regressive and exclusionary; and because the logic behind them is, some argue, flawed.

Back in 1979, Paul Cloke studied key settlement policies in England in some depth, in both more pressured parts of the country (tending to be closer to major urban centres) and more remote areas. He found that in both cases KSPs had to some extent worked – in more pressured areas settlements had been conserved, with their 'rural' character protected (see Woods, 2005 for a discussion of conservation and rural character in Britain); while in remoter areas there had been a reduction in depopulation to an extent. However, likewise in both cases the policies had led to problems – in more pressured areas viable settlements which had been designated non-key had been prevented from growing, and house price rises had forced out some 'local' people at the expense of in-migrants; and in remoter areas rural depopulation had been replaced by migration from hinterland to key settlements, 'thus exacerbating the crises in outlying small settlements' (Cloke, 1979: 200).

Despite these mixed outcomes, the KSP approach continued (and continues) to dominate rural planning policy in England. In 1996, Stephen Owen highlighted the findings of a series of reports looking at the problems of rural areas, including a lack of employment, reduction in (affordable) housing supply and a reduction in availability of services, and argued that KSPs and similar urban containment policies had contributed to these problems (Owen, 1996). Ten years later the problems, and their alleged causes, remained or had worsened (ARHC, 2006; Best and Shucksmith, 2006). Sturzaker and others have argued that the failure of policy-makers to address these issues suggests a fundamental problem with how planning for rural areas in England – and indeed elsewhere – is conceptualised (Sturzaker, 2010), and that this problem has been a feature of the English planning system since its inception in the 1940s (Hall et al., 1973).

As discussed above, the justification for KSPs has changed over the years, and is currently predicated on the notion that it is contrary to the principles of sustainable development to allow development in rural areas in all but a small number of tightly bound places. As Owen (1996) noted, the problem with concepts such as sustainable development is that, while they are sufficiently broad to be hard to disagree with, they are also sufficiently vague to allow particular perspectives to be privileged over others (Allmendinger and Haughton, 2010). A number of authors have argued that the definition of sustainable development as it applies to rural areas has increasingly narrowed, moving from the balancing of social, economic and environmental facets to an almost exclusive focus on the environment, and in turn an increasingly narrow focus on one aspect of the environment – the need to reduce CO_2 emissions by reducing private car use, and consequently not allowing housing in places with no public transport provision (see, among others, Owen, 1996; Scott and Murray, 2009; Sturzaker and Shucksmith, 2011). What this can result in is what has been termed the 'sustainability trap', wherein 'otherwise beneficial development can only be approved if the settlement is considered sustainable in the first place. Failure to overcome this hurdle essentially stagnates the settlement – freezing it in time . . . it makes such communities less, not more, sustainable' (Taylor, 2008: 45).

What Taylor (in a report commissioned by the then prime minister) and others argue is needed is a more balanced and nuanced understanding of sustainability, that: firstly, questions whether those living in rural areas *do* make more use of private cars than their urban counterparts (Champion, 2009); secondly, considers other ways in which the environmental sustainability of communities can be conceptualised (Breheny, 1993); and thirdly, brings back the considerations of equity and social justice which have been argued to be at the heart of the Brundtland Commission's definition of sustainability (Sturzaker and Shucksmith, 2011).

As noted above, until very recently national planning policy in England stubbornly stuck to the now dominant definition of sustainable development, which has also been taken up by planners and policy-makers in other contexts including Ireland and Finland. In March 2018, however, a draft revision to English national policy was published. This appeared to suggest a different position, including as it did the instruction that 'Plans should identify opportunities for villages to grow and thrive, especially where this will support local services' (MHCLG, 2018: 21) – that is, suggesting a way out of the 'sustainability trap'. At the time of writing the finalised policy is yet to be published, and the test of any change will be how it is reflected in local decision-making. In the next section existing examples of different practice at the local and community level are introduced to illustrate the possible results of such decision-making. These approaches can broadly be regarded as attempting to work within the formal structures of planning and development policy, or deliberately challenging/ignoring formal 'rules'.

Formal approaches

Some local planning authorities in England have, in the wake of the Taylor Review (see above), reflected more deeply on the role and purpose of KSPs. Winchester District Council, in the southeast of England, is one example. They made efforts to develop a more sophisticated KSP policy which did not rule settlements as inherently unsustainable and considered other factors including the desire of residents to see their villages grow. Ultimately they concluded that 'Small communities should not be regarded as unsustainable simply because of their size or location' (Winchester District Council, 2013: 59), and adopted a policy that would allow development in most settlements in the district –albeit in many cases for 'local needs' only.

Without a comprehensive survey of all rural local authorities it is hard to know how common this more sophisticated approach is, but evidence from Cumbria in the north west of England suggests frustration on the part of some local communities, resulting in direct action. Such action in plan-making and implementation was made substantially easier by the 2011 Localism Act, a piece of legislation which according to the Government that introduced it would offer communities unprecedented power to produce their own 'neighbourhood plans' and carry out development which met community needs (DCLG, 2011). The Upper Eden Neighbourhood Plan was the first to be 'made' (i.e. brought into use as planning policy). The Upper Eden Community Plan (UECP) group, formed in 2002, had been attempting to influence the local and regional planning policy which covered their area for a number of years in an attempt to avoid the 'sustainability trap' which they found various settlements in their area had fallen into as a result of the KSP promoted by the North West Regional Spatial Strategy and adopted in the Eden District Local Plan.

In the words of a local activist, the process through which such policy was derived was:

> a fairly closed shop . . . those at the top of the profession have a lot invested in the extant system . . . It felt like a closing of ranks by the Regional Planning Body and Local Planning Authority when offered an alternative view of, say, 'sustainability'.
>
> *(Sturzaker and Shaw, 2015: 601)*

The opportunities offered by the 2011 Localism Act were seized by the UECP group, who subsequently produced a Neighbourhood Plan which permitted single homes to be built outside of key settlements and within settlements which had formerly been designated as key but had lost an important service and so had been de-designated by Eden District Council (EDC). The latter was apparently a real fear for the UECP group, the key settlements in their area containing shops, pubs, etc. which were of questionable viability (UECP, 2013).

A similar approach has subsequently been adopted in the Neighbourhood Plan produced for the community of Matterdale, also within EDC but, due to the complexity of the English planning system, under the planning jurisdiction of a different local planning authority, the Lake District National Park (LDNP). The KSP adopted by the LDNP categorises Matterdale as 'open countryside' because of its dispersed settlement character, which is the lowest category of settlement in the KSP and means housing can only be permitted in Matterdale 'in limited circumstances' (Matterdale Parish Council, 2015: 7). The Matterdale Neighbourhood Plan is permissive of housing in a wider range of situations, including for conversions of buildings and for employees of existing businesses.

These examples, small in scale as they are, demonstrate that it is possible, albeit not easy, for communities to successfully establish an alternative to the mainstream discourse around sustainable development within English planning policy. This approach, however, does require a considerable amount of social capital within the community, and a local authority that is not actively hostile to the idea. In other locations, including outside England and its unique neighbourhood planning system, a different, more radical, approach may be required.

More radical alternatives

The KSP approach discussed in this chapter tends to mean that outside of key settlements, and particularly in what is often termed the *open countryside*, local authorities will not permit new housing. There are many instances of individuals or groups building, or attempting to build, housing or other forms of development without first obtaining planning permission for them, including in relation to agriculture. If such transgressions of planning law are identified by the local authority, then they are likely to pursue enforcement action, requiring those responsible to apply for planning permission, and ultimately demolish the construction if permission is not granted, directly by the local authority or subsequently on appeal. There are several instances where the discourse of sustainable development discussed above has been challenged as part of the application and appeal process.

Scott (2001) discusses two such developments in Pembrokeshire (Wales) and Somerset (south west England). In both cases the developments which had taken place were described as 'low impact', with the groups involved arguing that in fact they exemplify sustainable development, in contrast to the conventional forms of development which the mainstream discourse promotes. Low impact development goes beyond the built form and incorporates 'subsistence-based development managed, as far as practicable, in order to maximize environmental and community benefits and produce self-sufficiency in food' (Scott, 2001: 276). Whether or not one subscribes to the lifestyle thus described, this form of low impact development does appear to accord with the broader definitions of sustainability advocated by Owen (1996) and Taylor (2008). However, in these cases and others local planning officers and politicians have historically strongly resisted developments of this sort, citing primarily that as their location beyond key settlements has been defined as unsustainable, the detail of the development is irrelevant. It is hard to avoid returning to one author's description of County Durham's persistence with their Category D for over 20 years and reflect upon its similarity with today's approach in England and elsewhere: 'an approach to planning that pursued a single line of thought at the expense of other possibilities' (Pattison, 2004: 328). So what hope for change and a different approach?

Conclusion

Much as in the preceding section, we can consider informal and formal approaches to challenging the KSP, or more specifically the single-minded interpretation of sustainability which currently underlies it. We could, as some advocate, encourage more radical, community-led approaches that challenge the status quo. Such 'guerrilla warfare' (Adams et al., 2013) against the mainstream planning system is likely, however, to be an option for only a minority of communities, comfortable with the sort of antagonism towards authority it implies, and with the skills and social capital to pursue it. Change at the structural level, in terms of how planning policy is developed from the national to the local level, would be required for a wider range and number of communities to benefit. The Welsh case is an example of how this might happen, with current planning policy in Wales now supportive of 'One Planet Development' (Welsh Government, 2012) – a form of radically low impact development which can be permitted in the open countryside in Wales, subject to strict monitoring. The complexity of this monitoring, which includes scrutiny of records of food and water consumed, household income and trips to and from the development site, has led some to question its practicality (Harris, 2017).

An alternative approach might be as part of a move from discourses of 'sustainability' to 'resilience' in planning theory and practice. This shift, identified by Davoudi (2012) as being widespread, implies a concomitant change in approach to consider how resilient communities are – resilience in this sense being in some way about adaptability, for example to climate change. Scott and Gkartzios (2014) make use of the concept in a powerful way to consider how rural housing beyond key settlements (specifically in Ireland, but the lessons are transferable) may or may not be resilient. They specify a range of factors which must be part of any such consideration, which they conclude will require housing policy to become better integrated with community development policy, among other things and, crucially, a more sophisticated approach to determining whether particular forms of development, settlements, or individual family homes are resilient or can be made more so. There is a great deal to commend this approach. However, unfortunate as it is to finish this chapter on a pessimistic note, while resilience may be a more advanced concept than sustainability, the prescriptions Scott and Gkartzios identify are very similar to those put forward by Owen in 1996 and, indeed, Cloke in 1979. They may have more success in instigating a shift in approach in Ireland, with its distinctive history and pattern of land ownership, but experience suggests that in Great Britain, the discourse of 'urban good, rural bad' when it comes to the location of new housing will remain a powerful one, and may continue to exert a strong influence in other places in Europe and farther afield.

References

Adams, D., Scott, A.J. and Hardman, M. (2013) Guerrilla warfare in the planning system: revolutionary progress towards sustainability? *Geografiska Annaler. Series B, Human Geography* 95, 4, 375–387.

Allmendinger, P. and Haughton, G. (2010) Spatial planning, devolution, and new planning spaces, *Environment and Planning C: Government and Policy* 28, 5, 803–818.

ARHC (2006) *Final Report*, London: Affordable Rural Housing Commission and Department for the Environment, Food and Rural Affairs.

Best, R. and Shucksmith, M. (2006) *Homes for Rural Communities: Report of the Joseph Rowntree Foundation Rural Housing Policy Forum*, York: Joseph Rowntree Foundation.

Breheny, M. (1993) Planning the sustainable city region, in A. Blowers (ed.), *Planning for a Sustainable Environment*, London: Town and Country Planning Association, 150–189.

Champion, T. (2009) Urban–rural differences in commuting in England: a challenge to the rural sustainability agenda?, *Planning, Practice and Research* 24, 2, 161–183.

Cloke, P. (1979) *Key Settlements in Rural Areas*, New York: Methuen.

Cloke, P. and Shaw, D. (1983) Rural settlement policies in structure plans, *Town Planning Review* 54, 3, 338–354.

Davoudi, S. (2012) Resilience: a bridging concept or a dead end?, *Planning Theory and Practice* 13, 2, 299–307.

DCLG (2011) *A Plain English Guide to the Localism Act*, London: Department for Communities and Local Government.

DCLG (2012) *National Planning Policy Framework*, London: Department for Communities and Local Government.

Geyer, N.P. and Geyer, H.S. (2017) Counterurbanisation: South Africa in wider context, *Environment and Planning A* 49, 7, 1575–1593.

Guo, Y. and Zhong, S. (2016) Collaborative approaches for planning the rural areas of Chinese cities, in G. Verdini, Y. Wang and X. Zhang (eds), *Urban China's Rural Fringe: Actors, Dimensions and Management Challenges*, London: Routledge, 17–32.

Hall, P., Gracey, H., Drewitt, R. and Thomas, R. (1973) *The Containment of Urban England*, London: Allen and Unwin.

Harris, N. (2017) Planning for One Planet: a critical evaluation of Welsh Government's planning policy on One Planet development, UK–Ireland Planning Research Conference, Queen's University Belfast, 11–13 September.

Matterdale Parish Council (2015) *Matterdale Parish Neighbourhood Plan: Neighbourhood Plan Proposal for Referendum*, retrived from http://www.lakedistrict.gov.uk/__data/assets/pdf_file/0004/616315/Matterdale-Neighbourhood-Development-Plan-final.pdf (Accessed: 9 August 2017).

MHCLG (2018) *National Planning Policy Framework Draft Text for Consultation*, London: Ministry of Housing, Communities and Local Government.

Murray, M. (2010) *Participatory Rural Planning: Exploring Evidence from Ireland*, Burlington, VT: Ashgate.

Owen, S. (1996) Sustainability and rural settlement planning, *Planning, Practice and Research* 11, 1, 37–47.

Pattison, G. (2004) Planning for decline: the 'D'-village policy of County Durham, UK, *Planning Perspectives* 19, 3, 311–332.

Scott, A. (2001) Contesting sustainable development: low-impact development in Pembrokeshire Coast National Park, *Journal of Envrionmental Policy and Planning* 3, 4, 273–287.

Scott, L.F. (1942) *Report of the Committee on Land Utilisation in Rural Areas*, London: HMSO.

Scott, M. (2006) Strategic spatial planning and contested ruralities: insights from the Republic of Ireland, *European Planning Studies* 14, 6, 811–829.

Scott, M. (2012) Housing Conflicts in the Irish Countryside: Uses and Abuses of Postcolonial Narratives, *Landscape Research* 37, 1, 91–114.

Scott, M. and Gkartzios, M. (2014) Rural housing: questions of resilience, *Housing and Society* 41, 2, 247–276.

Scott, M. and Murray, M. (2009) Housing rural communities: connecting rural dwellings to rural development in Ireland, *Housing Studies* 24, 6, 755–774.

Sireni, M. (2016) When urban planning doctrine meets low density countryside, *European Countryside* 8, 3, 189–206.

Stockdale, A., Findlay, A. and Short, D. (2000) The repopulation of rural Scotland: opportunity and threat, *Journal of Rural Studies* 16, 2, 243–257.

Sturzaker, J. (2010) The exercise of power to limit the development of new housing in the English countryside, *Environment and Planning A* 42, 4, 1001–1016.

Sturzaker, J. and Mell, I.C. (2016) *Green Belts: Past; Present; Future?*, London: Routledge.

Sturzaker, J. and Shaw, D. (2015) Localism in practice – lessons from a pioneer neighbourhood plan in England, *Town Planning Review* 86, 5, 587–609.

Sturzaker, J. and Shucksmith, M. (2011) Planning for housing in rural England: discursive power and spatial exclusion, *Town Planning Review* 82, 2, 169–193.

Sturzaker, J. and Verdini, G. (2017) Opposition and resistance: governance challenges around urban growth in China and the UK, *Journal of Urban Management* 6, 1, 30–41.

Taylor, M. (2008) *Living Working Countryside: The Taylor Review of Rural Economy and Affordable Housing*, Wetherby: CLG.

Turnock, D. (1991) The planning of rural settlement in Romania, *The Geographical Journal* 157, 3, 251–264.

UECP (2013) *Upper Eden Neighbourhood Development Plan*, Kirkby Stephen: Upper Eden Community Plan.

van Dam, F., Heins, S. and Elbersen, B.S. (2002) Lay discourses of the rural and stated and revealed preferences for rural living: some evidence of the existence of a rural idyll in the Netherlands, *Journal of Rural Studies* 18, 4, 461–476.

Welsh Government (2012) *One Planet Development: Practice Guidance*, Cardiff: Welsh Government.

Winchester District Council (2013) *Winchester District Local Plan Part 1: Joint Core Strategy*, Winchester District Council, retrieved on 9 August 2017 from www.winchester.gov.uk/planning-policy/local-plan-part-1/adoption.

Woods, M. (2005) *Contesting Rurality: Politics in the British Countryside*, Aldershot: Ashgate.

World Commission on Environment and Development (1987) *Our Common Future*, Oxford: Oxford University Press.

33

The complementarity of participatory and strategic village planning

Michael Murray

Introduction

Recent years have borne witness to significant physical transformations of village character in areas within commuting distance of urban workplaces as well as in more remote, high amenity settings popular as second home and retirement destinations. In these varying contexts, village planning and development is frequently contested between pro-growth, slow growth and no growth advocates whose shared focus is largely on the quantum of new housing to be provided for within the development limits of the statutory planning framework. While the careful management of land for development is a necessary planning response, not least in facing-down the tyranny of unacceptably high over-zoning conventions, it nonetheless remains an essentially incomplete practice routine. Two considerations arise: firstly, planning process needs to give more attention to the participatory involvement of local people in a way that transcends consultation and bounded state co-option; and secondly, plan content needs to extend beyond the location, scale and design of physical development and embrace the social, economic and cultural dimensions of village vitality. Accordingly, the core component of that quest for a different and better future revolves around the art of citizen-engaged, wide-ranging and action-oriented strategic village planning that can move beyond the narrow policy calculus of regulatory land-use allocation. This chapter critically examines the application of that complementary trajectory of village planning. It explores the potential contribution of local stakeholders to the shaping of more complete village futures but also reveals the tensions, uncertainties and frustrations that can arise from a different form of planning praxis.

The chapter is structured as follows: an initial contextualisation is provided by drawing attention to the role of the professional planner in local government and to village resident diversity; this is followed by discussion of planner-led and community-led village plans that engage with the regulation of land use; finally, and by way of contrast, the potential and limitations of participatory and strategic village planning are explored. The geographical focus is mainly on the United Kingdom, but additional insight is derived from village and small-town planning in other parts of Europe and in the United States.

The village planning context

The literature on village personality is longstanding, not least in regard to the history and character of English villages. Much of this early writing seeks to describe their distinctive physical attributes and spatial variation using illustrated lists of notable villages (see for example: Pakington, 1945; Blunden, 1947; Bonham-Carter, 1952). It commonly portrays a visual romanticism linked to an arcadian interpretation of green and pleasant land which on occasion only hints at the challenges prompted by a wider restructuring of the rural economy and changing relationships across the settlement hierarchy, especially in regard to the availability of services (see the previous two chapters in this section). These tendencies have a wider geography and, as noted by Mak (1996) in his ethnographic case study of Jorwerd in the Netherlands, they are emblematic of a silent revolution that has swept through Europe over the period from 1945. And yet a formidable counterbalance to this modernity induced change remains a much vaunted neighbourliness and sense of community which can prompt local struggles to create alternative village realities. It is within this arena of transformational planning that some contextual comment is necessary on the role played by two key stakeholder groups: the professional planners in local government and village residents.

The day-to-day operation of the planning system is rooted in local government (Allmendinger et al., 2000) with decision-making in the main being the responsibility of elected representatives having regard to the technical input and advice of professional planners. In the United Kingdom these working relationships are simultaneously upwards to central government and its agencies in regard to legislation, planning policy and strategic investment and outwards to developers, landowners, pressure groups, community groups and the general public, all of which do have a significant bearing on the shaping and delivery of village plans. Decisions about how land should be used involves making choices which will affect the interests of these different stakeholders in different ways (Taylor, 2004). Central government, for example, may be pushing for land release for new village housing in or close to designated green belts while local communities on the other hand may be strongly resistant to expanded development limits in order to protect visual amenity. Sensitive planning responses to this village development-conservation dialectic do play out internationally. Research demonstrates that many rural areas across North America, for example, are experiencing unprecedented residential development that negatively impacts on the rural character that attracts new residents to these areas in the first place and which in the case of New England villages has identified the potential of innovative sub-division design as a way to protect open space, biodiversity and rural setting (Ryan, 2006).

Village residents are not a homogeneous bloc and as Pahl (1975: 45) notes 'there is no village population as such' but rather a series of social groups with specific problems and housing needs to which rural planning should be more closely aligned. In his seminal essay titled 'The social objectives of village planning', based on research in southeast England, Pahl categorises the following elements of village life: large property owners, salaried immigrants with some capital, spiralists, those with limited income and little capital, the retired, council house tenants, tied cottages and other tenants, local tradesmen and owners of small businesses. The important point Pahl makes is that dealing with the village 'as a sort of average of all such groups is extremely misleading and possibly accounts for much of the confusion in rural planning' (ibid.: 45). Understanding why people find themselves in a village and how they relate to where they live is key to understanding community involvement and the nature of the membership and role of voluntary associations which engage in planning related activities. Linked to this commentary on segmentation is the ongoing debate about the social worlds of established residents and newcomers with the latter, perhaps, perceived as living primarily in recently constructed edge-of-village housing developments and having a weaker attachment to place. There is research evidence, however, that throws interesting light on any such distinction between locals and newcomers, with Gieling et al. (2017) demonstrating in the Dutch context that origin and length of residence are not conclusive in predicting levels of village attachment. Indeed all residents are located along a spectrum of attachment with 'rural idyll seekers', for example, as equally likely to volunteer to be involved in village life as those in the 'traditionally attached' group, more than half of whom are village born residents. In short, this more nuanced appreciation of different types of village resident gives insight into the appropriateness of participation diversity in village planning practice.

Planner-led village plans

Planner-led approaches to village planning have historically operated at two interdependent levels: regional/sub-regional and local. The former, rooted in central place theory, service thresholds and economies of scale established the village mosaic as one layer in the settlement pattern. Thus, in the United Kingdom, post war development plans commonly offered a strategic framework for village planning through an explication of key settlement policy (see previous chapter) whereby housing, employment, services and infrastructure would be concentrated in selected centres in order to maintain a level of rural investment that might support both the key settlement and its hinterland villages (Cloke, 1983). A significant corollary to that enterprise was the desire to rationalise the rural settlement pattern which notably was taken to extreme in the case of Durham County Council (1951). It advanced a policy of planned decline for those settlements affected by the closing of coalmines. All settlements in the County were placed into one of four categories on the basis of population and employment potential, along with the proximity of industrial sites and the physical condition of property and services. A total of 121 villages out of a total of 357 towns and villages in the county were at some time placed in Category D, with the 1951 Written Analysis of the County Development Plan stating:

> Those from which a considerable loss of population may be expected. In these cases it is felt that there should be no further investment of capital on any considerable scale, and that any proposal to invest capital should be carefully examined. This generally means that when the existing houses become uninhabitable they should be replaced elsewhere, and that any expenditure on facilities and services in these communities which would involve public money should be limited to conform to what appears to be the possible future life of existing property in the community.
>
> *(County Council of Durham, 1951: 77–78)*

While the Plan was at pains to emphasise that there 'is no proposal to demolish any village, nor is there a policy against village life' this was not how the planning prescription was perceived and strong political and community dissent followed. Pattison argues that this process of planning attrition, nonetheless, did work itself through in subsequent years and in so doing a scathing reproach is directed at this manifestation of technocratic planning:

> From the 1930s until the early 1990s, Durham's mining communities were continually and overwhelmingly understood by planners in economic terms. No serious consideration was ever given to the social and cultural formations within the settlement structures.

> Planners invariably saw the built environments of mining settlements in negative terms. In the reporting of events by the local press and in the letter columns of local papers, policy makers seemingly held little regard for community commitments and concerns.
>
> *(Pattison, 2004: 329)*

More generally, key settlement policy has been critiqued regarding its limitations of discouraging beneficial forms of development in non-selected villages, incomplete control by planning authorities and the displacement of investment to replace or expand facilities that have closed elsewhere (Cloke, 1983). However, an adherence to the positivist planning traditions of key settlement policy continues to endure with the content of planner-led village plans nesting below the favoured status of higher order settlements and invariably reduced to the minimalist delineation of a tightly drawn development envelop that indicates the outer boundary of the village and white land opportunities within it for future house building. The case of Northern Ireland is illustrative in that regard.

In this region of the UK, the scope for the physical expansion of villages has been frequently constrained in development plans for those settlements close to the Belfast Urban Area in order to control the risk of sprawl and protect greenbelt. In more remote rural areas, in contrast, the past approach has tended to define more generous development limits in line with a wider practice to also over-zone in towns. However, over the period from 2001, when a new regional planning framework for Northern Ireland was endorsed by the devolved government (for a further discussion, see Murray, 2009), the technical preference for calculating area-based Housing Growth Indicators, and then allocating housing units to constituent elements of the settlement hierarchy within each area, has allowed for the application of a much more precise determination of housing land availability in each village. Across rural Northern Ireland the scope for future village growth has been reigned-in through tighter development limits, linked to the contentious de-zoning of land without planning permission, in new development plans. Additionally, a raft of development management policies has been set out in these plans to enable precautionary decision-making on planning applications. Clearly at work here is the need to plan at the village scale in line with infrastructure capacity, avoid excessive public sector investment costs and promote town-based living – key settlement policy *déjà vu*?

Community-led regulatory village plans

In Thorburn's (1971) classic textbook on village planning, the following advice is offered under the heading of 'public relations':

> Public or private discussion on objectives and approach are desirable and can help to reduce objections at a later stage. It is preferable if these discussions follow publication of survey information so that all present are equally well informed. In practice it is usually found that the public are unable to put forward any ideas which have not already occurred to the professional planners, probably because they are not experienced in initiating ideas of this kind. Hence these meetings should not be conceived as a search for ideas, but rather as an exchange of views educative to both sides.
>
> *(Thorburn, 1971: 113)*

The discussions that should take place during village plan preparation were further canvassed as giving recognition to the influence of 'prominent individuals' and those 'known to have an active interest in planning' (ibid.: 115).

Planning theory and practice in the interim has more generally moved well beyond that prescription. The involvement of local residents in the physical analysis of their villages to include the crafting of planning guidance around settlement structure, building form, design details, landscaping, street furniture and road layout is exemplified by *village design statements*, introduced in England in 1993 (Owen, 2002). They were proposed by the then Countryside Commission to help reverse the widespread erosion of village distinctiveness by becoming a supplementary development control tool. They would more fully involve local communities in assessing village character and giving voice to the type of design that they felt would be appropriate for new development. Chapter 34 of this Companion, by Arthur Parkinson, considers more fully the theme of village design statements drawing on the experience of Ireland. The important point in the context of the current chapter is that they demonstrate that local communities can and do act in the regulatory planning arena and which has gained momentum with the subsequent introduction in England of neighbourhood planning initiatives.

Neighbourhood planning emerged out of the Localism Act 2011 for England. The Department for Communities and Local Government explicates its remit as follows:

> Neighbourhood Planning gives communities direct power to develop a shared vision for their neighbourhood and shape the development and growth of their local area. They are able to choose where they want new homes, shops and offices to be built, have their say on what those new buildings should look like and what infrastructure should be provided, and grant planning permission for the new buildings they want to see go ahead. Neighbourhood planning provides a powerful set of tools for local people to ensure that they get the right types of development for their community where the ambition of the neighbourhood is aligned with the strategic needs and priorities of the wider local area.
>
> *(Department for Communities and Local Government, 2014)*

This initiative fits well with a public policy agenda for local empowerment in planning but, as noted by Archer and Cole (2014), it also needs to be appreciated as an attempt to liberalise access to land for the volume house-builders in the context of a mismatch between housing need and housing supply in many parts of England. It has been headlined: 'Fixing Our Broken Housing Market' (Department for Communities and Local Government, 2017). Accordingly, there are very specific rules to the game set from above which require initiation, for example, by a parish council or community organisation, approval for a defined neighbourhood boundary by the local planning authority, a schedule of consultation and research to inform the preparation of a draft neighbourhood plan, an independent examination of draft policies and site proposals (which have often been co-produced with guidance from local authority planners), and a neighbourhood planning referendum which brings the neighbourhood plan into force when more than half of those voting from within the designated area boundary vote in favour of the neighbourhood plan. The net result is that provided there is general conformity with local government plans, the neighbourhood plan will become part of the statutory development plan for the area. In 2017 around 2,200 groups across England as a whole have started a neighbourhood plan since 2012 and 410 successful referenda have taken place with an average 'Yes' vote of 88 per cent (Locality, 2017). In the context of this chapter, Parker and Salter (2016) indicate that the majority of plans have come from rural parish councils and market towns under pressure from housing development. Community-led regulatory village plans are a significant part of that output.

The key question which must be posed, however, is the extent to which these neighbourhood plans represent a sufficient response to local needs and challenges? Thus far the academic commentary here is varied with Parker and Salter (2017) identifying that opportunities have been taken to advance socially and environmentally sensitive housing solutions, while Vigar et al. (2017) conclude in their village case study that securing better local services or improved housing conditions have largely been framed out of consideration. This points towards a more circumscribed brief in regard to community-led place planning and where the primacy of concern rests with the devising of locally acceptable regulatory frameworks for housing provision. Meanwhile, proposals to introduce broadly comparable neighbourhood planning initiatives in Scotland (Local Place Plans) and Wales (Place Plans) are being brought forward at the time of writing (January 2018) and the Government has confirmed financial aid for a new 2018–2022 programme of neighbourhood planning in England which will bring additional communities into the participatory arena, as well as allowing for the replacement or modification of existing plans. Nonetheless, this must surely remain an incomplete harnessing of the potential that village communities possess to shape their own destinies. That capability rests with a complementary village planning approach which links citizen participation and broad-based strategic thinking and it is to this matter that the next section of this chapter now turns.

Participatory and strategic village planning

Over the past three decades village and small town residents in many parts of Europe and the United States have become involved in attempting to shape the future of their communities by drawing up their own plans that identify a range of social, economic and environmental projects (see for example: Murray and Dunn, 1996; Cavazzani and Moseley, 2001). This community-led local development approach is group-based and is often looking to take up tasks passed over by government and private enterprise but, in so doing, there remains a dependency on some public or philanthropic funding for, say, technical assistance. Moreover, these initiatives must connect upwards and outwards which requires the formation of strategic alliances and partnership working routines in order to maximise the chances of implementation. The management style of groups is important here and they may lack the hands-on control and business sensitivity of the skilled entrepreneur and be constrained by a weakly developed organisational capacity and a precarious financial structure. Additionally, community groups in their composition may not fully reflect the diversity of local interests and thus there can arise issues around their composition, mandate and purpose. Nonetheless what is very apparent is that rural residents are seizing opportunities to express mutual, locality-oriented interest in collective actions (Wilkinson, 1991) and, in short, people are connecting with place, not least in village settings.

A core component of that community quest for a different and better future is strategic planning and, simply defined, it is a structured way by which to analyse a local situation and to make provision for dealing with these circumstances. A strategic plan for a village looks forward over a period of years and seeks to incorporate both current and potential future activities into a positive framework for delivering and managing change. It can be single themed (for example, related to townscape improvements) or more commonly be multi-dimensional and bring together different stakeholders at the local scale to agree and implement an action-oriented agenda. A village development strategy should not, however, be presented or be regarded as a 'shopping list' of projects. It should be viewed as a useful tool to prioritise the things that must be done by the constituent civil society groups in the village, more likely in partnership with others. Accordingly, participatory and strategic village planning can enhance the delivery of pilot and mainstream funding programmes, and can stimulate the private, voluntary and public sectors to higher levels of performance by harnessing citizen enthusiasm for getting involved in making a contribution to these collective efforts (Murray and Greer, 2001). This is not to deny that practice can fall short of these aspirations and serious obstacles can arise in any local development context, for example, volunteer burnout and resident apathy, probity of action in the identification and advocacy of proposals, and tension between project workers and community group management as to who is in charge.

These development strategies with their combination of breadth, community provenance and action focus have been labelled 'village action plans' (Moseley, 2002) and are well captured in terms of their preparation and content by the comments of Action for Communities in Rural England (ACRE):

> Embarking on making a plan is the start of a journey that should bring a community closer together; realise the resources and skills available of members of that community and; take control of the community's future needs in terms of services, housing, health etc.
>
> *(ACRE, n.d.)*

The key to success in this endeavour is dialogue which requires reaching out to others, recognition that many people hold pieces of the desired way forward, and respectful listening by those steering the participatory and strategic village planning process. Hard-to-reach groups should be encouraged to attend events and conversations should be marked by creativity, openness and flexibility. At the same time it is important to appreciate that there are real limits to participatory engagement. A published village development strategy, for example, does not command legal status as a statutory document. And yet this may well be its singular strength in that it can act as an informal lever for change by transforming a deep sense of external dependency to a confident civic activism that is appreciative of wider connectedness. That engagement can extend beyond single village planning to encompass multi-community or village cluster participatory and strategic action which recognises that each place can contribute to the collective effort according to its strengths. The challenge here, however, is dealing with the varying capacity of village planning groups to transcend parochialism and sign up to a shared vision of what may be possible. Accordingly, any strategic planning process is unlikely to be short in terms of time and linear in the easy style of survey, analysis, plan which in turn underscores the contribution of an embedded community development perspective in participatory and strategic village planning (Murray and Greer, 1998). In the final analysis, the public value of this public involvement in strategic village planning may be regarded as a collective process of discovery to produce a transformative trajectory that is much more appropriate for local community and institutional realities.

Conclusion

This chapter has explored the rubric attached to three forms of village planning which move the loci of engagement from local authority planner to citizen and from land-use regulation to multi-dimensional local development. There is an international resonance to that analysis given the shared challenges of managing and facilitating change at this local scale. Quite clearly the emphasis remains firmly set on regulating land use through the codes of physical planning but, as argued above, the complementarity of a wider participatory, strategic and action oriented planning process and output deserves continued attention. Any perception that the latter be viewed as a minority activity is perhaps more deeply reflective of its separate academic provenance which champions broad based rural development over a land-use management paradigm cloaked with the language of environmental sustainability. This, perhaps, is *the* fundamental arena of debate that will continue to engage those whose brief or interest is village planning.

References

ACRE. n.d. Community planning. Retrieved on 5 January 2017 from www.acre.org.uk/rural-issues/community-planning.

Allmendinger, P., Prior, A. and Raemaekers, J. (2000) *Introduction to planning practice*. Chichester: John Wiley and Sons.

Archer, T. and Cole, I. (2014) Still not plannable? Housing supply and the changing structure of the house-building industry in the UK in 'austere' times. *People, Place and Policy*, 8(2), 97–112.

Blunden, E. (1947) *English villages*. London: Collins.

Bonham-Carter, V. (1952) *The English village*. Harmondsworth: Penguin Books.

Cavazzani, A. and Moseley, M. (2001) *The practice of rural development partnerships in Europe*. Soveria Mannelli: Rubbettino.

Cloke, P. (1983) *An introduction to rural settlement planning*. London: Methuen.

County Council of Durham. (1951) *County development plan: 1951 written analysis*. Durham: County Council of Durham.

Department for Communities and Local Government. (2014) *Guidance: neighbourhood planning*. London: Department for Communities and Local Government.

Department for Communities and Local Government. (2017) *Notes on neighbourhood planning*, 19th edition (March). London: Department for Communities and Local Government.

Gieling, J., Vermeij, L. and Haartsen, T. (2017) Beyond the local-newcomer divide: village attachment in the era of mobilities. *Journal of Rural Studies*, 55, 237–247.

Locality (2017) Neighbourhood Planning at Locality Convention '17. Retrieved on 13 December 2017 from https://mycommunity.org.uk/2017/11/24/neighbourhood-planning-locality-convention-17.

Mak, G. (1996) *Jorwerd: the death of the village in late 20th century Europe*. London: Harvill Press.

Moseley, M. (2002) Bottom-up 'village action plans': some recent experience in rural England. *Planning Practice and Research*, 17(4), 387–405.

Murray, M. (2009) Building consensus in contested spaces and places? The Regional Development Strategy for Northern Ireland. In S. Davoudi and I. Strange (eds), *Conceptions of space and place in strategic spatial planning*, 125–146. London: Routledge.

Murray, M. and Dunn, L. (1996) *Revitalizing rural America: a perspective on collaboration and community*. Chichester: John Wiley and Sons.

Murray, M. and Greer, J. (1998) Strategic planning for multi-community rural development: insights from Northern Ireland. *European Planning Studies*, 6(3), 255–269.

Murray, M. and Greer, J. (2001) *Participatory village planning: practice guidelines workbook*. Belfast: Rural Innovation and Research Partnership, Queen's University Belfast and Rural Development Council.

Owen, S. (2002) From Village Design Statements to Parish Plans: some pointers towards community decision-making in the planning system in England. *Planning Practice and Research*, 17(1), 81–89.

Pahl, R.E. (1975) *Whose city? And further essays on urban society*. Harmondsworth: Penguin Books.

Pakington, H. (1945) *English villages and hamlets*. London: B. T. Batsford.

Parker, G. and Salter, K. (2016) Five years of Neighbourhood Planning: a review of take-up and distribution. *Town and Country Planning*, 85(5), 181–188.

Parker, G. and Salter, K. (2017) Taking stock of Neighbourhood Planning in England 2011–2016. *Planning Practice and Research*, 32(4), 478–490.

Pattison, G. (2004) Planning for decline: the 'D'-village policy of County Durham, UK. *Planning Perspectives*, 19 (July), 311–332.

Ryan, R.L. (2006) Comparing the attitudes of local residents, planners and developers about preserving rural character in New England. *Landscape and Urban Planning*, 75(1–2), 5–22.

Taylor, N. (2004) *Urban planning theory since 1945*. London: Sage Publications.

Thorburn, A. (1971) *Planning villages*. London: The Estates Gazette Limited.

Vigar, G., Gunn, S. and Brooks, E. (2017) Governing our neighbours: participation and conflict in neighbourhood planning. *Town Planning Review*, 88(4), 423–442.

Wilkinson, K.P. (1991) Social stabilization: the role of rural society. Paper presented to the International School of Rural Development. Galway: University College Galway.

34

Village design and distinctiveness

Arthur Parkinson

Introduction

Concerns about the quality of design and loss of local distinctiveness in rural settlements that emerged from the middle of the twentieth century onwards (see, for example, Sharp, 1946) reflect the wider observation that contemporary shifts in society were having a profound impact upon the distinctive character of places (Relph, 1976; Norberg-Schulz, 1980). In relation to rural places, specifically, these changes were characterised by the emergence of standardised speculative housing layouts, building designs and construction techniques. Furthermore, within rural debates, design tended to be ignored, while 'the professional territory of design, architecture, even aesthetics and taste' was often separate and distinct, ignoring the 'social, economic and policy context into which building design fits' (Bishop, 1994: 259). Although concerns over loss of village distinctiveness are not new, they have become heightened in recent years in the context of globalisation, which is contributing to the homogenisation of places, while simultaneously driving the emergence of agendas seeking to protect and capitalise upon the distinctiveness of individual places for competitive advantage (Woods, 2007).

Loss of distinctiveness in rural settlements has alternatively been described as a loss of 'responsiveness' (Owen, 1995, 1998), founded on the notion that local distinctiveness evolved primarily as the result of local responses to local circumstances. A loss of responsiveness may result, for example, from the replacement of local vernacular building traditions by developer speculation and economies of scale in the construction industry, or through a disconnection from the natural environment by generic design solutions. In other words, historically, needs and problems resulted in distinctive change in the physical fabric of villages – a process that has relevance for contemporary village planning, in responding with sensitivity to local character and managing the pace and scale of physical change. On this basis, spatial planning is seen as having a role in fostering responsiveness to the loss of local physical character, through a bottom-up approach to village planning, involving the local community in the formulation of place-specific design solutions. This chapter examines spatial planning efforts to address these issues and, in particular, the tool referred to as village design statements (VDSs), which emerged in the UK from the 1990s onwards, and which was adopted in Ireland ten years later. In the Irish context, the unprecedented rate and extent of change during the so-called 'Celtic Tiger' boom period took place in a culture that has not traditionally placed a high value upon the designed historic built environment. Nevertheless, concerns over loss of local distinctiveness have been voiced as elsewhere internationally, and the application of the VDSs model is still emergent and evolving. This provides useful insights relating to the utility and applicability of the VDS tool internationally, as well as how future efforts can be further advanced to become more effective. The chapter begins by exploring the conceptual territory associated with character and place distinctiveness. Thereafter, different spatial planning approaches to maintaining small town character are very briefly introduced, before moving on to focus in detail on the key characteristics of the VDS approach in particular. This focuses first on the UK context, before moving on to examine the second (and most recent) phase of the VDS programme in Ireland, focusing on content and the ways in which local distinctiveness is represented within guidance and actual VDSs completed.

Conceptualising place distinctiveness

A range of related, overlapping and shifting terminology and concepts have been used over time in policy and the literature to describe the distinctive qualities of urban and rural places, including character (Jacobs, 1961; Worskett, 1969; Alexander, 1977), placelessness (Relph, 1976), sense of place (Steele, 1981), and *genius loci* (Norberg-Shultz, 1980). However, two ideas are central to each of these: firstly, that the distinctive character of a place is the sum of its physical attributes and visual appearance (tangible attributes), as was perhaps emphasised in the work of Cullen (1961); secondly, these texts also to varying degrees emphasise the relationship between people and place. Thus, places may be subject to the meanings people apply to them (e.g. collective memory, symbolism, nostalgia – intangible attributes), and the relative value that people assign to physical places as a result, 'So we take delight in physically distinctive, recognizable locales and attach our feelings and meanings to them' (Lynch, 1976: 24–25). This is closely related to the view of *genius loci* as 'the sense people have of a place, understood as the sum of all physical as well as symbolic values in nature and the human environment' (Jivén and Larkham, 2003: 70), and echoes Conzen's (1966) writings on the relationships between culture, history and the physical character of places. Implicit in these writings is the broader relationship between people, place, and identity. In this regard, Dixon and Durrheim (2000: 40) argue that people's sense of place-identity – their attachment to a place – is formed 'through individuals' transactions with their environment', and is 'constructed, produced and modified through human dialogue' (i.e. via discourse; see also Proshansky et al., 1970; Twigger-Ross and Uzzell, 1996). These concepts and relationships are also connected with the definition and conservation of heritage (dealt with more fully in Chapter 35 of this Companion). Moreover, what the public identifies as important to the character of a place may be quite different from the priorities embedded within professional discourses (Jones, 1995; Davison, 2017). Thus, the physical qualities of places cannot be considered in planning policy and practice without simultaneously also taking people-place relationships into account, particularly to ensure inclusion and equity in associated planning efforts. Although these relationships have been recognised within village planning efforts for some time (see, for example, Owen, 1998), Stephenson (2010) contends that planning theory has tended to almost exclusively focus on the physicality of places, and to largely ignore intangible qualities. Moreover, Jivén and Larkham (2003: 78) argue that:

> historical and theoretical perspectives on *genius loci* and spirit of place have become confused, particularly in the tendency of practical planning and design to work towards 'creating' a sense of place and the contributions of topography, natural conditions and variations, and symbolic meanings, tend to be given less weight than built form.

Rural settlement planning and design

The community-led approach that characterises VDSs, as practiced in the UK and Ireland, is part of a wider shift towards greater involvement of the public in planning across many European countries in recent decades, addressing social, economic and environmental improvement concerns (Campbell and Marshall, 2000; Albrechts, 2002) in addition to specific concerns around the erosion of the distinctive character of villages. Underpinning this trend is a theory of rural development that 'stresses the individuality of localities, the empowerment of local people and the need for development activists to work with local communities not simply in, for, or through them' (Moseley, 1997: 197). A number of factors have fuelled this trend, including a public perception of neglect, a desire to harness social capital, as well as the emergence of a realisation that public involvement is essential to the successful pursuance of sustainability. Within the field of rural planning, this wider shift has resulted in the emergence of a range of policy processes and responses, each of which can play an important role in capacity building and in closer coordination of strategic planning with community priorities (Gallent et al., 2008).

With respect to initiatives that address a loss of rural distinctiveness, specifically, there have been a number of planning responses, some of which are situated within the built heritage field, and are outside the scope of this chapter (see for example Williamson, 2010). In relation to the USA context, Arendt et al. (1994) discuss the loss of character resulting from the imposition of standard design requirements in land use zoning, and a response rooted in a new urbanist approach to planning, achieved through community-based guidelines, that draws on traditional townscape concepts while also emphasising the role of creative design. In relation to small towns in Australia, Green (1999, 2000) conceptualises town character in terms of shifting community values, and as multi-dimensional and non-static, encompassing 'Environmental aesthetics, community sentiment and identity, naturalness, change, uniqueness and affective responses to the environment' (Green, 2000: 84). Perhaps most notable is an emphasis on intangible attributes, such as the meanings people apply to place, as well as the relationship between place and the personal and collective identity of the people that live there – elements not as strongly or clearly considered within UK and Irish village design efforts, at least until more recently (see Owen et al., 2011). Similarly, Thorbeck (2012) emphasises community values in planning and design in the rural USA, through a new 'rural design' discipline, though Childs (2014) questions the need for a new discipline.

In the UK, a number of planning initiatives were instigated by the Countryside Commission from the 1980s onwards, focused on smaller rural settlements and understanding local needs and priorities (Owen, 2002). In the early 1990s, VDSs were one such project that examined public concerns around the impact of standardised suburban housing development on rural distinctiveness, which Owen describes:

> Typically such developments take the form of layouts designed around purpose-built access roads of regulation dimensions with groups of similar houses arranged in culs-de-sac. Suburban forms are transplanted into villages. Houses, usually detached or semi-detached, sometimes deliberately set back to give a spurious 'variety'. They have large windows and shallow roofs. Each has a garage, sometimes two, attached to the side of the house. They are transplanted with few adaptions because, being capable of replication, they can be built more cheaply if the basic design is repeated. They are popular with house buyers, but probably because there are few alternatives available.
>
> *(Owen, 1995: 145)*

This form of unsympathetic standardisation led over time to an ingrained view that any form of new development 'inevitably detracts from the character of villages' (Owen et al., 2011: 408). In turn, this led to the adoption of restraint policies relating to rural housing development in many development plans. However, blanket restraint policies are contrary to the sustainability of small rural settlements, particularly resulting from a focus largely upon reducing CO_2 emissions, without sufficient cognisance of other dimensions of sustainable development (Owen and Herlin, 2009; also Chapter 32 of this Companion). The VDS initiative therefore also sought to address these issues, and to foster a positive but discriminating attitude to development in small rural settlements, contributing to their social and economic vitality – and thereby, their sustainability. Over the course of the following two decades, VDSs became established in planning practice, sometimes in a systematic fashion through adoption and use as supplementary planning guidance, thereby becoming a material consideration in planning decisions. While the VDS model is community-led, the information gathered is framed by a professional – normally appointed by the local community, itself. In contrast with other types of community-led planning tool in the UK, VDSs were seen as fitting better with higher forms of strategic plan due to their clearer framing, and thereby came to be regarded as an effective means of addressing concerns about loss of local distinctiveness (Gallent et al., 2008).

In terms of their utility, VDSs can serve two purposes. Firstly, as a planning tool, VDSs are of use within the formal planning system to address the loss of local distinctiveness in villages through the following attributes (after Owen, 1999):

- VDSs provide design guidance for architects, builders and developers.
- More specifically, VDSs can provide guidance on how new development can be sensitive and appropriate to its local setting (though not whether or where development might be more appropriately located). In this regard, concerns around the landscape setting and its relationship with settlement form are emphasised.
- VDSs provide an insight into what makes a specific place distinctive.
- VDSs foster better understanding by local planning authority officers.
- VDSs provide planning authorities with a clearer and firmer basis on which decisions can be based and justified.
- VDSs add another layer of influence at a different level from the statutory process.

In contrast, as a tool for involving local communities in the design and setting of new development, VDSs have been found to be particularly useful in raising awareness of design issues among the public. However, while the form of public engagement in VDSs is situated at a higher level than 'tokenism' on Arnstein's (1969) ladder of participation – and thereby offers more substantive involvement than many other planning initiatives – they may be of less interest than alternative local planning tools with a broader scope (focused on responsiveness to social and economic needs as well as design issues). A related but key point in relation to VDSs is that, because of their narrow focus upon physical design and distinctiveness, they have been found to be most successful when used alongside other complementary forms of local planning instrument that deal with a broader range of issues.

Moreover, there is often a tension between top-down plans (which seek to meet strategic objectives and coordinate wider interests), and bottom-up community-led efforts that inevitably focus on the priorities, values, meanings, etc., of the local public, but which may also tend to unfairly reflect the interests of the most powerful and outspoken participants – reflecting a wider tension between representative and participative democracy, respectively. To overcome these tensions, Owen et al. (2007) point to the importance of effectively bridging top-down (strategic, development plans and local area plans) and bottom-up (VDSs etc.) forms of planning. More specifically, there needs to be an awareness among all of what the 'other level' can contribute, and a continuous flow of information (through deliberative or associative governance), for successful bridging to occur. In relation to VDSs, specifically, their content can often be easier to reconcile with more strategic local planning frameworks than other forms of local community-led plan, as they are narrower in scope and more consistent (Owen et al., 2011). However, VDSs can still be out of step with the objectives of broader design strategies, leading to a reluctance among local authorities in England to formally adopt them into strategic frameworks (Gallent et al., 2008). In relation to spatial planning more generally, Gallent et al. (ibid.) similarly argue for the necessity of an approach that mediates between different values, and for the bridging of community-led plans with strategic plans. This, they contend, is more in keeping with the principles of participative democracy, than attempting to frame and integrate (and even alter) community-led plans into top-down strategic plans, as has on occasion taken place. Exactly how this can be achieved is outside the scope of this paper, but the issues and possibilities have been discussed elsewhere (see, for example, Gallent, 2013), and hold potentially valuable lessons relating to the VDS model.

Erosion of village distinctiveness in Ireland

Although villages and small towns in Ireland are largely the result of successive waves of colonial settlement, much of the building stock within towns, often lacking in ornamentation and modest in scale, dates from the nineteenth century. Towns fall into a number of overlapping functional categories, and most had a market function, or formerly serviced an adjacent landlord estate, though rural settlements' traditional market and service functions have been substantially lost in recent decades. In contrast with England, where development in villages has often been characterised as harmful, Ireland's planning culture is characterised by relatively permissive *laissez-faire* planning policies (Donovan and Gkartzios, 2014). Furthermore, a high value has not traditionally been placed upon the designed historic built environment (see Chapter 35 of this Companion), and a national system for the statutory protection of built heritage was only put in place as late as 2000. These factors, together with shifts towards default generic suburban housing design, and an unprecedented rate of development during the so-called 'Celtic Tiger' boom period, have combined to erode the distinctiveness of Irish villages, and similar concerns around loss of distinctiveness have been expressed as elsewhere – including among the wider public (Owen et al., 2008; Owen et al., 2011; Parkinson et al., 2016; McManus, 2017). Certain civil society organisations, notably An Taisce (the Irish National Trust) and Friends of the Irish Environment, have been particularly vocal in campaigning against the erosion of the character of Irish towns and villages.

A system of development control (now management), operated by local authorities, has existed in Ireland since 1963, and local authorities have statutory responsibility for strategic planning and development management, though the Department of Housing, Community and Local Government has an overarching policymaking role. As part of local authorities' role, county, city, and large town development plans are updated once every six years; however, since 2000, local authorities have also been required to prepare local area plans (LAPs) for settlements with over 5,000 population (2,000 before 2010), and at the discretion of the local authority for smaller settlements. LAPs are the lowest level of the statutory planning system and, given their broader scope, are an important complementary form of plan alongside VDSs. Under the Planning and Development Act 2000, local authorities have statutory responsibilities with respect to built heritage. In particular, these involve the compilation of a Record of Protected Structures, and the power to designate Architectural Conservation Areas to preserve the character of a place, area, group of structures or townscape; however, the concept of 'character' is not clearly defined in legislation or related guidelines. The state-funded Heritage Council, a national body independent of central government, also has an advisory and promotional role with respect to built and natural heritage, and administers a number of heritage-related national programmes, including county heritage plans, public realm plans, VDSs, and grant funding for heritage projects. The council also has a partnership role alongside the DCHG and Fáilte Ireland (the national tourism development authority) in delivering the Historic Towns initiative, which seeks to promote heritage-led regeneration of towns through partial grant funding of regeneration projects.

Village design statements in Ireland

Bottom-up approaches to rural settlement planning in Ireland have had considerable success in meeting local needs, perhaps most notably through the EU's LEADER initiative. However, specifically in relation to the loss of local distinctiveness in rural settlements, VDSs were brought to Ireland by the Heritage Council in 2000, essentially by adopting the English approach, though with a stronger emphasis on loss of heritage as well as distinctiveness (Owen et al., 2011). The VDS model in Ireland was revised in 2006 to become more participative (Jordan and Harvey, 2010), and by 2008 around 40 VDSs had been prepared. However, a review of the programme conducted by CCRI and BDOR Ltd on behalf of the Heritage Council, concluded that it had on the whole not been successful in meeting its stated aims and objectives (Owen et al., 2008). The authors noted that few VDSs in Ireland had been initiated and produced by local communities, despite this being the Heritage Council's initial intention, and no VDSs were adopted as supplementary guidance by local planning authorities. Moreover, community-led instruments embracing broader social, economic and environmental matters may often have been more appropriate in many of the situations where a VDS had been prepared. More specifically, in certain cases, the scope of a VDS was extended to encompass broader planning concerns including employment and service provision. Some wider contextual issues were also identified as prejudicing the success of the programme, specifically: LAPs were a new instrument often not in place; low levels of commitment to the tool at local and national levels; the limited experience of the public in relation to design; and limited professional skills in facilitating this. In the small number of cases where a VDS was deemed in the report to be successful in meeting its objectives, this was seen as being because the local community valued the character of the town, local authority officials were actively engaged (resulting in its extensive use in planning decision-making), but also because an individual within the local authority took on a leadership role throughout the process.

Following the review of the first phase of the VDS programme, the Heritage Council commenced a second phase with two pilot projects, in Julianstown, Co. Meath, and Sandymount, an urban village within the boundaries of Dublin City. Building on these, the Heritage Council published a village design 'Toolkit' (Heritage Council, 2012), which includes a checklist of heritage and design elements that should be considered within the scope of the statement, as well as eight steps that aim to guide a community in preparation of a VDS. The overarching approach within the guidance clearly recognises and emphasises the role and importance of a community-led, collaborative and participative approach. Crucially, it also recognises the

Figure 34.1 Slane, Co. Meath, Ireland: VDS completed in 2008 as part of the first phase of the Heritage Council's VDS programme.

Figure 34.2 Slane's setting, adjacent to the River Boyne.

Figure 34.3 Julianstown, Co. Meath, Ireland: one of two locations chosen as a pilot for phase two of the Heritage Council's VDS Programme, completed in 2010.

Figure 34.4 Despite the 'Celtic Tiger' boom period, vacant and dilapidated historic buildings are not uncommon in the centre of Irish villages, as in this view of Julianstown.

relationship between the 'sense of place created by a village's cultural heritage' and 'a community's sense of identity' (Heritage Council, 2012: 1). Although funding for Heritage Council's grant schemes (including support for VDSs) was reduced dramatically (60% between 2010 and 2015) (DHCG, 2016), a number of VDSs have been prepared since the Heritage Council's Toolkit was published.

With respect to the relationship between VDSs and wider layers of strategic planning, while current national guidelines relating to the preparation of Local Area Plans refer to the role of VDSs and other supplementary planning guidance, they warn that these must be, 'consistent with the parameters laid down by statutory plans, to guide specific development proposals'. (DECLG, 2013: 3). Therefore, although there have been efforts to resolve tensions between, and to align, VDS and national policy, guidelines suggest a more top-down style than the more deliberative, associative, or collaborative approaches advocated by Owen et al. (2007) and others (Gallent et al., 2008; Gallent, 2013). However, a number of local authorities have actively taken the VDS approach on board as supplementary guidance, or have included objectives to prepare or support further VDSs (Laois, Limerick, Wexford, South Dublin and Cork County Councils). Notably, Limerick explicitly seeks to implement VDSs where no Local Area Plan exists, and to identify the need for social and community infrastructure though VDSs, contradicting lessons from elsewhere. Further, Cork County Council emphasises that VDSs should be consistent with Local Area Plan Policy, though it is unclear whether this might be achieved in a top-down or more collaborative manner. Nevertheless, although questions remain about the appropriateness of top-down instigation and the influence of statutory policy on VDS content, VDSs are becoming increasingly integrated into the statutory planning system.

Some of the most recently published VDSs have been instigated by the local community, are prepared on the basis of the Heritage Council's Toolkit, and feature a strong emphasis on the role of the community in their formulation. Looking specifically into the detail of how local distinctiveness is represented within these VDSs, the analysis in the VDS for Coachford lists physical elements of the villages that people like (e.g. school; church; sports fields; ACRCC, 2013a: 17), but does not elaborate on *why*, i.e. due to their tangible value, or also due to the intangible (memory, symbolism, etc.). The VDS for Rylane features a similar emphasis (ACRCC, 2013b). Preparation of the VDS for Cong (MCC, 2013) commenced shortly before the publication of the VDS toolkit, with the support of the Heritage Council. In relation to identity, it emphasises place-branding for tourism purposes (rather than people–place relationships) (ibid.: 28), aesthetics, and the identity *of the place* (rather than place attachment and community identity) (ibid.: 32). In contrast with these examples, the VDS for Marlfield (TCC, 2013) features no explicit community role in its formulation, and focuses on top-down guidance relating to architectural and townscape elements. While this is not a comprehensive analysis of the formulation of these VDSs, and cannot be taken to represent efforts under way in other counties, a greater emphasis could be placed on why specific elements of places are valued – which underpins the basis for VDSs, and has implications for the nature and detail of the resulting guidance.

In relation to the broader content of these recent VDSs, key ambitions and actions still include issues outside the normative scope of a VDS, e.g. increasing the provision of community facilities, or instigation of a community renewable power scheme (ACRCC, 2013b). However, not all recent VDSs are as broad in scope; some of the more recent VDSs focus more narrowly on design (MCC, 2013). While landscape, ecology, green infrastructure and biodiversity are listed as considerations within the Heritage Council's Toolkit, reflecting much earlier suggestions around the appropriate scope of VDSs (Owen, 1998), the extent to which these considerations feature in VDSs in Ireland varies. For example, ecology features prominently as a concern within the VDS for Mulranny (MCC, 2012), but does not feature at all within the VDS for Coachford (despite the village's landscape context featuring heavily).

The content of recent VDSs therefore varies, and the scope is sometimes still broader than has been argued as being appropriate. While this may to some extent reflect local priorities, previous lessons around the success of VDSs in both the UK and Ireland suggest that the broadening of VDSs' scope may not be appropriate or successful. Echoing Murray (2016), this implies a need for a more expansive approach to village planning in Ireland to encompass social, economic and cultural dimensions.

Conclusion

In early work relating to village distinctiveness in the UK, Owen (1995: 143) defined village distinctiveness as the 'physical character', *as seen* by inhabitants and other users. More recent work identifies the sense of attachment that people form with place, and the role of this attachment in heritage and local identity (Owen et al., 2011), though understanding of the relationships between people, place attachment, identity and local heritage are still unclear. Despite numerous efforts in the intervening years, this is an ongoing theme within both the literature and in practice.

Within community-led planning efforts that seek to provide place-specific guidelines to ensure the sensitivity of design and development to local distinctive character, it is incumbent upon professionals involved to ensure that the meanings and value the wider public assign to places are properly understood and meaningfully incorporated within resulting policy and guidance. This obviously means that the local public must continue to play a leading role in the formulation of VDSs and other related efforts. However, it is also essential that guidance and planning practices relating to design and local distinctiveness must seek to find and employ novel and appropriate tools to break down barriers between professionals and the public in conversations around the intangible qualities of places – rooted in collective memory, symbolism, and nostalgia. These qualities are central to people's sense of identity and crucial in understanding what people see important in the character of the places they inhabit. The choice of specific methods for public interaction through which this can be achieved is outside the scope of this chapter, but – above all else – these must seek to creatively break down barriers of knowledge and power between professionals and the public, to meaningfully represent public values relating to place in design and development, and thereby protect and enhance the distinctive character of rural villages.

References

ACRCC (2013a) *Coachford: Community-Led Village Design Statement*. Coachford: ACR Community Council.

ACRCC (2013b) *Rylane: Community-Led Village Design Statement*. Coachford: ACR Community Council.

Albrechts, L. (2002) The planning community reflects on enhancing public involvement. Views from academics and reflective practitioners. *Planning Theory & Practice*, 3(3), 331–347.

Alexander, C. (1977) *A Pattern Language: Towns, Buildings, Construction*. Oxford: Oxford University Press.

Arendt, R., Brabec, E.A., Dodson, H.L., Reid, C. and Yaro, F.R.D. (1994) *Rural by Design*. Chicago, IL: American Planning Association.

Arnstein, S. R. (1969) A ladder of citizen participation. *Journal of the American Institute of Planners*, 35(4), 216–224.

Bishop, J. (1994) Planning for better rural design. *Planning Practice and Research*, 9(3), 259–270.

Campbell, H. and Marshall, R. (2000) Public involvement and planning: looking beyond the one to the many. *International Planning Studies*, 5(3), 321–344.

Childs, M.C. (2014) Rural design: a new design discipline. *Journal of Urban Design*, *19*(2), 263–265.

Conzen, M.R. (1966) Historical townscapes in Britain: a problem in applied geography. In J. W. House (ed.), *Northern Geographical Essays in Honour of GHJ Daysh*, 56–78. Newcastle-upon-Tyne: Dept of Geography, University of Newcastle.

Cullen, G. (1961) *The Concise Townscape*. London: Butterworth.

Davison, G. (2017) The character of the Just City: the regulation of place distinctiveness and its unjust social effects. *Town Planning Review*, *88*(3), 305–325.

DECLG (2013) *Local Area Plans: Guidelines for Planning Authorities*. Dublin: Stationery Office.

DHCG (2016) *Review of the Heritage Council*. Retrieved on 1 December 2017 from www.chg.gov.ie/app/uploads/2017/10/2017-10-04_final_report.pdf.

Dixon, J. and Durrheim, K. (2000). Displacing place-identity: a discursive approach to locating self and other. *British Journal of Social Psychology*, *39*(1), 27–44.

Donovan, K. and Gkartzios, M. (2014) Architecture and rural planning: 'Claiming the vernacular.' *Land Use Policy*, *41*, 334–343.

Gallent, N. (2013) Re-connecting 'people and planning': parish plans and the English localism agenda. *Town Planning Review*, *84*(3), 371–396.

Gallent, N., Morphet, J., and Tewdwr-Jones, M. (2008) Parish plans and the spatial planning approach in England. *Town Planning Review*, *79*, 1–29.

Green, R. (1999) Meaning and form in community perception of town character. *Journal of Environmental Psychology*, *19*, 311–329.

Green, R. (2000) Notions of town character. *Australian Planner*, *37*(2), 76–86.

Heritage Council (2012) *Community-Led Village Design Statements in Ireland: Toolkit*. Kilkenny: Heritage Council.

Jacobs, J. (1961) *The Death and Life of Great American Cities*. New York: Vintage Books.

Jivén, G. and Larkham, P. J. (2003) Sense of place, authenticity and character: a commentary. *Journal of Urban Design*, *8*(1), 67–81.

Jones, O. (1995) Lay discourses of the rural: developments and implications for rural studies. *Journal of Rural Studies*, *11*, 35–49.

Jordan, D. and Harvey, A. (2010) *Village Design Statements (VDS) in Ireland: The Way Forward? VDS Questionnaire Analysis*. Kilkenny: Heritage Council.

Lynch, K. (1976). *Managing the Sense of a Region*. Cambridge, MA: MIT Press.

McManus, R. (2017) Identity crisis? Heritage construction, tourism and place marketing in Ireland. In M. McCarthy (ed.), *Ireland's Heritages: Critical Perspectives on Memory and Identity*, 235–250. London: Routledge.

MCC (2013) *Cong: Community-Led Village Design Statement*. Castlebar: Mayo County Council.

MCC (2012) *Mulranny Village Design Statement*. Castlebar: Mayo County Council.

Moseley, M.J. (1997) Parish appraisals as a tool of rural community development: an assessment of the British experience. *Planning Practice & Research*, *12*(3), 197–212

Murray, M. (2016) *Participatory Rural Planning: Exploring Evidence from Ireland*. London: Routledge.

Norberg-Schulz, C. (1980) *Genius Loci: Towards a Phenomenology of Architecture*. New York: Rizzoli.

Owen, S. (1995) Local distinctiveness in villages: overcoming some impediments to clear thinking about village planning. *Town Planning Review*, *66*(2), 143.

Owen, S. (1998) The role of village design statements in fostering a locally responsive approach to village planning and design in the UK. *Journal of Urban Design*, 3(3), 359–380.

Owen, S. (1999) Village design statements: some aspects of the evolution of a planning tool in the UK. *Town Planning Review*, *71*(1), 41–59.

Owen, S. (2002) From village design statements to parish plans: some pointers towards community decision making in the planning system in England. *Planning Practice and Research*, *17*(1), 81–89.

Owen, S., Bishop, J., Moseley, M., Boase, R. and Coffey, F. (2008) *Enhancing Local Distinctiveness: Evaluation of Village Design Statements (VDSs) in Ireland (2000 to date). Final Report to the Heritage Council*. Kilkenny: Heritage Council.

Owen, S., Bishop, J. and O'Keeffe, B. (2011) Lost in translation? Some issues encountered in transferring village design statements from England to Ireland. *Journal of Urban Design*, *15*(3), 405–424.

Owen, S. and Herlin, I. S. (2009) A sustainable development framework for a landscape of dispersed historic settlement. *Landscape Research*, *34*(1), 33–54.

Owen, S., Moseley, M. and Courtney, P. (2007) Bridging the gap: an attempt to reconcile strategic planning and very local community-based planning in rural England. *Local Government Studies*, *33*(1), 49–76.

Parkinson, A., Scott, M. and Redmond, D. (2016) Competing discourses of built heritage: lay values in Irish conservation planning. *International Journal of Heritage Studies*, *22*(3), 261–273.

Proshansky, H. M., Ittelson, W. H. and Rivlin, L. G. (1970) *Environmental Psychology: Man and His Physical Setting*. New York: Holt, Rinehart and Winston.

Relph, E. (1976) *Place and Placelessness*. London: Pion.

Sharp, T. (1946) *The Anatomy of the Village*. Harmondsworth: Penguin Books.

Steele, F. (1981) *The Sense of Place*. Boston, MA: CBI Pub. Co.

Stephenson, J. (2010) People and place. *Planning Theory & Practice*, *11*(1), 9–21.

TCC (2013) *Marlfield Local Area Plan 2013*. Clonmel: Tipperary County Council.

Thorbeck, D. (2012) *Rural Design: A New Design Discipline*. London: Routledge.

Twigger-Ross, C. L. and Uzzell, D. L. (1996) Place and identity processes. *Journal of Environmental Psychology*, *16*, 205–220.

Williamson, K., (2010) *Development and Design of Heritage Sensitive Sites: Strategies for Listed Buildings and Conservation Areas*. London: Routledge.

Woods, M. (2007) Engaging the global countryside: globalization, hybridity and the reconstitution of rural place. *Progress in Human Geography*, *31*(4), 485–507.

Worskett, R. (1969) *The Character of Towns*. London, Architectural Press.

35

Conserving rural heritage
The cases of England and Ireland

Arthur Parkinson and John Pendlebury

Introduction

Reflecting a wider cultural turn in the social sciences, the heritage field in recent decades has witnessed a shift away from positivist epistemological theories of heritage, stressing intrinsic universal value, towards a discursive understanding, whereby heritage value is seen as being socially constructed and plural, rather than intrinsic to the fabric and universal (Gibson and Pendlebury, 2009; Harrison, 2010). In other words, what is considered heritage and how it is valued is culturally constructed and varies over time and space. This inevitably affects the development and implementation of public policy for heritage protection as management decisions are made on the basis of a set of subjective and contestable values unique to a specific place and time (Graham, 2002). Moreover, people's interactions with the physical environment around them can also play a key role in the formation of identity, particularly in terms of the symbolism of built heritage, rooted in collective memories of past events (Halbwachs, 1950; Moore and Whelan, 2007; Graham and Howard, 2008). However, these relationships are not settled and fixed, but in constant flux (Agnew, 1991), and can be subject to manipulation by different policy actors in the pursuance of particular objectives (Archer, 1997, after Foucault).

Using this understanding of the nature of heritage, this chapter explores how ideas of the countryside, heritage and national identity have been articulated in England and Ireland and how these have, in turn, influenced the very different approaches to heritage protection adopted in the two countries. This is notwithstanding the similarities that exist between the countries in the way the respective countryside was produced within the context of a shared over-arching political system prior to the ceding of the Irish Free State from Britain in 1922. Put simply, both countrysides are heterogeneous with many landscape types and (historically) tenure patterns, but with a strong influence over their formation from large aristocratic estates, with small farmers tending to be pushed towards the productive margins.

In England, the landscape values produced have been normalised as something beautiful and important in constructions of national identity. These values have, in turn, fuelled a wider romanticisation of the rural and been important in the evolution of a planning approach of strict control in the open countryside. In Ireland the cultural context is very different. While the notion of a Gaelic rural idyll was invoked prominently in nation-building efforts in the early years after Irish independence, a low value has often been placed on rural built heritage, aspects of which have traditionally been linked with collective memory of colonial domination. This, in tandem with a contemporary political and planning culture that tends to celebrate development, leads to a quite different approach to management of rural heritage. The chapter explores the antecedents for these very different readings and imaginaries of rural heritage before considering how these have fed into rural planning and regulation and the different planning cultures that have developed in the two countries.

England

The imaginary of rural England

In this section we discuss the importance of the countryside in the construction of ideas of England very specifically; each of the nations of the United Kingdom has a somewhat different relationship between the rural and national identity. In short, rural England, as encapsulated in the phrase 'this green and pleasant land'[1] has, over an extended historical period, been closely associated with national identity; as something that is particular and precious to be protected. This is not accidental or inevitable but has been achieved by conscious processes of argumentation and reinforcement over many years. Equally, while the rural has endured as an axiomatic component of English identity for many people, its precise articulation has changed over time.

Darby (2000) discusses the modern formation of English rural mythology with, for example, reference to the aristocratic Grand Tour, the formation on the modern nation state, the disjuncts caused by the French Revolution, and the nineteenth century English Romanticism movement. Wiener (1981), in his classic polemic about anti-industrialism in the British ruling classes, assesses how Englishness became defined as the rural and, more specifically, in terms of metaphors of gardens, southern landscapes and an old country. He argues that in the nascent industrial country of the nineteenth century, a rentier aristocracy largely maintained cultural hegemony. New industrial capital bought social capital by buying into the cultural tastes of the existing elite and bought into their institutions, such as public schools, that reinforced these values in the (male) heirs of the industrialists, effecting a smooth transition to modernity. Howkins (2014) specifically discusses the significance of the period 1880–1920 (as part of a wider argument about the importance of these decades in defining Englishness; Colls and Dodd, 2014). Like Wiener, Howkins emphasises the increasing centrality of southern landscapes, or 'south country', in defining the English countryside. All these authors emphasise in different ways the class politics that underpin these representations of England. But Howkins also shows how by the First World War such images of England and her values had been normalised across class boundaries, with troops in the trenches sharing a romanticised vision of a bucolic rural England of which they frequently can have had no direct knowledge. During this period, a proto-conservation movement was developing that

> expressed a desire to maintain those continuities between past and present which valued landscapes were seen to embody. By doing so, it addressed a key problem of modernity: how to present a coherent and historically rooted sense of national identity at a time of change.
>
> *(Readman, 2008: 198)*

Villages and rural buildings were integral to the south country landscapes in particular. In parallel with efforts to protect the countryside, pioneers of building conservation John Ruskin and William Morris were also expressing ideas about English identity. Alongside their well-known

Figure 35.1   Painshill Park, Surrey: an eighteenth-century aristocratic vision of landscape.

concern for ecclesiastical buildings, they focused on older, often relatively modest country houses, which were seen as representing a kind of medieval yeoman England (Pendlebury, 2009a). Founded in 1895, the National Trust had an initial (and enduring) emphasis upon protecting scenic countryside but also the rural built heritage. The first building it acquired in 1896 was Alfriston Clergy House, a rural and relatively modest medieval timber framed hall house.

Protecting rural England

The inter-war period was a critical time when such romanticised visions of England became promulgated further; elegiac laments about the despoliation of rural England from modernisation and urban sprawl rubbed alongside mass movements to extend the enjoyment of and access to the countryside, along with halting efforts to construct the English countryside as something orderly and modern and to plan for and regulate rural development. In regards to planning and protection, or in the term adopted by Matless (1998) 'planner-preservationism', the period saw a burgeoning of activity including, for example, polemical works (e.g. Williams-Ellis, 1928) and a series of plans for rural areas commissioned by the newly established Council for the Preservation of Rural England (Sheail, 1981). The protection of the countryside, it was felt, could no longer be entrusted to traditional practices; the consequences of laissez-faire urbanisation dictated that modern planning was required if the specialness of the English countryside was to endure. Indeed, urban sprawl, ribbon development and the intrusion of urban activities and facilities into the countryside were particular concerns. Figures such as Thomas Sharp argued for a strict separation between town and country (Pendlebury, 2009b). Furthermore, the planner-preservationists were not bound into the south country vision of English landscape but differentiated between uplands and lowlands in a more multi-faceted Englishness (Matless, 1998).

The Second World War was the pivotal moment that translated wider agitations for developing an effective planning system into action. War time activity on planning included the Scott Report (Committee on Land Utilization in Rural Areas, 1942) on land utilisation in rural areas that had, at its heart, an assumption that rural land was a precious asset needing protection. Alongside the central post-war legislation of the 1947 Town and Country Planning Acts, a whole series of other relevant measures were introduced that collectively served to identify and protect large tranches of countryside. The net effect was to create a planning system that was restrictive in the countryside (apart from for agricultural operations) with the additional layering from 1949 of National Parks and Areas of Outstanding National Beauty (AONBs) and green belts (from 1955). National parks were concentrated in uplands (apart from the later Norfolk Broads designation), with a responsibility for promoting recreation as well as landscape protection. AONBs focused more narrowly on landscape protection and were mostly uplands or coastal. Green belts had the task of preventing urban sprawl (Cullingworth et. al., 2014). It is no coincidence that Hall and colleagues' epic study of the post-war planning system was entitled *The Containment of Urban England* (Hall et al., 1973).

This legislation focused upon the protection of the countryside, rather than historic rural buildings. However, in parallel, there was a new impetus for positioning the English stately home as the epitome of English civilisation (following an earlier wave of interest in the nineteenth century when, for a time, older stately homes were popularised as symbols of a shared common national heritage). In the post-war period many country houses were opened up to visitors. A significant part of this was the National Trust's country house programme that acquired significant historic buildings from an impecunious aristocracy (Mandler, 1997). There were also advocates for the value of humble rural buildings and anxieties about the unnecessary loss of old cottages to slum clearance (Godfrey, 1944).

Figure 35.2 An estate cottage in Northumberland with cottage garden. A bucolic vision of England.

As the role of heritage protection as part of a modern planning system reformulated over subsequent decades, national debate was firmly on urban areas, as the nature of the future evolution of towns and cities was contested. However, the stately home remained a significant focus for the conservation movement. Landmarks included *The Destruction of the Country House* exhibition at the Victoria and Albert Museum in 1974 and the campaigning of the new organisation SAVE Britain's Heritage (Mandler, 1997). Furthermore, there was a steady extension of heritage protection within rural areas. Archaeological sites were identified, the rapid growth of conservation area numbers following their introduction in 1967 was fuelled in part by systematic programmes of village designation in some areas of the country, particularly in the south, and the huge growth in the number of listed buildings in the 1980s was partly as a consequence of the resurvey of rural areas. Furthermore, the 1980s and 1990s saw English Heritage introduce new Registers for Historic Parks and Gardens and for Battlefields that were predominantly rural.

Subsequently the Countryside Commission in the 1990s attempted to move beyond designated areas to encompass the whole English landscape. Working closely with English Heritage and English Nature, the country was systematically mapped as part of a characterisation programme designed to highlight the qualities of all English countryside, not just the rather atypical landscapes of the National Parks and AONBs. As part of this work, English Heritage undertook landscape archaeology-based historic landscape assessments (Fairclough, 1999).

Management of rural heritage today

In the twenty-first century the English planning system has endured much turmoil and change. However, the planning policies restricting development in the countryside established in 1947 remain largely in place, with the main pressure being for housing development in some localities.[2] A vision of England underpinned the formation of this planning approach, attaching both cultural and economic value to the country cottage, with roses round the door, and a view out over open fields. Restriction on development has further enhanced the economic value of existing buildings, particularly in the housing market and the economic capital locked up in such rural idylls helps ensure there is usually a strong political voice for sustaining the status quo. Indeed, in reinforcing the spatial separation of urban and rural, the English planning system has contributed to a strong trend of counter-urbanisation through urban out-migrants leap-frogging the greenbelt into rural villages (Woods, 2011); these incomers often in turn work to reinforce this spatial divide in what Murdoch and Lowe (2003) term the 'preservationist paradox'. The Localism Act of 2011 has introduced a new dynamic into rural planning, albeit there are antecedents for more bottom-up processes, such as village design statements and parish plans, dating back to the 1990s. The Act enables the production of community-led Neighbourhood Plans and there has been a higher take-up in rural areas than in cities. The evidence on the plans emerging is mixed but does not seem to herald a radical departure from a protectionist approach to the countryside and issues of heritage and identity are often prominent (Parker and Salter, 2017). Indeed, Tait and Inch (2016) have shown how both the underlying ideologies of localism held by the Conservative Party and the way that localism was subject to an organised response in practice, are closely bound up with the romanticised view of rural England as at the heart of English identity.

Ireland

The protection of rural heritage in Ireland is in many ways characterised by similar themes as elsewhere. For example, competing professional and lay representations of heritage can lead to sometimes bitter conflict in heritage management, with dominant groups often tending to win out. However, the Irish case is marked out by its postcolonial context, unique in western Europe, the distinct regulatory history stemming both directly and indirectly from this, as well as the distinct way this has shaped subtly shifting public attitudes and official policy relating to heritage.

The imaginary of rural Ireland

Aalen et al. (1997) describe the diverse character of Ireland's cultural landscape, which exhibits considerable variation both across the country as a whole and at a local level, significantly as a result of environmental constraints. Notably, the absence of glacial drift along much of the western seaboard, historically led to the dominance of small farming in the west. In contrast, generally greater agricultural productive potential in the east has led to larger landholdings, more mixed farming and higher levels of affluence. Enclosed fields emerged largely in the postplantation period, and particularly during agricultural reform of the eighteenth century; these are dominant in the cultural landscape, and both their size, and forms of boundary construction, vary considerably. Each of these factors has a significant bearing upon geographical variation in the character of the landscape and buildings. However, the cultural landscape is also shaped by Ireland's colonial history: towns, landlord estates and houses (colloquially referred to as the 'big house'), the distribution of historical vernacular housing typologies (in particular byre dwellings in the west and labourers' houses in the east), and historical patterns of agricultural production, are all largely the result of successive waves of settlement, from Viking, to Anglo-Norman, to the Plantations of the sixteenth and seventeenth centuries when huge numbers of English and Scottish settlers were given lands confiscated from the native population.

A 'postcolonial' perspective in relation to Ireland, set out in greater detail elsewhere (Kearns, 2006), is particularly useful in understanding heritage value and management in Ireland, as colonial history is often central in framing meanings and debates (Howe, 2000). This history has led to a discourse wherein aspects of tangible and intangible cultural heritage originating before the sixteenth and seventeenth century plantation period, and rural society in general, can be represented as the cultural inheritance of the perceived 'native' (Gaelic) Irish, while built heritage and planning can be associated with the pre-independence ruling elite Ascendancy and British colonial power. This view of planning emphasises the relationship between the people and the land, and is deeply embedded in Ireland's political and social culture. In short, there are three key and inter-related collective memories of national events related to this history which are of particular significance. The first is a memory of poverty and famine, and an associated struggle for land rights (Kane, 2011). Second is a memory of the colonial power's role in creating both the rural and urban environment (Hood, 2002), which often resulted in indifference or outright hostility towards that heritage, both before and after independence (Bielenberg, 2013). This is closely connected with the landlord estates and houses that dominated both land ownership and the means of agricultural production, each of which were key contributors to poverty and famine (D'Arcy, 2010). The third memory is of the violent political campaign for independence, notably culminating in the 1922 Anglo-Irish Treaty and effective independence. This postcolonial experience has tended to shape various aspects of society – including planning.

In relation to Ireland's formal planning system, set against years of failed protectionist policies and resulting economic stagnation, Bannon (1989: 128) suggests that a peculiarly Irish approach to economic programming began to emerge from the late 1950s onwards, characterised by 'an abhorrence of planning in any rigid sense'. This has perhaps shaped Irish planning ever since, and reflects Ireland's 'post-colonial tradition of individualism' (ibid.: 148). The Local Government (Planning and Development) Act, 1963, which emerged in this context, and established the

Figure 35.3 Ruined vernacular house, Gap of Dungloe, Co. Kerry.

system of development control, was therefore aimed primarily at the stimulation of development rather than control. This has shaped a planning and political culture that has tended to celebrate development as an indicator of rural vitality (Scott and O'Neill, 2013). Moreover, Bunreacht na hÉireann (the Constitution of Ireland) protects the individual private property right, though this right may be curtailed where it conflicts with the public interest. With respect to heritage protection, economic stagnation in the decades after independence resulted in little environmental threat, and the conservation movement developed considerably later in Ireland than in many Western societies. An Taisce (the National Trust for Ireland) was not formed until 1948 (50 years after the equivalent in England), and the Irish Georgian Society (the principal civil society organisation in Ireland advocating for architectural heritage protection) was formed as late as 1958. From the late 1950s onwards, the new pro-development context also tended to overshadow heritage concerns, and the statutory measures for heritage protection that emerged at the start of the twenty-first century were perhaps more due to external shifts such as Ireland's ratification of the Granada Convention in 1997, than out of concerns emanating within Ireland. In particular, the Planning and Development Act 2000 introduced a national system for statutory protection of built heritage for the first time, as well as separate provisions for local authorities to preserve landscape areas through designation of Areas of Special Amenity (relating to natural beauty or recreational value) and Landscape Conservation Areas (LCA). Aside from protection within the planning system, Ireland's Heritage Council, a state-funded statutory body formed in 1995, has a significant role in provision of heritage advice, grants, and promotional efforts relating to both built and natural heritage.

Management of rural heritage today

Postcolonial representations of planning and heritage continue to frame planning and conservation debates in Ireland, resulting in challenges to their legitimacy and to conflict in both rural and urban locations (Scott, 2012; Moore-Cherry and Ó Corráin, 2016), reflecting a disrespect for authority observed in other postcolonial contexts (Keating and Martin, 2013). Heritage protection can also be seen as an unwanted burden, particularly given that conservation grant funding has tended to be modest (Pickard and Pickerill, 2002). Even vernacular buildings are typically not valued by their owners or rural communities, perhaps as a reminder of collective memories of the Great Famine and past oppression. It is of note that these attitudes are at odds with tourism policy from the mid-to-late twentieth century onwards, which has tended to romanticise and promote aspects of the Irish vernacular building tradition (Scott, 2012) and has attempted to market it to tourists as part of an Irish rural idyll. The bulk of Irish traditional rural housing stock can be characterised by its modest scale, and its architectural and technical simplicity. While there were some notable attempts to modernise the rural housing stock in the late nineteenth and early twentieth centuries, and again in the 1930s (Aalen, 1997), a substantial portion of the housing stock had remained largely unaltered until around 1960, primarily due to rural poverty, both before independence, and in the initial decades thereafter. However, the early 1960s marked the beginning of a period of modernisation that transformed the rural landscape dramatically. This was the result of a number of factors, including broader shifts in national economic policy mentioned above, increased affluence and car use, the resulting broadening of choice in residential location, as well as standardisation in housing design and construction exemplified in the widely popular and used pattern book, 'Bungalow Bliss' (Fitzsimons, 1971). Houses built from this period onwards were often grander in typology than previously, tended to echo the Irish 'Big House' – for example in the use of classical architectural language – and were signifiers of wealth and social status (Rowley, 2017). In the subsequent decades, many vernacular houses were abandoned or replaced with modern bungalows, dramatically changing the character of the landscape. Colloquially referred to as 'bungalow blight' (McDonald, 1997) and, later, 'bungalow blitz', this change was also facilitated to a degree by the absence of statutory heritage protection until after 2000. During the 'Celtic Tiger' boom period, the process of rural housebuilding continued apace, again facilitated by increased affluence, and a relatively permissive planning regime (Donovan and Gkartzios, 2014).

Despite the continuing existence of historically dominant and deeply rooted postcolonial narratives, an increasingly inclusive attitude towards the postcolonial architectural legacy has begun to emerge (Parkinson et al., 2017). This gradual change is characterised, firstly, by a view that many of the negatives associated with colonialism are now gone; secondly, a view that the modernity initially brought by imperialism is now an integral part of the richness of Irish society; and, thirdly, by an awareness of the economic benefits that heritage can bring through tourism. In recent years, official discourse has also sought to win public support for 'big house' heritage by stressing the involvement of Irish builders and craftspeople in the design and construction of these buildings (see for example Ahern, 2003: 2). However, perhaps more significant is the gradual drift – and occasional active shift – in the symbolic meaning of the historic environment. While the postcolonial legacy is now rarely the focus of antipathy or political struggle as in the past, neither has it been fully embraced by the wider public as an element of the Irish national 'patrimony', as suggested by others (e.g. Kearns, 1982; Negussie, 2001). Many Irish people feel no sense of collective ownership of the 'big house' or many other historic rural structures as *their* heritage, save for cases where colonial memory has been displaced by newer representations.

Figure 35.4   Bellinter House, Co. Meath. Designed by architect Richard Cassels around 1750
for Dublin merchant John Preston. Now a boutique hotel.

This 'cultural distance' perhaps makes it more difficult to seek public and political support for
conservation than elsewhere.

In more general terms conservation in Ireland tends to be concerned primarily with tangible
value and associated technical matters; intangible value (e.g. relating to cultural or social categories primarily), and its relevance in conservation decisions, are not dealt with in guidance
in the same level of breadth or detail – which raises a question mark over whether designation is as representative and inclusive as it should be (Parkinson et al., 2016). In particular, the
relevance of identity in the cultural construction of heritage value is inadequately expressed in
legislation, and is barely touched upon in built heritage guidelines, despite its frequent role in
framing both historical and more recent heritage debates – often leading to conservation conflict. O'Rourke (2005) has similarly raised concern around whether the top-down nature of
landscape and resource management in Ireland, and whether cultural as well as ecological factors
are adequately taken into account – the values of the local public in particular. She argues that
this reflects a tendency in Ireland for strong representative democracy, but weak participatory
democracy – itself arguably a legacy of pre-independence patterns of authority – and highlights
a need for closer coupling of natural and cultural systems to achieve effective management.

Conclusion

England and Ireland both have rich, heterogeneous rural landscapes and rural built heritage,
encompassing both aristocratic grandeur and humble vernacular. While both countries have

great variety encompassing distinct landscapes and typologies, not shared across the Irish Sea, there is a degree of shared history in the forces that formed the appearance of these places today and specifically in terms of an Anglo-Irish aristocracy. However, how these landscapes and their attendant buildings are understood is very different between England and Ireland due to profound differences in the cultural and political histories of these countries. An important manifestation of this that extends beyond the scope of this chapter is the very different planning cultures and traditions that developed in England and Ireland in the twentieth century. Put simply, a more laissez-faire approach has prevailed in Ireland, less conservative and more systemically pro-development than has prevailed in England. While there has been a long-standing discourse in England to liberalise planning, rural development planning remains restrictive in most circumstances (with the exception of agricultural operations).

It is also the case that countryside as manifest as rural heritage has played a very different role in the cultural imagination of the two different countries. In England, while the particular conception of countryside has changed over time, the notion that the rural heritage forms a key embodiment of English national identity has persisted over an extended historical period. In Ireland the legacies of colonialism and the hardships of famine have engendered an equally important but very different response to rural legacy.

While consideration of local cultural factors – and local public values, in particular – is crucial in heritage management, this need can be especially important in Ireland, due to the postcolonial experience, and the role this has played in shaping aspects of the culture of politics, planning, and of wider society. Despite this, in recent years, the Northern Ireland peace process, alongside generally improved relationships between the UK and Ireland, have helped engender a newfound willingness to reconsider issues of heritage and identity – thereby contributing to a revalorisation of the colonial legacy in Ireland. More recently, the ground has begun to shift again, however. The UK's vote to leave the European Union has created new tensions between the two countries, in particular over the future of the Irish border and possible impacts upon the Irish economy. These recent developments hold the potential in the coming years to at least slow the pace at which the aforementioned reconsideration and revalorisation continues to take place.

The English landscape of countryside and villages remains a powerful imaginary of England and Englishness. This has not been fixed or static and has had different discursive and political roles; elegiac/progressive, reactionary/reformist and so on. However, perhaps like all expressions of national identity, there is an 'othering' dimension; expressing Englishness denies other identities. Specifically, in the contemporary moment the 'other' might be seen as 'European' (although the English are of course European). While the vote to leave the European Union was by no means entirely an English rural phenomenon, we cannot help but feel that deeply embedded and recursive national mythologies centred upon the English countryside are inextricably bound into some of the mythologies mobilised by the campaigns to leave the EU. As a final irony, the form of that countryside is likely to change, in ways that are not yet predictable, as a consequence of Brexit.

Notes

1 'Green and pleasant land' is a well-known phrase taken from the last line of the preface to 'Milton', an epic poem by William Blake of $c.$1810, but immortalised when set to music by William Parry in 'Jerusalem', a major patriotic song, first performed during the First World War in 1916.
2 The rural heritage is not completely immune from liberalising reform, albeit changes have tended to be relatively minor. For example, at the time of writing, some extensions of permitted development rights to allow barns to be converted to residential use have been made. This has been accompanied by guidance from Historic England (2017) on sympathetic conversion.

References

Aalen, F. H. (1997) Buildings. In F. H. Aalen, K. Whelan and M. Stout (eds), *Atlas of the Irish Rural Landscape*, 145–179. Toronto: University of Toronto Press.

Aalen, F. H., Whelan, K. and Stout, M. (eds.) (1997) *Atlas of the Irish Rural Landscape*. Toronto: University of Toronto Press.

Agnew, J. (1991) Place and politics in Post-war Italy: a cultural geography of local identity in the provinces of Lucca and Pistoia. In K. Anderson and F. Gales (eds), *Inventing Places: Studies in Cultural Geography*, 52–71. London: Halsted Press.

Ahern, P. B. (2003) Foreword. In T. A. M. Dooley (ed.), *Historic House Survey*. Dublin: Irish Georgian Society/Department of the Environment, Heritage and Local Government.

Archer, J. (1997) Situating Australian national identity in theory and practice. In G. Stokes (ed.), *The Politics of Identity in Australia*, 23–37. Cambridge: Cambridge University Press.

Bannon, M. J. (1989) Irish planning from 1921–1945: an overview. In M. J. Bannon (ed.), *Planning: The Irish Experience, 1920–1988*, 38–40. Dublin: Wolfhound Press.

Bielenberg, A. (2013) Exodus: the emigration of southern Irish Protestants during the Irish War of Independence and the Civil War. *Past and Present*, 218(1), 199–233.

Colls, R. and P. Dodd (eds) (2014) *Englishness: Politics and Culture 1880–1920*, 2nd edition. London, Bloomsbury.

Committee on Land Utilization in Rural Areas (1942). *Report of the Committee on Land Utilisation in Rural Areas* [the 'Scott Report']. London: HMSO.

Cullingworth, J. B., Nadin, V., Hart, T., Davoudi, S., Pendlebury, J., Vigar, G., Webb, D. and Townshend, T. (2014) *Town and Country Planning in the UK*. London: Routledge.

Darby, W. J. (2000) *Landscape and Identity*. Oxford: Berg.

D'Arcy, A. (2010) The potato in Ireland's evolving agrarian landscape and agri-food system. *Irish Geography: Bulletin of the Geographical Society of Ireland*, 43, 119–134.

Donovan, K. and Gkartzios, M. (2014) Architecture and rural planning: 'Claiming the vernacular.' *Land Use Policy*, 41, 334–343.

Fairclough, G. (1999) Protecting the Cultural Landscape: National Designation and Local Character. In J. Grenville (ed.), *Managing the Historic Rural Landscape*, 179. London: Routledge.

Fitzsimons, J. (1971) *Bungalow Bliss*. Kells: Kells Art Studies.

Gibson L. and Pendlebury J. (2009) Introduction. In L. Gibson and J. Pendlebury (eds), *Valuing Historic Environments*, 1–18. Farnham: Ashgate.

Godfrey, W. H. (1944) *Our Building Inheritance: Are We to Use it or Lose it?* London: Faber & Faber.

Graham, B. (2002) Heritage as Knowledge: Capital or Culture? *Urban Studies*, 39: 1003–1017.

Graham, B. J. and Howard, P. (2008) *The Ashgate Research Companion to Heritage and Identity*. Burlington, VT: Ashgate.

Halbwachs, M. (1950) *The Collective Memory*. New York: Harper and Row.

Hall, P., Gracey, H., Drewett, R. and Thomas, R. (1973) *The Containment of Urban England* (2 volumes). London: George Allen & Unwin Ltd.

Harrison, R. (2010) *Understanding the Politics of Heritage*. Manchester: Manchester University Press in association with the Open University.

Historic England (2017) *The Adaptive Reuse of Traditional Farm Buildings*. London: Historic England.

Hood, S. (2002) The significance of the villages and small towns in rural Ireland during the eighteenth and nineteenth centuries. In P. Borsay, and L. Proudfoot (eds), *Provincial Towns in Early Modern England and Ireland: Change, Convergence and Divergence*, 241–262. Oxford: Oxford University Press.

Howe, S. (2000) *Ireland after Empire: Colonial Legacies in Irish History and Culture*. Oxford: Oxford University Press.

Howkins, A. (2014) The discovery of rural England. In R. Colls and P. Dodd (eds), *Englishness: Politics and Culture 1880–1920*, 85–112. London: Bloomsbury.

Kane, A. (2011) *Constructing Irish National Identity: Discourse and Ritual during the Land War 1879–1882*. Basingstoke: Palgrave Macmillan.

Kearns, G. (2006) Dublin, modernity and the postcolonial spatial fix. *Irish Geography*, 39, 177–182.

Kearns, K. C. (1982) Preservation and transformation of Georgian Dublin. *Geographical Review*, 72, 270–289.

Keating, M. A. and Martin, E. J. (2013) Leadership and culture in the Republic of Ireland. In J. S. Chhokar, F. C. Brodbeck and R. J. House (eds), *Culture and Leadership Across the World: the GLOBE Book of In-Depth Studies of 25 Societies*, 361–396. Hoboken, NJ: Taylor and Francis.

Mandler, P. (1997) *The Fall and Rise of the Stately Home*. New Haven, CT, Yale University Press.

Matless, D. (1998) *Landscape and Englishness*. London: Reaktion Books.

McDonald, F. (1997) Writing off the Irish countryside. *Irish Times*, 26 June, 13.

Moore, N. and Whelan, Y. (2007) *Heritage, Memory and the Politics of Identity: New Perspectives on the Cultural Landscape*. Aldershot: Ashgate.

Moore-Cherry, N. and Ó Corráin, D. (2016) 1916 then and now: reflections on the spatiality of the Rising's urban legacies. *Irish Geography*, 49(2), 117–124.

Murdoch, J. and Lowe, P. (2003) The preservationist paradox: modernism, environmentalism and the politics of spatial division. *Transactions of the Institute of British Geographers* 28(3): 318–332.

Negussie, E. (2001) Ireland. In R. Pickard (ed.), *Management of Historic Centres*, 133–161. London: Spon.

O'Rourke, E. (2005) Socio-natural interaction and landscape dynamics in the Burren, Ireland. *Landscape and Urban Planning*, 70(1–2), 69–83.

Parker, G. and K. Salter (2017) Taking stock of neighbourhood planning in England 2011–2016. *Planning Practice and Research*, 32(4), 478–490.

Parkinson, A., Scott, M. and Redmond, D. (2016) Defining 'official' built heritage discourses within the Irish planning framework: insights from conservation planning as social practice. *European Planning Studies*, 24(2), 277–296.

Parkinson, A., Scott, M., and Redmond, D. (2017) Revalorizing colonial era architecture and townscape legacies: memory, identity and place-making in Irish towns. *Journal of Urban Design*. 22(4), 502–519.

Pendlebury, J. (2009a) *Conservation in the Age of Consensus*. London: Routledge.

Pendlebury, J. (2009b) The urbanism of Thomas Sharp. *Planning Perspectives* 24(1): 3–27.

Pickard, R. and Pickerill, T. (2002) Conservation finance 1: support for historic buildings. *Structural Survey*, 20(2), 73–77.

Readman, P. (2008) Preserving The English Landscape, c.1870–1914. *Cultural and Social History* 5(2): 197–218.

Rowley, E. (2017) Housing in Ireland: 1740–2016. In E.F. Biagini and M.E. Dal (eds), *The Cambridge Social History of Modern Ireland*. Cambridge: Cambridge University Press.

Scott, M. (2012) Housing conflicts in the Irish countryside: uses and abuses of postcolonial narratives. *Landscape Research*, 37(1), 91–114.

Scott, M., and O'Neill, E. (2013) Displacing wind power across national boundaries or eco-innovation? Spatial planning implications of UK–Ireland renewable energy trading. *Planning Theory & Practice*, 14, 418–424.

Sheail, J. (1981) *Rural Conservation in Inter-War Britain*. Oxford: Clarendon Press.

Tait, M. and A. Inch (2016) Putting localism in place: conservative images of the good community and the contradictions of planning reform in England. *Planning Practice and Research*, 31(2), 174–194.

Wiener, M. J. (1981) *English Culture and the Decline of the Industrial Spirit 1850–1980*. Cambridge: University Press.

Williams-Ellis, C. (1928) *England and the Octopus*. Portmeirion: Council for the Protection of Rural England.

Woods, M. (2011) *Rural*. London: Routledge.

36

Contours and challenges of rural change in transition economies

The case of China

Karita Kan

Introduction: urbanisation and rural change in China

The advent of market reform in the late 1970s heralded a period of rapid, growth-driven transformation in post-Mao China. From the perspective of territorial change, one significant development is the rise and growth of Chinese cities through rapid urbanisation processes. Seen as engines propelling national economic growth, cities took on primary roles as nodes of economic organisation and centres of consumption in reforming China. Fresh injection of capital into city building and the decentralisation of authority to urban governments fuelled the expansive growth of cities, with important implications for rural–urban relations and rural planning. This introductory section places the physical transitions confronting rural places in context by outlining the political, fiscal and territorial drivers of urban expansionism in reform-era China.

Prior to reform, rural and urban planning under state socialism followed the strict *diktat* of economic and industrial planning and was largely a centralised and technical exercise (Xie and Costa, 1993). The downward transfer of power to local governments in the reform era greatly enhanced local initiatives in planning. The introduction of the City Planning Act in 1989 equipped municipal governments with the power to oversee land development projects, while sub-municipal officials at the district level were also granted the authority to prepare development control plans to regulate land use and development intensity (Yeh and Wu, 1999). The changing conception of cities as growth poles further provided capital for development. The opening up of China to foreign direct investment enabled the inflow of capital into real estate and infrastructure (He and Zhu, 2010; Kan, 2017). Local governments were given permission to raise funds through establishing local financing platforms, and the proliferation of these financing vehicles allowed city officials to vastly expand investment in city building efforts.

The commodification of land and the creation of an urban property market played a direct role in accelerating urbanisation processes. During the socialist era, land was treated primarily as a means of agricultural and industrial production. As socialist property, land was assigned free of charge to users through administrative allocation, with no cost and time limit imposed on land use (Lin and Ho, 2005). Revisions to the Land Administration Law in the late 1980s brought fundamental changes to this free land-use system. A paid land-use track was introduced by making the use right of urban land available for market circulation (Yeh and Wu, 1996). In China, urban land was placed under state ownership. By separating the ownership right of urban land from its use right and commoditising the latter, the reform leadership directly enabled the appropriation and extraction of rent from land. To obtain the leasehold right to develop urban land, commercial users must pay a premium, known as land conveyance fee, to the local government to purchase the use right to land plots. As it is itself the owner of urban land, the Chinese state has essentially created for itself a new stream of income. Over the course of the 1990s, land-derived revenue grew to become a major pillar of municipal finance, especially in coastal cities where prime urban land was highly sought-after by property developers (Hsing, 2010). Growth in the real estate economy was further fuelled by the extensive privatisation of housing in the late 1990s and early 2000s, a strategy adopted to stimulate consumption in the aftermath of the Asian financial crisis (Lee and Zhu, 2006).

What emerged from these developments was a gradual shift in the way Chinese cities were governed and planned. Paralleling the transition from managerialism to entrepreneurialism in western cities (Harvey, 1989; Hall and Hubbard, 1998), the literature on China has noted the rise of entrepreneurial strategies in urban governance and the new dominance of pro-growth economic considerations in political thinking. Researchers have observed the emergence of 'growth coalitions' between local states and property developers in land development (Duckett, 1998; Zhu, 1999; Zhang, 2002), as well as the rise of investments in the built environment and urban image projects that are designed to attract capital and the consumer dollar (Broudehoux, 2004; Gaubatz, 2005). Distinct from the private sector-led growth observed in the West, however, urban development in China is predominantly steered by the entrepreneurial arms of the state at the local level. Wu (2003) characterised the post-socialist entrepreneurial city as a 'state project', while Ong (2011) similarly noted how megaprojects in China showcase not only the power of transnational capital, but more importantly the sovereign power of the party-state. Indeed, the ability to deliver growth and promote development has become a key criterion by which local officials are appraised and evaluated (Chung, 2001; Edin, 1998). As succinctly put by Hsing (2010: 212), 'while local states build the city, the city builds the local state'.

The promotion of urbanism by entrepreneurial leaders has a marked impact on the Chinese landscape. While in city centres the built environment was transformed by the construction of skyscrapers and shopping complexes, the impact of rapid urbanisation on rural areas was no less profound. At the rural–urban interface, the expansion of cities into the countryside has led to the large-scale conversion of farmland to non-agricultural uses. An average of 208,000 hectares of cultivated land were lost per year to construction between 1985 and 1988 (Lin and Ho, 2005), while the first six years of the 1990s saw the further disappearance of 4.85 million hectares of farmland (Lin and Ho, 2003). Encroachment has resulted in dispossession and landlessness in some rural communities, while absorbing and incorporating others into urbanised economies and land-use regimes. In the more remote countryside, meanwhile, the impact of urbanisation is most keenly felt in the exodus of the able-bodied rural population to cities in search of wage employment. The issue of abandoned or 'hollow' villages has presented challenges to planners and policymakers as they seek to revive rural economies and rebuild rural communities.

The remaining sections of this chapter provide an overview of these developments by surveying the contours and challenges of rural change as observed in peri-urban areas, and in the more remote hinterlands.

Rural transformation in peri-urban areas

Rural places situated at the fringe of cities have experienced the most intensive processes of territorial change as a result of urban expansion. It is at the rural–urban interface that the majority of land conversions for construction purposes has taken place. The transfer of rural land has occurred through two main channels, namely expropriation and administrative conversion. As a legacy of state socialism, rural land in China is collectively owned by the village. Expropriation refers to the formal legal process by which collectively owned rural land is converted into state-owned urban land. After the state unit undertaking expropriation has compensated farmers for their lost land and crops, the acquired land can be used for construction and its use rights can be circulated in the urban land market. Another method of obtaining rural land is achieved through administrative means, whereby rural counties neighbouring large cities are converted into urban districts and placed under the municipal government's jurisdiction (Ma, 2005). This practice of *de facto* annexation is commonly used to facilitate the expansion of cities, with the effect of placing vast quantities of suburban land and rural resources under the control of urban officials (Yew, 2012).

In the first decades of reform, land conversions mainly took place as an outcome of rapid rural industrialisation. Market transition led to the influx of small- and medium-sized investment from places like Hong Kong and Taiwan, which provided important initial capital for the development of labour-intensive manufacturing industries in the form of township-village enterprises (Oi, 1992). In the Pearl River Delta, especially, geographic proximity and kinship ties helped accelerate industrialisation and contributed to the export-driven or 'exo-urbanisation' of rural areas (Sit and Yang, 1997; Smart and Lin, 2007). Such industrialisation-driven growth paved the way for the so-called 'zone fever' of the 1990s, during which farmland was converted on a large scale to make way for industrial development zones that have emerged in the thousands across China (Cartier, 2001).

The significant loss of arable land resulting from these developments prompted the Chinese government to impose a temporary moratorium on farmland conversion in the late 1990s. Concerns for environmental sustainability and food security aside, policymakers also saw the need to rein in illicit land transfer practices which have grown alongside the proliferation of black land markets (Lin and Ho, 2005). The ambiguity surrounding the ownership rights of rural land creates expansive room for rent-seeking and collusion by officials (Ho, 2001), and this means that villagers often receive only a fraction of the value of their land in compensation. Dispossession and the use of extra-economic means by state and parastatal agencies in expropriation processes further intensify conflicts surrounding rural land transfers. The relationship between land grabs and rural resistance in reform-era China has been extensively observed and analysed in the literature (e.g. Cai, 2010; Guo, 2001; Hsing, 2010; O'Brien and Li, 2006; Sargeson, 2013; So, 2007; Yep, 2013).

As rural land expropriation became increasingly delegitimised, planners and policymakers have turned to alternative approaches of land development. Xu et al. (2009: 905) noted that beginning in the 2000s, the state has begun to adopt an 'ad hoc approach toward bottom-up initiatives by individual land users' that allowed for greater flexibility. For instance, by law, rural land must first undergo expropriation by the state before its use rights can be traded on the land market. To facilitate development, some local governments now allow the direct trade of rural land-use rights by village collectives to commercial users. Another example is achieving urbanisation and land development without displacing indigenous villagers. Instead of expropriating rural land and resettling villagers in new residential areas, where they are confronted with loss of employment, territorial dislocation, community disintegration and social marginalisation, an increasingly popular alternative has been to regenerate rural areas by allowing villagers to stay and enabling their participation in the development process. This is most commonly practised in China's 'urban villages' (*chengzhongcun*); rural places that have become part of the urban fabric as cities expand, further discussed below.

Urban villages and the politics of redevelopment

Also known as villages in the city, urban villages have emerged in post-socialist China as a distinct spatial form that traverses rural–urban boundaries. Their formation has to do with the expansion of cities into the surrounding rural hinterland, which results in the latter's enclosure by urbanised areas. As they retain collective ownership over rural land, urban villages constitute neighbourhoods with unique institutional features. There are hundreds of urban villages in all major metropolises around China, from Beijing and Shanghai to Guangzhou and Shenzhen.

Urban villages can be seen as informal settlements serving similar functions to those found in developing countries, in that they provide low-cost accommodation for the city's transient population (Webster et al., 2016; Zhang, 2011). In the rural land regime, each household is allotted housing or homestead plots on which low-rise rural apartments can be built for villagers' accommodation. Over time, informal rental housing markets have evolved in urban villages as rural residents capitalised on their land and began renting out rooms to rural migrant workers (Zhang et al., 2003; Song et al., 2008; Wang et al., 2009). As the influx of migrants into cities continued to grow and demand for low-cost housing exceeded the state's supply, urban villages became transitional neighbourhoods for rural migrants as they sought to integrate into urban society (Liu et al., 2010). They furnish migrants with affordable facilities and provide job opportunities to those who choose to rent space and set up small businesses inside urban villages (Hao et al., 2011).

From the state's planning perspective, however, urban villages constitute disorderly spaces with poor land-use coordination and ambiguous property rights (Wu et al., 2013). Building intensity in urban villages is extremely high due to rural households' competitive bid to maximise rentable floor space, resulting in a general lack of public space and poor hygiene and safety conditions. As land in the village officially belongs to the collective, the state also lacks effective means of spatial intervention. Governance in urban villages is often fragmented between several interlocking organs of authority, including the villagers' committee, the village party branch and the village economic development company at the village level, as well as upper-level urban governments. The fact that urban villages often accommodate thousands or tens of thousands of migrant workers further presents policymakers with the added challenge of governing a highly mobile population.

Deemed spaces of blight and informality, urban villages have become prime targets of redevelopment projects. To increase villagers' incentive to consent to the wholesale redevelopment of their neighbourhoods, local governments have decentralised planning authority and allowed villagers to directly share in the monetary benefits of redevelopment. First, instead of state-led expropriation, village collectives are given the power to take the lead in redevelopment processes with the state taking up a mediating role. Village leaders can directly negotiate with property developers, set down criteria for the competitive bidding of projects, and sign joint development contracts with chosen private partners. Second, instead of being the passive recipient of state-determined compensation, villagers are given a direct stake in the redevelopment of their land. Rather than being excluded from the lucrative land-derived revenue that redevelopment would bring, villagers are given a share in future profits as they enjoy partial ownership rights over the developments to be constructed in the neighbourhood. Last but not least, instead of eviction and resettlement, villagers are given new apartments in the redeveloped neighbourhood where they can continue living.

One exemplary case of redevelopment took place in the urban village of Liede in Guangzhou (Chung and Unger, 2013; Li et al., 2014a). Between 2007 and 2012, the neighbourhood underwent wholesale demolition and reconstruction under the leadership of the village economic development company, in partnership with both local and transnational capital. Prior to redevelopment, Liede was a typical urban village with a high-density housing environment of low-rise rural apartments. Following redevelopment, villagers were re-accommodated in residential high-rises. The vacated space was used to develop office buildings, shopping complexes and luxury hotels, to become part of the new financial district of Guangzhou. Holding ownership rights to some of these new developments, the village company reaps a continuous stream of rental revenues which is either re-invested, used for the community's welfare provision, or directly distributed to members of the village in the form of dividends.

By giving villagers a stake in the speculative commodification of their land, therefore, the state has effectively enrolled fractions of the rural population in its pro-growth developmental agenda. In the literature, such developments have largely been viewed in a positive light. Villagers' participation in planning has been seen as a way of bottom-up empowerment (Po, 2008), and as representative of an emergent 'village corporatism' that sees villagers mobilising and self-organising in confronting extractive urban governments (Hsing, 2010). Nonetheless, other researchers have also pointed to the contentious and undemocratic aspects of these projects. Although planning authority is decentralised to villages, decision-making at the village level is often monopolised by cadres and rural elites without transparent consultation with villagers (Kan, 2016). After redevelopment, some urbanised rural residents struggle to find alternative employment in the urban economy and have to adjust to higher living costs. Furthermore, while indigenous villagers are re-accommodated in the redeveloped neighbourhood, the large numbers of migrant workers whose livelihoods had been dependent on urban villages are confronted with the reality of dislocation (Chung, 2012). Although the state's aim in undertaking redevelopment is to create 'governable spaces', informality is merely being displaced to the outskirts of cities as migrants seek alternative housing at the rural–urban periphery (Wu et al., 2013).

The incorporation of rural places into circuits of capital accumulation through the redevelopment and commodification of rural land thus engenders its own politics of distribution and geography of differentiation. The absorption of segments of the rural population into speculative rentiership has created an enriched class of villager-landlords, but this has also exposed entire communities to the instability and variability of property and rental markets. The associated risks were brought home in the city of Dongguan in the aftermath of the global financial crisis, when falling rental incomes brought severe fiscal crises to villages that relied on entrepreneurial land development as their main sources of revenue (Xue and Wu, 2015). Meanwhile, the large population of rural migrants that have not been included in regimes of land-based accumulation confront a different type of risk and instability as redevelopment and displacement force them to seek viable livelihoods elsewhere. The continued marginalisation of the housing needs of the transients brings to the foreground the issue of rural migrants' citizenship and right to the city (Chung, 2012).

Recently, planners and architects in China have contested the dominance of the redevelopment-by-demolition model by proposing alternative approaches that champion social equity and spatial inclusiveness. In Shenzhen, for example, urban planners have begun to take an active part in redevelopment projects by working with local communities to come up with creative designs that are socially focused (Kochan, 2015). Community-led development has been seen as a more democratic way to achieve planning and regeneration through local action (Gallent, 2013; Gallent et al., 2015).

How to harness community and societal resources in addressing planning challenges will be crucial for the equitable development of urbanising rural communities.

Development and change in the rural hinterlands

While rural communities in peri-urban areas have been subject to processes of encroachment and enclosure, those in the more remote countryside have predominantly experienced neglect and abandonment. As development in reform-era China focuses resources on urban growth in coastal areas, significant disparities between wealthy coastal cities and poor inland areas have persisted. In the countryside, de-collectivisation has led to a vacuum in political authority and economic decline (Unger, 2002). At the beginning, the contracting of farmland to individual households under the newly introduced household responsibility system contributed to a boost in rural productivity (Brandt et al., 2002). Over time, however, household-based farming became difficult to sustain due to prohibitive transaction costs and farmers' lack of capital, technology and access to information (Zhang and Donaldson, 2013). Farmers' economic woes were aggravated by excessive fee collections by predatory states, which have been a key source of rural grievances (Bernstein and Lu, 2003). Widespread poverty and the lack of opportunities in the countryside led to the mass exodus of the able-bodied rural population. As rules governing rural-to-urban migration were relaxed in the 1980s, hundreds of millions of rural residents have departed from their native villages in search of wage employment in towns and cities (Chan, 2012). This has resulted in the phenomenon of left-behind children and elderly, as well as the proliferation of 'hollow villages' (*kongxin cun*) – rural communities that have been emptied of their young working population (Jacka, 2012; Ye et al., 2013; Murphy, 2014).

Beginning in the mid–2000s, the Chinese government began to implement rural reform aimed at alleviating poverty and decline in the countryside. In 2005, the campaign to build a 'new socialist countryside' (NSC) was announced, a nationwide initiative focused on bolstering rural development (Ahlers and Schubert, 2009). A central feature of the NSC campaign was to achieve village modernisation through improving public infrastructure, sanitation and housing in rural areas. The campaign has been described as 'the extension of state-sponsored urban planning regimes into rural jurisdictions', where 'urban planning came to be seen as an appropriate tool for solving a range of intractable rural problems' (Bray, 2013: 53). Previously, systematic spatial planning was applied mainly in urban areas at provincial, municipal and district levels, but the NSC campaign extended such practices to towns and villages with the formal inclusion of 'village plans' into the new 2008 Urban and Rural Planning Law (Bray, 2013). The incorporation of rural areas into the same spatial planning regime as cities signified the state's modernist agenda in its attempt to reverse rural 'backwardness' (Looney, 2015).

The NSC campaign consists of two main aspects. The first involves basic infrastructural construction and improvement works on village sanitation. In Jiangsu province, the implementation of the village improvement programme brought integrated systems of water supply, waste disposal and sewage treatment to rural settlements, as well as the extension of urban transportation services to villages (Wu and Zhou, 2013). In Ganzhou city of Jiangxi province, the NSC helped supply 21,000 villages with clean water, electricity, flush toilets and garbage collection services (Looney, 2015).

The second and more controversial aspect is spatial restructuring through the concentration of rural settlements. This ranged from encouraging villagers to renovate and rebuild their old homes as new multi-storey apartments, to resettling entire rural communities in 'new villages' or 'new rural communities'. Some of the resettlements take place under the rubric of 'poverty alleviation resettlement', where poor households are relocated from marginal rural areas either to resettlement sites built in the outskirts of cities and towns, or to more accessible sites within or nearby the households' home villages (Lo et al., 2016; Xue et al., 2013). Other resettlements are performed in response to the national land consolidation initiative, where land is reclaimed to mitigate the prevalent phenomenon of farmland abandonment and rural hollowing (Li et al., 2014b). Some of the reclaimed land is returned to agricultural use, while others is developed into rural landscapes for tourism and recreation purposes (Wilczak, 2017).

The construction of new rural housing and resettlement of villagers have attracted controversy for their revenue-driven agenda and impact on rural livelihoods. As highlighted in the previous section, the strategy of elevating villagers into high-rises has been used to create more space for the accumulation of rent in peri-urban areas, and the move to concentrate rural settlements has been interpreted in a similar light. Due to rapid land loss, the Chinese state has tightened regulations on the conservation of arable land. In some cities, local governments can obtain a land quota for urban construction by reclaiming land released by village households in rural areas (Yep and Forrest, 2016). Through residential concentration, therefore, the stock of rural land released in the countryside effectively translates into more land for high-value development in urban areas (Wilczak, 2017). The construction of new rural housing is also seen as furthering the state's goal of stimulating domestic consumption through investment in infrastructure and housing (Looney, 2015). These are clear instances of the direct impact of urban governments' revenue imperative on the planning of rural places.

Peasant households resettled in near-urban areas face similar adjustment challenges as those who have lost their land to expropriation and are displaced to urban fringes by redevelopment projects. They must adapt to higher living costs, find non-farm employment, and rebuild community ties in new ways. Notably, however, some studies have found considerable support among rural households for resettlement. Villagers appreciated the improvement in sanitary conditions and saw themselves as achieving a more 'urbanised' style of living (Yep and Forrest, 2016).

The planning process and manner by which resettlement projects are carried out may be crucial in determining satisfaction. While there is general consensus on how improvements in infrastructure and sanitation have brought betterment in rural living conditions, researchers have made different observations on implementation processes and outcomes. In some projects, rural households were actively included in the planning process and their property rights were respected. The modernisation programme was described as 'the farmers' regeneration of their homes rather than an imposed activity or regulation from the government' (Wu and Zhou, 2013: 168). In other cases, however, the displacement and relocation of the rural poor showed 'strong bureaucratic mobilisation on the one hand and villagers' limited participation in the policy process on the other' (Looney, 2015: 926). The divergence in outcomes reflects significant variation in regional experiences and state–society dynamics, and calls for further research into institutional processes that would be more conducive to bottom-up empowerment.

Conclusions

The past four decades have witnessed the emergence of an increasingly differentiated countryside in post-socialist China. This chapter has examined the changes and challenges confronting rural places in near-urban areas and the more remote countryside. While the respective dynamics differ, both areas saw the clear impact of rapid urbanisation on political economies and land politics. Immense growth pressure has affected not only rural communities situated close to metropolitan centres, but also further reaches of the countryside through processes such as migration and arable land conservation. The increasing incorporation of rural places into urban regimes of development has also led to the extension of urban planning approaches and practices into rural areas. As highlighted in this chapter, there have been more experimentation and entrepreneurship in the redevelopment and modernisation of rural places in which planners take an active part in the design and development of new rural communities. More broadly, the growth and internationalisation of the planning profession in reform-era China and the decentralisation of authorities to local governments will continue to bring more opportunities for planners to directly participate in decision-making processes concerning rural spatial development.

Going forward, rural planning will have a crucial role to play in shaping discourses of rurality and fostering community development and participation. In urban village redevelopment, the involvement of design studios and architectural firms has contributed to more participatory design processes and facilitated the preservation of historic and cultural heritage. In remote rural areas, similar initiatives have emerged such as in Sichuan province in the aftermath of the 2013 earthquake. The 7.0-magnitude earthquake in Lushan County affected some two million residents and led to the destruction of 70,000 homes. Architects and designers were brought in to engage villagers in the post-disaster reconstruction of their homes, with a view to strengthening local community empowerment, sustainable development and the revitalisation of traditional cultures (Ku and Dominelli, 2017). Following decades of pro-growth development that prioritises economic efficiency and urban-centred regimes of accumulation that emphasise land rents, sustainability concerns and alternative visions of development have begun to challenge the hegemony of market-led growth. The re-valuation of rurality, community and local cultures will have profound implications not only for rural development but also for rural–urban relations in China.

References

Ahlers, A. L. and Schubert, G. (2009) Building A New Socialist Countryside–Only A Political Slogan? *Journal of Current Chinese Affairs* 38(4), 35–62.

Bernstein, T. P. and Lü, X. (2003) *Taxation Without Representation in Contemporary Rural China*, Cambridge: Cambridge University Press.

Brandt, L., Huang, J., Li, G. and Rozelle, S. (2002) Land Rights in Rural China: Facts, Fictions and Issues, *The China Journal* 47, 67–97.

Bray, D. (2013) Urban Planning Goes Rural: Conceptualising the New Village, *China Perspectives* 3, 53–62.

Broudehoux, A. M. (2004) *The Making and Selling of Post-Mao Beijing*, New York: Routledge.

Cai, Y. (2010) *Collective Resistance in China: Why Popular Protests Succeed or Fail*, Stanford, CA: Stanford University Press.

Cartier, C. (2001) 'Zone Fever', The Arable Land Debate, and Real Estate Speculation: China's Evolving Land Use Regime and Its Geographical Contradictions, *Journal of Contemporary China* 10(28), 445–469.

Chan, K. W. (2012) Migration and Development in China: Trends, Geography and Current Issues, *Migration and Development* 1(2), 187–205.

Chung, H. (2012) The Spatial Dimension of Negotiated Power Relations and Social Justice in The Redevelopment of Villages-In-The-City in China, *Environment and Planning A* 45, 2459–2476.

Chung, H. and Unger, J. (2013) The Guangdong Model of Urbanization: Collective Village Land and The Making of a New Middle Class, *China Perspectives* 3, 33–41.

Chung, J. (2001) Reappraising Central-Local Relations in Deng's China: Decentralization, Dilemmas of Control, and Diluted Effects of Reform, in C. Chao and B. J. Dickson (eds), *Remaking the Chinese State: Strategies, Society and Security*, New York: Routledge, 46–75.

Duckett, J. (1998) *The Entrepreneurial State in China: Real Estate and Commerce Departments in Reform Era Tianjin*, London: Routledge.

Edin, M. (1998) Why Do Chinese Local Cadres Promote Growth? Institutional Incentives and Constraints of Local Cadres, *Forum for Development Studies* 25(1), 97–127.

Gallent, N. (2013) Re-Connecting 'People and Planning': Parish Plans and The English Localism Agenda, *Town Planning Review* 84(3), 371–396.

Gallent, N., Hamiduddin, I., Juntti, M., Kidd, S. and Shaw, D. (2015) *Introduction to Rural Planning: Economies, Communities and Landscapes*, New York: Routledge.

Gaubatz, P. (2005) Globalization and The Development of New Central Business Districts in Beijing, Shanghai and Guangzhou, in L. J. C. Ma and F. Wu (eds), *Restructuring the Chinese City: Changing Society, Economy and Space*, London: Routledge, 98–121.

Guo, X. (2001) Land Expropriation and Rural Conflicts in China, *The China Quarterly* 166, 422–439.

Hall, T. and Hubbard, P. (1998) (eds) *The Entrepreneurial City: Geographies of Politics, Regime, and Representation*, Chichester: Wiley.

Hao, P., Sliuzas, R. and Geertman, S. (2011) The Development and Redevelopment of Urban Villages in Shenzhen, *Habitat International* 35(2), 214–224.

Harvey, D. (1989) From Managerialism to Entrepreneurialism: The Transformation in Urban Governance in Late Capitalism, in D. Harvey (ed.), *Spaces of Capital: Towards a Critical Geography*, New York: Routledge, 345–368.

He, C. and Zhu, Y. (2010) Real Estate FDI in Chinese Cities: Local Market Conditions and Regional Institutions, *Eurasian Geography and Economics* 51(3), 360–384.

Ho, P. (2001) Who Owns China's Land? Policies, Property Rights and Deliberate Institutional Ambiguity, *The China Quarterly* 166, 394–421.

Hsing, Y. (2010) *The Great Urban Transformation: Politics of Land and Property in China*, Oxford: Oxford University Press.

Jacka, T. (2012) Migration, Householding and The Well-Being of Left-Behind Women in Rural Ningxia, *The China Journal* 67, 1–22.

Kan, K. (2016) The Transformation of The Village Collective in Urbanizing China: A Historical Institutional Analysis, *Journal of Rural Studies* 47(B), 588–600.

Kan, K. (2017) The (Geo)politics of Land and Foreign Real Estate Investment in China: The Case of Hong Kong FDI, *International Journal of Housing Policy* 17(1), 35–55.

Kochan, D. (2015) Placing the Urban Village: A Spatial Perspective on the Development Process of Urban Villages in Contemporary China, *International Journal of Urban and Regional Research* 39(5), 927–947.

Ku, H. B. and Dominelli, L. (2017) Not Only Eating Together: Space and Green Social Work Intervention in A Hazard-Affected Area in Ya'an, Sichuan of China, *The British Journal of Social Work* 48(5), 1409–1431.

Lee, J. and Zhu, Y. P. (2006) Urban Governance, Neoliberalism and Housing Reform in China, *The Pacific Review* 19(1), 39–61.

Li, L. H., Lin, J., Li, X. and Wu, F. (2014a) Redevelopment of Urban Village in China–A Step Towards an Effective Urban Policy? A Case Study of Liede Village in Guangzhou, *Habitat International* 43, 299–308.

Li, Y., Liu, Y., Long, H. and Cui, W. (2014b) Community-Based Rural Residential Land Consolidation and Allocation Can Help to Revitalize Hollowed Villages in Traditional Agricultural Areas of China: Evidence from Dancheng County, Henan Province, *Land Use Policy* 39, 188–198.

Lin, G. C. and Ho, S. P. (2003) China's Land Resources and Land-Use Change: Insights from the 1996 Land Survey, *Land Use Policy* 20(2), 87–107.

Lin, G. C. and Ho, S. P. (2005) The State, Land System, and Land Development Processes in Contemporary China, *Annals of the Association of American Geographers* 95(2), 411–436.

Liu, Y., He, S., Wu, F. and Webster, C. (2010) Urban Villages Under China's Rapid Urbanization: Unregulated Assets and Transitional Neighbourhoods, *Habitat International* 34(2), 135–144.

Lo, K., Xue, L. and Wang, M. (2016) Spatial Restructuring Through Poverty Alleviation Resettlement in Rural China, *Journal of Rural Studies* 47, 496–505.

Looney, K. E. (2015) China's Campaign to Build a New Socialist Countryside: Village Modernization, Peasant Councils, and the Ganzhou Model of Rural Development, *The China Quarterly* 224, 909–932.

Ma, L. J. (2005) Urban Administrative Restructuring, Changing Scale Relations and Local Economic Development in China, *Political Geography* 24(4), 477–497.

Murphy, R. (2014) Study and School in The Lives of Children in Migrant Families: A View From Rural Jiangxi, China, *Development and Change* 45(1), 29–51.

O'Brien, K. and Li, L. (2006) *Rightful Resistance in Rural China*, New York: Cambridge University Press.

Oi, J. C. (1992) Fiscal Reform and The Economic Foundations of Local State Corporatism in China, *World Politics* 45(1), 99–126.

Ong, A. (2011) Worlding Cities, or the Art of Being Global, in A. Roy and A. Ong (eds), *Worlding Cities: Asian Experiments and the Art of Being Global*, Oxford: Blackwell, 1–26.

Po, L. (2008) Redefining Rural Collectives in China: Land Conversion and The Emergence of Rural Shareholding Co-Operatives, *Urban Studies* 45(8), 1603–1623.

Sargeson, S. (2013) Violence as Development: Land Expropriation and China's Urbanization, *Journal of Peasant Studies* 40(6), 1063–1085.

Sit, V. F. and Yang, C. (1997) Foreign-Investment-Induced Exo-Urbanisation in the Pearl River Delta, China, *Urban Studies* 34(4), 647–677.

Smart, A. and Lin, G. (2007) Local Capitalisms, Local Citizenship and Translocality: Rescaling From Below in the Pearl River Delta Region, China, *International Journal of Urban and Regional Research* 31(2), 280–302.

So, A. Y. (2007) Peasant Conflict and the Local Predatory State in the Chinese Countryside, *Journal of Peasant Studies* 34(3–4), 560–581.

Song, Y., Zenou, Y. and Ding, C. (2008) Let's Not Throw the Baby Out with the Bath Water: The Role of Urban Villages in Housing Rural Migrants in China, *Urban Studies* 45(2), 313–330.

Unger, J. (2002) *The Transformation of Rural China*, Armonk, NY: M. E. Sharpe.

Wang, Y. P., Wang, Y. and Wu, J. (2009) Urbanization and Informal Development in China: Urban Villages in Shenzhen, *International Journal of Urban and Regional Research* 33(4), 957–973.

Webster, C., Wu, F., Zhang, F. and Sarkar, C. (2016) Informality, Property Rights, and Poverty in China's Favelas, *World Development* 78, 461–476.

Wilczak, J. (2017) Making the Countryside More Like the Countryside? Rural Planning and Metropolitan Visions in Post-Quake Chengdu, *Geoforum* 78, 110–118.

Wu, F. (2003) The (Post-)Socialist Entrepreneurial City as A State Project: Shanghai's Reglobalisation In Question, *Urban Studies* 40(9), 1673–1698.

Wu, F. and Zhou, L. (2013) Beautiful China: The Experience of Jiangsu's Rural Village Improvement Program, in J. Colman and C. Gossop (eds), *Frontiers of Planning: Visionary Futures for Human Settlements*, Beijing: ISOCARP, 156–169.

Wu, F., Zhang, F. and Webster, C. (2013) Informality and the Development and Demolition of Urban Villages in The Chinese Peri-Urban Area, *Urban Studies* 50(10), 1919–1934.

Xie, Y. and Costa, F. J. (1993) Urban Planning in Socialist China: Theory and Practice, *Cities* 10(2), 103–114.

Xu, J., Yeh, A. and Wu, F. (2009) Land Commodification: New Land Development and Politics in China since the Late 1990s, *International Journal of Urban and Regional Research* 33(4), 890–913.

Xue, D. and Wu, F. (2015) Failing Entrepreneurial Governance: From Economic Crisis to Fiscal Crisis in the City of Dongguan, China, *Cities* 43, 10–17.

Xue, L., Wang, M. Y., and Xue, T. (2013) 'Voluntary' Poverty Alleviation Resettlement in China, *Development and Change* 44(5), 1159–1180.

Ye, J., Wang, C., Wu, H., He, C. and Liu, J. (2013) Internal Migration and Left-Behind Populations in China, *Journal of Peasant Studies* 40(6), 1119–1146.

Yeh, A. G. O. and Wu, F. (1996) The New Land Development Process and Urban Development in Chinese Cities, *International Journal of Urban and Regional Research* 20(2), 330–353.

Yeh, A. G. O. and Wu, F. (1999) The Transformation of the Urban Planning System in China from a Centrally-Planned to Transitional Economy, *Progress in Planning* 51(3),167–252.

Yep, R. (2013) Containing Land Grabs: A Misguided Response to Rural Conflicts Over Land, *Journal of Contemporary China* 22(80), 273–291.

Yep, R. and Forrest, R. (2016) Elevating the Peasants into High-Rise Apartments: The Land Bill System in Chongqing As A Solution For Land Conflicts in China? *Journal of Rural Studies* 47, 474–484.

Yew, C. P. (2012) Pseudo-Urbanization? Competitive Government Behavior and Urban Sprawl in China, *Journal of Contemporary China* 21(74), 281–298.

Zhang, L. (2011) The Political Economy of Informal Settlements in Post-Socialist China: The Case of Chengzhongcun(s), *Geoforum* 42, 473–483.

Zhang, L., Zhao, S. X. B. and Tian, J. P. (2003) Self-Help in Housing and Chengzhongcun in China's Urbanization, *International Journal of Urban and Regional Research* 27(4), 912–937.

Zhang, Q. F. and Donaldson, J. A. (2013) China's Agrarian Reform and the Privatization of Land: A Contrarian View, *Journal of Contemporary China* 22(80), 255–272.

Zhang, T. (2002) Urban Development and A Socialist Pro-Growth Coalition in Shanghai, *Urban Affairs Review* 37(4), 475–499.

Zhu, J. (1999) Local Growth Coalition: The Context and Implications of China's Gradualist Urban Land Reforms, *International Journal of Urban and Regional Research* 23(3), 534–548.

37

Planning strategically in light of rural decline

Experiences from Denmark

Anne Tietjen and Gertrud Jørgensen

Introduction: rural planning under new conditions

Rural decline is considered a major policy and planning issue in Denmark as population decline in peripheral rural areas jeopardises local economic development and the provision of services. While such peripheral areas suffer from a declining and ageing population (especially in the smallest villages and towns), lower levels of educational attainment, a high proportion of citizens living off public welfare, and a struggling property market, they also benefit from low unemployment rates and a high proportion of export-oriented industries (Regeringen Erhvervsministeriet, 2017). At the same time, sections of the population in such rural areas live an increasingly urbanised life (KL's Analyseenhed, 2015): they commute to work over long distances, ask for well-functioning service infrastructures and cultural events, and place great value on attractive built environments and accessible landscapes for recreation and outdoor activities.

Instruments such as tax redistribution and other subsidies play a considerable role in strengthening development in peripheral rural areas (Vejrup et al., 2017). However, new strategies for managing rural decline, and its social and spatial consequences, are now being developed at municipal and local level which adhere to the 'new rural paradigm' (OECD, 2006) and place greater emphasis on place-based development.

Following local government reform in 2007, rural municipalities in Denmark are now larger than they once were, with a minimum of around 30,000 inhabitants who often reside in extensive territories. Eight out of a total of 98 rural municipalities – including Thisted – cover more than 1,000 km². Municipalities have been the recipients of new planning responsibilities, covering not only urban areas, but also countryside planning. They have broad political powers, comprehensive responsibilities and corresponding administrative competencies, enabling them to instigate *strategic spatial development initiatives* for the municipal territory at large.

Such initiatives are generally associated with broader development policies; all Danish municipalities must prepare a political strategy for the municipal planning every fourth year (within the first two years of a new electoral term). There are no rules for thematic orientation of the strategy, but many municipalities now include rural development (which was explicitly identified as a new priority in the 2007 reform). In addition, many municipalities develop specific rural policies, often in collaboration with local action groups (LAG) and other actors, and in accordance with European rural development programmes (Salling Kromann and Just, 2009).

External grant funding plays an increasingly important role in rural planning in Denmark. European LEADER+ funding is one example, but applications are also made to national funding bodies and co-funding is sometimes sought from charitable foundations. Such investments featured in the spatial development projects examined in this chapter (see also Tietjen and Jørgensen, 2016, 2018).

Two cases – one planning innovation initiative

Two of the most peripheral municipalities in Denmark are Thisted, a predominantly rural area adjacent to the North Sea, and the island of Bornholm in the Baltic Sea. Both have strong traditions in agriculture and fishing, growing tourism sectors, and ambitions to develop new third sector businesses. They share the problem of population decline resulting in challenges around service provision and difficulties in attracting a highly skilled workforce. On the other hand, both Thisted and Bornholm have unique place-based qualities and an abundance of active local communities and citizen organisations. These place-based potentials became the basis for the planning innovation Land of Opportunities (2007–2012), initiated by the charitable trust Realdania with the goal of developing new place-based, collaborative and project-oriented approaches to rural planning in partnership with peripheral municipalities (Sloth Hansen et al., 2012). In terms of strategic outcomes, which are examined below, Thisted and Bornholm can be held up as relatively successful case studies. Hence they provide a focus for the rest of this chapter.

At the start of the initiative, both municipalities were relatively new and had existed in their current forms for only a short period of time. Thisted was the product of a merger of three small municipalities, following local government reorganisation in 2007. Bornholm had been

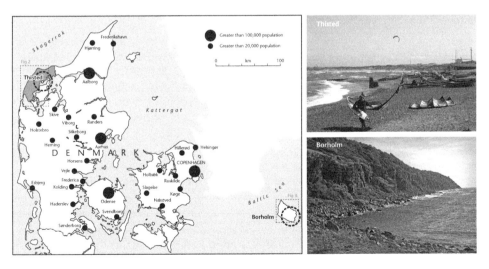

Figure 37.1 Locations of Thisted and Bornholm. Situated in the periphery of Denmark, these municipalities have significant landscape amenity characteristics, with unique wave conditions suitable for surfing in Thisted, and (unique in Denmark) granite rock formations in Bornholm.

created a few years earlier, in 2003, when five smaller municipalities were merged following local referenda that supported the proposal to create a larger, more integrative, authority. Neither of these new authorities was experienced in strategic planning and nor were they familiar with the particular challenges of countryside planning, which was a new municipal responsibility at that time.

In Thisted, Land of Opportunities resulted in six strategic projects, each including several physical interventions, with a total investment of 89 million Danish Crowns (€12 million). In Bornholm, seven projects were implemented with a total investment of 77 million Danish Crowns (€10 million). All projects were co-financed by the municipalities, the Realdania trust, and through other external funding (Sloth Hansen et al., 2012). Throughout the initiative, the Trust has monitored and actively advised the municipalities on planning decisions. These projects focused on developing the physical framework for daily life and tourism in order to improve the local quality of life and ultimately to attract new citizens and more tourists.

We began by studying the overall innovation initiative and the visions and planning processes in each of the cases, while over time we re-directed our focus to five of the projects, one in Thisted and four on Bornholm, that appeared to have particular strategic foci and seemed to align with core definitions of 'strategic spatial projects' set out in literature (e.g. see Kühn, 2010; Oosterlynck et al., 2011). These five projects were (1) the development of new infrastructure in three villages to support the ambition of turning coastal Thisted into a premier surfing destination dubbed 'Cold Hawaii'; and on Bornholm, (2) the development of recreational facilities in a former fishing harbour at Hasle; (3) a new beach promenade at Sandvig, (4) the restoration, re-use and celebration of industrial heritage at Hammer Harbour, incorporating new spaces for community use, and (5) the reclamation of a former granite quarry at Vang for outdoor recreation.

Between 2013 and 2017 we carried out documentary analysis, interviews with key actors from the municipal administration (alongside interviews with other local actors) and recurrent site visits. The relatively long investigation period afforded us the opportunity to trace the evolution of the projects, their implementation, and the immediate and derivative outcomes over a period of 10 years.

Actor–network theory (ANT) provided the framework to study both the strategic planning processes and their long-term strategic outcome. ANT proposes that people and things can be made to act together in socio-material actor-networks for common purposes (in this case strategic planning objectives) (Latour, 2005). This happens through *translation* processes in which human and non-human actors by interaction negotiate a shared problem-understanding, gather relevant actors to address this problem and, finally, enrol the actors necessary to work for a problem solution (for example a strategic spatial project) in a hybrid actor-network with clearly defined actor-roles (Tietjen and Jørgensen, 2016, 2018). Because ANT focuses on the capacity to induce change, things, such as particular wave conditions in Thisted or granite rock formations on Bornholm, are perceived to have agency too: 'The structure of the material world pushes back on people' (Yaneva, 2009: 277). At the same time people and things never act alone; action is always distributed in hybrid actor-networks and any change is therefore a 'network effect'. By following human and non-human actors through the planning processes and for some years after project implementation in Thisted and Bornholm, we studied (1) how planners acted to negotiate strategic planning objectives, formulate strategic spatial projects and enrol the actors necessary to implement these projects and (2) what were the strategic outcomes of the projects in terms of the network effects of the actor-networks of people and things assembled by the projects.

Two different processes

While the general context and project set-up of Land of Opportunities were common to the two cases, the planning processes proved to be very different, as were the relationships between the overall visions and the particular strategic spatial projects undertaken.

Thisted

Thisted Municipality was involved in the planning initiative from an early stage. The municipality was engaged in a pre-project dialogue with the charitable trust and had started mapping local development potential in Thisted's 50 villages in 2006. Land of Opportunities set out with two broad aims; to deliver population and tourism sector growth but through a relatively narrow focus on physical projects – because the purpose of the funding trust was to improve quality of life via the built environment. Thus, projects needed to develop the physical framework for daily life and tourism, focusing on the quality of buildings, public open spaces and cultural heritage. On a strategic level, all projects had to focus on the future role of villages in the municipality, for example through providing the infrastructure for the growth of community networks.

In light of these goals, the municipality invited local communities to propose project ideas based on what they perceived as place-specific qualities and potentials. More than 600 citizens participated in eight 'ideas-development' workshops, held in collaboration with a number of local rural development actors. Throughout the 'ideas-development' phase, the municipal planners remained in dialogue with citizens and civic associations. More than 100 project ideas were eventually submitted by 45 associations. These ideas were reviewed, sifted, and prioritised. Those to be taken forward were grouped together and further developed into six 'strategic projects' by the municipal planners, in collaboration with local experts, the charitable trust and key stakeholders from the local communities (Tietjen and Jørgensen, 2016).

Nineteen of the original 100 project ideas originated from the coastal villages of Klitmøller, Nr. Vorupør and Krik. These ideas were developed into one strategic spatial project, 'The Good Life at the Seaside', with interventions in all of the three villages. Members of the Klitmøller-based surf club, NASA, had proposed a 'Coastal Experience Centre' in Klitmøller and, together with local associations in Nr. Vorupør, beach promenades in both Klitmøller and Nr. Vorupør to connect the fishermen's landing places with other local activities. This was badged as the 'Association Path'. These projects aimed to support the existing surfing community and make better use of the unique wave conditions along the coast, with the broad goal of promoting the area as *the* centre for surfing in northern Europe, branded 'Cold Hawaii'. The municipal planners readily embraced this strategic vision and the idea for a community centre in Krik village was quickly brought into line with this vision and adapted to accommodate the needs of surfers and other outdoor activities' tourists. Together with the local project proposers and other stakeholders, the municipal planners developed an integrative project plan that precipitated a master-planning competition for the landing places and piers in the three villages which sought worked-up design proposals for the prioritised project ideas. The competition brief also asked for draft sketches for second-priority projects: among these was a sea bath in Klitmøller (later implemented in Nr. Vorupør). The winning plan/design was further developed and finalised with the input of future users and stakeholders and, in August 2011, 'The Good Life at the Seaside' was officially opened in Klitmøller. The initiative has informed several municipal policies and plans, most notably, 'place-based potential as a starting point for development', is the general theme for the municipal spatial development plan for 2010–2022 and for the 2011

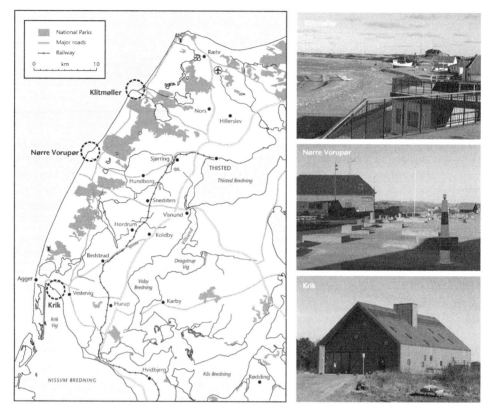

Figure 37.2 Thisted: 'The Good Life at the Seaside'. The map shows the location of the projects at the North Sea coast. The photographs show impressions from the projects: from top to bottom, a view from the public roof of the new surfers' club in Klitmøller, the Association Path in Vorupør, and the new community centre in Krik.

business and tourism policies, both of which make explicit reference to Land of Opportunities (Tietjen and Jørgensen, 2016).

Bornholm

In Bornholm, the process was very different. The municipality had not been involved in the pre-project dialogue but was asked by the Realdania trust, quite late in the initiation process, if it wished to participate in Land of Opportunities – which of course it did. The funding available for projects from the Trust was significant (30 million Danish Crowns, €4 million) with the municipality required to match fund that budget. But neither the politicians nor the planners were prepared for the challenge and did not, in the initial phase, come up with particularly convincing project ideas. The Trust required project development to focus on the shrinking villages in the interior of the island, which were thought to have the most pressing problems. But after two years, the planning process had resulted in only three small projects (and drawn on very little of the overall budget) each with very local foci and limited strategic ambition. Therefore, it was decided to develop additional new projects, based on existing ideas from local communities, existing development initiatives and, not least, existing possibilities for co-financing. All of these projects came to be located at the coast where several existing municipal initiatives were already up and running. With a total of 69.7 million Danish Crowns (€9.4 million), the four second-generation projects made up the major part of the investment and they all had strong strategic ambitions, hence this chapter's focus on these projects (see also Tietjen and Jørgensen, 2018).

All four projects focused on the transformation of post-industrial landscapes. They were developed in the space of just three years between 2010 and 2012, and they partly built upon ongoing projects, where project organisation, user involvement, stakeholder networks and trust were already established, and opportunities for co-financing were already in place.

Two of the projects were developed within the framework of on-going area renewal. Funding from the Trust allowed for additional and architecturally ambitious interventions; a harbour bath and a new beach island in the former fishing harbour in Hasle and the establishment of a new beach promenade in Sandvig.

In a third project, the redevelopment of the former granite shipping harbour, Hammer Harbour, was expanded to include the presentation of the industrial heritage of granite quarrying on North Bornholm which, during this project, was analysed and mapped by the local museum as offering place-based development potential. Following an architecture competition, the port facility was restored to reveal its original granite and timber constructions and a new jetty was constructed to create an up-to date recreational marina. A new multi-functional building, a café, and toilet facilities were established in three small timber buildings on a wooden deck on the large open space next to the harbour. With their characteristic architecture these

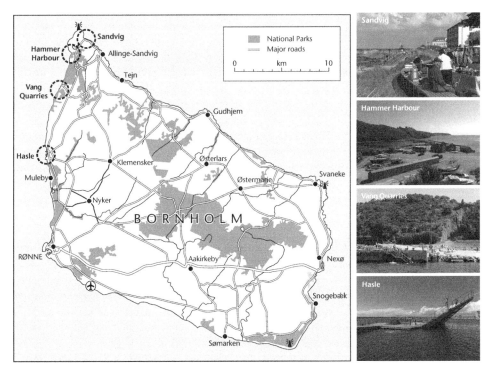

Figure 37.3 Bornholm: the map shows the location of the projects location at the Baltic Sea coast, and the photographs show impressions from the projects: from top to bottom, local market at the Sandvig beach promenade, view over Hammer harbour, the renovated Vang Pier and the new Hasle harbour bath.

buildings form a new landmark and provide a public space shared by locals and tourists for multiple uses.

In a fourth project, the municipality – stimulated by a proposal from the Bornholm Outdoor Council – took over the ownership of the recently abandoned Vang granite quarries and reclaimed this vast and dramatic rocky area as a multifunctional open space for nature and outdoor recreation (Tietjen and Jørgensen, 2018).

All four projects and their underlying strategic visions were created in collaboration with local residents, users and stakeholders. However, there was little strategic coordination across the individual projects despite geographic proximity, which was possibly due to the short implementation time frame. Three of the four projects, in Sandvig, Hammer Harbour and Vang Quarries, worked with the industrial heritage of granite quarrying in North Bornholm as an asset for future place-based development. While the conservation and reuse of heritage was developed as a rationale for each project individually it was only framed as a municipal strategic focus after implementation. Furthermore, the municipality conducted a collaborative initiative for a large-scale strategic plan, Future Landscape North Bornholm, to follow up on these and other place-based development projects in the area (for further details, see http://fremtidens landskaber.ku.dk). There was an element of post-hoc rationalisation of these individual projects, which sought to tie them together and lend them greater coherency.

The effects: creating, strengthening and developing actor-networks

The Land of Opportunities projects in Thisted and Bornholm were all completed about five years ago. The projects helped transform a larger coastal area (in the case of Thisted) and a beach promenade, two harbours and an abandoned quarry (in the case of Bornholm) for new uses, users and activities through focused interventions with a characteristic architectural design. These have had considerable impacts on the aesthetics and the performance of the particular places and also improved residents' and visitors' perceptions of these places; especially 'The Good Life at the Seaside', which has had a considerable branding effect on Thisted, as the 'Cold Hawaii' surfing destination.

Both in Bornholm and in Thisted, shared facilities and activities for tourists and local users have increased 'public life' in these sparsely populated areas. In Bornholm, new regional communities of interest have developed around new recreational activities. The run-down town of Hasle, formerly the least attractive coastal town in Bornholm, has become a popular place for people all over the island to come for a swim in the harbour bath. Small scale, locally based tourism businesses have developed in connection with all four projects and new events are being organised, for example by local sports associations or by native Bornholmers, who are now living and working in Copenhagen but come back to organise events on the coast (Tietjen and Jørgensen, 2018). Thisted, and especially the coastal village of Klitmøller, has experienced some population growth and a rise in house prices. Moreover, a new 'Cold Hawaii' college focused on surfing and business innovation has been co-developed by the surfer community in Klitmøller. This was further expanded in 2017 on the back of its success (see www.tvmidtvest. dk/artikel/hf-cold-hawaii-udvider-antallet-af-pladser).

A key observation from this research concerns the strength of new actor-networks created during the planning process and their capacity to catalyse further development after project implementation. In Thisted, the extremely well-orchestrated planning process spawned enduring active networks of people, places and things, particularly around and across the projects in Klitmøller and Nr. Vorupør. The case of Nr. Vorupør is particularly interesting. Following the project, local businesses, fishermen and civic associations came together to produce a strategy and master plan for the development of tourism in Vorupør, which drew heavily on the winning proposal from the competition for 'The Good Life at the Seaside'. To date, this has resulted in the renovation of the fishermen's buildings at the landing place, the renovation of the pier and the construction of a sea bath, which was originally proposed for Klitmøller and added late in the process to 'The Good Life at the Seaside' in Nr. Vorupør.

In Bornholm, both the residents in Sandvig and in Hasle are engaged in local follow-up projects to the completed area renewal and Land of Opportunities projects, while local participants from all four projects continue to engage in maintaining and developing activities in the project areas, such as mountain biking and events in the Vang quarries or flea markets and public gatherings in Sandvig. However, the actor-networks that were created based on the Land of Opportunities projects in Bornholm have a more local outlook, are less interconnected, and thus have less strategic impact than those developed in Thisted.

Land of Opportunities as strategic spatial planning

What broader lessons for strategic spatial planning in rural areas can be drawn from the case studies presented in this chapter? While strategic planning is not a fixed concept and can be seen as a 'weakly structured whole of different practises' (Sartorio, 2005: 26), some characteristics repeatedly appear in the literature: strategic plans are long term, and as such their implementation is likely (or bound) to be influenced by uncertainties which the strategic plan needs to deal with – not by reducing complexity, but rather by embracing it, exploring possible advantages of working with multiple actors in heterogeneous networks of people and things (Balducci et al., 2011). Planning, according to Hillier, is a kind of 'strategic navigation' (Hillier 2011:503) towards an uncertain future; It needs to be guided by a strategic vision while adapting to the winds and currents that it will inevitably encounter. Some authors (e.g. Kühn, 2010; Oosterlynck et al., 2011) stress that strategic spatial planning combines visions with key projects in recursive processes. In this context, strategic visions should be based on broad stakeholder participation and comprehensive analyses, while strategic projects steer development in the envisioned direction, addressing selected local areas (and needs) and generating immediate results for the actors involved.

The project in Thisted is almost a textbook example of this approach with its very broad initial vision, the comprehensive involvement of actors in a 'first ideas' creation phase, inclusion of new key project ideas in the re-formulation of the vision during the process (Cold Hawaii), and the limited number of key projects that embodies the vision and points towards new key projects and strategic re-visioning. The process in Bornholm was quite different: even though the second phase started with the selection of four key projects (chosen by opportunity rather than selected strategically), the projects did end up with strong strategic effects, supporting a common narrative (the Granite Adventure) and inspiring the municipality to new ways of co-creating strategic visions. This suggests that there may be more than one way to obtain the interaction between long-term visions and key projects, though the Thisted experience was one of intent rather than accident.

Bafarasat (2014) identifies three 'schools' of strategic spatial planning, defined by their main focus: the 'Performance School', focused on efficient planning and implementation centred on corporatist and representative actors; the 'School of Innovative Action', focused on bottom-up driven spatial innovation, and the 'School of Transformative Strategy Formulation', in which the approach to planning is an innovation itself and a collaborative, all-inclusive, co-production exercise (ibid.). This typology may be useful for understanding the planning processes in Thisted and Bornholm and their outcomes. While Thisted was clearly rooted in the school of transformative strategy formulation, with its open-ended initial vision and comprehensive participation, Bornholm straddled the boundary between the school of innovative action (with much inclusion of local actors and respect for their inputs into the single projects) but elements of the performance school, as the projects were selected in a top-down manner, as those which were likely to become a success within the short period of time available. The learning-oriented planning approach, based on adaptive, collaborative and transdisciplinary methods (Healey, 1997, 2007; Albrechts, 2004), seemed to work well in the Thisted case, making it possible to collectively re-imagine possible futures for a particular place and translate them into concrete priorities and coherent actions for implementation (Oosterlynck et al., 2011). This may also have shaped the outcomes in terms of creation of actor networks that were able to catalyse further development: this was patently more successful in Thisted than in Bornholm.

The scholarly discussion of strategic spatial planning juxtaposes strategic planning approaches to the 'traditional' statutory planning, which governs future land use and gives residents and entrepreneurs greater economic certainty but lacks the will and methods to explore visions beyond the mundane tomorrow. The need for balancing certainty and uncertainty in planning is being (re)introduced into the discussion today (Albrechts, 2017; Hillier, 2017; Kempenaar and van den Brink, 2018).

In both municipalities, the strategic plans and projects were made within the framework of the statutory planning system, but they introduced what some scholars call 'soft space' (Haughton et al., 2013), understood as imprecise spatial delineations or non-statutory planning themes, which can guide action and serve to build local consensus. In both cases, soft spaces were in focus as the visions and projects went across planning themes, transgressed urban and rural boundaries, and included design qualities as basic instruments in the plans and projects.

Some scholars have explored the extent to which the proponents of strategic, collaborative and communicative planning are naïve, well-meaning contributors to a basically neo-liberal agenda (Sager, 2005; Olesen, 2013), and even Albrechts, being one of the main proponents, discusses this dilemma in a recent book (Albrechts, 2017). While our cases certainly worked strategically and within soft-spaces, the planning was hardly neo-liberal, because clear public interests steered the implemented projects. However, the strong influence of an essentially private funding body and the need for further external co-funding might be seen as part of a neo-liberal trend, where municipalities are increasingly dependent on private finance (be it private trusts or public earmarked funding) to achieve public goals. Our cases show that the methods and processes of strategic spatial planning may be a way to handle this new situation and even utilise it to achieve enhanced participation and design quality.

Although strategic planning emphasises procedural aspects of planning, the actual outcome in terms of the specific design and quality of space should not be forgotten: according to van den Broeck (2011: 89), 'space is not merely derived from social processes, but has its own relative autonomy and, by its logic, also contributes to ordering, shaping and identifying human activities'. Kempenaar and van den Brink (2018) likewise underscore that regional strategic design should not only take a collaborative starting point, but also use design principles in the planning process. It is very clear from all our project cases that place and architectural quality is essential for the wider outcome of the projects – not only in terms of aesthetics, but also in terms of how the physical projects spatially connect people and non-human things in new ways. Several examples illustrate this: the Hammerhavn redevelopment gathers and concentrates visitors and local users in a characterful multi-functional building cluster which works as a recognisable landmark from the sea and from the land; it also tells the story of historical uses and users while inviting new users and uses in. The Association path in Nr. Vorupør re-orders a previously chaotic landing place into a shared space for fishermen, surfers and other users. In the reclaimed granite quarry in Vang, a new lake separates a nesting peregrine falcon from recreational users, while it also collects rainwater, thus enabling humans and non-humans to share an increasingly multifunctional area.

The problems that 'shrinking' rural areas face today are manifold and complex. Strategic spatial planning alone cannot solve them, but can help relieve some symptoms and contribute to restoring or enhancing place quality. Combining open-ended, long-term visions with strategic spatial projects can initiate actor-networks that empower local residents and point towards new developments and a more positive outlook on the future. Planning processes that include many actors in long processes are of course challenging and difficult to manage, but they also generate long term effects – partnerships, collaborations and networks – that provide a foundation for future rural action and development.

References

Albrechts, L. (2004) 'Strategic (spatial) planning reexamined', *Environment and Planning B: Planning and Design* 31(5), 743–758.

Albrechts, L. (2017) 'Some ontological and epistemological challenges', in L. Albrechts, A. Balducci and J. Hillier (eds), *Situated Practices of Strategic Planning*, 389–404, London: Routledge.

Bafarasat, A. Z. (2014) 'Reflections on the three schools of thought on strategic spatial planning', *Journal of Planning Literature* 30(2), 132–148.

Balducci, A. et al. (2011) 'Introduction: strategic spatial planning in uncertainty: theory and exploratory practice', *Town Planning Review* 82(5), 481–501.

Haughton, G., Allmendinger, P. and Oosterlynck, S. (2013) 'Spaces of neoliberal experimentation: soft spaces, postpolitics, and neoliberal governmentality', *Environment and Planning A: Economy and Space* 45(1), 217–234.

Healey, P. (1997) *Collaborative Planning: Shaping Places in Fregmented Societies*, London: Macmillan.

Healey, P. (2007) *Urban Complexity and Spatial Strategies: Towards a Relational Planning for Our Times*, New York: Routledge.

Hillier, J. (2011) 'Strategic navigation across multiple planes: Towards a Deleuzean-inspired methodology for strategic spatial planning', *TPR*, 82(5).

Hillier, J. (2017) 'Strategic spatial planning in uncertainty or planning indeterminate futures? A critical review', in L. Albrechts, A. Balducci and J. Hillier (eds), *Situated Practices of Strategic Planning*, 298–316, London: Routledge.

Kempenaar, A. and van den Brink, A. (2018) 'Regional designing: a strategic design approach in landscape architecture', *Design Studies*, 54, 80–95.

KL's Analyseenhed (2015) *Det lokale Danmark (Local Denmark)*, Copenhagen: KL's Analyseenhed.

Kühn, M. (2010) 'Strategic planning - approaches for the regeneration of shrinking cities in Eastern Germany', paper presented at Session K05: Strategic Planning and Spatial Development, RSA Annual International Conference, Pécs, 24–26 May.

Latour, B. (2005) *Reassembling the Social: An Introduction to Actor–Network Theory*, London: Routledge.

OECD (2006) *The New Rural Paradigm Policies and Governance*, Paris: OECD.

Olesen, K. (2013) 'The neoliberalisation of strategic spatial planning', *Planning Theory* 13(3), 288–303.

Oosterlynck, S., Albrechts, L. and van den Broeck, J. (2011) 'Strategic spatial planning through strategic projects', in S. Oosterlynck et al. (eds), *Strategic Spatial Projects, Catalysts for Change*, 1–14, London: Routledge.

Regeringen Erhvervsministeriet (2017) *Regional og landdistriktspolitisk redegørelse 2017*, Regeringen Erhvervsministeriet.

Sager, T. (2005) 'Communicative planners as naïve mandarins of the neo-liberal state?', *European Journal of Spatial Development* December, 1–9.

Salling Kromann, D. and Just, F. (2009) *Kommunalreformen og landdistriktspolitikken*, Odense: Syddansk Universitet. Institut for Forskning og Udvikling i Landdistrikter.

Sartorio, F. S. (2005) 'Strategic spatial planing. a historical review of approaches, its recent revival, and an overview of the state of the art in Italy', *disP*, 162(3), 26–40.

Sloth Hansen, S., Møller Christensen, S. and Skou, K. (2012) *Mulighedernes land: Nye veje til udvikling i yderområder* [*Land of opportunities: New approaches to rural development*], Copenhagen: Realdania.

Tietjen, A. and Jørgensen, G. (2016) 'Translating a wicked problem: a strategic planning approach to rural shrinkage in Denmark', *Landscape and Urban Planning*, 154, 29–43.

Tietjen, A. and Jørgensen, G. (2018) 'There is more to it than meets the eye: strategic design in the context of rural decline', *Transactions of the Association of European Schools of Planning*, 2(1), 9–31.

Van den Broeck, J. (2011) 'Spatial design as a strategy for a qualitative social–spatial transformation', in S. Oosterlynck et al. (eds), *Strategic Spatial Projects, Catalysts for Change*, 87–96, London: Routledge.

Vejrup, K., Gjermansen, H., Moltrup-Nielsen, B., Drescher, M., Andersen A. K. and Pedersen, N. (2017) *Udligning og tilskud udgør mere end 30 pct. af indtægterne i flere kommuner*, Copenhagan: DST Analyse.

Yaneva, A. (2009) 'Making the social hold', *Design and Culture*, 1(3), 273–288.

Part VII

Landscape, amenity and the rural environment

Protecting landscapes has been a longstanding feature of planning approaches to countryside management for much of the twentieth century, particularly in more urbanised societies. Concerns with competing demands on emblematic and highly prized landscapes, urban sprawl, loss of farmland and 'countryside character', and promoting access to the countryside have provided the backdrop to some of the most enduring legacies of 'town and country planning'. In North America, for example, National Parks were first established and protected by law in the late nineteenth century – here the emphasis was on wilderness protection, often through the public ownership of land resources as a means of preserving nationally symbolic landscapes. National parks were also widely established and institutionalised in many European countries in the post war era with an emphasis on protecting 'special places' often on the grounds of scenic value or providing access to the countryside for the urban working and middle classes. Within an era of rapid urbanisation in the twentieth century, cherishing landscapes and rural places became politically important reflecting the rise of a popular conservation movement motivated by defending the countryside from inappropriate development. Protecting the countryside from urban growth, sprawl and suburbanisation emerged as key planning concerns, leading to urban containment and the prevention of urban coalescence through establishing green belts to protect the rural hinterlands of cities in many countries.

As Selman (2006) argues, landscape needs little justification as a subject of importance for land use planning, as planning systems have sought to protect areas of exceptional scenic beauty and safeguard locally important landscapes while managing or accommodating sensitive new development. In this context, landscapes are recognised as important to national, regional and local identity; as a repository of cultural heritage; associated with nature and connected to habitats and biodiversity; as a social space where communities live and work; and a commodified space, with landscapes exploited for their potential for local economic development – therefore, the role of planning is to mediate and balance these competing demands, particularly as new drivers of landscape change emerge.

This part of the Companion examines the origins of landscape protection as a central planning concern and explores how landscape policy is being reimagined to balance competing demands for rural space and address twenty-first-century challenges. The section begins by critically examining the evolution of national parks and green belts for countryside protection and also evaluates how these traditional tools are being reinvented to balance new development pressures with conservation imperatives. While these traditional instruments have often been characterised as top-down interventions providing 'islands of protection' (Owens and Cowell, 2011), the section also goes on to consider more integrated and participatory approaches towards managing both highly prized and so-called ordinary landscapes, including chapters dealing with cultural heritage landscapes through participatory methods, green infrastructure approaches, wellbeing perspectives and rewilding as a management strategy. Therefore, this section aims to critically examine the legacy of traditional approaches while reimagining landscape and conservation planning for twenty-first-century needs. This includes the need to reframe landscape debates away from narrow approaches focused on high amenity and commodified landscapes towards understanding the role of effective landscape management in climate change mitigation and adaptation, capturing ecosystem services, flood risk management, maximising biodiversity gain, restoring damaged or abandoned landscapes, and health and wellbeing, alongside more traditional planning concerns with place-making and a sense of place. In other words, planning for multifunctional and connected rural landscapes.

The first chapter in Part VII, by Bell and Stockdale, charts the evolution of National Parks as a central approach towards protecting highly prized landscapes. As the authors note, the creation of national parks have a long history, often predating the wider institutionalisation of planning systems, notably with the establishment of Yellowstone National Park in 1872 in the US representing one of the earliest examples of environmental law-making. These early examples of national parks in the US (with similar designations in Canada and Australia) focused on conservation and the protection of wilderness, often of emblematic importance to national identity, created through public ownership of land. The idea of national parks gained further traction in post-war Europe, as many western European countries legislated for landscape protection through establishing national parks aiming to conserve beauty, wildlife and cultural heritage, while promoting opportunities for increased public access to the countryside and promoting opportunities for enjoyment. While drawing on international experiences, Bell and Stockdale focus on the UK experience of national parks, in particular how national parks face new challenges in the twenty-first century and a constant rebalancing of economic and environmental goals. The chapter highlights important shifts in the governance of national parks, moving from direct control and protection towards broader sustainable development principles underpinned by new stakeholder based governance approaches based on co-management principles and negotiation through the increased role of non-state actors. Bell and Stockdale also provide a critical review of the increased marketisation of national parks, including a greater emphasis on natural capital approaches and the designation of new national parks on economic potential rather than environmental grounds, suggesting new imperatives for national parks in quantifying natural assets and framing their contribution to society primarily in economic terms. Bell and Stockdale argue that this rebalancing towards an economic rationality may further entrench conflict between economic and environmental interests, suggesting that this emphasis on economic goals and valuation may strengthen the case for accommodating new development within national parks.

Issues of heritage and landscape are also explored in Chapter 39, which examines the increasing adoption of cultural heritage landscapes as a countryside management concept. In this chapter, Drescher, Feick, Degeer and Shipley define cultural heritage landscapes as 'distinct geographic areas that have been altered by human activity that can help us understand a people or place'. The chapter examines the emergence of cultural landscapes within high-level policy arenas and its adoption by UNESCO and application within the North American context and in Europe through the European Landscape Convention. For Drescher et al., the key challenge for managing cultural heritage landscapes is to fairly evaluate and regulate proposed developments in a way that accommodates change and enables economic prosperity while maintaining landscape values important to a community. To explore this challenge, the chapter evaluates the use of participatory planning tools to provide an evidence-base for decision-making through a case study of planning for cultural landscapes in Ontario Province, Canada. The case study area faces many of the typical demands and changes of near urban rural landscapes, balancing agricultural modernisation, urban expansion, new demands for renewable energy infrastructure, and a rising demand for mineral extraction – all within the context of a distinctive rural landscape and distinct local groups, such as religious communities and First Nations (Indigenous people). The chapter outlines a series of linked, low cost engagement tools, including web-based surveys and mapping tools, focus groups and interviews to create a visualised and digitised evidence base to identify and map the community's values. Notably, the authors illustrate that some highly valued areas emerged that were not identified by heritage committees or municipal officials, suggesting that a participatory approach to landscape planning enables additional richness and nuance to policy formulation, and to ascertain rather than approximate community values.

Alongside the establishment of national parks, green belts have remained an enduring planning legacy in many parts of the world. Taylor (Chapter 40) provides an overview of the evolution of green belts and the varying purposes of developing green belt policy. She reminds us that green belts are essentially an urban policy: that planning for green belts has been conceived first to manage urban growth and make *cities* more sustainable with the sustainability of rural areas a secondary consideration. She then goes on the critique three goals of green belt policy, namely growth management, the protection of agricultural land, and natural environment protection, arguing that green belts work best when they combine all three goals, creating multifunctional places rather than as simply a boundary between urban and rural, serving as an urban container. The chapter then explores the case of Toronto's green belt strategy, which is positioned as an exemplar of combining multifunctional goals: recreation access and tourism, heritage, climate action (e.g. carbon storage), flood mitigation, and biodiversity protection sit alongside more traditional goals of preventing urban sprawl and agricultural land protection. Increasingly, more emphasis has been placed on protecting water resources, coined the 'blue belt'. This integrative agenda, facilitated by a regional approach to planning, has ensured that multiple stakeholders have been able to buy-in and support the strategy and enabled a long established planning tool to be 'fit for purpose' for twenty-first-century challenges.

Understanding the spaces where city and country meet is further explored by Scott, who discusses the rural urban fringe (RUF) as a hybrid space that should command greater attention from planners to capture the opportunities that fringe places provide. Scott provides an extensive review of the literature relating to the RUF, providing a typology of concepts used to describe and analyse fringe spaces. In doing so, Scott highlights the potential for moving towards a systems-based approach towards understanding and planning for the RUF. Specifically, Scott proposes a greater integration of ecosystem approaches with spatial planning, suggesting that the commonalities between these two approaches offers much potential for more integrative planning across the built and natural environment. The chapter also provides three case studies of planning in the RUF (with additional contributions from Cash and Qvistrom): a countryside management approach in England, dealing with land-use and social justice in the RUF in Jamestown, South Africa, and sustainable planning innovation in the RUF of Scania, Sweden. Each of these case studies examines the planning approach, outcomes and key lessons to be developed with these different types of place-making experiments. Scott concludes by suggesting that the 'transitory and contested nature of RUF character presents significant challenges set within the increasing pace and scale of change'. While change is often dominated by urban

433

expansion, Scott also counters that a notable characteristic of twenty-first-century RUFs is the way these areas are also becoming increasingly 'transformed by traditional rural land uses associated with new spaces of woods, forest and urban agriculture'. Finally, a more rural-centric transformation provides important opportunity spaces for rural planners to help mitigate the worst excesses of past and present urban centric agendas and to influence the kind of RUF identity and character for the future.

The following chapter by Mell examines green infrastructure as an emerging framework for planning at the landscape scale. Green infrastructure (GI) is often promoted as a new conservation paradigm, moving away from a traditional 'islands of protection' approach towards a more integrated and proactive process of protecting, enhancing, creating and restoring landscapes. GI planning is more focused on landscape functions and services, further developing an ecosystem approach to spatial planning. In this regard, Mell identifies key principles for GI planning before outlining a series of illustrative case studies of GI in practice. The chapter emphasises that GI approaches have shifted landscape planning from its traditional focus on 'special landscapes' towards a more inclusive approach towards so-called 'ordinary landscapes' to ensure that landscape connectivity, functionality, ecosystem services and appropriate development are managed and debated. Therefore, rather than protecting specific sites, a GI approach aims to develop ecological networks to enhance connectivity of landscapes – addressing habitat fragmentation and enabling species movement – and multifunctional landscapes, which capture benefits relating to, for example, amenity, wellbeing, biodiversity, river catchment management, and carbon storage. Mell also suggests that GI frameworks provide a conduit between urban and rural planning, demonstrating the interconnectedness of urban centres to their surrounding landscapes – from providing accessible greenspace and greenways to the countryside, to examples of landscape based flood resilience interventions to protect cities from flood risk based on upstream measures.

Building on Mell's assessment of landscape as a multifunctional resource, Chapter 43 examines landscape from a health and wellbeing perspective. While approaching landscape from a wellbeing perspective first appears novel, Rogerson et al. remind us that the early designation of national parks and recreational use of coastal areas are early examples of linking environment and wellbeing, providing restorative environments to enhance individual health. The chapter provides an overview of landscape as a valuable health resource, providing opportunities for physical activity, to reduce stress, restore mental fatigue, and enhance mood and self-esteem. To understand the relationships between wellbeing and landscape more effectively, the chapter outlines three alternative perspectives to wellbeing that can assist with designing nature-based interventions: firstly, hedonic approaches are reviewed, drawing on the wider happiness and life satisfaction literature to understand the influence of contact with nature on self-reported wellbeing, including stress reduction. Secondly, eudaimonic approaches are discussed, stressing the importance of fulfilment, suggesting that landscapes 'afford' individuals with opportunities to behave autonomously upon their own agency. Thirdly, a positive psychological wellbeing approach is examined. This examines the benefits of natural environments to the 'pleasant life' (maximising desirable experiences), the 'good life' (developed from activities that an individual enjoys), and the 'meaningful life' (contributing to the greater good e.g. friendships, serving the community). Employing this latter approach, Rogerson et al. provide an empirical evaluation of a wellbeing programme centred on engagement with natural landscapes, concluding that participants with the lowest initial wellbeing scores benefited the most from engagement with nature and natural landscapes. From a planning perspective, this review of the evidence-base provides an important contribution to understanding the positive benefits of nature and natural environments for physical and mental health when managing pressure for development, new housing, infrastructure etc. in rural localities.

The final chapter of Part VII examines rewilding as a countryside management approach. While the opening chapters in this section address the challenges of maintaining or protecting highly prized and pressured landscapes, this final chapter examines landscape management approaches applicable to more marginal rural areas, particularly in regions faced with land abandonment following agricultural restructuring or depopulation. Following a discussion of the drivers of land abandonment, Lennon traces the origins and evolution of rewilding theory and practice. While rewilding concepts are often explored and debated within the ecological literature, this chapter attempts to explore the wider potential of rewilding (ecological, economic, social) and related governance issues and institutional barriers. Although rewilding ideas can be controversial – such as the (re)introduction of predatory species – Lennon examines the increasing application of 'passive rewilding' in European contexts. This refers to the withdrawal of human intervention from the management of the landscape, but not necessarily humans from that landscape. In this way, in addition to capturing additional ecosystem services (e.g. carbon sequestration), landscape rewilding can also be linked to supporting local economic development centred on eco-tourism, hunting or adventurous recreation. However, while offering potential, Lennon also notes the institutional barriers or competing policy agendas constraining rewilding strategies – in this context, how nature is valued and interpreted is deeply contested. The chapter explores these issues by examining efforts to implement rewilding approaches in Ireland, which have been framed as a potential tourism boost for rural economies. Lennon concludes that the interactions between social, economic and cultural dimensions of rewilding remain under-investigated in the literature and within policy debates.

References

Owens, S. and Cowell, R., 2011. *Land and limits: interpreting sustainability in the planning process*. Routledge, London.

Selman, P., 2006. *Planning at the landscape scale*. Routledge, London.

38

National parks as countryside management

A twenty-first-century dilemma

Jonathan Bell and Aileen Stockdale

Introduction

The concept of national park designation is based on the principle of preserving park areas for the nation. Since the designation of Yellowstone in 1872, national parks have evolved considerably in their type and form. Originally a wilderness and conservation concept, national parks now integrate a wider set of management objectives (conservation, recreation and economic development). As a result, 'there is no single model of national park' (Frost and Hall, 2009: 11). Their evolution has paralleled the changing role of the countryside and the emergence of the sustainability agenda. A New Rural Economy (Shucksmith, 2012) is evident, characterised by an increasingly consumptive and commodified rural space (Woods, 2011). In this context, meeting the challenges of sustainable development has proved problematic for national park management. Parks have been criticised in the past for prioritising conservation of the natural environment at the expense of rural development needs (Bishop et al., 1998; McCarthy et al., 2002). Today there are fears that rural development (such as, extractive industries) alongside market based conservation strategies are prioritised in the context of economic austerity (Cortes-Vasquez, 2017). Reconciling competing land uses and stakeholder interests within national park boundaries, therefore, continue to present challenges for contemporary rural planning and national park management. This chapter reports on the changing role of national parks in increasingly contested rural landscapes, and on the on-going and emerging dilemmas for rural planning. While multiple challenges could be reported, here we choose to focus on two: the installation of energy infrastructure and the emergence of a natural capital approach to countryside conservation.

National park – an evolving concept

The original nineteenth-century wilderness and conservation model of national parks in the US stemmed from a desire to protect 'natural monuments' (Dilsaver, 1997) from human influence. Wilderness based national park models were brought into state ownership and managed through a system of resource use restrictions. While intended to be enjoyed by wider society, this was only under controlled conditions. This model of national park corresponds to IUCN category II areas which are defined as:

> A natural area of land and/or sea, designated to (a) protect the ecological integrity of one or more ecosystems for present and future generations, (b) exclude exploitation or occupation inimical to the purposes of designation of the area, and (c) provide a foundation for spiritual, scientific, educational and visitor opportunities, all of which must be environmentally and culturally compatible.
>
> *(IUCN, 1994: 19)*

The legacy of this wilderness model has been the development of 'conservation islands' and a distorted appreciation of the relationship between nature and society (IUCN, 2002, 2008). These criticisms, and the core principle of managing land in the national interest, remain central to national park debates today, particularly in relation to the challenges of reconciling national/local and public/private interests.

In contrast to Runte's (1987) 'worthless lands' hypothesis (alleging that US parks were only designated because they had no alternative economic uses), the European countryside is both inhabited and cultivated. Many European landscapes are the product of centuries of resource use activity and encapsulate the co-evolution of nature and society (Phillips, 2005). Therefore, the wilderness model was deemed unsuitable (Barker and Stockdale, 2008) and a multi-purpose national park, recognising the importance of nature as well as the needs of local communities, was required. This heralded a 'new paradigm' for protected areas and led to the creation of the IUCN category V protected landscape/seascape (Locke and Dearden, 2005), where 'the interaction of people and nature over time has produced an area of distinct character with significant aesthetic, ecological and/or cultural value, and often with high biological diversity' (IUCN, 1994: 22).

Even though they are multi-purpose, European parks still perceivably adopt an 'overriding emphasis on environmental management, protection and enhancement' with social and economic issues being of secondary importance (McCarthy et al., 2002: 667). Zoning is commonly introduced to minimise land-use conflicts (Bishop et al., 1998; McCarthy et al., 2002), whereby a highly restrictive central protection zone is typically surrounded by peripheral zones where some development and recreational activity is permitted.

The UK has taken the evolutionary development of national parks a step further. Initially, the National Parks and Access to the Countryside Act (1949) designated national parks in England and Wales (the first in the Peak District in 1951) with two statutory aims similar to the early wilderness model: to conserve natural beauty, wildlife and cultural heritage and, to promote opportunities for understanding and enjoyment. Like in the US, the Act was introduced in response to concerns about unplanned rural development (in this case, forestry, housing, quarrying) and a lack of public access to scenic areas (MacEwen and MacEwen, 1982). Unlike in the US, lands within the designated national parks remained in private ownership reflecting past and intended future management practices. Each designation is managed by a National Park Authority that also operates as a planning authority for the park. Designation under the 1949 Act imposed national conservation priorities in the 'public interest'; this restricted development and arguably impacted negatively on local communities due to issues of housing affordability, employment availability, and population sustainability (Marshall and Simpson, 2009; Richards and Satsangi, 2004) – these issues are also examined by Gallent and Scott in Chapter 22 of this Companion. Then in the 1990s, owing to the influence of the sustainability agenda, a duty to foster the economic and social well-being of park residents was placed on national park authorities (but not as an additional statutory aim). The Environment Act (1995) and (English) Natural Environment and Rural Communities Act (2006) gave greater prominence to development within parks but not at the expense of conservation: in situations of conflict between the park's aims, greater weight is given to 'conserve natural beauty, wildlife and cultural heritage'. Nevertheless,

> the remit of the national parks – pursuing the economic and social wellbeing of communities alongside environmental management – is a thoroughly modern concept. . . . In this sense, the [National] Parks are 'exemplars of sustainable development'.
>
> *(Arup, 2013: 9)*

Sustainable development principles are even more firmly embedded in the model of national park introduced in Scotland by the National Park (Scotland) Act 2000. Two further statutory aims accompany those already associated with English and Welsh national parks: 'promotion of sustainable use of natural resources' and 'promotion of the sustainable economic and social development of the area's communities'. While '[t]his signalled a firm legislative commitment to sustainable development' (Bell and Stockdale, 2015: 216), like in England and Wales, greater weight is given to conservation in situations where irreconcilable conflict arises between the four Park aims. Moreover, in Loch Lomond and the Trossachs National Park, the Park Authority became the planning authority, while for the Cairngorms National Park planning powers remain with the local authorities (councils). Therefore, arrangements for the delivery of planning functions have also evolved.

Protected areas increasingly function as instruments for regional development (Mose, 2007). National park purposes have also evolved to the extent that many are now designated as much for their economic potential as for their conservation function (Fredman et al., 2007). This approach is typified in Scotland where national parks include an explicit socio-economic aim and potentially represent the starkest contrast yet to the original wilderness model of national park. A 'new conservation' paradigm has emerged representing a re-conceptualisation of conservation based on particular interpretations and applications of sustainable development; typically an approach which shifts from state to a community-led focus and the inclusion of neoliberal ideology to 'make conservation pay' (Brown, 2002, 2003). Tourism, for example, is a means to contribute towards the financial sustainability of protected areas (Spenceley et al., 2017). Evolution of the national park concept also reflects a 'new rural' paradigm (Horlings and Marsden, 2014) and the emergence of a New Rural Economy (Shucksmith, 2012) including closer links between the urban and rural economy, a multi-sector place-based approach to rural development, and an increasingly consumptive countryside. Indeed, 'national park' is now a globally recognised brand label and offers a structure through which commodified tourism can operate (Reinius and Fredman, 2007). However, national parks as popular tourist attractions, according to La Page (2010) and others, raise concerns about the natural environment.

'New conservation' recognises that the role of government has evolved from 'controller and provider' to 'facilitator and enabler' through the involvement of non-state actors in decision-making; participatory forms of governance are regarded as essential for effective protected area management (Lockwood, 2010; Worboys et al., 2015). Indeed, national park governance today involves co-management, negotiating diverse and frequently competing stakeholder interests, particularly public and private interests as well as local and national interests, and achieving an appropriate balance between centralised and decentralised decision making. Accordingly, partnership working has become a key feature of both rural development and national park management (Hamin, 2001; Blackstock et al., 2017; Tatum et al., 2017). However, within such governance structures, power dynamics often determine which stakeholder interests exert greatest influence. While the UK planning system is built around mediating public interests and private property rights (Cullingworth et al., 2015) and includes collaborative and participatory approaches to accommodate diverse stakeholder interests (Healey, 2006), modern-day multifunctional national parks present on-going – and mounting – planning and management challenges.

Contemporary planning challenges

Protected areas are facing profound challenges in the context of economic austerity promoted in the global North since the 2008 financial crisis. Specifically within a UK context, the challenge is potentially exacerbated by the decision to leave the European Union; for example, in the wake of 'Brexit', the UK government is proposing to boost the economy with large infrastructure projects in the countryside (Kentish, 2017; Phillips, 2017). Therefore, contextual circumstances (state divestment in conservation, reduced public funding and Brexit) have arguably furthered the potential for conflict between nature conservation and economic development. Squeezed public sector budgets means national parks are increasingly conceived of in terms of the services they provide to society (Tatum et al., 2017). The imperative to find new ways of making conservation 'pay its way' has resulted in the extractive economies and market-based conservation strategies taking on a new importance (Cortes-Vazquez, 2017). This is especially apparent in the UK, in the form of energy infrastructure and the growing prominence of the 'natural capital' agenda in public policy. Each of these issues are now discussed in turn.

National parks and energy infrastructure

The social acceptance of energy infrastructure in the countryside has been a topic of controversy for some time (Ellis, 2009). For example, the erection of wind turbines created 'green on green' tensions (Warren, 2009), whereby conservationists are divided between those who cite local landscape impact versus those who endorse the pursuit of global climate goals (Cowell, 2016). However, the use of the countryside for energy production has taken on a new dimension in the UK with the emergence of the fracking agenda. Attempts to exploit shale gas deposits for the production of gas and oil through the process of hydraulic fracturing or 'fracking' could be viewed as a controversial form of countryside commodification. Indeed, conflict rather than consensus has become a common feature of the countryside, as demonstrated during the late 1990s/early 2000s by the mobilisation of major protests around issues such as hunting, rural service provision and the future of farming (Woods, 2003). Over a decade later the countryside is still typified by conflict, protest and discontent, as evidenced by recent anti-fracking protests (Pidd, 2017; BBC, 2017). National parks are not immune from the fracking policy controversy. While UK legislation prohibits surface drilling within national park boundaries, conservationists have expressed concern at the Onshore Hydraulic Fracturing (Protected Areas) Regulations 2016 which enables horizontal drilling 'under' national parks in ground at least 1,200 metres below the surface (Delebarre and Smith, 2017; Gosden, 2015). Not only could fracking equipment positioned adjacent to national park boundaries disturb the unique landscape setting of these designated areas, but such a precedent could pave the way for future above ground operations within national parks.

The fracking debate is ensconced in a 'framing contest', whereby supporters and opponents vie for acceptance of their interpretative frames in the public consciousness and in public policy (Hilson, 2015). Recent analysis suggests that the positive government framing of fracking is overriding local environmental concerns and planning authorities in England continue to overlook the climate implications of embarking on further fossil fuel exploration in the form of fracking derived gas (Hilson, 2015; Short and Szolucha, in press). Short and Szolucha (2017) argue that post-Brexit, already 'austerity-ravaged councils' will be under increasing pressure to approve commercial fracking against the wishes of local people. If permitted on a commercial scale, fracking could dramatically and rapidly transform parts of rural Britain. Assessing the influence of power in shaping the articulation of local interests, examining how the inevitable conflict between rural stakeholders is managed and gauging the extent to which local interests get to determine how this issue progresses (against a backdrop of national economic prerogatives), represent emerging areas of inquiry for rural planning research.

Clearly, shale gas deposits within national parks are an exploitable resource which could contribute to national economic priorities. In the context of nationally imposed economic pressures, quantifying the long-term economic contribution of well maintained natural assets can help challenge 'exploitive' activities that typically carry a powerful short-term economic argument. Therefore, market-based conservation strategies can help support the case for policy and investment that protect the natural environment.

National parks and the natural capital approach

Market based conservation strategies emerged out of a neoliberal rationale which suggests that, in order to protect nature, it is necessary to quantify the economic value of protected nature and commodify the ecosystem services that flow from nature (Cortes-Vazquez, 2017). One such approach is based on the concept of 'natural capital', which is understood as the natural assets (including geology, soil, air, water) that provide the services upon which human life depends (World Forum on Natural Capital, 2017). It has been claimed that a natural capital approach, which quantifies natural assets and any changes to this capital in a systematic way, can contribute to better informed decisions about land use, conservation and human development (Bateman et al., 2013; Agarwala et al., 2014).

Natural capital assessments have been championed as an efficient, practical and readily understandable approach to supporting more effective policy and investment decisions. Dieter Helm, chair of the UK Natural Capital Committee, which advises the UK government, equates the natural capital contained within national parks to more widely recognised examples of economic development:

> The national parks are every bit as much a part of our economy as a jaguar car factory is, or a high speed rail route or a block of new houses; these [national parks] are part of the fabric of our economy and they need to be paid for.
>
> *(Helm, 2016)*

A natural capital approach is considered integral to the delivery of the UK's 25 Year Environment Plan, currently being developed by the Department for Environment, Food and Rural Affairs (DEFRA). In its advice to DEFRA on the 25 Year Plan, the UK Natural Capital Committee recommends assessment of natural capital assets within national parks to inform policy and investment decisions:

> England's National Parks contain very significant natural capital . . . where practical, each National Park should quantify and value the main natural capital assets in its area, using the accounting framework recommended by the Committee in its first term. Valuation should play a key part in the assessment of natural capital investment options.
>
> *(UKNCC, 2017: 18)*

The UKNCC also proposes that in order to further protect and enhance the UK's stock of natural capital 'consideration should be given to the creation of new national parks' (ibid.: 18).

Through the adoption of natural capital approaches in national planning and environmental policy it could be argued that the ethos underpinning protected area management in the UK is entering a new phase. If the guidance offered by the UKNCC (ibid.) is adhered to, and delivered through DEFRA's 25 Year Environment Plan, approaches to national park management could move beyond the conventional sustainable development approaches which have dominated the last two decades. A natural capital approach, which views the environment as a key economic asset, will result in existing, and potentially new, national parks attempting to frame their contribution to society in predominantly economic terms. For example, the Lake District National Park Authority has identified the opportunity to become an 'early adopter' of natural capital approaches to inform Local Plan development and the pursuit of strategic objectives. Meanwhile, the Royal Society for the Protection of Birds applied a natural capital approach to its English Nature Reserves. While recognising that care needs to be taken to ensure that the approach does not neglect the intangible benefits of nature, the report asserts that a 'natural capital approach needs to be at the heart of the way decisions are made by both the private and public sectors' (Bolt et al., 2017: 4).

UK National Parks are likely to embrace market based conservation strategies to varying degrees over the next decade. In some instances, this will lead to further entrenchment of the conflict between economic and environmental interests. A commitment to market based conservation may leave national parks vulnerable to the forces of the neoliberal economic agenda. For example, the natural capital approach can use economic arguments to justify investment in, and protection of the environment; however, financial calculations could have the opposite effect, making a stronger case for development rather than protection. The possibility of the latter is potentially heightened given that the economic case for environmental protection is strengthened by a longer-term assessment, while politicians typically work to shorter-term economic priorities. Furthermore, the natural capital approach adopts compensatory measures or off-setting in instances where natural asset loss cannot be avoided. On the one hand, this approach could benefit the environment by securing compensatory measures for developments that would not otherwise make a contribution to the environment. On the other hand, due to the implementation of compensatory measures, the case for development (which would not otherwise be permitted) could be strengthened. It is not inconceivable that justification for perceivably inappropriate or controversial forms of development (such as fracking) could be strengthened by a form of off-setting.

Conclusion

National parks have evolved in their purpose and management and now encapsulate a diversity of landscapes; they accommodate multiple land uses that reflect historical and contemporary interactions between society and nature. Traditionally, nationally defined understandings of the 'public interest' were deemed to favour conservation at the expense of the everyday needs of park residents. In the current context, national understandings of the 'public interest' have altered dramatically towards the pursuit of economic growth as embedded in national UK policy. These contemporary understandings of the 'public interest' therefore threaten the traditional purposes of national park management. As a result, Maidment (2016) detected National Park Authorities increasingly having to promote conservation purposes back to government (government was historically the main proponent of the conservation focussed public interest narrative). This is particularly pertinent within the context of contemporary energy infrastructure.

Recent policy developments, such as the adoption of the natural capital approach, suggest some National Park Authorities are also subscribing to economic understandings of the 'public interest', to justify investment in, and protection of, the environment.

Tensions around how to define, and negotiate between multiple and often opposing interpretations of, what constitutes the public interest will continue to challenge decision making approaches within UK National Parks (Maidment, 2016). For example, the issue of 'public interest' and the 'common good' has come to the fore in discussions about post-Brexit subsidy models to replace, for example, the Single Farm Payment. The idea of a subsidy model based on 'public money for public benefit' is gaining traction (WCL, 2017; Harrabin, 2017). Natural capital assessments represent one option for measuring existing levels, and any changes to the level of natural assets. National parks offer an opportunity to pilot a natural capital approach to inform future investment and policy decisions.

Regardless of the extent to which the 'language' of natural capital is embraced in UK and specifically national parks policy, it remains to be seen whether the 'approach' will be effectively adopted to influence decision making and ultimately support conservation objectives or add to the commodification of an already commodified national park concept.

References

Agarwala, M., Atkinson, G., Baldock, C. and Gardiner, B. (2014) Natural capital accounting and climate change, *Nature Climate Change* 4(7), 520–522.

Arup (2013) *Valuing Wales' National Parks*. Brecon: National Parks Wales.

Barker, A. and Stockdale, A. (2008) Out of the wilderness? Achieving sustainable development within Scottish national parks, *Journal of Environmental Management* 88, 181–193.

Bateman, I. J., Harwood, A. R., Mace, G. M., Watson, R. T., Abson, D. J., Andrews, B., Binner, A., Crowe, A., Day, B. H. and Dugdale, S. (2013) Bringing ecosystem services into economic decision-making: land use in the United Kingdom, *Science* 34(1), 45–50.

BBC (2017) Police plea over Kirby Misperton anti-fracking protests, retrieved on 14 November 2017 from www.bbc.co.uk/news/uk-england-york-north-yorkshire-41167673.

Bell, J. and Stockdale, A. (2015) Evolving national park models: The emergence of an economic imperative and its effect on the contested nature of the 'national' park concept in Northern Ireland, *Land Use Policy* 49, 213–226.

Bishop, K., Green, M. and Phillips, A. (1998) *Models of National Park*, Perth: Scottish Natural Heritage.

Blackstock, K., Dinnie, E. and Dilley, R. (2017) Governing the Cairngorms National Park – Revisiting the neglected concept of authority, *Journal of Rural Studies* 52, 12–20.

Bolt, K. Ausden, M., Williams, L. and Field, R. (2017) *Accounting for Nature: A Natural Capital Account of the RSPB's Estate in England*, Sandy: Royal Society for the Protection of Birds (RSPB).

Brown, K. (2002) Innovations for conservation and development, *Geographical Journal* 168(1), 6–17.

Brown, K. (2003) Three challenges for a real people-centred conservation, *Global Ecology and Biogeography* 12, 89–92.

Cortes-Vasquez, J.A. (2017) The end of the idyll? Post-crisis conservation and amenity migration in natural protected areas, *Journal of Rural Studies* 51, 115–124.

Cowell, R. (2016) Decentralising energy governance? Wales, devolution and the politics of energy infrastructure decision making, *Environment and Planning C, Government and Policy* 35(7), 1242–1263.

Cullingworth, B., Nadin, V., Hart, T., Davoudi, S., Pendlebury, J., Vigar, G., Webb, D. and Townshend, T. (2015) *Town and Country Planning in the UK*, London: Routledge.

Delebarre, E.A. and Smith, L. (2017) *Shale Gas and Fracking*, House of Commons Briefing Paper 6073, London: House of Commons Library.

Dilsaver, L.M. (1997) *America's National Park System: The Critical Documents*, Lanham, MD: Rowman & Littlefield.

Ellis, G. (2009) Wind power: Is there a planning problem? *Planning Theory and Practice* 10(4), 521–547.

Fredman, P.L., Hornsten, F. and Emmelin, L. (2007) Increased visitation from National Park designation, *Current Issues in Tourism* 10(1), 87–95.

Frost, W. and Hall, M. (2009) *Tourism and National Parks: International Perspectives on Development, Histories and Changes*, London: Routledge.

Gosden (2015) Fracking to be allowed beneath National Parks despite ban pledge, retrieved on 14 November 2017 from www.telegraph.co.uk/news/earth/energy/fracking/11408206/Fracking-to-be-allowed-beneath-national-parks-despite-ban-pledge.html.

Hamin, E.M. (2001) The US National Park Service's partnership parks: collaborative response to middle landscapes, *Land Use Policy* 18(2), 123–135.

Harrabin, R. (2017) Farm subsidies 'must be earned' – Michael Gove, retrieved on 14 November 2017 from www.bbc.co.uk/news/science-environment-40673559.

Healey, P. (2006) *Collaborative Planning: Shaping Places in Fragmented Societies*, Basingstoke: Palgrave Macmillan.

Helm, D. (2016) Natural capital and national parks, retrieved from www.dieterhelm.co.uk/natural-capital/environment/natural-capital-and-national-parks/

Hilson, C. (2015) Framing Fracking: Which frames are heard in English planning and environmental policy and practice? *Journal of Environmental Law* 27(2), 177–202.

Horlings, L. and Marsden, T. (2014) Exploring the 'New Rural Paradigm' in Europe: Eco-economic strategies as a counterforce to the global competitiveness agenda, *European Urban and Regional Studies* 21(1), 4–20.

IUCN (1994) *Guidelines for Protected Management Categories*, Gland: Switzerland.

IUCN (2002) *Management Guidelines for IUCN Category V Protected Areas*, Gland: Switzerland.

IUCN (2008) *Guidelines for Protected Areas Management Categories, Part II. The Management Categories*, Gland: Switzerland.

Kentish, B. (2017) Government's post-Brexit business plans are putting Engalnd's countryside at risk – campaigners warn, retrieved on 14 November 2017 from www.independent.co.uk/news/uk/home-news/government-investment-infrastructure-brexit-countryside-national-parks-high-weald-a7504846.html.

La Page, W. (2010) *Rethinking Park Protection: Treading the Uncommon Ground of Environmental Beliefs*, Wallingford: CAB.

Locke, H. and Dearden, P. (2005) Rethinking protected area categories and the new paradigm, *Environmental Conservation* 32(1), 1–10.

Lockwood, M. (2010) Good governance for terrestrial protected areas: a framework, principles and performance outcomes, *Journal of Environmental Management* 91(3), 754–766.

MacEwen, A. and MacEwen, M. (1982) *National Parks: Conservation or Cosmetics?* London: Allen & Unwin.

McCarthy, J., Lloyd, G. and Illsley, B. (2002) National Parks in Scotland: balancing environment and economy, *European Planning Studies* 10(5), 665–670.

Maidment, C. (2016) In the public interest? Planning in the Peak District National Park, *Planning Theory* 15(4), 366–385.

Marshall, A. and Simpson, L. (2009) Population sustainability in rural communities: the case of two British national parks, *Applied Spatial Analysis and Policy* 2(2), 107–127.

Mose, I. (2007) *Protected Areas and Regional Development in Europe: Towards a New Model for the 21st Century*, Aldershot: Ashgate Publishing Limited.

Phillips, A. (2005) Landscape as a meeting ground: category V protected landscapes/ seascapes and world heritage cultural landscapes, in J. Brown, N. Mitchell and M. Beresford (eds), *The Protected Landscape Approach: Linking Nature, Culture and Community*, Gland: IUCN.

Phillips, J. (2017) The why and wherefores of citizen participation in the landscapes of HS2, *Planning Theory and Practice* 18(2), 328–333.

Pidd, H. (2017) 'This has been my life for the past six years': on the anti-fracking frontline, retrieved on 14 November 2017 from www.theguardian.com/environment/2017/jul/16/lancashire-anti-fracking-protest-camp-cuadrilla.

Reinius, S., W. and Fredman, P. (2007) Protected areas as attractions, *Annals of Tourism Research* 34(4), 839–854.

Richards, F. and Satsangi, M. (2004) Importing a policy problem? Affordable housing in Britain's National Parks, *Planning Practice and Research* 19(3), 251–266.

Runte, A. (1987) *National Parks: The American Experience*, Lincoln, NE: University of Nebraska Press.

Short, D. and Szolucha, A. (2017) Fracking Lancashire: the planning process, social harm and collective trauma, *Geoforum* (online first) doi.org/10.1016/j.geoforum.2017.03.001.

Shucksmith, M. (2012) *Future Directions in Rural Development*, Dunfermline: Carnegie UK Trust.

Spenceley, A., Snyman, S. and Eagles, P (2017) *Guidelines for Tourism Partnerships and Concessions for Protected Areas: Generating Sustainable Revenues for Conservation and Development*, Montreal: Secretariat for the Convention on Biological Diversity and IUCN.

Tatum, K., Porter, N. and Hale, J. (2017) A feeling of what's best: Landscape aesthetics and notions of appropriate residential architecture in Dartmoor National Park, England, *Journal of Rural Studies* 56, 167–179.

UKNCC (2017) *Advice to Government on the 25 Year Environment Plan*, UK Natural Capital Committee.

Warren, C. (2009) Powering Scotland: windfarms and the energy debate, in C. Warren (ed.), *Managing Scotland's Environment*, 2nd edn, Edinburgh: Edinburgh University Press, 340–368.

WCL (2017) *A Future Sustainable Farming and Land Management Policy for England*, London: Wildlife and Countryside Link. www.nationaltrust.org.uk/documents/a-sustainable-farming-and-land-management-policy-for-england-2017.pdf

Woods, M. (2003) Deconstructing rural protest: the emergence of a new social movement, *Journal of Rural Studies* 19(3), 309–325.

Woods, M. (2011) *Rural*, London: Routledge.

Worboys, G., Lockwood, M., Kothari, A., Feary, S. and Pulsford, I. (2015) *Protected area governance and management*, Gland: IUCN.

World Forum on Natural Capital (2017) What is natural capital? Retrieved on 14 November 2017 from https://naturalcapitalforum.com/about.

39

Participatory methods for identifying cultural heritage landscapes

Michael Drescher, Robert Feick, Christopher DeGeer and Robert Shipley

Introduction

As seen elsewhere in this book, rural areas have planning challenges that differ materially from what is encountered in urban contexts. Planning approaches developed for urban complexes cannot be implemented uncritically in areas with more dispersed populations and distinct sociocultural and environmental needs and resources. This is especially true with regard to culture and heritage. Most of us have a strong sense of the cultural value of rural areas. In fact, those places can form a significant part of our personal and group identity. Unfortunately, these features are often taken for granted until some major change is proposed, for example: aggregate extraction, the erection of large buildings or the placement of infrastructure such as wind turbines. When that happens, we frequently do not know how to fairly evaluate and regulate the proposed developments in ways that accommodate change while maintaining values important to the community's culture. This chapter considers different methods for planners to engage community members in defining the landscapes that are particularly valuable and worthy of protection.

The relatively new concept of cultural landscape has evolved to (a) identify significant landscapes as distinct from the ambient environment, and (b) aid in decision-making about land use. These two points clearly lead to prioritising some landscapes over others because planning cannot protect everything from unsympathetic change. When difficult decisions have to be made, finding ways to systematically evaluate significance with fairness and transparency becomes necessary.

We begin by outlining some of the challenges of rural planning and exploring how cultural landscape definitions have evolved at the international, national, sub-national and local levels. We also consider the shortcomings of the current, expert-led, approach to identifying cultural landscapes and highlight the need for more public input. Then, we describe a case study in Ontario, Canada where one photo-based and three map-based approaches were used by community members to identify potential cultural landscapes. The approaches included interviews, focus groups, web based questionnaire and photo-voice. This study engaged knowledge brokers and stakeholders in the general population and also reached out to distinct local religious communities and two First Nations (Indigenous people). Finally, we discuss the findings from the perspective of different public participation methods[1] before developing conclusions and recommendations based on our community engagement exercises.

Landscape pressures, policy and practice

While a relatively new term in most planning frameworks, cultural landscapes have long been a focus of other academic disciplines, and indeed, of preservation practice itself. Since Carl Sauer introduced the term 'cultural landscape' in his 1925 publication *Morphology of the Landscape*, debates surrounding the meaning of cultural landscape have spread from the field of geography, and cast them into a variety of disciplines including planning. Interest in cultural landscape as a concept to aid in decision-making around proposed changes in rural areas is driven by new and more complex pressures on the countryside. These pressures are both direct and indirect in nature and result primarily from ongoing urban expansion, industrialisation of agriculture as well as emerging forms of energy production. Urban development creates an almost insatiable demand for sand, gravel and cement. At the same time, large numbers of city dwellers desire to move to rural areas, particularly within commuting distance to larger urban centres. These forces of change take the form of urban expansion onto rural lands and dispersed, ribbon-like settlements of houses and estates. Increasing urban populations and the ongoing global energy transition require new forms of energy, which increasingly comes from wind turbines and solar panels. All of these complications are set against the imperatives of the natural environment such as water, soil and air quality, place-related values of inhabitants, and the spectre of climate change.

To help decision-makers address the issues facing rural planning, policy has evolved at various levels. In 1972, the United Nations Educational, Scientific and Cultural Organization (UNESCO), adopted the World Heritage Convention in 1972 with the aim of identifying, protecting, conserving, and transmitting cultural and natural heritage (UNESCO, 2016). Cultural landscapes, however, were not included on the World Heritage List until 1992. The relatively late adoption of cultural landscapes appears to have resulted from a reworking of the definition of heritage in an effort to be more representative of 'universal value' and to reflect changes to the meaning of cultural heritage (Harrison, 2013: 114–140).

UNESCO acknowledges that cultural landscapes are the combined works of nature and humans and that they illustrate the evolution of society under constraints and opportunities presented by the natural environment. Three main categories of cultural landscape are identified by UNESCO: designed (e.g. a formal garden), associative (e.g. a battlefield), and organically evolved (e.g. farmland). The last category is sub-divided into two categories: fossil (or relict), and those that retain an active role in contemporary society (continuing). The UNESCO categories are not discrete, and a landscape may be part of all categories to various degrees. However, the definition developed by UNESCO continues to influence and shape policy at various political levels. For example, the International Council on Monuments and Sites (ICOMOS) as well as the International Union for the Conservation of Nature (IUCN) both follow the definition developed by UNESCO. At the national level, in Canada, the UNESCO definition is used in the *Standards and Guidelines for the Conservation of Historic Places*. The US Park Service guidelines for cultural landscapes, on the other hand, emphasises tangible associations to history and intangible aesthetic values through the definition: 'a geographic area (including both cultural and natural resources and the wildlife or domestic animals therein), associated with a historic event, activity, or person or exhibiting other cultural or aesthetic values' (US Parks Service, 2006).

An alternative definition of cultural landscapes is found in the European Landscape Convention (ELC, signed in 2000). Article 1a of the ELC defines a landscape as an 'area as perceived by people, whose character is the result of the action and interaction of natural and/or human factors'. That the ELC defines landscape as a perceived area is complemented by the explicit role of the public in identifying and conserving cultural landscapes. The Convention requires each state party to 'establish procedures for the participation of the general public, local and regional authorities, and other parties with an interest in the definition and implementation of the landscape polices . . .' (Article 5c). Policy that necessitates the inclusion of the general public and 'other interested parties' in the identification and implementation of landscape polices is not exclusive to the ELC. In Ontario, the term 'heritage' is added to 'cultural landscape' and Cultural Heritage Landscape (CHL) means:

> a defined geographical area that may have been modified by human activity and is identified as having cultural heritage value or interest by a community, including an Aboriginal community. The area may involve features such as structures, spaces, archaeological sites or natural elements that are valued together for their interrelationship, meaning or association.
>
> *(Ministry of Municipal Affairs and Housing, 2014)*

Although community involvement is, at least, an implied cornerstone of Ontario's policy, what constitutes a community and how planners ought to elicit community values remains unclear. Our experiment in identifying candidate CHLs through community engagement will hopefully help not only in Ontario but also in other planning jurisdictions that recognise the formative role of the public in landscape planning.

Participatory approaches to planning are desirable for several reasons, ranging from substantive rationales, such as developing better policy (Cass, 2006), to normative ideas of addressing democratic deficiencies and fostering locally informed place-governance systems (Healey, 1997). Yet, while the ideology of participatory planning has been adopted in policy documents almost universally, cultural landscape policy has not kept pace (Conrad et al., 2011). In much of Europe, the ELC is explicit in calling for a bottom-up, participatory approach but the Convention's own Explanatory Report recommends 'performing the evaluation according to objective criteria first (ELC, 2000: 10) and then using those results in comparison with assessments by the general public and interest groups. The explanation for this is the 'inevitably subjective and varying public perceptions of landscape' (Jones, 2007). The praxis of identifying candidate sites prior to, or in isolation from, community opinion occurs in Canada, as well.

In the Canadian Constitution, land-use planning is a provincial jurisdiction. In the province of Ontario, several statutes and policy documents allow municipalities to formally recognise and protect cultural landscapes. These include the Planning Act and the Ontario Heritage Act. Under the Planning Act, the Province publishes a statement of their goals for land-use planning matters, called the Provincial Policy Statement (PPS). Municipalities are required to create plans that are consistent with the PPS. That document states that 'significant cultural heritage landscapes shall be conserved' (Ministry of Municipal Affairs and Housing, 2014). Through these documents, the Province establishes the importance of ascertaining community values in CHL studies, and both empowers and conveys responsibility to municipalities for their protection.

When CHLs are identified by municipalities, the work tends to be done either in isolation from the general public, or with a candidate list developed by municipal staff often assisted by consultants and volunteer heritage committees. Some municipalities in Ontario have adopted a framework similar to the US Parks Service, while others have developed protocols for the inclusion of the general public through traditional planning forums such as open houses and meetings where planners can communicate findings and seek input. One exemplary study from the Regional Municipality of Waterloo did address the participatory deficit that characterises most CHL studies (Shipley and Feick, 2009). For a proposed CHL centred on the last wooden covered bridge in Ontario the researchers held two focus groups with local residents. Those meetings informed the development of a web-based survey. Their response to the survey was robust, the local Township designated a CHL with strong local support, and an application for aggregate extraction nearby was withdrawn.

Rural cultural landscapes in Waterloo region

To explore the relative advantages of different methods for soliciting public input on CHL identification, a case study was conducted in the Ontario Townships of Woolwich (population 22,000) and Wellesley (population 10,500). Woolwich and Wellesley cover more than 600 square kilometres of rural area and, together with two other rural townships and three urban municipalities, form the Regional Municipality of Waterloo (population 535,000) (Statistics Canada, 2016). Waterloo Region is located in south-central Ontario about one and half hour drive west of Toronto, Canada's largest city.

The rolling countryside of Waterloo Region has been occupied for many millennia by Indigenous people and has experienced two centuries of prosperous Euro-Canadian settlement. Today the Region is one of the fastest growing areas in North America. The Region's Official Plan establishes a framework and a set of policies for coordinated social, environmental and

Figure 39.1 View of the Conestogo River meandering through Mennonite agricultural properties in Wellesley Township, Ontario, Canada. Some of the character-defining elements discernable in this landscape photo include the relatively large number of barns and silos, the tradition of planting flowers in garden plots, and a horse drawn vehicle parked adjacent to the foregrounded house.

physical planning across all of the lower tier municipalities. After the adoption of the PPS in 2005, the Regional Municipality of Waterloo began the process of identifying Cultural Heritage Landscapes (CHLs).

The Townships of Woolwich and Wellesley occupy the northern sections of the Region. The main topographical features running through Waterloo are the Grand River and its tributaries. Typically these watercourses have a broad floodplain bordered by farmland, punctuated by villages that developed in the past around river crossings and mill sites. Claims to the land are somewhat complicated. Virtually all of the area in the two Townships is located on the traditional territory of the Mississauga of the New Credit First Nation. In 1784, the land for 10 kilometres on either side of the Grand River was deeded to the Six Nations of the Haudenosaunee Confederacy (previously called the Iroquois Confederacy) in return for their allegiance to the British Crown during the American Revolution. One of the problems is that the Mississauga only ceded the land to the Crown by the Between the Lakes Treaty of 1792. According to the treaty the land was opened to immigrants for settlement, farming and industry but was not sold outright. The First Nations retained rights such as hunting and fishing. The resolution of the situation is ongoing but there is now a clear understanding that First Nations have the right to be consulted about land-use decisions.

Another community makes up a large part of the rural population in the two Townships, 'Old Order Mennonites'. Old Order Mennonite is a colloquial term used to describe several related yet diverse Anabaptist religious groups whose ancestors settled here over 200 years ago. In accordance with their faith, they have rejected most aspects of modernity in favour of a simpler lifestyle. The picturesque nature of the area is often enhanced by the sight of their horses and buggies and their traditional mixed agricultural practices throughout the study area. In keeping with their beliefs, they do not generally participate in civic life or politics, living in but not being of the world. Because they are valued members of the larger community, there is a long tradition of local decision-makers carefully considering the interests of this religious community.

The term picturesque has already been used to describe the areas that comprise our case study. The land consists of undulating hills, rivers and streams, and wood lots, overlaid with the grid road pattern, farmsteads with garden plots, orchards and big barns that have largely disappeared from other parts of rural Ontario. It is easy to focus on the details of this arcadian idyll and to think about protecting it from change. But change is inevitable and necessary in contemporary economies. When considering what kinds of change to permit, planners try to focus not on physical detail but on the cultural values. If functioning agriculture is the value, then farmers may have to be allowed to build bigger barns, supplement their income with the tariffs from wind turbines and communication towers. The challenge lies in enabling economic prosperity while preserving the essence of local cultural heritage.

Consultation approaches

Four approaches to gathering community input for delineating CHLs were selected. These approaches have a history of use for various circumstances (see Aoki, 1999 and Shipley and Utz, 2012, for examples). The two objectives were (1) to identify candidate CHLs, and (2) to test different ways of gathering community input on candidate CHLs. The outcomes for the first objective relate directly to supporting heritage landscape planning in Waterloo Region. The second objective was cast more broadly with the intent to inform cultural heritage landscape planning in rural areas where (1) community input and local knowledge are needed and valued, (2) local education of residents is sought, and (3) local planning resources are modest.

We used passive and active recruitment strategies. The passive recruitment strategy included advertisements though various media. Sixty-five flyers advertising the study were placed in frequently used community locations around the Townships, such as cafes, general stores, and bakeries. Each flyer had 'tear-away' sections at the bottom with details on how to contact the research team. Two articles were published in local newspapers, which detailed the nature of the study and how people could get involved. Additionally, a live interview was given for local radio, a Facebook page was maintained, a presentation was made at a televised Township Council meeting, and flyers were handed out at community events. The active recruitment strategy relied on identification of key stakeholders through web searches and contacts relayed through the project Steering Committee. Stakeholders and other knowledge brokers were contacted by phone or email and invited to participate. To recruit focus groups, gatekeepers were identified and contacted for their group's participation.

The passive recruitment strategy prompted participants to self-select for the method of their choosing. In most cases, however, community members selected the web-based survey. The web-based survey and mapping tool was custom-developed by the researchers. This tool enabled participants to draw polygons around valued locations on a digital map of the Townships and respond to questions about why they valued these locations. We also developed a photo-voice approach that allowed participants to submit photos they had previously taken or they took travelling the Townships for the purposes of this study. Participants were supplied with booklets to facilitate data recording and were asked to indicate the locations in the booklet where photos were taken as well as why these locations were of value. This method was developed with the intention of community members self-selecting for participation, but those who participated were actively recruited.

Community members were actively recruited for participation in focus groups and interviews. Focus groups, ranging in size from 5 to 12 people, were divided into groups of 3 to 5 and each sub-group was provided a paper map of the Township of the their choosing. Participants were asked to draw boundaries around locations they valued on the map and then to fill out a discussion form for each location. Focus groups were developed so that they could be completed with or without the presence of a researcher. Lastly, semi-structured interviews were used to gain a richer understanding of geographical locations, values, stories, and history of particular landscapes. During each interview, the participants were provided a map of one or both Townships and were asked to identify places they valued by drawing a boundary around the geographic location and to explain why they valued the location. The maps of each Township displayed forested areas, waterways, waterbodies, roads with easily identifiable names, cycling routes, and place names of the various settlements.

The interviews were audio recorded, data was transcribed verbatim and analyzed using qualitative data analysis software. The discussion forms generated through the focus groups and the responses provided through photo-voice and web-survey were also included in the analysis, which enabled triangulation of CHLs. The maps created through the interviews, focus groups, and web-survey were digitised and overlaid using a geographic information system to visually display the areas people identified as valuable.

Results of study

The results of the study are presented in three parts, the first dealing with an evaluation of the consultation approaches, the second with the substantive results of cultural landscape identification, and the third concerning the values expressed.

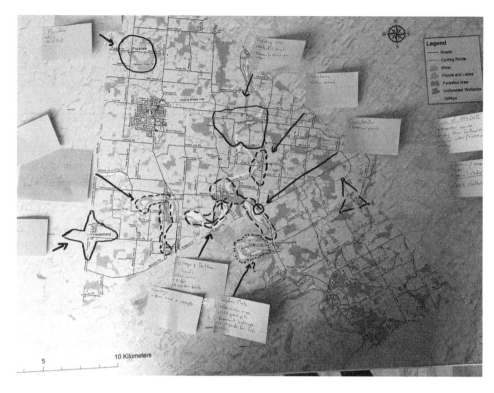

Figure 39.2 An example of a map of Woolwich Township made during an interview. The respondent marked specific areas and made notes on the importance of each.

During four months of public engagement, July to October 2017, 119 participants responded to the four consultation approaches. Fifty-eight maps were created with a total of 433 landscapes identified. Over 90 per cent of the participants were actively recruited through telephone calls, email, or snowball sampling. The remaining participants self-selected through passive recruitment. Where the participant's gender was known almost two thirds were male and a third female. The web-based survey garnered 10 valid submissions from nine participants, and, in most cases, only one candidate CHL was identified. All participants in this approach were passively recruited. Several participant comments suggested that the web-survey did not allow full expression of which landscapes were valued and why, and that it was challenging to use.

Table 39.1 Summary of results.

Instrument	Number of participants	Number of males to females	Participants actively recruited	Participants passively recruited	Number of maps created	Average number of landscapes identified per mapping exercise
Web-based survey	9	N/A	0	9	10	1
Photo-voice	3	0: 3	3	0	N/A	5
Focus group	72	44: 28	72	0	15	9
Interview	35	25: 10	34	1	33	9
Total	**119**	**69: 41**	**109**	**10**	**58**	

Data collected with the web-survey were neither rich nor nuanced, and did not elicit areas of overlap between participants to the method. The three participants in the photo-voice exercise were all actively recruited. The data recording booklets provided space for up to nine valued locations, but on average only five locations were identified. All submissions were geographically aggregated around relatively limited areas. There was no overlap between the photo-voice submissions.

Ten focus groups were conducted with 72 participants. These groups created 15 maps with an average of about 9 valued landscapes per map. Participants included two Township Councils, two Heritage Committees, one historical society, several Lions Clubs and other community organisations. Several areas of overlap were identified between the different groups but explanation of values was often limited. Nevertheless, the researchers witnessed fruitful discussions wherein participants shared stories and meaning of landscapes.

Thirty-five people participated in semi-structured interviews. All but one interview participant were actively recruited. On average, participants identified nine valued landscapes per map. This method reached landscape artists, religious leaders, historians, natural heritage experts, politicians, recreational group leaders, First Nations, Old Order Mennonites and other long-time residents. Numerous areas of overlap were identified through this method and it allowed for the collection of diverse values that coexist in the landscapes.

Figure 39.3 is a 'heat map' showing valued landscapes in the Township of Woolwich. This style of map represents an aggregation of all spatial data for valued landscapes. Stronger colour intensity indicates higher density of valued landscapes. A similar map was created for the Township of Wellesley. The heat map suggests that four areas or types of landscape are most valued: the major rivers and their tributaries, outdoor recreational amenities, the historic villages, and various natural or re-naturalised areas. Various watercourses were identified in 27 of the 29 maps that were created for the Township of Woolwich. Most of the settlements are located on the banks of rivers or streams.

While the mapped spatial data can serve to visualise locations of valued areas, the intangible values that were captured by the various approaches are more challenging to present. For example, several instances occurred where values conflicted. Some participants attached aesthetic value to seeing unfenced pastures adjacent to watercourses, while others regarded livestock access to streams a potential source of bacterial pollution. Another example is the dilemma concerning dams. Some participants wanted to preserve them as historic mill sites, while First Nations and other participants indicated they should be decommissioned to restore the aquatic environment for wildlife and fishing.

The interview and focus group approaches were best at eliciting and capturing tangible and intangible values, both historical and contemporary. Several participants identified river segments as especially aesthetic: 'the Grand River is absolutely beautiful when you get over into West Montrose'. Others valued the entirety of the river for recreational and historical purposes: 'You kayak in your modern, plastic, synthetic things, but you can't help but feel that these were major transportation conduits in years gone by.' For others, the Grand River is a source of inspiration:

I kept an eye on the rivers. Watched the light fall on them. I've painted – I think I've painted all of the rivers that I would come across. I like them in the wintertime where they would create abstract patterns.

One participant expressed how the Grand River contributed to a sense of self: 'I was born and raised in Poland, so water is very important . . . the Grand River and any waterways are very

Figure 39.3 Heat map analysis of valued landscapes in Woolwich Township. The curvilinear dark red area running north–south is the Grand River Valley and the adjacent communities. Linear hotspots running east to west represent a repurposed rail line now used for recreation and The Conestogo River until its confluence with the Grand River. The dark circles in the northeast and southwest overlay historic and traditional settlements.

important to keep in public access'. Another person enthused: 'We know that there are bald eagles all along the Grand now, yeah, it's pretty cool'. Importantly, many of these values are congruent with an interpretation of the Grand River as expressed by a First Nations respondent: 'So, it's all culturally significant to us. There's not one area where you can pick it out'. Similar statements were recorded concerning the importance of the traditional villages, their streets, houses, churches and views of them across the countryside.

Beyond the descriptions of places and the reasons for which they are valued, there was considerable agreement among participants concerning the threats and challenges faced in these rural areas. Aggregate pits were mentioned often: 'Aggregate extraction causes the destruction of the natural heritage of the area'. Other participants felt that aggregate extraction was an inappropriate land use 'right in the middle of a historically significant landscape'. Somewhat ironically, one of the areas that was valued for recreational opportunities contains a series of abandoned gravel pits that have filled with water and are now surrounded by new tree growth and prairie grasses. It seems clear that the perception of what is an appropriate land use can change with time.

Conclusions and recommendations

In liberal democracies where it is government policy to identify and conserve areas of cultural heritage significance – those places that help us understand ourselves and our histories – conceptual and practical challenges remain. In the matter of conserving cultural landscapes the research described in this chapter addressed two questions: first, how can we go about identifying significant landscapes as distinct from the ambient environment? Second, which approaches are most efficacious at identifying these landscapes?

First, most previous research in this field has concluded that the identification of significant cultural landscape works best when it is based on public consultation and is driven from the bottom up. Having undertaken extensive public engagement in the context of the current case study, we conclude that it is quite possible, at a modest cost, to engage citizens and to reach a broadly supported consensus on the identification of candidate areas of cultural and heritage landscape significance. While some people contend, for good reasons, that all of the cultural landscape is important, our current planning systems make difficult demands. These systems operate in support of making individual land-use decisions, which in turn necessitates having criteria for differentiation. Having good evidence for landscape significance provides a meaningful framework for fair and transparent decision-making.

We have also found that researchers approaching public consultation with sound methods can have confidence in their recommendations and planners can in turn be confident that their advice to decision-makers will stand the test of community acceptance. This study sought and received participation from a variety of stakeholders in the community, ranging from Old-Order Mennonites and conservation archivists to natural heritage experts, politicians, and avid cyclists. Striking similarities in terms of the geographic locations identified as valuable emerged between these respondents, clearly indicating some areas within the political boundaries of the Townships as being highly appreciated by the majority of the community.

Moreover, we found that some valued areas emerged that were not identified by heritage committees or municipal staff members. Therefore, by taking a participatory approach to landscape planning, we were able to add richness and nuance to policy recommendations, and to ascertain rather than approximate community values, which lends confidence to our recommendations. Confidence, however, is not the same as certainty and we must remember that not all land-use decisions will be popular with everyone. In this case study, there are two other factors

Figure 39.4 Aggregate extraction is a major industry that feeds urban growth and transportation infrastructure so decisions need to be made about where it will be allowed and where it will be off limits. Knowing what areas are most valued by the community helps in making those decisions.

that increase confidence in cultural landscape identification. First, the Township of Woolwich already has one of the first designated cultural heritage landscapes in the Province of Ontario. Recognition of the value of that landscape was based on previous research by two of the authors of this chapter (Shipley and Feick, 2009). Second, professional planners for both Townships included in this research, have been observers from the very beginning of the research and have endorsed the research process and outcomes.

The authors have no reluctance in recommending that local governments undertake robust public engagement exercises to identify their cultural landscapes. This is especially challenging on two counts. First, in many jurisdictions such as Ontario, junior levels of government must comply with policy authored by more senior levels of government. These policies are often somewhat vague by necessity since they must apply to many differing locales. Second, participatory planning is a notoriously tricky endeavour and there is no well-defined suite of tools that can be used in every situation (Shipley and Utz, 2012). Nevertheless, the current research does provide direction.

One challenge to participatory planning is recruiting participants. The current study was widely advertised through various medium (e.g. televised council meeting, live radio broadcast, multiple newspapers, flyers), yet only a small fraction of participants self-selected for participation. The vast majority of participants were actively contacted for participation through telephone or email. These findings lead to the recommendation, that to achieve sufficient participation, researchers may need to proactively contact potential participants.

We found that face-to-face engagement methods were most successful, not only for reaching participants, but also for their ability to elicit the identification of candidate sites. The majority of participants in this study were engaged through interviews or focus groups and it was in these environments that the greatest number of candidate sites were identified. Although many factors could contribute to these results, such as challenges of engaging with the web-based instrument, demographics of the study area, or recruitment strategy, the role of the researchers in empowering participants through reinforcing dialogue should not be neglected. The final recommendation, therefore, is that although pubic engagement exercises are trending towards greater use of technology, the more direct and interactive approaches such as interviews and focus groups continue to be well suited for meaningful public engagement.

Acknowledgements

The authors would like to thank students Shanqi Zhang (Ashley), Netzach Straker and Rebecca Korrol for their assistance and to recognize the support of the MITACS Accelerate funding programme, the Region of Waterloo and the Architectural Conservancy of Ontario. The research was a project of the University of Waterloo Heritage Resources Centre.

Note

1 For the present purpose we will use such terms as public engagement, participation and community consultation interchangeably although in specific contexts they may have slightly different meanings.

References

Aoki, Y. (1999). 'Review article: trends in the study of the psychological evaluation of landscape.' *Landscape Research*, *24*(1), 85–94.

Cass, N. (2006). *Participatory-Deliberative Engagement: A Literature Review*. Working Paper. Manchester: School of Environment and Development, Manchester University.

Conrad, E., Christie, M. and Fazey, I. (2011). 'Is research keeping up with changes in landscape policy? A review of the literature.' *Journal of Environmental Management*, *92*(9), 2097–2108.

ELC. (2000). European Landscape Convention. Retrieved on 21 December 2017 from https://rm.coe.int/1680080621.

Harrison, R. (2013). *Heritage: Critical Perspectives*. New York: Routledge.

Healey, P. (1997). *Collaborative Planning: Shaping Places in Fragmented Societies*. Vancouver: UBC Press.

Jones, M. (2007). The European Landscape Convention and the question of public participation. *Landscape Research*, *32*(5), 613–633.

Ministry of Municipal Affairs and Housing. (2014). *Provincial Policy Statement*. Toronto: Ministry of Municipal Affairs and Housing. Retrieved on 25 October 2018 from www.mah.gov.on.ca/AssetFactory.aspx?did=10463.

Sauer, C. O. (1925). *Morphology of the Landscape*. Berkeley, CA: University of California Press.

Shipley, R. and Feick, R. (2009). A practical approach for evaluating cultural heritage landscapes: lessons from rural Ontario. *Planning Practice & Research*, *24*(4), 455–469.

Shipley, R. and Utz, S. (2012). Making it count: a review of the value and techniques for public consultation. *Journal of Planning Literature*, *27*(1), 22–42.

Statistics Canada. (2016). Census profiles. Retrieved on 21 December 2017 from www12.statcan.gc.ca/census-recensement/2016/dp-pd/prof/index.cfm?Lang=E.

UNESCO. (2016). Operational guidelines for the implementation of the World Heritage Convention. Retrieved on 21 December 2017 from http://whc.unesco.org/en/guidelines.

US Parks Service. (2006). Guidelines for the treatment of cultural landscapes. Retrieved on 21 December 2017 from www.nps.gov/tps/standards/four-treatments/landscape-guidelines/index.htm.

40

The future of green belts

Laura E. Taylor

Introduction

A green belt is a deceptively simple idea: a belt of green space encircling a city to constrain urban sprawl. The area contained by a green belt is urban, or will be urban one day, while land inside the green belt itself remains permanently rural. From a rural planning perspective, the problem is that planning for green belts has been conceived first to manage urban growth and make *cities* more sustainable with the sustainability of rural areas a secondary consideration. The intent of green belt planning was originally to impose discipline on a city by stopping sprawl, but the contemporary goals of green belt planning are much more comprehensive.

Many cities today have green belts, the most famous of which encircles London, England. Today, London's iconic green belt is only one of many around cities in the United Kingdom and around the world, with the newest being Toronto's Green belt. Scholars have written extensively about London's Green Belt (Amati, 2016; Mace, 2016; Sturzaker and Mell, 2017). The *Greater London Plan 1944* is a work of art and a remarkable example of town and country planning survey methods influenced by Patrick Geddes, one of the first regional scholars (Abercrombie, 1945; Welter, 2002). Green belts in many other cities have been established over the years with varied success, including: Hong Kong (Tang et al., 2005); Ottawa, Canada (Gordon and Scott, 2016); Frankfurt and Berlin in Germany; Seoul, Korea; Melbourne, Australia; and many others but with few in countries of the global south (see Amati, 2016; Amati and Taylor, 2011). The scholarly literature on green belts, written mostly by planners, geographers, and urbanists, is generally positive about the idea of green belts although with little agreement on what the future holds.

London's Green Belt is an archetype of rational comprehensive planning created in the interwar years when the planning profession was emerging to understand and manage urban growth (Amati, 2016). It was designed to limit the size of the city and give city residents access to the countryside. But today, London's Green Belt is under political attack for being outdated and no longer fit for purpose (Mace, 2016; Sturzaker and Mell, 2017). Conversely, in Canada, Toronto's much newer green belt, established in 2005, is a more contemporary model and considered to be quite successful as it is better aligned with today's planning paradigms[1] (Ali, 2008; Amati and Taylor, 2010).

Are green belts still a good idea to deal with land-use conflicts in near-urban rural areas? This chapter argues that their potential lies in balancing traditional narrowly focused urban growth management goals with a sustainable approach to planning multifunctional peri-urban areas. Arguably no other planning intervention is better suited to be a comprehensive, flexible planning solution for the rural–urban periphery.

Green belt planning goals

Green belts have three main types of goals: growth management, or urban containment to create permanent urban boundaries; agricultural land protection; and natural environment protection, including source water protection, ecological conservation, parks and recreation, and climate action. The goals respond to city and regional planning politics when the green belt is created. Typical political issues include public health and safety, environmental degradation, availability of drinking water, air quality concerns, and urban sprawl. The goals are, of course, not mutually exclusive and more successful green belts may be those where are all goals are present.

Growth management

A green belt used for growth management is first and foremost an urban boundary designed to contain urban growth and stop urban sprawl. This founding vision of maintaining separation between town and country underpins much of land use planning in the Western world. As an urban boundary, it is meant to contain the future expansion of a central city. But in most places the green belt actually functions as an urban separator, separating the central city from existing or planned towns on the periphery. Most large urban regions are 'polycentric', meaning that there exist one or more important settlement areas in addition to the central city (Hodge and Robinson, 2001: 277). So, while a green belt may contain the growth of the central city, it will not stop growth itself, which should be expected to not only result in intensification of the central city but also disperse to other settlement areas in the region. The scholarly literature uses many synonyms with respect to growth management such as 'urban containment', 'urban separator', or 'urban boundary', but all concepts have the goal identifying where urban development is and is not planned.

From a land-use planning perspective, an urban boundary is a relatively straightforward policy intervention where a line mapped onto the landscape, usually following main roads or natural features, creates expectations for land use on both sides of the line. On the urban side: developable, serviced land. On the rural side: development is restricted, and municipal services are limited. Arguably, green belts can be seen to support development by 'fixing' where development can and cannot occur as the area outside the green belt is given tacit permission for development when a green belt is established (Macdonald and Keil, 2012; Sandberg et al., 2013).

From a rural planning perspective, green belts are an urban solution to an urban problem imposed on rural folks. Once a green belt is imposed, land use options are restricted to rural-type uses. But urban-type uses in the countryside (and speculative interest in future urban uses) are sometimes welcomed as economic investment.

Agricultural land protection

A second common green belt goal is agricultural land protection. Long-term protection is intended to support rural economies, to maintain rural cultures, to protect water quality, to ensure local food security, and to protect the scenic countryside. Different types of farms and farm practice will be privileged by the planning and policy tools used in green belt implementation, which tend to be lost in the politics of urban-imposed farmland protection (Bunce, 1994; Woods, 2005). Protecting farmland using a green belt may reduce urban development pressures but protection itself does not ensure the long-term economic viability of the farm nor the livelihood of the farmer (Cadieux et al., 2013).

How are farms within a green belt different from farms on the outside? Should green belt farms get a break on their property taxes? Should green belt farmers be compensated for the public good they are providing by maintaining the scenic landscape and for the ecosystem services their lands provide? For example, with carbon-trading, farmers can be compensated for carbon sequestration (Zasada, 2011).

Natural environmental protection

The third common goal is the protection of the natural environment. The near-urban rural area provides natural resources and ecological processes, which make city life possible. The protection of those resources and processes ensures the healthy ecological function of the urban region, the 'bioregion'. A bioregion is based on the way that water flows through a region as the presence (or absence) of water shapes a region's ecology. Our understanding of watersheds, rivers, streams, and headwaters, and related ecosystems is based on biophysical and ecological science, which in turn informs planning. Environmentally focused green belt planning may protect the source of drinking water (both above-ground natural or engineered lakes or groundwater aquifers), protect air quality, protect ecosystems, provide amenities for recreation, and/or take action against climate change.

A green belt created primarily for environmental planning purposes is a beguiling concept because it is supported by ecological science. Urban ecology has emerged over the past two decades, at least in North America, as the study of landscape ecology at the scale of the urban region (Forman, 2008, 2014). The label 'urban ecology' is slightly misleading as it includes not only urban areas but rural and peri-urban areas, too, as parts of the interrelated ecological function of the larger region.

Green infrastructure refers to the use of natural processes to provide services that conventionally have required engineered responses, largely brought on by increased interest in climate change mitigation and adaptation through planning. Green belts may include green infrastructure goals as protected areas are characterised by less development and therefore by fewer permeable surfaces, improving stormwater management as well as protecting environmentally sensitive land. The Toronto Green belt provides an example of an approach to green belt planning that tries to be inclusive of all three of these goals.

The case of Toronto's Greenbelt

The original vision for Toronto's Greenbelt was ambitious as it included multiple goals, extended existing protections for near-urban rural areas, and was part of an overall vision for the urban region. Toronto's Greenbelt is actually a bundle of three plans: the Niagara Escarpment Plan (Niagara Escarpment Commission, 2017), the Oak Ridges Moraine Conservation Plan, and the Greenbelt Plan (Ontario Ministry of Municipal Affairs, 2017a). The Greenbelt Plan designates the 'Protected Countryside'. Together with the Growth Plan (Ontario Ministry of Municipal Affairs, 2017b), the Greenbelt was designed to rein in urban sprawl in order to prevent the loss of farmland and green space, and to improve air and water quality in the region. The total

Greenbelt area is 730,000 hectares (almost 2 million square acres) and lies in an arc around the most urbanised area with the city of Toronto at its centre. The 'Greater Golden Horseshoe' region[2] that includes the city of Toronto and surrounding suburban cities and small towns and countryside is home to one-quarter of Canada's population and is one of the fastest-growing regions in North America due to international immigration: one in three immigrants to Canada settles here. The region lies at the southernmost edge of the vast province of Ontario.

The Greenbelt Plan was accompanied by the Growth Plan, which provides the policy framework for development within the region. The Greenbelt is a wide green arc containing the city of Toronto and several large suburban cities. There are some small towns in the Greenbelt (holes in the green fabric of the Greenbelt) and towns and cities beyond it, many within the Growth Plan area and subject to its policies. The Growth Plan requires intensification within existing built-up areas, planned densities for new development in greenfield areas, and density targets for 'urban growth centres' to concentrate the highest intensity of growth around existing suburban centres and major transit stations. The new planned urban structure is supported by population and employment projections and the distribution of growth over the 25-year life of the Plan. Growth distribution before the introduction of the plans was based on the accommodation of projected market demand by housing and employment type. A new transportation agency was created that devised a transportation plan called The Big Move (Metrolinx, 2008) to better link

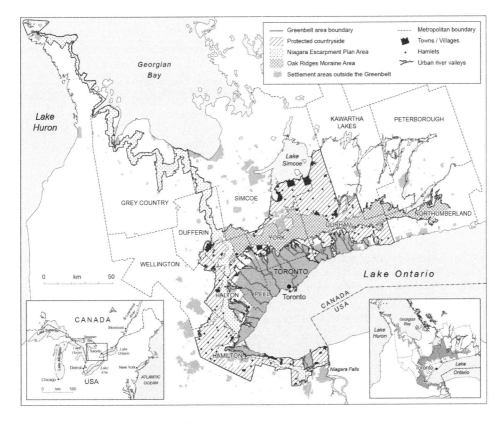

Figure 40.1 Toronto Greenbelt Plan 2017.

Source: Ontario Ministry of Municipal Affairs (2017a)

transit with land use patterns in the planned urban structure set out in the Greenbelt Plan and Growth Plan. A foundation was created and funded by the Province to support the Greenbelt through public education and research and to provide local economic support. The Greenbelt was but one of several interrelated initiatives to protect rural areas to manage sustained economic growth in the Toronto – or Greater Golden Horseshoe – region. The Province also changed planning processes around municipal decision-making and the arbitration of planning matters. Therefore, the Greenbelt Plan represents progressive attempts at landscape conservation over the past fifty years and continues to be part of an overall regional strategy to manage growth and conservation.

Regional planning has been a cornerstone of planning in southern Ontario since the birth of the organised planning profession in Canada at the beginning of the twentieth century (Caldwell, 2011; White, 2007, 2016). Since the provinces have the constitutional responsibility for land use, the Province of Ontario has shaped the urban structure of the Toronto region and retains jurisdiction over local decision-making. The Provincial Policy Statement (Ontario Ministry of Municipal Affairs, 2014), updated regularly, provides guidance to local municipalities on planning issues. Since the 1940s, the Province has created planning boards, offices, and secretariats to undertake public consultation and produce plans and reports to deal with growth management issues in the region, of which the Greenbelt is the most recent.

The Greenbelt shares not one but all of three green belt goals discussed above. Designed as a giant green arc containing the existing built-up area of Toronto (plus greenfield areas to accommodate several years of urban expansion), the Protected Countryside of the Greenbelt is first and foremost an urban containment boundary. Agricultural protection is achieved in the Protected Countryside through an agricultural system designation. The Protected Countryside includes a natural heritage system to protect biodiversity and ecological function over the long term. Compared globally to other jurisdictions, the Toronto region has a robust conservation regime (see Toronto and Region Conservation Authority, 2014) and the Greenbelt is an added layer of protection over and above other policies and regulations.

Recreational use of the Greenbelt is provided by a number of different public and private agencies and organisations, including the conservation authorities and the parks systems in local municipalities. The Friends of the Greenbelt Foundation provides funding for public and non-profit organisations to develop tourist amenities, and provides curated walking, hiking, and cycling routes, with points of interest clearly signed throughout.

Climate change mitigation and adaptation have become important issues for land use planning. The Greenbelt Plan was amended in 2017 to directly acknowledge the role that Greenbelt lands play in climate action, including carbon capture and storage and biodiversity protection. The natural heritage system is described as green infrastructure, improving the region's resilience to extreme weather events.

One of the most fascinating things about the Toronto Greenbelt is that it includes every conceptual type of green belt within it. Rather than imploding under the weight of the complexity of dealing with all of the expectations placed upon it, there seems to be something for everyone. The revised Greenbelt Plan (2017), which was subject to a mandated 10-year review, is even more ambitious than the original in 2005. A lot of effort went into the 10-year review, including wide public consultation. The report on the 10-year review was essentially a review of planning in the region itself (Crombie, 2015). All related plans were amended to fix kinks in the old plans and to adjust to new priorities. Climate change and the extension of the Greenbelt into river valleys within urban areas were added to the original goals for agricultural protection, environmental protection, culture, recreation and tourism, settlement areas, and infrastructure and natural resources.

In the next 10 years, the Greenbelt is expected to continue as planned with scenic open countryside protection, geologic landform protection, source water protection, urban growth boundaries, agricultural land protection, and natural heritage system protection, and climate action. The Province has completed work to extend Greenbelt-level protection of agricultural systems and natural heritage systems into the Growth Plan area. Work has also begun to extend Greenbelt protection to more areas for water resource protection, dubbed the 'Bluebelt'. Shifting politics and future elections will continue to shape implementation of Greenbelt policies.

Fit for purpose? Green belts and competing planning values

A green belt is a simple idea that sells the promise of permanent rural landscape protection. At the same time, it hides from most people the complexities of its designation. Green belts may have originally been based on an over-simplified understanding of life and landscapes both inside and outside of the city and may continue to run the risk of perpetuating urban–rural environmental imaginaries. But the simple rationale for a green belt – to protect rural areas from urbanisation – means that it is more likely to survive tests of fitness for particular purposes so that the green belt will be around to be fit for future, yet unseen purposes.

The green belt concept has expanded from the first designs for the London Green Belt and is continuing to evolve. Older ideas of urban separation based on ideal city form and public health were superseded by the desire to protect the ecology of rural environments and are today extended to climate change mitigation and adaptation. Nonetheless, a green belt because of the simplicity of the idea, embodies the 'precautionary principle' with respect to the protection of rural landscapes. We know that rural landscapes provide all kinds of benefits for both urban and rural residents and there are very compelling social and environmental reasons for their continued existence (and new reasons such as climate adaptation continue to emerge). Whether or not a green belt is fit for purpose depends on whether the originating 'fix' continues to be seen as in the public interest and whether outcomes conform to the plan's intent. Green belts that have been around for a long time tend to be accompanied by a founding mythology based on the nature of the planning crisis at the time. The danger becomes that the founding crisis becomes less and less relevant to the contemporary moment. For example, Sturzaker and Mell (2017: 150) make the case that 'greenways' are a compelling contemporary alternative to better respond to local social and natural ecologies compared to the original containment purpose of green belts in the UK. The reasons why a green belt is first established will affect the challenges of their long-term viability unless the plan and its implementation evolve along with contemporary circumstances.

Challenges to green belts deals most often with economic and competing planning values regarding the use of land in rural–urban periphery (Gallent et al., 2006; Scott et al., 2013; Taylor and Hurley, 2016; see also Chapter 41 of this Companion). Green belts are a major planning intervention in rural land economics as they either expropriate lands within the green belt, remove future development rights, and/or limit to rural-type uses the options property owners may have on their lands. A green belt is intended to make agricultural and other rural uses permanent (although governments may use green belts as land reserves for future public uses such as highways, transit, airports, and other public facilities) and is intended to end real estate market speculation.

Market forces bring much political pressure to bear on avoiding the establishment of new green belts and on modifying existing ones. In the UK especially, debates have raged over the market limitations created by existing green belts. Arguments in favour of rethinking existing green belts to allow development, especially new housing, include replacing the idea of the green belt as a 'blunt' urban separator with 'wedges' (Mace, 2016) or 'greenways' with the hope that a more focused parks-related purpose will provide a better political economic use-value argument in favour of creating or maintaining the protected (albeit much smaller) area. The increased financialisation of land also seems to be putting increased pressure on land in peri-urban areas as land is increasingly seen as a commodity with exchange value in global securities markets, encouraging property investors to seek profit through the planning process by redesignating land from rural to urban uses and increasing land value. Development may or may not occur as long as the commodity value increases.

Green belts are another layer of state land use policy and regulation. As a planning tool, a green belt can be an overlay, a designation, or a zone. A wide variety of planning regulations is available to jurisdictions depending upon the local land use planning regime. In most places, the responsibility for land use is written into constitutional documents as well as longstanding legislations that have structured decision-making and bureaucratic processes over the long term. For example, in Canada's case, the *British North America Act*, 1867 gives each province the responsibility for land use (Hodge and Robinson, 2001: 117–118). Each province creates local governments as it sees fit to deliver public services (Ontario Legislative Assembly, 2018). This state-level responsibility stands in contrast to the United Kingdom where the national government legislates planning and provides policy guidance (Cullingworth and Nadin, 1997) and the United States, where local jurisdictions retain much of the power over land use decision-making (Wheeler, 2013). So, the types of planning regulations available depend on the planning regime. Nonetheless, an overriding consideration is to determine the most appropriate level of government to have carriage of the green belt. The acceptance of state-led green belts is possible only in places where there is a societal culture that embraces more collective public values and that has upheld strong top-down planning.

The future of the green belt concept

Green belts are a part of planning history, especially in Anglo-American countries, and have been around for a long time. Is there the danger that green belts will be seen as anachronisms rather than useful to contemporary near-urban rural planning? They are under political attack in the UK as no longer 'fit for purpose', meaning that they are artefacts of a previous era of British planning and urban life, and have failed to visibly and evenly contain urban growth and protect the countryside and the natural environment. In contrast, Toronto's Greenbelt is a contemporary and so far successful example of a comprehensive yet flexible plan with wide public support.

Green belts differ from other forms of rural greenspace protection because of this holistic regional imagination. Unless part of a larger regional planning strategy, the establishment of a new green belt or the maintenance of an existing one is difficult to undertake. The green belt contains lands that are planned to remain rural in character (in all its complexity) and undeveloped for urban-type uses. A green belt is one layer of land-use protection, within which policies can be tailored to natural and cultural landscapes within the rural area. Ongoing public benefits have to be visible and accessible to the public to experience.

Planning for climate change is increasingly seen as a goal to which green belts are well-suited. A common conflict in rural landscape protection is between aspirations for farmland protection and natural environment protection. Cadieux et al. (2013) discuss the tensions between agricultural land protection and natural environment protection in the implementation of the Toronto Greenbelt where the environmental sustainability of farming practice is often questioned and ecologically protected areas are seen as infringing on farmers' property rights. With increasing focus on climate action such as carbon sequestration, the benefits provided by both farmland and natural heritage landscapes converge.

While the potential for green belts still exists in its comprehensiveness and adaptability, what other concerns might affect future discourse about green belts?

Reconciling green belts and settler colonialism

A discussion of the potential of green belt planning to advance truth, healing, and reconciliation between settler states and indigenous peoples is missing from the green belt literature. Green belts perpetuate settler colonialism (see Porter and Barry, 2016) as they are part of the rise of land-use planning to discipline the landscape. The opportunity to reshape the near-urban rural landscape to be more inclusive of indigenous rights and values with respect to those lands exists and seems in line with the spirit of landscape-scale protection offered by green belts.

Global South

The literature is also lacking narratives and evidence from the experience of green belt efforts in the global south. As a single layer land-use designation imposed from the national or provincial/ state level, a green belt plan requires a strong state planning apparatus and social acceptance of the state's authority to impose limits on land use and development. Green belt planning (at least as presented in this chapter) may not be possible or practical in many countries where planning itself is 'informal', blurring the boundaries between legal and illegal practices (Roy, 2009). Under what conditions and regimes will future green belts make sense?

The digital revolution

The way we understand green belts seems poised to become both more extensive and detailed if the promises of big and open data are realised. Environmental and other kinds of data have the potential to change the way we see, use, and understand green belts. Performance indicators for green belts may be derived from data on commuting, stormwater flows, farmland production, and home building. With the ease of cross-platform data sharing today, especially as software like Geographic Information Systems (GIS) becomes more user-friendly, all information has the potential to be public spatial information. Green belts have already become more readily 'visible' to residents of the urban region through the widespread accessibility of aerial landscape views through Google maps, for example. People are more easily able to see what is going on within green belt areas, which may shift the perception of the green belt as a sort of rural *tabula rasa* by making productive rural activities visible. This accessible perspective seems only set to improve with the convergence of big data and real-time mapping.

Sustainable livelihoods within the green belt plan area

Finally, one of the lessons from the Toronto Greenbelt has been the need to consider how to create and maintain rural areas. The aesthetic appeal of the scenic countryside exists largely because of the existence of working farms and other productive rural uses using environmental land management strategies that enhance the beauty of the scenic countryside. To support long-term livelihood strategies of those working there, state programmes and other means are needed. The danger exists that the Greenbelt will be designated as rural on paper but will not continue to be a functional rural system, instead owned by exurbanites and maintained as the pastiche of a working landscape (Abrams and Gosnell, 2012; Morse et al., 2014).

Conclusion

A green belt is a simple idea that sells the promise of permanent rural landscape protection. At the same time, it hides from most people the complexities of its designation. A single-goal green belt does not live up to the potential of the green belt concept. For example, a green belt that is just an urban boundary (where the green belt is a wide green line) is a central city containment strategy but is not a positive plan for the lands within it or beyond it. A green belt that is just a protected area designed for agricultural land protection is too narrowly defined as the value hinges on farming potentially at the expense of other rural landscape uses. A green belt that is meant for just natural environment protection – just as green infrastructure or just as a linear greenway – also runs the risk of being too narrowly defined; while environmental science supports boundaries and provides a rationale for protection seemingly away from political criticism, the lands must be managed as ecological or as recreational and may be at odds with other desirable rural uses. When designed to be more than a single-goal protected area designation, green belts can provide flexibility in the public interest over the long term. The rural–urban fringe is hybrid, multifunctional, contains multiple geographies and every fringe of every city is as unique as the landscapes and cultures and processes that have shaped the city's region. Green belts are a powerful planning tool as they transcend localised land-use conflicts and remain aligned with the persistent spatial imagination of an interconnected city and countryside.

Notes

1 Arguably, contemporary planning paradigms are the same as they always have been. Planning has been about public health and safety from the beginning and remain so today. The goals of planning are, perhaps, differently expressed today: food security; air quality; and equality of access to nature for physical and mental health.
2 'Toronto' is the label for many geographies and planning jurisdictions. While Toronto is a municipal jurisdiction, Toronto often refers to the entire urban region.
   Toronto is a city, a large municipality of about 2.8 million people. The present-day City of Toronto has the same jurisdiction as the former Regional Municipality of Metropolitan Toronto area, which is widely touted as one of the first regional governments in North America. Started in 1953, Metro Toronto had responsibility for coordinating major transportation and other infrastructure across the six cities of Toronto, Etobicoke, York, North York, East York, and Scarborough; each of those cities was itself an amalgamation of smaller, historic towns and their countrysides. The 'rural' near-urban countryside was already integrated into this metropolitan – regional – governance structure beginning in the post-war period. By 1970, regional municipalities (Halton, Peel, York, and Durham) had been created across the area known as the Greater Toronto Area (GTA). Each of these 'upper tier' regional municipalities included large expanses of countryside, in addition to cities and towns; the regional municipalities are so large that much of that countryside still exists as 'rural' today, either as part of the green belt or as yet undeveloped countryside. Metro Toronto became the City of Toronto in 1998, roughly coinciding with the 'full build-out' of the city's geography (i.e. very few rural lands remained following years of rapid suburbanisation).
   In 2004, with the creation of the Greenbelt Act and the Places to Grow Act, the Greater Golden Horseshoe region became the focus of the regional planning by the Province. At the time, the Greater Golden Horseshoe had a population of about 7.8 million people and has grown to about 9.2 million at the time of this writing with 13.5 million expected by 2041, the planning horizon of the Growth Plan (Ontario 2015). The Greater Golden Horseshoe is often colloquially referred to as just 'Toronto' when people are referring to the large urban region.

References

Abercrombie, P. (1945) *Greater London Plan 1944: A Report Prepared on behalf of the Standing Conference on London Regional Planning*. London: His Majesty's Stationery Office.

Abrams, J.B. and Gosnell, H. (2012) The politics of marginality in Wallowa County, Oregon: Contesting the production of landscapes of consumption. *Journal of Rural Studies* 28(1), 30–37.

Ali, A.K. (2008) Greenbelts to Contain Urban Growth in Ontario, Canada: Promises and Prospects. *Planning Practice and Research* 23(4), 533–548.

Amati, M. ed. (2016) *Urban Green Belts in the Twenty-first Century*, 2nd edition. New York: Routledge.

Amati, M. and Taylor, L.E. (2010) From Green Belts to Green Infrastructure. *Planning Practice and Research* 25(2), 143–155.

Amati, M. and Taylor, L.E. (2011) Conference report: The death and life of greenbelts: 'Local Solutions for Global Challenges', Toronto, 22–24 March 2011. *Town Planning Review* 82(6), 733–737.

Bunce, M.F. (1994) *The Countryside Ideal: Anglo-American Images of Landscape*. London: Routledge.

Cadieux, K.V., Taylor, L.E. and Bunce, M.F. (2013) Landscape ideology in the Greater Golden Horseshoe Greenbelt Plan: Negotiating material landscapes and abstract ideals in the city's countryside. *Journal of Rural Studies* 32, 307–319.

Caldwell, W.J. ed. (2011) *Rediscovering Thomas Adams: Rural Planning and Development in Canada*. Vancouver: UBC Press.

Crombie, D. (2015) *Planning for Health, Prosperity and Growth in the Greater Golden Horseshoe: 2015–2041: Recommendations of the Advisory Panel on the Coordinated Review of the Growth Plan, the Greenbelt Plan, the Oak Ridges Moraine Conservation Plan and the Niagara Escarpment, Toronto*. Toronto: Queen's Printer for Ontario. Retrieved on 25 October 2018 from http://ossga.com/multimedia/0/planning_for_health_prosperity_and_growth_-_expert_panel_report.pdf.

Cullingworth, J.B. and Nadin, V. (1997). *Town and Country Planning in the UK*, 12th edn. London: Routledge.

Forman, R.T.T. (2008) The urban region: natural systems in our place, our nourishment, our home range, our future. *Landscape Ecology* 23(3), 251–253.

Forman, R.T.T. (2014) *Urban Ecology: Science of Cities*. Cambridge: Cambridge University Press.

Gallent, N., Andersson, J. and Bianconi, M. (2006) *Planning on the Edge: The Context for Planning at the Rural–Urban Fringe*. London: Routledge.

Gordon, D.L.A. and Scott, R. (2016) Ottawa's Greenbelt evolves from urban separator to key ecological planning component. In M. Amati (ed.), *Urban Green Belts in the 21st Century*. New York: Routledge, 129–148.

Hodge, G. and Robinson, I.M. (2001) *Planning Canadian Regions*. Vancouver: UBC Press.

Macdonald, S. and Keil, R. (2012) The Ontario Greenbelt: shifting the scales of the sustainability fix? *The Professional Geographer* 64(1), 125–145.

Mace, A. (2016) The metropolitan green belt, changing an institution. *Progress in Planning* 121, 1–28.

Metrolinx (2008) *The Big Move: Transforming Transportation in the Greater Toronto and Hamilton Area*. Toronto: Queen's Printer.

Morse, C.E. et al. (2014) Performing a New England landscape: Viewing, Engaging, and Belonging. *Journal of Rural Studies*, 36, 226–236.

Niagara Escarpment Commission (2017) *The Niagara Escarpment Plan*. Toronto: Queen's Printer.

Ontario Legislative Assembly (2018) Municipal Act, 2001, S.O. 2001, c. 25. Retrieved on 26 May 2018 from www.ontario.ca/laws/statute/01m25.

Ontario Ministry of Municipal Affairs (2014) *Provincial Policy Statement*. Toronto: Queen's Printer.

Ontario Ministry of Municipal Affairs (2015) The Greater Golden Horseshoe Region: an international economic, agricultural and cultural hub. Retrieved from http://ontario.ca/landuseplanningreview.

Ontario Ministry of Municipal Affairs (2017a). *Greenbelt Plan*. Toronto: Queen's Printer.

Ontario Ministry of Municipal Affairs (2017b). *Growth Plan for the Greater Golden Horseshoe*. Toronto: Queen's Printer.

Porter, L. and Barry, J. (2016) *Planning for Coexistence? Recognizing Indigenous Rights Through Land-Use Planning in Canada and Australia*. London: Routledge.

Roy, A. (2009) Why India cannot plan its cities: informality, insurgence and the idiom of urbanization. *Planning Theory* 8(1), 76–87.

Sandberg, L.A., Wekerle, G.R. and Gilbert, L. (2013) *The Oak Ridges Moraine Battles: Development, Sprawl, and Nature Conservation in the Toronto Region*. Toronto: University of Toronto Press.

Scott, A.J. et al. (2013) Disintegrated development at the rural–urban fringe: re-connecting spatial planning theory and practice, *Progress in Planning* 83, 1–52.

Sturzaker, J. and Mell, I. (2017) *Green Belts: Past; Present; Future?* London: Routledge.

Tang, B., Wong, S. and Lee, A.K.W. (2005) Green belt, countryside conservation and local politics: a Hong Kong case study. *Review of Urban and Regional Development Studies* 17(3), 230–247.

Taylor, L.E. and Hurley, P.T. (2016) *A Comparative Political Ecology of Exurbia: Planning, Environmental Management, and Landscape Change.* London: Routledge.

Toronto and Region Conservation Authority (2014) *The Living City Policies for Planning and Development in the Watersheds of the Toronto and Region Conservation Authority.* Toronto: Toronto and Region Conservation Authority.

Welter, V.M. (2002) *Biopolis: Patrick Geddes and the City of Life.* Cambridge, MA: MIT Press.

Wheeler, S.M. (2013) *Planning for Sustainability: Creating Livable, Equitable and Ecological Communities.* 2nd edn. New York: Routledge.

White, R. (2007) *The Growth Plan for the Greater Golden Horseshoe in Historical Perspective.* Toronto: Neptis Foundation.

White, R. (2016) *Planning Toronto: The Planners, Their Plans, Their Legacies, 1940–80.* Vancouver: UBC Press.

Woods, M. (2005) *Contesting Rurality: Politics in the British Countryside.* London: Ashgate.

Zasada, I. (2011) Multifunctional peri-urban agriculture: a review of societal demands and the provision of goods and services by farming, *Land Use Policy* 28(4), 639–648.

41

Rediscovering the rural– urban fringe

A hybrid opportunity space for rural planning

Alister Scott

Introduction

> If we want to change the landscape in important ways we shall have to change the ideas that
> have created and sustained what we see.
>
> *(Meinig, 1979: 42)*

The spaces where town and countryside meet are dynamic, complex and 'messy', providing an arena within which urban and rural land uses, infrastructure(s), identity(ies) and values are negotiated and contested within wider place-making processes and outcomes (Gant et al., 2011; Hiner, 2016; Luka, 2017). Here competing agendas of sustainability, spatial planning, post-productivism, social and environmental justice, neoliberalism and economic growth clash (Gallent et al., 2006; Scott et al., 2013, 2018; Cash, 2016). The rural–urban fringe (RUF) is therefore schizophrenic; subject to a conflation of terms reflecting different conceptual priorities – e.g. RUF (Piorr et al., 2014), urban–rural fringe or rural urban fringe (Scott et al., 2013); Exurbia (Taylor and Hurley, 2016); Edgelands (Farley and Roberts, 2012); and the peri urban interface (McGregor et al., 2006). And also differing characterisations: a dominant space of the twentieth century (OECD, 2011); a misunderstood space (Gallent et al., 2006); a weed (Cresswell, 1997); a battleground for urban and rural uses (Hough, 1990); and a landscape out of order (Qvistrom, 2007). Cumulatively, this obfuscates and artificially restricts the existing/ potential nature and functions of the space itself (Scott et al., 2013). As a starting point, the chapter needs to adopt one identity, favouring the rural urban fringe (RUF) to reflect its rural planning primacy. To understand RUF characterisations and potential, a review of the literature will be undertaken to explore its past, present and future pathways and, with the addition of three case studies, help to illuminate RUF potential within a rural planning lens.

The RUF is now, arguably, the predominant land type globally (Mckenzie, 1997; OECD, 2011, 2013) if you accept definitions that go beyond simple land-use considerations to consider the intersection of urban and rural values lifestyle and interests. Yet perversely, most policy and academic narratives are still conceived and implemented within either urban or rural narratives;

each with their own associated paradigms (e.g. ecosystem approach and spatial planning) and increasingly complex governance frameworks, with considerable potential for academic and policy 'disintegration' (Scott et al., 2013). Consequently, different land-use interests contest RUF space within their own specialist sector silos and associated needs (e.g. housing, manufacturing, retail and other commercial businesses, infrastructure, gypsies, energy, farming, recreation, tourism, landscape and nature conservation) and on an ad hoc basis (Gallent et al., 2006). Consequently, problems are identified, diagnosed and treated, within different spatialities and temporalities, without reference to wider RUF interdependencies (Hodge and Monk, 2004). Interventions in the form of siloed planning policies and associated tools use ad hoc combinations of zoning, designations and incentive interventions (market-based instruments). These may regulate, constrain, exclude or allocate land for development but, within RUF spaces, they rarely provide the integration and multi-functionality which is increasingly demanded (Henderson, 2005; Westerlink et al., 2013). Hence policy and decisions for RUF are largely reactive and discordant, fuelling conflict and leading to sprawl, social fragmentation, pollution and waste (Taylor and Hurley, 2016; Qvistrom, 2007; Niner and Brown, 2009; Cash, 2016) and, in some cases, the approved policies actually conflict, as in Ilberry's (1991) study of Birmingham's RUF.

What is evident from the academic and policy literatures is that the RUF has received limited attention as a place in its own right; rather it has been bolted on to predominantly urban expansion/regeneration or rural preservationist discourses (Scott et al., 2013; Gallent et al., 2006). Such urban–rural polarisation is both unhelpful and oversimplistic, restricting the opportunities and benefits that may arise from more holistic considerations of the RUF space itself (Qvistrom, 2007, 2010). Furthermore, the compartmentalisation and classification of data into rural and urban domains hinders effective RUF diagnostics. Thus, we have a limited grasp of what is happening in the RUF. For example there is scant evidence about RUF quality and potential; limited evidence of the needs and priorities of the people who live and work there (e.g. see Luka, 2017 on contrasting narratives between second home owners and residents in Canada), and species and human/non-human interactions in shaping the RUF (e.g. see Connolly, 2016 on Swfitlet farming in Malaysia). This problem is exacerbated when the RUF crosses several municipal boundaries, making coherent programmes of delivery increasingly problematic, although recent development and application of landscape-scale thinking has helped this (Wyborn and Bixler, 2013).

Thus the RUF becomes a reactive space where things often happen as by-products of urban or rural policy priorities (Rauws and de Roo, 2011). Such un-coordinated planning interventions magnify disconnects and local community impotence thus fuelling conflict and protest (Luka, 2017). According to Ravetz (2011, 2013), this reflects multi-level, multi-sectoral, multi-functional and multi-scalar attributes, thereby rendering any generalities and one size fits all RUF solutions oversimplistic, dangerous and incomplete. This 'messy' state of RUF affairs is compounded by the lack of clarity of the identity, nature and extent of exactly what is and is not RUF with definitional variation, vagueness and obfuscation apparent in the academic and policy literatures (Scott et al., 2013). This disintegration narrative provides a starting point for more interdisciplinary/transdisciplinary explorations (Tress et al., 2005) from the literature to help unpack the complexity, character and potential of RUF space.

It is this theoretical and practical quest for improving our understanding of RUF identity and character, realising the associated opportunities and benefits, that this chapter addresses. First, the chapter explores and positions RUF literatures within a functional typology. This is then unpacked through three international case studies which collectively inform a final section considering the role for RUF space as a hybrid priority space for rural planning.

Towards a typology of RUF literatures

While seven RUF research themes are identified from the literature below, it is important to realise that these are not meant to be mutually exclusive. Furthermore, they are bounded within a rural planning lens that forms the basis of this book. So this is not meant to be an exhaustive sift of all relevant literatures. The typology embraces both functional and disciplinary components across groupings of writers and hopefully adds value to other extant RUF typologies based on analysis of RUF spaces themselves. For example, Zasada et al. (2013) delineated a method to identify degrees of peri-urbanity by using population density of particular classes of land use/land cover and logistic regression models; Gonçalves et al. (2017) used a transdisciplinary approach to define five distinctive peri urban spaces within the Lisbon metropolitan area of Portugal; and Hornis and Van Eck (2009) used travel patterns within and across peri urban areas in the Netherlands to identify distinctive peri urban characters highlighting the key role of networks and scales.

Perceptions of peri-urban landscapes

There is a consensus across these works highlighting the strong attachment and sense of place to these 'ordinary landscapes' surrounding major urban centres (Qvistrom, 2007; Scott, 2011; Roberts and Farley, 2012). However, there is no consensus about the kind of fringe spaces that are wanted reflecting the different urban and rural lifestyles that clash in these occupied spaces (Garner, 2017; Luka, 2017). Some studies have reported common concerns around the pace, scale and quality of land-use change resulting in the perceived homogenisation and 'placelessness' of new developments which fail to respect local landscape character (Scott et al., 2009). Qvistrom's (2007, 2010, 2012) work in Malmo, Sweden, has highlighted the way different perceptions of the fringe lead to its view as a landscape 'out of order'. Here the tension is between the professional planners with their fetish for order and citizens and business with innovation, informality and entrepreneurship which challenges this professional perspective (see also Adams et al., 2014). These different professional versus lay discourses provide important narratives of RUF opportunity and creativity; all of which are rarely heard and accounted for in policy and decision-making (Garner, 2017; Beilin et al., 2013).

Urban form (morphology) and the implications of urban expansion

There is a substantive and long established literature exploring the formation processes and later fate of relict urban fringes now embedded within built-up areas (Scott et al., 2013; Larkham, 2003). The overall development pattern is one of 'belts' (Whitehand, 1967, 1988; Whitehand and Gu, 2017). Outward expansion of cities is driven by regional and national economies and is not continuous. In periods of economic growth and high land values, development is dense and usually favours high-density housing, while in periods of stagnation, land values are low and large sites can be acquired for 'institutional' land uses. These belts are recognisable, albeit in modified form, from periods of growth and stagnation of even the medieval period (Conzen, 1960). Such low-density 'fringe belts' can be recognised in numerous countries and cultures (e.g. see Conzen, 1960 for Whitby, England; Whitehand and Gu, 2017 for China; Ducom, 2005 for France). They provide particular environments and locations for activity, with specific implications for planning (Ducom, 2005; Whitehand and Morton, 2006) and ecology/biodiversity (Hopkins, 2012).

Cultural geographers, in particular, have shown a richness of approaches to, and experiences of suburbia where there is a strong RUF theme associated with sprawl (see for example, Harris and Larkham, 1999; Silverstone, 1997; Phillips, 2010). Here the residential development at varying densities produced in the high-growth-period belts produces characterless and homogeneous suburbia (Bruegmann, 2005). However, this is an elite view, and an over-simplification, as detailed morphological studies demonstrate (Whitehand and Carr, 2001). Whitehand (2001: 108) observes that, in this historical context, fringe belts are not products of coherent plan-making or decision making:

> They are products of large numbers of separate decisions about individual sites. Indeed the decisionmakers frequently had no knowledge of one another and almost invariably no conception of the way in which their decisions and those of others would in combination have the effect that we refer to as a fringe belt.
>
> *(Whitehand, 2001: 108)*

Yet the potential of fringe belts in terms of strategic value, ecological significance and development potential is high, if only these sites can be identified and treated in policy terms as coherent wholes, rather than as discrete 'windfall' sites (Hopkins, 2012).

Environmental and social justice

Sharma-Wallace (2016) makes a powerful case for an extension of environmental justice research to the RUF through a more systematic exploration and synthesis of its edge characteristics and transitional character. This provides a fertile space for more theoretical investigation in light of the inequalities that can occur in practice. Here residents' differential experiences within rapidly expanding cities can provide important perspectives on spaces of marginalisation, exclusion and poverty (Scott and Oelofse, 2005). There are interesting perspectives exposed with limited access to key resources such as water (Mehta et al., 2014) and new community infrastructure developments such as schools, shops and medical services (Scott, 2013). Additionally, new undesired facilities such as waste treatment plants can be imposed without providing the 'promised' community and biodiversity benefits in landscapes of power and influence (Cash, 2014). Understanding how social justice is embedded within existing power relations and planning decisions provides a useful political economy perspective on the political dynamics of RUF spaces representing an important but neglected component of sustainability (Taylor and Hurley, 2016; Cash, 2016).

Land-use narratives, models and planning for urban growth

The contested and highly pressurised RUF spaces generate clashes of different values and aspirations of the places that are wanted, reflecting the simultaneous intersection of consumption and production functions within the same geographic space (Slee, 2005; Phillips, 2010). Consequently, this fuels a tension between the prevailing urban-centric and economic narratives of economic growth with more environmental and community based narratives stressing localism, localisation and natural capital (Scott et al. 2013; Kenter et al., 2015). Such narratives are often seen in opposition leading to disintegrated policy and development but in the context of competent spatial planning should form part of a theoretical fusion as suggested by Cowell and Lennon (2014). However, they do not and the management and mediation of this conflict is undertaken primarily through regional and local government operation of nationally legislated planning systems. Hitherto, assessments of planning policy and practice have received scant academic attention in RUF research (Qvistrom, 2007; Gant et al., 2011); but more research on planning policy and development applications is now evident with a peri urban flavour (Gonçalves et al., 2017; Hornis and Van Eck, 2009; Lui and Robinson, 2016).

Planning systems (both urban and rural) are, however, heavily shaped by traditional neoclassical economics thinking, associated with traditional land value models reflecting the priority for market exchange within which urban centric ideals associated with economic growth flourish (Longlands, 2013) and trickle down to the rural periphery. Consequently, most statutory planning interventions associated with the RUF are technocentric, driven by a fetish for zoning and order (Adams et al., 2014; Qvström, 2007). Within such models, economic growth trumps other social and environmental considerations which are not easily costed and valued (Cash, 2016; Davies, 2013). Thus contemporary policy and practice has increased polarisation between the protected RUF spaces (e.g. green belts) on the one hand (Elson, 1986) versus other greenfield sites and brownfield spaces which are subjected to mounting planning pressures for new urban developments on the other (Gant et al., 2011). Here the RUF is polarised between being a zone of stagnation or rapid change (Hiner, 2016; Verdú-Vázquez et al., 2017).

Exchange and hybrid spaces

Building on ideas first exposed by Hodge and Monk (2004), on urban–rural interrelationships, places become defined from the pattern of interrelationship and dependencies between them rather than simply conforming to urban or rural characterisations. Significantly, within RUF spaces we are seeing new flows of dependencies and relationships associated with rural land uses within urban spaces in contrast to the more established urban expansion flows as the city has moved inexorably into the countryside. Thus we see the rise of urban agriculture (Hardman et al., 2018) and urban forests (La Rosa et al., 2018; Tu et al., 2016) with important outcomes for optimising the delivery of ecosystem services benefits to citizens within RUF locations (UKNEAFO, 2014; Tu et al., 2016; Caro-Borrero et al., 2015). As already mentioned, the importance of networks and exchange was a key feature in Hornis and Van Eck's (2009) work on travel patterns in the Netherlands RUF.

Innovation and creativity spaces

Using more proactive, strategic and positive examinations of RUF identity (or identities), there is a set of literatures focused around the RUF as innovation and creativity spaces. This has two main foundations: multi-functional 'innovation', and 'service' interdependency (Ravetz, 2006, 2011). We can distinguish between:

- *Social innovation* – for example, social enterprise, cooperative business, socio-ecological communities, enabled by social learning and capacity building. See for example, Eakin et al.'s (2010) investigation into adaptive management contrasting resident and public officials' interpretations within a case study of flooding in the Upper Lerma River Valley, Mexico.
- *Policy innovation* – for example, devolution from central government, partnership, networked governance, decision making, enabled by new social media and new forms of governance. See for example, Scott et al.'s (2013) historical review of the countryside management approach as a hidden and unexploited innovation.
- *Economic innovation* – for example, ecosystems services, new models of stewardship, localisation, management of commons. See for example, Verdú-Vázquez et al.'s (2017) work on the characterisation, analysis and evaluation of the resilience and dynamic of the green belt and associated adaptive eco-design of green infrastructures.

- *Technical innovation* – for example, new transport networks, new concepts of local agri-food, low carbon transitions. See for example, Zasada's (2011) work on post-productive characteristics and potential for peri-urban agriculture.
- *Environmental innovation* – for example, landscape ecology and ecological systems thinking, market-based instruments and environmental planning. See for example, Radford and James's (2013) creation of a new analytical tool for the non-economic valuation of a number of ecosystem services in Manchester.

Holistic and systems-based approaches

These approaches generally accord with social ecological systems thinking (Bruckmeier, 2016) which has provided a wealth of new models, toolkits and even games to help understand and view processes of RUF dynamics (Piorr et al., 2011; Scott et al., 2013). Here the RUF is being framed within wider placemaking and place-based agendas (Hiner, 2016; Westerlink et al., 2013) championing more interdisciplinary and mixed methods approaches. There is some important conceptual development here given that the RUF can also represent the academic interface of built and natural environments with the intersection of spatial planning and ecosystem approach paradigms. Previously these have been developed and operationalised within separate disciplines, governance frameworks and tools (Scott et al., 2013, 2018). Significantly, Cowell and Lennon (2014) stress the importance of using social learning and methodological approaches that better incorporate and integrate competing theories and ideas rather than producing yet more complexity and competition through such creeping theoretical incrementalism. Thus by deconstructing their competing principles within a simple mapping exercise, we can start to identify areas of convergence which, offer potential hybrid policy pathways to turn conceptual theory into more effective RUF practice and delivery (Scott et al., 2013). This theoretical convergence helps signpost the three case studies that follow.

Case studies

From this typology of academic literatures, three RUF case studies are now considered reflecting different RUF characterisations and written specifically by the authors themselves for this chapter. These 'deep dive' case studies provide a fertile avenue to further illuminate the RUF opportunity space.

Rediscovering an asset-based framework for RUF areas in England (Alister Scott)

In the late 1970s in England, the Countryside Commission made a series of bold new policy interventions in the RUF based around a 'countryside management' approach (CMA). This was partly in response to local government reorganisation with the perceived need to provide a human 'bridge' between the more distanced local government structures and the needs of local communities. Countryside project officers were financed through grant-aid programmes within local authorities to implement community-based projects in response to challenges as cities and towns expanded into the RUF. The project officers acted as mediator, negotiator and enabler between the needs of the communities; the needs and capacities of the environment; and the delivery of plans and policies. In such aspects, the process of building community capacity and skills was deemed of equal importance as project outcomes. Significantly, CMA directly challenged sectoral thinking through its focus on integration, joining up different policy priorities and managing conflicting needs of the communities, environment and place. This shaped a

Table 41.1 Comparing spatial planning principles with the Convention on Biological Diversity ecosystem approach.

Spatial planning principles	Ecosystem approach principles
The Governance Principle (e.g. authority. legitimacy, institutions power; decision making) (e.g. Tewdwr Jones et al., 2010; Kidd, 2007)	1 The objectives of management of land, water and living resources are a matter of societal choice. 3 Ecosystem managers should consider the effects (actual or potential) of their activities on adjacent and other ecosystems. 9 Management must recognise the change is inevitable.
The Subsidiarity Principle (e.g. delegation to lowest level; shared responsibility; devolution) (e.g. Haughton and Allmendinger, 2014)	2 Management should be decentralised to the lowest appropriate level.
The Participation Principle (e.g. consultation; inclusion; equity; deliberation) (e.g. Albrechts, 2015; Gilliland and Laffoley, 2008)	11 The ecosystem approach should consider all forms of relevant information, including scientific and indigenous and local knowledge, innovations and practices. 12 The ecosystem approach should involve all relevant sectors of society and scientific disciplines.
The Integration Principle (e.g. holistic; multiple scales and sectors; joined up) (e.g. Low, 2002; Mommaas and Janssen, 2008)	3 Ecosystem managers should consider the effects (actual or potential) of their activities on adjacent and other ecosystems. 5 Conservation of ecosystem structure and functioning, in order to maintain ecosystem services, should be a priority target of the ecosystem approach. 7 The ecosystem approach should be undertaken at the appropriate spatial and temporal scales. 8 Recognising the varying temporal scales and lag effects that characterise ecosystem processes, objectives for ecosystem management should be set for the long term. 10 The ecosystem approach should seek the appropriate balance between, and integration of, conservation and use of biological diversity.
The Proportionality Principle (e.g. deliverable viability; pragmatism; best available information) (e.g. Nadin, 2007)	4 Recognising potential gains from management, there is usually a need to understand and manage the ecosystem in an economic context. 9 Management must recognise the change is inevitable.
The Precautionary Principle (e.g. adaptive management; limits; uncertainty; risk) (e.g. Counsell, 1998)	6 Ecosystem must be managed within the limits of their functioning, 8 Recognising the varying temporal scales and lag effects that characterise ecosystem processes, objectives for ecosystem management should be set for the long term. 10 The ecosystem approach should seek the appropriate balance between, and integration of, conservation and use of biological diversity,

Sources: Scott et al. (2018: 233); www.cbd.int/ecosystem

significant Countryside Agency research programme between 2001 and 2006 illuminating the RUF opportunity space and its potential (see Gallent et al., 2006). From this work emerged a positive and holistic vision set within ten core themes with creativity, innovation and subsidiarity at its core (Countryside Agency and Groundwork Trust, 2005):

(1) A bridge to the country.
(2) A gateway to the town.
(3) A health centre.
(4) A classroom.
(5) A recycling and renewable energy centre.
(6) A productive landscape.
(7) A cultural legacy.
(8) A place for sustainable living.
(9) An engine for regeneration.
(10) A nature reserve.

The 'Countryside Around Town' initiative was the delivery vehicle to achieve this vision but crucially was stalled and failed to become mainstreamed in policy and practice.

Hidden research and practice

Much of this Countryside Agency work now lies inaccessible to the public, buried and forgotten due to institutional change and merger (Countryside Commission to Natural England) with no electronic outputs. However, the CMA philosophy, experience and outcomes may provide a useful way forward today for realising the assets of RUF spaces, set within a critical review of the lessons learnt from earlier failures. Undertaking this type of reflective gaze is rare in research, policy and practice but the lessons from this failure could help us signpost a pathway to the integration we currently desire in RUF spaces today.

Lessons

- The launch of the 2005 vision coincided with the creation of Natural England, involving the merger of the Countryside Agency, English Nature and parts of the Rural Development Commission within a new government Non-Departmental Public Body. This involved significant re-structuring of staff with new functions and responsibilities resulting in a hiatus in existing programmes, including the RUF (Countryside Around Towns initiative).
- Many countryside managers were pioneers, with considerable flexibility and freedom to pursue their work with limited managerial interventions. They were located in various local authority departments across the UK (e.g. tourism, planning, recreation and environment). As these were new appointments, senior managers were ill-equipped to understand their work role, exacerbated by the rapid turnover of staff in these CMA positions.
- The influence of Countryside Commission grant aid budgets was crucial in driving appointments which provided a significant income stream to stressed local authority budgets. However, when grant aid was switched away from specific posts towards outputs, many countryside management projects in the RUF were phased out.
- The CMA projects were generally small scale and practical; they did build community capacity, capability and skills and as such dovetail with current localism initiatives associated with neighbourhood plans and green infrastructure.

- Although CMA was championed as a new model working across traditional boundaries and silos with active involvement of communities and stakeholders, this was increasingly seen as subsidiary and separate to statutory planning processes. There was limited contact with statutory policies and decision-making associated with planning, health, education, transport and social services. Thus, CMA interventions were being carried out separately to the statutory work of the local authorities, resulting in CMA being disconnected from conventional statutory functions and hence vulnerable to cutting when resources were scarce.
- The ten principles for RUF today map well into the current natural capital agendas which involve taking an asset based approach. The principles were built on solid research foundations and thus should be reinvigorated within a new round of CMA project officers to support RUF identity and character. Here, RUF woodlands and food growing offer significant potential.

Conclusion

The overriding lesson from this experience highlights the importance not only of embedding new approaches such as CMA into existing governance structures but also changing behaviours across all the key stakeholders themselves so that CMA can become mainstreamed (Scott et al., 2018). Because CMA was seen as peripheral and additional to policy and decision-making processes, it was ultimately vulnerable to cuts with all the loss of expertise and intelligence that entailed.

Managing social justice in Jamestown, South Africa (Corrine Cash)

Location

Jamestown is a part of Stellenbosch Municipality and reflects the race-class organisation of space typical of apartheid, with marginalised 'Cape Coloured' people occupying the outskirts of the urban centre. It is surrounded by attractive and highly valued countryside and is world famous for wine growing. In the post-apartheid era, the Municipality was tasked with delivering sustainable development across the newly demarcated urban area which also includes areas such as Jamestown and the high-density township of Kayamandi. The key delivery vehicle for this was the Integrated Development Plan.

Land use

Historically the area consisted of subsistence and small-scale agriculture. Jamestown was gazetted as a Group Area for 'Cape Coloured' people in 1944. It sits along a secondary highway connecting the wealthy historically white towns of Stellenbosch and Somerset West. Since 1993, land use has changed dramatically: adjacent to Jamestown proper are car dealerships and a shopping mall, and across the street is a golf course and gated community named De Zalze. The entire area is also part of the Cape Winelands Biosphere Reserve, so development is ostensibly subject to restrictions pertaining to the health of the natural environment.

Tensions

There are considerable tensions across the different communities, stakeholders and decision-makers, which all revolve around the pressures for development and how they are managed.

Here there is a significant disconnect between the theory of the plan as published and decision-making in practice. This creates winners and losers in a resource management process that has led to one of the highest gini coefficients in the world. Specifically:

- The housing market puts pressure for expansion rather than densification so Greenfield development is preferred to redevelopment of brownfield sites. This is driven by both students (Stellenbosch is home to an important South African university), and domestic and international buyers.
- Planners believe that housing development is an economic multiplier with trickle down effects to the rest of the region.
- A proposed golf course was presented as sustainable development in the biosphere plan (though this is clearly contentious particularly in the context of regular and increasing instances of severe drought in the Western Cape). The golf course created rifts within the Municipality itself where planners saw it as contravening the IDP but politicians were supporting it in spite of an approved plan.
- Developers promised Jamestown residents a Community Trust as part of the betterment process for developments in their community. This helped secure support. However, the Trust never materialised thereby creating mistrust among races and classes in post-apartheid Stellenbosch Municipality.

Outcomes

Far from achieving an integrated development outcome, what is becoming evident is community disintegration, where groups live together but separately with increasing tensions and inequalities. This contradicts the core goals of the IDP process. Hence, the RUF continues to be a place of contradicting and messy land-use practices that, in the South African case, stem from and reflect the continuing social and cultural disintegration in these spaces that was codified in law under apartheid.

Conclusions

Post apartheid South Africa has some of the most advanced legislation on the books. Adherence to the letter of the law, however, is another matter. What the Jamestown case makes clear is that you cannot legislate change where the change envisioned challenges existing social relations and configurations of power. So, despite shared physical space, much of the apartheid dynamics still remain 'hidden' in play.

Managing the rural urban fringe as a test bed for sustainable planning in Scania, Sweden (Mattias Qviström)

Rationale

In order to gain deeper understanding of the interaction between planning and land-use change at the RUF, a series of research projects were set up between 2004 and 2016, focusing on southern Scania and examining the evolution and character of the RUF of Malmö (e.g. Qviström 2007, 2013, 2018). The projects combined ethnographic methods (e.g. field studies and interviews) with landscape/planning history and policy analysis of contemporary planning, to illuminate everyday uses and incremental land-use changes, and cumulative impacts of a planning regime which is still predicated on a rural–urban divide. The studies position the RUF as a testbed for sustainable planning, and by revealing the intricate and deep-seated conflicts therein allow us to gain improved intelligence for planning beyond the RUF too, especially regarding managing rural–urban conflicts.

Background

The southern part of region of Scania, Sweden, has some of the most productive soils in the country. It is one of the most densely populated areas in Sweden, with expanding cities and increasing RUF development. The countryside is characterised by large plains dominated by farming, leaving very little space for recreation (beyond the coastal zones) or maintaining/enhancing the biodiversity of the region. Farming is done on an industrial scale, oriented towards a global market, with pockets of hobby farms and RUF settlements. At the inner urban fringe, land waiting to be developed offers a temporary zone of importance for everyday recreation and play. While the Planning and Building Act 2010 and the Environmental Code offers limited protection for farmland, the public debate on planning has nevertheless come to focus on the need for imposing growth-boundaries rather than on how to manage the RUF.

Lessons

The RUF is characterised by temporary uses, incremental processes, delayed projects and places where uncertainty prevails. This mix of temporalities is very difficult to handle within the planning system, which emphasises clearly defined (and thus static) categories of land-use inhibiting innovation. This presents a significant obstacle for planners and those interests which seek to optimise the assets of the RUF within wider placemaking initiatives.

The studies detected a remarkable number of institutional and policy silos standing in the way of an integrated approach to the RUF. Here Qviström has used landscape as a multifunctional and temporal framework for illuminating and demystifying the fringe beyond the usual administrative, conceptual, sectoral or legal silos.

By taking an historical perspective, a better understanding of RUF challenges and potential is captured. For instance, in Sweden land politics dating back to the 1960s still influences planning, and the conflicts it caused with farmers then still frames their engagement with planning and planners, which is based on mistrust and suspicion. Thus, innovative thinking easily gets stranded unless planning history (and the difficulties of silo thinking) is taken into consideration together with the micropolitics that exist. Relational geography – e.g. the notion of asymmetric analysis and of hybrids – has been an increasingly useful concept to grasp the potentials of the RUF moving beyond urban–rural divides (Qviström, 2013). For example, work has looked at its post-industrial nature developed in abandoned gravel pits, green houses and military grounds.

Conclusion

The case study of Malmö resulted in an applied project using game based approaches in which rural and urban stakeholders were invited to dialogues on developing a planning strategy for the RUF. Such a proactive planning approach has considerable potential. Despite its dynamic geography, a spatial planning strategy for the RUF offers a pathway to exploit the place based potential and significantly Malmö is now starting to develop such a strategy as a world first.

Effective rural planning in and for the RUF: future prospects and priorities for rural planning

The case studies and literature review collectively share common components that could form a fertile focus for future rural planning research and practice agendas. At the core of each are different notions and experiences of place-making. The transitory and contested nature of RUF character presents significant challenges set within the increasing pace and scale of change. For the most part, this change has been driven by urban centric ideas, associated with domination of nature (Adams et al., 2014) and SMART engineered cities, which have led to rapid urban expansion, heralding the arrival of a diverse range of social, economic and environmental challenges that characterise the RUF. However, a notable characteristic of twenty-first-century RUFs is the way these areas are also becoming increasingly transformed by traditional rural land uses associated with new spaces of woods, forest and urban agriculture (e.g. La Rosa et al., 2018). In addition, new market-based instruments associated with the newly emerging natural capital and ecosystem services agendas are beginning to feature (Caro-Borrero et al., 2015; Jiang and Swallow, 2015). This rural-centric transformation provides important opportunity spaces for rural planners to help mitigate the worst excesses of past and present urban centric agendas and to influence the kind of RUF identity and character for the future.

However, drawing from all three case studies and successful place-making initiatives, one core prerequisite to enable this is the development of a bold vision that can be created or co-designed to engage publics and other stakeholders as well as lever the necessary finance (La Rosa et al., 2018). It is the leadership and conflict management associated with the delivery of this vision that is crucial to the future success of RUF spaces as centres of creativity and innovation as stressed by Ravetz (2013, 2014). It is here that Qviström's (2007, 2010, 2018) contribution is most valuable in highlighting the need for planners and decision-makers to become more adaptable and less risk-averse in the RUF, moving away from their perceived fetish for order and to embrace more innovation and entrepreneurship. This requires a significant change in culture as noted by Adams et al. (2014), with a willingness to use more temporary/flexible planning permissions and agreements and to develop more innovative policies that support development as well as investing in the necessary resources (i.e. through a new breed of countryside managers) to support and deliver place-making initiatives.

Given the rapid pace of RUF dynamics and pressures, my own work on CMA suggests that this might offer a suitable support and delivery model (see the first of the three case studies in this chapter). However, in such aspects, it is necessary to learn the lessons from their past failures and to integrate project officers more effectively into statutory planning and decision-making processes. Working at, and crucially, across the RUF interface and mediating between the needs of place, people and the environment, arguably rediscovers a powerful model for rural planning practice and theory. Here the practice component builds on the Countryside Agency and Groundwork Trust's (2005) ten ingredients for RUF planning; while the theoretical component builds upon the merger of the ecosystem approach and spatial planning paradigms (Scott et al., 2018). This type of theoretical and practical convergence develops Cowell and Lennon's (2014) ideas for competing theoretical mergers with a shift away from the fringe as either an urban or rural led construct waiting for something better to come along.

Both Sharma-Wallace (2016) and Cash (2016) caution us that planning for the RUF can all too easily create spaces of exclusion and marginalisation with negative environmental and social justice outcomes. Exposing and understanding the different lived-in experiences of RUF inhabitants – for example, Kirkey and Forsyth's (2001) landmark study on gay men and Luka's (2017) work with amenity migrants and long term residents – provides an important reality check of the efficacy of past and present planning interventions. Through more effective evaluation of interventions (a notable gap in practice), we can start to capture and exchange institutional memory as part of a wider social learning and knowledge exchange agendas. Significantly, there are such studies emerging, as evidenced in Lui and Robinson's study (2016) of Adelaide. However, in general current literature and practice does not really address this.

Finally, the case studies reveal my fervent hope to move from studying rural and or urban spaces in isolation to start to move towards more integrated work studying the dependencies between places and spaces. It is ultimately from these new understandings and improved RUF diagnostics (ideally within data observatories) that we can start to make better interventions. This leads to changing the way we 'see' the RUF and I leave the final words to Farley and Roberts, who provide perhaps the most compelling call to action which requires us to rethink how we view these places and indeed challenge the conventional rural or urban labels we utilise that restrict or hinder RUF potential:

> Somewhere in the hollows and spaces between our carefully managed wilderness areas and the creeping, flattening effects of global capitalism there are still places where an overlooked England truly exists, places where ruderals familiar here since the last ice sheets retreated have found a way to live with each successive wave of new arrivals, places where the city's dirty secrets are laid bare, and successive human utilities scar the earth or stand cheek by jowl with one another; complicated, unexamined places that thrive on disregard, if we only could put aside our nostalgia for places we've never really known and see them afresh.
>
> *(Farley and Roberts, 2012: 10)*

Acknowledgements

The author gratefully acknowledges the additional input into the case studies section of this chapter, namely Corrine Cash (Assistant Professor at St Francis Xavier University, Canada) for the Jamestown (SA) contribution and Mattias Qviström (Professor in Landscape Architecture, chair in Spatial Planning, at the Department of Urban and Rural Development, SLU, Uppsala, Sweden) for the case study of Scania, Sweden.

References

Adams, D., Scott, A.J. and Hardman, M. (2014) Guerrilla warfare in the planning system: revolutionary progress towards sustainability? *Geografiska Annaler B. Series B, Human Geography*, 95(4), 375–387.

Albrechts, L. (2015) Ingredients for a more radical strategic spatial planning. *Environment and Planning B*, 42 (3), 510–525.

Beilin, R., Reichelt N. and Sysaket, T. (2013) Resilience in the transition landscapes of the peri-urban: from 'where' with 'whom' to 'what'. *Urban Studies*, 52(7), 1–17.

Bruckmeier, K. (2016) *Social-Ecological Transformation: Reconnecting Society and Nature*. London: Palgrave Macmillan.

Bruegmann, R. (2005) *Sprawl: A Compact History*. Chicago, IL: University of Chicago Press.

Caro-Borrero, A., Corbera, E., Neitzel, K.C. and Almeida-Lenero, L. (2015) 'We are the city lungs': payments for ecosystem services in the outskirts of Mexico City. *Land Use Policy*, 43, 138–148.

Cash, C. (2014) Towards achieving resilience at the rural–urban fringe: the case of Jamestown, South Africa. *Urban Forum*, 25, 125–141.

Cash, C. (2016) Good governance and strong political will: are they enough for transformation? *Land Use Policy*, 50, 301–311.

Connolly, C. (2016) 'A place for everything': moral landscapes of 'swiftlet farming' in George Town, Malaysia. *Geoforum*, 77, 182–191.

Conzen, M.R.G. (1960) *Alnwick, Northumberland: A Study in Town Plan Analysis.* London: Institute of British Geographers.

Counsell, D. (1998) Sustainable development and structure plans in england and wales: a review of current practice. *Journal of Environmental Planning and Management,* 41(2), 177–194.

Countryside Agency and Groundwork Trust. (2005) *The Countryside in and around Towns: A Vision for Connecting Town and Country in the Pursuit of Sustainable Development.* Cheltenham: Countryside Agency.

Cowell, R. and Lennon, M. (2014) The utilisation of environmental knowledge in landuse planning: drawing lessons for an ecosystem services. *Environment and Planning C,* 32(2), 263–282.

Cresswell, T. (1997). Weeds, plagues, and bodily secretions: a geographical interpretation of metaphors of displacement. *Annals of the Association of American Geographers,* 87(2), 330–345.

Davies, H. (2013) Environmental justice and equal protection: intent, motivation and cognition in decision making. *Journal of Human Rights and the Environment,* 4(2), 191–214.

Ducom, E. (2005) Fringe belts in French cities: comparative study of Rennes, Nantes, Tours. In M. Barke (ed.), *Approaches in Urban Morphology.* Newcastle upon Tyne: University of Northumbria.

Eakin, H., Lerner, A.M. and Murtinho, F. (2010) Adaptive capacity in evolving peri-urban spaces: responses to flood risk in the upper Lerma River valley, Mexico. *Global Environmental Change,* 20, 14–22.

Elson, M. (1986) *Green Belts: Conflict Mediation in the Urban Fringe.* London: Heinemann.

Farley, P. and Roberts, M.S. (2012) *Edgelands: Journeys into England's True Wilderness.* London: Vintage.

Gallent, N., Bianconi, M. and Andersson, J. (2006) Planning on the edge: England's rural–urban fringe and the spatial-planning agenda. *Environment and Planning B,* 33, 457–476.

Gant, R.L., Robinson, R.M. and Fazal, G. (2011) Land-use change in the 'edgelands': policies and pressures in London's rural–urban fringe. *Land Use Policy,* 28(1), 266–279.

Garner. B. (2017) 'Perfectly positioned': the blurring of urban, suburban, and rural boundaries in a southern community. *Annals of the American Academy,* APSS 672, 46–63.

Gilliland, P. and Laffoley, D. (2008) Key elements and steps in the process of developing ecosystem-based marine spatial planning. *Marine Policy,* 32(5), 787–796.

Gonçalves, J., Gomes, M.C., Ezequiel, S., Moreira, F. and Loupa-Ramos, I. (2017) Differentiating peri-urban areas: a transdisciplinary approach towards a typology. *Land Use Policy,* 63, 331–341.

Hardman, M., Chipungu, L., Magidimisha, H., Larkham, P.J. and Scott, A.J. (2018) Guerrilla gardening and green activism: rethinking the informal growing movement. *Landscape and Urban Planning,* 170, 6–14.

Harris, R. and Larkham, P.J. (eds) (1999) *Changing Suburbs: Foundation, Form and Function.* London: Spon.

Haughton, G. and Allmendinger, P. (2014) Spatial planning and the new localism. *Planning Practice & Research,* 28(1), 1–5.

Henderson, K. (2005) Tensions, strains and patterns of concentration in England's City Regions. In K. Hoggart (ed.), *The Citys Hinterland: Dynamism and Divergence in Europe's Peri Urban Territories,* 119–155. Ashgate: Aldershot.

Hiner, C.C. (2016) Beyond the edge and in between: (re)conceptualizing the rural–urban interface as meaning–model–metaphor. *The Professional Geographer,* 68(4), 520–532.

Hodge, I. and Monk, S. (2004). The economic diversity of rural England: stylised fallacies and uncertain evidence. *Journal of Rural Studies,* 20(3), 263–272.

Hopkins, M.I.W. (2012) The ecological significance of urban fringe belts. *Urban Morphology,* 16(1), 41–54.

Hornis, W. and Van Eck, J.R. (2009) A typology of peri-urban areas in the Netherlands. *Tijdschrift voor Economische en Sociale Geografie,* 99(5), 619–628.

Hough, G. (1990) *Out of Place: Restoring Identity to the Regional Landscape.* New Haven, CT: Yale University Press.

Ilberry, B.W. (1991) Farm diversification as an adjustment strategy on the urban fringe of the West Midlands. *Journal of Rural Studies,* 7(3), 207–218.

Jiang, Y. and Swallow, S.K. (2015) Providing an ecologically sound community landscape at the urban–rural fringe: a conceptual, integrated model. *Journal of Land Use Science,* 10(3), 323–341.

Kenter, J.O., et al. (2015) What are shared and social values of ecosystems? *Ecological Economics,* 111, 86–89.

Kidd, S. (2007) Towards a framework of integration in spatial planning: an exploration from a health perspective. *Planning Practice & Theory,* 8(2), 161–181.

Kirkey, K. and Forsyth, A. (2001) Men in the valley: gay male life on the suburban–rural fringe. *Journal of Rural Studies,* 17, 421–441.

Larkham, P.J. (2003) The teaching of urban form. In A. Petruccioli, M. Stella and G. Strappa (eds), *The Planned City.* Bari: Uniongrafica Corcelli Editrice.

La Rosa, D., Geneletti, D., Spyra, M., Albert, C. and Fürst, C. (2018) Sustainable planning for peri-urban landscapes. In A. Perera, U. Peterson, G. Pastur and L. Iverson (eds), *Ecosystem Services from Forest Landscapes*. Cham: Springer.

Longlands, S.L.J. (2013) Growing nowhere: privileging economic growth in planning policy. *Local Economy*, 28(7–8), 894–905.

Low, N. (2002) Ecosocialisation and environmental planning: a Polanyian approach. *Environment and Planning A*, 34(1), 43–60.

Lui, Z. and Robinson G. (2016) Residential development in the peri-urban fringe: the example of Adelaide, South Australia. *Land Use Policy*, 57, 179–192.

Luka, N. (2017) Contested periurban amenity landscapes: changing waterfront 'countryside ideals' in central Canada. *Landscape Research*, 42(3), 256–276.

McGregor, D., Simon, D. and Thompson, D. (eds) (2006) *The Peri-Urban Interface: Approaches to Sustainable Natural and Human Resource Use*. London: Earthscan.

McKenzie, F. (1997) Growth management or encouragement? A critical review of land use policies affecting Australia's major exurban regions. *Urban Policy and Research*, 15(2), 83–99.

Mehta, L., Allouche, J., Nicol, A. and Walnycki, A. (2014) Global environmental justice and the right to water: the case of peri-urban Cochabamba and Delhi. *Geoforum*, 54, 158–166.

Meinig, D.W. (ed.). (1979) *The Interpretation of Ordinary Landscapes*. New York: Oxford University Press.

Mommaas, H. and Janssen, J. (2008) Towards a synergy between 'content' and 'process' in Dutch spatial planning: the Heuvelland case. *Journal of Housing and the Built Environment*, 23, 21–35.

Nadin, V. (2007) The emergence of the spatial planning approach in England. *Planning Practice & Research*, 22(1), 43–62.

Niner, P. and Brown, P. (2009) First steps towards regional planning for Gypsy and Traveller sites in England: evidence based planning in practice. *Town Planning Review*, 80(6), 627–646.

OECD. (2011) *OECD Rural Policy Reviews*. Paris: OECD Publishing.

OECD. (2013) *Rural Policy Reviews: Rural–Urban Partnerships: An Integrated Approach to Economic Development*. Paris: OECD Publishing.

Phillips, M. (2010). Rural community vitality and malaise: moving beyond the rhetoric. In ESRC/The Scottish Government (ed.), *Rural Community Empowerment in the 21st Century: Building a 'Can-Do' Culture*, 9–12. Swindon: ESRC.

Piorr, A., Ravetz, J. and Tosics, I. (2011) *Peri Urbanisation in Europe: Towards European Policies to Sustain Urban Rural Futures*. Copenhagen: PLUREL.

Qvistrom, M. (2007) Landscapes out of order: studying the inner urban fringe beyond the rural–urban divide. *Geografiska Annaler*, 89B(3), 269–282.

Qvistrom, M. (2010) Shadows of planning: on landscape/planning history and inherited landscape ambiguities at the urban fringe. *Geografiska Annaler*, 92B(3), 219–235.

Qvistrom, M. (2012) Contested landscapes of urban sprawl: landscape protection and regional planning in Scania (Sweden), 1932–1947. *Landscape Research*, 37, 399–415.

Qvistrom, M. (2013) Searching for an open future: planning history as a means of peri-urban landscape analysis. *Journal of Environmental Planning and Management*, 56, 1549–1569.

Qvistrom, M. (2018) Farming ruins: a landscape study of incremental urbanization. *Landscape Research*, 43, 575–586.

Radford, K.G. and James, P. (2013) Changes in the value of ecosystem services along a rural–urban gradient: a case study of Greater Manchester, UK, *Landscape and Urban Planning*, 109(1), 117–127.

Rauws, W.S. and de Roo, G. (2011) Exploring transitions in the periurban area. *Planning Theory and Practice*, 12(2), 269–284.

Ravetz, J. (2006) Regional innovation and resource productivity – new approaches to analysis and communication. In S. Randles and K. Green (eds), *Industrial Ecology and Spaces of Innovation*, 45–76. Cheltenham: Edward Elgar.

Ravetz, J. (2011) Peri-urban ecology: green infrastructure in the twenty-first century metro-scape. In I. Douglas, D. Goode, M.C. Houck and R. Wang (eds), *The Routledge Handbook of Urban Ecology*. London: Routledge.

Ravetz, J. (2013) Introduction: from 'sustainable' city-regions to synergistic. *Town and Country Planning*, 8, 402–407.

Ravetz, J. (2014) Inter-connected responses for inter-connected problems: synergistic thinking for local urban development in a global urban system. *International Journal of Global Environmental Issues*, 13(2–4), 362–388.

Roberts, M. and Farley, P. (2012) *Edgelands: Journeys into England's True Wilderness*. London: Jonathan Cape.

Scott, A. (2011) Beyond the conventional: Meeting the challenges of landscape governance within the European Landscape Convention? *Journal of Environmental Management*, 92(10), 2754–2762.

Scott, A.J. (2013) Re-thinking English planning: managing conflicts and opportunities at the urban–rural fringe. In K. Shaw and J. Blackie (eds), *New Directions in Planning: Beyond Localism*, 20–27. Newcastle: University of Northumbria.

Scott, A.J., Shorten, J., Owen, R. and Owen, I.G. (2009) What kind of countryside do we want: perspectives from Wales. *Geojournal*, 76, 417–436.

Scott, A.J., et al. (2013) Disintegrated development at the rural urban fringe disintegrated Scott (2011) Re-connecting spatial planning theory and practice. *Progress in Planning*, 83, 1–52.

Scott, A.J., Carter C., Hardman, M., Grayson, N. and Slaney, T. (2018) Mainstreaming ecosystem science in spatial planning practice: exploiting a hybrid opportunity space. *Land Use Policy*, 70, 232–246.

Scott, D. and Oelofse, C. (2005) Social and environmental justice in South African cities: including 'invisible stakeholders' in environmental assessment procedures'. *Journal of Environmental Planning and Management*, 48(3), 445–468.

Sharma-Wallace, L. (2016) Toward an environmental justice of the rural–urban interface. *Geoforum*, 77, 174–177.

Silverstone, R. (ed.) (1997) *Visions of Suburbia*. London: Routledge.

Slee, R.W. (2005) From countrysides of production to countrysides of consumption. *The Journal of Agricultural Science*, 143(4), 255–265.

Taylor, L. and Hurley, P. (eds) (2016) *A Comparative Political Ecology of Exurbia: Planning, Environmental Management, and Landscape Change*. Cham: Springer.

Tewdwr-Jones, M., Gallent, N. and Morphet, J. (2010) An anatomy of spatial planning: coming to terms with the spatial element in UK planning. *European Planning Studies*, 18(2), 239–257.

Tress, B., Tress, G. and Fry, G. (2005) Integrative studies on rural landscapes: policy expectations and research practice. *Landscape and Urban Planning*, 70(1–2), 177–191.

Tu, G., Abildtrup, J. and Garcia, S. (2016) Preferences for urban green spaces and peri-urban forests: an analysis of stated residential choices. *Landscape and Urban Planning*, 148, 120–131.

UKNEAFO (2014) UK National Ecosystem Assessment Follow on: synthesis of the key findings. Retrieved from http://uknea.unep-wcmc.org/LinkClick.aspx?fileticket=5L6%2fu%2b%2frKKA%3d&tabid=82.

Verdú-Vázquez, A., Fernandez-Pablos, E., Lozano-Diez, R.V. and Lopez-Zaldivar, O. (2017) Development of a methodology for the characterization of urban and periurban green spaces in the context of supra-municipal sustainability strategies. *Land Use Policy*, 69, 75–84.

Westerlink, J., Lagendijk, A., Duhr, S., Van der Jagt, P. and Kempenaar, A. (2013) Contested Spaces? The use of place concepts to communicate visions for peri-urban areas. *European Planning Studies*, 21(6), 780–800.

Whitehand, J. W. R. (1967). Fringe belts: a neglected aspect of urban geography. *Transactions of the Institute of British Geographers*, 41, 223–233.

Whitehand, J.W.R. (1988). Urban fringe belts: development of an idea. *Planning Perspectives*, 3, 47–58.

Whitehand, J.W.R. and Carr, C.M.H. (2001) The creators of England's inter-war suburbs. *Urban History*, 28, 218–234.

Whitehand, J.W.R. and Gu, K. (2017) Urban fringe belts: evidence from China. *Environment and Planning B*, 44(1), 80–99.

Whitehand, J.W.R. and Morton, N. J. (2006) The fringe-belt phenomenon and socio-economic change. *Urban Studies*, 43(11), 2047–2066

Wyborn, C. and Bixler, R.P. (2013) Collaboration and nested governance: scale dependence, scale, framing and cross-scale interactions in collaborative conservation. *Journal of Environmental Management*, 123, 59–67.

Zasada, I (2011) Multifunctional peri-urban agriculture-a review of societal demands and the provision of goods and services by farming. *Land Use Policy*, 28, 639–648.

Zasada, I., Loibl, W., Berges, R., Steinnocher, K., Koestl, M., Piorr, A. and Werner, A. (2013) Rural–urban regions: a spatial approach to define urban–rural relationships in Europe. In K. Nilsson, S. Pauleit, S. Bell, C. Aalbers and T.A.S. Nielsen (eds), *Peri-urban Futures: Scenarios and Models for Land Use Change in Europe*, 45–68. Berlin: Springer.

42

Integrating green infrastructure within landscape perspectives to planning

Ian Mell

Introduction

How we manage our environmental resource base is subject to intense pressures. These include the need to ensure sufficient space is made available for housing and other built infrastructure, for food production, and to meet ecological system needs. However, we are currently in a position where our use of the landscape is undermining its functionality (Benedict and McMahon, 2006). Since the mid-1990s, and particularly over the last decade, green infrastructure has developed as the 'go-to' approach within landscape planning to moderate the impacts of development and manage the environment in a more sustainable manner.

A significant proportion of green infrastructure research focusses on its design, implementation and management within urban areas. However, the basic premise of green infrastructure planning is that is marries the urban and rural through an understanding of landscape connectivity and the promotion of supportive socio-ecological systems management, both of which promote landscape multi-functionality (Selman, 2006). The attainment of such goals has met with mixed success, as different planning and environmental stakeholders have engaged with the concept of green infrastructure in varying ways (Mell, 2016). What is apparent from a review of the research and practitioner literature is that an understanding of ecological systems, their spatial articulation within and across landscapes, and how humans interact, manage and value these resources is essential to long-term investment in sustainable green infrastructure planning at the landscape scale (Austin, 2014).

This chapter reflects upon how green infrastructure planning is being applied at the landscape scale, discussing how alternative thematic approaches to green infrastructure thinking can deliver a range of ecological, social and economic benefits. It goes on to examine how the policy of landscape planning facilitates this process and reviews examples of good practice in terms of investment and management. Finally, the chapter considers how green infrastructure provides planners with options for the development of more integrated and sustainable approaches to planning at the landscape scale.

What is green infrastructure?

Green infrastructure has been noted as being both a singular resource – one of a set of land-scape elements (Mell, 2010) – as well as a wider and connective network of landscape resources (Herzog, 2013). While such a broad description is, or can be, valuable to planners who can uti-lise a range of policy instruments and investment options to meet their specific planning needs, it introduces a vagueness to discussions of its utility and functionality within planning (Sturzaker and Mell, 2017). It is therefore important to fix the parameters of what green infrastructure is and what it does to contextualise its use within discussions of landscape planning. Natural England's Guidance set out a wide range of resources they considered 'green infrastructure'. This encompassed individual elements of a given landscape but also, importantly, resources that function at a landscape scale as key hubs, nodes or links (Natural England and Landuse Consultants, 2009). Benedict and McMahon (2006) in the USA and Boyle et al. (2013) in New Zealand present similar assessments of green infrastructure. Natural England's research helped to frame these debates from an ecological perspective, which could then be adapted by stakeholders, including the Town and Country Planning Association (2012) and the Landscape Institute (2009) in the UK, as well as the Conservation Fund and Chicago Wilderness in the USA (Chicago Wilderness, n.d.; Weber, 2007), and subsequently by practitioners in Germany, the Netherlands and Belgium in Europe (South Yorkshire Forest Partnership and Sheffield City Council, 2012).

Each of these stakeholders extended the elemental debate proposed by Natural England and aligned them with the conceptual discussions of Benedict and McMahon (2006), who also reflect upon the thematic articulations of green infrastructure. These discussions highlight several core principles used to review the development and management of green infrastruc-ture, including connectivity between people/resources, access to nature, supporting ecological networks, sustainable praxis, strategic investment/management, and the delivery of socio-economic benefits across urban/rural landscapes. Each of these principles is deemed central to how we develop green infrastructure as they offer *ecological* and *socio-economic dimensions* to its delivery (Sinnett et al., 2015).

These principles locate how we develop green infrastructure as they offer a geographical and thematic understanding for its delivery. Spatially, green infrastructure supports environmental resources at the site through to the landscape scale, and is particularly important in managing water and ecological elements. Furthermore, the promotion of connective networks of green infrastructure have been proposed as providing linear corridors across the landscape, drawing on the greenways and Countryside In and Around Towns (CIAT) mandates (Countryside Agency and Groundwork, 2005; Little, 1990). The alignment of scaled investment with connective landscape elements provides scope for people to access nature, for species movement, and for environmental resources to function according to landscape principles (Farina, 2006). This is especially relevant in water management, where supply and quality issues require landscape-scale management (Ahern, 2007). Similarly, biodiversity can be managed more effectively when ecological corridors are developed within the landscape to facilitate species movement and limit ecological island effects (Liquete et al., 2015).

The creation of green infrastructure networks at a regional scale has also been promoted as an effective mechanism to deliver economic growth. Again, the management of water resources is a critical aspect of green infrastructure planning as it promotes the management of landscapes that are more resilient to climatic change when compared to other forms of built infrastruc-ture (Young, 2010). Likewise, where landscape-scale nature reserves are developed, they can mitigate pollution and climatic variation caused by transport or domestic resource use, thus limiting the costs of control and/or adaptation (Weber et al., 2006). We can also identify where agriculture engages with green infrastructure through farm diversification or agri-environmental schemes that promote on-site biodiversity management (Kleijn et al., 2006). Although these actions may be incentivised, they illustrate how engaging with green infrastructure in landscape management can deliver financial benefits to the land owner.

All of these elements need to be planned in a strategic way to ensure that green infrastructure resources are developed in the right places, focus on pertinent issues, and form part of a wider discussion of how ecological systems provide added value to the landscape. This has proven hard to engineer (cf. Mell, 2016), as the protection of green infrastructure in the form of green belts, national parks, nature reserves or expanses of grassland are seen by commentators as limiting the economic value of these resources (Sturzaker and Mell, 2017). In places, this view has credence, but fundamentally the protection of the landscape remains integral to the delivery of an extensive range of rural and urban services (Gallent et al., 2015).

Green infrastructure and landscape policy

The employment of green infrastructure may appear to be focussed on urban planning issues; however, due to the spatial dynamism of the concept, there is a wealth of opportunity to employ it at the landscape scale. Moreover, in many areas of the UK and North America, green infrastructure is already embedded within landscape-focussed regional policy (Ahern, 2013). Examples include the now revoked Regional Spatial Strategies in England, which framed green infrastructure investment as a mechanism to support environmental protection at the landscape scale (Mell, 2010). Likewise, the signing into law of the Maryland Greenways Commission set in motion a management programme that protects critical habitats from urban expansion (Benedict and McMahon, 2006). Moreover, we also see the application of green belt policies in Canada, South Korea and Australia, bridging urban/rural planning and providing landscape-scale protection for environmental resources (Sturzaker and Mell, 2017). Recently, the designation of the major rivers of India as 'human entities' with civic rights proposed to protect them against pollution (Safi, 2017) is a further way in which green infrastructure can be managed from a landscape perspective.

Each of these examples illustrates how green infrastructure can be embedded within landscape focussed policy. However, to ensure that it is provided with an equal platform as other infrastructure within these discussions, we need to assess where good practice has occurred to date. For example, during the exploration (1995–2005) phase of green infrastructure development in Britain, we witnessed the establishment of its conceptual principles, and the identification of its elements within land–use classifications (Mell, 2016). This included the first era of green infrastructure strategies, for example the Cambridgeshire sub-region, Bedfordshire and Luton, and in the East Midlands (Mell, 2010), although these documents were simple spatial representations of where green infrastructure was located. The development of Regional Spatial Strategies (RSSs) in England presented the first extended discussion of green infrastructure as a multi-functional and connective form of landscape management. Located within what Mell (2015) describes as the 'expansion era' (2005–2010) of green infrastructure planning, the development of a spatially diverse policy reflected a growing understanding of its principles, which has been largely absent in previous policy.

This debate can be extended to two, highly pertinent international policies that have been developed to support landscape-scale green infrastructure, the European Landscape Convention (ELC, signed in 2000) and the Water Framework Directive (WFD, adopted in 2000).

Both were created to provide clarity on how landscape should be managed, placing specific emphasis on the protection of quotidian landscapes, planning for water resources at the catchment scale, and the promotion of a coordinated approach to landscape planning, which integrates multiple stakeholders in a cross-boundary form of management.

A key aspect of the ELC is its proposal that all landscapes should be embedded within environmental planning policy, not just those ascribed high value, to ensure that the roles of connectivity, functionality and appropriate management are debated (Roe et al., 2009). Compliance with this mandate remains variable though, as EU member states engage differently with the discretionary nature of the policy. This was a result of the ELC's proposal that landscapes are more than physical elements but are also imbued with socio-cultural identities (Olwig, 2007). This has enabled some EU governments to rethink how they locate 'landscape' within their broader development and management policy. Unfortunately, for some, this is simply rhetoric with little actual uptake of landscape understanding within policy and practice (Scott, 2011). Therefore, although the ELC has facilitated debate regarding the value of 'landscape' planning, there has remained a lack of compliance in some areas with the convention's mandates (Roe, 2013).

The WFD, in comparison, has been applied more frequently to facilitate effective catchment and river basin management across the EU, although barriers to its implementation remain. As a process of EU-wide monitoring, the WFD sets out an ambitious agenda to improve water quality within landscape management practices (Kallis, 2001). The key benefit of the WFD, according to Hering et al. (2010), has been the systematic collection of data, the development of water monitoring practices and a greater engagement by politicians and non-water specialists in water catchment discussions. However, developing a pan-European consensus on the methods of data collection, monitoring and evaluation has been difficult, leading some commentators to question the relevance of the WFD to real-world landscape management. This is, in part, a consequence of the inability of the framework to facilitate more effective collaborative practices between landscape practitioners and policy-makers/planners. This remains a multi-scalar issue as landscape resources and the policies or legal guidelines that control them are often unaligned. Furthermore, a proportion of stakeholders show little desire to bring policy/practice into closer equilibrium (Moss, 2008).

The utility of the ELC and WFD has been in focussing discussions of green infrastructure at the landscape scale, requiring advocates to adopt a purposefully spatial approach to management. This embeds a more cognisant view of ecologically centred systems within the planning process and provides the basis for the development of more appropriate green infrastructure policy. The role of policy in ensuring that green infrastructure principles are embedded within landscape practice has been evidenced in England within a series of regional and sub-regional developments.

Until their revocation in 2011, England had a structure of regionally evidenced development plans, or RSSs, which were used to shape the identification and allocation of strategic investment sites (Baker and Wong, 2013). This included the provision of housing, transport infrastructure and, significantly, green infrastructure. However, across England, there was variation in the approach to green infrastructure planning within this process. Where a strong advocacy environment existed, for example in the north-west, green infrastructure was embedded within the RSS. In contrast, Yorkshire and Humberside were less forward-thinking in their uptake of green infrastructure (Mell, 2010). Furthermore, although the RSS process was fraught with complexities and disagreements over where green infrastructure funding should be allocated, and what landscape-scale projects should be supported, where outcomes were agreed, they generally focussed most frequently on the linking of ecological networks and the integration of green infrastructure across the wider landscape (Horwood, 2011).

Following revocation of the RSSs in 2010/11, a more significant emphasis has been placed on sub-regional strategies to fill the sub-national policy gap. The second iteration of the Cambridgeshire Green Infrastructure Strategy (Cambridgeshire Horizons, 2011), the emerging Greater Manchester Spatial Framework (Greater Manchester Combined Authority, 2016), and the London Plan (Greater London Authority, 2016) all promote green infrastructure as an essential component in delivering landscape functionality. The adoption of these strategies remains variable, however, as they are subject to ongoing economic and political influences, which have been seen to dilute green infrastructure praxis in many locations (Mell et al., 2017). The lack of universal support for green infrastructure in local authority-led policy reflects the lack of value placed on landscape in the National Planning Policy Framework (NPPF) in the UK (Jorgensen, 2017). Although the NPPF offers protection to green belt designations, and promotes ongoing actions to manage protected and ecologically sensitive locations, its support of green infrastructure remains opaque. The Department for Communities and Local Government (2012) advises local authorities to plan for green infrastructure within their Local Development Plans, yet it appears reluctant to promote the creation, funding and management of green infrastructure networks across our wider landscapes (Mell, 2015).

The inclusion of green infrastructure in landscape policy is therefore subject to several political, legal and administrative barriers which need to be addressed to effect successful implementation. For example, political support is needed to ensure that green infrastructure is debated with the same level of detail and clarity as other built infrastructure (Benedict and McMahon, 2006; Lennon, 2014). Secondly, there is a need to negotiate what defines best practice for green infrastructure at the landscape scale (Austin, 2014). Finally, an emphasis should be placed on the formation of strategic and integrated policy that links green infrastructure with the provision of other engineered infrastructure and/or services, which can lead to greater buy-in and long-term cooperation (Ugolini et al., 2015), and help identify appropriate funding, from public, private or community sources to support long-term commitments for management (Natural England and Landuse Consultants, 2009).

Connectivity, access to nature and multi-functionality

Examples of large-scale investment in green infrastructure can be identified around the world, especially where forestry and conservation are embedded within development/management practices (Austin, 2014). For example, the Forestry Commission in the UK has been a prominent supporter of green infrastructure investment, noting its socio-economic and ecological value (England's Community Forests and Forestry Commission, 2012). Across many of their sites, they have used landscape-scale plantations for recreation, conservation and economic development activities. These plantations are landscape-scale interventions with the capacity to ensure multiple uses and users can engage with the landscape simultaneously without conflict. At a sub-regional scale, England's Community Forests has reinvigorated former post-industrial landscapes in northern England using the green infrastructure principles of connectivity, access to nature and multi-functionality (Kitchen et al., 2006). Their ongoing mandate has been to invest in landscape-scale tree planting, the conversion of derelict sites into functional country parks, and to facilitate engagement through health, wellbeing and conservation projects with these landscapes (Mell, 2011).

In the USA, a similar approach to connectivity is thriving in the ongoing investment in greenways. Greenways are linear or circular corridors that promote movement and interaction with the environment across landscape boundaries (Little, 1990). Originally conceived as parkways enabling people to drive to the countryside, greenways have transitioned from the car to the individual on foot or by bicycle. Many of the most successful greenways link urban and rural landscapes and act as key recreational and conservation resources in areas with little history of green infrastructure investment, e.g. in Indianapolis (Lindsey et al., 2001). However, greenways have been extended from this initial urban/rural interface to become prominent landscape-scale resources. This is linked directly to McHarg's (1969) proposals of functionality, which require planning and management to acknowledge the role of landscapes in the promotion of effective conservation strategies. Greenways are therefore seen as assisting the maintenance of landscape integrity, addressing fragmentation and establishing connected habitats and wildlife corridors (Hellmund and Smith, 2006). Moreover, the Florida Ecological Greenway Network (Fábos, 2004), the New England Greenway (Ryan et al., 2006) and the Southeastern Wisconsin Environmental Corridors (Hellmund and Smith, 2006), all illustrate the value of connecting green infrastructure resources at a landscape-scale to deliver ecological conservation and recreation mandates.

Green belts are a further example of planning for green infrastructure at the landscape scale. In the UK, they have been used to limit urban coalescence (Amati, 2008), but internationally they have also been employed to meet a range of ecological needs. For example, the Greater Golden Horseshoe Green Belt in Ontario (see also Chapter 40) aims to limit sprawl while protecting the sensitive ecological and hydrological landscapes of the Niagara Escarpment and Oak Ridge Moraine (Cadieux et al., 2013). The city of Milan in Italy has also used green belts to link urban and rural areas through the *Piano intercomunale milanese* (Milan Inter-Municipal Plan) (Sturzaker and Mell, 2017). Implemented in the 1960s, it has seen extensive investment in forestry linking the neighbouring agricultural land of Lombardy with Milan through landscape-scale urban parks, i.e. Parco Sud (Sanesi et al., 2007). Moreover, the *Ceinture verte* located around Paris manages the greater Île-de-France green belt, ensuring that approximately 300,000 hectares of land remain protected as natural and wooded areas as part of the regional-scale protected belt (Plan vert regional) (Laruelle and Leganne, 2008).

Water, ecosystem services and nature-based resource management

In addition to promoting connectivity and multi-functionality, green infrastructure planning at the landscape scale supports biodiversity and ecosystem services. Rouse and Bunster-Ossa (2013) present a detailed discussion of the intersections of policy, practice and green infrastructure in their analysis of how water, biodiversity and nature-based investment can lead to positive landscape management. They highlight how state-level Departments of Natural Resources (DNR), the Department of Housing and Urban Development (HUD) and Environmental Protection Agency (EPA) in the USA all act as key advocates for the development of landscape-scale green infrastructure (Mell, 2016). State DNRs and the EPA, in particular, work with landscape resources to support the regulation and function of cultural ecosystem services due to the range of temporal, scalar and elemental variables that influence how they can be managed. Through a connection with ecological resource management, green infrastructure planning can ensure that water systems, habitats, and ecological functions are maintained (Ahern et al., 2014). There is also a growing body of research within landscape planning that uses the principles of landscape ecology, namely: *links, hubs* and *nodes* to ensure that ecological resources form wider networks of connected and supportive systems (Jongman and Pungetti, 2004).

The development of a pair of landscape nature reserves in Cambridgeshire and Huntingdonshire: *Wicken Fen* and *Great Fen* are examples of such ideas. Wicken Fen is an ambitious 100-year landscape wetland restoration project located in high quality agricultural land in East Cambridgeshire (Hughes et al., 2016), while the Great Fen aims to covert 3,000 hectares of farmland into new wetland and other habitats to protect the National Nature Reserves at Woodwalton and Holme Fens (Rotherham, 2011). Both aim to establish landscape-scale resources that promote flood mitigation, climate change adaptation, biodiversity and public engagement with the landscape (Hodder et al., 2014). These aims have been achieved through the use of the waterway restoration to ensure that seasonal (and large-event) fluctuations can be managed. Moreover, through the inclusion of a diverse mosaic of landscape elements, species and management techniques, the site managers, the National Trust at Wicken Fen and the Wildlife Trust at Great Fen, have diversified what could have been considered a monofunctional landscape. The shift to a more biodiverse form of landscape management required a rethinking of the species mix of plants and animals, including the incorporation of native plant species more adaptive to fluctuating water levels. It also introduced bovine and equestrian species to facilitate animal and ecological diversity (Hughes et al., 2011). Both projects were included in the first and second iterations of the Cambridgeshire Green Infrastructure Strategy. In each document, the role of ecosystem services was promoted through the development of a connected network of landscape hubs linked through water and ecological corridors, creating a diverse habitat. These are complemented by a series of connecting greenways, waterways and footpaths, promoting species migration and diversification, as well as human uses of the site (Cambridgeshire Horizons, 2008, 2011).

The Chicago Wilderness and Forest Preserves projects are two examples where multi-agency buy-in has been generated to promote more effective landscape management on the Chicago River catchment. This has used forest preserves (sessional wetland woods) to mitigate fluctuations in peak rainfall flows in the area. The forest preserves enable the controlled collection and distribution of water resources, minimising the impacts of flooding in Chicago, and more widely in southern Illinois (Lake County Forest Preserve Department, 2014). The sites have also been located to provide key landscape resource hubs within the wider Chicago greenspace network, linking the city with water and terrestrial resources in the surrounding counties (Mell, 2016). The work of the Conservation Fund has focussed several projects on protecting the Chesapeake Bay area. This work brought together political leaders in Maryland, Pennsylvania, Virginia and Washington, DC to establish a collaborative approach to watershed protection and to ensure that the ecological integrity of the area is not undermined (Weber, 2004; Weber and Wolf, 2000).

A further example, the Columbia Basin Ecosystem Management Plan, was developed through a collaboration between USDA Forests and Bureau of Land Management, and manages over 145 million acres of the Columbia River Basin. The plan identifies ecosystem protection, nutrient cycling, key species protection, and environmental resource stability as key objectives that are delivered through a spatially diverse approach to landscape management. This project was reported by Benedict and McMahon (2006) as highlighting best practice in the integration of scientific evidence to underpin a more transparent and accountable form of landscape management. The project also promotes educational outreach to ensure that the key messages of landscape protection are disseminated.

Conclusions

Although the research evaluating green infrastructure planning focusses predominately on urban issues, there is a need to apply the same principles to the wider landscape. Ecological systems are not limited by legal or administrative boundaries, and as such should be managed through a greater understanding of ecological systems at multiple scales. Moreover, water systems and ecological habitats are not confined to urban areas and should be thought of as environmental features that provide a conduit between urban and rural planning: what happens to ecological systems in upland, heath or moorland, and in wider catchments has a direct influence on urban ecosystems. Therefore, failures to plan strategically to integrate green infrastructure thinking across the continuum of urban/rural policy and practice hinder both the functionality of these systems and the benefits they deliver. With the ever-changing demands on ecological resources, such strategic approaches are potentially more necessary now than in any other era of urban–rural planning.

References

Ahern, J. (2007). Planning and design for sustainable and resilient cities: theories, strategies and best practice for green infrastructure. In V. Novotny, J. Ahern, and P. Brown (eds), *Water-Centric Sustainable Communities* (pp. 135–176). Hoboken, NJ: Wiley-Blackwell

Ahern, J. (2013). Urban landscape sustainability and resilience: the promise and challenges of integrating ecology with urban planning and design. *Landscape Ecology, 28*(6), 1203–1212.

Ahern, J., Cilliers, S., and Niemelä, J. (2014). The concept of ecosystem services in adaptive urban planning and design: a framework for supporting innovation. *Landscape and Urban Planning, 125,* 254–259.

Amati, M. (2008). *Urban Green Belts in the Twenty-first Century.* Farnham: Ashgate.

Austin, G. (2014). *Green Infrastructure for Landscape Planning: Integrating Human and Natural Systems.* New York: Routledge.

Baker, M., and Wong, C. (2013). The delusion of strategic spatial planning: what's left after the Labour Government's English regional experiment? *Planning Practice and Research, 28*(1), 83–103.

Benedict, M. A., and McMahon, E. T. (2006). *Green Infrastructure: Linking Landscapes and Communities.* Washington, DC: Island Press.

Boyle, C., Gamage, G., Burns, B., Fassman, E., Knight-Lenihan, S., Schwendenmann, L., and Thresher, W. (2013). *Greening Cities: A Review of Green Infrastructure.* Auckland: University of Auckland.

Cadieux, K. V., Taylor, L. E., and Bunce, M. F. (2013). Landscape ideology in the Greater Golden Horseshoe Greenbelt Plan: negotiating material landscapes and abstract ideals in the city's countryside. *Journal of Rural Studies, 32,* 307–319.

Cambridgeshire Horizons. (2008). *Cambridgeshire Charter for Quality Growth and Cambridgeshire's Vision 2007–2021.* Cambridge: Cambridgeshire Horizons.

Cambridgeshire Horizons. (2011). *Cambridgeshire Green Infrastructure Strategy.* Cambridge: Cambridgeshire Horizons.

Chicago Wilderness. (n.d.). Chicago Wilderness. Retrieved on 1 September 2015 from www.chicagowilderness.org/index.php.

Countryside Agency and Groundwork. (2005). *The Countryside in and around Towns: A Vision for Connecting Town and County in the Pursuit of Sustainable Development.* Wetherby: Countryside Agency.

Department of Communities and Local Government. (2012). *National Planning Policy Framework.* London: DCLG.

England's Community Forests and Forestry Commission. (2012). *Benefits to Health and Wellbeing of Trees and Green Spaces.* Farnham: England's Community Forests and Forestry Commission. Retrieved from www.communityforest.org.uk/resources/case_study_health_and_wellbeing.pdf.

Fábos, J. G. (2004). Greenway planning in the United States: its origins and recent case studies. *Landscape and Urban Planning, 68*(2–3), 321–342.

Farina, A. (2006). *Principles and Methods in Landscape Ecology: Towards a Science of the Landscape.* London: Springer.

Gallent, N., Hamiduddin, I., Juntti, M., Kidd, S., and Shaw, D. (2015). *Introducing Rural Planning: Economies, Communities and Landsacpes.* London: Routledge.

Greater London Authority. (2016). *The London Plan: The Spatial Development Strategy for London Consolidated with Alterations since 2011.* London: Greater London Authority.

Greater Manchester Combined Authority. (2016). *Greater Manchester Spatial Framework: Draft for Consultation.* Manchester: Greater Manchester Combined Authority.

Hellmund, P. C., and Smith, D. (2006). *Designing Greenways: Sustainable Landscapes for Nature and People.* Washington, DC: Island Press.

Hering, D., Borja, A., Carstensen, J., Carvalho, L., Elliott, M., Feld, C. K., . . . de Bund, W. van. (2010). The European Water Framework Directive at the age of 10: a critical review of the achievements with recommendations for the future. *The Science of the Total Environment, 408*(19), 4007–4019.

Herzog, C. P. (2013). A multifunctional green infrastructure design to protect and improve native biodiversity in Rio de Janeiro. *Landscape and Ecological Engineering, 12*(1), 141–150.

Hodder, K. H., Newton, A. C., Cantarello, E., and Perrella, L. (2014). Does landscape-scale conservation management enhance the provision of ecosystem services? *International Journal of Biodiversity Science, Ecosystem Services and Management, 10*(1), 71–83.

Horwood, K. (2011). Green infrastructure: reconciling urban green space and regional economic development: lessons learnt from experience in England's north-west region. *Local Environment: The International Journal of Justice and Sustainability, 16*(10), 37–41.

Hughes, F. M. R., Adams, W. M., Butchart, S. H. M., Field, R. H., Peh, K. S.-H., and Warrington, S. (2016). The challenges of integrating biodiversity and ecosystem services monitoring and evaluation at a landscape-scale wetland restoration project in the UK. *Ecology and Society, 21*(3), article 10.

Hughes, F. M. R., Stroh, P. A., Adams, W. M., Kirby, K. J., Mountford, J. O., and Warrington, S. (2011). Monitoring and evaluating large-scale, 'open-ended' habitat creation projects: a journey rather than a destination. *Journal for Nature Conservation, 19*(4), 245–253.

Jongman, R., and Pungetti, G. (eds). (2004). *Ecological Networks and Greenways: Concept, Design and Implementation*. Cambridge: Cambridge University Press.

Jorgensen, A. (2017). What is happening to landscape? *Landscape Research, 42*(1), 1–5.

Kallis, G. (2001). The EU water framework directive: measures and implications. *Water Policy, 3*(2), 125–142.

Kitchen, L., Marsden, T., and Milbourne, P. (2006). Community forests and regeneration in post-industrial landscapes. *Geoforum, 37*(5), 831–843.

Kleijn, D., Baquero, R. A., Clough, Y., Díaz, M., Esteban, J., Fernández, F., . . . Yela, J. L. (2006). Mixed biodiversity benefits of agri-environment schemes in five European countries. *Ecology Letters, 9*(3), 243–254.

Lake County Forest Preserve Department. (2014). *100-Year Vision for Lake County*. Libertyville, IL: Lake County Forest Preserve Department.

Landscape Institute. (2009). *Green Infrastructure: Connected and Multifunctional Landscapes. Landscape Institute Position Statement*. London: Landscape Institute.

Laruelle, N., and Leganne, C. (2008). The Paris-Ile-de-France Ceinture Verte. In M. Amati (ed.), *Urban Green Belts in the Twenty-First Century* (pp. 227–241). Aldershot: Ashgate.

Lennon, M. (2014). Green infrastructure and planning policy: a critical assessment. *Local Environment, 20*(8), 957–980.

Lindsey, G., Maraj, M., and Kuan, S. (2001). Access, equity, and urban greenways: an exploratory investigation. *The Professional Geographer, 53*(3), 332–346.

Liquete, C., Kleeschulte, S., Dige, G., Maes, J., Grizzetti, B., Olah, B., and Zulian, G. (2015). Mapping green infrastructure based on ecosystem services and ecological networks: a Pan-European case study. *Environmental Science and Policy, 54*, 268–280.

Little, C. (1990). *Greenways for America*. Baltimore, MD: Johns Hopkins University Press.

McHarg, I. L. (1969). *Design with Nature (Wiley Series in Sustainable Design)*. New York: John Wiley and Sons.

Mell, I. C. (2010). Green infrastructure: concepts, perceptions and its use in spatial planning. PhD thesis, University of Newcastle.

Mell, I. C. (2011). The changing focus of England's Community Forest programme and its use of a green infrastructure approach to multi-functional landscape planning. *International Journal of Sustainable Society, 3*(4), 431–446.

Mell, I. C. (2015). Green infrastructure planning: policy and objectives. In D. Sinnett, S. Burgess, and N. Smith (eds), *Handbook on Green Infrastructure: Planning, Design and Implementation* (pp. 105–123). Cheltenham: Edward Elgar Publishing.

Mell, I. C. (2016). *Global Green Infrastructure: Lessons for Successful Policy-making, Investment and Management*. London: Routledge.

Mell, I., Allin, S., Reimer, M., and Wilker, J. (2017). Strategic green infrastructure planning in Germany and the UK: a transnational evaluation of the evolution of urban greening policy and practice. *International Planning Studies*, 1–17.

Moss, B. (2008). The Water Framework Directive: total environment or political compromise? *Science of The Total Environment, 400*(1–3), 32–41.

Natural England and Landuse Consultants. (2009). *Green Infrastructure Guidance*. Peterborough: Natural England.

Olwig, K. R. (2007). The practice of landscape 'Conventions' and the just landscape: the case of the European landscape convention. *Landscape Research, 32*(5), 579–594.

Roe, M. (2013). Policy change and ELC implementation: establishment of a baseline for understanding the impact on UK national policy of the European Landscape Convention. *Landscape Research, 38*(6), 768–798.

Roe, M., Selman, P., Mell, I. C., Jones, C., and Swanwick, C. (2009). Establishment of a baseline for, and monitoring of the impact of, the European Landscape Convention in the Bristol. Retrieved from http://landscapecharacter.org.uk/elc/baseline-monitoring-ELC.

Rotherham, I. D. (2011). Implications of landscape history and cultural severance for restoration in England. In *Human Dimensions of Ecological Restoration* (pp. 277–287). Washington, DC: Island Press/ Center for Resource Economics.

Rouse, D. C., and Bunster-Ossa, I. (2013). *Green Infrastructure: A Landscape Approach*. Chicago, IL: APA Planners Press.

Ryan, R. L., Fábos, J. G., and Allan, J. J. (2006). Understanding opportunities and challenges for collaborative greenway planning in New England. *Landscape and Urban Planning, 76*(1–4), 172–191.

Safi, M. (2017). Ganges and Yamuna rivers granted same legal rights as human beings. *The Guardian*, 21 March. Retrieved from www.theguardian.com/world/2017/mar/21/ganges-and-yamuna-rivers-granted-same-legal-rights-as-human-beings.

Sanesi, G., Lafortezza, R., Marziliano, P., Ragazzi, A., and Mariani, L. (2007). Assessing the current status of urban forest resources in the context of Parco Nord, Milan, Italy. *Landscape and Ecological Engineering, 3*(2), 187–198.

Scott, A. (2011). Beyond the conventional: meeting the challenges of landscape governance within the European Landscape Convention? *Journal of Environmental Management, 92*(10), 2754–2762.

Selman, P. (2006). *Planning at the Landscape Scale (RTPI Library Series)*. London: Routledge.

Sinnett, D., Smith, N., and Burgess, S. (eds). (2015). *Handbook on Green Infrastructure: Planning, Design and Implementation*. Cheltenham: Edward Elgar Publishing.

South Yorkshire Forest Partnership and Sheffield City Council. (2012). *The VALUE Project: The Final Report*. Sheffield: South Yorkshire Forest Partnership and Sheffield City Council.

Sturzaker, J., and Mell, I. (2017). *Green Belts: Past, Present and Future*. London: Routledge.

Town and Country Planning Association. (2012). *Planning for a Healthy Environment – Good Practice Guidance for Green Infrastructure*. London: Town and Country Planning Association.

Ugolini, F., Massetti, L., Sanesi, G., and Pearlmutter, D. (2015). Knowledge transfer between stakeholders in the field of urban forestry and green infrastructure: results of a European survey. *Land Use Policy, 49*, 365–381.

Weber, T. (2004). Landscape ecological assessment of the Chesapeake Bay Watershed. *Environmental Monitoring and Assessment, 94*(1–3), 39–53.

Weber, T. (2007). *Ecosystem Services in Cecil County's Green Infrastructure: Technical Report for the Cecil County Green Infrastructure Plan*. Annapolis, MD: Conservation Fund.

Weber, T., Sloan, A., and Wolf, J. (2006). Maryland's Green Infrastructure Assessment: development of a comprehensive approach to land conservation. *Landscape and Urban Planning, 77*(1–2), 94–110.

Weber, T., and Wolf, J. (2000). Maryland's Green Infrastructure - using landscape assessment tools to identify a regional conservation strategy. *Environmental Managment and Assessment, 63*(1), 265–277.

Young, R. F. (2010). Managing municipal green space for ecosystem services. *Urban Forestry and Urban Greening, 9*(4), 313–321.

43

Landscape and wellbeing

Mike Rogerson, Valerie Gladwell, Jules Pretty and Jo Barton

Introduction

Parks were designated and designed in the nineteenth century informed by a belief that they might provide health benefits (Maller et al., 2008). In the UK, during this time beach huts and promenades were also built in order to boost opportunity to spend time on the coast, for both the enjoyment and the wellbeing improvements on offer. Our landscape is a significant health resource that promotes physical, mental and social wellbeing. There are three main approaches to understanding wellbeing, namely those of hedonic wellbeing, eudaimonic wellbeing and positive psychological wellbeing, which emphasise the importance of different feelings, cognitions and experiences. In this chapter we draw on findings from a range of research approaches, to outline current understanding of how greenspaces and nature landscapes can influence different aspects of wellbeing, including the 'five ways to wellbeing' set out by the UK government. We then provide an overview of findings from our own evaluative research into environment-based wellbeing interventions for individuals with low wellbeing.

Hedonic wellbeing and nature

The hedonic perspective views 'subjective wellbeing' as a multi-dimensional construct, comprising moderately correlating, but independent components which include momentary feelings of positive affect and negative affect, and more considered cognitive components of life satisfaction and satisfaction within domains important to the individual (Diener et al., 2003; Carruthers and Hood, 2004; Diener et al., 2009; Gale et al., 2013). The UK Office for National Statistics reports that people living in predominantly rural areas have higher feelings of happiness, life satisfaction and feelings that the things that they do in life are worthwhile (Office for National Statistics, 2017). Although these associations may be linked with other socioeconomic variables such as income, epidemiological research has controlled for many of these and reported that individuals have higher wellbeing when living in urban areas with more greenspace compared to less greenspace (White et al., 2013). Several studies have highlighted the positive association between richness of wildlife and plant species within an environment and improved wellbeing (Huby et al., 2006; Fuller et al., 2007; Luck et al., 2011; Dallimer et al., 2012; Clark et al., 2014), with visits to these enriched landscapes providing emotional, social and psychological benefits, such as improvements in self-esteem and mood (Huby et al., 2006; Barton et al., 2009; Curtin, 2009; Lemieux et al., 2012; Clark et al., 2014). Wellbeing is increased in individuals who perceive themselves to be in areas more diverse in birds, butterflies and plants (Clark et al., 2014). There is also evidence for the importance of landscapes rich in nature close to the home, with the proximity and quantity of natural spaces in the local environment being related to fewer health outcomes and risk factors (Lovell et al., 2014). Degradation of the natural environment can exert adverse effects particularly mental wellbeing, greater than the adverse effects associated with economic decline, nutritional threats and pollution (Speldewinde et al., 2009; Lovell et al., 2014). Bluespace is an important element of nature much like greenspace; living near to the coast is also associated with better wellbeing (Wheeler et al., 2012). Coastal landscapes also cater for various therapeutic needs and promote wellbeing via emotional connections, which are often deeply embodied and shared (Bell et al., 2015).

Acute contact with natural environments has been shown to improve psychological affective states, whether contact is in the form of exercise or mere exposure to or views of natural landscapes (Teas et al., 2007; Berman et al., 2008; Abkar et al., 2010; Bratman et al., 2012; Gladwell et al., 2013). Stress reduction theory (SRT), partly derived from psycho-evolutionary theory (Plutchik, 1980a, 1980b), suggests that, based on historical human-environment interactions and relationships, when there is an absence of threat, nature environments invoke positive primary emotional responses, which promote recovery from stress (Ulrich, 1981, 1984; Ulrich et al., 1991; Herzog and Strevey, 2008; Ewert et al., 2011). This is important for wellbeing as primary emotions are related to secondary and tertiary emotions (Plutchik, 1980a), which can also be termed 'affect'. As previously outlined, within the hedonic perspective, affective state is an important component of wellbeing. Natural environments provide positive distractions from daily stresses and invoke feelings of interest, pleasantness and calm, thereby reducing stress symptoms and promoting positive affect (Ulrich, 1981, 1984; Ulrich et al., 1991; Herzog and Strevey, 2008; Ewert et al., 2011).

Many studies have shown that spending time relaxing or exercising in nature landscapes enhances individuals' attention levels (Berto, 2005; Berman et al., 2008; Rogerson and Barton, 2015; Rogerson et al., 2016). Attention restoration theory (ART) outlines that built environments contain more stimuli that humans are less efficient at processing, which together with daily tasks function to deplete one's attentional capacity. Natural landscapes, however, are easier for the human brain to process and respond to, serving as respite and an opportunity for restoration of depleted attentional resources and associated improvement of affective state, as the two are linked. Indeed, consistent with SRT and ART, research has shown that when walking from built urban areas into urban greenspaces, the brain activates patterns identified as representing lower frustration and arousal and a higher relaxed meditative state; and vice-versa when moving from greenspace to built areas (Aspinall et al., 2015). In this way, via improved affect and attentional capacity, nature landscapes function to enhance our temporary ability and likelihood to take notice of our surroundings, which in line with SRT, is in turn cyclically good for our hedonic wellbeing as influenced by affective state. Further to this, 'taking notice' is also identified as one of the public-health oriented 'five ways to wellbeing' outlined by the UK government and National Health Service (Aked et al., 2009).

Box 43.1 Five ways to wellbeing

The New Economics Foundation (NEF) identified five evidenced-based actions to improve wellbeing (New Economics Foundation, 2008):

(i) **Connect** – Social relationships, a sense of 'belongingness', social interaction and the support of family and friends are important for wellbeing and can also protect against becoming ill, while social isolation and exclusion are associated with poorer health (Tones and Green, 2004).

(ii) **Be active** – Engaging in physical activity promotes mental wellbeing, leading to improvements in self-esteem, mood and quality of life and a reduction in the risk of cardiovascular disease associated risk factors such as hypertension, high blood cholesterol and Type II diabetes.

(iii) **Take notice** – Studies have shown that being aware of what is taking place in the present directly enhances wellbeing and that 'savouring the moment' can help to reaffirm life priorities (Brown and Ryan, 2003). This ability to engage in 'mindfulness' enhances self-understanding, reduces stress and improves mental health and wellbeing (Feldman et al., 2007; New Economics Foundation, 2008; Chu, 2010; Williams, 2010; Howell et al., 2011; Keng et al., 2011).

(iv) **Keep learning** –For children and young people learning contributes to social and cognitive development, increases self-esteem and social interaction and encourages participation in physical activity (Hall-Lande et al., 2007). In adults learning is associated with wellbeing; life satisfaction, optimism and self-efficacy; self-esteem and resilience (Feinstein and Hammond, 2004; Hammond, 2004); a sense of purpose and hope; encouraging social interaction and enabling people to feel competent (Takano et al., 2002). Learning is also protective against depression and in older people in particular, work and educational opportunities can lift them out of a depressive state (Feinstein and Hammond, 2004; Steptoe et al., 2013).

(v) **Give** – Giving to others by volunteering, joining a community group or helping a friend or stranger can provide substantial wellbeing benefits. Mutual cooperation and working with others can increase neuronal responses in the reward areas of the brain, indicating that social cooperation is intrinsically rewarding (Rilling et al., 2007; New Economics Foundation, 2008). Helping and giving to others contributes to improvements to cognitive and social functioning, which is crucial to mental wellbeing. Feelings of life satisfaction and happiness are also strongly associated with engagement in community activities (New Economics Foundation, 2008).

Eudaimonic wellbeing

The eudaimonic perspective considers individuals' *psychological wellbeing* as involving one's perception of engagement with existential challenges of life (Keyes et al., 2002). Eudaimonic wellbeing is based on individuals' feelings and appraisals of meaning and virtue (Waterman,

1993; Ryan and Huta, 2009; Hasnain et al., 2014). The eudaimonic approach primarily views wellbeing as related to one's feelings of self-actualisation, that is, self-fulfilment as a functioning individual (Maslow, 1954; Maslow et al., 1970; Deci and Ryan, 2000; Keyes et al., 2002). Self-determination theory suggests that wellbeing is related to fulfilment of three domains: competency, autonomy and relatedness (Deci and Ryan, 2000).

Offering an approach to understanding behavioural relationships between the individual and the environment, the ecological dynamics perspective views both the individual and the environment as systems comprising a complex arrangement of factors (termed 'constraints') (Brymer and Davids, 2012). It considers how this complex system of factors can constrain or afford processes relating to human wellbeing, and describes that constraints act as boundaries, within which behavioural invitations or possibilities (termed 'affordances') exist, thereby shaping behaviour (Brymer and Davids, 2014; Brymer et al., 2014; Yeh et al., 2015). Natural landscapes often afford opportunities for engagement with challenge and behaviours that contribute to fulfilment of competency, autonomy and relatedness. Whereas more built (often urban) landscapes have more associated rules and set behavioural patterns, nature landscapes afford individuals with opportunities to behave autonomously upon their own agency, for example, by navigating a route through non-uniform terrain.

Natural landscapes also offer great affordance for individuals to develop skills and knowledge, and thereby gain a sense of enhanced competency. For example, conservation activities, or taking part in wilderness trails through natural landscapes often encompass elements of skill acquisition and development. This supports the idea that activities in natural landscapes often provide opportunities for learning, which is another of the 'five ways to wellbeing' described above. Learning can promote increases in self-fulfilment (Konu and Rimpelä, 2002). Landscapes can play an important role in childhood learning (Cooper, 2015), but can also continue to provide opportunities for learning through adulthood, for example, about ecosystems, and learning to recognise plant, bird, fish and animal species.

The eudaimonic domain of relatedness resonates with the 'Connect' way to wellbeing, and to this domain, spending time in and engaging with nature, for example through conservation activities, is associated with increases in feelings of relatedness to the natural world (Nisbet et al., 2009) – a metric which is itself associated with reported happiness (Zelenski and Nisbet, 2014). Place making, through physical and mental connection to the environment, can result from being and even further becoming a strong canvas for the creation of positive experiences and memories (Pretty et al., 2017). Time in natural landscapes is also often a social occasion, such as a walk with family, friends or other groups, and many activities also foster teamwork experiences, such as conservation activities or orienteering. This social characteristic of nature landscapes functions in line with the 'Connect' way to wellbeing.

Positive psychological wellbeing

The positive psychological approach to wellbeing draws on elements of both the hedonic and eudaimonic perspectives (Sin and Lyubomirsky, 2009). Authentic happiness theory suggests that wellbeing comprises components called the pleasant life (one which maximises pleasurable and positive experiences), the good life (individuals develop their strengths and virtues or signature strengths in activities that the individual enjoys and is passionate about), and the meaningful life (individuals apply their signature strengths in activities that contribute to the greater good, such as parenting, developing friendships or servicing the community) (Seligman, 2002; Norrish and Vella-Brodrick, 2008). The eudaimonic aspect of relatedness contributes to positive psychological wellbeing, but the component of the meaningful life is also contributed to by the five ways to wellbeing's facet of 'Give'. Being involved in community activities in urban or rural greenspaces, for example, volunteering projects to create or maintain community gardens, provides wellbeing benefits through altruistic behaviour, as well as through feelings of relatedness to community places and groups (Pretty et al., 2016). Additionally, nature-based landscapes are often venues for horticultural, social or physical activity hobbies, which resonates with the component of the good life.

As described by ecological dynamics perspective, natural landscapes afford physically active behaviours. Conservation work tends to be manual, from moving across the greenspace area to planting trees or food seeds. Urban and rural greenspaces, and coastal areas are often used for physical activity pursuits. A multi-study analysis found that walking, cycling, sailing, horse riding, and boating, all induced improvements in individuals' affective state, resonating with both the eudaimonic and positive psychological understandings of wellbeing (Barton and Pretty, 2010). Additionally to affective pleasure, as alluded to by the 'Be Active' facet of the Five Ways To Wellbeing, reported physical activity level is associated with wellbeing, thereby further linking landscape to wellbeing.

Landscape-based wellbeing programmes

Whereas built landscape and busy urbanised lifestyles are often not conducive to good wellbeing, nature landscapes can be a central part of the solution (Pretty et al., 2017). A growing number of wellbeing programmes are centred on engagement with natural landscapes and features. There are indications that individuals and vulnerable groups who are most in need, or who have lowest wellbeing, may benefit the most from natural landscapes, as having access to nature within neighbourhood landscapes can be equigenic for human wellbeing across populations. One study found that across 34 countries, compared to poor access to recreational greenspaces, good access was associated with 40 per cent narrower socioeconomic inequality in mental wellbeing (Mitchell et al., 2015). A recent evaluation of Wildlife Trust conservation programmes for vulnerable adults found that attendees with low wellbeing at the start of the programme reported greater improvements in wellbeing than did attendees who had average wellbeing at the start (Rogerson et al., 2017).

Findings from evaluative research

The Green Exercise Research Team has conducted research to evaluate the efficacies of wellbeing programmes for vulnerable groups, within which nature landscapes are a central facet of the programme. Many of the programmes are based on the promotion of the 'Five Ways to Wellbeing'. They encourage participants to 'connect' to their local community, environment and other participants through the sharing of knowledge, skills, experience and stories. The activities are designed to develop social cohesion and networking, thereby reducing social inclusion, which positively impacts on health and wellbeing. Programmes encourage participants to be physically 'active' and actively work with their communities, resulting in increased fitness, social cooperation and teamwork. They enable participants to 'take notice' of the environment by becoming more aware of their local environment and learning about conservation/actions to protect it. They provide opportunities for participants to 'keep learning' through assistance with gaining qualifications or enrolling on courses, skill and knowledge sharing and educational sessions to learn about nature and their local environment. Learning also enhances self-efficacy, self-esteem and overall wellbeing. The programmes also enable participants to 'give' to others by volunteering, listening and talking, helping the community and giving up their time.

Currently in the UK, the most universally used and recognised measure of mental well-being is the Warwick-Edinburgh Mental Wellbeing Scale (WEMWBS) (Stewart-Brown and Janmohamed, 2008). The positively focused design of the WEMWBS enables it to be used by mental health promotion initiatives (Parkinson, 2006). Its comprising items were selected with the intention of covering relevant aspects of positive affect, psychological functioning (autonomy, competence, self acceptance, personal growth) and interpersonal relationship (Tennant et al., 2007). Whereas the full scale comprises fourteen items, a short version (Short-form Warwick-Edinburgh Mental Wellbeing Scale; SWEMWBS) comprises only seven, and correlates strongly with the full version. The 14- item scale is scored by summing responses to each item answered on a five point Likert scale, from 1 (none of the time) to 5 (all of the time). The minimum scale score for the 14-item version is 14 and the maximum is 70, with a higher score representing a better wellbeing. Both the long form and short form WEMWBS have been validated for use in the UK with those aged 13 years and above (Stewart-Brown and Janmohamed, 2008; Stewart-Brown et al., 2009; Clarke et al., 2011). The scales show good content validity; with Cronbach's alpha scores ranging from 0.89 to 0.91 and high correlations with other mental health and wellbeing scales. Test-retest reliability at one week was high (0.83) and social desirability bias was lower or similar to that of other comparable scales.

For the WEMWBS, no official threshold points have been developed for categorisation purposes. However, some research classifies wellbeing as 'low', 'average' or 'high' using the mean and standard deviation (SD) of the data (Braunholtz et al., 2007), and other evaluations have compared to UK national values. The most recently surveyed national average wellbeing for adults in England is a score of 49.9, with a weighted-sample standard deviation of 10.8 (Morris and Earle, 2017). Average wellbeing can be considered to be within one standard deviation of the mean (39–61). A score greater than one standard deviation above the mean (62 or higher) can be categorised as high wellbeing, and a score one standard deviation below the mean (38 or lower) can be categorised as low wellbeing.

We now present collated findings from these evaluations in which data was collected using the WEMWBS and SWEMWBS (for the current purpose, SWEMWBS scores have been doubled to bring these in line with full WEMWBS data). Participants completed the questionnaire at the start and end of the programme. Data included is only that from participants who were new to the respective programme and reported low wellbeing scores at the start of the evaluation. Table 43.1 outlines the six evaluated wellbeing programmes and Figure 43.1 displays the average wellbeing scores reported at the start and the end of the respective programmes. A total of 87 participants are included in this analysis (male = 33; female = 45 undisclosed = 9). Their ages ranged from 17 to 80, (see Table 43.1 for details of ages by project).

The combined mean $\pm$ SD wellbeing score at the start of all programmes was 30.7 $\pm$ 6.1, representing a low wellbeing. After engagement with the landscape-based wellbeing programmes, the overall mean significantly improved to 43.4 $\pm$ 11.3 [t_{86} = 11.7; $p < 0.001$], representing average wellbeing and a shift in categorisation. The mean difference of 12.7 (95% CIs 10.5–14.8) indicates that scores rose by 22.7 per cent of the total score-span scope of the wellbeing measure. Figure 43.1 shows some variation in the starting wellbeing scores between the six programmes, but mean wellbeing score in each of the six had moved into the average wellbeing bracket at post-intervention. The biggest change was seen in Programme D, but this also had a very large standard deviation due to the particularly small sample size.

Figure 43.2 shows changes in wellbeing scores for the six programmes when split by sex and programme duration. Male and female wellbeing scores at the start of the programme were similar and improved by similar magnitudes (change for males was 14.0 $\pm$ 10.0 and

Table 43.1 Nature-based intervention programmes for vulnerable groups.

Programme	Cohort	Landscape subtype	Duration	Programme description
A	General population and adults with defined needs (21–57 years old; mean age 36.3; M = 8, F = 1)	Nature reserves	12 weeks	Volunteering in conservation and craft activities (multiple projects included)
B	Vulnerable adults (ranging in their defined needs) (34–52 years old; mean age 42.5; M = 2, F = 6)	Urban greenspaces such as community gardens and allotments	12 weeks	Community gardening and food growing projects (multiple projects included)
C	Adolescents (17–20 year olds; mean age 22.9; M = 3, F = 4)	Woodland	6 weeks (one day per week)	Personal development programme using bush craft for improving confidence and teamwork
D	Youth at risk, experiencing self-esteem and/or behavioural issues (17–18 year olds; M = 1, F = 2)	Mountains	6 months	Activities that centred around two week-long wilderness trails
E	Adults experiencing mental health difficulties (22–51 years old; mean age 38.5; M = 4, F = 2)	Countryside and mountains	6 months	Walking and outdoor-based therapy activities
F	General public, including asylum seekers, individuals experiencing physical or mental health disabilities, elderly, socioeconomically deprived (19–80 years old; mean age 42.6; M = 15, F = 30, Undisclosed = 9)	Urban greenspaces such as parks, allotments.	6 months	Provision of opportunities for participation in nature-based activities

change for females was 12.3 ± 11.0), so although there was a significant improvement over time $[F_{1,76} = 118.8, p < 0.001]$ there was no significant differences according to gender $(p > 0.05)$. Both male and female wellbeing scores moved into the average category after participating in the programmes. Both the starting wellbeing scores and the reported improvements in wellbeing were similar across programmes lasting 12 weeks and less, and those lasting longer than 6 months. This alludes to the efficacy of nature-based interventions for meaningfully

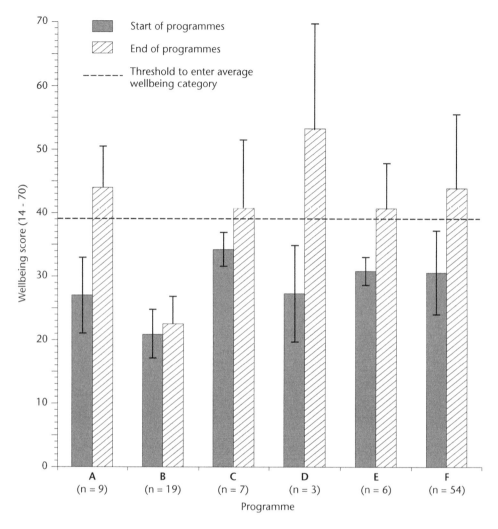

Figure 43.1   Self-reported changes in wellbeing from start to end of programmes.

improving individuals' wellbeing, and maintaining those improvements across many months. The landscape-based interventions enabled 63 per cent of participants to move out of the low wellbeing category and into the average and high categories; 93 per cent of participants reported better wellbeing at the end of the programmes than at the start. The data clearly illustrates the efficacy of harnessing the landscape for wellbeing benefits for vulnerable groups.

Conclusion

Major health and wellbeing issues face the UK (both at an individual and population level) including physical inactivity; obesity; mental ill-health; dementia; loneliness and continuing health inequalities. These issues have created real challenges for policy makers, public health and for statutory, voluntary and private sector organisations responsible for providing health and social care services. In addition, the pressure on natural landscapes in the UK continues to grow,

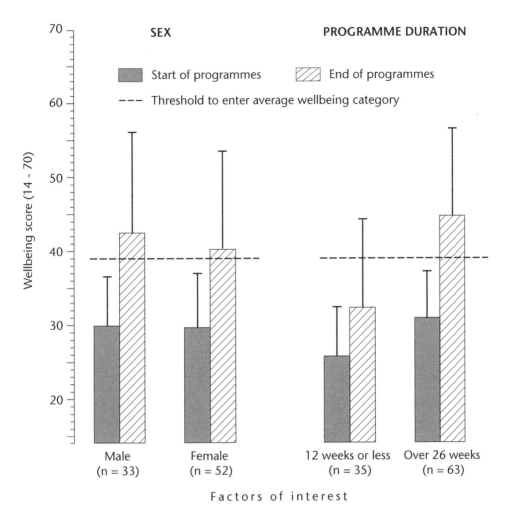

Figure 43.2 Changes in wellbeing scores in relation to sex and programme duration.

the need for housing and economic opportunities continues to rise and inevitably, development is likely to put greenspaces at risk. This chapter has outlined some of the many ways in which landscape and wellbeing are linked. We have focussed on the positive influences of nature landscapes and the multiple wellbeing and social inclusion outcomes that come from contact with nature. Research evidence of nature's potential to improve wellbeing has implications for not only the wellbeing and resilience of individuals but also for public health of communities and for the management of natural environments. Landscapes play an important role in providing a valuable public health resource.

We have illustrated this by presenting data collected from landscape-based interventions for individuals with low wellbeing. Such interventions are designed to enhance wellbeing, by enabling participants to connect to others, be active, take notice of the environment, learn and give. By participating in the five ways to wellbeing participants reported increased social cohesion and inclusion, reduced social isolation, and improved fitness, self-esteem and self-efficacy, all of which contributed to their improved mental wellbeing. Natural landscapes play a nuanced and instrumental role in adult daily life (Finlay et al., 2015; Pretty et al., 2016, 2017). It is important we enhance our understanding of adult health and landscape experiences to inform the development and design of nearby nature to improve the quality of life for ageing populations. The need for access to good quality nature has important policy implications for a wide range of sectors, including public health, mental health and social care, social inclusion, the management of natural places and urban planning.

References

Abkar, M., Kamal, M., Maulan, S. & Mariapan, M. 2010. Influences of viewing nature through windows. *Australian Journal of Basic and Applied Sciences*, 4, 5346–5351.

Aked, J., Marks, N., Cordon, C. & Thompson, S. 2009. *Five ways to wellbeing: A report presented to the Foresight Project on communicating the evidence base for improving people's well-being*. London: Ncf.

Aspinall, P., Mavros, P., Coyne, R. & Roe, J. 2015. The urban brain: analysing outdoor physical activity with mobile EEG. *British Journal of Sports Medicine*, 49, 272–276.

Barton, J., Hine, R. & Pretty, J. 2009. The health benefits of walking in greenspaces of high natural and heritage value. *Journal of Integrative Environmental Sciences*, 6, 261–278.

Barton, J. & Pretty, J. 2010. What is the best dose of nature and green exercise for improving mental health? A multi-study analysis. *Environmental Science & Technology*, 44, 3947–3955.

Bell, S. L., Phoenix, C., Lovell, R. & Wheeler, B. W. 2015. Seeking everyday wellbeing: The coast as a therapeutic landscape. *Social Science & Medicine*, 142, 56–67.

Berman, M. G., Jonides, J. & Kaplan, S. 2008. The cognitive benefits of interacting with nature. *Psychological Science*, 19, 1207–1212.

Berto, R. 2005. Exposure to restorative environments helps restore attentional capacity. *Journal of Environmental Psychology*, 25, 249–259.

Bratman, G. N., Hamilton, J. P. & Daily, G. C. 2012. The impacts of nature experience on human cognitive function and mental health. *Annals of the New York Academy of Sciences*, 1249, 118–136.

Braunholtz, S., Davidson, S., Myant, K. & O'Connor, R. 2007. *Well? What do you think? The third national Scottish survey of public attitudes to mental health, mental wellbeing and mental health problems*. Edinburgh: Scottish Government Social Research.

Brymer, E. & Davids, K. 2012. Ecological dynamics as a theoretical framework for development of sustainable behaviours towards the environment. *Environmental Education Research*, 19, 45–63.

Brymer, E. & Davids, K. 2014. Experiential learning as a constraint-led process: an ecological dynamics perspective. *Journal of Adventure Education & Outdoor Learning*, 14, 103–117.

Brymer, E., Davids, K. & Mallabon, L. 2014. Understanding the psychological health and well-being benefits of physical activity in nature: an ecological dynamics analysis. *Ecopsychology*, 6, 189–197.

Carruthers, C. P. & Hood, C. D. 2004. The power of the positive: leisure and well-being. *Therapeutic Recreation Journal*, 38, 225.

Chu, L. C. 2010. The benefits of meditation vis-à-vis emotional intelligence, perceived stress and negative mental health. *Stress and Health*, 26, 169–180.

Clark, N. E., Lovell, R., Wheeler, B. W., Higgins, S. L., Depledge, M. H. & Norris, K. 2014. Biodiversity, cultural pathways, and human health: a framework. *Trends in Ecology & Evolution*, 29, 198–204.

Clarke, A., Friede, T., Putz, R., Ashdown, J., Martin, S., Blake, A., Adi, Y., Parkinson, J., Flynn, P. & Platt, S. 2011. Warwick-Edinburgh Mental Well-being Scale (WEMWBS): validated for teenage school students in England and Scotland. A mixed methods assessment. *BMC Public Health*, 11, 487.

Cooper, A. 2015. Nature and the outdoor learning environment: the forgotten resource in early childhood education. *International Journal of Early Childhood Environmental Education*, 3, 85–97.

Curtin, S. 2009. Wildlife tourism: the intangible, psychological benefits of human–wildlife encounters. *Current Issues in Tourism*, 12, 451–474.

Dallimer, M., Irvine, K. N., Skinner, A. M., Davies, Z. G., Rouquette, J. R., Maltby, L. L., Warren, P. H., Armsworth, P. R. & Gaston, K. J. 2012. Biodiversity and the feel-good factor: understanding associations between self-reported human well-being and species richness. *BioScience*, 62, 47–55.

Deci, E. L. & Ryan, R. M. 2000. The 'what' and 'why' of goal pursuits: human needs and the self-determination of behavior. *Psychological Inquiry*, 11, 227–268.

Diener, E., Scollon, C. N. & Lucas, R. E. 2003. The evolving concept of subjective well-being: the multifaceted nature of happiness. *Advances in cell aging and gerontology*, 15, 187–219.

Diener, E., Scollon, C. N. & Lucas, R. E. 2009. *The evolving concept of subjective well-being: the multifaceted nature of happiness*. Berlin: Springer.

Ewert, A., Overholt, J., Voight, A. & Wang, C. C. 2011. Understanding the transformative aspects of the wilderness and protected lands experience upon human health. Retrieved from www.fs.fed.us/rm/pubs/rmrs_p064/rmrs_p064_140_146.pdf.

Feinstein, L. & Hammond, C. 2004. The contribution of adult learning to health and social capital. *Oxford Review of Education*, 30, 199–221.

Feldman, G., Hayes, A., Kumar, S., Greeson, J. & Laurenceau, J.-P. 2007. Mindfulness and emotion regulation: the development and initial validation of the Cognitive and Affective Mindfulness Scale-Revised (CAMS-R). *Journal of Psychopathology and Behavioral Assessment*, 29, 177.

Finlay, J., Franke, T., McKay, H. & Sims-Gould, J. 2015. Therapeutic landscapes and wellbeing in later life: impacts of blue and green spaces for older adults. *Health & Place*, 34, 97–106.

Fuller, R. A., Irvine, K. N., Devine-Wright, P., Warren, P. H. & Gaston, K. J. 2007. Psychological benefits of greenspace increase with biodiversity. *Biology Letters*, 3, 390–394.

Gale, C. R., Booth, T., Mõttus, R., Kuh, D. & Deary, I. J. 2013. Neuroticism and Extraversion in youth predict mental wellbeing and life satisfaction 40 years later. *Journal of Research in Personality*, 47, 687–697.

Gladwell, V. F., Brown, D. K., Wood, C., Sandercock, G. R. & Barton, J. L. 2013. The great outdoors: how a green exercise environment can benefit all. *Extreme Physiology & Medicine*, 2, 1–7.

Hall-Lande, J. A., Eisenberg, M. E., Christenson, S. L. & Neumark-Sztainer, D. 2007. Social isolation, psychological health, and protective factors in adolescence. *Adolescence*, 42, 265.

Hammond, C. 2004. Impacts of lifelong learning upon emotional resilience, psychological and mental health: fieldwork evidence. *Oxford Review of Education*, 30, 551–568.

Hasnain, N., Wazid, S. & Hasan, Z. 2014. Optimism, hope, and happiness as correlates of psychological well-being among young adult assamese males and females. *IOSR Journal Of Humanities And Social Science (IOSR-JHSS)*, 19, 44–51.

Herzog, T. R. & Strevey, S. J. 2008. Contact with nature, sense of humor, and psychological well-being. *Environment and Behavior*, 40, 747–776.

Howell, A. J., Dopko, R. L., Passmore, H.-A. & Buro, K. 2011. Nature connectedness: associations with well-being and mindfulness. *Personality and Individual Differences*, 51, 166–171.

Huby, M., Cinderby, S., Crowe, A., Gillings, S., McClean, C., Moran, D., Owen, A. & White, P. 2006. The association of natural, social and economic factors with bird species richness in rural England. *Journal of Agricultural Economics*, 57, 295–312.

Keng, S.-L., Smoski, M. J. & Robins, C. J. 2011. Effects of mindfulness on psychological health: a review of empirical studies. *Clinical Psychology Review*, 31, 1041–1056.

Keyes, C. L., Shmotkin, D. & Ryff, C. D. 2002. Optimizing well-being: the empirical encounter of two traditions. *Journal of Personality and Social Psychology*, 82, 1007.

Konu, A. & Rimpelä, M. 2002. Well-being in schools: a conceptual model. *Health Promotion International*, 17, 79–87.

Lemieux, C. J., Eagles, P. F., Slocombe, D. S., Doherty, S. T., Elliott, S. J. & Mock, S. E. 2012. Human health and well-being motivations and benefits associated with protected area experiences: an opportunity for transforming policy and management in Canada. *Parks*, 18, 71–85.

Lovell, R., Wheeler, B. W., Higgins, S. L., Irvine, K. N. & Depledge, M. H. 2014. A systematic review of the health and well-being benefits of biodiverse environments. *Journal of Toxicology and Environmental Health, Part B*, 17, 1–20.

Luck, G. W., Davidson, P., Boxall, D. & Smallbone, L. 2011. Relations between urban bird and plant communities and human well-being and connection to nature. *Conservation Biology*, 25, 816–826.

Maller, C., Townsend, M., Henderson-Watson, C., Pryor, A., Prosser, L., Moore, M., Leger, L. S. & Victoria, P. 2008. *Healthy parks healthy people: The health benefits of contact with nature in a park context*. Melbourne: Deakin University and Parks Victoria.

Maslow, A. 1954. *Motivation and personality*. New York: Harper & Row.

Maslow, A. H., Frager, R., Fadiman, J., McReynolds, C. & Cox, R. 1970. *Motivation and personality*. New York: Harper & Row.

Mitchell, R. J., Richardson, E. A., Shortt, N. K. & Pearce, J. R. 2015. Neighborhood environments and socioeconomic inequalities in mental well-being. *American Journal of Preventive Medicine*, 49, 80–84.

Morris, S. & Earle, K. 2017. *Health survey for England 2016: well-being and mental health*. London: Health and Social Care Information Centre/NHS Digital.

New Economics Foundation. 2008. *Five ways to wellbeing: a report presented to the Foresight project on communicating the evidence base for improving people's wellbeing*. London: New Economics Foundation.

Nisbet, E. K., Zelenski, J. M. & Murphy, S. A. 2009. The nature relatedness scale linking individuals' connection with nature to environmental concern and behavior. *Environment and Behavior*, 41, 715–740.

Norrish, J. M. & Vella-Brodrick, D. A. 2008. Is the study of happiness a worthy scientific pursuit? *Social Indicators Research*, 87, 393–407.

Office for National Statistics. 2017. Measuring national well-being: Life in the UK. London: Office for National Statistics.

Parkinson, J. 2006. *Measuring positive mental health: Developing a new scale*. Glasgow: NHS Health Scotland.

Plutchik, R. 1980a. A general psychoevolutionary theory of emotion. *Emotion: Theory, Research, and Experience*, 1, 3–33.

Plutchik, R. 1980b. *Measurement implications of a psychoevolutionary theory of emotions: Assessment and modification of emotional behavior*. Berlin: Springer.

Pretty, J., Barton, J., Pervez Bharucha, Z., Bragg, R., Pencheon, D., Wood, C. & Depledge, M. H. 2016. Improving health and well-being independently of GDP: dividends of greener and prosocial economies. *International Journal of Environmental Health Research*, 26, 11–36.

Pretty, J., Rogerson, M. & Barton, J. 2017. Green mind theory: how brain-body-behaviour links into natural and social environments for healthy habits. *International Journal of Environmental Research and Public Health*, 14, 706.

Rilling, J. K., Glenn, A. L., Jairam, M. R., Pagnoni, G., Goldsmith, D. R., Elfenbein, H. A. & Lilienfeld, S. O. 2007. Neural correlates of social cooperation and non-cooperation as a function of psychopathy. *Biological Psychiatry*, 61, 1260–1271.

Rogerson, M. & Barton, J. 2015. Effects of the visual exercise environments on cognitive directed attention, energy expenditure and perceived exertion. *International Journal of Environmental Research and Public Health*, 12, 7321–7336.

Rogerson, M., Barton, J., Bragg, R. & Pretty, J. 2017. The health and wellbeing impacts of volunteering with The Wildlife Trusts. University of Essex.

Rogerson, M., Gladwell, V. F., Gallagher, D. J. & Barton, J. L. 2016. Influences of green outdoors versus indoors environmental settings on psychological and social outcomes of controlled exercise. *International Journal of Environmental Research and Public Health*, 13, 363.

Ryan, R. M. & Huta, V. 2009. Wellness as healthy functioning or wellness as happiness: The importance of eudaimonic thinking (response to the Kashdan et al. and Waterman discussion). *The Journal of Positive Psychology*, 4, 202–204.

Seligman, M. E. P. 2002. *Authentic happiness: Using the new positive psychology to realize your potential for lasting fulfillment*. New York: Free Press.

Sin, N. L. & Lyubomirsky, S. 2009. Enhancing well-being and alleviating depressive symptoms with positive psychology interventions: A practice-friendly meta-analysis. *Journal of Clinical Psychology*, 65, 467–487.

Speldewinde, P. C., Cook, A., Davies, P. & Weinstein, P. 2009. A relationship between environmental degradation and mental health in rural Western Australia. *Health & Place*, 15, 880–887.

Steptoe, A., Shankar, A., Demakakos, P. & Wardle, J. 2013. Social isolation, loneliness, and all-cause mortality in older men and women. *Proceedings of the National Academy of Sciences*, 110, 5797–5801.

Stewart-Brown, S. & Janmohamed, K. 2008. *Warwick–Edinburgh mental well-being scale: User guide, version*, 1. Glasgow: NHS Health Scotland.

Stewart-Brown, S., Tennant, A., Tennant, R., Platt, S., Parkinson, J. & Weich, S. 2009. Internal construct validity of the Warwick-Edinburgh mental well-being scale (WEMWBS): a Rasch analysis using data from the Scottish health education population survey. *Health and Quality of Life Outcomes*, 7, 15.

Takano, T., Nakamura, K. & Watanabe, M. 2002. Urban residential environments and senior citizens' longevity in megacity areas: the importance of walkable green spaces. *Journal of Epidemiology & Community Health*, 56, 913–918.

Teas, J., Hurley, T., Ghumare, S. & Ogoussan, K. 2007. Walking outside improves mood for healthy postmenopausal women. *Clinical Medicine: Oncology*, 1, 35–43.

Tennant, R., Hiller, L., Fishwick, R., Platt, S., Joseph, S., Weich, S., Parkinson, J., Secker, J. & Stewart-Brown, S. 2007. The Warwick–Edinburgh mental well-being scale (WEMWBS): development and UK validation. *Health and Quality of life Outcomes*, 5, 63.

Tones, K. & Green, J. 2004. *Health promotion: planning and strategies.* Oxford: Sage.

Ulrich, R. S. 1981. Natural versus urban scenes some psychophysiological effects. *Environment and Behavior*, 13, 523–556.

Ulrich, R. 1984. View through a window may influence recovery. *Science*, 224, 224–225.

Ulrich, R. S., Simons, R. F., Losito, B. D., Fiorito, E., Miles, M. A. & Zelson, M. 1991. Stress recovery during exposure to natural and urban environments. *Journal of Environmental Psychology*, 11, 201–230.

Waterman, A. S. 1993. Two conceptions of happiness: Contrasts of personal expressiveness (eudaimonia) and hedonic enjoyment. *Journal of Personality and Social Psychology*, 64, 678.

Wheeler, B. W., White, M., Stahl-Timmins, W. & Depledge, M. H. 2012. Does living by the coast improve health and wellbeing? *Health & Place*, 18, 1198–1201.

White, M. P., Alcock, I., Wheeler, B. W. & Depledge, M. H. 2013. Would you be happier living in a greener urban area? A fixed-effects analysis of panel data. *Psychological Science*, 24, 920–928.

Williams, J. M. G. 2010. Mindfulness and psychological process. *Emotion*, 10, 1.

Yeh, H.-P., Stone, J. A., Churchill, S. M., Wheat, J. S., Brymer, E. & Davids, K. 2015. Physical, psychological and emotional benefits of green physical activity: an ecological dynamics perspective. *Sports Medicine*, 1–7.

Zelenski, J. M. & Nisbet, E. K. 2014. Happiness and feeling connected the distinct role of nature relatedness. *Environment and Behavior*, 46, 3–23.

44

Rewilding as rural land management

Opportunities and constraints

Mick Lennon

Introduction

Agricultural and silvicultural land abandonment represents a significant challenge for the future sustainability of rural areas. Indeed, the extent of this challenge is reflected in the calculations of Ramankutty and Foley (Ramankutty and Foley, 1999), who estimate that between 1700 and 1992 roughly 1.5 million square kilometres of arable land was abandoned globally. The ecological consequences of such abandonment are complicated, as some species associated with agricultural landscapes decline and others associated with successional vegetation increase (Queiroz et al., 2014). These alterations to the land-use mosaic also impact the social and economic fabric of rural societies as the demographic imbalances generated by rural–urban migration throw many areas into a spiral of decline characterised by underinvestment and a perceived lack of opportunity. Although tackling these issues has been a constant presence in rural planning activities since they first emerged, the acceleration of European land abandonment during the second half of the twentieth century brought this matter to the fore of policy debates (MacDonald et al., 2000). The drivers of such land abandonment are complex and often contextualised, but nevertheless interact with the globalisation of agricultural production characterised by increasing competition and decreasing financial margins (Lasanta et al., 2017). In this context, Navarro and Pereira (2015a) identify three prominent land-use policy responses that have emerged in recent years: intensification, extensification and afforestation. In their analysis, 'intensification' of agricultural practices reflects a perceived need to maximise productivity and is most frequently pursued in fertile areas with topographical conditions conducive to mechanisation. Concurrent in time but not space, is a process of extensification, whereby improved productivity is sought through diversifying agricultural land uses and/or increasing farm sizes via land consolidation. Finally, afforestation is increasingly practiced in peripheral lands of comparatively poor soil fertility. However, less attention has been allocated to a fourth option, that of 'rewilding' (Jepson, 2016; Navarro and Pereira, 2015b; Schnitzler, 2014). Normally the preserve of ecological debates, this chapter thereby explores the broader potential benefits and constraints of rewilding as a holistic response to the multifarious challenges presented by agricultural and silvicultural land abandonment in the twenty-first century. To effectively achieve this, it is first necessary to examine the origins and evolution of the concept.

Origins and evolution of the rewilding concept

The term 'rewilding' first emerged in North America from the collaborative work of environmental activist Dave Foreman and biologist Michael Soulé, which resulted in the creation of The Wildlands Project in 1991 (subsequently renamed the Wildlands Network) (Sandom et al., 2013). In essence, this initiative seeks to 'recreate' extensive natural landscapes throughout North America by reintroducing predators. The work of the Wildlands Network is neatly summarised in the 3Cs of 'cores, corridors and carnivores'. The underlying logic of this initiative is to create large core areas free from human interference that are linked by corridors of emerging or regenerating habitats. The ecological dynamics of this landscape are managed through the predation activities of carnivores. While the temporal referent for this proposal is not specified, it appears to predate the modern period as 'Under this earliest rewilding concept, the wild is a time when large carnivores were abundant in North America' (Jørgensen, 2015: 483). In this context, Soulé and Noss (1998) reference the reintroduction of wolves into Yellowstone National Park in the USA as an example of their 3Cs concept in action. Here, wolves were progressively reintroduced following an absence from the park for several decades due to their eradication by humans. These carnivores subsequently began predating the park's population of elk, resulting in the behaviour modification of park herbivores, which in turn had a series of cascading effects on local ecology. For example, large herbivores altered their patterns of grazing such that less time was spent in comparatively exposed river valley areas. This facilitated the regeneration of riparian willow that caused silting and ensuing river realignment, which in turn facilitated an increase in local biodiversity, with bird, fish, mammal and insect species taking advantage of regenerated habitat areas wherein new foodwebs emerged. Nevertheless, as noted by Jørgensen (2015), the 3Cs concept has enjoyed only limited purchase within the literature on rewilding (Fraser, 2009).

Receiving much more attention is the rewilding concept advanced of Donlan et al. (2006) in the years following the initial work of The Wildlands Network. Ostensibly motivated by what they view as the denuding of ecological complexity, the advocates of this form of rewilding expand the visions of The Wildlands Network by proposing the reintroduction of proxies for extinct species into large protected areas of North America. This is to be undertaken with the aim of mimicking the ecological dynamics present at the close of the Pleistocene, which ended about 13,000 years ago in North America (often referred to the 'Ice Age'). Thus, in contrast to the vague temporal referents of The Wildlands Network, advocates of 'Pleistocene Rewilding' propose recreating the past landscape of a specific period. Supporters of this approach view it as necessary to seek the regeneration of Pleistocene ecological dynamics, and not just pre-colonial landscapes, as it is contended that the degree of megafaunal extinction was greater following initial human migration into North America via the Bering Strait, than was the case following the rediscovery of the continent by Columbus. From this perspective, currently extant species of cheetah, lion, elephants, camels and tortoises should be translocated to North America and used as surrogates to re-engineer the landscape. Unsurprisingly, the rewilding proposals advanced by both The Wildlands Network and promoters of Pleistocene Rewilding have proved controversial. Specifically, those involved in nature conservation question the scientific wisdom of such approaches given that the reintroduction of predators or their proxies into landscapes in which they have been absent for a considerable period may result in unintended consequences as prey species may not adjust easily to their presence, with potentially catastrophic implications for foodweb dynamics (Buchholz, 2007; Caro, 2007). However, more influential are the normative dimensions of reintroducing predators into rural landscapes to facilitate the conversion of such spaces into eco-centric wilderness areas. Here, concern regarding the presence of large predators in the environment reveals the limits of acceptable rewilding as issues surrounding agricultural conflict and personal safety come to the fore. As noted by a proponent of Pleistocene Rewilding, 'what we did is expose the real reason why this will not happen in the United States- because we are mostly not willing to put up with dangerous animals' (Harry Green quoted in Carey, 2016: 807).

A different approach to animal-led rewilding has been adopted in Europe. Similar to the geography of the 3Cs approach, this model focuses on ecological networks of spatially arranged cores, corridors, buffers and restoration areas. However, in contrast to prominent North American rewilding proposals, greater emphasis is placed in European rewilding on naturalistic grazing regimes. Essentially, these involve using robust herbivores outside of field-based agricultural systems to 'manage' the landscape. This approach is rooted in a paleoecological understanding of European landscapes in the middle Holocene period, approximately 7,000–5,000 years ago (Vera, 2002). It posits that European landscapes of the mid-Holocene were continually shifting between open canopy mixed grassland and forested ecosystems maintained by the grazing activities of large herbivores. Hence, proponents of naturalistic grazing seek to establish guilds of large herbivores (e.g. bison, horses, cattle) on substantial sites free from human interference wherein complex biodiversity rich landscapes can be recreated. Such sites are generally identified as unprofitable agricultural and forestry areas (Van Wieren, 1995). Nevertheless, similar to other forms of animal-led rewilding proposals, the theories underpinning naturalistic grazing regimes are not without their scientific detractors (Mitchell, 2005). However, as with predator based rewilding proposals, it is normative questions that have proved most controversial. For example, in the case of the Oostvaardersplassen Nature Reserve outside Amsterdam in The Netherlands, it was not the science of naturalistic grazing that posed the greatest threat to the continuation of this rewilding experiment. Instead, significant public unease regarding animal welfare found expression in political debate consequent on substantial animal mortalities in the harsh winters of 2005 and 2010 (Lorimer, 2015). This emotive issue wound its way through protests and into the legal system, with a compromise eventually being reached in which a professional hunter would cull enfeebled animals unlikely to survive harsh weather. As such, a less interventionist form of rewilding not reliant on the controversial reintroduction of animal species may prove more normatively acceptable. It is this form of 'passive rewilding' that is quietly gaining ground as an approach in marginalised European regions experiencing agricultural and silvicultural decline.

Rewilding marginalised rural environments

Instead of focusing upon animal-led landscape rewilding, those advocating passive forms of rewilding most frequently concern themselves with the successionary vegetative dynamics in the context of land abandonment. Here, less stress is placed on pre-human referents, with a greater emphasis laid on 'the passive management of ecological succession with the goal of restoring natural ecosystem processes and reducing human control of landscapes' (Pereira and Navarro, 2016: v). Working in landscapes with long histories of human modification, those attempting such passive forms of rewilding in Europe are generally aware that nature is co-produced through changing configurations of culture, politics, economics and the non-human environment. Hence, while North American perspectives often seek to extract culture from nature, in Europe the aim is frequently to inject nature into culture (Hall, 2014). Although natural processes may be initially restored via human intercession, the objective is to withdraw human intervention from the management of the landscape, but not necessarily humans from that landscape. It is in this sense that those advocating the passive rewilding of non-profitable agricultural and silvicultural areas often support local economic development centred on eco-tourism, hunting and adventure recreation (see www.rewildingeurope.com). With vegetative succession and complex forest re-establishment frequently seen as consequences of such rewild-ing, additional ecosystems services are understood to accrue than would otherwise be the case should landscapes continue to be dominated by economically unprofitable agriculture or com-mercial forestry monoculture. For example, supporters of such rewilding initiatives contend that forest rejuvenation assists the sequestration of atmospheric carbon (Poulton et al., 2003), a processes widely deemed essential to tackling climate change. Likewise, it is asserted that flood risk management benefits from such forms of rewilding. For example, the freedom given to alter course to the River Liza at Wild Ennerdale in England's Lake District has been shown to help mitigate flooding (Oyedotun, 2011). Furthermore, cost savings ensuing from less expenditure on traditional labour intensive modes of biodiversity management can be redirected into local community development and other forms of nature conservation initiatives.

Hence, it is ironic that perhaps the main impediment to the realisation of passive rewilding in Europe is not so much local community opposition as the institutional configuration of nature conservation regulation at the supranational level. Specifically, while the provisions of European Union (EU) policy facilitate agri-environmental schemes (EEC Regulation no. 2078/92) that are a cornerstone of both agricultural and environmental policy across Europe, they concurrently constrain rewilding as a feasible conservation approach. This is because such agri-environmental schemes provide farmers with payments for environmental improvement and maintenance activities grounded in a 'compositional' ecological perspective. From this perspective, certain constellations of species such as meadows and heathland are valued above others. The species configurations associated with these habitats thus form the end-point of the financial incen-tives to farmers, such that fostering and maintaining these ecological configurations through targeted vegetation clearances and planting schemes become requisites for subsidy payments (EEA, 2009). However, rewilding promotes a 'functionalist paradigm' (Jepson, 2016: 121) of ecosystem self-direction wherein ecological configurations such as meadows and heathlands are but points in the continuum of successional processes characterising the ecological dynamics of a self-willed landscape. It is against this backdrop of institutionalised nature conservation policy that the abandonment of farmland to passive rewilding is perceived as a threat to biodiversity (MacDonald et al., 2000; Merckx and Pereira, 2015). Furthermore, many involved in nature conservation fret that passive rewilding increases the risk of unwelcome ecological features such as zoonotic diseases, invasive species spread, forest fires, as well as unchecked population explosions of previously controlled and undesirable animals (Goulding and Roper, 2002; Pausas et al., 2009). At heart here are contending understandings of the very nature of nature itself, as fundamentally the principles of rewilding's 'functionalist' ecological perspective that promotes a 'hands-off' approach to landscape runs counter to the 'compositional' paradigm underpin-ning nature conservation at the EU level. This thereby constrains the prospects of rewilding consequent on: (a) how nature is interpreted and valued; (b) the personal investment costs of time, money and identity made by those schooled in compositional modes of assessment and management; and (c) the institutional arrangements embedded within the machinery of EU and national policy resultant from (a) and (b), and given most salient expression in the compositional audit culture of the EU Birds (ECC, 1979) and Habitats (ECC, 1992) Directives. Consequently, even though rewilding is growing as a nature conservation discourse in Europe, it lacks the overarching institutions and policy architecture necessary for pan-European coherence as an approach to managing rural land abandonment across the continent. Accordingly, landscape level rewilding in Europe is currently characterised by an array of different approaches. These range from those based on naturalised grazing regimes, such as the Oostvaardersplassen Nature

Reserve initiative, to those that actively seek to minimise grazing in an effort to protect the initial vegetative compositions of rewilded woodlands, such as the Carrifran Wildwood Project in Scotland (Ashmole and Ashmole, 2009), to those that promote a retreat from intervention more broadly, such as in the forest of the Ardèche in France (Schnitzler, 2014). It is against this backdrop that competing perspectives on what rewilding may mean and how it should be implemented can be observed. Identifying and examining such contending viewpoints thereby proffers greater understanding as to the issues associated with advancing rewilding as an approach, or suite of approaches, to addressing land abandonment in rural environments. Hence, the next section explores the case of a rewilding initiative in the peripheral west coast of Ireland. This is undertaken as a vehicle for investigating the issues encountered when seeking to rewild modified and marginalised rural landscapes.

Wild Nephin

In March 2013, the National Parks and Wildlife Services (NPWS) and Coillte, the body that manages Ireland's state-owned forests, signed a Memorandum of Understanding formalising a shared aim to create a 'Wild Nephin' area by combining and rewilding lands under their respective administrations on Ireland's western seaboard. In an effort to progress the project, a number of stakeholders deemed relevant to the initiative were also invited to become involved via membership on the project's steering committee. These included Mayo County Council, in whose administrative area both these areas are located, and Fáilte Ireland (the national tourism development organisation). The initiative seeks to unite the existing Ballycroy National Park with the contiguous Nephin forest to create Ireland's first wilderness area. The Ballycroy National Park is managed by the NPWS and is Ireland's most recent addition to the state's national park system. Adjoining the eastern boarder of the national park are the Coillte forestry lands, amounting to 4,606 hectares, which are 'currently managed as a production forest with the predominant species being Sitka spruce (*Pices sitchensis*) and lodge pole pine (*Pinus contorta*)' (Murphy et al., 2011: 2:2). However, the feasibility study produced as part of the Wild Nephin project concluded that 'An analysis of yield and net present value demonstrated that harvesting the Nephin forest will result in a negative outcome for the company' (ibid.: 1:7). Thus, the forestry plantation is commercially unviable. The prospect of rewilding these lands thereby offers an opportunity to minimise financial losses while concurrently maximising public goods provision from the forest, such as biodiversity enhancement, climate change mitigation, recreation and eco-tourism based local economic development. The provision of these public goods part-fulfils Coillte's mandate (see www.coillte.ie). Most attention in the project is focused on rewilding this commercial forestry plantation, with a fifteen-year timeline proposed for its conversion to wilderness. While there is general agreement that the Wild Nephin project should primarily centre on landscape scale vegetative regeneration rather than animal-led rewilding, interviews conducted with project participants reveal differences of opinion regarding the most appropriate way to implement such rewilding. Much of this stems from a lack of national referents on what rewilding may entail in a landscape heavily modified by commercial forestry. Hence, different objectives for rewilding have emerged as project participants reflectively interpret the meaning of rewilding relative to the specific cultural, economic and ecological contexts of the forest. This is undertaken against a backdrop in which no officially institutionalised national or pan-European framework for rewilding exists.

Central to this issue are diverging opinions regarding the role of conventional interventionist approaches focused on 'compositional' ecological understandings of nature conservation and more passive 'functionalist' perspectives on how to rewild a landscape. Here, a key fault line is the position of invasive species. This is particularly important in the context of managing the ecological conditions of the adjacent Ballycroy National Park, where EU nature conservation designations permitting inclusion in the Natura 2000 network require a management regime part-focused on the eradication of flora and fauna deemed invasive. To the fore of such debates is the spread of rhododendron (*Rhododendron ponticum*) and laurel (*Prunus laurocerasus*), as 'Where these species invade, they often kill both ground vegetation layers and prohibit natural regeneration of woody plants' (Murphy et al., 2011: 8:6). Controlling the range of these species is thereby seen as fundamental to successful rewilding. The standing of this issue is reflected in a commitment to limiting the range of these plants despite acknowledgment that 'the eradication of these species will require a long-term commitment and experience to date indicates that the cost of removal is high' (ibid.: 8:6). However, this view is not unanimous as others central to the project's conception question the premise of seeking to promote and maintain a particular species composition. As expressed by an interviewee deeply involved in the project:

> Habitats are very dynamic. They're constantly changing and this idea of a climax habitat that I would've learned in college, this in an oak forest, that's it, it's all over – rubbish! It's constant change. You might get new species coming in. You might get sycamore coming in. And we had a lot of discussions in the wilderness project about rhododendron. Why do anything with it? Why? Why that particular species? Now I know that there are various reasons, you know, it tends to crowd out. Well we know it crowds out because we are constantly interfering with it. But what happens if we let it go?
>
> *(Interviewee A1)*

Despite such disagreement, it is probable that in the case of the Wild Nephin project, the 'compositionalist' view will win out through alignment with hierarchies of policy embedded within national land-use management regimes that are profoundly influenced by EU nature conservation regulation (Bastmeijer, 2016; EC, 2013). Indeed, many promoting rewilding as a viable nature conservation approach to addressing issues of land abandonment contend that the power of such embedded policy hierarchies is the principle obstacle to implementing the functionalist ecological paradigm on which much rewilding activity is based (Jepson, 2016; Merckx and Pereira, 2015).

Economic and cultural issues concerning the attributes and benefits of a rewilded landscape have also emerged as matters of debate in the Wild Nephin initiative. Indeed, the project feasibility report concluded that substantial economic advantage would accrue from rewilding the commercial forest, both in terms of discrete visits to the converted forest area and in terms of tourist revenue generated in the wider local economy. As such, rewilding of this commercial forest is viewed by some as presenting a potential income stream for a marginalised rural area. It is therefore unsurprising that Mayo County Council, which is mandated to enhance tourism in the area is supporting the rewilding of the Coillte lands. However, concerns have arisen on how this agenda may impact on the subjective experience of somebody engaging with the landscape. Specifically, fear has been voiced that the insensitive exploitation of a tourism opportunity threatens the very foundation on which such tourism is based. From this perspective, it is not compositional or functionalist ecologies that generate unease. Instead, what matters is the sense of remoteness, challenge and minimal degree of human influence that rewilding is seen to promote. Indeed, some involved in the project believe that it is these attributes that should determine the rewilding activities to be undertaken. In this context, several interviewees expressed the view that the greatest threat to the project is on-going human intervention in the landscape. As asserted by one interviewee,

Unless you're guided by that very strong vision for what the wilderness area is, there's a danger of creeping infrastructure, recreation infrastructure creeping back in. And that's something that we need to be mindful of. I would suspect that whatever motivation, well whatever involvement Mayo County Council has, I would suspect that they're largely motivated by the potential of this project from a tourism perspective. And I think tourism pressure is going to challenge the core proposition of wilderness.

(Interviewee A3)

Thus, there is a strong aesthetic dimension to rewilding, particularly when it is seen as a process that can facilitate the conversion of a marginalised but heavily modified area into a space with a self-willed appearance that is perceived as relatively devoid of human influence (Prior, 2012). Accordingly, debate on how to provide access to rewilded areas, yet safeguard a perception of wildness, also reveals diverging viewpoints not usually addressed in literature on rewilding

Hence, the Wild Nephin project throws into relief the difficulties that may be encountered in attempting to rewild unproductive rural landscapes in the absence of agreement on what 'rewilding' means. While the institutionalisation of compositional ecological approaches within the architecture of EU nature conservation policy may prove an impediment to the realisation of functionalist ecological perspectives, the interactions between the social, cultural and economic dimensions of rewilding remain a topic still comparatively neglected by academic study. Consequently, there is a risk that as the science of rewilding advances, it will be the neglect of the rewilding's social dimensions that limit progress in policy and planning. Ultimately, this rests on the contention that rewilding is as much as complex mix of cultural and economic issues as it is a matter of ecological debate.

Conclusion

Many forms of rewilding emphasise a particular temporal reference and thereby position it within the broader field of restoration ecology. However, more recent developments in the field have construed rewilding as not so much 'looking back' as 'looking forward'. This turn towards the self-willed wild through greater 'non-human autonomy' (Prior and Ward, 2016: 133) challenges traditional modes of practice while concurrently providing an opportunity for debate on how nature is conceived. In doing so it forces to the fore discussion on the perceived services supplied by nature and how to tackle the complex intertwining of ecological conservation and the marginalisation of rural areas. Hence, rewilding presents an opportunity to holistically re-evaluate assumptions on the best way to manage our interactions with the rural landscape. Nevertheless, for this critical reflection to be successful, it will need to broaden the scope of planning, implementation and assessment beyond the narrow confines of ecological science. In particular, should advocates of rewilding seek to overcome the obstacle of institutionalised nature conservation policy, they will have to devise new modes of practice that work with those conventional forms of biodiversity conservation that are socio-economically and culturally embedded in the functioning of many marginalised rural societies. The task of doing so involves complementing rather than challenging such conventional thinking, while not replicating the compositional approaches underpinning it. This probably entails greater focus on collaboratively formulating objectives, practices and regulatory tools with those most likely to be affected by rewilding practices, be they farmers whose livestock and crops are potentially threatened by animal introductions, or local communities who stand to win or loose in the contest to cater for eco-tourism. Ultimately it involves generating and realising new narratives of hope and opportunity that help society re-read the landscape in ways informed by new understandings of what nature may mean in the context of a world pervasively transformed by humanity.

References

Ashmole, M. & Ashmole, P. (eds). 2009. *The Carrifran Wildwood story: ecological restoration from the grass roots.* Ancrum: Borders Forest Trust.

Bastmeijer, K. 2016. Natura 2000 and the protection of wilderness in Europe. *In*: Bastmeijer, K. (ed.) *Wilderness protection in Europe: the role of international, European and national law*, 177–198. Cambridge: Cambridge University Press.

Buchholz, R. 2007. Behavioural biology: an effective and relevant conservation tool. *Trends in Ecology & Evolution*, 22, 401–407.

Carey, J. 2016. Core concept: rewilding. *Proceedings of the National Academy of Sciences*, 113, 806–808.

Caro, T. 2007. The Pleistocene re-wilding gambit. *Trends in Ecology & Evolution*, 22, 281–283.

Donlan, C. J., Berger, J., Bock, C. E., Bock, J. H., Burney, D. A., Estes, J. A., Foreman, D., Martin, P. S., Roemer, G. W. & Smith, F. A. 2006. Pleistocene rewilding: an optimistic agenda for twenty-first century conservation. *The American Naturalist*, 168, 660–681.

EC. 2013. *Guidelines on wilderness in Natura 2000: management of wilderness and wild areas within the Natura 2000 Network*. Technical report 2013-069. Brussels: Commission of the European Communities.

ECC. 1979. *Directive 79/409/EEC of the European Parliament and of the Council on the conservation of wild birds*. Brussels: Commission of the European Communities.

ECC. 1992. *Council Directive 92/43/EEC on the conservation of natural habitats and of wild fauna and flora*. Brussels: Commission of the European Communities.

EEA. 2009. *Distribution and targeting of the CAP budget from a biodiversity perspective*. Technical report 12/2009. Luxembourg: European Environmental Agency.

EEC. 1992. *Council Regulation (EEC) No 2078/92 of 30 June 1992 on agricultural production methods compatible with the requirements of the protection of the environment and the maintenance of the countryside*. Brussels: European Commission.

Fraser, C. 2009. *Rewilding the world: dispatches from the conservation revolution*. New York: Metropolitan Books.

Goulding, M. & Roper, T. 2002. Press responses to the presence of free-living wild boar (*Sus scrofa*) in southern England. *Mammal Review*, 32, 272–282.

Hall, M. 2014. Extracting culture or injecting nature? Rewilding in transatlantic perspective. *Old World and New World Perspectives in Environmental Philosophy*. London: Springer.

Jepson, P. 2016. A rewilding agenda for Europe: creating a network of experimental reserves. *Ecography*, 39, 117–124.

Jørgensen, D. 2015. Rethinking rewilding. *Geoforum*, 65, 482–488.

Lasanta, T., Arnáez, J., Pascual, N., Ruiz-Flaño, P., Errea, M. & Lana-Renault, N. 2017. Space–time process and drivers of land abandonment in Europe. *Catena*, 149, 810–823.

Lorimer, J. 2015. *Wildlife in the Anthropocene: conservation after nature*. Minneapolis, MN: University of Minnesota Press.

MacDonald, D., Crabtree, J., Wiesinger, G., Dax, T., Stamou, N., Fleury, P., Lazpita, J. G. & Gibon, A. 2000. Agricultural abandonment in mountain areas of Europe: environmental consequences and policy response. *Journal of Environmental Management*, 59, 47–69.

Merckx, T. & Pereira, H. M. 2015. Reshaping agri-environmental subsidies: From marginal farming to large-scale rewilding. *Basic and Applied Ecology*, 16, 95–103.

Mitchell, F. J. 2005. How open were European primeval forests? Hypothesis testing using palaeoecological data. *Journal of Ecology*, 93, 168–177.

Murphy, B., Clarke, T., Tiernan, D., Strong, D., Neville, P., Murphy, P., Cregan, M., Bullock, C., Kavanagh, T., Keane, M. & Quinn, L. 2011. The Wild Nephin Project: an examination of the feasibility of setting aside the Nephin Forest area as a wilderness area. Unpublished report.

Navarro, L. & Pereira, H. M. 2015a. Rewilding abandoned landscapes in Europe. *In*: Navarro, L. & Pereira, H. M. (eds) *Rewilding European Landscapes*, 3–24. London: Springer.

Navarro, L. & Pereira, H. M. (eds). 2015b. *Rewilding European Landscapes*. London: Springer.

Oyedotun, T. 2011. Long-term change of the River Liza, Wild Ennerdale, England. *In*: Brebbia, C. A. (ed.) *River Basin Management*, VI. Southampton: WIT Press.

Pausas, J. G., Llovet, J., Rodrigo, A. & Vallejo, R. 2009. Are wildfires a disaster in the Mediterranean basin? A review. *International Journal of Wildland Fire*, 17, 713–723.

Pereira, H. M. & Navarro, L. 2016. *Rewilding European Landscapes*. London: Springer International Publishing.

Poulton, P., Pye, E., Hargreaves, P. & Jenkinson, D. 2003. Accumulation of carbon and nitrogen by old arable land reverting to woodland. *Global Change Biology*, 9, 942–955.

Prior, J. 2012. The roles of aesthetic vales in ecological restoration: case studies from the United Kingdom. PhD thesis, University of Edinburgh.

Prior, J. & Ward, K. J. 2016. Rethinking rewilding: a response to Jørgensen. *Geoforum*, 69, 132–135.

Queiroz, C., Beilin, R., Folke, C. & Lindborg, R. 2014. Farmland abandonment: threat or opportunity for biodiversity conservation? A global review. *Frontiers in Ecology and the Environment*, 12, 288–296.

Ramankutty, N. & Foley, J. A. 1999. Estimating historical changes in global land cover: Croplands from 1700 to 1992. *Global Biogeochemical Cycles*, 13, 997–1027.

Sandom, C., Donlan, C. J., Svenning, J.-C. & Hansen, D. 2013. Rewilding. *In*: Macdonald, D. W. & Willis, K. (eds) *Key topics in conservation biology*, 430–451. Oxford: Wiley.

Schnitzler, A. 2014. Towards a new European wilderness: embracing unmanaged forest growth and the decolonisation of nature. *Landscape and Urban Planning*, 126, 74–80.

Soulé, M. & Noss, R. 1998. Rewilding and biodiversity: complementary goals for continental conservation. *Wild Earth*, 8, 18–28.

Van Wieren, S. 1995. The potential role of large herbivores in nature conservation and extensive land use in Europe. *Biological Journal of the Linnean Society*, 56, 11–23.

Vera, F. W. 2002. The dynamic European forest. *Arboricultural Journal*, 26, 179–211.

Part VIII

Energy and resources

Part VIII of the Companion focuses on rural resources and resource extraction. Rural regions have traditionally been perceived as 'resource banks' based on the withdrawal or extraction of resources for the benefits of the wider economy (Markey et al., 2008). Often the benefits of the resource base is captured outside of the rural economy – for example, the added value to agricultural products often arises from global supply chains, while mining extraction in remote rural regions is increasingly on the basis of 'fly-in, fly-out' operations (Cheshire, 2009) with limited impact on local economies. Similarly, the 'rural' is often the backdrop or site of controversial new infrastructure (often relating to renewable energy), whereby the benefits of new development are often perceived to be captured elsewhere with the costs associated with rural communities. Rural resource management has a long tradition within the wider rural studies literature. However, in this section, the various chapters present new challenges to managing resources in the context of twenty-first-century challenges, from climate change to the influence of financialisation of rural resources and globalisation.

The part begins with Phillips and Dickie (Chapter 45) examining post-carbon ruralities, providing a comprehensive account of the rural dimension of transitioning to a low or post-carbon future. While this issue will be of critical importance to rural economies and lifestyles, the authors firstly note that engagement with debates on climate change and energy issues has been quite limited among rural scholars (certainly compared to the expansion of literature on low carbon cities). Phillips and Dickie address this deficit by examining John Urry's concept of a post-carbon sociology, arguing for a greater focus on space and place (and rurality) in developing a post-carbon geography. The paper then examines the challenges facing rural areas in transitioning to a low or post-carbon future, drawing on the UK experience. This includes analysis of energy consumption patterns across rural households, noting the higher per capita electricity use in rural areas compared to urban counterparts, highlighting challenges related to an older housing stock in rural areas that is often less energy efficient and located off the gas grid. The chapter also identifies longer commuting distances to work among rural households – 33 per cent higher for people living in rural areas, with a greater reliance on private car travel due to sparse public transport networks. Significant greenhouse gas emissions arising from the agricultural sector are also identified. Taken together, these factors suggest that rural areas face significant challenges in transitioning to the imperative of post-carbon futures – however, this also provides a pathway to identifying possible policy actions that may have benefits, such as retrofitting older housing stock to become more energy efficient or developing enhanced public transport networks possibly using new smart technologies. As well as being important points of origin for carbon emissions, Phillips and Dickie also identify rural areas as significant sites for the production of renewable energy including biomass, waste, hydro-electric, solar and wind. The authors explore some of the contested nature of land-use change arising from new renewable energy landscapes, including issues related to environmental 'land grabbing', competition between food production and the shift to biofuels, and the visual change to the countryside emerging within this transition. The chapter focuses on the controversial and multiscalar issues related to the deployment of wind energy and related infrastructure and the rural as a site of opposition to low carbon transitions. Within this context, the authors note the mobilisation of 'green' arguments by both proponents and opponents to wind energy development, as biodiversity concerns and landscape impacts are often used to counter the benefits of renewable energy and wider climate change debates.

These themes are further developed in the following chapter by Natarajan, which explores planning for major renewable energy infrastructure and engagement with rural communities. Natarajan examines planning decision-making around renewable energy infrastructure from the perspective of rural communities, particularly the balance of power between urban and rural, which parallels long established debates surrounding rural resource extraction. Whereas Phillips and Dickie highlight the mobilisation of alternative green narratives to support or oppose renewable energy deployment, Natarajan takes as a departure point the gap between national support for renewable energy and local resistance to the development of renewable energy infrastructure. Empirically, the chapter draws on a case study of planning decisions related to renewable energy infrastructure in Wales, which were made under the UK's Nationally Significant Infrastructure Projects regime. This centralising tendency in decision-making has prioritised carbon reduction in consenting to new renewable energy infrastructure, with Natarajan arguing that decisions have been driven by higher tier or national concerns. Instead, Natarajan suggests that there should be a balance with local needs, including conservation and alternative land-uses. In this context, the Nationally Significant Infrastructure Projects regime has repositioned 'the rural' within the planning process by the upscaling of decision-making. Moreover, within this decision-taking arena rural planning actors have limited capacities to engage, and their voice is a secondary consideration, particularly in relation to national policy priorities. Instead, local interests have to fight to establish the importance of their concerns, and seek mitigation for impacts. Local individuals also encounter well-resourced developers who are more equipped to deliver evidence of the sort that finds traction within regulatory processes.

The following chapter, by Köhne and Rasch, also examines energy related issues, this time exploring the impacts of hydraulic fracturing ('fracking') – unconventional gas extraction – on rural communities. Köhne and Rasch begin the chapter by assessing the impacts of fracking, including the potential economic benefits (local employment growth, construction etc.), but also the negative externalities associated with environmental and health impacts and erosion of local place attachment. The chapter then goes on to examine opposition to fracking, drawing on the rapidly growing number of studies on fracking conflicts in the US, Europe and Australia. Köhne and Rasch identify parallels with other extraction or energy conflicts where 'meaning-making' around nature, rurality and place have been central to local opposition campaigns. For Köhne and Rasch, fracking, although often framed as a resource, greener than coal and a step towards the transition to renewable energy, has many negative impacts on rural communities.

The way these impacts are experienced and perceived, is mediated by the meaning that people attach to land and natural resources and their place attachment. The chapter also notes how fracking causes tensions *within* communities and engender feelings of exclusion and disempowerment because people in communities near extraction sites are often excluded from decision-making, raising issues of social and environmental justice.

In Chapter 48, Fazito, Scott and Russell build on ideas of environmental justice as a framework to examine mineral extraction conflicts in the Espinhaço Range Biosphere Reserve, located with the State of Minas Gerais, Brazil. Firstly, the chapter provides an overview of the literature on mining and rural studies, identifying some of the key themes explored by researchers since the 1970s, including benefits and costs to host areas – such as 'resource boomtown' studies and, more negatively, the so-called resource curse at both national and local levels – and new modes of governance between mining operations and local communities. Fazito et al. propose an environmental justice framework to examine recent conflicts between proposals for a new mining operation and plans to promote tourism based in a biosphere reserve. While much work on environmental justice employs a Rawlsian perspective for examining environmental 'bads' (i.e. *how* and *what* gets distributed) this chapter also draws on the work of Amartya Sen, Iris Young, Nancy Fraser and Axel Honneth to develop a three-pronged approach to justice based on: (1) *distribution* of environmental goods and bads; (2) *recognition* to examine the *context* of oppression, which underpins maldistribution; and (3) *capabilities* to focus on how distribution of goods and bads affect our well-being and how we 'function'. Fazito et al. apply this framework to examine the process through which consent was given for a new mining operation within a fragile environmental context, which displaced attempts to develop a sustainable tourism strategy for the region. Through adopting a multidimensional approach to environmental justice, Fazito et al. demonstrate that a sense of injustice among locally affected people did not arise simply from the distribution of new mining projects (recognised as potentially bringing local benefits), but from the exclusion of local values, norms and knowledge from decision-making arenas, the ability of the mining companies to fragment and undermine decision-making, and the sense of loss following the commencement of mining activities (due to environmental degradation and land enclosure). In this regard, including a capabilities perspective in environmental justice-framing provides insights into how mining can undermine local 'freedoms', traditional practices, alternative livelihoods, and a sense of place attachment that formal decision-making failed to take into account.

The next chapter, by Butt, explores food security and the role of planning. Protecting and maintaining rural farmland has been a longstanding goal of many planning systems in the global north; indeed as argued in other chapters, the primacy of agricultural preservation has often been to the detriment of other policy goals related to affordable housing, economic diversification and social equity. Butt explores these issues by examining the evolution of agricultural and planning policy and examining the peri-urban context as one of the most contested arenas between farming and alternative uses for rural space. However, Butt begins by placing these land-use conflicts in the wider context of global food security and wider notions of food systems (e.g. food access, equity and quality) to propose a multidimensional role for planning policy in supporting food security priorities. This approach moves beyond the rural, requiring planning practice 'to reflect on issues as diverse as farmland preservation, urban food 'deserts', productive urban greenspace and industrial agriculture'. In relation to rural planning practice, Butt identifies critical issues surrounding the changing nature of rural land uses driven by industry restructuring, land and water ownership, globalised food systems, increasing interests in 'niche' and source-identified food in developed economies, and 'tensions in much of the world relating to livelihoods and self-determination, to land grabbing and to contested modes of agricultural modernisation'. Following a wide ranging discussion on food security and land-use regulation,

Butt then moves on to provide a detailed account of Australia's rural planning and agriculture regimes, with a specific focus on conflicts within peri-urban areas. Butt uses this arena to explore rapid rural change based on agricultural restructuring, demands for rural lifestyle housing, energy production and the emergence of alternative farming practices – all of which combine to produce a contested rural identity. Within this context, the chapter suggests that planning practice has kept its traditional focus on farmland preservation; however, Butt calls for planning to embrace food security to move beyond a one dimensional productivist focus, to 'one that considers food security as a process from production to consumption, and one that considers the opportunities of multi-scalar and diverse agricultural systems as a resolution to the challenges of global trade, climate change and diverse rural land values'.

The final chapter in this section, by Visser and Spoor, focuses on land as a fundamental rural resource and explores processes of land grabbing in the former Soviet Union. Land grabbing has increasingly been explored in the global south context particularly in relation to the growing influence of state interests (such as China and oil-rich Gulf states) and corporate investment funds. As Visser and Spoor argue, Russia also provides an insightful case, which includes the added complexity of transitioning from socialist governance of land to the concentration of land ownership in new large agricultural holdings. The chapter provides an overview of global land grabbing trends, motives and local impacts, charting the rise of land grabbing by state and corporate interests in the aftermath of the global financial crisis in 2008, when agricultural and food markets were perceived as less risky investments. The downsides of this trend are considerable: Visser and Spoor highlight, for example, dispossession of land from smallholders, a decline of *local* food security, the precariousness of the labour created, and a tendency of shifting to unsustainable mono-cropping. Moreover, large-scale land deals frequently generate conflict through, for instance, direct displacement of smallholders or indirect dispossession of smallholders through driving up land prices and making land inaccessible for local farms.

Within this context, the authors position the former Soviet Union as a frontrunner within the global trend of the expansion of large-scale farms as it is the region with the largest share of farmland controlled by large-scale corporate farms (only parts of Brazil and Argentina come close to such dominance of corporate farms). This has been enabled through 'the most drastic and widespread overhaul of land governance in recent history', with rural areas within the former Soviet Union's vast land mass experiencing a rapid shift from central state land governance to the privatisation of farms, the liberalisation of land and lease markets, and shift in the state's social functions in the countryside. Within this transition, the chapter notes concerns with widespread rent-seeking behaviour, corruption and the inability of small farmers and farm employees to maintain opposition in the face of more powerful interests.

References

Cheshire, L., 2009. A corporate responsibility? The constitution of fly-in, fly-out mining companies as governance partners in remote, mine-affected localities. *Journal of Rural Studies*, 26(1), 12–20.
Markey, S., Halseth, G. and Manson, D., 2008. Challenging the inevitability of rural decline: Advancing the policy of place in northern British Columbia. *Journal of Rural Studies*, 24(4), 409–421.

45

Post-carbon ruralities

Martin Phillips and Jennifer Dickie

Introduction

As Woods (2012: 131) has remarked, engagement with debates over climate change and energy issues have been quite limited by exponents of rural studies, with Dickie and Phillips (forthcoming) also noting that energy has rarely figured in rural textbooks or indeed, as indicated in Figure 45.1, in articles in rural journals, at least up until the last decade. However, the last decade has seen a significant rise in the number of studies, arguably reflective of increasing wider concerns over energy availability, accessibility and its environmental impacts. The latter concerns often revolve around the use of carbon-based energy sources (coal, oil, gas) and the impacts of their combustion of the climate through the release of CO_2 and other greenhouse gases such as nitrous oxide.

The effects of climate change, and of policies and actions designed to prevent or mitigate its occurrence, have since the 1990s become a subject of concern, debate and action among policy-makers and researchers. Much of this debate has been focused on urban areas, with concerns being expressed both about the impacts of climate change on urban life (see Schreider et al., 2000; London Climate Change Partnership, 2002; Bulkeley and Betsill, 2003; Lindley et al., 2006; McEvoy, 2007; Wilby, 2007; Bicknell et al., 2009; Hunt and Watkiss, 2011) and the role that urban areas play in fostering climate change through the through the use of hydrocarbons (e.g. Hunt, 2004; Dodman, 2009; World Bank, 2010; Hoornweg et al., 2011; Rutherford and Coutard, 2014).

Urry (2012: 534) has argued that oil 'fuels almost all movement of people and objects, and is also central in manufacturing industry, in almost all agriculture and in distributing water worldwide'. It should be acknowledged that this viewpoint may exhibit something of a bias towards the Global North, as it is clear that many areas in the Global South make considerable use of organic forms of energy (i.e. animal and human labour power, the burning of biomass, water, wind and solar power (see Nansaior et al., 2011; Guta, 2014; Hiemstra et al., 2018), although oil and other hydrocarbons (e.g. coal and natural gas) also play a major role in these areas as well. For Urry, this reliance on carbon fuels is unsustainable and there is a real need for social theories to engage with 'catastrophic' lines of thought related to a series of interlocking forms of

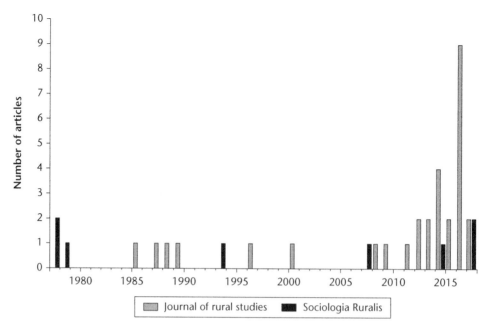

Figure 45.1 Energy articles in two leading international rural journals.

resource depletion and environment change, including 'climate change . . . [and] the probable decline in the availability of energy' (ibid.: 43). He argues for the development of a 'post-carbon sociology' that seeks to both highlight the 'carbon underpinning' (ibid.: 16) of social life and the possibilities of transitioning to a society that has lost this dependency.

While Urry's arguments are, we feel of critical importance, they can be criticised for neglecting geographic differentiation, as indicated by his neglect of differentiations between the Global North and Global South, and also by the almost complete neglect of rural areas in his account, beyond brief references to the potential and impacts of agro-fuel development, community-based car sharing and the planning challenges of creating eco-communities (Dennis and Urry, 2009). In this Urry is far from alone, with there being a clear urban-centricity in discussions of post-carbon transitions with cities being widely viewed as both a crucible in the formation of the 'contemporary high carbon world' (ibid.: 25) and as a key 'instrument' (Rutherford and Coutard, 2014: 1354) for transitioning from this (see Phillips and Dickie, forthcoming), with little research exploring 'the "rural" dimension of "low-carbon" transitions' (Markantoni and Woolvin, 2015: 202). Hence, while we feel that Urry's concept of a 'post-carbon sociology' is valuable, we wish to add a stronger geographical dimension to it, such that this present work might contribute to calls to establish a 'post-carbon geography' (Matthews and Morgan, 2013; Hicks, 2013). Specifically, we wish to outline some of the work that has been done examining the rural dimensions to low-carbon transition, prior to examining some of the challenges that such a transition faces in relation to both energy consumption and energy production.

Sketches of low-carbon rural transition

While there has been explicit, and a wider implicit, emphasis on the role of urban space in both the formation and mitigation of carbon dependency, the neglect of these issues in rural space has not been total. There has, for instance, been a long-standing strand of rural transport studies that make reference to higher levels of energy consumption, and associated CO_2 emissions, among rural populations (e.g. Banister, 1992; Banister and Banister, 1995; Breheny, 1995; Commission for Rural Communities, 2010; Fahmy et al., 2011; Hall and Woolvin, 2012; Minx et al., 2013; Baiocchi et al., 2015). More recently, and particularly in association with the up-turn in rural energy studies identified in Figure 45.1, there has been a growth in studies exploring the employment of non-carbon based forms of energy production and/or considering rural community engagement with, or resistance to, such energy developments (e.g. Phillips and Dickie, 2014, 2015; Jolivet and Heiskanen, 2010; Markantoni and Woolvin, 2015; van der Schoor and Scholtens, 2015). Many of these studies focus on localised rural spaces, or networks of such spaces, drawn into energy production, but with limited engagement with notions of broader transitions.

One exception to this is Huber and McCarthy's (2017: 657) discussion of energy regimes, which argues that energy production 'requires space', with much of that space historically and, potentially into the future, being rural in character. Drawing on historical studies that have highlighted transitions from an organic to a minerals, or hydro-carbon, based economy (e.g. Wrigley, 1988), Huber and McCarthy (2017: 659) suggest that the emergence of industrial capitalism involved a transition from a 'horizontal surface-based energy regime' – where energy was largely derived from solar energy captured at the surface of the Earth in plants by photosynthesis, 'and the muscle power that comes from animals that feed off those plants' – to a vertical energy regime centred around subterranean energy extraction. In a manner similar to Urry (2012: 534), they highlight the resource dependency of modern capitalism, but in line with the concept of a 'post-carbon geography' seek to draw out its spatial dimensions as well. In particular they argue that this energy regime overall made less horizontal surface demands – or demands on land – than did its predecessor: compared to, say charcoal burning, a coal mine exhibited much higher 'power density' (Smil, 2015) in that it would produce much more energy from a relatively small patch of the land at the surface. They also consider the spatialities implied by a post-carbon transition, suggesting that not only might this be viewed as entailing a 'return to a surface energy regime', albeit one where energy production would not rely on the 'biological capture and metabolism' of an 'organic economy' but instead on 'highly advanced technologies that would be more labour, material and capital intensive, but would 'place intense pressures on rural lands and people because much of the energy production 'would almost certainly take place in rural areas' (Huber and McCarthy, 2017: 665). They add that transitioning to such an energy economy would pose enormous challenges, and also produce a series of social and environments effects, including 'contestation, conflicts and resistance' and 'many instances of dispossession' (ibid.: 666).

Huber and McCarthy's work very much frames these issues as an emergent possibility, but as we shall detail in the following sections, evidence can be found to demonstrate that rural space and low-carbon transitions are already facing profound challenges, not least in relation to the possible futures of each other.

The challenges of a rural post-carbon transition: illustrations from the UK

Energy consumption in the UK countryside

Reductions in energy reduction has been viewed as a key component of addressing carbon emissions, and in the UK there have been falls in this over the last decade (Figure 45.2), although such figures obscure significant spatial variations, with human life in many rural areas exhibiting heavy energy dependencies. Figure 45.3, for example, shows per-capita domestic electricity

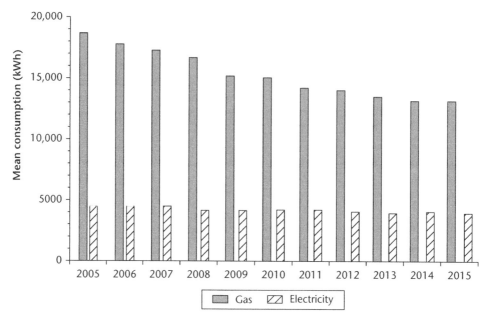

Figure 45.2   Mean gas and electricity consumption in England, 2005–2010, and England and Wales, 2011–2015.

consumption in England and Wales in 2016, for all areas and for areas classified as rural. It shows clearly that rural areas often have well-above average levels of domestic electricity consumption, a feature that is also clearly demonstrated in Table 45.1. Average rural domestic electricity

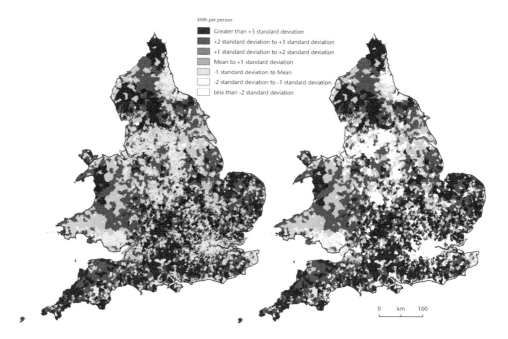

Figure 45.3   Electricity consumption per capita, England and Wales, 2016.

Table 45.1 Electricity consumption, fuel poverty, off-gas-grid and low-energy-efficiency dwellings in rural and urban areas of England and Wales, 2015–2016.

Area characteristic	Mean annual electricity consumption 2016 (kWh per household)	Mean annual electricity consumption 2016 (kWh per person)	Percentage of household off-gas-grid 2016	Percentage of household in fuel poverty 2015	Percentage of dwellings with energy efficiency certificates F & G
Rural village and isolated dwellings in sparse setting	5912.4	2564.3	27.7	18.5	0.32
Rural village and isolated dwellings	6037.0	2467.2	33.1	13.6	0.18
Rural town and fringe in sparse setting	4499.1	2071.2	9.5	11.6	0.13
Rural town and fringe	4408.9	1875.4	8.7	9.4	0.05
Urban city and town in sparse setting	3911.7	1735.1	7.3	11.4	0.06
Urban city and town	3917.4	1651.6	6.0	10.4	0.03
Urban minor conurbation	3542.5	1500.8	5.0	12.6	0.04
Urban major conurbation	3883.1	1592.3	9.1	11.5	0.03
All rural areas	5144.7	2155.5	19.5	11.5	0.12
All urban areas	3887.7	1621.6	7.2	10.9	0.03
All areas	4112.1	1716.9	9.4	11.0	0.05

Sources: see Figures 45.2–45.5.

consumption for 2016 is still almost 12 per cent higher than the 2005 national mean electricity consumption figures, and 25 per cent above the mean urban figures for 2016.

Reasons given to account for the high levels of rural per-capita electricity consumption include high relative numbers of off-gas-grid properties and so-called 'hard-to-heat' and 'hard-to-treat' homes (BRE Housing, 2008). As Figure 45.4 and Table 45.1 indicate, rural areas tend to have a higher proportion of households who have no mains gas supply, with areas of villages and hamlets and isolated dwellings being particularly likely to be impacted. In many of these areas, over eighty per cent of households are off the gas-grid, while in many areas of north and central Wales, Cornwall and Norfolk, this rises to almost all households.

Households which have no access to mains gas supply have to rely on electricity for cooking and heating, or else make use of some other source of energy such as oil, wood or bottled gas. Not only will such areas tend to have high per-capita electricity use, but the extensive use of alternative energy sources tends to mean that these areas are also among those with high levels of so-called fuel poverty. In the case of Figure 45.5 and Table 45.1, fuel poverty figures are modelled estimates of the proportion of households who appear to have fuel costs that are above the national average and whose cost would leave the household with a 'residual income below the official poverty line' (BEIS, 2017: 1). As with off-gas-grid figures, areas classified as areas of villages and hamlets and isolated dwellings have the highest proportion of households in fuel poverty, although the distribution is far from identical. Areas of villages and hamlets and isolated dwellings which are relatively remote from built up areas (i.e. which are in so-called 'sparse settings') are, for instance, more likely to have high proportions of households in fuel poverty, while in relation to households with no access to gas, less remote areas of villages and hamlets and isolated dwellings have the highest figures. The differences indicate that not only is access to the gas-grid not simply a function of geographical remoteness, but also that fuel poverty levels are not simply a reflection of accessibility to gas supplies. Important influences on the distribution of fuel poverty levels include variations in rural income levels – many of the areas

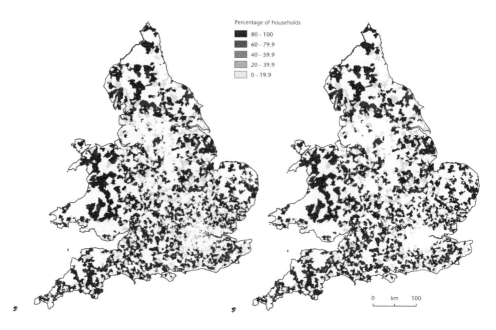

Figure 45.4   Percentage of off-gas-grid households, England and Wales, 2016.

of fuel poverty are areas of marginal agriculture, and hence, low agricultural incomes – and the distribution of the 'hard to heat'/'hard to treat' residential properties which have low energy inefficiency. As Figure 45.5 and Table 45.1 indicate, such properties have a stronger relative presence in rural than urban areas and particularly in areas classified as village and hamlets and isolated dwellings.

Among the features seen to account for the relative concentration of such homes in rural areas are the greater relative presence of older properties which may have less insulation and may have solid as opposed to cavity walls into which insulation can be inserted (BRE Housing, 2008), and higher proportions large and detached properties which may take more energy to heat and may lose relatively more heat to the atmosphere than semi-detached, terraced and apartment properties (see Jones and Lomas, 2015; Schubert et al., 2013; Viggers et al., 2017). It has, however, also been noted that there may be cross-correlations between some of these property characteristics and socioeconomic and demographic variables such as social class, income, household size and tenure (e.g. Commission for Rural Communities, 2007; DEFRA, 2008; Büchs and Schnepf, 2013; Huebner et al., 2015). Studies have shown, for instance, that domestic energy consumption tends to increase with income levels, which may also tend to be higher in gentrified rural areas (e.g. Druckman and Jackson, 2008; Büchs and Schnepf, 2013; Phillips and Dickie, 2015), and indeed that increases in property sizes may be more than off-setting gains in the energy efficiency of residential property through improved construction technologies and retro-fitting of energy conservation measures (Brecha et al., 2011; Viggers et al., 2017).

Energy is not just consumed in the home, but is often consumed in rural business and in transportation, including in commutes to and from work and in journeys to access a series of educational, welfare and retailing services. Anable et al. (1997), for instance, claimed that average distances travelled by rural residents in the UK is about twice that of urban dwellers, while the Commission for Rural Communities (2010) calculated that people living in areas of villages and dispersed settlements travelled 42 per cent further than those in England as a whole.

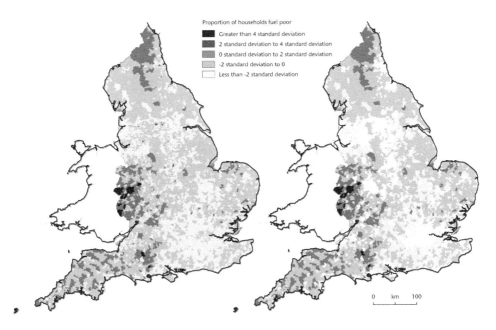

Figure 45.5 Proportion of households in fuel poverty in England, 2017.

Analysis of travel to work distances recorded in the 2011 Census, suggests that these journeys were 33 per cent higher for people living in rural areas than the overall average, and were, perhaps unsurprisingly, significantly higher across all categories of rural areas within sparse settings and also increased progressively down the settlement hierarchy (see Figure 45.6). As shown in Table 45.2, there was also heavy reliance on the use of private vehicles to undertake these journeys across rural areas, a feature that also been demonstrated in a series of localised case studies. Roberts and Henwood (2016), for example, highlights the travel patterns of a resident in the Llŷn Peninsula in northwest Wales, who used a car to journey over 50 miles a day in relation to their work plus also regularly travelled over 12 miles to access childcare services and leisure activities for their children. In a more extensive study of eight villages across five Districts in England, Phillips and Dickie (forthcoming) document that residents on average travelled almost 22 kilometres (over 13 miles) into work, with some travelling over 100 miles (161 kilometres). Extensive travel was also recorded for visits to supermarkets, cash and clothes shopping, with 97 per cent of the journeys to supermarkets being undertaken by car, 25 per cent of respondents making more than one visit to a supermarket per week, and almost 91 per cent of respondents never having made use of public transport from their place of residence. Such figures highlight the heavy mobility and carbon dependency of contemporary rural lives, with the former being well recognised by rural residents who made frequent comments about the importance of mobility to the performance of everyday rural lives. There were, however, significantly fewer references to environmental dimensions of this mobility, despite 80 per cent of respondents expressing acceptance that the world's climate was changing.

High levels of domestic energy use, long transport journeys and high reliance of the private car can all be seen to be significant contributors to climate change through carbon dioxide (CO_2) emissions. Given the evident prevalence of these features in rural areas in countries such as the United Kingdom, it is unsurprising that a range of studies have identified the presence of higher per capita CO_2 emissions in rural areas over urban ones (Commission for Rural Communities, 2010; Fahmy et al., 2011; Minx et al., 2013; Baiocchi et al., 2015; Phillips and Dickie, forthcoming). Figure 45.7 and Table 45.3 show per capita emissions in England and Wales and confirms the picture of higher rural emissions, although rural villages, hamlets and isolated dwelling in sparse contexts had the highest per capita emission levels.

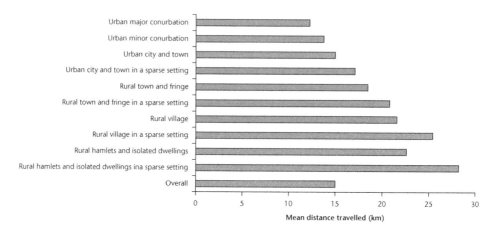

Figure 45.6 Travel to work distances, England and Wales, 2011 Census.

Table 45.2 Modes of travel to work, England and Wales, 2011.

Area characteristic	Mode of transport							Work mainly at or from home
	Driving car or van	On foot or bicycle	Bus, taxi or underground, metro, light rail, tram	Train	Passenger in car or van	Motorcycle, scooter or moped	Other mode of travel	
Rural hamlet & isolated dwellings in sparse setting	61.7	9.2	1.3	0.7	3.8	0.5	1.1	21.6
Rural hamlet & isolated dwellings	66.1	8.4	1.9	3.3	3.7	0.6	1.0	15.0
Rural village in sparse setting	67.3	10.6	2.2	0.8	4.9	0.6	1.0	12.6
Rural village	70.4	8.2	2.3	3.0	4.2	0.7	0.8	10.4
Rural town & fringe in sparse setting	58.2	24.4	2.6	0.7	5.5	0.6	0.9	7.2
Rural town and fringe	69.7	10.7	3.8	3.1	5.2	0.8	0.6	6.1
Urban city & town in sparse setting	52.6	29.9	3.6	0.9	6.3	0.5	0.9	5.3
Urban city & town	61.8	16.0	6.2	3.9	6.0	0.8	0.6	4.5
Urban minor conurbation	59.8	13.8	14.0	1.5	6.2	0.7	0.5	3.5
Urban major conurbation	45.2	12.3	24.1	8.5	4.0	0.9	0.7	4.3
All rural areas	68.7	9.9	2.9	2.9	4.6	0.7	0.7	9.5
All urban areas	54.9	14.5	13.8	5.7	5.2	0.8	0.6	4.4
All areas	57.5	13.6	11.7	5.2	5.1	0.8	0.6	5.4

Source: 2011 Census

Figure 45.7 CO$_2$ emissions, England and Wales, 2012.

Table 45.3 CO_2 emissions in urban and rural output areas within England and Wales, 2013.

Area characteristic	Mean CO_2 emissions (tonnes per annum)		
	Direct Consumption	Transport	Total
Rural hamlets & isolated dwellings in sparse setting	1.14	1.48	12.54
Rural hamlets & isolated dwelling	1.20	1.39	12.36
Rural village in sparse setting	1.33	1.47	13.33
Rural village	1.45	1.35	13.09
Rural town & fringe in sparse setting	1.55	1.16	12.91
Rural town & fringe	1.53	1.17	12.66
City & town in sparse setting	1.44	0.99	11.77
City & town	1.41	1.07	12.00
Urban minor conurbation	1.49	0.97	11.59
Urban major conurbation	1.34	0.77	11.03
All rural	1.44	1.27	12.75
All urban	1.39	0.95	11.71
All areas	1.39	1.01	11.89

Source: Experian GreenAware Dataset 2012, QS702EW. Office for National Statistics licensed under the Open Government Licence v.1.0

Emissions of CO_2 clearly do not reflect simply domestic and travel energy consumption, with agricultural and other rural businesses also being important emission sources. With respect to agriculture, it is important to recognise that this is also a significant source of other 'greenhouse gas' emissions, including methane and nitrous oxide, as well as an important 'carbon sink', as the crops, pasture and other plants grown within agriculture absorb or 'sequestrate' CO_2 as part of the process of photosynthesis. Estimates of the proportion of global greenhouse gases emissions that stem from agriculture have varied considerably, although it is widely recognised that livestock agriculture is a major contributor (e.g. see Bellarby et al., 2013; Herrero et al., 2011). Estimates of its impact again range widely (Herrero et al., 2011), largely due to variations in the methodologies and assumptions built into the calculations. The UN's Food and Agriculture Organization (2006), for instance, calculated that 8 per cent of global greenhouse gas emissions stem from livestock agriculture while Goodland and Anhang (2009) put the figure at 51 per cent. The Food and Agriculture Organization's low estimate stems in part because they assume that the amount of CO_2 emitted via livestock respiration equate to the amount of carbon that they consumed, which was itself was created through the sequestration of CO_2 through photosynthesis. Goodland and Anhang question these assumptions, arguing that livestock respiration accounts for 21 per cent of global greenhouse gas emissions connected to human activity. They also seek to quantify the amount of carbon sequestration lost through the use of land for grazing pasture and livestock feed as opposed to forestry, suggesting that just within the tropics, this represents at least half of existing levels of anthropogenic related CO_2 emissions. They also consider the impact of converting land used for livestock rearing for growing crops for direct human consumption and biofuels, suggesting this could reduce global CO_2 emissions by 4.2 per cent.

Although Herrero et al. (2011) have questioned many of the arguments advanced by Goodland and Anhang (2009), they do serve to highlight how transformations in agricultural land-use as well as practices could significantly impact greenhouse gas emissions hence potentially mitigate global climate change. These changes in land-use would, however, entail changes along the agricultural commodity chains, including so-called agri-food networks, not least because movement away from livestock production would require significant transformation in consumption habits. Furthermore, as Accorsi et al. (2016) note, greenhouse gas emissions, and associated mitigation activities, can be identified across all stages of the commodity chain, although attention often focuses at localised sites of production (i.e. at the scale of the farm or the local landscape).

The countryside as a site of energy production and location of social resistance to low-carbon transition

As well as being important points of origin of carbon emissions and sinks for carbon seques-tration, it is also important to recognise that rural areas have become significant sites for the production of renewable, non-hydrocarbon based, forms of energy, including biomass, waste, hydro-electric, solar and wind based sources of energy (see Figures 45.8 and 45.9). Frantal et al. (2014) remark that recent years have seen the rise of 'brightfield' initiatives, whereby brownfield sites are used as locations for renewable energy developments. However, as Table 45.4 indicates, within the UK many renewable developments are located in areas classified as rural.[1] Indeed, in Scotland over 93 per cent of the units and approaching 97 per cent of the installed capacity of renewable energy production above 1 MW has been located in areas classified as rural, while in Northern Ireland the figures are almost 82 per cent and 97 per cent respectively. The figures for England and Wales are lower, although there is still a clear rural focus in renewable energy production, with 77.5 and 74.9 per cent of operational units respectively being located in areas classified as rural, while figures for rural capacity are slightly higher at 82.9 and 82.1 per cent respectively. Reasons that might be given to account for this rural focus include the use of agri-cultural products and waste in biomass production, although, as Table 45.4 indicates, anaerobic,

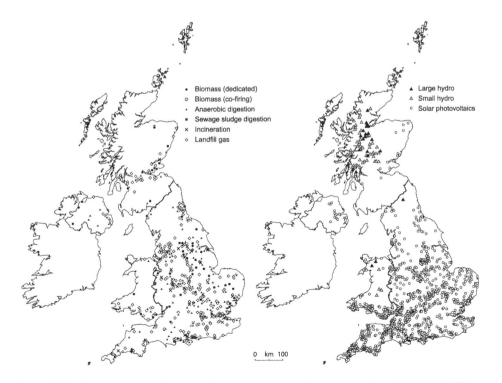

Figure 45.8   Biomass, waste, hydro and solar energy developments in the UK, 1990–2017.

biomass and waste production sites are among the most urban focused of the renewables. This might reflect the significance of population density in the supply of waste products (see Comber et al., 2015), although it has also become evident that biomass plants have drawn extensively on purpose grown crops which has led to concerns about potential shortfalls in feedstock supplies (Tranter et al., 2011), as well as concerns over its potential impacts on food production and knock on social and environmental impacts, including greenhouse gas emissions because some of the land uses being displaced by the biofuel energy production might have greater carbon sequestration capabilities that the energy crops being produced (see Howard et al., 2009; Lovett et al., 2009; Haughton et al., 2009; Rowe et al., 2008). Particular concerns have been expressed about the growth of biofuels for vehicle transportation, such as ethanol which can be produced from cereals such as wheat, barley, corn and rye, as well as sugar crops, and biodiesel that can be produced from oilseed crops such as rape, sunflowers and soybeans. Huber and McCarthy (2017) note that on a global scale there has been large scale conversion of land into the production of biofuels, which has often associated with processes of 'land-grabbing' by private and state interests, although in the EU there has been much lower levels of liquid biofuel production and associated land conversion (see Ajanovic, 2011; Mol, 2007, 2014).

A second identified locational influence on renewable energy production has been population sparsity, it being claimed that renewable energy developments, particularly large-scale units, have tended to be located to minimise the number of people impacted by their operation (e.g. Hanley and Nevin, 1999). In Scotland, for instance, just over half of on-shore wind power units come from sites located in areas classified as remote or very remote rural areas, with the latter contributing significantly in relation to capacity, reflecting the construction of more extensive

Figure 45.9   On-shore wind developments in the UK, by units and production capacity, 1990–2017.

Table 45.4 Operational renewable energy developments within the UK, 1990–2017.

Location	Biomass, Anaerobic Digestion, & Waste		Hydro-electric		Solar		Wind	
	Units	Capacity	Units	Capacity	Units	Capacity	Units	Capacity
England								
Rural hamlets & isolated dwellings in sparse setting	7	14.9	1	6.1	27	182.9	17	295.8
Rural hamlets & isolated dwellings	112	2519.3	0	0.0	409	3147.7	100	1149.9
Rural village in sparse setting	0	0.0	0	0.0	13	141.9	8	48.7
Rural village	74	285.5	0	0.0	272	2253.1	71	770.4
Rural town & fringe in sparse	1	1.3	0	0.0	2	6.2	0	0.0
Rural town & fringe	46	289.0	0	0.0	108	754.6	18	171.3
Urban city & town in sparse setting	0	0.0	0	0.0	0	0.0	2	14.2
Urban city and town	121	820.1	0	0.0	123	707.4	36	261.3
Urban minor conurbation	12	95.1	1	1.7	4	11.3	3	13.7
Urban major conurbation	47	444.6	0	0.0	12	33.7	12	78.6
All areas England	420	4469.8	2	7.8	970	7238.8	267	2803.9
Northern Ireland								
Mixed urban/rural	1	3.2	0	0.0	5	29.1	8	103.5
Rural	8	28.7	0	0.0	7	90.9	66	928.3
Urban	1	2.1	0	0.0	3	24.3	0	0.0
All areas Northern Ireland	10	34.0	0	0.0	15	144.3	74	1031.8
Scotland								
Accessible Rural Areas	34	188.8	1	20.0	11	56.7	130	3501.0
Accessible Small Towns	4	11.4	0	0.0	0	0.0	2	42.8
Large Urban Areas	4	19.1	0	0.0	0	0.0	2	16.0
Off-shore	0	0.0	0	0.0	0	0.0	0	0.0
Other Urban Areas	6	78.9	0	0.0	1	5.0	6	52.7
Remote Rural Areas	4	20.5	21	130.0	3	7.4	85	2509.5
Remote Small Towns	1	5.0	1	1.9	0	0.0	0	0.0
Very Remote Rural Areas	0	0.0	53	292.2	0	0.0	67	1187.7
Very Remote Small Towns	1	3.5	1	1.9	0	0.0	1	48.3
All areas Scotland	54	327.2	77	446.0	15	69.1	293	7358.0
Wales								
Rural hamlets & isolated dwellings in sparse setting	1	2.4	5	22.8	16	124.8	15	192.3
Rural hamlets & isolated dwellings	4	8.8	0	0.0	31	233.1	5	66.5
Rural village in sparse setting	1	3.2	0	0.0	5	20.6	15	194.6
Rural village	7	87.2	1	32	20	114.9	7	266.6
Rural town & fringe in sparse	0	0.0	0	0.0	2	20.8	0	0.0
Rural town & fringe	3	11.8	0	0.0	6	38.3	11	181.4
Urban city & town in sparse setting	0	0.0	0	0.0	0	0.0	0	0.0

(continued)

Table 45.4 (continued)

Location	Biomass, Anaerobic Digestion, & Waste		Hydro-electric		Solar		Wind	
	Units	Capacity	Units	Capacity	Units	Capacity	Units	Capacity
Urban city and town	11	87.8	0	0.0	33	206.6	8	58.9
Urban minor conurbation	0	0.0	0	0.0	0	0.0	0	0.0
Urban major conurbation	0	0.0	0	0.0	0	0.0	0	0.0
All areas Wales	27	201.2	6	54.8	113	759.1	61	960.3

Table 45.5 Solar and wind power related planning applications and rejections, England and Wales, 1990–2017.

Locations	Solar			Wind			Wind & Solar
	Number of applications submitted	Number of applications rejected	Percentage rejected	Number of applications submitted	Number of applications rejected	Percentage rejected	Percentage rejected
Rural village and isolated dwellings in sparse setting	110	16	14.5	150	58	38.7	28.5
Rural village and isolated dwellings	892	170	19.1	393	178	45.3	27.1
Rural town and fringe in sparse setting	9	3	33.3	4	1	25.0	30.8
Rural town and fringe	346	55	15.9	173	56	32.4	21.4
Urban city and town in sparse setting	3	0	0.0	3	0	0.0	0.0
Urban city and town	369	44	11.9	146	38	26.0	15.9
Urban minor conurbation	12	1	8.3	4	1	25.0	12.5
Urban major conurbation	33	7	21.2	24	4	16.7	19.3
All rural areas	1357	244	18.0	720	293	40.7	25.9
All urban areas	417	52	12.5	177	43	36.8	16.0
All areas	1774	296	16.7	897	336	40.1	23.7

wind farms with larger turbines (see also Figure 45.9). Table 45.4 confirms the strong focus in Scotland on on-shore wind energy production observed by Cowell et al. (2017) who also observe that onshore wind has been the predominant renewable energy technology employed in Wales and Northern Ireland, a situation confirmed by Table 45.4. These figures also suggest that sparsely populated rural areas in Wales similarly appear as areas of significant wind power developments, having significantly higher numbers of sites and levels of production capacity than their less sparse counterparts in both the village and rural hamlet and isolated dwelling categories. On the other hand, accessible rural areas in Scotland and rural town and fringe areas in Wales have also seen extensive wind power developments, and in England it is the non-sparse categories of rural areas that figure most strongly in wind power production, with areas classified as villages and rural hamlets and isolated dwellings figuring most strongly. In the case of Northern Ireland, there is a strong predominance of units and production capacity within rural areas, although the classification does not enable differentiation within this category.

The development of wind power within rural areas has attracted a lot of attention in countries such as the United Kingdom, often focused around its environmental impacts and their social acceptability (e.g. Bristow et al., 2012; Toke, 2005; van der Horst, 2007; van der Horst and Vermeylen, 2011; Woods, 2003). As Munday et al. (2011: 1) note, these have often been associated with considerable planning conflicts (see also Toke, 2005; Cowell, 2007), although there have also been calls to connect their construction to concepts of sustainable rural development (Hain et al., 2005; Stevenson and Richardson, 2003), eco-economies (Kitchen and Marsden, 2009) and community empowerment (Walker et al., 2007) as well as low-carbon transitions. As highlighted in the work of Woods (2003: 284), there can be a complex braiding of such discourses with notions of rurality, with for instance, both proponents and protestors employing notions of sustainability but employing it, in the former case, in association with a 'discourse of the rural as a productive space' as well as within global environmental framings, while the latter group frequently emphasised a 'discourse of the rural as a space of consumption', as well as a concern to retain spaces of wild, unspoilt, nature (see also van der Horst and Vermeylen, 2011). As Warren et al. (2005: 854) observe, contestations over wind power development often involve the mobilisation of 'green arguments on both sides', with Cowell (2007: 293) observing that planning decisions often requiring 'sharp trade-offs involving landscape aesthetics, place-based identities and deeply held values about nature'.

It has also been argued that reactions to the construction of wind turbines have been conditioned as much by the practices and relations surrounding their development than by their impacts. Bristow et al. (2012: 1108), for example, argue that in the UK public attitudes to wind power development relate in part to the 'great majority' of them being 'driven by major, commercial energy utilities with few other economic ties to development locations' (see also Warren and McFadyen, 2010). This stands in some contrast to countries such as Denmark and Germany where, as Munday et al. (2011: 4) note, government policies have sought to 'encourage significant financial participation and co-ownership from farmers and local citizens' (see also Breukers and Wolsink, 2007; Szarka and Bluhdorn, 2006; Szarka, 2007; Toke, 2005). In the UK, government policies have been market-orientated, which has created, among other features, 'higher up-front cost for potential investors' and thereby favoured the involvement of 'larger wind-farm developers and utilities, with a lower cost base and easier access to finance and contracts' (Munday et al., 2011: 4), although there has been a tempering of this, to a degree, in the devolved governments (Cowell, 2007; Cowell et al., 2017). There has also been, as Cowell (2007) notes, a tendency for local land-use planning to become identified by both government and business as a barrier to the growth of wind generation. As Cowell (2007) and Cowell et al. (2017) have remarked, this has encouraged practices of central direction and a streamlining or by-passing of local planning processes by successive UK Governments, at least up until 2010 General Election, when electoral concerns within a Conservative led coalition government produced a heightened willingness to respond to localised protests over wind energy developments via affording 'local planning authorities more control over renewable energy applications' (Cowell et al., 2017: 175).

Planning has also, conversely, often been viewed by opponents of wind power as working to favour its development, both through acting as a decision-making process mechanism where 'public involvement or "say" . . . may be minimal' (Cass and Walker, 2009: 62) and also providing 'an arena where unfair accusations of NIMBYism proliferate' (Lennon and Scott, 2015a: 593). Wind power development has indeed led to the emergence of a series of studies exploring the notion of NIMBYism (e.g. Wolsink, 2000; Devine-Wright, 2005; van der Horst, 2007). Much of this research has been highly critical of the concept, suggesting that while the term has 'had a strong influence in shaping how industry, policy-makers and media commentators' (Devine-Wright, 2011b: xxiii) respond to developments such as wind energy, and indeed on the responses of people seeking to object to their development (see Warren et al., 2005; McClymont and O'Hare, 2008; Cass and Walker, 2009; Burningham et al., 2015), it is actually empirically, conceptually and normatively problematic (e.g. see Burningham et al., 2007; Wolsink, 2006; Devine-Wright, 2007, 2009, 2011a). Empirically, the term is often used to characterise the attitudes of people seeking to resist developments within the localities in which they reside or have an interest – their 'backyards' – it being claimed that people may be supportive of developments in general but come to resist it when it impacts them directly. Studies have questioned the empirical presence of such attitudes, suggesting that resistance often draws upon and expresses social concerns such as equity and justice as opposed to self-interest (e.g. Gross, 2007; Simcock, 2014), as well as being influenced by a range of personal, social and contextual factors (e.g. Devine-Wright, 2008). Linked to these criticisms, there has been a conceptual questioning of some of the assumptions being made within the concept, including whether attitudes to developments are conditioned by spatial proximity/distance and cost/benefit reasoning (Jones and Eiser, 2009, 2010). Finally, there have also been criticisms of the normative use of the concept, it being argued by Wolsink (2000, 2006) among others that the term is used in a pejorative way to devalue the presence of development resistance. This devaluation can take a variety of forms, including characterising NIMBYism as stemming from people being emotional rather than rational (Cass and Walker, 2009; Burningham et al., 2015), or lacking in knowledge and understanding of the need for change (McClymont and O'Hare, 2008; Aitken, 2010; Burningham et al., 2015), or 'being overly selfish, focusing on private disbenefits and over-looking public goods' (Devine-Wright, 2011a: 61), or even cloaking this self-interest behind espousal of public good (Wolsink, 2006).

While these criticisms raise significant points and have fostered useful avenues for further research, including more symmetrical examination of the arguments and interests of proponents and objectors to wind power developments (Walker et al., 2011) and the presence and use of emotion by both groups (e.g. Cass and Walker, 2009; Lennon and Scott, 2015a; Simcock, 2014), it is also important to note that a value-neutral approach as seemingly envisaged by Wolsink (2006) is untenable (e.g. Phillips, 1994) and that as Hubbard (2006: 94) has remarked, campaigns of resistance may well involve a 'proxy politics' in which particular interests are rationalised and expressed in a range of 'different languages and registers'. It is also evident from empirical studies that public acceptability has formed, and still remains, a significant challenge to the expansion of wind power across many rural areas (e.g. Mason and Milbourne, 2014; Ogilvie and Rootes, 2015; van der Horst and Toke, 2010).

Table 45.6 Areas of England and Wales licensed for petroleum exploration and drilling, 14th round.

Location	Percentage of licence blocks falling within location
Rural village and dispersed settlement in sparse setting	4.33
Rural village and dispersed settlement	50.55
Rural town and fringe in sparse setting	0.58
Rural town and fringe	16.65
Urban city and town in sparse setting	0.07
Urban city and town	19.94
Urban minor conurbation	3.70
Urban major conurbation	4.17
All rural areas	72.12
All urban areas	27.88
All areas	100.00

Such challenges are not restricted just to wind power, with resistance to a series of renewable energy technologies having also been identified within studies of rural localities (e.g. Upreti, 2004; Upreti and van der Horst 2004; Röder, 2016). Table 45.1 indicates that in rural areas of England, the total number and capacity of anaerobic, biomass and waste production sites as well as solar photovoltaic sites exceed those of wind power, while in Wales solar photovoltaics exceeds wind in number of sites but not production capacity. In Northern Ireland and Scotland, wind power remains pre-eminent in both frequency of sites and production capacity. While such differences can in part be accounted for by the environmental conditions, they arguably have been conditioned as much by governmental responses in England to wind power protests and the reliance of market mechanisms. There would certainly be evidence to support Markantoni and Woolvin's (2015: 202) suggestion that there could potentially be emerging quite 'distinct trajectories of transition across the UK', distinctions that, as they argue, might also involve geographically differentiations in 'capacities and motivations' for community participation, as well as resistance, to energy transitions, some of which may involve rural and urban differentiations. Table 45.6, for instance, highlights how rural areas were not only locations with high numbers of applications for wind or solar energy developments, but also had higher levels of planning proposal rejections than found in areas classified as urban, with levels of rejections being particularly high for wind power in areas classified as rural village and isolated dwellings.

Unconventional carbons in the UK countryside: a bridge or challenge to a low-carbon future?

Hitherto we have been considering the transition to low-carbon societies as involving reductions or movement away from carbon-based energies, but there has been a set of energy developments that has both involved a heightened use of carbon and been linked to a transition to a low-carbon society, namely the exploitation of so-called 'unconventional hydrocarbons'. These are sources of oil and hydrocarbon gas that remain embedded within rocks rather than accumulating in porous and permeable rocks, and include shale gas and oil, tight gas from sandstones, and coal bed methane. These oils and gases are extracted by a range of methods (see Bradshaw et al., 2015), including heating at high temperatures (for oil shales), the injection of steam, water and solvents (for oil sands), and fracturing rock through pumping water or other liquids and propellants such as sand at high pressure underground, a process widely known as 'fracking' (for shale gas and, in some cases, coal bed methane; see also Chapter 47 of this Companion). Many of these methods have been employed across the oil and gas industries for many years, but the scale extended from the 1980s both as a consequence of technological innovation in their execution and changes in oil and gas prices which made their employment increasingly viable.

In countries such as the USA, Canada and Australia there has been extensive unconventional hydro-carbon development, much of it within rural areas (see Eaton and Kinchy, 2016; Morrone et al., 2015; Schafft et al., 2013; de Rijke et al., 2016; Sherval and Hardiman, 2014), while in many other countries, including the United Kingdom, there have been frequent calls for its construction. In 2014, for example, the UK prime minister, David Cameron, declared that the government was 'going all out for shale gas' (UK Government, 2014), arguing that the exploitation of this energy source would enhance economic growth and job opportunities, as well improve energy security. He also remarked that it held the potential to drive down domestic energy prices and contribute to transitioning to a low-carbon future. The extent to which these claims can be substantiated is debated (Bradshaw et al., 2015; McGlade et al., 2015), not least with respect to the role that unconventional oil and gas can act as a 'bridging fuel' to a low-carbon society, with Stephenson et al. (2012) drawing on research in Canada to argue that such labels simply provide a 'green wash' that merely provides a rhetorical justification for continuing carbon energy reliance. Despite such arguments, the UK Government continues to advance political support for unconventional carbon development in the UK, with Greg Clark, the Secretary of State for Business, Energy and Industrial Strategy, for example, recently reiterating the Government's view that 'there are potentially substantial benefits from the safe and sustainable exploration' of unconventional oil and gas and that 'every scenario proposed by the Committee on Climate Change setting out how the UK could meet its legally binding 2050 emissions reduction target includes demand for natural gas' (Department for Business Energy and Industrial Strategy, 2018: n.p.).

Despite the UK government expressing support for unconventional carbon developments, the country has, as yet, not become a location for their commercial extraction. While in England, central government direction has been in favour of their development, in 2015 all three devolved governments adopted significantly less enthusiastic approaches, with the Scottish Government announcing a moratorium on hydraulic fracturing, the Welsh Government issuing a Direction preventing local planning authorities from approving fracking developments and the government in Northern Ireland outlining a Strategic Planning Policy Statement that included a presumption against fracking until its benefits and risks were more fully understood (Delebarre et al., 2017). More recently, following a public consultation in 2017, members of the Scottish Parliament voted to extend the moratorium indefinitely and create what was described as an effective ban on fracking, although this action has been legally challenged by two petrochemical companies.

While the policy landscapes are differentiated, and in the Scottish case, currently uncertain, it is clear that across all parts of the UK there is considerable, and seemingly increasing, public opposition to the development of unconventional carbon (Andersson-Hudson et al., 2016; O'Hara et al., 2016; Bradshaw and Waite, 2017). Studies and debates within and beyond the UK have evidenced many similarities between resistance to fracking and contestations over renewable energy development, including the significance on people's attitudes of levels of knowledge (e.g. Stedman et al., 2016; Whitmarsh et al., 2016), notions of social equity and the right to participate in decision-making processes (Beebeejaun, 2017; Cotton et al., 2014; Williams et al., 2015; Whitmarsh et al., 2016; Whitton et al., 2017), as well as the significance of emotional attachments to place (Cotton, 2015; Sangaramoorthy et al., 2016).

Reference to place highlights the significance of location to considerations of responses to unconventional hydrocarbon developments. At the time this chapter was written, there are no locations in the UK where commercial unconventional extraction of hydrocarbons is being conducted, although there are three locations where exploratory drilling has commenced, namely near Balcombe in Sussex, at Kirby Misperton in North Yorkshire and at Preston New Road in Lancashire (for details of this last development see Bradshaw and Waite, 2017). However, Petroleum Exploration and Drilling Licences have been offered for a series of blocks of land across the United Kingdom (see Figure 45.10), and as Table 45.6 demonstrates, of the license blocks in England, it appears that 72 per cent encompass land classed as 'rural', with just over 50 per cent falling into the rural village and dispersed settlement category. Areas classified as sparse form a noticeably small proportion of the areas licensed for exploration, a feature that may be accounted for in part by the blocks' geological foundations, as many of the areas where suitable rocks for unconventional carbon extraction are areas that historically were used for conventional carbon extraction, and in and around which settlements emerge. The extent to which the licenced blocks will become sites for further exploration and exploitation remains very much an open question, it being widely argued that fracking currently lacks a 'social license to operate' in the UK (see House, 2013; McGlade et al., 2015; Bradshaw and Waite, 2017).

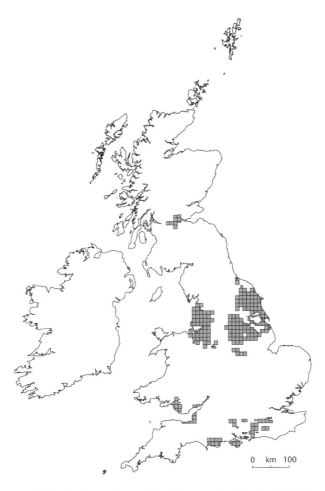

Figure 45.10   Licensed blocks for on-shore oil and gas exploration in the UK.

However, should this block on development be transformed or neglected, it would seem from the analysis of Table 45.6 that rural areas might once again become a major location for energy related developments, even if ones that might once again operate largely in the vertical energy regime identified by Huber and McCarthy (2017).

Conclusion

This chapter has explored the notion of a post-carbon transition through the concept of a post-carbon geography, focusing on the challenges such a transition might pose to the people living in rural spaces. It has focused principally on the UK, and there clearly is much work beyond this context that is of considerable value in developing post-carbon geographies of rural space. There is, for instance, extensive work on renewable energy developments within countries in Europe and North America (e.g. González et al., 2016; Jacquet, 2015; Jacquet and Stedman, 2013; Mayer et al., 2018; Moss et al., 2015; Phadke, 2011; Walker and Baxter, 2017), and also a growing number of studies of the rural Global South and Global East (e.g. Baurzhan and Jenkins, 2016; Love and Garwood, 2011; Mengistu et al., 2015; Shaaban and Petinrin, 2014; Song et al., 2014; Ulsrud et al., 2018). There would seem to be scope for a 'comparative ruralism' (Phillips and Smith, 2018) that might not only consider the similarities and differences in energy systems, discourses and practices in different countries and regions of the world, but also consider how these come to exert influences on one another. Lennon and Scott (2015a, 2015b), for example, have highlighted how wind energy developments in the 'Irish Midlands' were stimulated by the carbon reduction commitments of the UK government, and how these developments raised local and national concerns in Ireland. Reference has also been made in this chapter to concerns over land-grabbing in the Global South related to biofuels, which further illustrate the potential value of 'relational comparisons' in the development of post-carbon geographies.

This chapter has sought to highlight the significance of energy to contemporary rural life in the UK, and also how rural areas are, and have long been, important sites of energy production. Much of this energy is destined for consumption in urban areas, or in rural areas distant from their point of production, and one issue that has not been considered in this chapter is the impact that energy distribution infrastructure can have on rural areas. There is an extensive literature on this issue (e.g. Batel et al., 2013; Batel and Devine-Wright 2015; Cotton and Devine-Wright, 2010, 2013; Devine-Wright, 2013; Lienert et al., 2017), much of it focused on how the development of such infrastructure may be an issue of concern and contestation among rural residents. Much of this literature make use of, or seeks to critique, notions of NIMBYism, and hence many of the issues raised in this chapter in relation to this concept are also of relevance to this area of study.

This chapter has sought to highlight the challenges of low-carbon transitioning within rural areas, highlighting the significance of rural space within both energy consumption and production. In relation to both, rural space can be seen to impact material, symbolically and emotionally upon efforts to transition to a low-carbon society, and as we have argued elsewhere (Phillips and Dickie, 2014, 2015, forthcoming), rural residents are often highly conscious of such efforts and the challenges that surround them. In many instances they may distance themselves from such efforts, although this should not be viewed as necessarily expressing resistance but may reflect a complex range of feelings, including uncertainty, anxiety, incapacity and loss. Many of the studies of protest over renewable energy developments discussed in this chapter have also highlighted the role of emotional and affective responses to change, and the need for planners and others to engage with such responses. Such work clearly points to an important dimension of low-carbon transition, although it also important to consider the material challenges of this transition as highlighted by the work of Huber and McCarthy (2017) which, as discussed in this chapter, highlights how this transition may make extensive spatial demands, particularly of areas that are rural in character. This chapter has indeed detailed how a range of different forms of renewable energy production in the UK are exhibiting clear rural foci in their developments, although may be focused on different types of rural areas. The analysis of Huber and McCarthy (2017) suggests that the scale of rural, and indeed urban, land conversion to energy use will need to be on a much greater scale than is often envisaged by advocates of a low-carbon transition. Consequently planning, which often acts as a mechanism for arbitrating between competing land users, may be expected to be drawn yet further into social controversies through low-carbon transitions, notwithstanding the neo-liberal favouring of market distribution mechanisms that has been clearly evident in many parts of the UK.

Note

1 The classifications used in Table 45.1 are derived from governmental classifications employed in each of the component nations of the United Kingdom. See for example, Bibby and Brindley (2013) and Northern Ireland Statistics Research Agency (2015).

References

Accorsi R, Cholette S, Manzini R, Pini C and Penazzi S. (2016) The land-network problem: ecosystem carbon balance in planning sustainable agro-food supply chains. *Journal of Cleaner Production* 112: 158–171.

Aitken M. (2010) Why we still don't understand the social aspects of windpower: a critique of key assumptions within the literature. *Energy Policy* 38: 1834–1841.

Ajanovic A. (2011) Biofuels versus food production: Does biofuels production increase food prices? *Energy* 36: 2070–2076.

Anable J, Boardman B and Root A. (1997) *Travel Emissions Profiles*. Environmental Change Unit, Research Report 17. Oxford: University of Oxford.

Andersson-Hudson J, Knight W, Humphrey M, et al. (2016) Exploring support for shale gas extraction in the United Kingdom. *Energy Policy* 98: 582–589.

Baiocchi G, Creutzig F, Minx J, et al. (2015) A spatial typology of human settlements and their CO2 emissions in England. *Global Environmental Change* 34: 13–21.

Banister D. (1992) Energy use, transport and settlement patterns. In: Breheny M (ed) *Sustainable Development and Urban Form*. London: Pion, 160–181.

Banister D and Banister C. (1995) Energy consumption in transport in Great Britain: macro level estimates. *Transportation Research Part A: Policy and Practice* 29: 21–32.

Batel S and Devine-Wright P. (2015) A critical and empirical analysis of the national-local 'gap' in public responses to large-scale energy infrastructures. *Journal of Environmental Planning and Management* 58: 1076–1095.

Batel S, Devine-Wright P and Tangeland T. (2013) Social acceptance of low-carbon energy and associated infrastructures: a critical discussion. *Energy Policy* 58: 1–5.

Baurzhan S and Jenkins GP. (2016) Off-grid solar PV: Is it an affordable or appropriate solution for rural electrification in sub-Saharan African countries? *Renewable and Sustainable Energy Reviews* 60: 1405–1418.

Beebeejaun Y. (2017) Exploring the intersections between local knowledge and environmental regulation: a study of shale gas extraction in Texas and Lancashire. *Environment and Planning C* 35: 417–433.

BEIS. (2017) *Fuel Poverty: Methodology Handbook*, London: Department for Business, Energy and Industrial Strategy.

Bellarby J, Tirado R, Leip A, et al. (2013) Livestock greenhouse gas emissions and mitigation potential in Europe. *Global Change Biology* 19: 3–18.

Bibby P and Brindley P. (2013) *Urban and Rural Area Definitions for Policy Purposes in England and Wales: Methodology (v1.0)*, London: Government Statistical Service.

Bicknell J, Dodman D and Satterthwaite D. (2009) *Adapting Cities to Climate Change: Understanding and Addressing the Development Challenges*. London: Earthscan.

Bradshaw M, Chindo M, Dutton J, et al. (2015) Unconventional fossil fuels and technological change. In: Ekins P, Bradshaw M and Watson J (eds) *Global Energy: Issues, Potentials and Policy Implications*. Oxford: Oxford University Press, 268–290.

Bradshaw M and Waite C. (2017) Learning from Lancashire: Exploring the contours of the shale gas conflict in England. *Global Environmental Change* 47: 28–36.

BRE Housing. (2008) *A Study of Hard to Treat Homes Using the English House Condition Survey*, London: Defra & Energy Saving Trust.

Brecha RJ, Mitchell A, Hallinan K, et al. (2011) Prioritizing investment in residential energy efficiency and renewable energy – a case study for the U.S. Midwest. *Energy Policy* 39: 2982–2992.

Breheny M. (1995) Counterurbanisation and sustainable urban forms. In: Brotchie J, J. B, Batty M, et al. (eds) *Cities in competition: the emergence of productive and sustainable cities for the 21st century*. Melbourne: Longman Cheshire, 402–429.

Breukers S and Wolsink M. (2007) Wind power implementation in changing institutional landscapes: an international comparison. *Energy Policy* 35: 2737–2750.

Bristow G, Cowell R and Munday M. (2012) Windfalls for whom? The evolving notion of 'community' in community benefit provisions from wind farms. *Geoforum* 43.

Büchs M and Schnepf SV. (2013) Who emits most? Associations between socio-economic factors and UK households' home energy, transport, indirect and total CO2 emissions. *Ecological Economics* 90: 114–123.

Bulkeley H and Betsill M. (2003) *Urban Sustainability and Global Environmental Governance*, London: Routledge.

Burningham K, Barnett J and Thrush D. (2007) The limitations of the NIMBY concept for understanding public engagement with renewable energy technologies: a literature review. Beyond NIMBYism Project: Summary of Findings from Work Package 1.3. Exeter: University of Exeter.

Burningham K, Barnett J and Walker G. (2015) An array of deficits: unpacking NIMBY discourses in wind energy developers' conceptualizations of their local opponents. *Society and Natural Resources* 28: 246–260.

Cass N and Walker G. (2009) Emotion and rationality: the characterisation and evaluation of opposition to renewable energy projects. *Emotion, Space and Society* 2: 62–69.

Comber A, Dickie J, Jarvis C, et al. (2015) Locating bioenergy facilities using a modified GIS-based location–allocation-algorithm: Considering the spatial distribution of resourcesupply. *Applied Energy* 154: 309–316.

Commission for Rural Communities. (2007) *The State of the Countryside 2007*, Wetherby: Countryside Agency Publications.

Commission for Rural Communities. (2010) *The State of the Countryside 2010*, Wetherby: Countryside Agency Publications.

Cotton M. (2015) Stakeholder perspectives on shale gas fracking: a Q-method study of environmental discourses. *Environment and Planning A* 47: 1944–1962.

Cotton M and Devine-Wright P. (2010) NIMBYism and community consultation in electricity transmission network planning. In: Devine-Wright P (ed) *Renewable Energy and the Public: From NIMBY to Participation*. Abingdon: Earthscan, 115–128.

Cotton M and Devine-Wright P. (2013) Putting pylons into place: a UK case study of public perspectives on the impacts of high voltage overhead transmission lines. *Journal of Environmental Planning and Management* 56: 1225–1245.

Cotton M, Rattle I and van Alstine J. (2014) Shale gas policy in the United Kingdom: an argumentative discourse analysis. *Energy Policy* 73: 427–438.

Cowell R. (2007) Wind power and 'the planning problem': the experience of Wales. *European Environment* 17: 291–306.

Cowell R, Ellis G, Sherry-Brennan F, et al. (2017) Energy transitions, sub-national government and regime flexibility: how has devolution in the United Kingdom affected renewable energy development? *Energy Research and Social Science* 23: 169–181.

Dennis K and Urry J. (2009) *After the Car*. Cambridge: Polity.

De Rijke K, Munro P and Zurita MDLM. (2016) The Great Artesian Basin: a contested resource environment of subterranean water and coal seam gas in Australia. *Society and Natural Resources* 29: 696–710.

DEFRA. (2008) *Distributional Impacts of Personal Carbon Trading*. London: HMSO.

Delebarre J, Ares E and Smith L. (2017) Shale gas and fracking. House of Commons Library Briefing Paper No. 6073. London: House of Commons Library.

Department for Business Energy and Industrial Strategy. (2018) Energy Policy:Written statement – HCWS690. Retrieved on 23 May 2018 from www.parliament.uk/business/publications/written-questions-answers-statements/written-statement/Commons/2018-05-17/HCWS690.

Devine-Wright H. (2011a) From backyards to places: public engagement and the emplacement of renewable energy technologies. In: Devine-Wright P (ed) *Renewable Energy and the Public: From NIMBY to Participation*. Abingdon: Earthscan, 57–70.

Devine-Wright H. (2011b) Public engagement with renewable energy technologies: introduction. In: Devine-Wright P (ed) *Renewable Energy and the Public: From NIMBY to Participation*. Oxford: Earthscan, xxi–xxx.

Devine-Wright P. (2005) Beyond NIMBYism: towards an integrated framework for understanding public perceptions of wind energy. *Wind Energy* 8: 125–139.

Devine-Wright P. (2007) Reconsidering public attitudes and public acceptance of renewable energy technologies: a critical review. Manchester: School of Environment and Development, University of Manchester.

Devine-Wright P. (2008) Reconsidering public acceptance of renewable energy technologies: a critical review. In: Grubb M, Jamasb T and Pollit M (eds) *Delivering a Low-carbon Electricty System: Technologies, Economics and Policy*. Cambridge: Cambridge University Press, 443–461.

Devine-Wright P. (2009) Rethinking NIMBYism: the role of place attachment and place Identity in explaining place-protective action. *Journal of Community and Applied Social Psychology* 19: 426–441.

Devine-Wright P. (2013) Explaining 'NIMBY' objections to a power line: the role of personal, place attachment and project-related factors. *Environment and Behavior* 45: 761–781.

Dickie J and Phillips M. (forthcoming) Energising rural studies. *Journal of Rural Studies*.

Dodman D. (2009) Blaming cities for climate change? An analysis of urban greenhouse gas emissions inventories. *Environment and Urbanization* 21: 185–201.

Druckman A and Jackson T. (2008) Household energy consumption in the UK: a highly geographically and socio-economically disaggregated model. *Energy Policy* 36: 3167– 3182.

Eaton E and Kinchy A. (2016) Quiet voices in the fracking debate: ambivalence, nonmobilization, and individual action in two extractive communities (Saskatchewan and Pennsylvania). *Energy Research and Social Science* 20: 22–30.

Fahmy E, Thumim J and White V. (2011) The distribution of UK household CO2 emissions: interim report. York: Joseph Rowntree.

Food and Agriculture Organization. (2006) *Livestock's Long Shadow: Environmental Issues and Options*. Rome: Food and Agriculture Organization of the United Nations.

Frantal B, Pasqualetti M and van der Horst D. (2014) New trends and challenges for energy geographies: introduction to the special issue. *Morovian Geographical Reports* 22: 2–6.

González A, Daly G and Gleeson J. (2016) Congested spaces, contested scales: a review of spatial planning for wind energy in Ireland. *Landscape and Urban Planning* 145: 12–20.

Goodland R and Anhang J. (2009) Livestock and climate Change. What if the key actors in climate change were pigs, chickens and cows? *Worldwatch* November/December: 10–19.

Gross C. (2007) Community perspectives on wind energy in Australia: the application of a justice and community fairness framework to increase social accpetance. *Energy Policy* 35: 2727–2736.

Guta DD. (2014) Effect of fuelwood scarcity and socio-economic factors on household bio-based energy use and energy substitution in rural Ethiopia. *Energy Policy* 75: 217–227.

Hain J, Ault G, Galloway SJ, et al. (2005) Additional renewable energy growth through small-scale community oriented policies. *Energy Policy* 33: 1199–1212.

Hall C and Woolvin M. (2012) What are the implications for rural Scotland of a low-carbon future? In: Skerrat S (ed) *Rural Scotland in Focus 2012*. Edinburgh: Rural Policy Centre, Scottish Agricultural College, 86–101.

Hanley N and Nevin C. (1999) Appraising renewable energy developments in remote communities: the case of the North Assynt Estate, Scotland. *Energy Policy* 27: 527–547.

Haughton A, Bond A, Lovett A, et al. (2009) A novel, integrated approach to assessing social, economic and environmental implications of changing rural land-use: a case study of perennial biomass crops. *Journal of Applied Ecology* 46: 315–322.

Herrero M, Gerber P, Vellinga T, et al. (2011) Livestock and greenhouse gas emissions: The importance of getting the numbers right. *Animal Feed Science and Technology* 166–7: 779–782.

Hicks D. (2013) A post-carbon geography. *Teaching Geography* 38: 94–97.

Hiemstra G, van der Horst H and Hovorka A. (2018) Reassessing the 'energy ladder': household energy use in Maun, Botswana. *Energy Policy* 36: 3333–3344.

Hoornweg D, Sugar L and Trejos Gomez C. (2011) Cities and greenhouse gas emissions: moving forward. *Environment and Urbanization* 23: 207–227.

House E. (2013) Fractured fairytales: the failed social license for unconventional oil and gas development. *Wyoming Law Review* 13: 5–67.

Howard DC, Wadsworth RA, Whitakera JW, et al. (2009) The impact of sustainable energy production on land use in Britain through to 2050. *Land Use Policy* 26: S284–S292.

Hubbard P. (2006) NIMBY by another name? A reply to Wolsink. *Transactions of the Institute of British Geographers* 31: 92–94.

Huber MT and McCarthy J. (2017) Beyond the subterranean energy regime? Fuel, land use and the production of space. *Transactions of the Institute of British Geographers* 42: 655–668.

Huebner G, Hamilton I, Chalabi Z, et al. (2015) Explaining domestic energy consumption – the comparative contribution of building factors, socio-demographics, behaviours and attitudes. *Applied Energy* 159.

Hunt A and Watkiss P. (2011) Climate change impacts and adaptation in cities: a review of the literature. *Climate Change* 104: 13–49.

Hunt J. (2004) How can cities mitigate and adapt to climate change? *Building Research and Information* 32: 55–57.

Jacquet J. (2015) The rise of 'private participation' in the planning of energy projects in the rural United States. *Society and Natural Resources* 28: 231–245.

Jacquet J and Stedman RC. (2013) Perceived impacts from wind and natural gas development in Northern Pennsylvania. *Rural Sociology* 78: 450–472.

Jolivet E and Heiskanen E. (2010) Blowing against the wind: an exploratory application of actor network theory to the analysis of local controversies and participation processes in wind energy. *Energy Policy* 38: 6746–6754.

Jones C and Eiser JR. (2009) Identifying predictors of attitudes towards local on shore wind development with reference to an English case study. *Energy Policy* 38: 3106–3117.

Jones C and Eiser JR. (2010) Understanding 'local' opposition to wind development in the UK: how big is a backyard? *Energy Policy* 38: 3106–3117.

Jones R and Lomas K. (2015) Determinants of high electrical energy demand in UK homes: socio-economic and dwelling characteristics. *Energy and Buildings* 101: 24–34.

Kitchen L and Marsden T. (2009) Creating sustainable rural development through stimulating the eco-economy: beyond the eco-economic paradox? *Sociologia Ruralis* 49: 273–294.

Lennon M and Scott M. (2015a) Contending expertise: an interpretive approach to (re)conceiving wind power's 'planning problem'. *Journal of Environmental Policy and Planning* 17: 593–616.

Lennon M and Scott M. (2015b) Opportunity or threat: dissecting tensions in a post-carbon transition. *Sociologia ruralis* 57: 87–109.

Lienert P, Sütterlin B and Siegrist M. (2017) The influence of high-voltage power lines on the feelings evoked by different Swiss surroundings. *Energy Research and Social Science* 23,: 46–59.

Lindley S, J. H, Theuray N, et al. (2006) Adaptation strategies for climate change in the urban environment: assessing climate change risks in UK urban areas. *Journal of Risk Research* 9: 543–556.

London Climate Change Partnership. (2002) *London's Warning: the Impacts of Climate Change on London. Final Technical Report*, London: Entec UK Ltd.

Love T and Garwood A. (2011) Wind, sun and water: omplexities of alternative energy development in rural northern Peru. *Rural Society* 20: 294–307.

Lovett A, Sünnenberg G, Richter G, et al. (2009) Land use implications of increased biomass production identified by GIS-based suitability and yield mapping for Miscanthus in England. *BioEnergy Research* 2: 17–28.

Markantoni M and Woolvin M. (2015) The role of rural communities in the transition to a low-carbon Scotland: a review. *Local Environment* 20: 202–219.

Mason K and Milbourne P. (2014) Constructing a 'landscape justice' for windfarm development: the case of Nant Y Moch, Wales. *Geoforum* 53: 104–115.

Matthews S and Morgan J. (2013) The post-carbon challenge for curriculum subjects. *International Journal of Educational Research* 61: 93–100.

Mayer A, Malin SA and Olson-Hazboun SK. (2018) Unhollowing rural America? Rural human capital flight and the demographic consequences of the oil and gas boom. *Population and Environment* 39: 219–238.

McClymont K and O'Hare P. (2008) 'We're not NIMBYs!' Contrasting local protest groups with idealised conceptions of sustainable communities. *Local Environment* 13: 321–333.

McEvoy D. (2007) Climate change and cities: Special issue. *Built Environment* 33.

McGlade C, Ekins P, Bradshaw M, et al. (2015) *Conditions for Environmentally-Sound Shale Gas Development.* UK Energy Research Centre: Policy Briefing. London: UK Energy Research Centre.

Mengistu MG, Simane B, Eshete G, et al. (2015) A review on biogas technology and its contributions to sustainable rural livelihood in Ethiopia. *Renewable and Sustainable Energy Reviews,* 48: 306–316.

Minx J, Baiocchi G, Wiedmann T, et al. (2013) Carbon footprints of cities and other human settlements in the UK. *Environmental Research Letters* 8: 1–10.

Mol A. (2007) Boundless biofuels? Between environmental sustainability and vulnerability. *Sociologia Ruralis* 47: 297–315.

Mol A. (2014) Bounded biofuels? Sustainability of global biogas developments. *Sociolgia Ruralis* 54: 1–20.

Morrone M, Chadwick AE and Kruse N. (2015) A community divided: hydraulic fracturing in rural Appalachia. *Journal of Appalachian Studies* 21: 207–228.

Moss T, Becker S and Naumann M. (2015) Whose energy transition is it, anyway? Organisation and ownership of the Energiewende in villages, cities and regions. *Local Environment* 20: 1547–1563.

Munday M, Bristow G and Cowell R. (2011) Wind farms in rural areas: how far do community benefits from wind farms represent a local economic development opportunity? *Journal of Rural Studies* 27: 1–12.

Nansaior A, Patanothai A, Rambo AT, et al. (2011) Climbing the energy ladder or diversifying energy sources? The continuing importance of household use of biomass energy in urbanizing communities in Northeast Thailand. *Biomass and Bioenergy* 35: 4180–4188.

Northern Ireland Statistics and Research Agency. (2015) *Review of the Statistical Classification and Delineation of Settlements.* Belfast: Office of National Statistics.

O'Hara S, Humphrey M, Andersson-Hudson J, et al. (2016) *Public Perceptions of Shale Gas Extraction in the UK: From Positive to Negative.* Nottingham: University of Nottingham.

Ogilvie M and Rootes C. (2015) The impact of local campaigns against wind energy developments. *Environmental Politics* 24: 874–893.

Phadke R. (2011) Resisting and reconciling Big Wind: middle landscape politics in the New American West. *Antipode* 43: 754–776.

Phillips M. (1994) Habermas, rural studies and critical social theory. In: Cloke P, Doel M, Matless D, et al. (eds) *Writing the Rural: Five Cultural Geographies.* London: Paul Chapman, 89–126.

Phillips M and Dickie J. (2014) Narratives of transition/non-transition towards low-carbon futures within English rural communities. *Journal of Rural Studies* 34: 79–95.

Phillips M and Dickie J. (2015) Climate change, carbon dependency and narratives of transition and stasis in four English rural communities. *Geoforum* 67: 93–109.

Phillips M and Dickie J. (forthcoming) Moving to or from a carbon dependent countryside. *Journal of Transport Geography.*

Phillips M and Smith D. (2018) Comparative ruralism and 'opening new windows' on gentrification. *Dialogues in Human Geography* 8: 51–58.

Roberts E and Henwood K. (2016) Exploring the everyday energyscapes of rural dwellers in Wales: putting relational space to work in research on everyday energy use. *Energy Research and Social Science* 36: 44–51.

Röder M. (2016) More than food or fuel. Stakeholder perceptions of anaerobic digestion and land use: a case study from the United Kingdom. *Energy Policy* 97: 73–81.

Rowe R, Street N and Taylor G. (2008) Identifying potential environmental impacts of large-scale deployment of bioenergy crops in the UK. *Renewable and Sustainable Energy Reviews* 13: 271–290.

Rutherford J and Coutard O. (2014) Urban energy transitions: places, processes and politics of socio-technical change. *Urban Studies* 51: 1353–1377.

Sangaramoorthy T, Jamison AM, Boyle MD, et al. (2016) Place-based perceptions of the impacts of fracking along the Marcellus Shale. *Social Science and Medicine* 151: 27–37.

Schafft KA, Borlu Y and Glenna L. (2013) The relationship between Marcellus shale gas development in Pennsylvania and local perceptions of risk and opportunity. *Rural Sociology* 78: 143–166.

Schreider S, Smith D and Jakeman A. (2000) Climate change impacts on urban flooding. *Climate Change* 47: 91–115.

Schubert J, Wolbring T and Gill G. (2013) Settlement structures and carbon emissions in Germany: the effects of social and physical concentration on carbon emissions in rural and urban residential areas. *Environmental Policy and Governance* 23: 13–29.

Shaaban M and Petinrin JO. (2014) Renewable energy potentials in Nigeria: meeting rural energy needs. *Renewable and Sustainable Energy Reviews* 29: 72–84.

Sherval M and Hardiman K. (2014) Competing perceptions of the rural Idyll: responses to threats from coal seam gas development in Gloucester, NSW, Australia. *Australian Geographer* 45: 185–203.

Simcock N. (2014) Exploring how stakeholders in two community wind projects use a 'those affected' principle to evaluate the fairness of each project's spatial boundary. *Local Environment* 19: 241–258.

Smil V. (2015) *Power Density: A Key to Understanding Energy Sources and Uses.* Cambridge, MA: MIT Press.

Song Z, Zhang C, Yang G, et al. (2014) Comparison of biogas development from households and medium and large-scale biogas plants in rural China. *Renewable and Sustainable Energy Reviews* 33: 204–213.

Stedman RC, Evensen D, O'Hara S, et al. (2016) Comparing the relationship between knowledge and support for hydraulic fracturing between residents of the United States and the United Kingdom. *Energy Research and Social Science* 20: 142–148.

Stephenson E, Doukas A and Shaw K. (2012) 'Green washing gas': might a 'transition fuel' label legitimize carbon-intensive natural gas development? *Energy Policy* 46: 452–459.

Stevenson R and Richardson T. (2003) Policy integration for sustainable development: exploring barriers to renewable energy development in post devolution Wales. *Journal of Environmental Policy and Planning* 5: 95–118.

Szarka J. (2007) *Wind Power in Europe: Politics, Business and Society.* Basingstoke: Palgrave Macmillan.

Szarka J and Bluhdorn I. (2006) *Wind Power in Britain and Germany: Explaining Contrasting Development Paths.* London: Department of Food, Environment and Rural Affairs.

Toke D. (2005) Explaining wind power planning outcomes: some findings from a study in England and Wales. *Energy Policy* 33: 1527–1539.

Tranter RB, Swinbank A, Jones PJ, et al. (2011) Assessing the potential for the uptake of on-farm anaerobic digestion for energy production in England. *Energy Policy* 39: 2424–2430.

UK Government. (2014) Local councils to receive millions in business rates from shale gas developments. News release from the Prime Minister's Office. Retrieved on 23 May 2018 from www.gov.uk/government/news/local-councils-to-receive-millions-in-business-rates-from-shale-gas-developments.

Ulsrud K, Rohracher H and Muchunku C. (2018) Spatial transfer of innovations: South-South learning on village-scale solar power supply between India and Kenya. *Energy Policy* 114: 89–97.

Upreti B. (2004) Conflict over biomass energy development in the United Kingdom: some observations and lessons from England and Wales. *Energy Policy* 32: 785–800.

Upreti B and van der Horst D. (2004) National renewable energy policy and local opposition in the UK: the failed development of a biomass electricity plant. *Biomass and Bioenergy* 26: 61–69.

Urry J. (2012) Changing transport and changing climates. *Journal of Transport Geography* 24: 533–535.

Van der Horst D. (2007) NIMBY or not? Exploring the relevance of location and the politics of voiced opinion in renewable energy siting controversies. *Energy Policy* 35: 2705–2714.

Van der Horst D and Toke D. (2010) Exploring the landscape of wind farm developments: the local area characteristics and planning process outcomes in rural England. *Land Use Policy* 27: 214–221.

Van der Horst D and Vermeylen S. (2011) Local rights to landscape in the global moral economy of carbon. *Landscape* 36: 455–470.

Van der Schoor T and Scholtens B. (2015) Power to the people: local community initiatives and the transition to sustainable energy. *Renewable and Sustainable Energy Reviews* 43: 666–675.

Viggers H, Kealla M, Wickens K, et al. (2017) Increased house size can cancel out the effect of improved insulation on overall heating energy requirements. *Energy Policy* 107: 248–257.

Walker C and Baxter J. (2017) 'It's easy to throw rocks at a corporation': wind energy development and distributive justice in Canada. *Journal of Environmental Policy and Planning* 19: 754–768.

Walker G, Devine-Wright P, Barnett J, et al. (2011) Symmetries, expectations, dynamics and contexts: a framework for understanding public engagement with renewable energy projects. In: Devine-Wright P (ed) *Renewable Energy and the Public.* Oxford: Earthscan, 1–14.

Walker G, Evans B, Devine-Wright P, et al. (2007) Harnessing community energies: explaining community based localism in renewable energy policy in the UK. *Global Environmental Politics* 7: 64–82.

Warren C and McFadyen M. (2010) Does community ownership affect public attitudes to wind energy? A case study from south-west Scotland. *Land Use Policy* 27: 204–213.

Warren CR, Lumsden C, O'Dowd S, et al. (2005) 'Green on green': public perceptions of wind power in Scotland and Ireland. *Journal of Environmental Planning and Management* 48: 853–875.

Whitmarsh L, Nash N, Upham P, et al. (2016) UK public perceptions of shale gas hydraulic fracturing: the role of audience, message and contextual factors on risk perceptions and policy support. *Applied Energy* 160: 419–430.

Whitton J, Brasier K, Charnley-Parry I, et al. (2017) Shale gas governance in the United Kingdom and the United States: opportunities for public participation and the implications for social justice. *Energy Research and Social Science* 26: 11–22.

Wilby R. (2007) A review of climate change impacts on the built environment. *Built Environment* 33: 31–45.

Williams L, Macnaghten P, Davies R, et al. (2015) Framing 'fracking': exploring public perceptions of hydraulic fracturing in the United Kingdon. *Public Understanding of Science* 1–17.

Wolsink M. (2000) Wind power and the NIMBY-myth: institutional capacity and the limited significance of public support. *Renewable Energy* 21: 49–64.

Wolsink M. (2006) Invalid theory impedes our understanding: a critique on the persistence of the language of NIMBY. *Transactions of the Institute of British Geographers* 31: 85–91.

Woods M. (2003) Environmental visions of rural: windfarm development in mid Wales. *Sociologia Ruralis* 43: 271–288.

World Bank. (2010) *Cities and Climate Change: An Urgent Agenda.* Washington, DC: The World Bank.

Wrigley EA. (1988) *Continuity, Chance and Change: The Character of the Industrial Revolution in England.* Cambridge: Cambridge University Press.

46

Planning for rural communities and major renewable energy infrastructure

Lucy Natarajan

Introduction

This chapter considers the implications of major renewable energy infrastructure (REI) development for rural planning. It takes the view that rural planning should be community-oriented, and that it will involve contested decisions. It must grapple with conflictual issues associated with new developments in rural areas as major renewable energy infrastructure developments, which have proven controversial, become increasingly common. The UK is a particularly useful case to study as it has witnessed rapid expansion of REI roll-out (Huddleston, 2010). It is the detail of where and how such development is constructed, which will surely be the concern of rural planning and, as discussed below, such matters are opened up through the regulatory processes for 'Nationally Significant Infrastructure Projects' or NSIPs. This chapter draws on a recent study[1] of NSIPs in England and Wales, and particularly the data collected from the national infrastructure planning portal (available at https://infrastructure.planningin spectorate.gov.uk).

This assessment of NSIPs is rooted in an understanding of 'the rural' as a human construct (Marsden et al., 1993), where the uses of land are deeply entwined with societal notions of value. As such it sees rural planning as less oriented towards achieving an end-state as towards fulfilling the need 'to ensure sustainable and equitable use of resources in the countryside and to optimize the welfare produced from them over time for all stakeholders' (ibid.: 17). It recognises, the deep tension between development of national resources and the protection of local communities, which is writ large across the range of concerns and aspirations around the use of rural land negotiated within planning. The renewable energy infrastructure literature suggests this is most noted in relation to the 'social gap' (Bell et al., 2005, 2013) where there is national acceptance of wind farm developments but communities protest the appearance of them in their local countryside.

Given the intrinsic contestation of rural value, it is useful for any instance of planning, and particularly the REI development, to examine who speaks for a locality, how, and to what effect (Marsden et al., 1993). Knowledge claims over landscapes within NSIPs need to be 'carefully layered' to provide legal reasoning (Lee, 2017). Mardsen et al. (1993) also suggest local actors both struggle to find representation within networks of power relations and might be 'mobilised' to support the purposes of actors outside their locality. For this reason the representations that local rural communities provide on their own behalf, through participation in NSIP processes, as well as the overarching tension manifest within the regulatory stage of major REI, is of particular interest. This chapter therefore first assesses the 'NSIP regime' itself and then examines the concerns of rural communities, paying attention to how they are constructed and attributed significance within planning, drawing on the UK experience. Empirically, the research examined 12 NSIP cases that were complete by September 2015. These are:

- Kentish Offshore Wind Farm Extension;
- Galloper Offshore Wind Farm;
- Burbo Bank Offshore Wind Farm Extension;
- Rampion Offshore Wind Farm;
- Walney Offshore Wind Farm Extension;
- Triton Knoll Offshore Wind Farm;
- Navitus Bay Offshore Wind Farm;
- Brechfa Forest West Wind Farm;
- Clocaenog Forest Wind Farm;
- Swansea Bay Tidal Lagoon;
- North Blyth Biomass Plant; and
- Rookery South Energy from Waste Plant.

The reports from their examinations were revisited selecting those concerns that had been articulated explicitly by local people within rural communities within the NSIP regulation, and tracing their resolution or otherwise within the recommendations for mitigation or refusal of consent (a rare occurrence, only seen in Navitus Bay within the UK study). References from the reports are given with the name of the development and the paragraph number. Findings show the important role of local planning authorities in establishing the importance of rural concerns, particularly in the face of strong policy narratives in favour of major REI. In the concluding section, implications for planning for rural communities are discussed.

The NSIP regime

The NSIP regime provides a strong presumption in favour of consenting major REI through policy, and expediting it through planning control at the national level. It is doubtful whether major infrastructure developments are in fact brought from inception through to consent any quicker (Marshall and Cowell, 2016), however, the NSIP regime has certainly provided a powerful mechanism for delivering REI consent as set out by the Planning Act 2008 (TSO, 2008). Energy infrastructure proposals are determined 'NSIP' according to generating capacity thresholds – this included those with a maximum generating capacity of over 50 MW onshore or 100 MW offshore within England and Wales, although onshore wind farms in England have recently been removed from these provisions and are handled by local planning authorities, and those up to 350MW in Wales by the Welsh Assembly Government., Their 'promoters' prepare a draft development consent order as part of their application for consent. Applications must undergo a planning examination, which is conducted in an inquisitorial manner by an examining authority (ExA), either an individual or a panel appointed by the national planning inspectorate, within a six month period. Representations are primarily given in writing, although there are hearings where statutory bodies and local people (including individuals, businesses and groups) who have registered as Interested Parties by the allotted deadline can make representations orally, and site visits where local people can attend. All representations and procedural documentation such as ExA questions and developer application materials are made available online at the planning inspectorate's website. Subsequently, the ExA produces a report for the relevant Secretary of State who takes the decision, and these last reporting and decision-making stages are to be completed within 6 months. At the time of the UK study, 15 major REI cases had been through this new consenting system, and of these only one, Navitus, had been refused consent.

How then were such infrastructure projects deemed significant? The most prominent reason given is the need to meet the UK's legal obligations under the European Union's (EU) Renewable Energy Directive (2009/28/EC), and transposed into UK law via 'The Promotion of the Use of Energy from Renewable Sources Regulations 2011' (SI 2011/243). Current national policy in the UK leaves little room for doubt that REI is a high priority in two key National Policy Statements (NPS): EN-1 on energy and EN-3 on renewable energy (DECC, 2011a, 2011b), which set out the need, objectives and guidance for transitioning to renewables. These are key references within NSIP examinations. As EN-3 states: 'a significant increase in generation from large-scale renewable energy infrastructure is necessary to meet the 15% renewable energy target' (DECC, 2011a: section 1.1.1). Given the need for open space, and in the case of solar and wind power also uninterrupted sunlight and wind, these were mainly sited in rural areas and in the case of offshore wind farms and the Tidal Lagoon they had grid connections works running from the on-shoring point on the coast through rural areas to national electricity grid connection points. As such, the NSIP regime is not only infused with legal arguments for consenting major REI, but it also implies that REI should be considered a priority over possible rural concerns, such as tranquility and landscape views. As stated in EN-3, accepting large onshore wind farms 'will inevitably have some visual and/or noise impacts, particularly if sited in rural areas' (DECC, 2011a: section 2.7.2). Hence such local impacts may be deemed acceptable in rural areas, subject to tests such as conformity to ETSU-R-97 guidance on the assessment and rating of noise from wind farms (Meir et al., 1996) as stipulated in EN3.

Alongside the RE targets that underpin the NSIP regime itself, the expansion of renewable energy infrastructure appears to have been encouraged for economic reasons. In the UK, central government has emphasised the economic value of such development to rural farmers and landowners, as part of a drive to increase economic productivity in rural areas (DEFRA, 2015). As argued by the Department for Environment, Food and Rural Affairs (DEFRA) and Department for Communities and Local Government (DCLG), 'Better connectivity has led to farmers and landowners diversifying into renewable energy, such as wind turbines, solar panels and anaerobic digesters, producing energy for themselves and to sell, to provide an additional income stream. Farmers also utilise their buildings to provide storage facilities or office space for local businesses, providing much needed business accommodation' (DCLG and DEFRA, 2016). Such claims implicate NSIP in the longstanding tension between industrialising and conserving the countryside.

Critiques of industrialisation of rural land are rooted in the refutation of 'nature as a free gift', an idea that is attributed to Marx despite the argument that the interpretations are not as intended (Burkett, 1999). In any case, the key observation was that the (over)use of nature undermines the very conditions needed for production. 'Free nature' rests on a self-defeating logic in the face of the 'irreconcilable contradiction between use value and exchange value' for any commodity (Apostolopoulou and Adams, 2015: 17). However, arguments for REI suggest these can both use and conserve the environment, and thus go beyond notions of conflicting demands on rural land. Such narratives are highly contested, and popularly termed 'green grabbing' (Vidal, 2008) with the implication that 'green' arguments, where environmental value is attributed to developments and agricultural land uses such as bio-fuel crops, green infrastructure or ecotourism, are simply a cover to use rural land for private gain. They are seen as part of a set of planning discourses that can obscure local concerns by being vague about 'wide benefits' (Lennon, 2015), and thus smooth the path to 'mis-use' of the rural.

The NSIP regime was criticised for having a 'democratic deficit' in relation to onshore wind farms decision-making, which has been devolved back to local planning since the time of the UK study. Debates on rural governance lend some support to such a move, as they suggest that rural planning ought to err on the side of directly empowering local communities (Dalal-Clayton et al., 2002). Indeed in the UK, a new legitimacy has been given to arguments for local control (Wargent and Parker, 2018) by the introduction of statutory neighbourhood planning in England by the Localism Act (TSO, 2011). Nonetheless, rural planning scholars argue that 'despite an apparent localisation of rural policy delivery, the design of policy – and the framing of its delivery – is occurring at numerous levels and within a variety of different bodies above the point of delivery' (Gallent et al., 2015: 55). Focusing on major REI demonstrates this point, as both local and neighbourhood planning still operate in the context of decisions and policy made at national and Welsh scales.

While the introduction of the NSIP regime centralised decisions on REI and the lower tiers of planning, which are apparent within it, these have less discretion and authority than the ExAs. In Wales, the Welsh Assembly Government has arguably sought to bridge its own ambition for major REI even more firmly into local scale rural planning for onshore wind farms. Its Technical Advice Note 8 states that 'Local planning authorities should seek to maximise the potential of renewable energy by linking the development plan with other local authority strategies including the community strategy' (Welsh Government, 2005), and sets out designated Strategic Site Areas (SSAs) preferred for the construction of wind farms. Indeed, energy studies have criticised the means of constructing of SSAs for reducing the scope for reflexivity in Welsh energy policy (Cowell, 2010).

Local authorities are involved in NSIP examinations in an advisory capacity, as they are required to produce a local impact report (LIR) for the NSIP examination. LIR represent rural concerns and these are tested, as are all representations, through questioning by the ExA. Local plans are a material consideration in the NSIP examinations but throughout the cases studied, the NPSs may overrule them. This is clearly demonstrated in the case of Brechfa, in relation to the issue of minimum separation of turbines from rural properties. The Carmarthenshire unitary development plan (UDP) was being replaced by an LDP (local development plan), which was on deposit at the time of the exam. As the ExA noted, the deposit LDP contained a new policy that 'large-scale wind power proposals should be located a minimum of 1500m away from the nearest residential property' (Brechfa 3.12). However the ExA did not give weight to Carmarthenshire's proposed policy, citing the guidance in TAN8 of 'a normal minimum separation distance' of 500m (Brechfa 4.112).

Concerns of rural communities

As noted above, there are opportunities for local people to participate in NSIP processes. Applicants must undertake pre-application consultations in the local areas, and local people are included as Interested Parties (IP) in the examination. A survey of local participants across the UK study cases found that 'the strength of critique from participants in the NSIPs regulations serves to warn against any assumptions of procedural inclusiveness' (Natarajan et al., 2018: 209), although pro-active efforts of individual ExAs and developers that helped in particular episodes were acknowledged. The 12 cases studied were diverse, including onshore wind farms in Wales, as well as offshore wind farms, biomass and energy from waste plants, and a tidal lagoon. Broadly speaking, concerns identified by rural communities (including businesses and local interest and amenity groups as well as individual residents) related to tranquility, landscape, agriculture, tourism and ecology. Issues of tranquility, landscape and agriculture were particular to rural areas, whereas issues of local tourism and ecology were seen in both urban and rural areas, across the 12 cases. Therefore, while tourism and ecology are pertinent matters for planning for rural communities in relation to REI, in the interests of succinctness, this section focuses on how 'rural concerns' over tranquility, landscape and agriculture that were raised by IPs were substantiated and contested within the examination.

Tranquility

Looking first at the issue of noise as raised by rural communities, the representations predominantly cast their localities as quiet and tranquil by drawing on personal experience. For instance two IP in Triton Knoll raised concerns over 'noise and disruption from the construction of overhead cables, substations and converter stations' (Triton Knoll 5.1.15). The ExA framed these impacts as 'amenity impacts' but noted that 'the potentially serious harm to their own health that would arise from noise, there was no independent evidence on this point submitted to the Panel' (ibid). Similarly, in Brechfa IP representations regarding a 'swishing noise' of wind through turbine blades was related to experiences in local weather conditions; 'residents suggests that the concerns arise most frequently when the wind is from the south-east, and when the weather is damp. Mist, drizzle or light rain are seen as particularly associated with adverse noise conditions' (Brechfa 4.106). Again the ExA was unable to verify this on their site visits, and the information they requested on complaints over this type of noise coming from a nearby existing wind farm was not supplied by the local authority. The applicant argued that the issues experienced would not necessarily be replicated, and the ExA concluded on the evidence before them that 'the project could meet relevant standards and thus accord with policy' (Brechfa 4.110).

In some instance, IP concerns over the disruption of rural tranquility were given support by the LIR that local authorities supplied. Some LIRs argued that noise was a threat to the local economy. For instance in Rampion, the ExA reported how the South Downs National Park authority had 'highlighted a need to consider the impact of "loss of amenity and tranquillity in areas immediately adjacent to the cable route", suggesting that these concerns should not be considered lightly, given the marginal nature of many rural businesses dependent upon visitors' (Rampion 4.505). In that case, the ExA concluded the disruption would be temporary coming only from construction. Other times, LIRs and local policy helped articulate the 'residential amenity' aspect. For instance in Clocaenog, local councils reported that the cumulative background noise from all existing and permitted development would increase by 8 decibels, and noted that the ETSU-R-97 guidance indicated this was a major impact. The significance of this noise effect on rural communities was not contested by the ExA (Clocaenog 4.202) and the harm to residential amenity was acknowledged as being in conflict with local policies, i.e. NTE/7d of the Conwy local development plan and Policy VOE 9ii of the Denbighshire local development plan. Nevertheless following the guidance of EN3, the ExA attributed little or no weight to this impact as 'the correct methodology has been followed and a wind farm is shown to comply with ETSU-R-97 recommended noise limits' (Clocaenog 8.22).

Landscape

The second area of particular rural concern was landscape. This was an area where there was frequently difficulties in providing evidence that could hold weight in the planning examination.

The value of landscape was mainly described in terms of an intrinsic visual character of rural land, and for offshore REI, the value of landscape could also related to views from rural land to the seascape. In Burbo for instance, the local groups Wirral Society and Hoylake Village Life, as well as several local individuals, raised concerns about the impact of the REI developments on both aspects. Wirral Council was also concerned that the effects of the proposed offshore development on so-called 'receptors' within the designated Areas of Special Landscape Value, had only been given a 'moderate' impact rating by the developer. There were further hearings and ExA questions on this issue; however, several parties did not attend or follow up with any written representations (Burbo 4.124). However, the developer reported that there were discussions on mitigation that would proceed 'in their own time [and] at their own pace and we are not suggesting that any of these discussions should be taken into account by you [the ExA]' (Burbo 4.124). Thus the ExA concluded with reference to the NPSs that such local landscape impacts were not a reason to refuse consent.

Elsewhere, members of the public were at pains to present not only written material but also photomontages that visualised potential impacts. However, as demonstrated in Navitus Bay, visualisations from any party could be highly contested, and since the collection of industry guidance on assessing landscape impacts on 'visual receptors' (e.g. residential properties and people visiting viewpoints or using trails) was diverse, it could not easily resolve matters. The only consensus was that visualisations of landscape impacts should be used as tools (Navitus Bay 7.1.65), reaffirming their use as 'artefacts' (Rydin et al., 2017). Nonetheless, issues of landscape were evaluated and they proved decisive in the eventual refusal in this case. Counting against the development were the 'unique and outstanding qualities of the areas likely to be harmfully affected by the visual, intrusive presence of the turbine array and the offshore substations' (Navitus Bay 21.2.25), across several nationally designated areas (the Dorest and Isle of Wight AONBs Purbeck and Tennyson Heritage Coasts and New Forest Park). The ExA found that this – together with the 'less than substantial harm' to both the World Heritage Site and to the significance of designated heritage assets, and the point that those issues would 'preclude a favourable conclusion in terms of design quality' (Navitus Bay 21.3.19) – outweighed the benefits of the proposed wind farm.

Landscape concerns were not limited to views *per se*, but also extended to a more general rural character. In the Energy from Waste (EfW) case, local people and planning authorities indicated concerns about the industrial look of the facility and sought mitigation through redesign. As reported by the ExA, they argued that the proposed site was 'an area which is now changing its function and turning away from its historic role as an area where clay is extracted, in turn leaving large holes in the ground to be filled with waste from other parts of the country. Rather, it is now a rural, peaceful landscape, deserving to be left that way. The intrusion of the proposed EfW development would mean a return to the past' (Rookery 5.42). The ExA concluded that this weighed significantly against the proposal, although it did not outweigh the benefits, and no redesign would change this: 'Inevitably, the plant would be seen from many of the more distant viewpoints in the surrounding landscape as an essentially industrial plant in a rural location' (Rookery 6.18). Similarly, in Rampion, the energy substation, an associated works included in the application, was a concern in relation to rural vernacular design traditions. IP argued that 'the height of any substation buildings should be restricted to a single storey and that their design should echo agricultural buildings and thereby be appropriate to their rural setting' (Rampion 4.272). In this case, however, at the ExAs discretion a mitigation provision to protect existing hedgerows from removal was added into the developer's draft DCO, with the explanation that 'this provision is important given the maturity of trees and hedgerows in the location of the proposed substation and their importance in providing potential screening and the value attached to these landscape features by interested parties' (Rampion 4.288).

Agriculture

The third area of concern specific to rural communities was the impact of tracts of cable-laying that are necessary for energy transmission. While these types of works are not exclusive to REI, uncertainty is introduced where grid connections are not included as part of the applications. Such exclusions are relatively common in major REI, at least in part due to their scale and technological complexity. In the UK study, three of the 12 cases did not include associated grid connection in the application (Brechfa, Kentish and Triton Knoll), Burbo did not include those cables that were in Wales (with local authorities in Wales determining associated works for off-shore wind farms at that time), and Navitus Bay had not yet determined which of three sites to use. For agriculture, uncertainty threatened to impact seasonal operations and good soil practice, as well as income streams from crops. In Triton Knoll for instance, the ExA reported that there was 'uncertainty about the ability of farmers in the cable corridor to plan investment in facilities such as new agricultural buildings, drainage or irrigation, due to ongoing uncertainty about the location and width of the cable corridor or the depth at which cables would be buried' (Triton Knoll 5.1.21) and over 'how construction might be managed in a period of high rainfall, to avoid damage to soil structure and fertility' (Triton Knoll 5.1.12). The National Farmers' Union also provided representations on those matters and 'added concerns over the long-term impact on farmers' ability to reuse affected areas for crops through effects such as heating or drying' (Triton Knoll 5.1.21). In response to those concerns, combined with other points raised that related to cable laying including issues of tranquillity and other landowners' interests, the ExA recommended 'that no works shall commence until the SoS [Secretary of State] had confirmed in writing that all necessary consents for the connection have been granted' (Rampion 5.1.34).

Conclusions

The rural landscape has become a prominent 'lens' within which to examine energy debates internationally. Renewable energy infrastructure is a physical manifestation of rural change and a symbolically visible marker of the provision and societal uses of energy (Nadaï and van der Horst, 2010a, 2010b). In the UK, the reasoning for increasing renewable energy generating capacity is infused with a powerful narrative of helping to fight climate change. The use of rural land for major REI is thus framed as an alternative means to 'high carbon' forms of production. The section on the NSIP regime highlights strong narratives of carbon reduction, and potential for 'green-grabbing'. Planning that would consider equity in uses of the countryside must be able to learn about and consider rural concerns, not simply be driven by national or other higher tier concerns. This is not to suggest some type of automatic privilege of local rural concerns over wider need; on the contrary it invokes the argument that *both* conservation and use of rural land are crucial. This raises the question of how to plan for rural communities such that the impacts on rural localities are given adequate weight. To shed light on this, the previous section unpacked how the concerns of rural communities might be articulated within national regulatory processes of planning.

   The study of representations made by rural communities in view of proposed NSIPs, demonstrate how hard IPs have to fight to establish the importance of their concerns, and seek mitigation for impacts. Local individuals encounter a powerful national voice in the NPS and well-resourced developers who are more equipped to deliver evidence of the sort that finds traction within regulatory processes. When local communities were aligned with a civic organisation their points were more clearly heard within the planning examination; however, concerns were most effectively raised where they could establish a connection either to local planning policy or protection of areas of national designation. Thus, local authorities had an important role in helping support or refute rural communities concerns and in negotiating mitigation with developers. This suggests that, while active citizenry is essential to articulating community views, rural planning is important in ensuring 'upward recognition' of rural concerns over REI within the hierarchy of decision-making, especially when decisions are made at the national tier of statutory planning. As such, there is a reliance on local communities and local authorities to give voice to rural concerns within the NSIP examinations.

The position of 'the rural' within the planning processes for REI has been fundamentally shifted by the upscaling of decision-making. Identifying local concerns and relevant current rural plans and policies when a development is proposed appear to be the key means to keeping rural planning in the frame. However, in the NSIP regime rural planning actors have limited capacities to engage, and their voice is a secondary consideration, particularly in relation to national policy. Thus the centralisation of decision-making on major REI presents a serious challenge for rural planning. This remains true despite the devolution of certain aspects (as noted earlier), and is also an important consideration for other types of infrastructure such as transport and water works. These too are most likely to be developed in rural settings yet justified through regional and national priorities that, at least in the UK, are determined 'upstream' of regulation. This is important, since the impacts of major infrastructure development on the countryside are most likely to be considered when deciding development applications. There are two possible responses to such a situation. Firstly, local rural planning actors, can seek to engage with the higher tier arenas of decision-making. However such action does not ensure that the use of countryside land is strategically assessed, i.e. the value of finite rural resources and the costs of their uses are not an explicit consideration in planning at all (or in the case of REI bracketed out under the reasoning of 'mitigating climate change') and likewise the totality of the impacts on the rural population remain unknown. Therefore as a second response, rural planners might seek *post hoc* aggregate assessment and monitoring of rural impacts. In the present context, it appears that this will be critical to ensuring that rural planning concerns are not overlooked.

Note

1 This work was conducted at UCL and ESRC-funded. It was completed in December 2017, and findings and outputs are available through the project archive at www.ucl.ac.uk/nsips.

References

Apostolopoulou, E. and Adams, W. M. (2015). Neoliberal Capitalism and Conservation in the Post-crisis Era: The Dialectics of 'Green' and 'Un-green' Grabbing in Greece and the UK. *Antipode*, 47(1), 15–35.

Bell, D., Gray, T. and Haggett, C. (2005). The 'Social Gap' in Wind Farm Siting Decisions: Explanations and Policy Responses. *Environmental Politics*, 14(4), 460–477.

Bell, D., Gray, T., Haggett, C. and Swaffield, J. (2013). Re-visiting the 'Social Gap': Public Opinion and Relations of Power in the Local Politics of Wind Energy. *Environmental Politics*, 22(September), 115–135.

Burkett, P. (1999). Nature's 'Free Gifts' and the Ecological Significance of Value. *Capital and Class*, 23(2), 89–110.

Cowell, R. (2010). Wind Power, Landscape and Strategic, Spatial Planning: The Construction of 'Acceptable Locations' in Wales. *Land Use Policy*, 27(2), 222–232.

Dalal-Clayton, B., Dent, D. and Dubois, O. (2002). *Rural Planning in Developing Countries: Supporting Natural Resource Management and Sustainable Development*. London: Earthscan.

DCLG and DEFRA. (2016). *Rural Planning Review: Call for Evidence*. London: The Stationery Office.

DECC. (2011a). *National Policy Statement for Renewable Energy Infrastructure (EN-3)*. London: Her Majesty's Stationery Office.

DECC. (2011b). *Overarching National Policy Statement for Energy (EN-1)*. London: Her Majesty's Stationery Office.

DEFRA. (2015). *Towards a One Nation Economy: A 10-Point Plan for Boosting Rural Productivity*. London: The Stationery Office.

Gallent, N., Hamiduddin, I., Juntti, M., Kidd, S. and Shaw, D. (2015). *Introduction to Rural Planning*, 2nd edn. London: Routledge.

Huddleston, J. (Ed.). (2010). *Understanding the Environmental Impacts of Offshore Windfarms*. Oxford: COWRIE.

Lee, M. (2017). Knowledge and Landscape in Wind Energy Planning. *Legal Studies*, *37*(1), 3–24.

Lennon, M. (2015). Green Infrastructure and Planning Policy: A Critical Assessment. *Local Environment*, *20*(8), 957–980.

Marsden, T., Murdoch, J., Lowe, P., Munton, R. and Flynn, A. (1993). *Constructing the Countryside: An Approach to Rural Development*. London: UCL Press.

Marshall, T. and Cowell, R. (2016). Infrastructure, Planning and the Command of Time. *Environment and Planning C: Government and Policy*, *34*(8), 1843–1866.

Meir, R., Legerton, M. L., Anderson, M. B., Berry, B., Bullmore, A., Hayes, M., . . . Warren, J. (1996). *The Assessment and Rating of Noise from Wind Farms*. Didcot: ITSY.

Nadaï, A. and van der Horst, D. (2010a). Introduction: Landscapes of Energies. *Landscape Research*, *35*(2), 143–155.

Nadaï, A. and van der Horst, D. (2010b). Wind Power Planning, Landscapes and Publics. *Land Use Policy*, *27*(2), 181–184.

Natarajan, L., Rydin, Y., Lock, S. J. and Lee, M. (2018). Navigating the Participatory Processes of Renewable Energy Infrastructure Regulation: A 'Local Participant Perspective' on the NSIPs Regime in England and Wales. *Energy Policy*, 114, 201–210.

Rydin, Y., Natarajan, L., Lee, M. and Lock, S. (2017). Artefacts, the Gaze and Sensory Experience: Mediating Local Environments in the Planning Regulation of Major Renewable Energy Infrastructure in England and Wales. In M. Kurath, M. Marskamp, J. Paulos and J. Ruegg (eds), *Relational Planning: Tracing Artefacts, Agency and Practices*, 51–74. New York: Springer.

TSO. (2008). *Planning Act 2008*. London: The Stationery Office.

TSO. (2011). *Localism Act 2011*. London: The Stationery Office.

Vidal, J. (2008). The Great Green Land Grab. *The Guardian*, 13 February, 1–7.

Wargent, M. and Parker, G. (2018). Re-imagining Neighbourhood Governance: The Future of Neighbourhood Planning in England. *Town Planning Review*, 89(4), 379–402.

Welsh Government. (2005). *Technical Advice Note 8: Planning for Renewable Energy*. Cardiff: National Assembly for Wales.

47

Hydraulic fracturing in rural communities

Local realities and resistance

Michiel Köhne and Elisabet Dueholm Rasch

Introduction

This chapter examines impacts of hydraulic fracturing on rural communities. Hydraulic fracturing, commonly referred to as 'fracking', is a technique for recovering shale gas or oil from underground shale rock layers using pressurised water loaded with chemicals and sand to 'stimulate' the underground to release its resources (Norris et al., 2016). While hydraulic fracturing as an extraction technique has been used for decades to extract oil and gas, it was only in the late 1990s that it was combined with horizontal drilling, opening up the exploitation of deep shale rock layers. In 2004 shale gas was first extracted in Pennsylvania, soon becoming an epicentre of the American shale boom (Willow, 2016). In Europe, exploration licenses have been granted in several countries (France, UK, Poland, Romania and the Netherlands, among others) from 2010 onwards, but following environmental protests most governments have recently put a moratorium on hydraulic fracturing (e.g. Vesalon and Crețan, 2015) in contrast to the United States, Canada and Queensland, Australia, where shale oil and gas development has proceeded relatively unchallenged, despite ongoing protests (Eaton and Kinchy, 2016; Simonelli, 2014; Australian Government, 2015). In the United Kingdom, initial policies protecting local communities from seismic harm have now been replaced by more pro-industry policies curtailing community empowerment in fracking decision-making, increasing environmental risks to communities, and transferring powers from local to central government (Cotton, 2017).

Hydraulic fracturing as an extraction method has far-reaching impacts on the lived environment in terms of water, traffic, landscape and health. These impacts, together with the meaning that people give to natural resources shape people's perceptions of fracking, their positioning and how they may mobilise against it. These local impacts will be discussed in the first section. As plans for hydraulic fracturing are often resisted by people living near (proposed) extraction sites, these social dynamics and their relation to democracy is the focus of the second section. In a short third section, we discuss how impacts of, and social mobilisation against, shale gas extraction can be compared to the social dynamics and politics of renewable energy projects.

Arguments and impacts

This section discusses, first, the arguments that are used in support of hydraulic fracturing and, second, the negative impacts of this extraction method for rural communities. How these impacts are experienced, is shaped by how residents of (proposed) extraction areas relate to place and natural resources.

Hydraulic fracturing as an economic resource

The shale gas industry and its supporters draw primarily on a neoliberal framework to justify shale gas extraction. It is argued that shale gas extraction brings direct and indirect (local) economic benefits, such as the construction of superior roads, employment, increased tax revenues and improved education (Willow, 2016). A second argument is that the development of domestic shale resources is essential for energy independence of Europe as well as the United States (Bomberg, 2017). In addition, natural gas is presented as a cleaner alternative to coal (Finewood and Stroup, 2012) and as such seen as a way to facilitate the transition towards renewable energy. In Europe, it is argued that shale gas extraction will be safer than in the United States as it would be based on caution, sound science and robust regulation (Metze, 2017). Employment opportunities and wider economic benefits from associated economic activities, land lease income or infrastructure development, and royalties have been listed as possible local economic benefits (Hirsch et al., 2018; Fleming and Measham, 2015).

Industry and policymakers often underline the monetary value of natural resources to justify extraction processes. This emphasis can also be found at the local level. People see working in industries as a source of income, as a way of sustaining their livelihoods (Willow et al., 2014). In our own research in the Noordoostpolder (the Netherlands), some people would for example say: 'It's there, if we can find a way to get it out in a clean way, then there is no reason why we should not do so, to make a profit, as a way forward' or 'It will just bring some activity, . . . and that is something that just needs to be well organised.' This clearly resonates the neoliberal framework that is used to justify hydraulic fracturing on the national level (Willow, 2016).

Negative impacts

Studies show that economic benefits associated with fracking are often overstated (Barth, 2013), that greenhouse gas emissions are higher than originally presumed (Howarth et al. 2011) and that fracking significantly compromises the quality and availability of water (Entrekin et al., 2011). In recent publications, health effects of fracking are highlighted (see, among others, Hirsch et al., 2018). For example, many mobile gas/oil workers, as well as residents of affected communities, develop neurological and neuropsychological symptoms as a result of exposure to toxic chemicals and contaminated water. Furthermore, inhabitants of fracking communities experience a wide range of illnesses (e.g., increased rates of cancer) and physical symptoms, such as fatigue, headaches and dermatologic irritation (Hirsch et al., 2018).

Fracking also affects people's mental health. Residents worry about lifestyle, health, safety, and financial security, as well as about changes to the landscape. Such fears often become manifest through feelings of anger, anxiety and helplessness. Several studies have demonstrated that collective trauma may arise as a result of how industries impinge on community lives and intensify local divisions (Hirsch et al., 2018). In a case study of Lancashire (UK), a respondent is cited saying:

It's been really, really hard here. We have had 17 months of intense pressure as a Community as a result of this fracking application. Many of my team are now under the Doctor for anxiety stress, sleep issues and it's having such a detrimental effect on our quality of life.

(Short and Szolucha, 217: 6)

Mental effects are not limited to the actual exploration and extraction phase. Several studies also show that in the planning phase community members report an increased sense of powerlessness, fear, betrayal, anger, and stress among other symptoms (ibid.). These relate to the emotional ups and downs experienced in the various stages of the decision-making process, together with feelings of undue corporate influence and disillusionment with the political process. Community members fear the insecure: the unknown effects of chemicals on soil and water safety, the unknown effect of living on fracked land on land and housing prices, jobs and future livelihoods and the contested political process. During our research in the Noordoostpolder, people often highlighted that not knowing about the consequences of fracking was their prime motivation to organise against shale gas.

Fracking as a mining technique requires a dense grid of exploration-sites across property boundaries – as a result, the impacts of fracking do not discriminate in how they affect the lives of local people (Willow, 2014). However, in the United States, fracking often does occur in poorer, rural areas (Castelli, 2015). More generally, fracking can affect social cohesion because conflicts may arise between opponents and proponents of shale gas developments (Grubert and Skinner, 2017). In a community in Appalachia (Unites States) for instance, pre-existing tensions between long-time residents and newcomers exacerbated over the arrival of fracking to their community (Morrone et al., 2015). In other cases, the influx of (often male) workers can cause tensions because of real or perceived increases in crime, prostitution and alcoholism. At the same time, workers that come from outside are often isolated from and not supported by local residents (Hirsch et al., 2018).

Place attachment

The impacts of fracking described above affect people's quality of life which partly determines their views on fracking (Hirsch et al., 2018). Perceptions of fracking are further shaped by demographic factors (e.g. Brasier et al., 2011; Jacquet, 2012), environmental attitudes (e.g. Davis and Fisk 2014), political party preference and media effects (e.g. Boudet et al., 2014). In addition, the meaning that people give to land, landscape and natural resources – that is their 'place attachment' and belonging – mediates the way they experience (possible) effects of shale gas developments.

A good example of how this works can be found in our research in the Noordoostpolder. The Noordoostpolder was reclaimed from the sea in the 1940s and transformed into productive farmland by workers that hoped to improve their chances to be selected for a farm. These first inhabitants of the polder were called pioneers: a powerful figure that is associated with entrepreneurship, heroism, rationality and a strong voice. Both the risks of pollution associated with fracking and the experiences of political exclusion are considered as incompatible with this pioneer past. As John, a pioneer-son, explained to us, while looking out over the polder: 'our parents have reclaimed this land from the sea with their bare hands, so how can you even think of placing a drilling rig here?' Similar dynamics have also been reported from case studies in Ohio's shale gas country, where people feared losing their connections with the immediate environment as the land they had learned to appreciate for recreational and aesthetic reasons became a harbinger of illness, traumatic stress, and anxiety (Willow et al., 2014).

Resistance to fracking

Plans for hydraulic fracturing are often met with resistance from communities nearby (proposed) extraction sites because of the negative impacts discussed above (see, among many others, the work of de Rijke, 2013a, 2013b; Hudgins and Poole, 2014; Rasch and Köhne, 2016). In cases where fracking has not been resisted actively, this can often be attributed to a lack of organisational capacity and political opportunities, or to a choice to engage in individualist efforts to stop fracking that have failed, rather than to consent (Olive and Valentine, 2018). In what follows we discuss the most important characteristics of resistance to hydraulic fracturing developments.

Actors and alliances

Resistance to hydraulic fracturing is characterised by the involvement of many different actors and unexpected alliances that work together against shale or coal seam gas (Mercer et al., 2014; Colvin et al., 2015; Sherval et al., 2018). In the Noordoostpolder, for example, the partnership against shale gas seeks to represent all social layers of the municipality: it consists of local entrepreneurs, a left-wing environmentalist and a more conservative opposition group, village representatives, farmers and horticulturalists, the housing corporation, the municipal government and the health sector (see Rasch and Köhne, 2016 for further analysis). This way of organising blurs the lines between professionals, entrepreneurs and activists, as well as between local governments and social movements. In addition, local (activist) groups as well as (inter)national environmentalists link local concerns to cosmopolitan movements working on energy justice and climate change (Hopke, 2016; Rice and Burke, 2018).

In many cases, people that organise against fracking do not identify as activists, but consider themselves to be people who 'just want to solve a problem'. They engage in acts of resistance, because they see this as the only possible course of action left. In so doing, they often – but not always - continue to view activism as opposite to their own (often conservative) beliefs (Willow, 2014; Gullion, 2015). They do, however, work together with activist groups. In Australia, for example, landholders of the Tara Estate worked together with Friends of the Earth researching fracking in Australia and with the national activists group Lock the Gate in order to connect local community groups (Lloyd et al., 2013).

Social justice and local voice

Recent studies in Europe show that citizens that live near (proposed) fracking sites not only experience threats to the environment and their community, but also to democracy in general. The decision-making process related to fracking often results in a loss of trust in both industry and politicians, because they exclude the affected communities from participation (Thomas, 2017). Land use conflicts related to fracking are thus not only about the negative environmental and health related impacts, but also about who gets to decide about land use. As a consequence, resistance groups often formulate claims that are related to social justice and participation (see, for example, Thomas et al., 2017); opposition is not only directed at fracking itself (substantial values), but also at the process surrounding its regulation (procedural values) (Bomberg, 2015). Acts of resistance therefore often claim a local voice in decision-making (Rasch and Köhne, 2017) or at least a dialogue (Whitton et al., 2017).

Acts of resistance

Acts of resistance towards fracking can be subdivided into different categories: playful actions and media attention, political lobbying and legal resistance, knowledge production and knowledge dissemination.

The production of locally trusted knowledge about (possible) impacts of fracking is an important act of resistance (Rasch and Köhne, 2017) because the lack of information makes people feel excluded (Devey et al., 2014; Bec et al., 2016). Whereas the partnership against shale gas in the Noordoostpolder hired a consultancy firm to study the possible economic consequences of fracking for the region, communities themselves may also engage in different ways of knowledge production about the consequences of fracking. This citizen science approach includes local monitoring of, for example, water streams. Such knowledge could challenge the fracking boom (Matz et al., 2017), but channelling people's activities towards possible controversies over data, may also distract from political action (Kinchy, 2017). Opposition groups also engage in knowledge production by way of undertaking internet research, reading (scientific) articles and engaging with social media. In the Noordoostpolder, this production of knowledge, together with the dissemination of it, was the most important act of resistance.

Closely related to the production of knowledge as an act of resistance, is the dissemination of it through local information meetings (Lis and Stasik, 2017), songs (Highby, 2014) or film screenings (Vasi et al., 2015). In the Dutch case of Noordoostpolder, the partnership against shale gas organised information meetings in collaboration with village representatives. The knowledge presented was to be as neutral, measurable, scientific, and objective as possible. Such 'objective' knowledge, preferably expressed in numbers, is considered more effective to show how 'bad' or wrong fracking is, than intangible knowledges about place attachment and social meanings of natural resources. The rationale behind the focus on neutral knowledge and information is to enable people to make sound judgements on possible shale gas developments (Rasch and Köhne, 2017).

Different studies show that resistance groups often engage in 'playful acts' to draw attention of the public and to get media attention. A well-known playful form of resistance in many locations in Australia is 'the knitting nanna's': women sit down and knit at protests and blockades as a form of productive, peaceful protest (Larri and Newlands, 2017). In our research in the Noordoostpolder, we also witnessed a myriad of playful actions, among which the construction of a symbolic drill tower out of wood and participation in the 10 villages run in support of the partnership against shale gas.

Another important act of resistance is the lobbying of local and national politicians. In the Noordoostpolder, the partnership against shale gas not only worked extensively to influence politicians in their own local and regional networks, but also paid visits to both ministry and parliament. On several occasions, national politicians were invited to the Noordoostpolder and given a tour along what the partnership considered 'green projects', the message being: 'don't invest in shale gas, but in renewable energy'. Opposition groups also engage in what can be called 'legal resistance' (Whitton et al., 2017; Rice and Burke, 2018), such as challenging governmental and company actors before the court.

In some cases, acts of resistance are directly related to the fixed location of shale and coal seam gas and its extraction; people have resisted fracking physically in different ways. Landowners in Australia simply locked their gates in order to frustrate company access (de Rijke, 2013a, 2013b). In other cases, communities block access to fracking sites by way of building camps in front of access gates (Lis and Stasik, 2017; Steger and Milicevic, 2014).

Finally, a way of claiming participation in the decision-making process, is by municipalities and towns declaring themselves shale gas free. In 2013, the Noordoostpolder municipal council declared the municipality 'shale gas free'. In the Netherlands, one third of all municipalities declared themselves shale gas free. At a more local level, villages and towns also voted against shale-gas; the information evenings described above were always closed by a vote in favour or against shale gas. Although such self-organised consultations and declarations are not legally binding, they are used to legitimate participation in the decision-making process.

Hydraulic fracturing and renewable energy projects

The contestations and conflicts related to fracking in rural communities cannot be isolated from broader discussions about (large-scale) renewable energy projects and their impacts on rural communities (see also Chapter 46 of this Companion). The way in which these debates relate is threefold. On the one hand, as we discussed earlier this chapter, fracking is resisted because it is considered opposite to the transition to renewable energy. On the other hand, renewable energy projects are often resisted because they are perceived to have the same negative impacts on rural communities as fracking does, notwithstanding the fact that renewable energy technologies are often idealised as green or environmentally innocent alternatives to fossil fuels (Ottinger, 2013). Finally, engaging in anti-fracking resistance may evolve into a broader reflection about sustainability and renewable energy.

Negative impacts of renewable energy projects

Wind farms, solar parks and hydroelectric power dams can have several negative impacts on rural communities that resemble the consequences of fracking that we discussed above. As is the case with fracking, the (monetary) benefits of such projects mostly go to national governments and companies, while the communities suffer the negative impacts of such projects (Pepermans and Loots, 2013). Wind farms have further been resisted because of health effects (Pedersen and Waye, 2007; Schmidt and Klokker, 2014) and its effects on surrounding flora and fauna (Pasqualetti, 2011). In addition, renewable energy projects may also impact on the social cohesion of communities. A case study on Mexico for example shows how resistance against wind parks led to a violent conflict as the government used not only promises but also, violence and divisionary tactics to gain control over land for the development of a wind park (Dunlap, 2018). Finally, people fear the impact of wind parks on their experience of the landscape; wind parks often do not fit with the meaning people attach to their surroundings and the rural landscape as beautiful and restorative (Devine-Wright and Howes, 2010; Pasqualetti, 2011).

Local ownership and voice

Negative impacts of large-scale renewable energy projects also include complaints about the decision-making process, just as in fracking. In the case of wind parks, communities have been excluded from participation in decision-making processes as well, resulting in a sceptical attitude towards such projects (see, for example, Ottinger, 2013). At the same time, studies show that the building of wind parks is more easily accepted when communities engage in decision-making processes about wind energy (McLaren Loring, 2007) or when the wind park becomes a community endeavour (Walker, 2008).

From anti-fracking resistance to renewable energy production and awareness

Engagement in resistance against fracking may also result in reflections about renewable energy and the production of it. In the Noordoostpolder, some people started to become more reflective about their own personal consumption of energy because of their involvement with the shale gas resistance. Others started to develop activities in the realm of producing renewable energy, like putting solar panels on the roofs of their barns. In these cases processes of reflection, rooted in resistance to fossil fuel extraction, produce new norms and ideas of what are 'good' and 'just' way of producing energy.

Conclusion

The impacts of fracking on rural communities as well as the character of the resistance against it should be understood in the broader context of (resistance against) large-scale renewable energy projects. Fracking, although often framed as a resource, greener than coal and a step towards the transition to renewable energy, has many negative impacts on rural communities. The way these impacts are experienced and perceived, is mediated by the meaning that people attach to land and natural resources; their place attachment. In addition, fracking causes tensions within communities and engender feelings of exclusion and disempowerment because people in communities near (proposed) extraction sites are not allowed to participate in decision making processes related to resource extraction. Resistance towards fracking, then, is not only about the right to a clean and healthy environment, it is also about social justice and local ownership, about who gets to decide on land use.

References

Australian Government. (2015). *Review of the Socioeconomic Impacts of Coal Seam Gas in Queensland*. Canberra: Office of the Chief Economist, Department of Industry, Innovation and Science.

Barth, J. M. (2013). The Economic Impact of Shale Gas Development on State and Local Economies: Benefits, Costs, and Uncertainties. *New Solutions: A Journal of Environmental and Occupational Health Policy* 23(1), 85–101.

Bec, A., B. D. Moyle and J. M. Char-lee (2016). Drilling into Community Perceptions of Coal Seam Gas in Roma, Australia. *The Extractive Industries and Society* 3(3), 716–726.

Bomberg, E. (2017). Shale We Drill? Discourse Dynamics in UK Fracking Debates. *Journal of Environmental Policy & Planning* 19(1), 72–88.

Boudet, H., C. Clarke, D. Bugden, E. Maibach, C. Roser-Renouf and A. Leiserowitz (2014). 'Fracking' Controversy and Communication: Using National Survey Data to Understand Public Perceptions of Hydraulic Fracturing. *Energy Policy* 65, 57–67.

Brasier, K. J., M. R. Filteau, D. K. McLaughlin, J. Jaquet, R. C. Stedman, T. W. Kelsey, and S. J. Goetz. (2011). Residents' Perceptions of Community and Environmental Impacts from Development of Natural Gas in the Marcellus Shale: A Comparison of Pennsylvania and New York Cases. *Journal of Rural Social Science* 26(1), 32–61.

Castelli, M. (2015). Fracking and the Rural Poor: Negative Externalities, Failing Remedies, and Federal Legislation. *Indiana Journal of Law and Social Equality* 3(2), 6.

Colvin, R. M., G. B. Witt and J. Lacey (2015). Strange Bedfellows or an Aligning of Values? Exploration of Stakeholder Values in an Alliance of Concerned Citizens against Coal Seam Gas Mining. *Land Use Policy* 42, 392–399.

Cotton, M. (2017). Fair Fracking? Ethics and Environmental Justice in United Kingdom Shale Gas Policy and Planning. *Local Environment* 22(2), 185–202.

Davis C. and Fisk, J. (2014) Energy abundance or environmental worries? Analyzing public support for fracking in the United States. *Review of Policy Research* 31, 1–16

De Rijke, K. (2013a). The Agri-Gas Fields of Australia: Black Soil, Food, and Unconventional Gas. *Culture, Agriculture, Food and Environment* 35(1), 41–53.

De Rijke, K. (2013b). Hydraulically fractured: Unconventional gas and anthropology. *Anthropology Today* 29(2), 13–17.

Devey, S., V. Goussev, B. Schwarzenburg and M. Althaus (2014). Shale Gas U-Turns in Bulgaria and Romania: The Turbulent Politics of Energy and Protest. *Journal of European Management and Public Affairs Studies* 1, 47–60.

Devine-Wright, P. and Y. Howes (2010). Disruption to Place Attachment and the Protection of Restorative Environments: A Wind Energy Case Study. *Journal of Environmental Psychology* 30(3), 271–280.

Dunlap, A. (2018). Counterinsurgency for Wind Energy: The Bíi Hioxo Wind Park in Juchitán, Mexico. *The Journal of Peasant Studies* 45(3), 630–652.

Eaton, E. and A. Kinchy (2016). Quiet Voices in the Fracking Debate: Ambivalence, Nonmobilization, and Individual Action in Two Extractive Communities (Saskatchewan and Pennsylvania). *Energy Research & Social Science* 20, 22–30.

Entrekin, S., Evans-White, M., Johnson, B. and Hagenbuch, E. (2011). Rapid Expansion of Natural Gas Development Poses a Threat to Surface Waters. *Frontiers in Ecology and the Environment* 9(9), 503–511.

Espig, M. and K. de Rijke (2016). Unconventional Gas Developments and the Politics of Risk and Knowledge in Australia. *Energy Research & Social Science* 20, 82–90.

Finewood, M. H. and L. J. Stroup (2012). Fracking and the Neoliberalization of the Hydro-Social Cycle in Pennsylvania's Marcellus Shale. *Journal of Contemporary Water Research & Education* 147(1), 72–79.

Fleming, D. A. and T. G. Measham (2015). Local Economic Impacts of an Unconventional Energy Boom: The Coal Seam Gas Industry in Australia. *Australian Journal of Agricultural and Resource Economics* 59(1), 78–94.

Grubert, E. and W. Skinner (2017). A Town Divided: Community Values and Attitudes towards Coal Seam Gas Development in Gloucester, Australia. *Energy Research & Social Science* 30, 43–52.

Gullion, J. S. (2015). *Fracking the Neighborhood: Reluctant Activists and Natural Gas Drilling*. Cambridge, MA: MIT Press.

Highby, W. (2014). Critiquing and Conveying Information about Fracking through Song Parody: The Annotated Libretto of Frackville, the Horizontally Drilled Musical, with Interpolated Bibliography. *Progressive Librarian* 43, 81.

Hirsch, J. K., et al. (2018). Psychosocial Impact of Fracking: A Review of the Literature on the Mental Health Consequences of Hydraulic Fracturing. *International Journal of Mental Health and Addiction* 16(1), 1–15.

Hopke, J. E. (2016). Translocal Anti-Fracking Activism: An Exploration of Network Structure and Tie Content. *Environmental Communication* 10(3), 380–394.

Howarth, R. W., A. Ingraffea and T. Engelder (2011). Natural Gas: Should Fracking Stop? *Nature* 477(7364), 271.

Hudgins, A. and A. Poole (2014). Framing Fracking: Private Property, Common Resources, and Regimes of Governance. *Journal of Political Ecology* 21(1), 303–319.

Jacquet, J. B. (2012). Landowner Attitudes toward Natural Gas and Wind Farm Development in Northern Pennsylvania. *Energy Policy* 50, 677–688.

Kinchy, A. (2017). Citizen Science and Democracy: Participatory Water Monitoring in the Marcellus Shale Fracking Boom. *Science as Culture* 26(1), 88–110.

Larri, L. J. and M. Newlands (2017). Knitting Nannas and Frackman: A Gender Analysis of Australian Anti-Coal Seam Gas Documentaries (CSG) and Implications for Environmental Adult Education. *The Journal of Environmental Education* 48(1), 35–45.

Lis, A. and A. K. Stasik (2017). Hybrid Forums, Knowledge Deficits and the Multiple Uncertainties of Resource Extraction: Negotiating the Local Governance of Shale Gas in Poland. *Energy Research & Social Science* 28, 29–36.

Lloyd, D. J., H. Luke and W. E. Boyd (2013). Community Perspectives of Natural Resource Extraction: Coal-Seam Gas Mining and Social Identity in Eastern Australia. *Coolabah* 10, 144.

Matz, J. R., S. Wylie and J. Kriesky (2017). Participatory Air Monitoring in the Midst of Uncertainty: Residents' Experiences with the Speck Sensor. *Engaging Science, Technology, and Society* 3, 464–498.

McLaren Loring, J. (2007). Wind Energy Planning in England, Wales and Denmark: Factors Influencing Project Success. *Energy Policy* 35(4), 2648–2660.

Mercer, A., K. de Rijke and W. Dressler (2014) Silences in the Midst of the Boom: Coal Seam Gas, Neoliberalizing Discourse, and the Future of Regional Australia. *Journal of Political Ecology* 21, 279–302.

Metze, T. (2017). Fracking the Debate: Frame Shifts and Boundary Work in Dutch Decision Making on Shale Gas. *Journal of Environmental Policy & Planning* 19(1), 35–52.

Morrone, M., A. E. Chadwick and N. Kruse (2015). A Community Divided: Hydraulic Fracturing in Rural Appalachia. *Journal of Appalachian Studies* 21(2), 207–228.

Olive, A. and K. Valentine (2018). Is Anyone Out There? Exploring Saskatchewan's Civil Society Involvement in Hydraulic Fracturing. *Energy Research & Social Science* 39, 192–197.

Ottinger, G. (2013). The Winds of Change: Environmental Justice in Energy Transitions. *Science as Culture* 22(2), 222–229.

Pasqualetti, M. J. (2011). Opposing Wind Energy Landscapes: A Search for Common Cause. *Annals of the Association of American Geographers* 101(4), 907–917.

Pedersen, E. and K. P. Waye (2007). Wind Turbine Noise, Annoyance and Self-Reported Health and Well-being in Different Living Environments. *Occupational and Environmental Medicine* 64(7), 480–486.

Pepermans, Y. and I. Loots (2013). Wind Farm Struggles in Flanders Fields: A Sociological Perspective. *Energy Policy* 59, 321–328.

Rasch, E. D. and M. Köhne (2016). Hydraulic Fracturing, Energy Transition and Political Engagement in the Netherlands: The Energetics of Citizenship. *Energy Research & Social Science* 13, 106–115.

Rasch, E. D. and M. Köhne (2017). Practices and Imaginations of Energy Justice in Transition: A Case Study of the Noordoostpolder, the Netherlands. *Energy Policy* 107, 607–614.

Rice, J. L. and B. J. Burke (2018). Building More Inclusive Solidarities for Socio-Environmental Change: Lessons in Resistance from Southern Appalachia. *Antipode* 50(1), 212–232.

Schmidt, J. H. and M. Klokker (2014). Health Effects Related to Wind Turbine Noise Exposure: A Systematic Review. *PLoS One* 9(12), e114183.

Sherval, M., H. H. Askland, M. Askew, J. Hanley, D. Farrugia, S. Threadgold and J. Coffey (2018). Farmers as Modern-Day Stewards and the Rise of New Rural Citizenship in the Battle Over Land Use. *Local Environment* 23(1), 100–116.

Short, D. and Szolucha, A. (2017). Fracking Lancashire: The Planning Process, Social Harm and Collective Trauma. *Geoforum* (in press), doi.org/10.1016/j.geoforum.2017.03.001.

Simonelli, J. (2014). Home Rule and Natural Gas Development in New York: Civil Fracking Rights. *Journal of Political Ecology* 21(1), 258–278.

Smith, M. F. and D. P. Ferguson (2013). Fracking Democracy: Issue Management and Locus of Policy Decision-Making in the Marcellus Shale Gas Drilling Debate. *Public Relations Review* 39(4), 377–386.

Steger, T. and M. Milicevic (2014). One Global Movement, Many Local Voices: Discourse (s) of the Global Anti-Fracking Movement. In L. Leonard and S. B. Kedzior (eds), *Occupy the Earth: Global Environmental Movements*, 1–35. Bingley: Emerald Group Publishing.

Thomas, M., N. Pidgeon, D. Evensen, T. Partridge, A. Hasell, C. Enders, B. Harthorn and M. Bradshaw (2017) Public Perceptions of Hydraulic Fracturing for Shale Gas and Oil in the United States and Canada. *WIRES Climate Change* 8(3), e450.

Vasi, I. B., E. T. Walker, J. S. Johnson and H. F. Tan (2015). No Fracking Way! Documentary Film, Discursive Opportunity, and Local Opposition against Hydraulic Fracturing in the United States, 2010 to 2013. *American Sociological Review* 80(5), 934–959.

Vesalon, L. and R. Crețan (2015). 'We Are Not the Wild West': Anti-Fracking Protests in Romania. *Environmental Politics* 24(2), 288–307.

Walker, G. (2008). What are the Barriers and Incentives for Community-Owned Means of Energy Production and Use? *Energy Policy* 36(12), 4401–4405.

Whitton, J., K. Brasier, I. Charnley-Parry and M. Cotton (2017). Shale Gas Governance in the United Kingdom and the United States: Opportunities for Public Participation and the Implications for Social Justice. *Energy Research & Social Science* 26, 11–22.

Willow, A. J. (2014). The New Politics of Environmental Degradation: Un/expected Landscapes of Disempowerment and Vulnerability. *Journal of Political Ecology* 21(1), 237–257.

Willow, A. J. (2016). Wells and Well-being: Neoliberalism and Holistic Sustainability in the Shale Energy Debate. *Local Environment* 21(6), 768–788.

Willow, A. J., R. Zak, D. Vilaplana and D. Sheeley (2014). The Contested Landscape of Unconventional Energy Development: A Report from Ohio's Shale Gas Country. *Journal of Environmental Studies and Sciences* 4(1), 56–64.

48

Mineral extraction and fragile landscapes

Mozart Fazito, Mark Scott and Paula Russell

Introduction: mining, resources and rural regions

There is a long tradition of examining the relationship between mining, extractive industries and communities within the rural studies literature. For example, Tonts et al. (2012) provide a useful overview of the various waves of studies examining the economic and social challenges faced by remote, resource rich/dependent localities since the 1940s, with research intensifying in the 1970s within the context of increasingly volatile global markets. Many of these studies were motivated by understanding the combined effects of finite local resources, rising international competition and global economic crises on resource dependent towns and regions, focusing on the negative impacts of the extractive industry on local communities and economies. Within the global north, studies focused on the boom and bust nature of the mining industry and its impacts on single purpose resource communities, notably the North American 'boomtown' literature (e.g. Kassover and McKeown, 1981; Brown et al., 1989). These studies highlighted the longer-term vulnerability of mining and resource dependent communities in the face of diminishing resources and global market volatility. Subsequent research also investigated the impacts on local communities and economies of mine closures (Neil et al., 1992; McDonald et al., 2012) and of scarred or degraded landscapes following mining activities (Collier and Scott, 2010). Within the global south (and later inspiring research in more developed countries), research often centred on the so-called resource-curse – examining the relationships between resource abundance, economic underperformance and large-scale extractive industries dominated by international corporations (Auty, 1993; Ballard and Banks, 2003; Papyrakis, 2017). This literature draws our attention to the apparent contradiction of resource rich regions with mining complexes that generate huge wealth for owners and national governments, marked by high levels of poverty, social and economic marginalisation, political instability, negative local environmental impacts, and displacement of locals and/or indigenous peoples. While the early 'resource-curse' literature is dominated by economists focused on the macro level, increasingly a micro resource-curse literature has emerged (largely from within anthropology), with a closer focus on the impacts of the extractive industry on individual agency and community relationships and how resource extraction can exacerbate poverty for nearby communities (Hilson, 2010, 2012). As noted by O'Faircheallaigh (2013), local people obtain few of the benefits created by mining (with value being added elsewhere in the supply chain), and the environmental, social and economic effects of extractive industries can threaten existing, viable livelihoods, with indigenous peoples especially susceptible to marginalisation due to reliance on land resources which are often damaged in resource extraction.

With global demand for minerals rising in the 2000s, often related to the rapid expansion of the Chinese manufacturing sector and its export markets, the last 15 years has also witnessed an expanding literature focused on providing new understandings of mining and the extractive sector. Some of this research has continued to shed new light on the impacts of mining in rural regions and communities. This includes new pressures on local housing markets and service provision, particularly common within Australian studies; conflicts between mining and other rural sectors, such as agriculture and tourism (McManus, 2008; Everingham et al., 2015); competition with ecological services (Miller et al., 2012); the effects of transient populations and

Table 48.1 Key themes in literature on mining and rural regions.

Theme	Illustrative examples
Resource boomtowns	Examining the impacts on remote settlements of rapid growth in terms of benefits to local economies (e.g. Measham and Fleming, 2014), shifting community relations (e.g. loss of community control, e.g. Miller et al., 2012) and decline following resource depletion. Much attention focused on the long term vulnerability of resource-dependent localities (e.g. Tonts et al., 2012).
The 'resource curse'	Exploring and explaining the negative impact of natural resource abundance on long-term economic growth at macro scale (e.g. Auty, 1993).
	Increase in studies focused on the local scale resource-curse, and impacts of mining on local livelihoods, poverty and environmental degradation (e.g. Papyrakis, 2017; Hilson, 2010).
Indigenous peoples' rights	Investigating conflicts between mining operations and indigenous peoples' livelihoods (often based on land resources) (e.g. Hilson, 2002).
	Methods to address impacts on indigenous communities through so-called mining payments, employment schemes, and impact-benefit agreements (e.g. O'Faircheallaigh, 2004, 2013; Caine and Krogman, 2010).
Post-mining landscapes and economies	Examining the environmental degradation of mining operations and landscape rehabilitation following closure of mining works (e.g. Collier and Scott, 2008, 2009, 2010).
	Exploring the search for a post-mining viable economy (Neil et al., 1992; McDonald et al., 2012) and transitions to more sustainable economic pathways (e.g. Fazito et al., 2016).
From local to global environmental opposition	Examining local environmental degradation resulting from mining operations (e.g. Conde, 2017).
	Upscaling of environmental concerns to global issues, particularly related to carbon intensive activities and climate change (e.g. coal or new fracking operations) (e.g. Connor et al., 2009).
New governance modes and power	Charting new modes of governance between mining companies and local communities, including the formation of new private-community partnerships (e.g. Cheshire et al., 2014) and models of corporate responsibility (e.g. Cheshire, 2009).
	Models of decentralised decision-making, but also highlighting the emergence of conflicts among local elites (e.g. Arellano-Yanguas, 2011).

new models of 'fly-in, fly-out' mining companies (Cheshire, 2009); and the upscaling of environmental conflicts, from opposing new mining on the basis of local environmental degradation to global concerns of carbon intensive industries and climate change (Connor et al., 2009). Within this more recent literature, researchers have also moved beyond focusing solely on the negative impacts of mining to question the *inevitability* of conflicts and to highlight new governance approaches to mitigate local impacts. For example, Cheshire (2009) highlights new modes of corporate responsibility within mining companies in Australia, almost as a form of local patronage. This has led to the formation of new local partnerships, funding for local and regional projects, the donating of equipment to local schools and hospitals, and indigenous employment initiatives. Also based on Australian research, O'Faircheallaigh (2013) highlights the role of local action in influencing more positive outcomes, highlighting recent experiences of indigenous peoples in Western Australia in mobilising to shape local impacts. However, Arellano-Yanguas (2011) also notes the limitations of decentralised governance modes in relation to resolving mining conflicts in Peru, which resulted in aggravating local political conflicts. These studies also highlight the need to take a more nuanced view of actors involved in resource rich localities, with Cheshire urging researchers to move beyond depicting the mining industry as one dimensional, 'self-interested, [and] greedy' (2009: 12). In this context, local, regional and transnational actors may have complex motivations that shift over time (see also Ballard and Banks, 2003).

In this chapter, we seek to contribute to these debates through examining new mining conflicts in the Espinhaço Range Biosphere Reserve located within the state of Minas Gerais, Brazil. While this region is an historic mining area, during the 1990s to mid-2000s, local and regional policy increasingly promoted sustainable tourism as an alternative for economic growth. This was largely based on a re-valorisation of the region's unique landscape, historic mines and colonial era towns as key resources for tourism exploitation. While tourism was supported by local politicians and environmental actors, by the mid-2000s with the rising global demand for iron ore, mining was again profitable in the region and a mining corporation was successful in displacing tourism to win political support and approval for new mining and related developments. In this chapter, we explore the re-emergence of mining through the lens of environmental justice to examine the *framing* of environmental conflicts, which often centred on competing interpretations of 'development' and 'sustainability', and the legitimacy of different forms of knowledge within decision-making arenas. Empirically, the chapter is based on semi-structured in-depth interviews with 41 stakeholders (families affected by the mine, community leaders, policy-makers, politicians, journalists, environmental activists and consultants), a thorough documentary analysis, and the analysis of the recordings and reports of the State Environmental Policy Council meetings that approved the environmental licences for the mine to operate. The next section contextualises the case study material within the environmental justice literature.

Environmental justice and mineral extraction

Over the last three decades, it has become increasingly apparent that the issue of environmental quality is inextricably linked to that of human equality, equity and social justice, rights and people's quality of life. Agyeman et al. (2003) identify two dimensions of this relationship: firstly, globally, countries with a more equal income distribution, greater civil liberties and political rights tend to have higher environmental quality (e.g. air and water pollution, access to clean water). Secondly, environmental problems bear down disproportionally upon the poor – both within countries and globally. Within this context, environmental justice, both as a movement and theoretical lens, has emerged to link environmental issues to concerns with social justice. As a social movement, environmental justice has its origins in the USA, with prominent campaigns against the siting of toxic, hazardous and polluting industries in areas inhabited by poor and ethnic minority populations (see Taylor, 2000; Agyeman, 2002) – communities perceived to offer the path of least resistance. As recorded by Walker (2012), environmental justice movements quickly spread beyond the USA in the 1990s, and the scope of environmental justice campaigns widened considerably to frame debates surrounding a diverse range of topics from local to global issues – these include, for example: air pollution, waste facilities, contamination, flooding, transport, the distribution of environmental goods (such as urban greenspace), and increasingly climate change campaigns. These ideas have also been applied to mining conflicts around the globe with Urkidi and Walter (2011) providing a useful review of studies relating to mining activities from a spatial and economic distribution perspective.

Early examples of environmental justice campaigns and theoretical development of the concept have generally been framed within a traditional Rawlsian perspective of justice. They have been concerned with the distribution of environmental 'bads' and procedural fairness in this distribution through a focus on governance, 'rules' or standards producing just spatial outcomes – in other words, *how* and *what* gets distributed. Within the context of extraction industries, a focus on distributional aspects tends to consolidate research relating to the negative aspects of resource dependence (the resource-curse) and also studies demonstrating the negative economic, social and environmental impacts at the local level (Urkidi and Walter, 2011).

To advance these debates, environmental justice theory has moved beyond a focus on distribution to develop more holistic conceptual frameworks. For example Schlosberg (2004, 2007, 2013), drawing on the work of Amartya Sen, Iris Young, Nancy Fraser and Axel Honneth, develops a three-pronged approach to examine environmental justice (see Figure 48.1). The three elements are:

- **Distribution**: as outlined above, the distribution of environmental goods and bads has been the traditional way of examining justice – therefore, the central question is: how and to what end, should a just society distribute the various benefits it produces and the burdens required to maintain it? Drawing on John Rawls (1971), this approach encompasses both examining the procedural fairness (*how* goods and bads gets distributed) and outcomes (*what* gets distributed). For Schlosberg, key to procedural fairness is participation referring to political inclusion, empowerment and giving people voice, suggesting that procedural fairness is the most likely path towards fair distribution. Schlosberg emphasises 'authentic participation' enabling the breaking of silence, confronting and talking back to oppressors.
- **Recognition**: drawing on the work of Young (1990), Fraser (1997) and Honneth (1995), Schlosberg emphasises the need to examine the *context* of oppression, which underpins *maldistribution*. This suggests that distributional patterns happen for a reason, and the 'reality of domination and oppression must be taken as a starting point for any thorough and pragmatic theory of justice' (Schlosberg, 2007: 4). In this context, a lack of *recognition* in the social and political realm inflicts damage on oppressed individuals and communities, demonstrated by forms of insults, degradation, and devaluation at both individual and cultural level. Injustice, therefore, is viewed as the outcome of various social and institutional relations, with recognition framed as key to self worth and human dignity, but undermined by disrespect, denial of rights and denigration of ways of life (Honneth, 1995) or through different forms of institutional practice (Fraser, 1997). Whether, and how, individuals and communities are recognised is crucial in this regard.

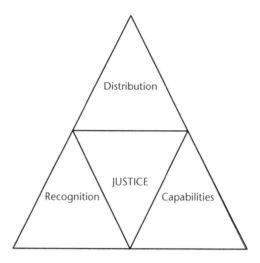

Figure 48.1   Towards an environmental justice framework.

Source: developed from Schlosberg (2007)

- **Capabilities**: drawing on Amartya Sen's (1999) *'Development as Freedom'* approach, Schlosberg proposes moving beyond simple distribution concerns, to focus on how distribution of goods and bads affect our well-being and how we 'function'. Sen (1999) argues that an alternative to the income-oriented view of development is the freedom-oriented view of development. He generalises the concept of development as the process of expansion of real freedoms that people enjoy, with attention paid particularly to the expansion of individuals' capabilities to lead the kind of lives they value. In this regard, Sen develops the concept of 'capabilities' to describe a person's opportunities to do and to be what they choose in the context of a given society, focusing on individual agency, functioning and well-being. In other words, for Sen, we need to move beyond a sole concern with the amount of goods we get, and to examine what those goods do for us and to translate basic goods into the functioning of human life – to examine what environmental goods do for us rather than simply focusing on their distribution.

This three-pronged approach to understanding environmental justice and claims-making will be applied to understanding mining conflicts in a specific global south case introduced in the next section.

Case study: Espinhaço Range Biosphere Reserve

To explore these themes within an emerging economy context, this chapter will examine the case of shifting development priorities within the Espinhaço Range Biosphere Reserve (ERBR) located within the state of Minas Gerais, Brazil. The ERBR has an area of 3,076,457.80 ha, crossing 53 municipalities and hosting 69 conservation units. This is an historic mining region within Brazil, which has hosted two of the most important cycles of economic development in colonial Brazil: the cycle of gold and the cycle of diamonds. Former mining cities, such as Ouro Preto and Diamantina, became historic sites. Colonial mining towns and the natural environment of mountains, forests, and waterfalls and natural springs comprise the attractive landscapes of the ERBR, and therefore its tourist potential. In the 1990s as the mining industry was declining, the state government increasingly sought to exploit this heritage and landscape potential and positioned 'sustainable tourism' as a means to generate economic growth within a framework of environmental conservation. At this time, many former mining sites were reimagined and commodified as 'heritagescapes' (cf. Wheeler, 2014) within state policy and incorporated into promoted tourism trails, such as the 'Circuit of Gold' and 'Circuit of Diamonds' (Minas Gerais, 2011). The objective was to utilise tourism as a means to modernise regions that were no longer reached by traditional industries (Fazito, 2015).

This shift from extractive industries towards amenity and environmental based tourism is particularly marked in our case study area, where a growing environmental awareness increasingly framed debates on future development pathways for the local and regional economy. Throughout the 1990s and into the 2000s, the central region of the Espinhaço Range increasingly focused on tourism as a key development policy, with the town of Conceição do Mato Dentro celebrated as the state 'capital of ecotourism'. The history of Conceição do Mato Dentro (the focus of this study) can be traced to the early eighteenth century, when the first deposits of gold were discovered, and a town was settled around the mining sites. After mining, agriculture and livestock production developed, but only for subsistence purposes among its inhabitants. With the decline of mining, Conceição do Mato Dentro became isolated and less developed than the neighbouring municipalities (Becker, 2009). However, the lack of economic development of Conceição do Mato Dentro led the town to maintain the old colonial buildings and preserve its natural landscapes and cultural traditions. The municipality has a population of 17,908 inhabitants, of which 12,269 (68.51%) are located in the town and 5,639 (31.49%), in the surrounding rural areas.

The promotion of eco-tourism followed re-valorisation of environmental capital in the locality and the establishment of an environmental movement committed to protecting nature and culture, mostly waterfalls and the distinctive landscape. Key environmental actors established the Society of Friends of Tabuleiro (named after a local waterfall) in 1998, which guided the establishment of an ecotourism programme. According to the local commercial association, from 2000 to 2006, the number of small accommodation enterprises (hotels and lodges, mostly) rose from 4 to 27. After getting some international support, environmental actors sought the recognition of UNESCO and the wider region became an internationally recognised biosphere reserve due to its high natural and cultural diversity and being a source of high quality fresh water (Minas Gerais, 2005). It hosts three typical Brazilian ecosystems: *cerrado* (savannah), *caatinga* (semi-arid) and *mata atlântica* (rain forest). *Cerrado* and *mata atlântica* are listed as two of the 25 world 'hotspots' for conservation priority (Myers et al., 2000).

However, in the 2000s, with the growing Chinese demand for iron, mining regained primacy, causing conflict between the newly established sustainable tourism policies and mining in the region. Among the various conflicts caused by the 'new mining wave' of the mid-2000s in the region, the one located in the town of Conceição do Mato Dentro attracted significant attention within the local media due to the size of the mining operation and the fragility and international recognition of the ecosystem. In 2006, transnational mining companies with the support of the State Government of Minas Gerais, launched the Project Minas-Rio, comprising a large iron ore mine, mostly located within Conceição do Mato Dentro, a power line, and a pipeline to transport the iron ore to the state of Rio de Janeiro and the port of Açu. As a result, mining displaced eco-tourism, both as a policy priority for local and regional development among local political elites (Fazito, 2012; Pereira and Becker, 2011) and also through physical displacement due to the impacts of mining on the local landscape. The process of establishing the Project Minas-Rio in Conceição do Mato Dentro has been an opaque process. Evidence

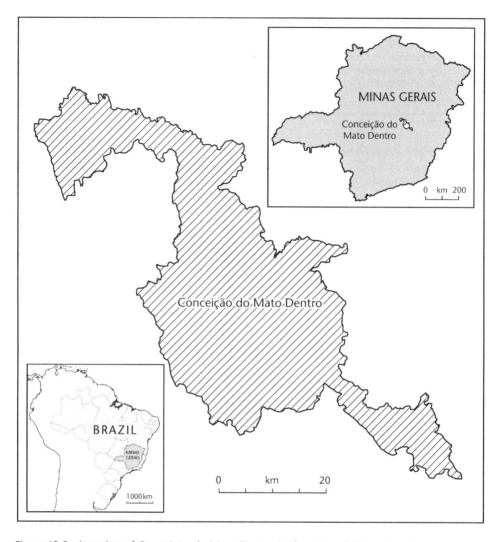

Figure 48.2 Location of Conceição do Mato Dentro in the state of Minas Gerais.

of manipulation, co-optation and coercion of people for the interests of the mining companies have been identified by NGOs and the local Public Prosecutor's Office, while complaints of violation of human rights have mushroomed since the first stages of the implementation of the mine. For example, in a public meeting on 17 April 2012, organised by the Public Prosecutor's Office of Minas Gerais, the Republic Prosecutor stated that violation of human rights occurred as a result of the strategy of fragmentation of the environmental licensing process for approving new mining operations, which is a common practice in Brazil, and which often benefits mining companies. According to the Prosecutor, the violations of human rights in Conceição do Mato Dentro are related to the negotiations between the mining company and affected families about the relocation of these families. Such violations are detailed by local opposition actors including the lack of water supply, denial of transit on public feeder roads, pollution of water springs, houses damaged by the traffic of heavy trucks and spread of dust, among others (MOVSAM et al., 2012). In the following sections we examine the growing opposition to the mining operations through Schlosberg's three-pronged approach to environmental justice.

Distributional justice

Within the case study, distributional justice provided a key frame to understand the conflicts between the mining company along with the state government and the local community. This focused primarily on *procedural* distribution and fairness, rather than the actual approval of mining operations in the area. On one hand, community interests recognised that there were potential local benefits derived from mining, from possible employment or spin-out benefits for existing businesses – in other words, it would be simplistic to suggest that there was a singular community response to the mining proposals with supporters as well as opposition. On the other hand, there was also a sense of powerlessness or resignation that the mine 'would happen'. However, the process of approving the mining operation was a source of considerable conflict.

Organised civil organisations and environmental organisations often have the ability to disrupt, delay or modify the actions of the extraction sector; however, in Conceição do Mato Dentro the role of these organisations was actively undermined. Within the context of an active local environmental movement, the mining company adopted a 'strategy of fragmentation' (Pereira and Becker, 2011: 255) to achieve its objectives, thus reducing transparency within local decision-making arenas. This refers to a series of actions leading to the fragmentation of variously: the environmental movement, the affected community, the environmental licensing process and the fragmentation of local politics.

The first tactic in this approach of fragmentation is the use of a shell company to purchase land before the mining project was launched. This acts as a way of diminishing the reparation and relocation costs allocated to the affected communities (Diversus, 2011). The use of a shell company to purchase land takes the mining companies out of the debate, and removes their legal responsibilities while avoiding local conflicts (Bachrach and Baratz, 1962). It prepares the ground for later decision-making and closes down opposition to the creation of the mine, even before the planning process has begun. Furthermore, before it was clear that the purpose of buying land was for mining, the shell company had been negotiating individually (rather than collectively) with landowners, often exploiting a lack of knowledge and experience concerning land values. This resulted in completely different prices being paid for land and created damaging, divisive resentment within the community. An anthropologist and consultant who coordinated a research project with the community affected by the mine in 2010 declared:

> Individual negotiations disrupt the society. Collective negotiation of resettlement . . . enforces that the links between people are kept . . . Community members [of the affected districts] are relatives and neighbours. In rural communities, neighbours help each other. Individual negotiations make this community context fragile. Nevertheless, individual negotiations prevailed [in the resettlement process of CMD].
>
> *(Interview 20, 29 March 2012)*

Another side of the strategy of fragmentation relates to the environmental licensing process itself. The project Minas-Rio comprises a mine, a port, a transmission line and a pipeline to transport minerals from the mine to the port. When the environmental licensing process for the mine was launched, the pipeline was already partially licensed. However, if the mine was not implemented, the pipeline would have no function. Nevertheless, there is a licensing process for the pipeline, another for the mine, and a third for the power line, each process substantially influencing the outcome of the subsequent stage in an accumulation of approvals. Moreover, the licence for the mine installation was also fragmented into two phases, because the company was not able to accomplish the mitigating actions required by the previous licence within the expected time limits. The fragmentation of the environmental licensing process, which was only possible through an active role of the state government, has caused confusion with the inclusion of different institutions, from different levels, and different information. This confusion was used by the company to build legal statements to defend their position. It has made the subsequent process opaque, lacking transparency, and therefore difficult to dispute.

The third dimension to fragmentation relates to fostering divisions within the local environmental movement. As outlined above, the environmental movement and the Society of Friends of Tabuleiro (SAT) were central in efforts to establish sustainable tourism as a priority in Conceição do Mato Dentro – indeed, some members of SAT were also elected to local government on this platform. However, with the launch of the mining project, the environmental movement held different standpoints concerning the way the mine proposal should be dealt with. While some activists completely opposed the mining project, SAT became a 'service provider' to the mining company, employed as consultants to contribute to the environmental impact assessment of the project and working to address the mitigating actions during the environmental licensing process. Furthermore, two SAT members occupied seats in the Environmental Policy Council, the body responsible for the licensing process for the mining project, with both SAT representatives voting to approve the licence. In the following quote an activist involved with the opposition movement complains about the practices of SAT members, illustrating his concerns: 'How come? . . . He favoured the mining company in the licensing process and made money for his own organisation [referring to SAT]' (Interview 17, 27 March 2012).

In this context, the environmental movement seemed to occupy multiple roles: an NGO, members of the Environmental Policy Council, members of local government working with the mining company, and providing consultancy services to the mining company – creating a lack of transparency within decision-making. Furthermore, the state government had entered into a formal partnership with the mining company to launch the mining operation (Pereira and Becker, 2011), and the majority of members of the Environmental Policy Council who were members of state organisations, were mandated to coordinate with state goals.

The fragmentation of the environmental movement that, until then, had been the strongest political force in the town of Conceição do Mato Dentro, helped the cause of the mining sector significantly through undermining organised civil society. The local political arena became increasingly marked by a struggle between local elites to benefit from the economic gains that the mining project would bring to the municipality. But more importantly for the mining companies, the political crisis generated the perfect environment for the negotiation of compensation due to the municipality as a consequence of the negative environmental and social impacts caused by the mine project. A weak and disorganised municipality has less bargaining power to negotiate and demand compensation. The influence of the proposed mine acted to fragment local politics with local elites competing to establish agendas and power, while decision-making appeared opaque.

Recognition

Reinforcing the fragmentation of local politics, the primary arena for decision-making – the environmental licensing for the project – was reduced to a technical exercise, proving to be exclusionary of many opponents to the mine. Within this process, the economic benefits and environmental impacts of the mining project were assessed through scientific and technical discourses, which served to exclude local knowledge and more informal experiences reported by local people affected by the mine. We argue that the exclusion of local values, voices and local knowledge and the absence of opportunities for lay-knowledge to be assessed is a form of misrecognition, silencing and devaluing local and cultural norms from decision-making. Indeed, the reduction of decision-making to simply an environmental licence process privileged scientific and technical discourses over wider political debates, enabling decision-making along narrow lines focused on impacts and possible mitigation measures. Instead the debate concerning the more fundamental question of the future of Conceição do Mato Dentro and whether mining should be favoured over other development pathways, such as tourism, was not debated. This technical interpretation favours well-resourced parties with access to legal and technical teams, while excluding individuals and groups affected.

However, following the approval process and failure of the mining company to comply with mitigation measures, local people along with activist organisations and the Public Prosecutor's Office in the local administration increasingly challenged official and technical discourses supporting the mine and provided new arenas to support local concerns. Key to this was enabling local people to speak to power though their experiential knowledge, particularly through public meetings organised by the Prosecutor's Office in 2012. These alternative arenas recognised the traditional social practices that exist in the locality and enabled local people to voice concerns through their actual experiences. In the words of a community member: 'We [community members] don't have psychological conditions to negotiate with them [the mining company]. This is why we will try to get organised into the association [of the Sapo district] in order to defend ourselves together' (Interview 14, 19th July 2010).

Capabilities

As highlighted, the actual presence of the mining operation has not been central in the environmental justice framing of resistance or opposition – itself reflective of a lack of transparency in decision-making. Instead, opposition grew after mining operations began due to two factors: (1) to ensure that the mining company met its obligations to mitigate the environmental impacts of its activities; and (2) the affects of the mining operation on existing or alternative livelihoods and its impact on community well-being, particularly due to the privatisation of land. In this regard, the opposition focused not on the siting of the mine (i.e. its distribution), but on its impact on eroding well-being, a sense of place and community, and livelihoods, suggesting that a loss of *functioning* provides a key frame in opposition discourses.

One example relates to the environmental degradation of local water quality, especially local springs and rivers, provided a key focal point for a profound sense of loss – ranging from impacts on small scale farming to local identity and cohesion. In relation to environmental impact on local livelihoods, during one public meeting when technical experts suggested that water quality had remained unchanged after mining started, one community member promptly replied:

'If you want, I can show you a water source in my community. Water was transparent, now we cannot see the bottom of the stream'.

To the idea that the mine would expand and distribute income, another community member states in the public meeting of 17 April 2012:

> No one buys my products in the market because they know my land is poisoned by the company. I used to earn 2500 reais [Brazilian currency] a month, and now I cannot make 500 reais. I had to take my children out of school . . . Water is poisoned in my land. Chickens are dead, pigs are dead and cows and steers are sick.

Demonstrating that this is not simply about economic impacts, the interview extract below with a farmer illustrates the impact of displacement from spaces thought of locally as communal meeting places affected by the mining project. It is possible to see how the water spring he mentions is viewed as a public space where the members of his community and other people used to meet, enjoy and socialise. It was a place to celebrate freedom. The fact that the community lost this important recreation site and informal meeting place has caused some of them to reflect upon the situation, and see that this was undermining support for the mine:

> There was, near my plot of land, a water spring known to have medicinal properties, very fresh and pure water. We used to meet our friends there, during our free time. Now it is located in their [the company's] land and we have no longer access to the water spring. I do not even know if it is still there or if it has been destroyed.
>
> *(Interview 12, 19 July 2010)*

Talking about another similar site used by locals and which was starting to be visited by tourists before the mining project, a community leader said: 'The place is still very beautiful, but we know that sooner or later it will be poisoned' (Interview 18, 29 March 2012).

In the statement above, it is possible to see the erosion of place-attachment and community bonds through a sense of loss following the mining operations. Local people were no longer able to access sites that had become important places and memories of a happier past framed local anger:

> I have roots in this region . . . I have a wonderful farm, where I grow all my food. I just have to buy salt, sugar and rice in the market . . . I had a dream to live a rural life, but now I cannot access my farm because of the dust and noise. The mining company ruined my dream . . . What am I going to do?
>
> *(Interview 14, 19 July 2010)*

It is remarkable that since mining operations began, there has been a growing feeling of indignation among the communities that are affected by the mine. The main causes of this are the violation of human rights (often related to the use of private security firms), lack of compliance with agreed mitigating actions, the initial approval of the environmental licences, and the ineffectiveness of judicial power concerning actions against the company. A document produced by the growing local opposition movement identified human rights violations highlighted in Box 48.1 (MOVSAM et al., 2012).

Box 48.1 Rights claims by local opposition movement

1   Right to information and participation.
2   Right to freedom of meeting, association and speech.
3   Right to work and to a dignified standard of living.
4   Right to adequate housing.
5   Right to education.
6   Right to health and a healthy environment.
7   Right to a fair negotiation, isonomy, according to transparently and collectively constructed criteria.
8   Right to come and go.
9   Right to the traditional practices and life-style as well as access and preservation of cultural, material and immaterial heritage.
10 Rights of the indigenous people, *quilombola* (communities formed by runaway slaves during slavery in Brazil) and traditional communities.
11 Right of vulnerable groups to special protection.
12 Right to access to Judge and to a reasonable duration of the judiciary trial.
13 Right to protect the family and the social and community solidary linkages.

These justice claims combine aspects of procedural justice and recognition (and dignity), but also express functioning, freedoms and capabilities perspectives as the opposition articulated the impact of the mine on traditional local practices, access to land, lifestyles and quality of life. These ideas formed the basis for local demands at public meetings organised by the Public Prosecutor's Office on 17 April 2012, when people had their voices listened to by members of the Public Prosecutor's Office and of the Public Defender's Office as well as by journalists, researchers and social activists.

Conclusion

The management of traditional rural resources for extraction has been a longstanding challenge for rural land-use governance, with mineral rich rural regions often positioned as a 'resource bank' for the wider regional, national and global economy, based on the withdrawal of wealth to meet ever fluctuating global demand (Markey et al., 2008). However, as many rural regions increasingly shift from natural resource extraction to consumption-centred economies based on amenities to drive rural transitions, mining operations represent a competing demand for rural land with alternative rural resource priorities (e.g. protecting fragile landscapes, ecosystem services) and alternative development pathways. This competition is particularly sharp in landscapes with a settled population (such as recent conflicts in the Hunter Valley, Australia), or where mining comes into conflict with traditional dominant rural interests, such as farming. In the case study presented in this chapter, mining competed with sustainable tourism to establish a narrative concerning positive local benefits – but ultimately a development discourse based on a finite resource for extraction that is incompatible with the future development of the tourism sector. The governance of mining and rural resource use varies considerably across the globe, with consequences for how mining corporations engage with governments and local stakeholders. As illustrated in the literature review, this includes new models of engagement, inclusion of indigenous peoples and a commitment to corporate responsibility. However, in emerging countries, mining corporations have also sought to fragment local politics and the ability of affected communities to participate in decision-making may be limited through the absence of transparent policy arenas and dominance of local elites.

In this chapter, we have employed an environmental justice framework to highlight the multiple claims in framing mining disputes. Adopting a multidimensional approach to environmental justice demonstrated that a sense of injustice did not arise simply from the distribution of new mining projects (recognised as potentially bringing local benefits), but from the exclusion of local values, norms and knowledge from decision-making arenas; the ability of the mining companies to fragment decision-making; and the sense of loss following the commencement of mining activities. In this regard, including a capabilities perspective in environmental justice-framing is important in understanding how mining can undermine local freedoms, traditional practices, alternative livelihoods, and a sense of place attachment that formal decision-making failed to take into account.

References

Agyeman, J. (2002). Constructing Environmental (In)ustice: Transatlantic Tales. *Environmental Politics*, 11(3), 31–53.

Agyeman, J., Bullard, R.D. and Evans, B. (eds). (2003). *Just Sustainabilities: Development in an Unequal World*. Cambridge, MA: MIT Press.

Arellano-Yanguas, J. (2011). Aggravating the Resource Curse: Decentralisation, Mining and Conflict in Peru. *The Journal of Development Studies*, 47(4), 617–638.

Auty, R.M. (1993). *Sustaining Development in Mineral Economies: The Resource Curse Thesis*. London: Routledge.

Bachrach, P., and Baratz, M. (1962). Two Faces of Power. *The American Political Science Review*, 56(4), 947–952.

Ballard, C. and Banks, G. (2003). Resource Wars: The Anthropology of Mining. *Annual Review of Anthropology*, 32(1), 287–313.

Becker, L. (2009). Tradição e Modernidade: o desafio da sustentabilidade do desenvolvimento da Estrada Real. PhD thesis, Federal University of Rio de Janeiro, Rio de Janeiro.

Brown, R.B., Reed-Geersten, H., et al. (1989). Community Satisfaction and Social Integration in a Boomtown: a Longitudinal Analysis. *Rural Sociology*, 54, 568–586.

Caine, K.J. and Krogman, N. (2010). Powerful or Just Plain Power-full? A Power Analysis of Impact and Benefit Agreements in Canada's North. *Organization and Environment*, 23(1), 76–98.

Cheshire, L. (2009). A Corporate Responsibility? The Constitution of Fly-in, Fly-out Mining Companies as Governance Partners in Remote, Mine-Affected Localities. *Journal of Rural Studies*, 26(1), 12–20.

Cheshire, L., Everingham, J.A. and Lawrence, G. (2014). Governing the Impacts of Mining and the Impacts of Mining Governance: Challenges for Rural and Regional Local Governments in Australia. *Journal of Rural Studies*, 36, 330–339.

Collier, M. and Scott, M. (2008). Industrially Harvested Peatlands and After-use Potential: Understanding Local Stakeholder Narratives and Landscape Preferences. *Landscape Research*, 33(4), 439–460.

Collier, M. and Scott, M. (2009). Conflicting Rationalities, Knowledge and Values in Scarred Landscapes. *Journal of Rural Studies*, 25(3), 267–277.

Collier, M. and Scott, M. (2010). Focus Group Discourses in a Mined Landscape. *Land Use Policy*, 27(2), 304–312.

Conde, M. (2017). Resistance to Mining: A Review. *Ecological Economics*, 132, 80–90.

Connor, L., Freeman, S. and Higginbotham, N. (2009). Not Just a Coalmine: Shifting Grounds of Community Opposition to Coal Mining in Southeastern Australia. *Ethnos*, 74(4), 490–513.

Diversus. (2011). *Diagnóstico Socioeconômico: Área Diretamente Afetada (ADA) e Área de Influência Direta (AID) da Mina da Anglo Ferrous Minas-Rio Mineração S/A – Conceição do Mato Dentro, Alvorada de Minas e Dom Joaquim.* Belo Horizonte: GESTA – UFMG.

Everingham, J.A., Devenin, V. and Collins, N. (2015). 'The Beast Doesn't Stop': The Resource Boom and Changes in the Social Space of the Darling Downs. *Rural Society, 24*(1), 42–64.

Fazito, M. (2012). Competing Rationalities of Tourism Development in the Espinhaco Range Biosphere Reserve, Brazil. Thesis, University College Dublin, Dublin.

Fazito, M. (2015). Modernização Turística: o papel do turismo nos discursos dominantes de desenvolvimento. In S. Figueiredo, F. Azevedo, and W. Nóbrega (eds), *Perspectivas Contemporâneas de Análise em Turismo, 108–126.* Belém: NAEA.

Fazito, M., Scott, M. and Russell, P. (2016) The Dynamics of Tourism Discourses and policy in Brazil. *Annals of Tourism Research, 57,* 1–17.

Fraser, N. (1997). *Justice Interruptus: Critical Reflections on the 'Postsocialist' Condition.* New York: Routledge.

Honneth, A. (1995). *The Struggle for Recognition: The Moral Grammar of Social Conflicts.* Cambridge, MA: MIT Press.

Hilson, G. (2002). An Overview of Land Use Conflicts in Mining Communities. *Land Use Policy, 19*(1), 65–73.

Hilson, G. (2010). 'Once a Miner, Always a Miner': Poverty and Livelihood Diversification in Akwatia, Ghana. *Journal of Rural Studies, 26,* 296–307.

Hilson, G. (2012). Poverty Traps in Small-Scale Mining Communities: The Case of Sub-Saharan Africa. *Canadian Journal of Development Studies, 33,* 180–197.

Kassover, J. and McKeown, R.L. (1981). Resource Development, Rural Communities, Rapid Growth: Managing Social Change in the Modern Boomtown (USA). *Minerals and the Environment, 3*(2), 47–54.

Markey, S., Halseth, G. and Manson, D. (2008). Challenging the Inevitability of Rural Decline: Advancing the Policy of Place in Northern British Columbia. *Journal of Rural Studies, 24*(4), 409–421.

McDonald, P., Mayes, R. and Pini, B. (2012). Mining Work, Family and Community: A Spatially-Oriented Approach to the Impact of the Ravensthorpe Nickel Mine Closure in Remote Australia. *Journal of Industrial Relations, 54*(1), 22–40.

McManus, P. (2008). Mines, Wines and Thoroughbreds: Towards Regional Sustainability in the Upper Hunter, Australia. *Regional Studies, 42*(9), 1275–1290.

Measham, T.G. and Fleming, D.A. (2014). Impacts of Unconventional Gas Development on Rural Community Decline. *Journal of Rural Studies, 36,* 376–385.

Miller, E., Van Megen, K. and Buys, L. (2012). Diversification for Sustainable Development in Rural and Regional Australia: How Local Community Leaders Conceptualise the Impacts and Opportunities from Agriculture, Tourism and Mining. *Rural Society, 22*(1), 2–16.

Minas Gerais. (2005). Reserva da Biosfera da Serra do Espinhaço: proposta de criação. Secretaria de Estado de Meio Ambiente de Minas Gerais. Retrieved in September 2012 from www.unesco.org/mabdb/br/brdir/directory/biores.asp?mode=all&code=BRA+06.

Minas Gerais. (2011). *Listagem dos Circuitos Turísticos.* Belo Horizonte: SETUR.

MOVSAM, UNICOM, AMD, ASCOR, Labcen/PUC Minas, CPT-MG . . . Associacao Comunitaria Sao Sebastiao do Bom Sucesso e Regiao. (2012). *Violações de Direitos Humanos na ADA e AID – Projeto Minas-Rio.* April. Conceição do Mato Dentro: MOVSAM.

Myers, N., Mittermeier, R., Mittermeier, C., Fonseca, G., and Kent, J. (2000). Biodiversity Hotspots for Conservation Priorities. *Nature, 403,* 853–858.

Neil, C., Tykkyläinen, M. and Bradbury, J. (eds) (1992). *Coping with Closure: An International Comparison of Mine Town Experiences.* London: Routledge.

O'Faircheallaigh, C. (2004). Denying Citizens Their Rights? Indigenous People, Mining Payments and Service Provision. *Australian Journal of Public Administration, 63*(2), 42–50.

O'Faircheallaigh, C. (2013). Extractive Industries and Indigenous Peoples: A Changing Dynamic? *Journal of Rural Studies, 30,* 20–30.

Papyrakis, E. (2017). The Resource Curse: What Have We Learned from Two Decades of Intensive Research: Introduction to the Special Issue. *The Journal of Development Studies, 53*(2), 175–185.

Pereira, D., and Becker, L. (2011). O Projeto Minas-Rio e o Desafio do Desenvolvimento Territorial Integrado e Sustentado: a grande mina em Conceição do Mato Dentro. Paper presented at the Seminario Recursos Minerais e Sustentabilidade Territorial, Brasília.

Rawls, J. (1971). *A Theory of Justice.* Cambridge, MA: Harvard University Press

Schlosberg, D. (2004). Reconceiving Environmental Justice: Global Movements and Political Theories. *Environmental Politics, 13*(3), 517–540.

Schlosberg, D. (2007). *Defining Environmental Justice: Theories, Movements, and Nature*. Oxford: Oxford University Press.

Schlosberg, D. (2013). Theorising Environmental Justice: The Expanding Sphere of a Discourse. *Environmental Politics*, 22 (1), 37–55.

Sen, A. (1999). *Development as Freedom*. Oxford: Oxford University Press.

Taylor, D.E. (2000). The Rise of the Environmental Justice Paradigm: Injustice Framing and the Social Construction of Environmental Discourses. *American Behavioral Scientist*, 43(4), 508–580.

Tonts, M., Plummer, P. and Lawrie, M. (2012). Socio-Economic Wellbeing in Australian Mining Towns: A Comparative Analysis. *Journal of Rural Studies*, 28(3), 288–301.

Urkidi, L. and Walter, M. (2011). Dimensions of Environmental Justice in Anti-gold Mining Movements in Latin America. *Geoforum*, 42(6), 683–695.

Walker, G. (2012). *Environmental Justice: Concepts, Evidence and Politics*. London: Routledge.

Wheeler, R. (2014). Mining Memories in a Rural Community: Landscape, Temporality and Place Identity. *Journal of Rural Studies*, 36, 22–32.

Young, I. (1990). *Justice and the Politics of Difference*. Princeton, NJ: Princeton University Press.

49

Food security and planning

Andrew Butt

Introduction

In many jurisdictions, rural spatial planning policy and practice is focused firmly on maintaining rural farmland, often in the face of persistent pressure to convert farms to urban, or urban-generated, land uses. The maintenance of *food security* has become central to discourses of farmland protection, and in recent decades this concept has seen renewed policy and scholarly interest. No longer confined to a focus on hunger and aggregate national food production, the framing, research and policy surrounding food security increasingly includes broader issues of the accessibility, equity and quality of food as central to debates and discussions in this field. For spatial planning, these debates extend to issues of the viability, diversity and vulnerability of food *systems*, extending to distribution and consumption, not simply food production, particularly in relation to land use conflict and transition in urban, peri-urban and rural areas.

Although international policy concerns regarding global hunger and food production emerged after 1945, the contemporary breadth of this agenda is evident in the most commonly accepted definition of *food security* that emerged from the 1996 World Food Summit; that food security exists when 'all people, at all times, have physical and economic access to safe nutritious food that meets their dietary needs and food preferences for an active and healthy life' (FAO, 1996, para. 1: 1). This in itself suggests roles for rural planning that include not only support for food and agricultural production, but also implicate notions of rural community development and locally derived food solutions.

Within the emerging discourses on food security, notions of food access, equity and quality have each been interwoven with previously settled understandings of hunger, aggregate food supply, national self-sufficiency and productivist rural policies. For land use planning systems, and for planning research, the increased scope of the food security agenda has allowed an interrogation of a range of issues including urban and rural food systems and the modes of production, distribution, consumption and spatial differentiation at the local and global scale.

Specifically, the changing nature and possibilities of food systems require planning practice to reflect on issues as diverse as farmland preservation, urban food 'deserts', productive urban greenspace and industrial agriculture. In rural planning practice, critical issues include the changing nature of rural land uses driven by industry restructuring, land and water ownership, globalised food systems, increasing interests in 'niche' and source-identified food in developed economies, and tensions in much of the world relating to livelihoods and self-determination, to land grabbing and to contested modes of agricultural modernisation.

This chapter will explore how notions of food, farm-scapes, and agricultural systems relate to the practices of rural planning, particularly in relation to tensions between local food, commodity-scale framing and post-agricultural rural land use. Using examples from Australian peri-urban regions, the chapter will identify these tensions, and frame them within broader narratives of food security and food economies at a local and global scale. In this example, the ways in which food security is conceptualised and operationalised within planning systems is shaped by broader notions of what food security is and can be, despite this being largely limited, in practice, to notions of food production at a commodity, export scale.

Food security as a policy agenda: production, access and equity

Since the 1970s, and specifically after the 1996 Rome Declaration (FAO, 1996), broadening notions of food security have emerged as increasingly vital areas of policy concern and public discourse. Mooney and Hunt (2009) describe this resurgent interest as a 'master frame' for twenty-first century public policy, although they identify that the meanings and policy actions associated with *food security* remain fluid and, akin to the concept of *sustainability*, retain contested meanings with unsettled definitions, operations and boundaries. This fluidity in meaning and purpose at once enlarges the scope of food security, while also requiring engagement with broad, diffuse and potentially contradictory definitions and concepts.

Distinctive discourses have extended beyond simply aggregate food production, to include notions of food sovereignty at a national level, and of the social and spatial justice issues of land, access to nutritious food, food 'localism', among other concerns. Each has become particular elements of resurgent discourses of food security, especially after global food price shocks during 2007–2008 (De Castro et al., 2013). As the critical issues of food scarcity and food pricing instability were revealed starkly in the events of the 2007–2008 food and oil 'price shocks', so too were new sets of concerns at what food security could mean, within a 'more volatile context for agri-food scholarship' and an 'explosion of interest in food studies' (Moragues-Faus and Morgan, 2015: 275). At this time the confluence of natural yet unseasonal events, increased commodity speculation, rising input costs and new market entrants such as bio-fuels resulted in new concerns and widening agenda for food policy.

However, the path to this expansion in food policy and scholarship emerges from a long-standing policy agenda. The formation of the Food and Agriculture Organization (FAO) after the Second World War, with its focus on hunger and improved aggregate production, coupled with the emergence of 'agricultural exceptionalism' within the General Agreement on Tariffs and Trade (GATT) led to a period where production improvements and national self-sufficiency (and trade protection) were considered to be key pathways to reducing hunger (Farsund et al., 2015). After the food price crises of 1972–1974, the focus of the FAO, and its conceptualisation of food security, pivoted away from national self-sufficiency towards the ability of individuals and households to access food (Jarosz, 2009: 51–52), a distinctly trade-focused approach that aligned with the emergent aims of the World Trade Organization (WTO) to remove trade distorting subsidies in the agricultural sector (Farsund et al., 2015). The ongoing international policy project of agricultural trade liberalisation has continued to influence discourses on food security as a production-focused task.

At a global level, a trade liberalisation discourse has largely been maintained, particularly as it supports the position of large exporting economies, such as the Australian case study described below.

However, the received stance has been increasingly challenged by discourses of food sovereignty that support continued differentiation of agricultural and food within trade policy, identifying the serious consequences of market failure when it comes to food in the developing world (De Schutter, 2011). The divergent rules and norms associated with global food trade 'remain a source of conflict and fragmentation' in food security and trade policy (Margulis, 2013: 65), especially between export and import economies, and between developed and developing nations.

Extending this concern, a wider, and typically more localised conceptualisation and operationalisation of food security has emerged which explores issues at the community level, and eschews oppositional narratives of production and consumption, of urban and rural, of local and global. Wider concerns of food justice (Cadieux and Slocum, 2015; Moragues-Faus, 2017) as a distributive and productive process, of the recognition of difference and diversity in production, and of the roles of urban and rural food systems as varied and essential ingredients in resolving socio-economic differences in food access at the sub-national, or community, level have each become part of policy and scholarship.

Such a broadening of the working definitions of food security are not without critique or disputed understandings, particularly when contrasting export-focused economies with highly reliable, and generally abundant food supply with poorer countries where politically and culturally contested food economies persist (Mooney and Hunt, 2009; Jarosz, 2014). Mooney and Hunt (2009) identify three modes of conceptualising contemporary *food security*: hunger, community food security (and food sovereignty) and vulnerability and risk. Each extends a genealogy of policy that has clear implications for food production systems, typically as a national public policy concern. Sonnino (2016) describes this as an increasingly 'bimodal' perspective of food security which recognises and encompasses both scaler and system-wide concerns raising significant challenges for understanding approaches to policy, particularly land use policy, as a context-specific and locally differentiated process in both urban and rural environments.

As a consequence of the breadth of definitions and issues now associated with food security, spatial planning scholarship and planning systems policy and research interests increasingly intersected with issues of planning for food accessibility in urban areas (and food deserts, see Wrigley, 2002), access to healthy food as a spatial justice issue (Cadieux and Slocum, 2015), urban foodscapes (Moragues-Faus and Morgan, 2015; Sonnino, 2016) offering a 'ruralisation' of urban spaces (Dandekar and Hibberd, 2016: 226) and a revived interest in the relationship between food, design and the lived environment (Barthel et al., 2015). Nonetheless, research emphasises the enduring potency of neoliberal, and global, trade-focused notions of food security (Jarosz, 2011), with productivist logics dominating in international agreements. The roles of international agencies, producer/exporter nations and of a globalised agricultural industry serve to maintain the prominence of this discourse, despite the increasingly disparate scholarly interest and local policy operationalisation of food security objectives.

For rural planning the challenge of this increasing breadth in understanding includes a renewed consideration of land use objectives, agricultural restructuring and community life. Largely, the core questions of farmland planning and protection sit comfortably with the original, singular, and production-focused conception of food security, particularly in its guise as a mechanism to increase net production and reduce hunger. These are particularly acute in agricultural economies focused firmly on scale, efficiencies and export potential. But the emergent notions of food security as responsive to issues such as food quality, local food systems and the political ecology of production and consumption also demand responses from rural planning. Despite a longstanding concern with agriculture, the increasing interest in reliability and vulnerability of food systems within these widening definitions have made *food security* a potent policy signifier with considerable relevance to rural land use planning, and demand its continued adaptation to emerging needs.

Rural planning practice and food systems

Contemporary rural planning, particularly rural planning practiced through land use regulation, has a fraught relationship with exiting an ideated agricultural production and farming systems in many locations globally. Within many rural planning systems, food production, security and vulnerability are at once central concerns, but also often seen through a lens of land use conflict, developmentalist agendas and agri-business transition.

In many regions, rural planning systems are removed from direct concern for food production systems, rather, focusing on land management and agribusiness objectives with planning regimes that seek multiple objectives and operate within a neoliberal framework. Particularly, the ideals of rural planning for farmland protection sit within a broader framework of objectives relating to ordered urbanisation, the management of rural landscapes and support for rural community life. Productivist logics of agriculture retain power in rural planning systems, acting as a basis for a focus on rural resource management in planning practice. Despite this, the realities of increasingly multi-functional landscapes, at least in those areas where rural planning tensions are most acute, align with the complexities of those definitions of food security that seek diversity in the face of vulnerability, emphasise a production-consumption nexus in multi-scalar food systems.

Typically, the roles of rural planning are considered to emanate historically from twin objectives; first, of managing urban expansion, and second recognising the multiple roles of rural place and landscape (Hall, 2014; Dandekar and Hibbard, 2016). Traditions of early rural planning include the 'anti-urban' objectives of the late nineteenth century, such as those exhibited in Howard's (1898) vision for the agrarian urbanism of 'garden cities' and similar reactions to contemporary urbanisation, or the fear of the 'urban octopus' (Williams-Ellis, [1928]1975) consuming the English rural landscape. Likewise, Patrick Abercrombie's (1926), concerns for the problem of 'weekend-shanties' or second homes for 'the owner of a car', of sprawl, 'ribbon' development and formation of dormitory satellite towns leading to excessive urbanisation were framed in response to seemingly ceaseless urban growth engulfing a countryside of both tradition and production. Similar themes are evident in Cole and Crowe (1937), whose survey of approaches in Europe and North America revealed increasing, yet often ineffective, concern for the protection of a rural land resource in the face of urbanisation.

The second emergent thread of rural planning – *rural development* – was more readily aligned with objectives of food production, first manifested in directive planning of agricultural expansion, especially in export-focused settler societies in North America and Australasia. While the utilitarian purposes of rural land, beyond managing sprawl were well considered in rural planning – such as Abercrombie noting that 'the work-a-day English countryside: this is really an *industrial* zone given over entirely to farming' (Abercrombie, 1926: 17, original emphasis) – examples such as irrigation colonies, (Kershner, 1953) and the significant advancement in watershed-scale rural planning evident in the Tennessee Valley Authority from the 1930s, and subsequently when the 'model and promise of American agricultural modernism was absolutely hegemonic in the three decades from 1945 to 1975' (Scott, 1998: 270).

This framework of productivity-focused, developmentalist and expert-driven rural planning for food production became a template for rural planning and agricultural development after 1945 as a state-directed project, one that retains influence in developing economies

(Planel, 2014) and remains criticised for rationalising practices that neglect local need and support investment and development techniques of alleged land (and water) grabbing (Cotula et al., 2009; Borras et al., 2011).

A decline in this form of production-focused rural planning after 1975 aligns with broader changes in agri-environmental governance, as well as the declining appetite for direct state-support to agriculture. This signalled a retreat for agricultural fundamentalism, at least as a state-directed project, and a consequent reframing of agricultural policy, whether as supporting 'multi-functionality' and an 'environmental case for continued farming' (Potter, 2010: 28), for instance in the EU, or a 'hands-off' hyper-competitive policy environment, such as in New Zealand, or a model seeking developmentalist goals through Foreign Direct Investment, as is presently occurring in East Africa (Cotula et al., 2009; Nolte and Väth, 2015), leads to a different set of possibilities for rural planning.

Consequent modes of rural spatial planning do, however, reveal a continuity in problem-setting around both agricultural protection and urban containment – both framed around ideals of farmland preservation. Since the 1970s, the dual pressures on effective rural planning have been the declining state-directed support for settlement and community support as an objective of agricultural development, and from the emerging and expanding field of multi-functional transition, particularly in peri-urban and amenity-rich regions.

In practice, rural planning systems have generally sought to take the lessons and techniques of urban planning to order land use and manage land use transition, particularly in contested peri-urban regions. Within this, a focus on the protection of farming soils, agricultural systems, and viable farming units in the face of land use transition has become a central feature. For planners, in turn, such responses reflect an increasingly diverse set of rural land uses – most evidently the growth in various forms of rural 'lifestyle' development where landscape amenity, proximity to urban areas and relative affordability are each drivers of an exurban, peri-urban and amenity-led expanding urban field into a landscape that 'looks like the country but often thinks like the city' (Goddard, 2009: 413).

For example, Angelo (2017) describes what is ostensibly rural land being, although not explicitly, categorised as *urban*, or more likely as 'urban in waiting' within planning systems. Similarly, 'greensprawl' (Cadieux and Taylor, 2013) is offered to describe an area of transition where urbanisation has occurred, yet remains seemingly concealed within an ostensibly non-urban landscape of farmlets, and 'leafy' exurbia, as in Rowe's (1991) 'middle landscape'. In these locations, the space were many of rural planning's challenges are most acute and visible, key challenges relate not only to notions of protecting farmland loss, but to understanding the fluid categories of farming in contested environments. In this regard, Antrop (2004) suggests the presence of both *extensification* and *intensification* of agriculture, for example through the increasing scale of intensive animal industries, the continued presence of diffuse, small-scale and often sub-commercial farms, and the consolidation, and geographical movement (including internationally) of expansive industries, such as cereal cropping.

Despite planning policy objectives to prevent the loss of farmland (and farming soils) in many jurisdictions, the connections between fluid definitions of *farming* and its emergent categories (Butt and Taylor, 2017), and the overall commitment to these policies (Slätmo, 2017) are often less than clear. Within planning, regulation for agricultural support, or conversely to prevent farmland loss, is often centred on spatial policy aimed at maintaining rurality, landscapes and opportunities for agricultural expansion.

In Europe, Australasia and North America this regulation has focused on preventing the fragmentation of land into small units, as well as the identification of valuable agricultural soils. In areas experiencing contestation, such as those areas undergoing peri-urbanisation

(Taylor et al., 2017; Lapping, 2006; Houston, 1994) increasingly, the complexity of agricultural production systems, including the intensification of production systems, render land fragmentation an insufficient measure of current or future productive agricultural potential. Likewise, approaches duffer to the ways in which agricultural and environmental goals can be considered concurrently (Marr et al., 2016). Nonetheless, as Lapping (2006: 118) identifies, 'policy frameworks all around the world continue to emphasise agriculture as the key sector for rural regions', and spatial planning for such regions seeks to 'support agriculture at all costs' (Curry and Owen, 2009: 575) despite examples of negative social and economic implications. Consequently, enacted policy is largely about an orthodoxy of farm protection, leading Lapping (2006: 118) to suggest that in the face of rural restructuring and differentiation, European and North American planning systems show an 'amazing consistency and lack of imagination' in relation to modes and approaches to agricultural protection, describing orthodoxy in responses to what are actually locally contextualised land use problems.

Contemporary rural planning practice operates within an increasingly diverse rural economy and environment, with contrasting objectives for rural futures. Models of rural planning as farming, settlement and agricultural development projects are arguably untenable in an era where multi-functional land use is anticipated, and where direct farm production support is at odds with global trade policy.

Rural planning, food and farming: an Australian story

Contemporary Australian land use planning in peri-urban and rural regions is firmly framed within a presumed primacy of commercial agricultural activities and an orthodoxy of assumptions about agriculture as the social and economic underpinning of rural land use. Australia has been active, and enthusiastic in agricultural trade liberalisation since the initial stages of the Uruguay Round of agricultural trade negotiations in the 1980s and 1990s (Kenyon and Lee, 2006; Botterill, 2016; Potter, 2010), resulting in a significant pivot from an earlier policy stance that sought to nurture agricultural development through developmentalist policy.

For almost two centuries, Australian agriculture has been considered an export industry, firstly as an imperial farm (Alomes, 1981), a part of Europe's global 'empire of grass' (Pawson and Brooking, 2008), and later as a 'food bowl' to Asia (DAFF, 2013). The developmentalism of the twentieth century remained predicated on advantageous trading arrangements, particularly with the UK, but became focused increasingly on higher value exports. Within these systems, highly centralised approaches to export quotas, to pricing and direct industry support were normalised, predicated on agricultural exceptionalism (Aitken, 1985) in public policy and the social imaginary at the local and national scale.

The land use implications of this strongly state-directed approach to both agriculture and regional development included the long-standing, and continuing, preference for a family-farming models, and most specifically through examples such as 'closer-settlement' schemes (often linked to returned soldier land grant programs), the development of rail infrastructure and localised agricultural extension and training in various forms, all typified policy approaches from the mid-nineteenth to mid-twentieth centuries. In a continent of climate uncertainty, the limits to productivity in a low rainfall environment were quickly uncovered, despite developmental and settlement objectives, often with significant economic and personal costs (Meinig, 1962; Lake, 1987).

By the 1970s, these arrangements gave way to open markets and a globalised export-orientation (Cockfield and Botterill, 2013), with production efficiency as the central goal (Potter, 2010).

Through this period, the declining terms of trade, the loss of preferential markets and the reduced subsidies and import protection resulted in a significant process of restructuring in Australian agriculture with farm numbers decreasing from over 200,000 in the mid-1960s to below 90,000 by 2015, coupled with substantial increasing in total farm productivity and overall agricultural exports (Sheng et al., 2011). Today, less than 3 per cent of the Australian workforce is employed in agriculture and less than 15 per cent of Australia's export income comes from agricultural products; this compares with close to half of export income in the 1970s (ABARES, 2012). Australian agriculture is export focused, with about 75 per cent of gross farm production for external markets. It is estimated that Australia grows enough food for 60–80 million people (PMSEIC, 2010) under current practice; three times the food needs of the existing domestic population. Continued productivity growth has been supported by specialisation and intensification in high-input agricultural systems, but barriers to this continuing include the constraints on water and soils on a dry and environmentally degraded continent.

Dominant trends towards fewer, larger farms have not been uniform. In peri-urban areas, small-scale farms have grown in number with production concentrated in few industries, notably cattle production and intensive animal operations, as well as some potentially tourism-focused activities such as wine grape growing. The share of peri-urban farm production to overall national agricultural output has not changed dramatically (Houston, 2005; Budge et al., 2012; Butt, 2013a). However, an increasing bifurcation is apparent between large-scale operations in industries such as intensive poultry raising and the many small-scale sub-commercial farms in industries such as cattle farming (Butt, 2013b).

Policy responses to the transition in Australian agriculture are influenced by notions of self-reliance in agriculture that have emerged since trade liberalisation, but also by underlying objectives of 'hyper-productivism' (Dibden et al., 2009) in national policy formation that eschew directive policy over rural place and industry, but retain a 'residual agrarianism' (Cockfield and Bottreill, 2013: 138) when regions and industries appear threatened in this competitive environment. Within the federal system, land use decisions are rarely the responsibility of national government, and tellingly Australia's first ever 'national food plan' aimed to foster a sustainable, globally competitive, resilient food supply that supports access to nutritious and affordable food, and to maintain and improve the natural resource base underpinning food production in Australia (DAFF, 2013). No concrete consideration of planning systems was included. The closest this policy came was to identify a planning dimension was to recognise that the effect of urban displacement of food production is not well understood, further suggesting that it is too early to assess the effect if any on the food sector or consumers (DAFF, 2013).

Despite a lack of national land use direction, concern at the local level for farmland loss and the consequences for food production (and food security) remains high in many productive agricultural regions, particularly in areas of higher quality soils and those close to metropolitan consumers. In these regions, land fragmentation and non-farming land use are critical concerns for planners, and rather than being seen to offer rural development and agrarian settlement, small farms have generally become an undesirable symbol of crypto-suburbia and sub-commercial pseudo-agriculture. Consistent with Lapping's (2006) international examples, Australian farmland protection strategies now centre around proscribing land fragmentation and non-agricultural land uses, particularly non-farm 'lifestyle' housing in rural areas (Taylor et al., 2017). Similar concerns also exist in areas where non-agricultural rural land uses conflict (or simply complement) ongoing agricultural production, such as eco-system services, energy production or mining. However, the scale of change and the distance of these transformations from larger populations often reduce conflict.

Contemporary rural planning for food security in Australia

Rural, spatial and land use planning regulations in Australia have only emerged as a comprehensive system of regulation in the 1970s, and most prominently in the past 30 years. These traditions and logics are developed from specific understandings of the nature of urban development and land use change and from the mechanisms and techniques of *urban* planning, although they also encompass the ideas and traditions of the long thread of rural land and agricultural industry planning policy. Essentially, they attempt to both prevent urban incursion, and to 'save' farmland in peri-urban regions – a task made difficult in the absence of complementary macro-scale social and economic policy.

Australian rural planning continues largely as a land use regulation activity, and within the Australian system this is chiefly the role of each state government, and usually delegated to local government planners. Strategic objectives and decision-making systems for rural land use typically include objectives for the protection of high-quality soils and the retention of farming systems in the face of encroachment from competing uses. They also typically recognise a range of other values for rural land – environmental, for mining, for wind-energy production and other uses. Land, including rural land is typically zoned to prescribe, or proscribe uses, including the subdivision of existing landholdings. With the exception of the vast areas of 'pastoral leases' in inland Australia, and the considerable areas of public land, this is considered to be an effective response to existing or potential demand for urban-generated land uses such as 'hobby' farms or rural lifestyle housing.

In practice, peri-urban regions are the key field of contestation in rural planning. Despite recent examples of concern at the loss of soil and water to mining (Turner et al., 2017; Ker, 2017) and an increasing interest in the impacts of coal-seam gas, most rural planning practice focuses on the prevention of land fragmentation to sizes deemed unviable for ongoing commercial-scale farming, or the prevention of non-farm dwellings being constructed in farming areas. Both of these issues are most focused in peri-urban and amenity landscapes, where commerciality, viability and their relationship to land in dynamic and competitive farming systems are complex. Agricultural technologies including intensive animal raising and protected horticulture are two examples of the types of changes that create tensions in planning systems predicated on largely uniform land use responses in policy and regulation. The orthodoxy of decision-making at the local level appears certain but uncertainty about productivity and viability in fluid and restructuring agricultural economies, and the essential categories of farming (and what is not farming) make land use planning for food production problematic.

Addressing diversity, vulnerability and resilience in rural food systems

The idea that farming could be more than a commodity-scale commercial activity in Australia is still often met with the same resistance (including by planners) as when Callaghan (1955: 12) identified that a 'psychological resistance to smaller types of production has been built up in Australia'. Despite this, the restructuring of agriculture, and the emergence of a range of alternative rural land uses (rural lifestyle housing, energy production and 'alternative' farming systems) has resulted in spaces of rapid change and locations of contested rural identity.

The bifurcation of farming scale in peri-urban regions is one example of agricultural and rural change that presents a challenge to rural planning practice that presumes an orthodox model of productive farming. It is a model that seems at odds with an increasingly diverse approach to food production and rural land use, especially in Australia's peri-urban regions. Critical concerns of non-farm land use and declining viability belie approaches to nurturing resilience through varied models, scale, markets and networks of food production and resilience. For example, James (2016) identifies the failure of planning systems to identify the cultural diversity evident in production systems on Sydney's fringe, with specific reference to small-scale horticultural operations among new immigrant communities. Likewise, Taylor et al. (2017) recognises the inherent tensions in supporting at-scale intensive animal industries in regions where more diverse, place-based farming systems rely on emerging networks of consumption, direct consumer relationships and 'alternative' models of food systems are growing components of the economies of land use and communities.

Specialisation, intensification and economic concentration of farming in Australia are key outcomes of the competitive global trade environment. While bio-economic models are apparently decoupled from place, conflict occurs and context still matters. The risks and vulnerabilities of these approaches are apparent factors in ongoing agricultural restructuring that seeks increasingly distant locations and relies on high inputs, transport costs and homogeneity in production. Australian rural planning systems enthusiastically support these models, despite the evident need to consider future food security through a lens of resilience and vulnerability.

Conclusions: rural planning in complex rural landscapes

Australian rural planning practice is predicated on the protection of farming systems, specifically through the protection of farming land from fragmentation and the encroachment of non-farm land uses. This focus reflects a continuity in purpose for rural planning at the local scale, despite the removal of complementary policies at the national scale that once supported agrarian exceptionalism, and despite the evident breadth of objectives for rural landscapes, such as environmental management, energy production and amenity-led housing.

Australia remains a large global food exporter, and Australian agriculture operates within a policy environment that supports efficiency and generally eschews trade distorting subsidies. Increasingly, the concentration and intensification of Australian farming is viewed as the basis of future productivity growth, but in many places these trends are the central focus of planning conflict. Despite Australia's large land mass and apparent need for agricultural expansion, the reality is that the limits to intensive agriculture are being reached, particularly limits to reliable water. In this context, the permanent removal of productive land is increasingly being seen as inappropriate; nonetheless, there is an enduring belief that there will always be more land available and that technology will continue to deliver in the pursuit of greater levels of production efficiency.

The capacity of Australian rural planning systems to address a *food security* agenda has been largely predicated on understandings of food security as a concern of global production and trade, and largely from the perspective of contributing as an export economy. However, broader understandings of food security require consideration of the sustainability and vulnerability of agriculture and the adaptability of food systems to their local context. In this regard, the production-focused agenda of Australian rural planning, including at the local level, insufficiently considers models of agricultural diversity, and multi-functionality in rural landscapes. The contributions made by diverse farming systems, and by diversity in rural land use and community life, are significant for addressing broader concerns of food security, especially at a local scale, where local food networks, alternative production systems and land uses that address the environmental externalities of high-input agriculture are each important.

The scalar issues of understanding food security matter in rural planning. With the demise of state-directed support for the explicit development of agricultural regions, the fortunes of Australian agricultural industries at an aggregate scale rely on global prices and, increasingly, global investment. However, the vulnerability of this model is apparent, whether in areas heavily reliant on variable water supply, or in areas where scaled-up intensive agriculture creates conflicts with other rural land uses. The response of Australian rural planning systems has largely been predicated on an understanding of directions for agriculture that preclude support for varied and localised farming systems in areas deemed 'lost' to agriculture, although perversely, attempts to transition to more explicitly multi-functional rural landscapes are often hampered by regulatory frameworks that presume singular models of future farming.

It is within this context that the significant concern of land use as spatial planning for rural land emerged in Australia. These were increasingly coupled with an agenda of environmental management and the emergent *transactional* or *consumption* value of rural place. Despite these changes, cultural significance remains in relation to agriculture. Productivist imperatives continue, and this results in 'winning' and 'losing' regions and industries. For rural planning practice and policy, the challenge is to recognise the broader conceptions of food security, beyond an aggregate production model, to one that considers food security as a process from production to consumption, and one that considers the opportunities of multi-scalar and diverse agricultural systems as a resolution to the challenges of global trade, climate change and diverse rural land values.

References

ABARES (2012). *The agricultural commodities statistics report*, Australian Bureau of Agriculture and Resource Economics and Sciences, Canberra.

Abercrombie, P. (1926). The preservation of rural England. *Town Planning Review*, 12(1), 5–56.

Aitken, D. (1985). Countrymindedness – the spread of an idea, *Australian Cultural History*, 4(1), 34–40.

Alomes, S (1981). The satellite society, *Journal of Australian Studies*, 5(9), 2–20.

Angelo, H. (2017). From the city lens toward urbanisation as a way of seeing: Country/city binaries on an urbanising planet, *Urban Studies*, 54(1) 158–178.

Antrop, M. (2004). Landscape change and the urbanization process in Europe. *Landscape and Urban Planning*, 67(1–4), 9–26.

Barthel, S., Parker, J. and Ernstson, H. (2015). Food and green space in cities: a resilience lens on gardens and urban environmental movements. *Urban Studies*, 52(7), 1321–1338.

Borras S., Hall, R., Scoones, I., White, B. and Wolford, W. (2011). Towards a better understanding of global land grabbing: an editorial introduction, *The Journal of Peasant Studies*, 38(2), 209–216.

Botterill, L. C. (2016). Agricultural policy in Australia: deregulation, bipartisanship and agrarian sentiment, *Australian Journal of Political Science*, 51(4), 667–682.

Budge, T., Butt, A., Chesterfield, M., Kennedy, M., Buxton, M. and Tremain, D. (2012). *Does Australia need a national policy to preserve agriculture land?* Australian Farm Institute, Sydney.

Butt, A. (2013a). Functional change and the peri-urban region: food systems and agricultural vulnerability, *Economic Papers*, 32(3), 308–316.

Butt, A. (2013b). Functional change and farming in the peri-metropolis: what does it really mean for agriculture? in Farmar-Bowers, Q. Higgins, V. and Miller, J. (eds.), *Food security in Australia: challenges and prospects for the future*, Springer, New York, 425–441.

Butt, A. and Taylor, E. (2017). Smells like politics: planning and the inconvenient politics of intensive peri-urban agriculture, *Geographical Research*, 45(2), 206–218.

Cadieux, K. and Slocum, R. (2015). What does it mean to do food justice?, *Journal of Political Ecology*, 22(1), 1–25.

Cadieux, K. and Taylor, L. (eds). (2013). *Landscape and the ideology of nature in exurbia: green sprawl*, Routledge, New York.

Callaghan, A. (1955). Development of old and new rural lands, *National Development*, No. 13 (September), 8–14.

Cockfield, G. and Botterill, L. (2013). Rural and regional policy: a case of punctuated incrementalism? *Australian Journal of Public Administration*, 72(2), 129–142.

Cole, W. and Crowe, H. (1937). *Recent trends in rural planning*, Prentice-Hall, New York.

Cotula, L., Vermeulen, S., Leonard, R. and Keeley, J. (2009). *Land grab or development opportunity? Agricultural investment and international land deals in Africa*, IIED/FAIO, Rome.

Curry, N. and Owen, S. (2009). Rural planning in England: a critique of current policy, *The Town Planning Review*, 80(6), 575–596.

DAFF (2013). *National food plan: our food future*, Department of Agriculture, Fisheries and Forestry, Canberra.

Dandekar, H. and Hibbard, M. (2016). Rural issues in urban planning: current trends and reflections, *International Planning Studies*, 21(3), 225–229.

De Castro, P., Adinolfi, F., Capitanio, F., Di Falco, S. and Di Mambro, A. (2013). *The politics of land and food scarcity*, Earthscan, London.

De Schutter, O. (2011). *WTO defending an outdated vision of food security*, Office of the United Nations High Commissioner for Human Rights, New York.

Dibden, J., Potter, C. and Cocklin, C. (2009). Contesting the neoliberal project for agriculture: productivist and multifunctional trajectories in the European Union and Australia, *Journal of Rural Studies*, 25(3), 299–308.

FAO (1996). *Declaration on world food security*, UN Food and Agriculture Organization, Rome.

Farsund, A. A., Daugbjerg, C. and Langhelle, O. (2015). Food security and trade: reconciling discourses in the Food and Agriculture Organization and the World Trade Organization, *Food Security*, 7(2), 383–391.

Goddard, J. (2009). Landscape and ambience on the urban fringe: from agricultural to imagined countryside, *Environment and History*, 15, 413–439.

Hall, P. (2014). *Cities of tomorrow: an intellectual history of urban planning and design since 1880*, 4th edn, Wiley-Blackwell, Chichester.

Houston, P. (1994). Planning for agriculture and agricultural land in Australia: observation on the state-of-the-art, in *Prime agricultural land and rural industries on the urban–rural fringe*, Australian Rural and Regional Planning Network, Bendigo, 103–124.

Houston, P. (2005). Revaluing the fringe: some findings on the value of agricultural production in Australia's peri-urban regions, *Geographical Research*, 43(2), 209–223.

Howard, E. (1898) *Garden cities of to-morrow*, Swan Sonnenschein & Co., London.

James, S. (2016). *Farming on the fringe: peri-urban Agriculture, cultural diversity and sustainability in Sydney*, Springer, Dordrecht.

Jarosz, L. (2009). The political economy of global governance and the world food crisis: the case of the FAO, *Review (Fernand Braudel Center)*, 37–60.

Jarosz, L. (2011). Defining world hunger: Scale and neoliberal ideology in international food security policy discourse, *Food, Culture & Society*, 14(1), 117–139.

Jarosz, L. (2014). Comparing food security and food sovereignty discourses, *Dialogues in Human Geography*, 4(2), 168–181.

Kenyon, D. and Lee, D. (2006). *The struggle for trade liberalisation in agriculture: Australia and the Cairns Group in the Uruguay Round*, Commonwealth of Australia, Canberra.

Ker, P. (2017). Shenhua to continue with Liverpool Plains coal project despite $262m NSW buy-back, *Australian Financial Review*, 12 July, retrieved from www.afr.com/business/mining/coal/nsw-buys-back-shenhua-coal-tenements-20170712-gx9glp.

Kershner, F. (1953). George Chaffey and the irrigation frontier, *Agricultural History*, 27(4), 115–122.

Lake, M. (1987). *The limits of hope: soldier settlement in Victoria, 1915–38*, Oxford University Press, Melbourne.

Lapping, M. (2006). Rural policy and planning, in Cloke, P., Marsden, T. and Mooney, P. (eds) *Handbook of rural studies*, Sage, London, 104–122.

Margulis, M. E. (2013). The regime complex for food security: implications for the global hunger challenge, *Global Governance: A Review of Multilateralism and International Organizations*, 19(1), 53–67.

Marr, E. J., Howley, P. and Burns, C. (2016). Sparing or sharing? Differing approaches to managing agricultural and environmental spaces in England and Ontario. *Journal of Rural Studies*, 48, 77–91.

Meinig, D. W. (1962). *On the margins of the good earth: the South Australian wheat frontier, 1869–1884*, Rigby, Adelaide.

Mooney, P. and Hunt, A. (2009). Food security: the elaboration of contested claims to a consensus frame, *Rural Sociology*, 74(4), 469–479.

Moragues-Faus, A. (2017). Problematising justice definitions in public food security debates: towards global and participative food justices, *Geoforum*, 84, 95–106.

Moragues-Faus, A. and Morgan, K. (2015). Reframing the foodscape: the emergent world of urban food policy, *Environment and Planning A*, 47(7), 1558–1573.

Nolte, K., and Väth, S. (2015). Interplay of land governance and large-scale agricultural investment: evidence from Ghana and Kenya, *The Journal of Modern African Studies*, 53(1), 69–92.

Pawson, E. and Brooking, T. (2008). Empires of grass: towards an environmental history of New Zealand agriculture, *British Review of New Zealand Studies*, 17(1), 95–114.

Planel, S. (2014). A view of a bureaucratic developmental state: local governance and agricultural extension in rural Ethiopia, *Journal of Eastern African Studies*, 8(3), 420–437.

PMSEIC. (2010). *Australia and food security in a changing world*, Prime Minister's Science, Engineering and Innovation Council, Canberra, Australia.

Potter, C. (2010). Agricultural liberalisation, multifunctionality and the WTO: competing agendas for the future of farmed landscapes, in Primdhal, J. and Swaffield, S. (eds) *Globalisation and agricultural landscapes: change patterns and policy trends in developed countries*, Cambridge University Press, Cambridge, 17–30.

Rowe, P. G. (1991). *Making a middle landscape*, MIT Press, Cambridge, MA.

Scott, J. (1998). *Seeing like a state: how certain schemes to improve the human condition have failed*, Yale University Press, New Haven, CT.

Sheng, Y., Mullen, D. and Zhao, S. (2011). *A turning point in agricultural productivity: consideration of the causes*, ABARES Research Report 11.4 for the Grains Research and Research and Development Corporation, Canberra.

Slätmo, E. (2017). Preservation of agricultural land as an issue of societal importance, *Rural Landscapes: Society, Environment, History*, 4(1), 1–12.

Sonnino, R. (2016). The new geography of food security: exploring the potential of urban food strategies, *The Geographical Journal*, 182(2), 190–200.

Taylor, E., Butt, A. and Amati, M. (2017). Making the blood broil: conflicts over imagined rurality in peri-urban Australia, *Planning Practice and Research* 32(1), 85–102.

Turner, G., Larsen, K., Seona Candy, S., Ogilvy, S., Ananthapavan, J., Moodie, M., James, S., Friel, S, Ryan, C. and Lawrence, M. (2017). Squandering Australia's food security: the environmental and economic costs of our unhealthy diet and the policy path we're on, *Journal of Cleaner Production*.

Williams-Ellis, C. ([1928]1975). *England and the octopus*, Robert MacLehose & Co., Glasgow.

Wrigley, N. (2002). 'Food deserts' in British cities: policy context and research priorities, *Urban studies*, 39(11), 2029–2040.

50

Land grabbing and rural governance in the former Soviet Union

Oane Visser and Max Spoor

Introduction

Large scale land acquisitions or 'land grabbing', whether by foreign states or private companies and investment funds, have become a global phenomenon in recent years, in particular after the food price hikes in 2007 and the global economic crisis in 2008–2009. Global capital has swiftly moved from more risky areas into 'safer' agricultural and food markets, while – at least initially – focusing on acquiring large tracks of land, for food, cash and biofuel crops, or for speculative motives. While advocates of such farmland investments have stressed (potential) benefits such as technology transfers, increased productivity and export earnings and creation of employment, critical studies have stressed downsides such as dispossession of land from smallholders, a decline of *local* food security, the precariousness of the labour created, and a tendency of shifting to unsustainable mono-cropping. A wide number of empirical studies have highlighted the problems these large-scale land acquisitions often pose to inclusive and sustainable land governance (White et al., 2012). Large-scale land deals frequently generate conflict through, for instance, direct displacement of smallholders or indirect dispossession of smallholders through driving up land prices and making land inaccessible for local farms (ibid.).

While the challenges that the current large-scale land rush poses for rural governance have widely been studied in the Global South, such challenges have remained understudied with regard to the vast agricultural space of the former Soviet Union (FSU). This region – with major (re)emerging agricultural powerhouses such as Russia, Ukraine and Kazakhstan – constitutes a particularly insightful setting to study large-scale land acquisitions, rural governance, and their conjunction.

Regarding *large-scale* land deals ('land deals' in short) the FSU is analytically interesting as it is the region with the largest share of farmland controlled by large-scale corporate farms (only parts of Brazil and Argentina come close to such dominance of corporate farms). Whereas countless news articles have reported on mega land deals in Africa, many of these announced deals never actually materialised. Thus, the FSU can be seen as a frontrunner within the global trend of the expansion of large-scale farms, with some of them applying the latest technologies in farming. In terms of land governance, the FSU region, constitutes a valuable research setting, as it features the most drastic and widespread overhaul of land governance in recent history, with rural areas from the Baltic sea to the Bering street, and from the arid steppes South in Central Asia to the far North all having experienced a drastic change from central state governance of land, to privatisation of farms, the liberalisation of land and/or lease markets, and the reshuffling of the state's social functions in the countryside. With variations across the various post-Soviet countries, overall these changes have brought about, as will be shown in this chapter, a drastic weakening of rural governance by the state and a parallel increase in power of large-scale, corporate farms and their financial investors. This shifting balance of power is further facilitated, as will be shown, by the near absence of effective collective action by the rural population. This chapter is based on earlier, largely empirical work conducted in various regions of Russia (and Ukraine) (Visser and Spoor, 2011; Visser et al., 2012, 2014, 2015).

Historical changes in land governance at local levels

During the Soviet period, land ownership/governance was formalised with a primordial role of the state, which owned all the land. Land use was managed by *kolkhozes* (collective farms) and *sovkhozes* (state farms), with tiny 'subsidiary' plots (0.10–0.25 ha) 'owned' by employees' households, mainly for subsistence purposes (Visser, 2008). Nevertheless, households had also informal claims on part of the *kolkhoz/sovkhoz* land and produce, mutually agreed upon (e.g. *sovkhoz* land at roadsides), but also through pilfering of fields and farm storages. Land institutions were represented by a *troika* of authorities: (1) the farm enterprise (*kolhoz/sovkhoz*) chairman; (2) the head of local state government; and (3) the local party secretary. As all the land was state-owned, and could not be transferred or sold, land registration was weakly developed, and land conflicts were rare.

Privatisation of land and farms in the 1990s

Land ownership/governance changed radically during the Yeltsin period, with reform undertaken through a 'share-based' rather than 'plot-based' land distribution. This meant that the 'formal' land ownership of *kolkhoz/sovkhoz* land went to the employees who became shareholders. This land share could be taken out as land for private farms, but in practice this was complicated, and often the collective farm remained the same, under another name. However, household plots became individual property, freely transferable. This land reform led to a new agrarian structure in the early to mid-1990s. Most *kolkhoz/sovkhoz* land came into the hands of various types of Large Farm Enterprises (LFEs), such as Joint Stock Companies, and Limited Liability Companies. The subsidiary plots became slightly larger (and remained very productive), in the hands of households, while a new sector of private family farms emerged, albeit representing not more than 5–10 per cent of agricultural land and production. Former *sovkhoz/kolkhoz* households, now employees of LFEs, continued to feel entitled to part of the LFE produce, and in a situation of crisis and deterioration of farm-centred social services, pilfering and steeling increased.

Not only land ownership and land use changed, but also land institutions. The Soviet *troika* disbanded; the party secretary disappeared, the influence of the head of local government declined due to the abolishment of the system of state-led production plans and the weak finances of the state. The chairman or corporate director of the LFE became the main player. As mentioned before, land registration was weakly developed, and personnel at the local level had no experience with land deals. Field visits by the first author over various years showed that the land registration office of the local state (called the *selsovet*), in most localities consists of at maximum 2–3 people, with limited means for land measurement. The privatisation of

LFEs in the early 1990s was often conducted by a travelling group of mostly young and urban economist/lawyers, trained by Western advisors. Local and regional authorities (with a few exceptions in regions with pro-reform governors) were not in favour of the emergence of private family farms, and often questioned the idea of privatisation at large.

After privatisation was largely completed, these specialists did not go to work in local state land institutions, which are very low-paid, but went to (agro)consultancy companies (e.g. Yugagrofond), large agrobusiness, or left the agro sector. Hence, in terms of the land governance there was hardly any continuity, and a lack of institutional memory. Land conflicts started to emerge between the LFEs and rural dwellers in two ways (Nikulin, 2003; Visser et al., 2012). Firstly, *LFE management vs family farmers*: In the 1990s, conflicts over land consisted mainly of conflicts between farmers willing to take out land counter to the wish of the LFE management (and sometimes the local community as a whole). Land plots which were contested, were rather small. Local authorities largely took the side of LFE management (due to close ties between LFE management and local authorities, and a remaining belief in large-scale farming at the side of local authorities). Secondly, *LFE management vs farm workers*: Conflicts over farm resources (equipment, inputs and agricultural produce) between LFE management and employees increased. Land was not a major issue in most cases, as it was abundant, outside investors virtually non-existent, and land value often negative (the land tax was higher than its market value).

The rise of large-scale land deals in the 2000s

There was an overall economic recovery of the Russian economy since 1999 (due to high oil-prices, political stabilisation after the chaotic Yeltsin years, and a positive influence of the dramatic devaluation of the Rouble for the agricultural sector, as imports became more expensive). In the early years of the Putin regime, in 2002 and 2003 a new land code was introduced (for non-agricultural and later for agricultural land), allowing land sales (with the exception of foreigners). This caused a sharp rise of large-scale land acquisitions by outside investors (including foreigners, working through proxies in joint ventures). While local governance has been somewhat stimulated in the 1990s and 2000s (partly with the aim of transferring financial responsibilities, or problems, downward), the overarching trend in the 2000s has been increasing re-centralisation and increasing the power of the state. As agriculture has become more profitable and land rapidly more valuable, local state administrations in particular became more interested in controlling land deals. This meant that personnel within the land offices (*Selsovets*) did the technical registration, while the power, control, and oversight over land deals was predominantly in the hands of the head of the agricultural department or the head of the municipality and district. The land offices are underfunded, and often lacked proper technical means for land administration. With the emergence of outside investors interested in acquiring LFEs and their landholdings, conflicts over land became more widespread, consisting of conflicts between agro-holdings (often in conjunction with LFE management and local authorities) versus local employees-shareholders. As these agro-holdings became bigger and bigger, and very much supported by the Putin administration, in these conflicts the latter party more than often lost out.

The relation between local state and agricultural producers

Control over the productive use of land

Although since the early 1990s the local state has lost its administrative power to set production goals for farm enterprises, informally state officials can still force, or convince, part of the LFEs to produce certain crops. The majority of the LFEs were in dire straits and needed commodity credits by parastatals to start off their production cycles. Local authorities used this situation to force LFEs to produce certain products which were not, or even marginally, profitable for LFEs, with goals of keeping the local processing industry functioning or keeping up rural employment with labour-intensive branches such as flax and livestock production. Only LFEs with sufficient financial resources and business connections to start off the production without the help of such parastatals were able to take production decisions without the local state's influence (Visser, 2008; Nikulin, 2011).

In the 2000s, various trends with opposing influences have played into the power balance between state and the private/corporate farm sector in Russian villages. On the one hand, with the increasing profitability of agriculture, part of the LFEs have become more economically independent of the state, which potentially undermines the control of the local state at the village level over production. Furthermore, many LFEs have been bought by outside investors and merged into agro-holdings, which operate beyond the level of the municipality, incorporating various LFEs within one district or even within various districts (or even various regions). In the case of the larger agro-holdings, the power balance of local (municipal) authorities towards these enterprises has markedly declined. On the other hand, in the 2000s under Putin, a process of recentralisation of the state, known as the establishment of the 'power-vertical' in Russia, and the growth of the state in terms of personnel, funding and influence over business has taken place. This trend, to some extent, has counterbalanced the increased influence of agro-business. As a result, the actual power balance of the state vis-à-vis agribusiness strongly depends, in particular, on the size, and finances of the agro-holding and the financial situation of the authorities, and their level in the state hierarchy (local, district or regional level).

The local state also had, and has, at least one legal tool regarding the productive use of land. Legislation exists that gives the state the right to dispossess agricultural land which is not used for agricultural purposes for a couple of years. In the 1990s, due to financial difficulties many LFEs left part of their land unused for many years. As local authorities understood that it was impossible to force LFEs to cultivate all their land, and land value was low or even negative, this law was hardly enforced. Enforcement of this legislation was different with respect to the newly emerged private family farms. In many localities, LFE farm management and local authorities seemed to work together to dispossess land of private farms which were unable to cultivate it (Visser, 2008; Visser and Spoor, 2011).

During fieldwork, the first author came across cases where the district authorities took away land from private farmers. In the Pskov region, one village was visited where villagers went to court to try and get the land they were entitled to from the LFE. One farmer even went to court several times but, at the moment of research in 2010, no decision had been taken. The emergence of such private family farms *was* (and often still *is*) seen as undermining the continuity of the LFEs, while in reality such farms account for only a small percentage of the agricultural land, and few other villagers are prepared to follow the risky endeavour of such private farmers. With the sharp increase of large-scale land acquisitions, partly aimed at speculative purposes or conversion of land into construction sites, land-use legislation has become much more important than in the 1990s. In 2016, the Federal Law on improvement of the order of land expropriation was adopted, which gave more power to the state to dispossess land which is not used according to its legally established purpose. Such a strict legislation in theory might limit land use for speculation, but is more likely to lead to control over land by the state *vis-à-vis* agricultural producers. In Russia and the large countries of the FSU about 40 million hectares of land have been abandoned in the 1990s or marginally used (Visser et al., 2014). Mostly, LFEs use only part of their land for various (non-speculative) reasons, land is retained or bought with a view on expansion of production in the coming years, land surface is temporarily reduced in years of economic difficulty (such as the years when the global financial crises that hit Russia in 1998 and 2008/2009), or land is fallow due to crop rotation schemes. Thus, a strict legislation on land use might, on the one hand, result in just giving local authorities more leeway to pressure LFEs into following their broader policies, or on the other hand stimulate the directors of LFEs and agro-holdings to give bribes to the authorities in order to keep their land.

Control over land in connection to social obligations

In the Soviet and post-Soviet rural context, land governance is strongly interwoven with social and economic policy in the villages. One cannot understand the rights and control over land without taking into account obligations and rights concerning agricultural production and the social infrastructure. Except from agricultural land in suburban areas, land in a land-abundant and sparsely populated country as Russia, the *de facto* value of agricultural land is strongly dependent on the social infrastructure in place. Attracting sufficient employees (and especially qualified and young people) is a major problem for agricultural companies (Visser, 2008). Pallot and Nefedova (2007) demonstrated that in the Soviet era, the social infrastructure was a key component in determining the level of agricultural production. Whereas the marginal returns of all kinds of capital investments in agriculture were extremely low and disappointing, the level of investments in social infrastructure such as housing, electrification, cultural and sports facilities had a clear positive impact on the returns in terms of increased agricultural production. The directors of LFEs, especially in the early 2000s, expressed their social support to villagers. A director of an agro-holding operating in two regions, was very open about his motivations, and gave the following answer on the question why they invested in social infrastructure:

> . . . it is a means to maintain good, favourable relations with the local authorities. That is very important. For example, to avoid ending up at the level of bribes, it is better to solve their problems in the social sphere. The other side of the topic is that there are, of course, the local villagers.
>
> *(Field notes, Moscow October 2010)*

First of all, as this director indicated, investing in social infrastructure 'is a way of protecting your business administratively'. Property rights to land were far from secure in that era. For domestic as well as foreign investors, this brought both problems as well as apparent opportunities. On the one hand, as stated by the American-Russian owner of the agro-holding 'Russian Farms': 'each single hectare of my land holdings can be disputed, and is potentially insecure', including for foreign investors (Visser and Spoor, 2010). As this farm director added, good ties with local and regional authorities were (and still are) indiscernible to ensure the security of your land claims. On the other hand, weak law enforcement also gives the agro-holdings opportunities to arrange land deals 'under the radar', and quickly acquire large tracts of land, through bypassing official regulations, exploiting the knowledge gap vis-à-vis the local population regarding land regulations, and infringing on the rights of the local landholders, when conducting land transactions (further discussed below).

Control over land transactions

The new land code itself left considerable power over the practical implementation of the regulations to regional and local authorities. In some regions, authorities have created extra rules which hinder the emergence of private farms. In the southern Krasnodar region aspiring private farmers should have no less than 300 hectares of land to be allowed to start a farm, whereas the average farm is only some 80 hectares large. Such rules curtailed the rights of the rural population and left them little choice than to rent or sell their shares to LFEs (large farm enterprises), while also making them more disadvantaged in transactions with foreign or domestic investors. Citing Wegren (2002), Visser and Spoor (2010) stated that local authorities have the right of first refusal for any land sale. The selling process is very cumbersome and prone to abuse (Wegren, 2002: 659). Thus, it is likely that the farm managers or wealthy investors will (continue to be) the winners of property reform. Indeed, 'it is not difficult to imagine land committee officials, who are not well-paid, being approached (paid) to exercise the right of first refusal to some land deemed desirable, but not other land, on behalf of hidden investors' (ibid.: 658).

Already, in the last years preceding the introduction of the land code, local former Communist Party bureaucrats were assigning themselves plots of land in expectation of legalisation by the land code (Visser and Spoor, 2011; citing Nikulin, 2003). A director of farm enterprise in Rostov, that was visited during fieldwork, was buying up shares from farm employees. This was a long process, and he regretted not choosing a 'smarter' way to obtain a majority ownership (ibid.). Since the introduction of the land code, the process of land concentration has been accelerating, also stimulated by worldwide developments such as rising food prices (in 2007–2008, and 2011) and transnational land acquisitions or 'land grabbing', as mentioned in the introduction of this chapter. In Russia, increasingly land shares formerly owned by the farm employees became concentrated in the hand of LFE management and/or outside investors. This process of concentration is especially strong in the Moscow region that surrounds the rapidly expanding capital, and in the most fertile lands in the south (such as the Krasnodar region). In these regions, the land acquisition process was also most concealed and proceeds predominantly through illegal means. The land market in the Moscow region has been characterised as the most non-transparent branch of all business spheres in Russia. Currently, the land market in the Moscow region also seems to be one of the most (if not the most) profitable branch of business in Russia. Once one knows this context, in hindsight the fall of the Moscow mayor Luzhkov in his struggle with the Kremlin comes as no surprise, centred on the conflict over the forestland designated for the construction of the highway between Moscow and St Petersburg.

When discussing the nexus between the state and the private/corporate sector in agriculture in more recent land acquisitions, it is important to recognise that the LFE director, as successor of the kolkhoz chairman, is still seen as the main representative of the state in the village by the rural population (Nikulin, 2011). This relates to not only labour issues on the farm, but also with regard to land property issues, with the farm director seen as the 'face' of the state. This gives these directors a strong power over the farm employees as well as the other villagers. Tamara Semenova, vice-president of the rural movement *Krestyansky Front* (Peasant Front, which later disbanded), stated:

> In the Soviet-era they [the rural population] were accustomed to believe the chairman of their kolkhoz. The chairman of the board of directors of the renewed farm or chairman of the joint stock company was elected from these kolkhoz chairmen. The peasants used to believe their chairman, as their own father. Where did they go in the Soviet time to complain? To the chairman of kolkhoz! For support, help and so on.
>
> *(Field notes, Moscow, October 2010)*

Furthermore, when the farm chairman and the formal state representative of the municipality joined forces, it was relatively easy to convince or manipulate the rural population regarding any sales of land. As Feifer (2003: 1) stated (cited in Visser and Spoor, 2010): 'few landowners understand their legal rights. In many, cases, regional and local officials have been able to keep land in the hands of collective-farm managers and other cronies'. Indeed, our evidence suggests that local authorities often support LFE management and/or agro\-holdings in appropriating land, at the cost of farm employees and rural dwellers more generally. As the previously cited vice-president of the 'Peasant's Front' then stated:

> . . . when the raiders came and started buying land, they came not just from a street. It was by prior arrangement with the regional and district authorities. Who will let the strangers to do this business in the Moscow region!? Well, we suppose they got the possibility to do this business for *otkati* ('kickbacks' – money or other gifts to authorities).
>
> *(Field notes, Moscow, October 2010; published in*
> *Visser et al., 2012; cf. Danizewski, 2001)*

In some cases the contacts between state and business are not only tight, but fully overlapping as, for example, in the case of the successful agro-holding 'Yuzhny Put' in the Southern region Krasnodar, whose head in the early 2000s was minister of agriculture of the region and at the same time director of this agro-holding (Danizewski, 2001). It is particularly the rank and file workers who have lost their shares. The chairman of a farm enterprise in Krasnodar stated: 'I could have made all into my own property, but to do it one must have no conscience at all' (Nikulin, 2003, cited in Visser and Spoor, 2010). Other farm directors, however, set out to concentrate farm ownership in their hands (Dvornik, 2000; Pallot and Nefedova, 2007: 117). In addition to the conjuncture of private farm business and the local authority, courts often decide in favour of the state and large business when rural dwellers and farmer associations start a court case to reclaim their farmland or receive compensation (Feifer, 2003).

> Until 2002 no one disputed the rights of peasants to land shares. But everything has changed in 2002 when the State law about turnover of agricultural lands went into power and land became a commodity. Since that moment illegal court practice became popular in the Moscow region . . . before 2002 the land rights were given to peasants, after 2002 – the other way around.
>
> *(Tamara Semenova, Peasant Front, interview,*
> *October 2010; also in Visser et al., 2012)*

Land conflicts and social movements

Since the mid-2000 some generally overlooked, but important changes have taken place. Various new rural movements have emerged in Russia and Ukraine. Moreover, some of these movements are clearly grassroots movements defending the rights of the rural population (Mamonova and Visser, 2014; Visser et al., 2012). They were and are defending land rights of peasants, promoting rural development and protesting against land speculation and environmental degradation. One such movements is the already cited Peasants' Front. Emerging from the cooperation of several rural dwellers deprived of their land shares in 2005, the *Krestyansky Front* now accounts for about 25,000 members in Russia. This organisation was focused mainly on the fight for land rights and against illegal land acquisition. The leaders organise public events such as meetings, pickets, large mass rallies, in order to draw the attention of federal authorities to these issues. The Front represented its members in courts, protesting against land 'raiders' by demonstrating in front of their offices, organising road blockades, requiring statutory documents and public explanations. Due to the efforts of *Krestyansky Front*, peasants in various regions got compensation for lost land plots. Nevertheless, the pressure from the government on these new social movements was large, the willingness 'from below' to become organised was minimal, and in the end the Peasant Front disbanded itself (Visser et al., 2015).

The decline of organised rural movements does not mean that rural collective action and protest around land issues has disappeared fully, rather it means that it has become more dispersed and *ad hoc*. Often these *ad hoc* movements are sparked by excessive acts of injustice. In the Krasnodar region, following the killing of 10 villagers in an attempt to grab land of a family farmer, farmers organised a 'march' of tractors to Moscow. Although they were arrested by the police in the neighbouring region, the tractor march did help to bring the violent land grab case to the attention of the public media, which made a court case unavoidable. While the agroholding involved in the land grab lost its illegally acquired land, the land was not returned to the family farmer, but handed over to another agroholding. Rural movements thus clearly face an uphill battle in their contacts with agroholdings and the state.

Conclusion

The relation between local government and the 'new' private/corporate farm owners shows elements of both continuity as well as substantial change. Currently local government in Russia still has substantial powers to influence land deals, to confiscate underutilised land, and, to some extent, to influence production decisions of agricultural producers on their land. Normally land governance organisations are heavily underpaid and weak, while the emerging large-scale corporate farms have become powerful and sometime omnipotent. The latter also re-invest in resolving some social problems (such as social infrastructure) in order to keep 'good relations' with local governments. However, rent-seeking is also widespread, and bribes are frequently paid in land deals. These have often turned out to be to the detriment of the private farmers and farm employees who received land (or at least land shares) in the re-distributive land reform of the 1990s. Since the mid-2000 new rural social movements have emerged, such as the Peasant Front, which became for a while a new actor regarding land deals, defending the interests of private farmers and farm employees against the agro-holdings and outside investors, who are facilitated in their land acquisition by local governments. However, even this movement could not resist the mounting pressure of the state and large-scale land owners against them, and combined with the hesitance of rural dwellers to get organised, it disappeared after only a few years. As a result an alliance of state and large farm enterprises governs rural areas without much interference from rural dwellers themselves, with large farms becoming increasingly more powerful with the ongoing expansion of agroholdings.

References

Danizewski, J. (2001) Betting the farm to save it, *The Moscow Times*, May, 12–13.

Dvornik, B.Z. (2000) Razvitie zemelnykh otnoshenii na regional'nom urovne: opyt, problemy, in: Agrarian Institute and All-Russian Institute of Agrarian Problems and Informatics (eds), *Rynochnaya transformatsiya sel'skovo khozyaistva*, Moscow: Agrarian Institute, 63–64.

Feifer, G. (2003) Russia: farmland reform may prove Putin's lasting legacy, retrieved from www.cdi.org/russia/johnson.

Mamonova, N. and O. Visser. (2014) State marionettes, phantom organisations or genuine movements? The paradoxical emergence of rural social movements in post-socialist Russia, *Journal of Peasant Studies*, 41 (4), 491–516.

Nikulin, A. (2003) The Kuban Kolkhoz between a holding and a hacienda: contradictions of post-Soviet rural development, *Focaal*, 41, 137–152.

Nikulin, A. (2011) *The Oligarkhoz as a successor of the post-Kolkhoz*, working paper Max Planck Institute for Social Anthropology, Halle/Saale, Germany.

Pallot, J. and T. Nefedova. (2007) *Russia's unknown agriculture: household production in post-socialist rural Russia*, Oxford: Oxford University Press.

Visser, O. (2008). Household plots and their symbiosis with large farm enterprises in Russia. In M. Spoor (ed.), *The Political Economy of Rural Livelihoods in Transition Economies*, 88–110, London: Routledge.

Visser, O. and M. Spoor. (2010) Land grabbing in Eastern Europe: Global food security and land governance in post-soviet Eurasia, paper presented at the 118th EAAE Seminar, Ljubljana, Slovenia, 25–27 August.

Visser, O. and M. Spoor. (2011) Land grabbing in post-Soviet Eurasia: the world's largest agricultural land reserves at stake, *The Journal of Peasant Studies* 38 (2), 299–323.

Visser, O., N. Mamonova and M. Spoor. (2012) Oligarchs, megafarms and land reserves: understanding land grabbing in Russia, *Journal of Peasant Studies* 39 (3–4), 899–931.

Visser, O., M. Spoor and N. Mamonova. (2014) Is Russia the emerging breadbasket? Re-cultivation, agroholdings and agricultural production, *Europe-Asia Studies* 66 (10), 1589–1610.

Visser, A., N.V. Mamonova, M. Spoor and A. Nikulin. (2015) Quiet food sovereignty as food sovereignty without a movement? Insights from post-socialist Russia, *Globalizations* 12 (4), 513–528.

Wegren, S. (2002) Observations on Russia's new agricultural land legislation, *Eurasian Geography and Economics* 43 (8), 651–660.

White, B., S.M. Borras Jr., R. Hall, I. Scoones and W. Woolford. (2012) The new enclosures: critical perspectives on corporate land deals, *Journal of Peasant Studies* 39 (1), 619–647.

Part IX

Reflections and futures

The four chapters in this final part of the Companion offer reflections on reframing rural planning in response to emergent challenges; these include the existential threat of climate change and the basic task of delivering effective planning within a rural governance context that must reconcile the sometimes opposing logics of external interests and local actors. The broader future of rural places is also considered, with globalisation, intensifying urbanisation and environmental change viewed as forces that could drive a range of rural outcomes, which will generate their own governance and planning challenges. Finally, the editors extract some of the key messages from this companion and outline some of the ways in which multi-scale and multi-sectoral planning in the countryside might respond to the wide variety of challenges and opportunities that the contributors have identified.

Homsy and Warner begin by claiming that 'rural planning requires local government leadership as well as collaboration across agencies and across communities to address climate change. This multisector and multilevel approach is the future of rural planning'. The importance of 'horizontal' and 'vertical' relationships, supplying/exchanging knowledge and resources, is an emergent theme from previous contributions, but critical given the scale of the climate challenge. Rural areas are also significant producers of greenhouse emissions – directly from forestry and agriculture and because of the lower densities of 'rural cities' and the longer commutes of rural residents. The likely impacts on rural places will be complex and rural planners will need to 'navigate this complex world', dealing with multiple issues – mineral extraction, commodity production, property rights and environmental protection. In the US, 'hazard mitigation' is a priority for many rural authorities and this can be seen as a first step 'towards sustainability', but 'it does not lead to the robust multisector and multilevel planning required to address climate change'. Such planning needs to link community-level collaboration with multi-level coordination and resourcing. The authors illustrate the impossibility of communities tackling climate change mitigation on their own, using two local case studies. These show the limits of independent action and unsupported/unresourced voluntary projects – which are too easily relegated behind other priorities. The Environmental Protection Agency (EPA) was established as a well-funded top down 'command and control' model and has achieved significant successes.

But such models are weakened by a lack of local knowledge and buy-in. However, if they can become part of multi-sectoral and multi-level partnerships then resources can be brought together with requisite knowledge. The means of achieving genuine partnership – and workable multi-level planning – then provides Homsy and Warner's main focus. They give the example of the Rouge River, Michigan. This waterway had diffuse sources of pollution. Without local knowledge and buy-in, the EPA would have been unable to coordinate its clean up. But by building local partnerships, it was able to work with different groups, understand sources of pollution and coordinate different actions that eventually resulted in the return of lost wildlife to the river. In relation to climate change, local partners must reconcile the competing goals of economic prosperity and environmental benefits. How this can be achieved – by pursing defined 'co-benefits' – is reviewed in another two case studies. These illustrate how 'officials in rural communities must tie climate change and other sustainability issues to economic development' in order to secure buy-in. Approaches must be 'creatively crafted' in order to bring together different interests, whose priorities might otherwise diverge. Rural planners must operate in a revised framework – one that coordinates local interest while working with a central authority that respects local action. Finally, climate change is presented as one of the most complex challenges that confronts the planet. It is also a challenge that usefully exposes the way in which rural planning must balance competing interests and work in a broader governance framework. That framework is explored in more detail in the second contribution to this section, which looks also at the asymmetry of power relations between local and external actors.

In Chapter 52 McMillan Lequieu and Bell ask how 'rural people interact with and mediate both horizontal and vertical power structures'. How do they deal with dis-connections and 'out-of-touch' political priorities? In response to these questions, the authors examine three pathways – examples of rural people 'opting out' of formal structures, 'opting in' and confronting/challenging them, and 'merging' – where 'merging represents a mutually beneficial, multi-scalar combining of power structures in pursuit of a locally-valued goal'. These responses are explored through case studies. The first example is of farmers 'opting out' of 'untenable market pressures, vacillating commodity prices, and environmentally damaging practices' – and into different methods of production and marketing. The originators of the 'Shepherd's Grain' project, in Washington, recognised the advantages of a high-quality wheat production method that reversed soil erosion, but needed to make it economically viable by organising into a collective which built strategic partnerships with restaurants and wholesalers. Opting out, in this example, required 'actors to build alliances outside of the rural to secure buyers who are willing to pay price premiums for social values, such as environmental and economic sustainability for the producers'. In the second example, of 'opting in', a community organisation formed to challenge a major mining application in Pennsylvania. They 'opted in' to legal processes, challenging the mining company's alleged lack of concern with pollution mitigation and seeking to address power asymmetry and defend recent gains in water quality – achieved through remediation following the defeat of a similar proposal 20 years before. The third example looks at land disputes in Uganda, where formerly displaced people are trying to re-establish customary land tenure rights. Those rights have hitherto been 'oral', passed down from one generation to the next along patriarchal lineage. War and displacement has disrupted that process. Local leaders are trying to establish customary rights by working with relatively new legal frameworks. The 'merging' here involves the codifying of the land tenure system, which had long been resisted, but which might resolve – or worsen – future conflicts. In all three cases, these different pathways connect communities to broader power structures but those communities are not 'cohesive units with unified desires'. There are internal divisions and competing goals, which may produce new conflicts and local power struggles. McMillan Lequieu and Bell shine a light on the complexity of the 'governance' challenge in rural areas – 'opting out' or going it alone presents its own opportunities and challenges, some of which were previously flagged by Homsy and Warner. 'Opting in' and challenging power asymmetries may be the only ways forward in some circumstances. Merging, or working with external power structures while representing local actors and their needs, seems to offer a means of mitigating the challenges of going it alone/opting out. The case studies presented in the opening chapters of this section clearly deal with very different challenges – but they both point to ways in which complexity and vertical/ horizontal linkages must be confronted.

The third contribution to Part IX, from Woods, stands on its own as an account of how current and emergent trends – in urbanisation, globalisation environmental change, commodification of the countryside, technological change and political/ideological pressures – may shape the future of rural places. Its focus is on the first three trends/drivers and also on the role of land use planning in mediating change – can planning work with the forces of change or is it too rigid: a constraint on the adaptive capacity of rural places? Three scenarios are presented, underpinned by the three main drivers – 'the urban countryside, the global countryside and the post-carbon countryside'. The first of these is conditioned by the increasing concentration of population in urban areas and therefore by urban encroachment, greater demands placed on rural resources (by urban society) and by the challenge of sustaining depopulating peripheral areas. Moreover, the idea of 'planetary urbanisation' – recently popularised by the work of Neil Brenner – suggests that interstitial rural regions are increasingly locked into urban economies and urban social relations. The countryside is 'de-ruralised' and part of an extended urban space. This current/future reality suggests a different approach to planning and development. Woods references authors who flag the dissolution of rural/urban boundaries and offer visions of neat and orderly rural development, much akin to Ebenezer Howard's Garden City prescription, which foresaw a coming together of town and country in a new hybridised, well-planned space. Such ideas find modern expression in different parts of the world – including in China's 'Building a New Socialist Countryside' policy, reviewed earlier in this Companion by Kan (Chapter 36). However, Woods argues that although 'planetary urbanisation may explain the economic processes underlying [new] dynamics, it arguably downplays the enduring cultural significance of rural identity' and co-exists with '*ruralisation* as a countervailing dynamic'. Broad acceptance of the 'urban countryside' thesis, but alongside additional rural dynamics – expressed, for example, through 'back to the land' migration – results in a 'global countryside' thesis that does not privilege 'the city as the driver of change'. Woods's global countryside is one that reflects the dynamic of globalisation; but that globalisation does not imply the imposition of urban power on rural space but rather new 'interactions in both urban and rural places'. These interactions produce new conflicts to be mediated; outcomes that are still shaped by local agency, resulting in places that struggle and places that prosper; a contested space in which alternate visions compete; and a space in which the path of globalisation, and the investment that it may bring, is uncertain. For rural planning, globalisation brings both new complexity and new uncertainty.

Woods's third scenario – of a 'post-carbon countryside' – supposes a radical break from current trends. Notions of the urban and global countryside are strongly rooted in present conditions. The post-carbon countryside, in contrast, breaks from the past and responds in radical ways to the challenges of climate change and peak oil. On the one hand, there may be no future justification for expending declining resources on servicing the needs of a dispersed population. That population may vacate rural areas, accentuating urban concentration and leaving behind an empty space for renewable energy or re-wilding. On the other hand, a different answer to peak oil may be a return to the land – to self-sufficient sustainable lives played out in eco-villages. Woods notes that these are strikingly different post-carbon futures – of a depopulated or repopulated countryside, and of cities as either the answer to, or generator of, excessive resource use – which would fundamentally challenge existing planning orthodoxies. Whether or not they come to pass, planning should concern itself with thinking about the future and with preparing itself for different eventualities. While rural futures are unlikely, in their entirety, to conform to any of the visions offered, 'elements of each are likely to feature in differing combinations in different locations'. And 'rural planning will play a critical role in negotiating these diverse pressures, seeking to develop and implement policies that will mediate their impacts on rural localities'.

Finally, in Chapter 54, Scott and colleagues seek to tie up any remaining loose ends following the three summative contributions to this section. The importance of thinking through alternative futures, addressing the big planetary challenges, and ensuring that planning enhances rather than hinders the adaptive capacity of rural areas – through multilevel and multisector actions that connect expertise with experiential knowledge – are all cues for a final consideration of future planning for the countryside. This final chapter tracks what planning should seek to achieve, and seek to avoid, going forward.

51

Reframing rural planning

Multilevel governance to address climate change

George C. Homsy and Mildred E. Warner

Introduction

Rural communities contribute significant amounts of greenhouse gases to the atmosphere and, at the same time, are highly vulnerable to the impacts of climate change. Around the world, droughts, fires and flooding transform the rural landscape and challenge traditional livelihoods. In parts of the Global South, 'climate refugees' flee rural communities and head into cities that are ill-equipped to receive them (Beard et al., 2016). In the US, greenhouse gases produced by agriculture and forestry make up as much as 31 per cent of total emissions (Satterthwaite, 2008). At the same time, rural cities and towns have lower residential and commercial densities and are more likely to contain older buildings and use energy technologies which make them less carbon efficient. In the US, rural residents have longer commutes (Renkow and Hoover, 2000) and less public transit, making them less carbon efficient (Glaeser and Kahn, 2010).

Rural planning must link together the environmental and economic needs of rural areas, as climate change threatens the resource base, economic stability, and quality of life within these communities. Some rural places and sectors will be hurt by climate change, while others may benefit. For example, corn yields in the US could fall from 21 to 34 per cent with soybean yields plummeting more than 50 per cent (Keane and Neal, 2017), while timber yields are expected to rise with warmer temperatures (Sohngen and Tian, 2016). Coastal fishing communities are vulnerable to economic and social dislocation due to changes in sea level, ocean temperature, and acidification (Colburn et al., 2016). Tourism will be hard hit as ski resorts lose tens of millions of visitors due to warmer temperatures and reduced snowfall (Wobus et al., 2017).

Rural planners must navigate this complex world, which includes challenges to the natural resource base as well as to land use, transportation, and economic development, which are issues more typically found in a planner's domain. Unfortunately, sustainability planning is lacking in most rural communities (Morrison et al., 2015). Rural municipalities often struggle to bring together the sometimes conflicting demands of resource extraction, commodity production, property rights, and environmental protection without the necessary knowledge, structural, fiscal, or political resources (Wolf, 2011). Indeed, while a majority of farmers from around the world believe that climate change is occurring, a significant portion do not connect the change to human activity (Houser, 2016), though government officials in rural communities more readily associate changes in local weather to human-induced climate shifts (Millar et al., 2015). As young people migrate to the cities, rural communities have to deal with increasing proportions of older residents who are increasingly vulnerable to flooding and other disasters (Krawchenko et al., 2016).

In the United States, a national survey of local government sustainability action in 2015 found that while 49 per cent of cities had sustainability plans, only 28 per cent and 29 per cent of suburbs and rural areas respectively did (Homsy et al., 2016). Suburbs can 'free ride' on city sustainability plans and actions, but rural areas cannot (Homsy and Warner, 2015). The survey also revealed that rural local governments are less likely to report having environmental, energy or climate change goals, but more commonly report economic development and disaster mitigation goals. While a third of urban communities had conducted greenhouse gas inventories and set reduction targets, less than 3 per cent of rural communities had. Both rural (73%) and urban (75%) communities reported economic development goals as part of their sustainability plans, and rural areas (58%) surpassed urban (38%) in incorporating disaster mitigation into their plans (Homsy et al., 2016). The survey also found that hazard mitigation is one area where rural planning is closer to urban planning with almost 90 per cent of respondents reporting disaster preparedness or hazard mitigation plans – increasingly important as climate change instigates more weather extremes. So while hazard mitigation may be a first step toward sustainability, it does not lead to the robust multisector and multilevel planning required to address climate change (Homsy et al., 2017).

A multilevel, multisectoral approach to climate change

Climate change requires a multilevel approach that links bottom-up efforts with top-down coordination and sanctioning power. By linking across environmental and economic sectors, rural governments can create sustainability solutions that fit local circumstances. This approach requires strong internal motivation. Based on research in rural communities around the world, Elinor Ostrom (2009, 2010) hypothesised that multisectoral collaborative efforts offer the best governance model for global commons issues, such as climate change. Community-level collaboration can build local solutions to environmental problems in a manner best suited to a particular environment and culture. The problem is that such approaches fail to address free rider and spillover concerns. For example, rapid industrialisation of many developing countries, such as China, has led to overwhelming water and air pollution as smog suffocates regions. Tight budgets, and local fragmentation and competition hinder cleanup. Air and water pollution do not respect jurisdictional boundaries, and building voluntary collaboration for environmental protection is hard. Local governments, especially rural ones, cannot control pollution by well-connected industries, particularly in the absence of strong national policy. At the same time, bottom-up efforts fail to account for gaps in fiscal and technical capacity that make planning and implementation for climate change in rural communities fall short.

Two case examples illustrate how rural communities struggle when they seek to undertake climate change mitigation on their own. The rural town of Enfield, New Hampshire (2010 population: 4,582) garners little help or coordination from either the state or federal governments. Without financial assistance or rules from above, little gets done in this poor town. 'It's a money thing,' the town manager reports. 'We need to replace the windows in town hall to save a lot of money and usually that is the first thing that gets cut [by the budget committee] because it's not an essential service'. Enfield does have an active citizens committee for sustainability and it has been a driving force bringing local and technical knowledge, political influence, and some capacity through grant writing. The committee's efforts have resulted in an energy audit for town hall, which also had its heating system replaced. In addition, they pushed the town to upgrade streetlights throughout the community. Despite the small successes, the chair of the sustainability committee laments that their bottom-up effort has trouble gaining traction because it is all volunteer-based, faces funding problems, and works within a political context that is not conducive to climate change mitigation policies.

Another rural community, Homer, Alaska (2010 population: 5,003) tried to band together in a horizontal collaboration with four other Alaskan municipalities (three urban: Juneau, Fairbanks, Anchorage, and a second rural: Sitka) to share knowledge and reduce costs after they found little support from the state or federal government for climate change mitigation efforts. The officials met at conferences and regularly connected over the telephone to discuss strategies, successes, and pitfalls. However, this voluntary network was short-lived. Other priorities distracted officials in the communities and, without top-down coordination, the collaboration ended when one local official, who had been organising the effort, retired.

In these cases, we see the limits of independent local action. Such a lesson is well-known to US environmentalists as decentralised and voluntary efforts through most of the twentieth century failed to clean up devastating air and water pollution. The federal government stepped in the 1970s with a series of major environmental laws. The newly created Environmental Protection Agency was charged with enforcement. This top-down governance framework made the United States the leader in environmental protection, by cleaning up the worst environmental problems (Fiorino, 2006). Some see such a top-down, command and control system as the necessary model of action on issues of the global commons, such as climate change mitigation (Stavins, 2010).

However, we argue that such a system lacks the local knowledge and on-the-ground buy-in to tackle more complex problems, such as climate change. Instead a multisectoral, multilevel approach is needed. Multilevel governance is best known among scholars of the European Union (Bulkeley and Betsill, 2003). This framework engages multiple tiers of government in a communicative process; authority becomes diffuse, not only across various public actors, but also among private entities (Kern and Bulkeley, 2009). Local knowledge and expertise become part of the discourse, not as a public participation exercise, but as a partnership between citizens, policymakers, and technical analysts (Homsy and Warner, 2013a). Such democratisation of knowledge and power offers the most effective path for avoiding disasters and addressing climate change, especially in rural areas. Knowledge and policy are embedded in customs, identities, institutions, and other components of society (Funtowicz and Ravetz, 1993). A multilevel governance framework must address these specific local conditions as well as differences in capacity between rural and urban governments. Rural municipalities often face weaker economic development and smaller tax bases or technical know-how and fiscal resources to act on climate change mitigation (Carter and Culp, 2010). Governing climate change requires a multilevel approach because climate change sits at the intersection of environmental protection, economic vitality, and social justice. Also, actors must break through administrative silos and geographical boundaries to tackle the complexity of reducing greenhouse gas emissions.

Homsy and Warner (2013a) set out a framework for multilevel planning around climate change mitigation, which involves co-production of knowledge with rural communities, reframing planning around co-benefits, engaging civil society, providing technical support and sanctioning and coordinating power from higher levels of government. A watershed management case illustrates the power of this framework for planning (ibid.). The Rouge River watershed, in the US state of Michigan, consists of four waterways, a 44-mile-long main stream and three major branches.

More than 1.5 million people live in 48 communities including rural, suburban, and urban places, within the 467-square mile watershed. The Rouge River was once the dirtiest in Michigan, and one of the worst in the nation, due to waste from auto companies and steel mills. After the passage of the Clean Water Act in 1972, the direct discharge of toxic pollution ended, but non-point source pollution, a more complex concern, continued to afflict the waterway.

Since non-point source pollution consists of runoff from, for example, all of the various roads, farms, and lawns in a region, it is trickier to manage in a one-size-fits-all, top-down manner. To tackle this more diffuse source, the US Environmental Protection Agency wanted the input and support of local communities, and it undertook what evolved into a multilevel approach. The federal government used the threat of a top-down, expensive court-ordered cleanup to press 45 municipalities to sign on to a collaborative effort. Regulators divided the watershed into seven sub-areas, each with different physical and cultural characteristics. Advisory groups in each sub-watershed developed plans for controlling non-point source pollution based on the local conditions and local economies; rural areas crafted much different plans than urban ones. The process in each place engaged citizens, activists, and experts as well as local officials and business owners. Over decades, the federal government provided hundreds of millions of dollars to fund different projects in the watershed. The multilevel approach worked; wildlife is returning to the area and fish are now safe to eat.

Translating multilevel governance to climate change mitigation would require that national (or state/regional) governments coordinate local efforts, insure that all participated, set goals, and provide capacity for initiatives tailored to regions or communities. For example, some rural communities might tackle agricultural emissions while others might find the biggest reductions can come from upgrading municipal operations or revising building codes or adding transit options. The key is the construction of both vertical and horizontal collaborations that use capacity at higher levels in conjunction with local expertise and knowledge.

Reframing climate action as a local concern

The discourse in rural communities, when dealing with sustainability issues, such as climate change, focuses on the tension between economic prosperity and environmental protection (Kessler et al., 2016). Therefore, rural planners often try to localise climate change by bundling it with other issues or reframing it in terms of other benefits (Mayrhofer and Gupta, 2016). A co-benefit approach allows rural officials to locally pursue sustainability actions with long-term and distant concerns by reframing it in terms of more immediate local gains (Spencer et al., 2017). This reframing also helps broaden the constituencies for and boost the effectiveness of sustainability actions (Puppim de Oliveira et al., 2013).

In the following two case examples, we found that officials in successful rural communities and small towns sought to reframe global climate change in terms of local economic development and undertake a variety of strategies to reduce greenhouse gas emissions. In each, the leaders recognised the importance of climate change to local and global environmental health as well as a responsibility to future generations. Yet, these officials were also adamant that the initial framing of the issue had to be along local and economic lines.

Homer, Alaska started developing greenhouse gas reduction strategies in 2006. A state-wide conference that year made it clear to local leaders that their rural city faced dual threats with its location on the coast and at a high latitude, where climate change's impacts will be stronger. The city manager reported that Homer's position on Kachemak Bay in southern Alaska means that tourism, fishing, and logging developed as important economic sectors and that business leaders and residents saw the negative impact of climate change damage – and its threat to these sectors. 'We are definitely seeing changes in the climate, more frequent storms and pests, like the spruce bark beetle, that haven't happened for many years, but are devastating forests around here'.

Such local impact to their economic foundation sparked action. Homer drafted a climate change plan in 2007 that set a 20 per cent target in greenhouse gas emissions (from 2000 levels) by 2020. A separate economic development plan for the city addressed both the negative economic impacts of the changing climate as well as some potential positive ones. For example, the city took steps to more sustainably manage fisheries given the rising temperatures and increased acidification caused by climate change and mitigate the increased pest attacks on the forests. At the same time, the report recognised that warming temperatures may mean a longer growing season and open agricultural opportunities. Therefore, the plan recommended appointing a food commission to investigate ways to boost local farming. The community also seeks to implement 'Smart Growth' land use and transportation policies to reduce the reliance on cars; support programs to improve energy efficiency in homes and businesses; and develop renewable energy sources, such as tidal power.

Homer mandates that all new employees undergo sustainability training, one of the first communities in the nation to do so. New municipal employees received a 13-page energy policy guide that emphasises the connection between energy conservation and saving money. The booklet is a model for other local governments with similar efforts. To finance environmental improvements in municipal operations, the city created a revolving energy fund. Every new project must have a plan to conserve energy and the projected savings are put back into the fund. Using high efficiency lights in the harbour saved $150,000 annually, and another project cut energy costs at the sewerage treatment plant by 20 per cent. The city manager calls the reframing a success:

> Our economic development commission is pretty darn green. They're pushing things like sustainable agriculture and farmers' markets and buying local and eating organic produce is a really big deal here. It started out as a philosophy, but now it's becoming a major part of the economy.

Rather than just saving money, Columbus, Wisconsin (2010 population: 4,991) reframed sustainability as a strategy for attracting economic development. The state limits the ability of municipalities to use policy tools, such as property tax abatements, to entice companies to locate there. So, in 2007 the city manager used a grant to create a new staff position with dual responsibilities: sustainability and economic development. The new person sought to market an identity for Columbus as a green community. The municipality was a pioneer in converting street lights to high-efficiency LED fixtures. They added electric car charging stations and purchased hybrid vehicles for the government fleet. Every purchasing decision had to consider Columbus's growing green reputation. In addition to upgrading public buildings, educational programs and incentives sought to encourage energy audits as well as conservation upgrades to homes and the purchase of energy-efficient appliances.

State-wide economic development and construction magazines picked up on the sustainability story and that publicity generated many economic development leads. Local officials report that in 2012, about $30 million dollars of capital investment flowed to Columbus including a new housing development, an assisted living center, and an arts incubator. A packaging plant and a plumbing manufacturer, already in the community, expanded their operations. Other factors certainly play into the location decisions, such as Columbus's proximity to the state capital and its access to a major highway. Still, the city manager estimates that 50 per cent of his community's recent success is attributable to its sustainability marketing program. Columbus is saving money as well. For example, municipal electric use dropped by 15.4 per cent from 2007 to 2012.

We can see in these case examples that officials in rural communities must tie climate change and other sustainability issues to economic development. Whether the community is rich or poor, this framing appears to be the main way forward. This link between environmental sustainability and economic development has been found more broadly in national surveys of local economic development policy (Zhang et al., 2017). Other cases find that local planners can leverage water crises or rising fuel prices to push climate change mitigation strategies (Homsy and Warner, 2013b). This co-benefits approach allows planners to appeal to some in a community with concerns about global warming; while reaching out to others on the grounds that a policy saves money, attracts investment, or lessens the risk of hazards. Rural planners, especially facing scepticism about climate change, need to reframe such controversial issues so that they can appeal to a broad audience.

Rural planning for climate change

Planners and officials in small cities and rural towns seeking to reduce greenhouse gas emissions have an important, but complex task. Not only must they plan land use and built environments, but they also need to manage their natural resource economies. Many rural communities have farming, forestry, mining, tourism, fishing, industrial, and other operations that both contribute greenhouse gases to the climate change problem and will suffer the impacts of a warming planet. Rural communities typically must tackle these challenges with fewer resources than urban or suburban places and in political environments often less open to such issues. Still, as our cases illustrate, small towns and rural communities can creatively craft policies to reduce greenhouse gas emissions. Similar examples in the literature demonstrate that planners must think broadly about the governmental and citizen audiences for proposed action and build constituencies of support. Other research has similarly found that farmers embrace climate change adaptations based on their local experience with weather even if they do not connect shifts in temperature or rain to climate change (Yung et al., 2015; Takahashi et al., 2016). Another study has found that rural residents are more likely to take part in community energy programs than their urban peers (Kalkbrenner and Roosen, 2016). This re-framing strategy is important as communities build internal motivation for action on global and regional commons issues. Reframing broadens the buy-in of citizens and increases the likelihood of implementation success.

However, we have also shown that individual action by local governments is not enough. Rural planners, who find ways to create internal motivation for action, still face commons challenges. Horizontal and vertical collaboration within a multilevel and multisectoral framework is a crucial part of policy co-production and successful implementation. But such collaboration is difficult without a central authority that respects local action and has the power to sanction local governments that do not act. Such an entity can also be a source of technical or fiscal capacity to rural governments and can leverage the internal goals of a community into action on global issues, such as climate change.

Climate change is one of the most complex challenges facing the planet. As such, it cannot be tackled simply by top-down, one-size-fits-all mandates or by bottom-up individual action by local governments. Planning for climate change in rural communities requires creativity of local officials in driving appropriate policies and the collaboration that comes within a multisectoral, multilevel framework. Rural governments are crucial to climate change mitigation efforts, but local efforts, especially in rural communities, need multiple levels of support to be successful.

References

Beard, V. A., Mahendra, A. and Westphal, M. I. (2016) *Towards a More Equal City: Framing the Challenges and Opportunities*. Working paper. Washington, DC: World Resources Institute.

Bulkeley, H. and Betsill, M. (2003) *Cities and Climate Change: Urban Sustainability and Global Environmental Governance*. London: Routledge.

Carter, R. and Culp, S. (2010) *Planning for Climate Change in the West*. Policy Focus Report Code PF024. Cambridge, MA: Lincoln Institute of Land Policy.

Colburn, L. L. Jepson, M., Weng, C., Seara, T., Weiss, J. and Hare, J. A. (2016) 'Indicators of climate change and social vulnerability in fishing dependent communities along the Eastern and Gulf Coasts of the United States', *Marine Policy*, 74(Supplement C), pp. 323–333. doi: 10.1016/j.marpol.2016.04.030.

Fiorino, D. J. (2006) *The New Environmental Regulation*. Cambridge, MA: The MIT Press.

Funtowicz, S. O. and Ravetz, J. R. (1993) 'Science for the post-normal age', *Futures*, 25(7), pp. 739–755. doi: 10.1016/0016-3287(93)90022-L.

Glaeser, E. L. and Kahn, M. E. (2010) 'The greenness of cities: carbon dioxide emissions and urban development', *Journal of Urban Economics*, 67(3), pp. 404–418.

Homsy, G. C., Liao, L. and Warner, M. E. (2017) 'Sustainability and disaster planning: what are the connections?', *Working paper*.

Homsy, G. C. and Warner, M. E. (2013a) 'Climate change and the co-production of knowledge and policy in rural USA communities', *Sociologia Ruralis*, 53(3), pp. 291–310. doi: 10.1111/soru.12013.

Homsy, G. C. and Warner, M. E. (2013b) *Defying the Odds: Sustainability in Small and Rural Places*. Briefing Paper. Washington, DC: ICMA Center for Sustainable Communities. Retrieved on 16 June 2014 from www.nado.org/wp-content/uploads/2013/10/defying_the_odds_briefing_paper.pdf.

Homsy, G. C. and Warner, M. E. (2015) 'Cities and sustainability: polycentric action and multilevel governance', *Urban Affairs Review*, 51(1), pp. 46–73. doi: 10.1177/1078087414530545.

Homsy, G. C., Warner, M. E. and Liao, L. (2016) 'Sustainability and local governments: planning helps balance environmental, economic, and social equity priorities', *Local Government Review*, December, pp. 5–14.

Houser, M. (2016) 'Who framed climate change? Identifying the how and why of Iowa corn farmers' framing of climate change', *Sociologia Ruralis*, p. n/a-n/a. doi: 10.1111/soru.12136.

Kalkbrenner, B. J. and Roosen, J. (2016) 'Citizens' willingness to participate in local renewable energy projects: the role of community and trust in Germany', *Energy Research & Social Science*, 13, pp. 60–70. doi: 10.1016/j.erss.2015.12.006.

Keane, M. and Neal, T. (2017) *The Impact of Climate Change on US Agriculture: New Evidence on the Role of Heterogeneity and Adaptation*. Oxford: Economics Group, Nuffield College, University of Oxford.

Kern, K. and Bulkeley, H. (2009) 'Cities, Europeanization and multi-level governance: governing climate change through transnational municipal networks', *JCMS: Journal of Common Market Studies*, 47(2), pp. 309–332. doi: 10.1111/j.1468-5965.2009.00806.x.

Kessler, A., Parkins, J. R. and Huddart Kennedy, E. (2016) 'Environmental harm and "the good farmer": conceptualizing discourses of environmental sustainability in the beef industry', *Rural Sociology*, 81(2), pp. 172–193. doi: 10.1111/ruso.12091.

Krawchenko, T., Keefe, J., Manuel, P. and Rapaport, E. (2016) 'Coastal climate change, vulnerability and age friendly communities: linking planning for climate change to the age friendly communities agenda', *Journal of Rural Studies*, 44(Supplement C), pp. 55–62. doi: 10.1016/j.jrurstud.2015.12.013.

Mayrhofer, J. P. and Gupta, J. (2016) 'The science and politics of co-benefits in climate policy', *Environmental Science & Policy*, 57(Supplement C), pp. 22–30. doi: 10.1016/j.envsci.2015.11.005.

Millar, J. E., Boon, H. and King, D. (2015) 'Do wildfire experiences influence views on climate change?', *International Journal of Climate Change Strategies and Management*, 7(2), pp. 124–139. doi: 10.1108/IJCCSM-08-2013-0106.

Morrison, T. H., Lane, M. B. and Hibbard, M. (2015) 'Planning, governance and rural futures in Australia and the USA: revisiting the case for rural regional planning', *Journal of Environmental Planning and Management*, 58(9), pp. 1601–1616. doi: 10.1080/09640568.2014.940514.

Ostrom, E. (2009) *A Polycentric Approach for Coping with Climate Change*. Policy Research Working Paper 5095. Washington, DC: World Bank.

Ostrom, E. (2010) 'Polycentric systems for coping with collective action and global environmental change', *Global Environmental Change*, 20(4), pp. 550–557. doi: 10.1016/j.gloenvcha.2010.07.004.

Puppim de Oliveira, J. A., Doll, C. N., Kurniawan, T. A., Geng, Y., Kapshe, M. and Huisingh, D. (2013) 'Promoting win–win situations in climate change mitigation, local environmental quality and development in Asian cities through co-benefits', *Journal of Cleaner Production*, 58, pp. 1–6. doi: 10.1016/j.jclepro.2013.08.011.

Renkow, M. and Hoover, D. (2000) 'Commuting, migration, and rural–urban population dynamics', *Journal of Regional Science*, 40(2), pp. 261–287. doi: 10.1111/0022-4146.00174.

Satterthwaite, D. (2008) 'Cities' contribution to global warming: notes on the allocation of greenhouse gas emissions', *Environment and Urbanization*, 20(2), pp. 539–549. doi: 10.1177/0956247808096127.

Sohngen, B. and Tian, X. (2016) 'Global climate change impacts on forests and markets', *Forest Policy and Economics* (New Frontiers of Forest Economics: Forest Economics beyond the Perfectly Competitive Commodity Markets), 72(Supplement C), pp. 18–26. doi: 10.1016/j.forpol.2016.06.011.

Spencer, B., Lawler, J., Lowe, C., Thompson, L., Hinckley, T., Kim, S. H., Bolton, S., Meschke, S., Olden, J. D. and Voss, J. (2017) 'Case studies in co-benefits approaches to climate change mitigation and adaptation', *Journal of Environmental Planning and Management*, 60(4), pp. 647–667. doi: 10.1080/09640568.2016.1168287.

Stavins, R. N. (2010) 'The problem of the commons: still unsettled After 100 years', *National Bureau of Economic Research Working Paper Series*, No. 16403. Available at: www.nber.org/papers/w16403.

Takahashi, B., Burnham, M., Terracina-Hartman, C., Sopchak, A. R. and Selfa, T. (2016) 'Climate change perceptions of NY State farmers: the role of risk perceptions and adaptive capacity', *Environmental Management*, 58(6), pp. 946–957. doi: 10.1007/s00267-016-0742-y.

Wobus, C., Small, E. E., Hosterman, H., Mills, D., Stein, J., Rissing, M., Jones, R., Duckworth, M., Hall, R., Kolian, M. and Creason, J. (2017) 'Projected climate change impacts on skiing and snowmobiling: a case study of the United States', *Global Environmental Change*, 45(Supplement C), pp. 1–14. doi: 10.1016/j.gloenvcha.2017.04.006.

Wolf, S. A. (2011) 'Network governance as adaptive institutional response: the case of multifunctional forested landscapes', *Journal of Natural Resources Policy Research*, 3(3), pp. 223–235. doi: 10.1080/19390459.2011.591760.

Yung, L., Phear, N., DuPont, A., Montag, J. and Murphy, D. (2015) 'Drought Adaptation and Climate Change Beliefs among Working Ranchers in Montana', *Weather, Climate, and Society*, 7(4), pp. 281–293. doi: 10.1175/WCAS-D-14-00039.1.

Zhang, X., Warner, M. E. and Homsy, G. C. (2017) 'Environment, equity, and economic development goals: understanding differences in local economic development strategies', *Economic Development Quarterly*, 31(3), pp. 196–209. doi: 10.1177/0891242417712003.

52

Rural governance and power structures

Strategies for negotiating uneven power between local interests and external actors

Amanda McMillan Lequieu and Michael M. Bell

Introduction

Rural communities have their own internal logics of power – cultural, economic, and political – which can exist in some isolation from the external logics of the market, state, or legal power structures. When the interests of external social structures impinge upon the internal logics of rural places, the internal may well be overwhelmed by the external, which is by and large more potent and impactful. Yet rural communities can also intentionally, and sometimes successfully, engage with the resulting conflicts of power.

This chapter focuses on rural governance processes and how local actors can contend with and mediate external power dynamics. We animate the concept of rural governance, that process through which rural people and their institutions make decisions about the social organisation of their collective wellbeing. Crucially, we extend governance as being more than a matter of government or decision-making, traditionally conceived. Both within and without institutions, rural people engage formal and informal responses to power differentials and mismatched interests and cope with the results of decisions made throughout the structures of governance. We propose a distinction for this engagement that accounts for differences in conflicts of power: horizontal and vertical power structures. By *vertical power structures*, we mean forces that stratify social relations within a place, such as class, patriarchy, racism, and homophobia. By *horizontal power structures*, we mean forces that subjugate one place to another such as imperialism, colonialism, core-periphery relations, the distribution of hazards, and the sense of some regions as culturally advanced and others as culturally backward. Vertical and horizontal power structures typically mutually support each other, combining in the maintenance of kyriarchy. As well, vertical power structures in one place may depend upon horizontal subjugation of other places, such as economic exploitation of rural places supporting class relations in urban places (Bell, 2018), and vice versa.

Through both formal and informal means of rural governance, however, rural people are not helpless in the face of horizontal power. They can resist. There is power in the rural itself, in its people, and in their social organisations (Bell et al., 2010). We distinguish between three general options for resistance: *opting out, opting in*, and *merging*. Opting out refers to a process of systematic removal of ones' local governance structures from the most oppressive or troubling of horizontal or vertical power structures. Opting in is characterised by intentional engagement in conflict with power structures through formal and informal means. Merging represents a mutually beneficial, multi-scalar combining of power structures in pursuit of a locally valued goal.

In what follows, we offer a case study of each option. First, we consider a grain farming cooperative in Washington which emerged as a response to increasing marketisation pressures by opting out of dominant market structures and pursuing alternative markets for direct-market, value-based flour. Next, in the foothills of the Appalachian Mountains, a small watershed organisation responded to a proposed coal mine by opting into legal appeal processes. Finally, Ugandan national leaders successfully merged oral, customary land tenure into their constitution to both protect traditional leadership structures and expand federal jurisdiction.

In each case, resistance emerges from conflict about the future of a place and between structural forces. Opting in, opting out, and merging could be understood as expressions of otherness, extensions of Castell's (1997) 'resistance identity'. Resistance identities are individual or collective response to the hegemonic power of top-down policymaking, typically generated by actors that are in positions or conditions often devalued, stigmatised, or overlooked by logics of political or economic domination (Zimmerbauer and Paasi, 2013: 33). Strategies for resistance are strategies for governance. Which strategies are selected by key actors in these case studies reflect the different solutions offered by opting out, opting in, and merging. Through opting out of existing power structures, individuals and institutions can pursue alternative sources of economic power for rural economies. Opting in attempts to mitigate a lack of access and efficacy of local groups in local, state, and federal bureaucracies. Merging accounts for a lack of vertical legitimisation of locally valued rights.

Through each typological category – opting out, opting in, and merging – we take care to describe the requirements of each form of response to external power structures in order to illuminate the context-based desirability and limitations of that form of response. While doubtless limited, as is any typology, our hope is this clear articulation of forms and requirements of grassroots response to external interests may aid policymakers, activists, and engaged community members in not only empowering local communities to more cohesively and creatively engage in power struggles, but recognise existing, and perhaps understated, responses.

Opting out: Shepherd's Grain

Opting out of an unjust, unsustainable, or otherwise undesirable situation is a common response of rural actors. Consider the familiar case of farmers turning from conventional, high-input production of plants and animals to alternative, value-added, and value-based growing methods. By opting out of untenable market pressures, vacillating commodity prices, and environmentally damaging practices and into different methods of production and marketing, even mid-sized farmers can often remain on their farms and in their preferred occupation across eras of economic insecurity.

The case of the Shepherd's Grain cooperative in the Palouse region of eastern Washington is a good example. In the mid-1980s, two wheat farmers, Karl Kupers and Fred Fleming, shared a concern for the ecological and economic unsustainability of conventional dryland wheat farming in the American northwest. They envisioned an alternative to what was a high-input and high-risk agricultural project: a method of high-quality wheat production that could actually reverse soil erosion while keeping its farm financially viable enough to reject federal commodity subsidies. In cooperation with local university scientists, they began using direct seed, no-till cropping systems that minimised soil turning, and rotational cropping with legumes, barley and alfalfa. The farmers were pleased with how this low-input farming system improved soil quality, decreased pesticide use, and promised to provide significant carbon sequestration from reduced tillage and soil disturbance. However, low-input grain production requires more intensive labour and upfront costs than continuing high-input, traditional production. Kupers and Fleming identified farmers in their region who were similarly concerned about environmental consequences of grain production and established the Shepherd's Grain collective to coordinate their alternative agricultural efforts.

To sustain the cooperative, however, Kupers and Fleming had to develop a similarly alternative marketing model that allowed them to earn consistently higher prices per bushel of grain. They opted out of traditional farming methods for wheat; they then opted out of typical market mechanisms as well. First, the cooperative developed strategic partnerships with local restaurants willing to pay a premium for 'regionally identified flours'. Shepherd's Grain uses strategic supply chain partnerships to replace the capital and expertise that otherwise would be required to handle grain milling and distribution. Nearly all sales are direct wholesale, which means that Shepherd's Grain depends on its customers to preserve its brand identity. Strategies for maintaining brand identity include recognition on partner websites and farmer visits to partner enterprises like the Bon Appétit cafés.

Next, Shepherd's Grain decided to set stable, year-long prices for their wheat based on cost of production plus a reasonable rate of return rather than charge a premium above commodity wheat prices. Cost of production is calculated as the sum of on-farm production expenses, transportation costs, Shepherd's Grain administrative fees and milling fees. Their aim was to disconnect the price received by Shepherd's Grain producers from commodity wheat prices so their farmers could receive a more stable and equitable return. By 2011, Shepherd's Grain was processing and marketing more than 500,000 bushels of wheat for 33 growers. Shepherd's Grain emphasises its story of *opting out* of the conventional grain production approaches by utilising environmentally friendly methods on multigenerational, family farms, and emphasising regional identity throughout the processing and consumption of the wheat. *Opting out* requires actors to build alliances outside of the rural to secure buyers who are willing to pay price premiums for social values, such as environmental and economic sustainability for the producers.

Opting in: the Mountain Watershed Association

Second, rural community organisations often selectively *opt in* to some power structures to resist other ones, particularly when there is a clear and viable mechanism for protest and resistance built into those structures. Consider the example of a certain western Pennsylvania watershed organisation attempting to resist state permitting of a new coal mine. In early 2017, the Mountain Watershed Association filed an appeal of the issuance of a permit for LCT Energy 'Rustic Ridge #1' mine with the Environmental Hearing Board. The proposed Rustic Ridge deep coal mine, in the Indian Creek sub-basin of the Youghiogheny River watershed of western Pennsylvania, would span nearly 3,000 underground acres with a 67-acre surface facility. This rural region of Pennsylvania is familiar with abandoned mine runoff impacting waterways and wells. While several decades of remediation efforts have improved water quality in the Youghiogheny River, the region remains economically reliant on rural resources, ranging from farming, to existing extraction, to tourism.

The Watershed's legal team filed suit against both the company pursuing mining and the local branch of government in charge of environmental regulation, arguing that neither group of actors had done their due diligence to protect the environment surrounding the proposed mine.

Permitting activity that would cause harm to waterways is illegal under Pennsylvania law. The mining company, the suit argued, offered insufficient preventative strategies to limit polluting discharges into the watershed, and that the Pennsylvania Department of Environmental Protection (DEP) did not adequately challenge the environmental gaps in the company's proposal.

Through this legal action, Watershed pointed out that twenty years ago, a proposed deep coal mine was proposed in the same area. At that time, the regional DEP office predicted significant water pollution and denied that mine proposal. 'As there have been no major geologic events in the permitted area since 1994 that would alter the potential for harms previously identified by the DEP, in issuing this permit the DEP has ignored their own findings from 1994,' stated Beverly Braverman, Executive Director of the Mountain Watershed Association. 'In appealing this decision, we hope the DEP's decision will be voided and that LCT will be required to permanently and immediately cease mine construction'.

The Mountain Watershed Association has faced resistance to their appeal. At an event sponsored by the company proposing the new coal mine, a spokesman for LCT Energy stated that a two million dollar 'community trust' would be established to support area projects only if residents did not exercise their legal right to bring the appeal. The company spokesman explained: 'If we spend a couple million dollars defending the permit, which we are prepared to do, it will be very difficult to provide funding for the community fund'.

Formally filing a legal appeal is a form of *opting in* to a formalised process of protest and resistance. Opting in requires resources – social capital, financial support, and knowledge about the appropriate and myriad ways to respond through the legal and other forma, extrajudicial systems. The organisation claims an active membership of fifty people and benefits from the pro bono aid of a local environmental justice firm. While not claiming a large standing budget, the organisation has some leverage in the region as the acting grant-disburser for waterway reclamation funds donated by formerly active mines in the region. Since opting in to formal protest involves fewer actors (organisation leadership, expert witnesses, and lawyers), members continue to be involved in more traditional forms of resistance – letter writing, fund raising, and on-site protests.

Merging: the Acholi of Uganda

The third strategy is *merging*: integrating formerly opposed power structures to pursue some common, or at least agreed upon, goal. To explicate this strategy, consider the case of customary land tenure in northern Uganda. According to customary law, land is *accessed*, rather than *owned*. In Uganda, like much of Africa, access rather than ownership means that land cannot be privatised and sold. All customary land in Uganda is collectively owned by the clan, with access being granted based on patriarchal lineage. The Acholi, of northern Uganda, exemplify this pattern of clan-based, customary land tenure. Access to land depends on intergenerational continuity, oral tradition, and social relationships legitimised by clan members and patriarchal hierarchies. This land tenure model was threatened during the war and internal displacement in the region between 1994 and 2006. After two decades of war, family and clan members reconvened from internal displacement camps at overgrown homesteads, struggling to recall forgotten boundaries. Violence frequently erupted between junior and senior men, squatters and returnees, and between widows and their late husbands' families.

To mitigate this unrest, some Acholi chiefs have been drawing upon the 1998 Land Act. This statutory law formalised legal pluralism by explicitly recognising customary rights to property. The Land Act 1998 defines customary tenure rights and lays out a process for registration and administration of customary rights. Local leaders are actively merging oral traditions with this process of written registration to manage conflicts over access, belonging, and lineage in the wake of such prolonged, regional displacement. According to a legal officer working in Acholiland: 'We take their law books, we examine their customs and cultures. We intend to marry this together – because the law cannot function solely, it must function within the fabric of the customs'.

This pattern of merging is new, as the Acholi have long resisted privatising land, or even codifying their land tenure system based on oral tradition, due to fears that such formalisation of ownership may contribute to their loss of access to their land. Ugandans are familiar with the impact of privatisation of land. Nineteenth and twentieth century colonisation threatened and, in southern and central Uganda in particular, usurped customary land tenure. Indeed, land tenure issues hint at the tangle of power conflicts facing Uganda today. On one hand, the government of Uganda prefers freehold, or private title, land ownership. Privatisation allows development of rural areas through agribusiness, commerce, and other corporate investment in land. However, indigenous people prioritise collective ownership through customary land tenure, and any land sale requires agreement of the clan leadership. After all, even if residents were compensated for their land, they were displaced from a landscape that provided material, structural, and symbolic meanings to their social worlds. According to Otim (not his real name), a middle-aged Acholi man, 'home is where ancestors live, where your god dwells, where you were born, and where your kids live'.

Merging required two stages. First, the Ugandan government provided legitimisation, and legal protection, of customary land tenure by incorporating Acholi customary land tenure into the Ugandan constitution through the 1998 Land Act. Second, local, Acholi leaders legitimised the Land Act, and in the process, their own decisions regarding land access in a time of crisis. Both national and local leaders can benefit from *merging* as a strategy for managing power differentials, regional upheaval, and threats to traditional methods of conflict management. Incorporating community-based or regionally scaled customs into national jurisdiction may institutionally protect those regionally specific approaches while further legitimising the state. Long-term impacts are hard to predict, though. Merging can exacerbate power differentials, or the state may be legitimised on paper but impotent to make real change on the ground.

Caveats: navigating changing times

The long-term outcomes of each of these cases remain unclear, contingent on variable agricultural policies, legal outcomes, and hyperlocal power dynamics. Strategies of choice – opting in, opting out, or merging – may not fully address the changing power structures or vulnerabilities of rural people. The viability of seeking and achieving more appropriate power structures depends not only on the political efficacy of actors, but on the timeliness of external factors, the unanimity of local communities, political transparency of local and regional bureaucracies.

Moreover, at times, efforts to redistribute power within rural governance processes and between local interests and external actors are contingent upon volatile, external factors. For instance, in 2007, rapidly increasing commodity wheat prices challenged the year-long cost-of-production-based pricing model used by Shepherd's Grain. As the stable price that had been set for the year by Shepherd's Grain was well below the market price for commodity wheat, the cooperative had to renegotiate its strategies of opting-out of open-market pricing. It now takes into account the relationship between Shepherd's Grain wheat prices and the commodity market when setting its twice-yearly membership prices.

Second, organisational representation in formal processes of resistance, protest, or appeal, may occur concurrently with individual actions of acquiescence, compromise, or neutrality.

In Pennsylvania, some residents are trying to mitigate possible property value loss due to water pollution from the proposed deep coal mine by increasing value of their properties. These informal strategies range from planting mushrooms and ginseng to sell at nearby town farmers market to developing non-natural-resource based rural businesses. While these forms of compromise – or perhaps, acquiescence due to a sense of inevitability – complement the Watershed's formal engagement in the legal appeal, other individuals in the region maintain a neutral or even pro-mine stance, and either passively or actively critique the Watershed's approaches.

Third, intention and application of policies may not be implemented by the people closest to the conflict situation. In Uganda, the Ker Kwaro Acholi – the highest level of customary leaders among the Acholi people – are in the process of recording their rules governing customary tenure in the form of the Principles, Practices, Rights and Responsibilities. However, some Acholi chiefs are resisting the burden of formal, written laws in order to strengthen their own control of their villages. Although the formal system for administrating land tenure, carrying out transactions and settling disputes in Uganda is quite clear, the practice is less so.

Each of these cases blend formal and informal strategies of resistance to counter unbalances in both vertical and horizontal power structures. On one hand, horizontal forces subjugate markets in Washington state to core pricing structures, distribute risks of water pollution due to mining to a particular watershed in Pennsylvania, or trouble hierarchies in Acholiland. On the other hand, vertical forces of internal conflict stratify social relations and complicate actors' goals and responses to conflict. After all, it is key to keep in mind that communities are not cohesive units with unified desires. Communities are divisive to the individual level, fields of interaction, characterised by diversity within their geographically local landscapes. Communities are complex, with solidarity disintegrating on the individual level due to conflicts of values, histories, and power.

Conclusion

Rural people interact with and mediate horizontal and vertical power structures in myriad ways. Tensions between external structures and local actors can perpetuate political, economic, and legal disconnections between rural communities and their broader socioeconomic and political contexts. Rural communities facing their own power struggles often pursue different local outcomes than those proposed or pursued by power structures. Some may proceed by *opting out* of undemocratic, vertical or horizontal power dynamics by pursuing non-traditional markets. Others may face undesirable rural policies head-on by *opting into* legal due process. Still other rural communities may find solutions to the uneven distribution of power by *merging* formal and informal political authority.

This chapter utilised the cases of how members of three rural communities – a mid-sized grain cooperative in Washington, USA, a coal mining community in Pennsylvania, USA, and a region of northern Uganda facing challenges to their customary land tenure – engaged in alternative forms of resistance, protest, or compromise to power structures. The complexities of these specific cases, but briefly illuminated in this short space, emphasise the challenges of successful reorganisation of power.

Indeed, the viability of seeking and achieving more appropriate power structures depends on the social capital, economic savvy, and political efficacy of actors, as well as the political transparency of top-down bureaucracies. Thus, trying to change policy might not be the only or best ways for rural communities to pursue different local outcomes than those proposed or pursued by vertical and horizontal power structures.

References

Bell, M. M. (2018). *City of the Good: Nature, Religion, and the Ancient Search for What Is Right*. Princeton, NJ: Princeton University Press.

Bell, M. M., Lloyd, S. and Vatovec, C. (2010). Activating the Countryside: Rural Power, the Power of the Rural, and the Making of Rural Politics. *Sociologia Ruralis*, 50(3), 205–224.

Castells, M. (1997) *Power of Identity: The Information Age: Economy, Society, and Culture*. Oxford: Blackwell Publishers.

Zimmerbauer, K. and Paasi, A. (2013). When Old and New Regionalism Collide: Deinstitutionalization of Regions and Resistance Identity in Municipality Amalgamations. *Journal of Rural Studies*, 30, 31–40.

53

The future of rural places

Michael Woods

Introduction

In 2012, the noted architect and urban theorist Rem Koolhaas delivered a lecture at the Stedelijk Museum in Amsterdam, in which he argued that 'the countryside is now the frontline of transformation' (Koolhaas, 2012). The lecture, subsequently published as an essay in *ICON* magazine, described changes and trends that Koolhaas had observed in the Dutch and Swiss countrysides, and challenged architects to move beyond their urban fixation and pay more attention to rural landscapes. Noting that the corollary of global urbanisation is the depopulation of rural areas, Koolhaas suggested that, 'you could therefore see the countryside as a place where people are disappearing from. In this void, new processes are taking place and new experiments and developments are being made' (Koolhaas, 2014: n.p.). In spite of the tone of discovery in the lecture and essay, Rem Koolhaas was not the first to make these observations, and much of the content discussed changes in the countryside that had already been painstakingly documented and analysed by rural geographers, sociologists and planning researchers over the last few decades. However, Koolhaas's celebrity gave the issue of rural change sudden new prominence, and his application of an architect's eye to the countryside contrasted with the tendency of social science researchers to focus on past change and present experiences by emphasising future trajectories and possibilities, and particularly the question of how the future countryside might be planned and designed.

It is not only architects who have started thinking about the future of rural places. Policy-makers too have begun to debate the future of rural areas, as the assumptions and principles that underpinned the long period of stable rural policy in the post-Second World War era have been challenged by social, economic, environmental and political pressures. The European Commission published a discussion document titled *The Future of Rural Society* as early as 1988 and has repeatedly returned to the question in debates around reforms to the Common Agricultural Policy. In the United States, the five-yearly renewal of the Farm Bill have increasingly become sites of struggle over competing alternative visions of the rural future. In China, the policy of 'Building a New Socialist Countryside', or rural reconstruction, has been a priority for the government for over a decade now, as it negotiates rapid urbanisation and deepened rural–urban inequalities. In Britain, questions about the future of the countryside fed into intermittent efforts to reshape rural policy by the Conservative government in the 1990s and the Labour government in the 2000s and have resurfaced with the prospect of Britain's withdrawal from the European Union.

These debates and speculations are taking place in different geographical and political-economic contexts, with different participants, and different objectives, but they all draw on shared patterns and experiences of rural change, in which several key drivers can be identified:

Urbanisation – The trajectory of a shift in the balance of the population from rural to urban areas has intensified in Africa, Asia and Latin America, and persists in parts of Europe, North America and Australia, presenting challenges around urban encroachment into adjacent rural areas; demands on rural land for food and energy and water resources by the expanded urban population; and the viability of depopulating rural communities.

Globalisation – Rural areas around the world are increasingly connected to each other and to cities through social, economic and cultural flows and networks, leading to the restructuring of rural economies under changed market conditions and re-composition of rural populations, as well as challenges to traditional rural cultural practices.

Environmental change – Climate change has both direct and indirect impacts on rural areas: directly, through shifts in the geography of agricultural production and of rural tourism; indirectly, through changes in rural lifestyles and land uses to adapt to a post-carbon future.

Commodification of the countryside – The value of the rural environment increasingly resides in its cultural or recreational appeal than in its use for production, yet the shift towards a consumption-based economy involves different priorities for land management and planning, leading to localised conflicts.

Technological change – New digital and communications technologies are creating new economic opportunities in rural areas and reconfiguring rural service delivery and the practice of everyday life in rural communities, as well as reshaping agricultural practice and geographies.

Political and ideological pressures – Neoliberalism has questioned the legitimacy of established rural policies and promoted the deregulation of markets for agri-food commodities, and the privatisation of state-owned land, natural resources and rural service providers.

This chapter focuses on the first three of these drivers – urbanisation, globalisation and environmental change – to explore in more detail the challenges, opportunities and policy and planning approaches involved in visions of the urban countryside, the global countryside and the post-carbon countryside. In so doing, it highlights the complex and sometimes contradictory position of rural planning in shaping the future of rural places. Systems of land-use planning and development control are on the one hand called on to protect rural landscapes and regulate urban encroachment, but on the other hand have also been critiqued as too rigid to permit necessary change, with arguments advanced for more flexible approaches to facilitate changes in rural land uses. More broadly, spatial planning is critical to the reconfiguration of rural–urban relations, and to questions of rural economic development, infrastructure and service delivery, and the provisioning of new resource demands, including for ecosystem services.

Planning is therefore expected to react to observed challenges and anticipated future trends, but in doing so there is also scope for innovation, for the exploration of completely new ways of thinking about the future of rural spaces and landscapes, as hinted by Koolhaas. In another contribution from architecture, members of the ARENA Rurality Network, have imaginatively explored possibilities for future rural landscapes and communities, considering questions that ask:

> Are there aspects of the rural that can be re-inscribed within the urban habitat? What will rural territories become in the future? Can we rediscover rural qualities, or potentials, or values, or stakes when exploring new ways to organize human habitat? What is it that organizes rurality? In an increasingly humanist age of sustainability and facing the limitations of an urban/economic system inherited from industrial revolution, can we learn from rurality in order to rethink our way of inhabiting earth? Can we suspend the question of how to urbanize rural space?
>
> *(Versteegh, 2014: 4)*

The ideas that have emerged range across suggestions for combining rural living and work spaces; adopting dynamic approaches to rural infrastructure; finding new uses for vernacular rural buildings; designing sustainable buildings that blend with the landscape; planning settlement forms that respect traditional morphologies; and promoting small-scale food production in urban and rural settings (see Versteegh and Meeres, 2014). It is striking, however, that while this work seeks to disrupt established tropes, it continues to be conditioned by the three key drivers of urbanisation, globalisation and environmental change (see also Stringer, 2017), and thus to imagine options for rural futures that are located in (or indeed, may cut across) the broad visions of the urban countryside, the global countryside and the post-carbon countryside, which the remaining sections of this chapter discuss in more detail.

The urban countryside

At some point in the early 2000s, the world passed an historic threshold with more than half the global population living in cities and towns for the first time. The increasing concentration of the population in urban areas has been driven in part by differential population growth, but also by rural to urban migration, notably in Africa, Asia and Latin America, where the rates of recent urbanisation are most rapid. As such, global urbanisation raises issues for rural areas, including the loss of rural land to construction; the demands on remaining rural space to provide food and other resources for urban populations; and the challenge of sustaining depopulating rural communities, especially in peripheral regions. Moreover, with a majority urban population now established, the question arises of whether global society is in effect becoming completely urbanised? Expanses of rural land may still exist between the physical extents of urban areas, but the communities that occupy these spaces will be intrinsically locked into, and subject to, urban economies, social relations, culture and political structures. As such, the rural future is as an urban countryside.

This is the thesis of planetary urbanisation, which has emerged as a vibrant idea in urban studies in recent years (see Brenner, 2014; Buckley and Strauss, 2016), building on theoretical foundations developed by Henri Lefebvre from observations of rural change in south-west France in the 1960s and 1970s (Lefebvre, 1996, 2003). Planetary urbanisation emphasises the dissolution of the urban/rural boundary as urban processes extend their reach into rural spaces and commandeer rural resources for urban populations. Rural land is enclosed and cultivated to produce food for urban consumers; rural resources are mined or harnessed to supply raw materials or energy to urban industry; rural landscapes are commodified for urban leisure; rural workers are proletarianised to service these activities, or as migrant labour imported into cities;

rural consumers are enrolled into urban-centred markets and engaged in urban culture. The countryside hence is 'de-ruralised' as it loses its autonomy and cultural distinction, yet the urban form is also changed. The city is recognised as just one product of urbanisation, requiring 'a new vision of urban theory *without an outside*' (Brenner, 2014: 15), embracing 'the evolving, mutually recursive relations between agglomeration processes and their [rural] operational landscapes' (ibid.: 21).

Planetary urbanisation challenges planning theory and practice by undermining conventional assumptions about spatial order. Principles of the separation of rural and urban space and land uses would need to be abandoned and replaced by integrative models incorporating concepts such as fuzzy boundaries and city-regions. Moreover, Ajl (2014) extends the logic of planetary urbanisation to argue for more radical rethinking of planning: ditching suburban expansion for the resettlement of the countryside in 'agropolitan' districts of small cities linked to agricultural production; centralising agricultural in urban and peri-urban settings; ending development aid that encourages rural-to-urban migration in the global south and reforming rural property relations.

Thus, while planetary urbanisation is the outcome of urbanisation processes that extend back to the English enclosure movement and beyond (Monte-Mór, 2014; Sevilla-Buitrago, 2014), and authors such as Ajl draw inspiration from early twentieth century planners such as Ebenezer Howard and Lewis Mumford, the emphasis on dissolving the rural–urban boundary means that the vision for the future countryside articulated in planetary urbanisation differs significantly from that which guided early- and mid-twentieth-century planning responses to urbanisation in Europe and North America. These aimed to balance the twin principles of the protection of rural space and the modernisation of rural society and economies, laying the foundation for the separation of rural and urban that became central to planning thought. Various combinations of development controls, greenbelts, zoning, farmland preservation schemes, and the designation of national parks and other protected landscapes enforced the former, while the latter was promoted through agricultural mechanisation, economic diversification, electrification and the building of new infrastructure. As Matless (1994: 12) observed, this vision of the future modern countryside embodied a distinct landscape aesthetic, 'of planned compact, nucleated villages; modern, yet at the same time grounded in a particular notion of tradition'. Modern houses and school buildings, new roads, reservoirs, electricity pylons and industrial farm buildings could all be accommodated, so long as they were neat and orderly.

Contemporary urbanisation in Asia, Africa and Latin America has many parallels with the earlier urbanisation of Europe and North America, and there are similarly strong resonances of early and mid-twentieth century planning in the policy responses of countries such as China. The 'Building a New Socialist Countryside' policy in China is a modernisation programme that aims both to protect agricultural land for food security purposes, and to raise rural living standards to deincentivise out-migration (Looney, 2015). A core component is the reconstruction of rural villages, with populations concentrated into nucleated settlements with modern facilities and new, often urban-style housing such as villas and apartment blocks (Long et al., 2012) (Figure 53.1). As in mid-twentieth-century Britain, the vision is a modern, ordered countryside combining urban infrastructure with stylised elements of rural tradition.

In western countries, however, the modernist vision was confronted in the late twentieth century by a renewed fashion for rural nostalgia among urban middle classes, whose increased wealth, leisure time and relative residential freedom was converted into the commodification of the countryside for tourism, recreation and amenity migration. In-migrants and visitors have on the one hand valorised rural heritage and rural landscapes, yet on the other hand themselves imported urban values and lifestyles into rural communities. As a result, planners have been

Figure 53.1   Village modernisation in Tengtou, China.

Source: photograph by M. Woods

faced with divergent visions for the future countryside between production and consumption uses (Blunden and Curry, 1988), conflicts around new construction or changes in land use (Mormont, 1987; Woods, 2005), and challenges of managing exurban development (Taylor and Hurley, 2016), or the conundrum of squaring demand for rural housing with local opposition to new house-building (Satsangi et al., 2010; Woods, 1998).

Indeed, in many parts of the global north, the future of the countryside is likely to involve a dynamic interplay of urbanisation and counter-urbanisation processes, each with specific spatial expressions that will involve migration to and from both rural and urban areas. Although planetary urbanisation may explain the economic processes underlying these dynamics, it arguably downplays the enduring cultural significance of rural identity and the appeal of the 'rural idyll' as a driver of change in the countryside, and thus the continuation of 'ruralisation' as a countervailing dynamic articulated in trends such as back-to-the-land migration, repeasantisation and urban agriculture.

The global countryside

The figure of the 'global countryside' follows the model of the 'urbanised countryside' in emphasising the inter-connection and inter-dependency of rural and urban places, but it differs sharply in not privileging the city as the driver of change. Patterns of economic restructuring and population redistribution are products not of a logic of agglomeration, but of a logic of

globalisation – the creation, multiplication, stretching and intensification of social and economic relations between places, rural and urban (Woods, 2007; see also Steger, 2003). Thus,
for example, the multiplication and intensification of trade relations between countries, and the
stretching of these relations over longer distances, has changed market conditions in agriculture,
destabilised the viability of individual farms, and led to changes in land use in rural areas. By
extrapolating such trends, Woods (2007) posited several characteristics of the global countryside
(Box 53.1), noting that these described a hypothetical space that does not exist in totality, and
may never exist. Furthermore, the partial manifestation of the global countryside reflects the
dynamic of globalisation, which Massey (2005) argues is reproduced through localities. As such,
globalisation is not imposed on the countryside by an urban power centre but is the product of
everyday interactions in both rural and urban places. This observation has several implications.

First, rural institutions, residents and other actors are actively engaged in negotiating,
resisting, manipulating and responding to globalisation processes, and thus have capacity to
shape their own futures, albeit within structural constraints (Woods, 2013). Constraints may
include geographical location (on a coastline or a mountainous area) or natural resources
(minerals, forest) that will determine interactions with transnational tourism or extractive
industries, as well as human capital and institutional capacity. Second, as the outcomes of
globalisation are at least influenced by local actors, they will vary between places. As such,
the global countryside is not a homogeneous space, but a differentiated patchwork of hybrid
places, in which some localities prosper and others struggle. Aguayo (2008) makes a distinction between 'global villages' that have the agency to be proactive in seizing opportunities
from globalisation, and 'globalised villages' whose futures are determined by others. Third,
the global countryside is consequently a contested space, with competition between contrasting visions for the future – for example over deforestation, mining, biofuel or infrastructure projects. This may take the form of local resistance to globalisation processes, but
transnational actors are frequently drawn into conflicts on both sides, collapsing distinctions
between local and global politics (Magnusson and Shaw, 2003). Fourth, given these uncertainties, globalisation is not necessarily linear: global capital can withdraw from rural localities
as well as invest, transport connections can be cut as well as created.

Globalisation hence presents challenges for rural planning, not only in the form of pressures
on rural land use, but also because of the complicated transnational matrix of actors involved.
The resort area of Queenstown in New Zealand is a case in point. Its transformation from a
small farming town to a major site for international tourism and lifestyle migration has seen its

---

Box 53.1   Characteristics of the global countryside

1   Primary sector and secondary sector economic activity in the global countryside feeds, and
    is dependent on, *elongated yet contingent commodity networks*, with consumption distanced
    from production.

2   The global countryside is the site of increasing *corporate concentration and integration*, with
    corporate networks organised on a transnational scale.

3   The global countryside is both the *supplier and the employer of migrant labour*.

4   The globalisation of mobility is also marked by the *flow of tourists* through the global coun-
    tryside, attracted to sites of global rural amenity.

(continued)

(continued)

5   The global countryside attracts *high levels of non-national property investment*, for both commercial and residential purposes.

6   It is not only social and economic relations that are transformed in the global countryside, but also the *discursive construction of nature and its management*.

7   The landscape of the global countryside is *inscribed with the marks of globalisation*.

8   The global countryside is characterised by *increasing social polarisation*.

9   The global countryside is associated with *new sites of political authority*.

10   The global countryside is always a *contested space*.

Source: Woods (2007)

population increase from under 10,000 people in 1991 to over 28,000 in 2013, a third of whom were born outside New Zealand, creating pressure for new amenities, infrastructure, tourist accommodation and housing. In the late 1990s, the planning policies of the pro-development local council were contested by conservationists and residents concerned at the impact on the landscape and environment, with the conflict attracting global coverage as high-profile international supporters were drawn into the campaign. At the same time, local politicians defended their stance by arguing that development was necessary to allow local landowners and investors to benefit from international tourism, and to prevent the district becoming an exclusive enclave for a global elite (Woods, 2011).

The post-carbon countryside

While the models of the urbanised countryside and the global countryside both anticipate the future from the continuation of observed trajectories, the third perspective, the post-carbon countryside, envisages a radical break with the past brought about by the twin challenges of climate change and peak oil. Climate change directly presents challenges to rural places as altered climatic conditions lead to shifts in the geography of agricultural production and threaten the viability of some rural tourism centres, such as mountain ski resorts. At the same time, actions that many argue are required to mitigate or respond to climate change – from expanding renewable energy sources to reducing dependency on fossil fuel powered vehicles – will mean fundamental changes for rural land and rural lifestyles. Yet, there are two contrasting visions of how the post-carbon future might unfold for rural areas.

In the first scenario, switches away from oil (both as a response to climate change, and as oil reserves are depleted as 'peak oil' production is passed) will severely impact on rural lifestyles that are heavily dependent on transport – both car-based commuting to work or to access services in towns, and distribution of goods to shops and homes via trucks. In this post-oil future, it is argued, cities will be more sustainable places to live, accelerating urbanisation further, with an accompanying relocation of technology-enabled industrial farming in urban and periurban areas (Grewal and Grewal, 2012; Morgan and Sonnino, 2010; Zasada, 2011). Vacated rural land will be put to new uses, supplying cities with renewable energy (from wind, solar or hydro resources), biofuels, recreational sites, and ecosystem services (such as carbon sequestration and flood alleviation), or re-wilded, reconstructing past landscapes. For instance, the Zero Carbon Britain project, which has set out a blueprint for transitioning Britain to a zero-carbon economy, has proposed a reduction in grazing land from 11 million hectares at present to under

2 million hectares, with an 80 per cent reduction in the number of sheep and dairy cattle, and a 90 per cent reduction in the number of beef cattle. Overall, only 29 per cent of land currently used for food production would be retained for this function, with the remaining land converted to energy crops such as *Miscanthus*, short rotation coppice willow and energy silage, to nitrogen-fixing legumes, or to land uses that support carbon sequestration such as woodland and the restoration of peat moors (CAT, 2010).

In the second scenario, however, it is cities that are regarded as unsustainable due to their oil-based economy, and the future is projected to lie in smaller rural towns and communities, producing their own food and using local resources (Farinelli, 2008; Menconi et al., 2013; North, 2010). This alternative future has been pioneered by individuals who have returned to small-scale sustainable farming in the 'back-to-the-land' movement and, in more communal form, by ecovillages pursuing self-sufficient sustainable lifestyles (Forde, 2017; Halfacree, 2006, 2007; Pickerill and Maxey, 2009; Wilbur, 2014). It also has more mainstream variants in the transition town movement, that seeks to build community resilience for a post-carbon future, with rural towns such as Totnes in England among the most notable examples (Kenis and Mathijs, 2014; Mason and Whitehead, 2012); as well as in trends observed in countries such as Greece for people to move to rural areas in times of urban economic crisis in order to adopt aspects of self-sufficiency (Anthopoulou et al., 2017; Remoundou et al., 2016).

Both scenarios involve rethinking rural places in ways that are controversial, dissent from established discourses, and test the boundaries of existing planning policies and procedures.

Figure 53.2 Opposition to wind farm developments, Wales.

Source: photograph by M. Woods

Ideas of shifting rural land use from food production to energy resources or ecosystem services, for example, tend to be proposed from the top-down as technocratic solutions that fail to take account of the strength of people's emotional attachment to agrarian rural landscapes. A foretaste of likely conflicts can be found in opposition to wind farms, which are justified by supporters as necessary contributions to tackling climate change by increasing renewable energy generation, but which are characterised by protesters as despoiling the landscape by introducing alien, urban elements (Bell et al., 2005; Phadke, 2011; Woods, 2003; Zografos and Martinez-Alier, 2009). Accommodating these competing perspectives in the planning process for renewable energy projects has required adjustments in planning thinking and approaches, engaging with ideas of environmental justice and 'landscape justice' (Cowell, 2010; Mason and Milbourne, 2014; Stevenson, 2009). Ecovillages, although designed to be low impact on the landscape and environment, can also present difficulties and provoke conflicts when located in settings in which construction is normally restricted. In Britain, the Lammas project in Pembrokeshire and Tinker's Bubble community in Somerset have both been subjects of prolonged planning disputes after being built outside the settlement envelopes prescribed in local plans and have only recently been legitimised by the adoption of new planning guidance to permit low impact developments in certain circumstances.

Conclusion

Forecasting the future is an uncertain science, and rural researchers who have ventured into projecting future trends have tended to do so with trepidation. Yet, anticipating future developments is essential for rural planning, both reactively to prepare for challenges to existing rural structures and policies, and proactively to try to manage change in the countryside and achieve desired outcomes. There is little question that the social and economic character of rural areas around the world has changed substantially in recent decades, and that rural policy has had to adapt in response; nor that further significant changes lie ahead. What is less certain, is precisely how future changes will unfold, and how various people affected – from policymakers to rural residents – will respond. This chapter has discussed three articulations of the future countryside, extrapolated from current patterns. In practice, elements of each are likely to feature in differing combinations in different locations. Rural planning will play a critical role in negotiating these diverse pressures, seeking to develop and implement policies that will mediate their impacts on rural localities, and influenced by various stakeholders and interest groups. As such, the future countryside is a space of possibility, not fully shaped, and open to the imaginations of planners, architects, entrepreneurs, campaigners and lifestyle pioneers projecting their dreams and fears for rural places and competing to turn vision into material reality.

References

Aguayo, B. E. C. (2008) Global villages and rural cosmopolitanism: exploring global ruralities, *Globalizations*, 5, 541–554.

Ajl, M. (2014) The hypertrophic city versus the planet of fields, in Brenner, N. (ed.) *Implosions/Explosions*, Berlin: Jovis, 533–550.

Anthopoulou, T., Kaberis, N. and Petrou, M. (2017) Aspects and experiences of crisis in rural Greece: narratives of rural resilience, *Journal of Rural Studies*, 52, 1–11.

Bell, D., Gray, T. and Haggett, C. (2005) The 'social gap' in windfarm siting decisions: explanations and policy responses, *Environmental Politics*, 14, 460–477.

Blunden, J. and Curry, N. (1988) *A Future for our Countryside*, Oxford: Blackwell.

Brenner, N. (ed.) (2014) *Implosions/Explosions: Towards a Study of Planetary Urbanization*, Berlin: Jovis.

Buckley, M. and Strauss, K. (2016) With, against and beyond Lefebvre: Planetary urbanization and epistemic plurality, *Environment and Planning D: Society and Space*, 34, 617–636.

CAT (2010) *Zero Carbon Britain*, Machynlleth: Centre for Alternative Technology.

Cowell, R. (2010) Wind power, landscape and strategic, spatial planning – the construction of 'acceptable locations' in Wales, *Land Use Policy*, 27, 222–232.

Farinelli, B. (2008) *L'Avenir est à la Campagne*, Paris: Sang de la Terre.

Forde, E. (2017) The ethics of energy provisioning: living off-grid in rural Wales, *Energy Research and Social Science*, 30, 82–93.

Grewal, S. S. and Grewal, P. (2012) Can cities be self-sufficient in food? *Cities*, 29, 1–11.

Halfacree, K. (2006) From dropping out to leading on? British countercultural back-to-the-land in a changing rurality, *Progress in Human Geography*, 30, 309–336.

Halfacree, K. (2007) Back-to-the-land in the twenty-first century? Making connections with rurality, *Tijdschrift voor Economische en Sociale Geografie*, 98, 3–8.

Kenis, A. and Mathijs, E. (2014) (De)politicising the local: the case of the Transition Towns movement in Flanders (Belgium), *Journal of Rural Studies*, 34, 172–183.

Koolhaas, R. (2012) Countryside, lecture presented Stedelijk Museum, Amsterdam, 25 April, retrieved on 30 January 2018 from http://oma.eu/lectures/countryside.

Koolhaas, R. (2014) Rem Koolhaas: Countryside architecture, ICON, 23 September, retrieved on 30 January 2018 from www.iconeye.com/architecture/features/item/11031-rem-koolhaas-in-the-country.

Lefebvre, H. (1996) *Writing on Cities*, Oxford: Blackwell (translated and edited by E. Kofman and E. Lebas).

Lefebvre, H. (2003) *The Urban Revolution*, Minneapolis, MN: University of Minnesota Press (translated by R. Bonnono).

Long, H., Li, Y., Liu, Y., Woods, M. and Zou, J. (2012) Accelerated restructuring in rural China fueled by 'increasing vs decreasing balance' land use policy for dealing with hollowed villages, *Land Use Policy*, 29, 11–22.

Looney, K. E. (2015) China's campaign to build a New Socialist Countryside: Village modernization, peasant councils, and the Ganzhou model of rural development, *China Quarterly*, 224, 909–932.

Magnusson, W. and Shaw, K. (eds) (2003) *A Political Space: Reading the Global through Clayoquot Sound*, Minneapolis, MN: University of Minnesota Press.

Mason, K. and Milbourne, P. (2014) Constructing a 'landscape justice' for windfarm development: the case of Nant y Moch, Wales, *Geoforum*, 53, 104–115.

Mason, K. and Whitehead, M. (2012) Transition urbanism and the contested politics of ethical place-making, *Antipode*, 44, 493–516.

Massey, D. (2005) *For Space*, London: Sage.

Matless, D. (1994) Doing the English village, 1945–90: an essay in imaginative geography, in P. Cloke, M. Doel, D. Matless, M. Phillips and N. Thrift (eds), *Writing the Rural*, London: Paul Chapman, 7–88.

Menconi, M, Stella, G. and Grohman, D. (2013) Revisiting the food component of the ecological footprint indicator for autonomous rural settlement models in Central Italy, *Ecological Indicators*, 34, 580–589.

Monte-Mór, R. L. (2014) What is the urban in the contemporary world?, in Brenner, N. (ed.) *Implosions/Explosions*, Berlin: Jovis, 260–267.

Morgan, K. and Sonnino, R. (2010) The urban foodscape: world cities in the new food equation, *Cambridge Journal of Regions, Economy and Society*, 3, 209–224.

Mormont, M. (1987) The emergence of rural struggles and their ideological effects, *International Journal of Urban and Regional Research*, 7, 559–575.

North, P. (2010) Eco-localisation as a progressive response to peak oil and climate change – a sympathetic critique, *Geoforum*, 41, 585–594.

Phadke, R. (2011) Resisting and reconciling big wind: middle landscape politics in the new American West, *Antipode*, 43, 754–776.

Pickerill, J. and Maxey, L. (2009) Geographies of sustainability: low impact developments and radical spaces of innovation, *Geography Compass*, 3, 1515–1539.

Satsangi, M., Gallent, N. and Bevan, M. (2010) *The Rural Housing Question*, Bristol: Policy Press.

Sevilla-Buitrago, A. (2014) Urbs in rure: historical enclosure and the extended urbanization of the countryside, in Brenner, N. (ed.) *Implosions/Explosions*, Berlin: Jovis, 236–259.

Steger, M. (2003) *Globalization: A Very Short Introduction*, Oxford: Oxford University Press.

Stevenson, R. (2009) Discourse, power and energy conflicts: understanding Welsh renewable energy planning policy, *Environment and Planning C: Government and Policy*, 27: 512–526.

Stringer, B. (2017) Villages and urbanization, *Architecture and Culture*, 5, 5–20.

Taylor, L. and Hurley, P. T. (eds) (2016) *A Comparative Political Ecology of Exurbia*, Dordrecht: Springer.

Versteegh, P. (2014) Introduction, in P. Versteegh and S. Meeres (eds) *AlterRurality*, Fribourg: ARENA, 3–6.

Versteegh, P. and Meeres, S. (eds) (2014) *AlterRurality*, Fribourg: ARENA.

Wilbur, A. (2014) Cultivating back-to-the-landers: networks of knowledge in rural Northern Italy, *Sociologia Ruralis*, 54, 167–185.

Woods, M. (1998) Advocating rurality? The repositioning of rural local government, *Journal of Rural Studies*, 14, 13–26.

Woods, M. (2003) Conflicting environmental visions of the rural: windfarm development in mid Wales, *Sociologia Ruralis*, 43, 271–288.

Woods, M. (2005) *Contesting Rurality*, Aldershot: Ashgate.

Woods, M. (2007) Engaging the global countryside: globalization, hybridity and the reconstitution of rural place, *Progress in Human Geography*, 31, 485–507.

Woods, M. (2011) The local politics of the global countryside: boosterism, aspirational ruralism and the contested reconstitution of Queenstown, New Zealand, *Geojournal*, 76, 365–381.

Woods, M. (2013) Economic entanglements and the reshaping of place in the global countryside, in C. Tamásy and J. Diez (eds) *Regional Resilience, Economy and Society: Globalising Rural Places*, London: Taylor and Francis, 11–32.

Zasada, I. (2011) Multifunctional peri-urban agriculture – a review of societal demands and the provision of goods and services by farming, *Land Use Policy*, 28, 639–648.

Zografos, C. and Martinez-Alier, J. (2009) The politics of landscape value: a case study of wind farm conflict in Catalonia, *Environment and Planning A*, 41, 1726–1744.

54

Planning rural futures

Mark Scott, Nick Gallent and Menelaos Gkartzios

Introduction

Rural planning has, over the last 50 years, been relegated to the margins of planning theory and practice (see Lapping, 2006; Gallent and Scott, 2017). Yet, despite growing urbanisation, rural populations remain sizeable and politically important. Predominantly rural regions account for a quarter of the population across OECD countries, while globally the rural population stands at 46 per cent (UN, 2018). Moreover, a high proportion of land, even in advanced and otherwise urbanised countries, hosts rural land uses and is characterised by its openness. For example, while Europe is one of the most urbanised continents on the globe, agricultural land and forests represent around 85 per cent of land cover across the 28 EU Member States (CEC, 2018). At this basic, but fundamental level, rural regions provide the backcloth for a range of crucial planning issues (Frank and Hibbard, 2017) such as renewable energy deployment, the management and enhancement of ecosystem services, food production and security, and natural resource exploitation and management. Alongside their traditional primary sector functions, these regions are increasingly shaped by new consumption patterns and the commodification of the countryside (through heritage, tradition, lifestyle, landscape, etc.), generating a range of threats but also opening up opportunities to transform rural economies. As Woods highlighted in the previous chapter, the 'value' of the rural environment is increasingly underpinned by its cultural or recreational appeal rather than in its use for production; yet the shift towards a consumption-based economy involves different priorities for land management and planning. These competing functions, uses and demands for rural space and resources, which overlap the distinctive and multiple social representations of the rural, present unique planning challenges that are deserving of critical planning attention as rural localities become subject to contentious and controversial change processes, transforming them into arenas of deeply contested planning decision-making.

Changing contexts for rural planning practice

The contributions to this Companion have sought to challenge common misconceptions that have often become entrenched in framing debates on the scope, purpose and limits of rural planning. In turn, these narratives have often shaped one-dimensional planning responses to complex rural problems or resulted in a minimal engagement with rural issues within planning policy and practice. For example, in the post-Second World War era, the rural as an agricultural space has been a dominant discourse within planning policy (Chapter 3, Lapping and Scott), leading to a primacy of agricultural interests in shaping rural futures (Curry and Owen, 2009; Gallent et al., 2017). This viewpoint has led to the preservation of farmland as an overarching policy goal, which has resulted in a neglect of the wider rural economy and the marginalisation of socially progressive rural planning actions. While the demise of agriculture reported in the 1990s may have been overstated (Chapter 49, Butt), rural places are clearly much more diverse in terms of functions, economies, the labour force, and society, than they once were. Similarly, in more industrialised countries, rural planning has been a story of rural landscape preservation, whereby the rural is conceived as a picturesque backdrop to urban development. This perspective has often resulted in a narrow view of the rural environment, prioritising aesthetic qualities over ecological integrity, and has led to rural landscape preservation becoming an instrument of rural social exclusion under a 'no development' ethic (Gallent et al., 2017). In contrast, in other rural areas (particularly more remote regions), the countryside is framed as lacking development pressures (and opportunities), leading to a *laissez-faire* approach to development in rural localities, particularly facilitative of development that supports local elites (Chapter 2, Gallent and Gkartzios). In the present chapter, we draw out some of the key themes emerging from the Companion in relation to the changing context for rural planning.

Understanding diverse ruralities

Contributors to the Companion have highlighted both the nature of differentiated rural space and also a growing diversity within the countryside. For example, Smith et al. (Chapter 21) highlight the diverse nature of rural population change across different types of countryside, which in turn has implications for housing markets and the supply of housing examined by Gallent and Scott (Chapter 22). Internationally, we have witnessed many rural areas experiencing a population turnaround in recent decades, particularly in accessible rural regions or in highly prized landscapes and coastal areas, characterised by their attraction to in-migrants and new entrepreneurs (Chapter 18, Herslund) and by their capacity to develop the conditions for a diverse rural economy (Chapter 17, Cowie et al). However, more remote rural places continue to experience a rural exodus, undermining service provision and the local economy (Chapter 9, Bock). Rural areas are often on very different trajectories, changing rapidly or slowly, and have different capacities to act in the face of growth or decline. From a planning perspective, policy and practice has too often been characterised by a 'one size fits all' approach, suggesting the need for a more sophisticated understanding of the rural condition and a nuanced planning response that is spatially differentiated.

Overlapping these material processes of rural change, much research has also highlighted the importance of 'the rural' as a social construction (Chapter 2, Gallent and Gkartzios), specific to different groups with their competing visions of acceptable rural change and development. Social representations of rural space provide insights into what and for whom rural planning is for, illustrated by land-use conflicts surrounding the siting of renewable energy infrastructure (Chapter 46, Natarajan) or housing construction (Chapter 22, Gallent and Scott; Chapter 23, Paris), for example. Moreover, the social representation of rural places also frames rural change processes – for example, rural gentrification driven by 'idyllic' images of rural living.

While research has emphasised that 'the rural' comprises highly differentiated rural places and experiences, various chapters in the Companion also draw attention to greater diversity *within* rural places. This includes the influx of affluent middle class residents (with the potential for gentrification) (Chapter 21, Smith et al.), but also contributions calling for planning to represent and recognise gendered (Chapter 28, Shortall) and heteronormative (Chapter 29, Doan and Hubbard) ruralities as well as inequalities experienced by ethnic minority communities (Chapter 27, Satsangi and Gkartzios), and the growing importance of addressing an ageing countryside (Chapter 30, Bevan).

The new rural economy and rural planning

As outlined in Chapter 20 (Scott), contemporary rural economies are characterised by diversity and variable outcomes, reflecting place-specific assets, local political capacity to realise rural potential and the availability of social and human capital, which combine and interact to produce differentiated patterns of change across rural regions. Part III of the Companion explored various dimensions of the rural economy, from models of development (e.g. exogenous, endogenous and regional) to potential development pathways, including rural enterprise innovation, the potential offered by 'payments for ecosystem services' for some rural economies, and the role of the so-called 'creative class' as a catalyst of rural growth. These contributions to the Companion challenge the notion of an agriculture-based economy, and instead emphasise the importance of spatial and territorial approaches to support further economic diversification. However, in practice, planning has often been marginal in enabling rural economic growth and diversification (Scott and Murray, 2009; Curry and Owen, 2009) with embedded land-use planning practices undermining sustainable economic development in some rural localities. In order to play a more proactive role within rural economies, planning interventions need to take account of how rural economies have evolved beyond agriculture and how rural populations are now dependent on an array of different activities. Critical in such understandings is to move beyond academic attempts to theorise models of sometimes dichotomous narratives of economic/social development, and to proceed to place-based accounts of how rural development works in practice, how stakeholders and lobby groups participate in policy formation and for what purpose – a 'reflexive turn' in line with 'neo-endogenous thinking' as suggested in Chapter 14 (Gkartzios and Lowe).

Reconceiving the rural environment

Protecting rural landscapes and farmland preservation have been longstanding features of rural planning (Gallent et al., 2017). Early examples of planning interventions to 'protect the rural' included the establishment of national parks (see Chapter 38, Bell and Stockdale) and greenbelts (Chapter 40, Taylor), both of which have proved enduring and popular legacies of planning and countryside management. However, these traditional approaches to rural planning are limited by their narrow perspective on the rural environment, limited to 'landscape' and protecting the picturesque, or by their tendency to frame the 'countryside' as an antidote to urban development and sprawl. Moreover, these approaches tend to result in 'islands of protection' rather than holistic approaches to managing environmental change. Part VII examined how traditional methods of countryside management are being reinvented for the twenty-first century: greater emphasis is now being placed on multifunctional landscapes, ecosystem services and green infrastructure which all focus on connectivity and multidimensional approaches towards landscape governance. Rather than simply protecting the countryside from development threats, these approaches also represent a shift towards restoring, enhancing and creating new ecological networks with multifunctional benefits, from wellbeing and health to climate change adaptation.

Shifting land ownerships and markets

As noted by Hetherington (Chapter 7), land is a fundamental resource that underpins the rural economy, providing food, water, timber, energy, minerals, and space for housing and recreation. Land reform has been a politically contentious issue for centuries, often tied into wider political struggles (e.g. post-colonialism, indigenous rights). In the contemporary countryside, land ownership is complex and often poorly documented. In parts of the UK, large estate (gentry) landowners dominate, while in other countries land ownership is often fragmented, with small-holdings more prevalent.

In transition and emerging economies, land ownership is becoming increasingly concentrated in large agricultural holdings. For example, in Chapter 50, Visser and Spoor demonstrate the extent and process of 'land grabbing' in the former Soviet Union in the wake of the collapse of collective farms. Moreover, recent research has highlighted how agricultural lands in Sub-Saharan Africa are being acquired by both foreign state and corporate interests, illustrating the importance of international flows of capital and their engagements with recipient state actors (especially those tasked to capture foreign direct investment) and agri-businesses. A range of countries are implicated in such processes, including the oil-rich Gulf States and China (Borras and Franco, 2012). In relation to corporate interests, as recorded by White et al. (2012), some land deals (for example those made by hedge funds and pension funds) may be purely speculative, betting on rising global land values; investors put their money in land, as they might put it in gold, works of art, or blue-chip shares. However, the purpose of the great majority of corporate land grabs is to establish agricultural production on a large scale, and to guarantee access to its products. While land grabbing has a long history, more recently commentators have noted the rise of so-called green-grabbing (in both the Global North and South) – the appropriation of land for environmental ends (Fairhead et al., 2012) with the aim of extracting profit (Apostolopoulou and Adams, 2015). This involves, for example, applying market instruments to mitigate environmental degradation, such as land-based carbon off-setting or biodiversity offsetting schemes, or the privatization of publicly owned environmental assets, such as state-owned woodlands.

The emerging picture of landownership is complex with traditional patterns of ownership overlapping with new flows of capital and international investment. In this regard, Gallent et al. (2018) note (in a European context) how the profile of rural landowners has shifted markedly in recent decades, with family farmers and life-style buyers being joined by institutional investors concerned primarily with the extraction of financial income through rural land assets – farmland, woodland and wineries – but extending also to leisure uses such as golf courses and theme parks, and to renewable energy. However, Chapter 12 (Moore) outlines alternative models of ownership. Moore considers the role of community ownership of rural land assets (through land trusts), which have been used to deliver affordable housing and renewable energy projects, providing more equitable and sustainable outcomes than private speculation. Land-use planning is tied up with land ownership and private property rights. From a rural planning perspective, new patterns of ownership translate into shifting demands for the use of rural land and raise questions around how best to regulate land-use change in the face of local needs, but also market demands to extract profit from land resources.

Globalisation, urbanisation and the rural

Capital movement and patterns of investment in rural land evidence the globalised processes that are today an important driver of change in many rural places. Rural economies are being reshaped by a mix of local and extra-local interests, with decisions taken thousands of miles away impacting on patterns of land and environmental change. While rural scholars have long recognised the influence of globalisation, rural planning practices have often been framed as 'local' planning with less critical attention given to understanding how rural localities are shaped by globalising tendencies and are connected to regional and global circuits of capital.

Moreover, in an increasingly urbanising society, understanding rural–urban relationships is of critical importance as it becomes more difficult to discuss rural and urban change without acknowledging crucial linkages (Lichter and Brown, 2011). Within this context, the traditional opposition of city and countryside that has dominated planning theory and practice is not only largely obsolete but ideological and often key to the capital accumulation process (Wachsmuth, 2014) – for example, representation of rural places as 'traditional', 'green', 'close to nature', 'slower paced' and so on, are central to the development process in rural localities and uneven development across rural space. Wachsmuth also questions the notion that, within a neo-liberalised and globalised world order defined by economic competition, there is seemingly no alternative to the city as a growth engine: the corollary being that rural decline is a 'natural process'. Within this broader context, Woods's discussion of the 'global countryside' (Chapter 53) provides a critical framework for understanding how rural actors negotiate top-down globalising pressures to reconstitute rural space. While recognising the inter-connection of urban and rural places, Woods focuses on the capacity of rural places to influence their own futures, which may include developing new functions within global networks, such as amenity spaces, ecosystem services and sites of low carbon transition.

Challenging rural planning as practice

Building upon the discussion of the shifting and varied context for rural planning, in this final part we briefly consider the prospects and scope for reinventing or reframing rural planning for the twenty-first century. In doing so, we seek to foster critical debates on the role of rural planning in realising the potential of rural people and places as part of a more territorially balanced, integrative, progressive and sustainable trajectory. This includes fostering development outcomes tailored to local contexts, aspirations and needs, while recognising environmental limits.

Rural planning mobilised to tackle global challenges

While rural planning has often been interpreted as *local* planning, a recurring theme within the Companion has been the potential of rural planning to contribute to global challenges. Issues surrounding biodiversity and enhancing ecosystem services were discussed in Chapter 19 (Juntti) on payment for ecosystem services, Chapter 42 (Mell) on green infrastructure, and Chapter 44 (Lennon) on re-wilding, while Chapter 41 (Scott) called for greater integration of ecosystem approaches into spatial planning frameworks, particularly in developing more holistic approaches to planning across natural and built environments. By more fully embracing ecosystem approaches within spatial planning frameworks, rural planners should also be at the forefront of managing growing environmental risks, often related to climate change. In this context, Lennon and Scott (2014) call for an 'ecological turn' in planning to place ecology, ecosystem services and environmental risks as central concerns of rural planning practice. Similarly, Gawith and Hodge (2017) argue for an ecosystem approach to rural land policy to address the fragmentation of land governance across planning, conservation and agri-environmental policies. Within this approach, the policy priority is the conservation of ecosystem structure and functioning, to maintain ecosystem services.

In the context of climate change, a number of chapters examine the potential for post carbon transitions. Phillips and Dickie (Chapter 45) examine the prospects for a post-carbon countryside, presenting a challenging analysis of carbon consumption and the roll-out of low carbon technologies and renewables. So while rural areas offer potential as locations of renewable energy infrastructure, dispersed rural geographies and car-dependency remain problematic for reducing carbon emissions. The challenge of planning for renewable energy infrastructure is further explored by Natarajan (Chapter 46), outlining the conflict between local opposition and the national priority of further renewable energy deployment.

As highlighted by Homsy and Warner (Chapter 51), while there is now an extensive literature on climate change governance and low carbon transitions at an urban scale, the rural literature is less developed. Homsy and Warner address this deficit by examining different spheres of climate change facing rural planning, while also noting the lower capacity to act in rural areas due to smaller and often less-resourced local government structures. They argue for a local (re)framing of climate change – currently perceived as involving short-term costs while the benefits are diffuse and perceived as distant in time and place – towards emphasising local gains linked, for example, to economic development. It is argued that this requires citizen engagement, local leadership and multi-level governance to coordinate local actions.

Planning for the just and equitable countryside

Accounts of social inclusion, diversity and equality are abundant in urban planning discussions and frame diverse areas of intervention; however, the transferability of these debates outside the metropoles is less obvious. Building on Shucksmith's (2018) notion of the 'good countryside', Part V primarily aims to unpack selective experiences of inequalities in the countryside. Satsangi and Gkartzios in Chapter 27 review the literature internationally to show how race and ethnicity contribute to distinctive rural geographies and therefore planning concerns centred on residential location, housing quality and mobility. Gender bias is recognised in planning (Burgess, 2008), and Shortall (Chapter 28) demonstrates the impact of gendered constructions regarding agriculture, particularly for health and safety, by considering a space hardly ever explored by planners: the farmyard – a hybrid social, family and employment space which demonstrates assumptions about gender roles and the treatment of agriculture in planning. Similar concerns are discussed in the urban planning literature with regards to the LGBTQ community (Frisch, 2015). While many countries in the Global North appear to have delivered advances in LGBTQ rights, Doan and Hubbard (Chapter 29) point to bias within both rural planning academia (for example, when considering the intersection of queerness and rurality) and practice (for example, in services that target the needs of rural queers). The authors draw attention, in particular, to the need for investment in HIV/AIDS care and education in rural areas. Age is critically considered in planning debates (regarding both the needs and engagement of young people and older people) and Chapter 30 (Bevan) flags the actions that can be used to support rural people in later life while moving beyond ageist views of the countryside. Questions of social and spatial justice of course intersect with class inequalities, which may be accentuated through legal and zoning practices which exclude low income groups (e.g. Chapter 8, Ashwood et al.). The Companion does not provide comprehensive treatment of otherness and marginalisation – disability is an obvious gap – but points to a general need for future rural planning research to focus on groups with potentially less power and their treatment within planning theory and practice.

Engaged, interactive and reflexive modes of planning

Both the politics of planning and the governance context are distinctive in rural settings. For example, formal rules and institutions for planning often overlap with informal rules and local political cultures operating in different types of rural places (see Part II). Planning actors in rural contexts also interact with distinct policy networks shaping rural futures, including extensive land owning interests, agricultural interests and food producers, amenity groups, conservation actors, and rural development networks. At the same time, planning in rural localities shares common trajectories with wider planning and development processes, similar to urban areas but manifested in different ways. For example, neoliberalism dominates rural as well as urban space; indeed, 'roll back' neoliberalism has been a longstanding feature of rural life as public services are perceived by central government as inefficient and expensive, having to work with thin markets and dispersed geographies. Similarly, financialisation and the penetration of global flows of capital has been an emerging theme within the rural studies literature (discussed above in relation to land acquisition). Together, these formal institutions and rules (laws and regulations), local norms and political culture, networks and actors, neoliberalism and globalisation, overlap with contested and competing social imaginaries of the rural to create distinct '*rural planning assemblages*'.

Formulating agreed rural sustainable development policies at a local level remains an elusive and contested policy goal. As observed in many of the Companion's chapters, rural planning processes are often framed by competing narratives or beliefs that relate to contested social constructions of rurality. As Woods (2003: 287) comments, in analysing rural conflicts 'the researcher needs to understand the complex negotiation of discourses of nature, landscape, environment and rurality which frame collective and individual actions'. In this context, it may be wise to reconsider a key question raised by Rydin (2003): how can an integrated and holistic sustainable development narrative be generated, developed and embedded so that it supports policy and planning practice? Rural planning conflicts often revolve around specific and controversial episodes of rural change (e.g. new housing, wind turbines etc.). Therefore, actors have limited opportunities to participate in a wider discussion on the future of rural communities or rural transformations, instead finding themselves involved in micro-level debates which do not extend to an exploration of the values underpinning preferred policy directions.

As Healey (1999) argues, place-making should consciously explore 'new story-lines' and nurture a new discourse for broad based rural policy through deliberative action. For Murray and Greer (1997), this represents an opportunity for planners to engage in a more interactive style of plan-making involving partnerships with rural communities and involving interest group mediation and the building of trust-relations. While rural development and countryside management initiatives have increasingly been advanced in partnership with rural communities – involving a power shift towards community action and bottom-up processes – land-use planning in rural areas is primarily characterised as top-down and procedural. Partnership based local planning processes would enable the exploration of competing rural narratives re-orientated towards local needs, capacities and perspectives of local people and the adoption of cultural, environmental and community values within the policy process. This requires developing novel and reflexive approaches to creating a shared understanding of rurality and its dynamic nature. The role of planners and academics in the co-production of such shared understandings of rurality is essential, and as suggested in Chapter 14 (Gkartzios and Lowe), layers of reflexivity are facilitated by the creation of networks across the academy and policy, interdisciplinary research that acknowledges the complex nature of rural planning questions, as well as international comparative perspectives. Furthermore, effective rural planning is a multi-level activity, involving the blending of the local with globalisation and the blurring of urban/rural boundaries. There are opportunities, through regional development strategies, to foster such effective planning (Chapter 15, Tomaney et al.).

'Smart' rural planning

While much attention has been devoted to the impact and potential of smart cities, limited research and practice has focused on exploring the potential of new technologies for enhancing rural planning. To date, the literature on the role of ICT in relation to rural planning is often limited to two research areas: (1) to drawing attention to the need for effective broadband roll-out and the urban–rural digital divide (Malecki, 2003); and (2) the potential of good quality ICT networks in attracting new rural residents, specifically the young and the so-called 'creative class' (Herslund, 2012). However, there is significant scope for a much broader research agenda to encompass the intersection of knowledge communication and social infrastructure.

ICT is now an integral part of our daily lives, whether for communications, for leisure, retail, or accessing public services. ICT is embedded in the commercial, social and cultural aspects of places with online shopping, digitisation of culture and heritage, social media intersecting with place-based social networks, and many municipalities across the global north are turning to ICT to aid their work. Specifically, Geospatial ICT – ICT that incorporates geospatial location as an inherent feature – has increasingly come to the fore to facilitate data analytics, data visualisation, and data mapping, including traditional tools: GIS; remote sensing; GPS; Web 2.0 tools e.g. web-mapping, geo web tools, Volunteered Geographic Information; social media tools e.g. smart phones, crowdsourcing, geo-tagging, geospatial analytics; sensor based tools e.g. airborne sensors, remote sensors, sensor networks. In this context, smart rural planning may harness technology to monitor, manage and regulate rural places in real-time using ICT infrastructure and ubiquitous computing. As discussed in Part IV (e.g. Chapter 25, Oliva and Camarero), these real time systems could potentially enable the efficient control of rural services (e.g. public transport in remote places, new social care models), promote smart mobility (e.g. real-time technologies and geospatial tools to encourage carpooling, car-sharing and other 'sharing' formulae in rural locations), management of ecological resources (e.g. real-time management of visitor numbers), and responses to environmental shocks (e.g. citizen responses to flooding). Moreover, new technologies can also be applied to enhancing planning governance in rural localities, such as new methods of public engagement between policy-makers and rural citizens and the adoption of a citizen science approach to utilise local knowledge in co-design of rural development pathways, or participatory e-platforms to overcome distance barriers. Therefore, broadband and mobile phone enabled internet access becomes essential rural infrastructure (Chapter 31, Gallent).

Quality of life, wellbeing and place-making

Part VI of the Companion examined many of the traditional elements of rural settlement planning, including infrastructure, village planning, local design distinctiveness, heritage and local place-making, and managing settlement growth and decline. These chapters raise issues concerning the appropriate toolkits available to planners to manage rural settlement change, and the connections between built and natural environmental quality, place-identity and distinctiveness, which in turn often underpin successful rural economies.

Moreover, a number of chapters also draw our attention to the health and wellbeing dimensions of rural settlement planning, including Kilpatrick et al.'s discussion of rural community health planning challenges (Chapter 24), Rogerson et al.'s exploration of the relationship between landscape and wellbeing (Chapter 43), and the implications of an ageing rural population for health care provision (Bevan, Chapter 30). Giving increased attention to place-based and rural wellbeing (including physical and mental health wellbeing) within spatial planning and rural development is an emerging research area (e.g. Scott et al., 2018; OECD, 2018) and requires careful consideration as to how rural residents can avail themselves of wellbeing opportunities. Dimensions of rural wellbeing, linked to spatial planning, include: health service spatial distribution, transport links to health care centres, the rural landscape as a health promoting environment, the wellbeing benefits of exposure to nature, and walkable rural settlements. In this context, greenspace access provides an illustrative example: while rural residents are often surrounded by greenspaces, this is sometimes inaccessible agricultural land. A key 'paradox of rural living' (Douglas et al., 2018) is that some rural populations may have more limited access to open space, parks and greenspaces than the residents of urban or suburban locations. Therefore, ensuring adequate access and opportunities for physical activities may require more nuanced solutions, such as linear greenways or river corridor walks.

Sustainable rural livelihoods – linking spatial planning with rural development

Rural planning and rural development (in both theory and practice) have remained largely separate and distinct in their approaches (in contrast to urban planning and urban regeneration). To oversimplify, planning and rural development processes and goals often appear to be poorly integrated and a policy disconnection has emerged between environmental / landscape objectives and goals, which support the social, economic and cultural imperatives of rural communities. Rural economic development and environmental planning have often developed in isolation, or are framed as opposing rationalities in developing rural policy. But in reality, economic development and environmental quality are mutually dependent: the latter attracts inmigrants and investment; similarly, environmental goods provide a critical pathway for green development, for example in the form of renewable energy or ecosystem services. These linkages need to be central to the rural narrative, showing not only how rural places can be sustainable, but how they can contribute to bigger environmental goals at a planetary scale.

While planning is often portrayed as a barrier to economic development in the countryside, Chapter 20 (Scott) identified a key role for spatial plans and practices in diversifying the rural economy. This involves positioning planning as an *enabler* (rather than barrier) of rural economic development based on engaging and understanding the changing nature of the economy across different types of rural areas and seeking mutually reinforcing relationships between environmental, economic and social objectives, and bridging the traditional gap between environment and economy.

Fostering new urban–rural relationships

New explorations of dynamic and co-dependant urban–rural relationships have figured prominently in academic and policy literature (Davoudi and Stead, 2002; CEC, 1999) in keeping with the spatial, multi-sectoral and multi-scalar governance approaches to planning. While original models of policy interventions in the rural viewed these areas as culturally, technically and economically dependent on urban centres (Chapter 14, Gkartzios and Lowe), more nuanced understandings have emerged, recognising material and immaterial flows across increasingly networked urban and rural spaces. The 'mobility turn' (Cresswell, 2006; Sheller and Urry, 2006; Urry, 2007) has offered a useful way to examine these flows and most counterurbanisation and rural gentrification debates currently recognise the fusion of urban and rural realities of increasingly mobile households (Chapter 21, Smith et al.). This new urban–rural relationship does not reject differentiation across and between urban and rural spaces; but, fundamentally, it recognises that rural planning research and policy is not just about 'the rural'; and *vice versa*, urban planning research is not just about the metropolis.

For Cresswell and others, a mobility perspective is essentially relational: it moves beyond narrow and distinct fields, such as transport or migration studies, to embrace all forms of movement, inclusive of global flows and movements of capital and labour (Cresswell, 2010). This implies also a 'mobility of knowledge', a shift in the way we think about what constitutes urban planning and rural planning knowledge. In fact inherent in the emergent urban–rural relationship is the acknowledgement that urban studies research offers critical insights for rural and regional policy making (Gkartzios, 2013; Remoundou et al., 2016); and, likewise, rural research experiments can offer insightful and original perspectives for planning epistemologies (see for example Chapter 26 by Crawshaw).

Final thoughts

This chapter has not set a prescriptive agenda for the future of rural planning, but rather drawn attention to the range of pressures and realities that will frame planning practice in the coming century. Planning in rural areas has, for too long, grappled with seemingly conflicting and competing priorities around landscape protection and economic development. These priorities can be brought together. Research has shown that sustainable development pathways are only possible when environmental and economic goals are viewed as inseparable and where exploitative activities give way to an understanding of the complex ecosystems of which human economy and habitation are crucial parts. Planning research and practice have a key role to play in developing this understanding in different locations, thereafter development strategies, plans and policies that work with local opportunities and realities. The goal of this Companion has been to contribute to that understanding, in broad terms; to illustrate the breadth of challenges faced by different rural places; and also to pinpoint the opportunities that exist to collectively design more sustainable rural futures.

In practice, this will require that policy-makers and planners begin by reimagining the opportunities that exist in rural places, beyond the narrow agendas that have shaped past intervention and led to limited engagement with the full range of rural interests and possibilities. Thinking on the countryside cannot simply be an extension of urban debates (Gallent et al., 2015). Rather, it needs to focus on inter-dependences – urban–rural, rural–rural, local–global, and environmental-economic – and the particular opportunities and challenges that the rural presents. Critical social science perspectives, of the type contained in this Companion, provide a platform for this thinking and, we hope, will help students and researchers grasp what it takes to 'do' rural planning: to put the principles of social, spatial and environmental justice, sustainability, and economic development into practice in complex political and social contexts. We also hope that the Companion will provide some impetus for planning schools worldwide to embrace rural planning agendas within their curricula, recognising the diversity of places with which researchers and students of planning must engage.

References

Apostolopoulou, E. and Adams, W.M. 2015. Neoliberal capitalism and conservation in the post-crisis era: the dialectics of 'green' and 'un-green' grabbing in Greece and the UK. *Antipode*, 47(1), 15–35.

Borras Jr, S.M. and Franco, J.C. 2012. Global land grabbing and trajectories of agrarian change: a preliminary analysis. *Journal of Agrarian Change*, 12(1), 34–59.

Burgess, G. 2008. Planning and the gender equality duty: why does gender matter? *People, Place and Policy Online*, 2–3, 112–121.

CEC. 1999. *ESDP-European Spatial Development Perspective: Towards Balanced and Sustainable Development of the Territory of the European Union*. Office for Official Publications of the European Communities, Brussels.

CEC. 2018. *Land Cover and Land Use*. Office for Official Publications of the European Communities, Brussels. Retrieved on 8 June 2018 from https://ec.europa.eu/agriculture/sites/agriculture/files/statistics/facts-figures/land-cover-use.pdf.

Cresswell, T. 2006. *On the Move: Mobility in the Modern Western World*. Routledge, New York.

Cresswell, T. 2010. Towards a politics of mobility. *Environment and Planning D: Society and Space*, 28, 17–31.

Curry, N. and Owen, S. 2009. Rural planning in England: a critique of current policy. *Town Planning Review*, 80(6), 575–596.

Davoudi, S. and Stead, D. 2002. Urban–rural relationships: an introduction and a brief history. *Built Environment*, 28(4), 269–277.

Douglas, O., Russell, P. and Scott, M. 2018. Positive perceptions of green and open space as predictors of neighbourhood quality of life: implications for urban planning across the city region. *Journal of Environmental Planning and Management*, (online first), 1–27, DOI: 10.1080/09640568.2018.1439573.

Fairhead, J., Leach, M. and Scoones, I. 2012. Green grabbing: a new appropriation of nature? *Journal of Peasant Studies*, 39(2), 237–261.

Frank, K.I. and Hibbard, M. 2017. Rural planning in the twenty-first century: context-appropriate practices in a connected world. *Journal of Planning Education and Research*, 37(3), 299–308.

Frisch, M. 2015. Finding transformative planning practice in the spaces of intersectionality. In P. Doan (ed.), *Planning and LGBTQ Communities: the Need for Inclusive Queer Space*. Routledge, London, 129–146.

Gallent, N., Hamiduddin, I., Juntti, M., Kidd, S. and Shaw, D. 2015. *Introduction to Rural Planning: Economies, Communities and Landscapes*. Routledge, London.

Gallent, N. and Scott, M. (eds) 2017. *Rural Planning and Development*. Routledge, London.

Gallent, N., Tewdwr-Jones, M. and Hamiduddin, I. 2017. A century of rural planning in England: a story of fragmentation, contestation and integration. *Planum: Journal of Urbanism*, 35(2), 91–104.

Gallent, N., Hamiduddin, I., Juntti, M., Livingstone, N. and Stirling, P. 2018. *New Money in Rural Areas: Land Investment in Europe and its Place Impacts*. Palgrave, London.

Gawith, D. and Hodge, I. 2017. *Envisioning a British Ecosystem Services Policy*. Policy Brief. University of Cambridge Department of Land Economy and Cambridge Centre for Science and Policy, Cambridge.

Gkartzios M. 2013. 'Leaving Athens': narratives of counterurbanisation in times of crisis. *Journal of Rural Studies*, 32, 158–167.

Healey, P. 1999. Institutionalist analysis, communicative planning and shaping places. *Journal of Planning Education and Research*, 19, 111–121.

Herslund, L. 2012. The rural creative class: counterurbanisation and entrepreneurship in the Danish countryside. *Sociologia Ruralis*, 52(2), 235–255.

Lapping, M.B. 2006. Rural policy and planning. In P. Cloke, T. Marsden and P. Mooney (eds), *Handbook of Rural Studies*, 106–107. Sage Publications, London.

Lennon, M. and Scott, M. 2014. Delivering ecosystems services via spatial planning: reviewing the possibilities and implications of a green infrastructure approach. *Town Planning Review*, 85(5), 563–587.

Lichter, D. and Brown, D. 2011. Rural America in an urban society: changing spatial and social boundaries. *Annual Review of Sociology*, 37, 565–592.

Malecki, E.J. 2003. Digital development in rural areas: potentials and pitfalls. *Journal of Rural Studies*, 19(2), 201–214.

Murray, M. and Greer, J. 1997. Practice forum planning and community-led development in northern rural Ireland. *Planning Practice and Research*, 12(4), 393–400.

OECD. 2018. Edinburgh policy statement on enhancing rural innovation. The 11th OECD Rural Development Conference, Edinburgh. Retrieved from www.oecd.org/cfe/regional-policy/Edinburgh-Policy-Statement-On-Enhancing-Rural-Innovation.pdf.

Remoundou, K., Gkartzios, M. and Garrod, G. 2016. Conceptualizing mobility in times of crisis: towards crisis-led counterurbanization? *Regional Studies*, 50(10), 1663–1674.

Rydin, Y. 2003. *Conflict, Consensus, and Rationality in Environmental Planning: An Institutionalist Discourse Approach*. Oxford University Press, Oxford..

Scott, K., Rowe, F. and Pollock, V. 2018. Creating the good life? A wellbeing perspective on cultural value in rural development. *Journal of Rural Studies*, 59, 173–182.

Scott, M. and Murray, M. 2009. Housing rural communities: connecting rural dwellings to rural development in Ireland. *Housing Studies*, *24*(6), 755–774.

Sheller, M. and Urry, J. 2006. The new mobilities paradigm. *Environment and Planning A*, *38*, 207–226.

Shucksmith, M. 2018. Re-imagining the rural: from rural idyll to good countryside. *Journal of Rural Studies*, *59*, 163–172.

UN. 2018. *World Urbanization Prospects: The 2018 Revision*. United Nations Department of Economic and Social Affairs, Population Division, New York.

Urry, J. 2007. *Mobilities*. Polity Press, Cambridge.

Wachsmuth, D. 2014. City as ideology: reconciling the explosion of the city form with the tenacity of the city concept. *Environment and Planning D: Society and Space*, *32*(1), 75–90.

White, B., Borras Jr., S.M., Hall, R., Scoones, I. and Wolford, W. 2012. The new enclosures: critical perspectives on corporate land deals. *Journal of Peasant Studies*, *39*(3–4), 619–647.

Woods, M. 2003. Conflicting environmental visions of rural: windfarm development in mid Wales. *Sociologia Ruralis*, *43*, 272–288.

Index